MODERN METHODS IN ANALYTICAL ACOUSTICS
ACOUSTICS
Lecture Notes

D.G. Crighton, A.P. Dowling,
J.E. Ffowcs Williams, M. Heckl and
F.G. Leppington

MODERN METHODS IN ANALYTICAL ACOUSTICS

Lecture Notes

With 175 Figures

Springer-Verlag
London Berlin Heidelberg New York
Paris Tokyo Hong Kong
Barcelona Budapest

Cover illustration: Ch. 10, Fig.8. Sound rays in to deep ocean with a Munk-profile for a source at a depth of 1000 m. (After Porter & Bucker, 1987.)

ISBN 3-540-19737-0 Springer-Verlag Berlin Heidelberg New York
ISBN 0-387-19737-0 Springer-Verlag New York Berlin Heidelberg

British Library Cataloguing in Publication Data
Modern methods in analytical acoustics: lecture notes.
  I. Crighton, D. G., *1942-*
  620.2
ISBN 3-540-19737-0

Library of Congress Cataloging-in-Publication Data
Modern methods in analytical acoustics: lecture notes / D.G. Crighton
  ... [et al.].
  p. cm.
  Includes index.
  ISBN 0-387-19737-0. – ISBN 3-540-19737-0
  1. Sound-waves. 2. Vibrations. I. Crighton, D.G., 1942-
QC243.M63   1992   91-41977
620.2 – dc20   CIP

© Springer-Verlag London Limited 1992
Printed in Great Britain
2nd printing, 1994
3rd printing 1996

The use of registered names, trademarks etc. in this publication does not imply, even in the absence of a specific statement, that such names are exempt from the relevant laws and regulations and therefore free for general use.

The publisher makes no representation, express or implied, with regard to the accuracy of the information contained in this book and cannot accept any legal responsibility or liability for any errors or omissions that may be made.

Typesetting: Camera ready by authors
Printed and bound at the Athenæum Press Ltd., Gateshead, Tyne and Wear
69/3830-5432 Printed on acid-free paper

D.G. Crighton
Fellow of St John's College, Cambridge and
The Professor of Applied Mathematics,
Department of Applied Mathematics and Theoretical Physics
University of Cambridge

A.P. Dowling
Fellow of Sidney Sussex College, Cambridge and
Professor of Mechanical Engineering,
Department of Engineering
University of Cambridge

J.E. Ffowcs Williams
Fellow of Emmanuel College, Cambridge, and
Rank Professor of Engineering (Acoustics),
Department of Engineering
University of Cambridge

M.Heckl
Professor of the Institut für Technische Akustik,
Technische Universität Berlin

F.G. Leppington
Professor of Applied Mathematics,
The Imperial College of Science, Technology and Medicine,
London.

CONTENTS

26 Anti-Sound
J.E. Ffowcs Williams 705

PREFACE

This book has grown out of lecture notes for a course specially constructed for the Admiralty some 25 years ago. The then Head of Sonar at Portland, Mr S. D. Mason, knew the importance of keeping abreast of fundamental advances in the subject and recognized the value of understanding the ideas and mastering the techniques used to develop new knowledge. Thirty scientists in his charge attended an eight hour session once a month in which four lecturers presented the analytical techniques of selected topics relevant to modern SONAR, illustrating their material with reference to particular recently published scientific papers. The course evolved over twelve months to cover noise and unsteady flow and was immediately repeated to a general extra-mural audience at Imperial College; that five day course was oversubscribed with professionals coming from several centres, mainly from the aeronautical and naval industries of Europe and the U.S.A. Since then the course has been given many times in Europe and the U.S.A., the scope has been expanded and the lecturing team has grown, but the primary aim of teaching the analytical techniques alongside specific areas of wave motion and unsteady fluid mechanics has remained. Sadly Charles Ellen, a founder member of the group, is no longer with us and is much missed. Each time the course was given the notes developed, evolving out of their lecture note style towards something different. But they remain essentially notes on which lectures are based that we now wish to put on a more permanent record. Some topics are more mature than others and some did not appear in particular courses.

The lecturers enjoyed the courses very much, speaking always on topics they felt important to the fundamental understanding of noise, vibration and fluid mechanisms. They felt the audience enjoyed them too. We hope the readers of this collection of notes will find them both instructive and fun to read.

There are many who have helped with the production of these notes to whom we owe a great debt. In particular, Nigel Peake made valuable suggestions on a number of chapters; and Naomi Coyle, who has typed the notes in camera ready form, deserves our special thanks for her calm efficiency.

November 1991   JEFfW

ACKNOWLEDGEMENTS

We are grateful for permission to reproduce various figures from the authors named in each caption, and from publishers as follows:

Academic Press (Figure 16.3) – Journal of Sound and Vibration (Figure 15.4); American Institute of Aeronautics & Astronautics (Figure 15.3); American Institute of Physics – Journal of the Acoustical Society of America (Figures 10.8, 25.15, 25.16, 25.17, 25.18); Annual Reviews Inc. (Figures 21.1, 21.3); Cambridge University Press – Journal of Fluid Mechanics (Figures 12.5, 25.10, 25.11, 25.12, 25.13); Editions de Physique (Figure 3.5); Elsevier (Figure 9.2); Institute of Electrical and Electronics Engineers, Inc. (Figure 19.3); McGraw-Hill (Figure 25.1); "Nature" (Figure 25.14); North-Holland (Figures 25.4, 25.5, 25.6, 25.7, 25.8, 25.9); von Kármán Institute for Fluid Dynamics (Figure 15.5).

Chapter 6 derives from lecture notes first given in a course at David Taylor Research Center, Md. in 1976. DGC gratefully acknowledges permission of the Center for reproduction of these notes here.

Chapter 16 has been prepared with the support of Ferranti-Thomson Sonar Systems UK Ltd. and the Ministry of Defence, which is gratefully acknowledged.

We are also indebted to the United States Office of Naval Research for support in the preparation of some parts of the material presented in these lecture notes.

PART I
The Classical Techniques of Wave Analysis

1. COMPLEX VARIABLE THEORY

Some techniques of complex variable theory, particularly contour integration, provide useful methods for dealing with certain problems of wave theory and are outlined in the present chapter. It is assumed that the reader will be reasonably familiar with the basic ideas of complex numbers, which are therefore described only briefly. The book by Carrier, Krook & Pearson (1966) provides a good account of basic theory with applications.

1.1 COMPLEX NUMBERS

The need to extend our ideas of numbers beyond the set of ordinary, or "real", numbers is seen by consideration of elementary quadratic equations. For there is not always a real solution. The simplest example is the equation

$$z^2 + 1 = 0 \ ,\tag{1.1}$$

which is clearly not satisfied by any real number z.

The theory of complex numbers is initiated by introducing a new number, usually denoted by i, with the defining property that

$$i^2 = -1 \ ,\tag{1.2}$$

and the quadratic equation (1.1) has solutions $z = i$ and $z = -i$. The element i is sometimes called an "imaginary" number since it does not belong to the more familiar set of ordinary, or "real", numbers. A new number system can now be generated by taking combinations of real and imaginary elements. Thus a *complex number z* has the form

$$z = x + iy \ ,\tag{1.3}$$

where x and y are both real; x is called the *real part* of z, written as $\Re z$ and y is called the *imaginary part* of z, written as $\Im z$. A real number x can be regarded as a complex number with zero imaginary part; thus $z = x$ is the complex number $z = x + i0$.

Addition and multiplication of complex numbers are defined so that the algebra of complex numbers is formally the same as that for ordinary numbers, provided that i^2 is always interpreted as -1 in accordance with (1.2). Thus if $z_1 = x_1 + iy_2$ and $z_2 = x_2 + iy_2$, then we have

Definition:

$$z_1 \pm z_2 = (x_1 \pm x_2) + i(y_1 \pm y_2) \; ; \tag{1.4}$$

$$z_1 z_2 = (x_1 x_2 - y_1 y_2) + i(x_1 y_2 + x_2 y_1) \; . \tag{1.5}$$

The operation of division follows from the natural definition that $1/z$ or z^{-1} is the number which, when multiplied by z, gives unity; thus

$$z^{-1} = \frac{x}{x^2 + y^2} - i \frac{y}{x^2 + y^2} \; , \tag{1.6}$$

provided x and y are not both zero. Further, z_1/z_2 is defined to be z_1 times z_2^{-1}.

It is easy to verify that

$$z_1 z_2 = z_2 z_1$$

and

$$z_1(z_2 z_3) = (z_1 z_2) z_3 \; .$$

In view of the latter result, the product may be written without ambiguity as $z_1 z_2 z_3$.

Complex Conjugate

Associated with any complex number $z = x + iy$ is its complex conjugate, obtained by changing the sign of the imaginary part. Denoted by $\bar{z}$ or z^*, we have

$$z^* = x - iy \; . \tag{1.7}$$

It is easy to show that *any* quadratic equation with real or complex coefficients can be solved in terms of complex numbers; no new numbers are needed apart from the element i. More generally, the fundamental theorem of algebra, whose proof is developed later in this chapter, states that a polynomial equation of any degree n has a solution in terms of complex numbers. In particular, if the coefficients of the polynomial are all real, then solutions can easily be shown to be either real or else occurring in complex conjugate pairs.

Argand Diagram

A useful geometric interpretation can be assigned to a complex number $z = x + iy$ by representing it as a point in a Cartesian reference frame $0xy$, with coordinates (x, y). There is clearly a one–to–one correspondence between complex numbers and points in this plane, called the *complex z–plane* or *Argand diagram*.

An alternative representation for a complex number is obtained by the use of polar coordinates (r, θ) to specify a point in the Argand diagram. If r and θ are defined in the usual way, namely $x = r \cos \theta$, $y = r \sin \theta$, with $r \geq 0$, then

$$z = x + iy = r(\cos\theta + i \sin \theta) \ . \tag{1.8}$$

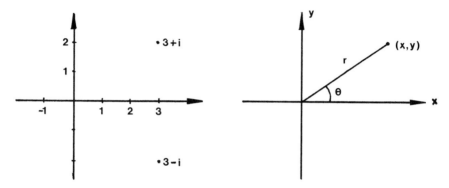

Figure 1.1 Argand Diagram   **Polar Coordinates**

The nonnegative number r is called the *modulus* or absolute value of a complex number z and is written variously as

$$r = \text{mod } z = |z| = +(x^2 + y^2)^{1/2} \ . \tag{1.9}$$

If $z = 0$, then $r = 0$, otherwise $r > 0$. The angle θ is called the *argument* or phase of z, written as

$$\theta = \arg z = \tan^{-1}(y/x) \ . \tag{1.10}$$

The multi–valued nature of θ is obvious: any multiple of 2π could be added to θ without changing the value of z. The argument can be made single–valued by defining the principal value of the argument (denoted by $\text{Arg}z$) to be such that

$$-\pi < \text{Arg}z \leq \pi \ . \tag{1.11}$$

De Moivre's Theorem

An interesting and suggestive consequence of the multiplication law (1.5) arises by considering numbers written in polar form $z = \cos\theta + i\sin\theta$, taking the modulus to be unity for simplicity. Thus

$$
\begin{aligned}
(\cos\theta_1 + i\sin\theta_1)(\cos\theta_2 + i\sin\theta_2) &= (\cos\theta_1\cos\theta_2 - \sin\theta_1\sin\theta_2) \\
&\quad + i(\sin\theta_1\cos\theta_2 + \cos\theta_1\sin\theta_2) \\
&= \cos(\theta_1 + \theta_2) + i\sin(\theta_1 + \theta_2) \quad .
\end{aligned}
\tag{1.12}
$$

If θ_1 and θ_2 both equal θ, for example, then

$$
(\cos\theta + i\sin\theta)^2 = \cos 2\theta + i\sin 2\theta \quad ,
$$

and a simple induction argument shows that

$$
(\cos\theta + i\sin\theta)^n = \cos n\theta + i\sin n\theta
\tag{1.13}
$$

for any positive integer n. Further, in view of the identity

$$
(\cos\theta + i\sin\theta)^{-1} = \cos\theta - i\sin\theta = \cos(-\theta) + i\sin(-\theta) \quad ,
$$

it follows that (1.13) holds for any integer, positive or negative. This result bears the name of de Moivre's theorem.

Multiplications of any two complex numbers z_1 and z_2 can be expressed in terms of the polar coordinate representation (1.8) together with (1.12). Evidently

$$
z_1 z_2 = r_1 r_2 \left\{ \cos(\theta_1 + \theta_2) + i\sin(\theta_1 + \theta_2) \right\} \quad ,
$$

which shows that

$$
|z_1 z_2| = |z_1||z_2| \quad \text{and} \quad \arg(z_1 z_2) = \arg z_1 + \arg z_2 \quad .
\tag{1.14}
$$

Any integral multiple of 2π could be added to the latter expression. It is not true in general that $\operatorname{Arg}(z_1 z_2) = \operatorname{Arg} z_1 + \operatorname{Arg} z_2$, as is seen by the example $\theta_1 = \theta_2 = 3\pi/4$. More generally,

$$
|z^n| = |z|^n \quad \text{and} \quad \arg(z^n) = n \, \arg \, z \quad .
$$

A more important application of de Moivre's theorem is that it paves the way for definitions of exponential and trigonometric functions of a complex variable z.

1.2 EXPONENTIAL, HYPERBOLIC AND TRIGONOMETRIC FUNCTIONS

The form of the multiplication formula (1.12) is similar to that of exponentials: the effect of multiplication is to add arguments. Guided by this idea, it is possible to define the exponential of an imaginary number $i\theta$ in such a way that the function has many of the properties familiar in the real variable case.

Definition. The exponential of an imaginary number $i\theta$ is defined as

$$\exp(i\theta) = e^{i\theta} = \cos\theta + i\sin\theta \ , \tag{1.15}$$

where θ is supposed real for the moment.

Example. $\exp(i\pi/2) = i, \ \exp(i\pi) = -1$.

Example. Show from the definition (1.15) that

(i) $e^{i\theta_1}e^{i\theta_2} = e^{i(\theta_1+\theta_2)}$;

(ii) $e^{i\theta} = 1 + (i\theta) + (i\theta)^2/2! + (i\theta)^3/3! + ...$.

Proof. Property (i) is simply formula (1.12). Property (ii) is verified by collecting together real and imaginary parts of its right–hand side to get

$$(1 - \theta^2/2! + \theta^4/4! + ...) + i(\theta - \theta^3/3! + ...) = \cos\theta + i\sin\theta = e^{i\theta} \ .$$

Definition. If $z = z(t)$ is a complex function of a *real* parameter t, its derivative can be defined as

$$\frac{dz}{dt} = \frac{dx}{dt} + i\frac{dy}{dt} \ . \tag{1.16}$$

Example. Show that $(d/d\theta)e^{i\theta} = ie^{i\theta}$. By definition $(d/d\theta)e^{i\theta} = (d/d\theta)(\cos\theta + i\sin\theta) = -\sin\theta + i\cos\theta = ie^{i\theta}$.

It is now a simple matter to take the important step of defining $\exp(z)$ for any complex number $z = x + iy$. A sensible definition should be such that $\exp(x+iy) = \exp(x)\exp(iy)$, in line with the result for real exponentials. Accordingly, we are led to:

Definition of Exponential Function: If $z = x + iy$,

$$e^z \equiv \exp z = e^x(\cos y + i\sin y) \ . \tag{1.17}$$

Clearly (1.15) is a special case with z purely imaginary.

Example. $\exp(1 + i\pi/2) = ie$, $\exp\left(\frac{\pi}{4}(1+i)\right) = e^{\pi/4}\left(\frac{1}{\sqrt{2}} + \frac{i}{\sqrt{2}}\right)$.

Example. Show that $\exp(z_1)\exp(z_2) = \exp(z_1 + z_2)$.

Proof. By definition (1.17), the left–hand side, is

$$e^{x_1}e^{x_2}(\cos y_1 + i\sin y_1)(\cos y_2 + i\sin y_2) = e^{(x_1+x_2)}(\cos(y_1+y_2) + i\sin(y_1+y_2))$$

by the rules of real algebra and formula (1.12). Hence the result, by (1.17).

It is easy to show that $e^{-z} = 1/e^z$, and that e^z can be written in power–series form

$$e^z = 1 + z + z^2/2! + \cdots .$$

Several further functions can now be defined.

Definition of Hyperbolic Functions.

$$\cosh z = \tfrac{1}{2}(e^z + e^{-z}); \sinh z = \tfrac{1}{2}(e^z - e^{-z}); \tanh z = \sinh z / \cosh z . \qquad (1.18)$$

Definition of Trigonometric Functions.

If (1.15) is added and subtracted from its complex conjugate, one finds

$$\cos\theta = \tfrac{1}{2}(e^{i\theta} + e^{-i\theta}) \quad \text{and} \quad \sin\theta = (e^{i\theta} - e^{-i\theta})/2i .$$

At this stage θ is supposed real, but the following generalization for trigonometric functions is immediately suggested:

$$\cos z = \frac{e^{iz} + e^{-iz}}{2}; \quad \sin z = \frac{e^{iz} - e^{-iz}}{2i}; \quad \tan z = \frac{\sin z}{\cos z} . \qquad (1.19)$$

Our ideas of trigonometric and hyperbolic functions are greatly extended by definitions (1.18) and (1.19), and the similarity between these functions is striking. It is readily verified that all the familiar addition laws are valid; thus

$$\sin(z_1 + z_2) = \sin z_1 \cos z_2 + \cos z_1 \sin z_2 ,$$
$$\cos(z_1 + z_2) = \cos z_1 \cos z_2 - \sin z_1 \sin z_2 ,$$
$$\cos^2 z + \sin^2 z = 1 ,$$
$$\cosh(z_1 + z_2) = \cosh z_1 \cosh z_2 + \sinh z_1 \sinh z_2 .$$

On the other hand, some inequalities that hold for real variables do not hold in the complex case. For example, it is easy to see that $|\sin iy| \to \infty$ as $y \to \infty$, so $\sin z$ can have absolute value greater than unity.

The procedure of writing $\sin z$, $\sinh z$, $\cos z$ and $\cosh z$ in terms of real and imaginary parts, and the various relationships between these functions, is greatly helped by the special results

$$\left.\begin{array}{ll} \cos iz = \cosh z, & \cosh iz = \cos z \\[2mm] \sin iz = i \sinh z, & \sinh iz = i \sin z \end{array}\right\} \tag{1.20}$$

for any complex z; each of these formulae follows directly from the definitions (1.18) and (1.19).

Example. $\sin z = \sin(x + iy) = \sin x \cosh y + i \cos x \sinh y$. Hence

$$|\sin z|^2 = \cosh^2 y - \cos^2 x \quad .$$

Logarithmic Function

One expects that the definition of $\log z$ should have the property, analogous to that for real variables, that

$$\log(z_1 z_2) = \log z_1 + \log z_2 \quad .$$

In particular the variable $z = re^{i\theta}$, in polar form, can be regarded as a product of r times $\exp(i\theta)$. Further, the logarithm of $e^{i\theta}$ is sensibly defined as $i\theta$, again by analogy with the real variable case. This suggests the following.

Definition. If $z = re^{i\theta}$ is any complex number, then define

$$\log z = \log r + i\theta \quad , \tag{1.21}$$

where $\log r$ is the natural logarithm of the positive number r. It is clear that $\log z$ is not single–valued, without further specification, on account of the ambiguity of the angle θ to which any integral multiple of 2π may be added. A particular value, or *branch* of the logarithmic function can be specified by arbitrarily choosing the angle θ to lie within the range $-\pi < \theta \leq \pi$. Denoting the function so defined by Log, we have

$$\mathrm{Log}\, z = \log r + i\theta, \quad -\pi < \theta \leq \pi \quad , \tag{1.22}$$

and this *principal value* of the logarithmic function is uniquely specified for any value of z in the complex plane. The single–valued nature of $\mathrm{Log}\, z$ has been achieved at some cost, however, since the function is necessarily discontinuous along the negative real axis; thus

$$\lim_{z \to re^{i\pi}} \text{Log} z = \log r + i\pi \text{ and } \lim_{z \to re^{-i\pi}} \text{Log} z = \log r - i\pi \;,$$

with a discontinuity of $2\pi i$. The idea of introducing a *branch cut*, in this case along the negative real axis, in order to render unique an inherently multi–valued function of z, plays an important role in the sequel and is accompanied by corresponding discontinuities across the cut.

Example. $\exp(\log z) = z$ for any branch of $\log z$

Proof.

$$\exp(\log z) = \exp(\log r + i\theta + 2n\pi i), \; n \text{ any integer}$$
$$= \exp(\log r) \exp(i\theta) \exp(2n\pi i) = re^{i\theta} = z \;.$$

Example. $\log(\exp z) = z$ for a suitably chosen branch of $\log z$.

Proof:

$$\log(\exp z) = \log e^x + iy + 2n\pi i : \; n \text{ any integer}$$
$$= x + iy \text{ choosing } n = 0 \;.$$

Similarly, $\log(z_1 z_2) = \log z_1 + \log z_2 + 2n\pi i$ for some appropriate value of the integer n (depending on the branches for $\log z_1$ and $\log z_2$).

1.3 MANY–VALUED FUNCTIONS

The idea of multi–valued functions has already been encountered through $\log z$, and is discussed here in further detail. As a simple prototype it is convenient to deal with a function that has just two values for a given z, namely the square–root function $z^{1/2}$.

Square Root

The function $w = z^{1/2}$ is the inverse of the function z^2 by definition. That is, $w = z^{1/2}$ is a number satisfying the equation $z = w^2$; there are obviously two solutions, since if w is a solution, then so is $-w$. When $z = 1$, for example, $z^{1/2}$ could mean either $+1$ or -1, with a similar ambiguity for all values of z other than $z = 0$.

The function $z^{1/2}$ can be rendered as single–valued, however, by suitably choosing one member of the two roots in the following way. At a particular point $z = z_0$, assign a definite value to $z^{1/2}$ from the pair of possibilities: to be specific, one might choose $z^{1/2} = +1$ at $z = 1$. Then take $z^{1/2}$ to be the function that varies *continuously* as z varies, i.e. $z^{1/2}$ does not jump about from one root to the other at neighbouring points.

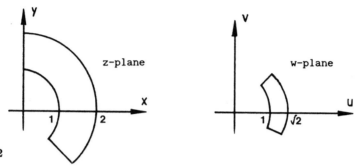

Figure 1.2

If $z^{1/2} = 1$ when $z = 1$, then as the variable z changes continuously from $1 \rightarrow e^{i\pi/2} \rightarrow 2e^{i\pi/2} \rightarrow 2e^{-i\pi/4} \rightarrow e^{-i\pi/4} \rightarrow 1$, round the circuit in the diagram, the variable $w = u + iv = z^{1/2}$ varies continuously from $1 \rightarrow e^{i\pi/4} \rightarrow \sqrt{2}e^{i\pi/4} \rightarrow \sqrt{2}e^{-i\pi/8} \rightarrow e^{-i\pi/8} \rightarrow 1$, and it is noted that $w = z^{1/2}$ reaches its starting value on completion of the tour. It is found that the function $z^{1/2}$ defined in this manner will always return to its original value for any such circuit, *provided that the circuit does not enclose the origin*. If z follows the circular path $z = e^{i\theta}$, for example, with θ varying continuously from 0 to 2π, then the finishing value of the function is $e^{i\pi} = -1$, and a similar property is readily demonstrated for any circuit enclosing the origin once: the origin is therefore called a *branch point* for the function $z^{1/2}$. Thus $z^{1/2}$ is continuous but not single–valued without some further constraint, since its value at point z depends on the path followed in going from z_0 to z. Two possible lines of approach are available to counter this difficulty, as follows.

Branch Cuts

The simplest idea, and the approach found most useful in the present context, is to prohibit paths from encircling the origin, with the device of "cutting" the complex plane by an arbitrarily chosen line from the origin to infinity (arbitrary except the line should not cross itself). To be definite one might make such a *branch cut* from the origin along the negative real axis. The function $z^{1/2}$, defined to take the value

+1 when $z = 1$, and varying continuously as z travels any path that does not cross the forbidden branch cut, is readily seen to be continuous and single–valued in the whole complex plane excluding the negative real axis. Specifically, if $z = re^{i\theta}$ $(-\pi < \theta < \pi)$, then with this definition if $w = z^{1/2}$

$$w = r^{1/2} e^{i\theta/2} \ (-\pi < \theta < \pi) \ .$$

In order to specify $z^{1/2}$ in this way, it is necessary and sufficient to assign a definite value to $z^{1/2}$ at a particular point $z \neq 0$ and to specify the branch cut from 0 to ∞.

Example. With $w = z^{1/2}$ defined as above, find $w(i)$, $w(-i)$, $w(1+i)$ and the limits of $w(-4+iy)$ as $y \to +0$ and $y \to -0$.

Solution. Writing $\quad z = re^{i\theta}(-\pi < \theta < \pi), \quad w = r^{1/2} e^{i\theta/2}$.

Thus

$$w(i) = e^{i\pi/4}, \ w(-i) = e^{-i\pi/4}$$

$$w \to \pm 2i \ \text{ as } \ y \to \pm 0, \ x = -4 \ .$$

Riemann Surfaces

It is readily seen that the function $z^{1/2}$ defined above is such that $-\pi/2 < \arg z^{1/2} < \pi/2$. If we took the other value $z^{1/2} = -1 = e^{i\pi}$ at $z = 1$, with the same branch cut along the negative real axis, then this would define the other *branch* of the function $z^{1/2}$ in which $z^{1/2}$ is in the left half of the complex plane. Thus the simple expedient of the "cut plane" defines a single–valued function at the expense of restricting the possible values of $z^{1/2}$. Evidently a single complex z–plane is not sufficient to contain both the possible branches of the function $z^{1/2}$. The idea of a *Riemann Surface* is to imagine fastened together several complex planes (two in the case of the function $z^{1/2}$) called *sheets* of the Riemann Surface, the sheets being thought to be stitched together along assigned lines (the negative real axis, for example). Thus instead of prohibiting z from crossing a branch cut, one imagines z transferring from sheet I to sheet II, with one branch of $z^{1/2}$ described on sheet I and the other on sheet II. Thus the value of the function at a point $z = 1 + i$, for example, depends on whether this is the point $z = 1 + i$ on sheet I or the point $z = 1 + i$ on sheet II, the change from one to the other occurring when z crosses the negative real axis.

For example, the function $w = z^{1/2}$ may be defined by choosing $w = 1$ at $z = 1$ on sheet I, with the two sheets of the Riemann Surface joined along the negative real axis. Consider the value of w as z travels twice round the unit circle, i.e. $z = e^{i\theta}$ where θ changes from 0 to 4π. When θ changes from 0 to π, we are on sheet I with $w = w_1 = e^{i\theta/2}$ changing from 1 to i; when θ increases beyond π through 2π up to 3π, z changes to sheet II and $w = w_2 = e^{i\theta/2}$ changes from i through -1 to $-i$; θ increasing beyond 3π up to 4π brings z on to sheet I again, with $w = w_1$ changing from $-i$ to 1. When $z = i$, for example, the two values $w_1 = \frac{1}{\sqrt{2}}(1+i)$ and $w_2 = -\frac{1}{\sqrt{2}}(1+i)$ depend upon the particular sheet of the Riemann Surface on which the variable z is situated.

The function $z^{1/2}$ evidently requires only two sheets for its Riemann Surface; $z^{1/3}$ has three possible values for any given z, so has three branches and requires a Riemann Surface of three sheets. The function $z^{1/3}(z-1)^{1/2}$ requires six sheets and functions like $\log z$ and $\tan^{-1} z$ have infinitely many possible values and require an infinite number of sheets.

This device, which involves the complication of several complex planes imagined to be fastened together in a prescribed manner, enjoys the advantage of including all possible branches of a multi–valued function. Henceforth, however, we shall be content with the simpler idea of a cut plane to specify a single branch of a many–valued function. In the language of Riemann Surfaces, the variable z will be confined to occupy just one sheet of the surface.

Branch Points and Branch Cuts

In the prototype problem $w = f(z) = z^{1/2}$ discussed above, the branch cut was required to travel from the origin to infinity. The special nature of the origin is made obvious by the observation that the function $z^{1/2}$ that varies continuously along a closed path P (say a circle) enclosing the origin once does *not* return to its starting value, no matter how small the radius of P. This property is possessed by no point other than the origin which is therefore accorded the special name of branch point. A more general definition is as follows.

Definition. If a function $w = f(z)$, varying continuously as z moves in a full circle of nonzero radius centred at z_o does not return to its starting value, no matter how small the radius, then z_0 is called a branch point of the function $f(z)$.

If $f(z)$ does not return to its starting value when z moves around a full circle of arbitrarily *large* radius, one says that "the point at infinity" is a branch point.

Evidently a single–valued continuous function has no branch points since f

14

always returns to its starting value as z travels any closed circuit.

In general, a particular branch of a many–valued function is specified by assigning a particular value of the function at a fixed point z_n (not a branch point) and by choosing suitable *branch cuts* joining the branch points. The function f is then single–valued if z is prohibited from crossing any branch cut.

Example. $w = f(z) = (z^2 - 1)^{1/2} = (z - 1)^{1/2}(z + 1)^{1/2}$.

This function has branch points at $z = +1$ and $z = -1$, but not at $z = \infty$. furthermore, $f(z)$ is double–valued. A single–valued function can be defined by choosing $w = +i = e^{i\pi/2}$ when $z = 0$, and by choosing cuts as shown in the diagram.

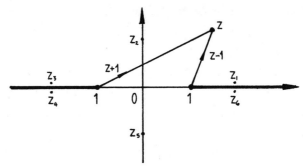

Figure 1.3

Let us calculate the value of w at each of the points

$$z_1 = 2 + i0, \quad z_2 = i, \; z_3 = -2 + i0 \; ,$$
$$z_4 = -2 - i0, \; z_5 = -i, \; z_6 = 2 - i0 \; .$$

An effective procedure is to give attention to the factors $(z - 1)^{1/2}$ and $(z + 1)^{1/2}$ of $f(z)$. At the origin $z = 0$, we may consider $z + 1$ to have argument 0 and $z - 1$ to have argument $+\pi$, whence $(z + 1)^{1/2}(z - 1)^{1/2} = e^{i\pi/2} = i$ has the correct starting value. For any other point z, the values of $(z - 1)^{1/2}$ and $(z + 1)^{1/2}$ can be obtained by allowing their arguments to vary continuously as z follows a path that does not cross either branch cut. This constraint is satisfied if we insist that $0 < \arg(z - 1) < 2\pi$ and $-\pi < \arg(z + 1) < \pi$, whence

$$0 < \arg(z - 1)^{1/2} < \pi, \; \text{and} \; -\pi/2 < \arg(z + 1)^{1/2} < \pi/2 \; .$$

Thus when

$$z = 2 + i0, \; (z + 1)^{1/2} = 3^{1/2}, (z - 1)^{1/2} = 1 \; \text{and} \; f(z_1) = 3^{1/2} \; .$$

When

$$z = 2 - i0, \ (z+1)^{1/2} = 3^{1/2}, \ (z-1)^{1/2} = e^{i\pi} \text{ and } f(z_6) = -3^{1/2} \ .$$

When

$$z = i, (z+1)^{1/2} = 2^{1/4}e^{i\pi/8}, (z-1)^{1/2} = 2^{1/4}e^{3i\pi/8} \text{ and } f(z_2) = 2^{1/2}i \ .$$

Similarly

$$f(z_3) = -3^{1/2}, f(z_4) = 3^{1/2} \quad \text{and} \quad f(z_5) = 2^{1/2}i \ .$$

As expected, there is a discontinuity across the branch cuts.

Example. In certain circumstances (i.e. when the point at infinity is not a branch point), the cuts can be confined to a finite region. Taking $f(z) = (z^2 - 1)^{1/2}$ again, we could choose $f = +\sqrt{3}$ when $z = 2$ with a cut passing along the real axis from $z = -1$ to $z = +1$.

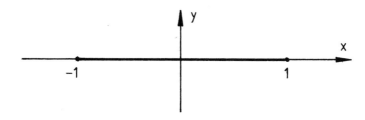

Figure 1.4

This can be thought of as being generated from the pair of cuts from $z = -1$ to $z = +\infty$ and from $z = +1$ to $z = +\infty$. If z crosses the real axis, with $x > 1$, say, then both these cuts are crossed, whence both factors $(z-1)^{1/2}$ and $(z+1)^{1/2}$ change sign simultaneously; although each factor is discontinuous, their product $f = (z-1)^{1/2}(z+1)^{1/2}$ is clearly continuous. Such a manoeuvre is not possible for the function $g(z) = (z-1)^{1/2}(z+1)^{1/3}$, for example, since the discontinuities would not annihilate each other in this case. The crucial difference is that $g(z)$ has a branch point at infinity, whilst $f(z)$ does not. For large z, $f(z) \sim z$, while $g \sim z^{5/6}$.

1.4 DIFFERENTIATION

The idea of differentiating a function $f(z)$ with respect to a complex variable z follows by direct analogy with real variable theory. Thus we have

Definition:
$$\frac{df}{dz} = f'(z) = \lim_{h \to 0} \frac{f(z+h) - f(z)}{h} \,, \qquad (1.23)$$

provided the limit exists. Since h is itself a complex number, there are infinitely many paths through which the number h can tend to zero: h could approach zero through real values, for example, or through imaginary values, or along a spiral path, and so on. The *derivative df/dz* is defined only if the limit (1.23) has the same value for all possible ways in which $h \to 0$, and this places very severe restrictions upon the function $f(z)$.

Example. If $f(z) = z^2$, then

$$\frac{f(z+h) - f(z)}{h} = \frac{2zh + h^2}{h} = 2z + h \to 2z \quad \text{as} \quad h \to 0 \ .$$

Example. If $f(z) = z^*$ (complex conjugate of z), then with $h = h_1 + ih_2$,

$$\frac{f(z+h) - f(z)}{h} = \frac{h_1 - ih_2}{h_1 + ih_2} \ .$$

This tends to $+1$ if $h_2 = 0$ and tends to -1 if $h_1 = 0$. Thus the limit (1.23) does not exist and z^* does not have a derivative.

Cauchy–Riemann Equations

The derivative of $f(z)$ exists only if the limit process (1.23) is independent of the way $h \to 0$. Thus necessary conditions for differentiability are obtained by equating the limits obtained by taking purely real and purely imaginary values. Write $f(z) = u + iv$ as a sum of real and imaginary parts, where u and v are of course functions of x and y. Taking $h = \delta x$ to be real, we have

$$\lim_{\delta x \to 0} \frac{u(x + \delta x, y) - u(x, y) + iv(x + \delta x, y) - iv(x, y)}{\delta x} = \frac{\partial u}{\partial x} + i\frac{\partial v}{\partial x} \qquad (1.24)$$

Similarly, taking $h = i\delta y$ to be imaginary, we have

$$\lim_{h \to 0} \frac{f(z+h) - f(z)}{h} = \frac{1}{i}\left(\frac{\partial u}{\partial y} + i\frac{\partial v}{\partial y}\right) = \frac{\partial v}{\partial y} - i\frac{\partial u}{\partial y} \ . \qquad (1.25)$$

Now these two limits must certainly be equal if $f(z)$ is differentiable. Equating real and imaginary parts leads to the celebrated *Cauchy–Riemann equations*

$$\frac{\partial u}{\partial x} = \frac{\partial v}{\partial y} \quad \text{and} \quad \frac{\partial u}{\partial y} = -\frac{\partial v}{\partial x} \ . \tag{1.26}$$

It has thus been shown that the Cauchy–Riemann equations provide *necessary* conditions for differentiability of $f(z)$, though they are not quite sufficient. However, the further weak conditions that u and v have continuous first derivatives at a point is now shown to ensure differentiability at that point. In this situation, the equality of the limit (1.23) as $h \to 0$ from the two special directions at right angles implies that the same limit obtains as $h \to 0$ in any manner:

Theorem. If the functions u and v have continuous first derivatives with respect to x and y at a point (x_0, y_0), then the function $f(z) = u(x, y) + iv(x, y)$ is differentiable at $z_0 = x_0 + iy_0$ if and only if the Cauchy–Riemann equations (1.26) hold at that point.

Proof. It has already been shown that differentiability implies that the Cauchy–Riemann equations (1.26) must hold, and it remains to prove the converse. This is established by writing $h = h_1 + ih_2$, whence

$$\lim_{h \to 0} \frac{f(z+h) - f(z)}{h}$$

$$= \lim_{\substack{h_1 \to 0 \\ h_2 \to 0}} \frac{h_1(\partial u/\partial x) + h_2(\partial u/\partial y) + ih_1(\partial v/\partial x) + ih_2(\partial v/\partial y) + o(h)}{h_1 + ih_2}$$

(using the continuity of the first derivatives)

$$= \lim_{\substack{h_1 \to 0 \\ h_2 \to 0}} \frac{h_1(\partial u/\partial x) + h_2(\partial u/\partial y) - ih_1(\partial u/\partial y) + ih_2(\partial u/\partial x)}{h_1 + ih_2}$$

(using the Cauchy–Riemann equations and neglecting the $o(h)$ terms in the limit)

$$= (\partial u/\partial x) - i(\partial u/\partial y) \tag{1.27}$$

and the proof is complete, since this is independent of the way $h_1 \to 0$ and $h_2 \to 0$. Formulae (1.26) and (1.27) show that $f'(z)$ may be expressed variously as

$$f'(z) = \frac{\partial u}{\partial x} + i\frac{\partial v}{\partial x} = \frac{\partial u}{\partial x} - i\frac{\partial u}{\partial y} = \frac{\partial v}{\partial y} - i\frac{\partial u}{\partial y} = \frac{\partial v}{\partial y} + i\frac{\partial v}{\partial x} \ .$$

Example. Show that $f(z) = \exp(z)$ is differentiable at all points, with $f' = f$.

Solution. Writing f in terms of its real and imaginary parts, we have $f = u + iv$ with $u = e^x \cos y$ and $v = e^x \sin y$. The Cauchy–Riemann equations are easily seen to be satisfied and are continuous for all (x, y). Thus $f'(z)$ exists and equals

$$f'(z) = \frac{\partial u}{\partial x} + i\frac{\partial v}{\partial x} = u + iv = f \ .$$

Example. If $f = z^*$, then $u = x$ and $v = -y$. Since $\partial u/\partial x = 1$ and $\partial v/\partial y = -1$, the Cauchy–Riemann equations are not satisfied for any (x, y) and f is nowhere differentiable.

Similarly, one can show that $\sinh z$ is differentiable at all points and that its derivative is $\cosh z$; $\sin z$ has derivative $\cos z$; $\cosh z$ has derivative $\sinh z$; $\cos z$ has derivative $-\sin z$.

Analytic Functions and Singular Points

A function is said to be analytic at a point z_0 if it is single–valued and differentiable at all points z within a neighbourhood of z_0: that is, within a region $|z - z_0| < \delta$ for some strictly positive number δ. A function is said to be analytic in a region R of the complex plane if it is analytic at each point of R.

A point z_1 at which $f(z)$ is not analytic is called a *singular point*, or singularity, of the function. Within its region of definition a function $f(z)$ may have several singularities that may be finite or infinite in number and may be arranged at either discrete intervals or clustered together.

Example. $f = |z|^2$ is analytic at no point. For although it is differentiable at $z = 0$, it does not have this property within any neighbourhood of $z = 0$.

Example. $f = e^z$ is analytic in the whole complex plane, with derivative e^z.

Example. $f = z^{-1}(z - 1)^{-2}$ is analytic in the whole plane except for the singular points $z = 0$ and $z = 1$.

Example. $f = 1/\sin(1/z)$ is analytic except for the singular points at $z = 0$ and at $z = \pm 1/n\pi$ for integer n. These singular points have a "limit point" at $z = 0$. That is, there are infinitely many of them within any arbitrarily small neighbourhood of $z = 0$, where their separation is decreasingly small.

Example. $f(z) = z^{1/2}$, with $f(1) = 1$ and a branch cut along the negative real axis: f is analytic with derivative $\frac{1}{2}z^{-1/2}$ except for points along the branch cut, which forms a line of singular points.

Properties of Analytic Functions

The following properties are consequences of the above definitions. Suppose f and g are analytic functions of z in a region R of the complex plane. Then

(i) $f + g$ is analytic in R, with derivative $f' + g'$;

(ii) fg is analytic, with derivative $fg' + f'g$;

(iii) f/g is analytic, if $g(z) \neq 0$, with derivative $(gf' - fg')/g^2$;

(iv) if $z = z(\zeta)$ is analytic for z in region R, ζ in region R_1, then f is an analytic function of ζ in R_1 with $df/d\zeta = (df/dz)(dz/d\zeta)$.

1.5 SERIES EXPANSIONS

Power Series

As in the theory of real variables, it is sometimes convenient to represent a given function by means of a power series (i.e. a Taylor series) in the form

$$f(z) = a_0 + a_1(z - z_0) + a_2(z - z_0)^2 + \cdots ,$$

provided the series converges. It will be shown later than, for an analytic function at $z = z_0$, the coefficients a_n are given by

$$a_n = \frac{1}{n!}f^{(n)}(z_0) , \tag{1.28}$$

as in the real variable case. The following properties, given here without proof, can be established and are important.

(i) The region of convergence is circular. That is, there is a real nonnegative number R_c, called the "radius of convergence", such that the series converges for $|z - z_0| < R_c$ and diverges for $|z - z_0| > R_c$. Convergence properties on the circle $|z - z_0| = R_c$ depend on the particular case in question. If $R_c = 0$, the series diverges except for the single point $z = z_0$. If the series converges for all z, one can say that $R_c = \infty$;

(ii) the coefficients a_n of the series are uniquely determined and given by (1.28);

(iii) convergent power series may be formally added, subtracted and multiplied to obtain another convergent series. Thus

$$\sum_0^\infty a_n(z - z_0)^n + \sum_0^\infty b_n(z - z_0)^n = \sum_0^\infty (a_n + b_n)(z - z_0)^n$$

$$\{a_0 + a_1(z - z_0) + a_2(z - z_0)^2 + \dots\}\{b_0 + b_1(z - z_0) + b_2(z - z_0)^2 + \dots\}$$

$$= a_0 b_0 + (a_0 b_1 + a_1 b_0)(z - z_0) + (a_0 b_2 + a_1 b_1 + a_2 b_0)(z - z_0)^2 + \dots \quad ;$$

(iv) convergent series may be differentiated and integrated term by term.

The property (iv) ensures the analytic nature of a power series within its circle of convergence; indeed the power series could be regarded as the basic definition of an analytic function if we insist that $R_c > 0$. Conversely it will be shown later in the present development that a function analytic at $z = z_0$ may be written as a series in powers of $(z - z_0)$, with radius of convergence R_c equal to the distance of z_0 from the nearest singularity.

Example. $f(z) = 1/(1 - z)$ may be expanded as the series

$$f(z) = 1 + z + z^2 + z^3 + \dots \quad \text{for } |z| < 1 \ . \tag{1.29}$$

Series in Inverse Powers

An alternative series representation for the function $f = 1/(1 - z)$ is initiated by the fact that f is analytic for all $|z| > 1$, hence is an analytic function of the variables $\zeta = 1/z$ for $|\zeta| < 1$, (i.e. $|z| > 1$). Thus

$$\frac{1}{1 - z} = \frac{-1}{z(1 - 1/z)} = -\frac{1}{z} - \frac{1}{z^2} - \frac{1}{z^3} \dots \quad \text{for } |z| > 1 \ . \tag{1.30}$$

A combination of the two types of series, with both positive and negative powers, leads to a much more general result than that of the Taylor series. To reach this goal in a natural way, consider the prototype example:

Example.

$$f(z) = \frac{1}{(1 - z)(2 - z)} = \frac{1}{1 - z} - \frac{1}{2 - z} \ , \quad z \neq 1; \ z \neq 2 \ .$$

The function f is analytic except at $z = 1$ and $z = 2$. It may be expanded in the powers of z if $|z| < 1$, in inverse powers for $|z| > 2$; in the region $1 < |z| < 2$, on the other hand, we may write the first term as

$$(1 - z)^{-1} = -1/z - 1/z^2 - 1/z^3 \dots \quad \text{for } |z| > 1 \ , \tag{1.31}$$

while the second term can be written as

$$(2 - z)^{-1} = \tfrac{1}{2} + z/2^2 + z^2/2^3 + \dots \quad \text{for } |z| < 2 \quad . \tag{1.32}$$

Thus

$$f(z) = \dots - \frac{1}{z^3} - \frac{1}{z^2} - \frac{1}{z} + \tfrac{1}{2} + \frac{z}{2^2} + \frac{z^2}{2^3} + \dots \quad \text{for } 1 < |z| < 2 \quad . \tag{1.33}$$

A generalization of this elementary example states that any function $f(z)$, analytic in an annulus $R_1 < |z - z_0| < R_2$ may also be so represented, as follows.

Theorem: Laurent Series

If $f(z)$ is analytic in the annulus $R_1 < |z - z_0| < R_2$, then it can be expressed as

$$f(z) = \dots + \frac{b_3}{(z - z_0)^3} + \frac{b_2}{(z - z_0)^2} + \frac{b_1}{(z - z_0)} + a_0 + a_1(z - z_0) + a_2(z - z_0)^2 + \dots , \tag{1.34}$$

which converges for $R_1 < |z - z_0| < R_2$.

A proof of this important result is deferred until the concept of complex integration is developed; the analysis will then provide the means to prove Laurent's theorem and give explicit formulae for the coefficients a_n and b_n.

The Laurent series expansion (1.34) implies that a function analytic in an annulus $R_1 < |z - z_0| < R_2$ can be expressed as a *sum* of two functions, the one involving positive powers being analytic inside the circle $|z - z_0| < R_2$, and the other analytic outside the circle $|z - z_0| > R_1$.

Singular Points

Any point where $f(z)$ is not analytic is called a singular point of the function $f(z)$. It is called *isolated* if there is no other singularity in its close neighbourhood. To be precise, a singular point z_0 is called isolated if there is a positive number δ such that the circle $|z - z_0| < \delta$ contains no other singularity except z_0. Otherwise there are infinitely many singularities within any arbitrary small neighbourhood of z_0. Included in this latter category are branch points and every point on a branch cut.

Example. $f(z) = z^{-1}(z - 1)^{-3}$ has isolated singularities at $z = 0$ and $z = 1$. Isolated singularities can be further classified with reference to the nature of their Laurent series (1.34), in the (degenerate) annulus $0 < |z - z_0| < R$, for some positive R which is the distance from z_0 to the nearest other singularity.

Pole of Order m

If the terms of (1.34) involving inverse powers of $(z - z_0)$ are finite in number, then there is an integer m such that $b_m \neq 0$ and $b_n = 0$ for all $n > m$, thus

$$f(z) = b_m(z - z_0)^{-m} + b_{m-1}(z - z_0)^{-m+1} + \ldots + b_1(z - z_0)^{-1} + a_0 + a_1(z - z_0) + \ldots$$
(1.35)

for $0 < |z - z_0| < R$, with $b_m \neq 0$. In this case z_0 is called a *pole of order m*. (The particular case where $m = 1$ is usually referred to as a *simple pole* as an alternative to a "pole of order one".) Evidently, $f(z) = 0(z - z_0)^{-m}$ as $z \to z_0$, and the function $g(z) = (z - z_0)^m f(z)$ remains finite as $z \to z_0$ and has a straightforward Taylor series expansion.

If the Laurent series has $b_n = 0$ for all n, in the expansion for $0 < |z - z_0| < R$, then it specifies a function that can be redefined to be analytic even at $z = z_0$, this point being a "singularity" only in a technical sense. For example, $f(z) = \sin z/z$ is undefined, in the first instance, at $z = 0$ where the denominator vanishes. But this is cancelled by the simultaneous vanishing of the numerator, so that its Laurent series is

$$f(z) = \frac{1}{z}\left(z - \frac{z^3}{3!} + \frac{z^5}{5!} \ldots\right) = 1 - \frac{z^2}{3!} + \frac{z^4}{5!} - + \ldots, \quad 0 < |z| \quad . \tag{1.36}$$

Evidently $f(z) \to 1$ as $z \to 0$, so that if $f(z)$ is extended by the additional definition $f(1) = 1$, then the "singularity" at $z = 0$ is removed, with f analytic at this point; the point $z = 0$ is called a *removable singularity*.

If the coefficients b_n of (1.34) have an infinite number of nonzero members in the expansion for $0 < |z - z_0| < R$, then the point z_0 is called an *isolated essential singularity*.

Example. $z^{-1}(z - i)^{-3}$ has a simple pole at $z = 0$ and a pole of order 3 at $z = i$; $\exp(1/z)$ has an isolated essential singularity at $z = 0$; $(z - 1)/(z^2 - 1)$ has a simple pole at $z = -1$ and a removable singularity at $z = +1$.

Any singularity other than a pole (of some order m) is called an essential singularity.

Example. $\operatorname{cosec}(1/z)$ has an essential singularity at $z = 0$, and simple poles at $z = \pm 1/n\pi$, n integral; $\operatorname{Log} z$, the principal value of the logarithmic function, has essential singularities of the origin and every point of the negative real axis.

Zeros

A point z_0 at which $f(z)$ vanishes is called a zero of f. If f is analytic within a neighbourhood of such a point, its Laurent series has the form

$$f = a_m(z - z_0)^m + a_{m+1}(z - z_0)^{m+1} + \dots \text{ for } |z - z_0| < R \quad , \qquad (1.37)$$

where the first nonzero coefficient a_m must have $m \geq 1$ in order that $f(z)$ vanish at z_0. This point z_0 is called a *zero of order m*.

Example. $\sin^2 z$ has a zero of order 2 at $z = \pm n\pi$, where n is an integer or zero.

1.6 INTEGRATION

It is required to extend our ideas of real integrals to the complex domain, and to give a suitable interpretation to the integral

$$I = \int_P f(z)dz \quad , \qquad (1.38)$$

where the function f and the path P are suitably smooth. Suppose firstly that f is continuous on P which is given in parametric form by

$$z = x(t) + iy(t), \ t_0 < t < t_1 \quad , \qquad (1.39)$$

where the derivatives of $x(t)$ and $y(t)$ are continuous for $t_0 < t < t_1$.

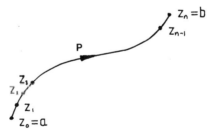

Figure 1.5

By direct analogy with real variable theory, suppose the path P to be subdivided into intervals $(z_0, z_1) \ (z_1, z_2) \dots (z_{n-1}, z_n)$ where $z_0 = a$ is the starting point of P, $z_n = b$ is its finishing point, and each point $z_0, z_1, \dots, z_n$ lies on P as shown in the diagram. In a typical interval (z_{r-1}, z_r) choose any point ζ_r on P, between z_{r-1} and z_r inclusive. We then define

$$\int_P f(z)dz = \lim \sum_{r=1}^{n} f(\zeta_r)\delta z_r \quad , \qquad (1.40)$$

where $\delta z_r = z_r - z_{r-1}$ and the limit is taken as $n \to \infty$ in such a way that the greatest arc length of the intervals (z_{r-1}, z_r) tends to zero. It is not obvious but true (as in the real case) that the limit so defined is independent of the choice of ζ_r in the intervals (z_{r-1}, z_r).

In terms of the parametric representation (1.39), definition (1.40) is found to be equivalent to the formula

$$\int_P f(z)dz = \int_{t_0}^{t_1} f(z(t))(\dot{x} + i\dot{y})dt \qquad (1.41)$$

whose real and imaginary parts are ordinary integrals and where $\dot{x} = dx/dt$ and $\dot{y} = dy/dt$. Similarly, on writing $f = u(x,y) + iv(x,y)$ in terms of its real and imaginary parts, one gets

$$\int_P f(z)dz = \int_P (udx - vdy) + i\int_P (vdx + udy) \qquad (1.42)$$

since $f\delta z = (u+iv)(\delta x + i\delta y) = (u\delta x - v\delta y) + i(v\delta x + u\delta y)$, and formula (1.42) expresses the complex integral in terms of two real line integrals.

An obvious extension of the definition occurs for the case of *piecewise smooth paths* and/or *piecewise continuous functions* f. Thus if f is continuous on each of a finite set of smooth sections $P_1, P_2, \ldots P_n$, we simple define

$$\int_P f(z)dz = \int_{P_1} f(z)dz + \ldots + \int_{P_n} f(z)dz \quad .$$

Example. $I = \int_P z^{-1}dz$, where P is the unit circle taken anticlockwise. Using the parametric representation $z = \exp(it)$, $0 \le t \le 2\pi$, $dz = ie^{it}dt$ and

$$I = \int_0^{2\pi} ie^{it}e^{-it}dt = 2\pi i \quad .$$

Note that the same result is obtained for a circle of any radius, and centre at 0.

Example. $I = \int_P \exp(z)dz$ where P is the straight line from 0 to $1+i$. Using the parametric form $z = (1+i)t$, $0 \le t \le 1$, we have

$$I = \int_0^1 \exp\{(1+i)t\}(1+i)dt = [\exp\{(1+i)t\}]_0^1 = \exp(1+i) - 1 \quad .$$

Example. $I = \int_P zdz$, along each of the three paths from 0 to $(1+2i)$

(i) straight line, $z = (1 + 2i)t, 0 \leq t \leq 1$;

(ii) parabola, $z = t + i2t^2, 0 \leq t \leq 1$;

(iii) pair of straight lines, $z = it, 0 \leq t \leq 2$ then $z = t + 2i, \ 0 \leq t \leq 1$. We have

$$I_1 = \int_0^1 (1 + 2i)^2 t\, dt = \tfrac{1}{2}(1 + 2i)^2 = -\frac{3}{2} + 2i \quad ,$$

$$I_2 = \int_0^1 (t + i2t^2)(1 + 4it)\, dt = \int_0^1 (t - 8t^3 + 6it^2)\, dt = -\frac{3}{2} + 2i \quad ,$$

$$I_3 = \int_0^2 it i\, dt + \int_0^1 (t + 2i)\, dt = -\frac{3}{2} + 2i \quad .$$

Note that the integral has the same value for each path from 0 to $(1 + 2i)$.

The following properties follow from definition (1.40):

$$\text{I.} \qquad \int_P (\alpha f + \beta g)\, dz = \alpha \int_P f\, dz + \beta \int_P g\, dz \qquad (1.43)$$

for complex constants α and β;

$$\text{II.} \qquad \int_{-P} f\, dz = -\int_P f\, dz \quad , \qquad (1.44)$$

where $-P$ denotes the path P taken in the opposite direction;

$$\text{III.} \qquad \left| \int_P f(z)\, dz \right| \leq \int_P |f| |dz| \quad , \qquad (1.45)$$

where $|dz| = ds$ denotes an element of arc length on P. In particular,

$$\text{IV.} \qquad \left| \int_P f(z)\, dz \right| \leq ML \quad , \qquad (1.46)$$

where M is the maximum value of $|f|$ on the path P of length L.

Cauchy's Theorem

The last worked example, on the integral of z along several paths from $z_1 = 0$ to $z_2 = (1 + 2i)$, produced the same answer $\tfrac{1}{2}(1 + 2i)^2$ for each path. This prompts the question whether the same value is obtained for *any* path from z_1 to z_2. This possibility is rendered more plausible by the observation that the integrand z is the derivative of the function $\tfrac{1}{2} z^2$, and one might anticipate by analogy with the theory of real integration that

$$\int_{z_1}^{z_2} z\, dz = \left[\tfrac{1}{2} z^2 \right]_{z_1}^{z_2} = \tfrac{1}{2}(1 + 2i)^2 \quad .$$

In a more general case, one is led to consider the integral $\displaystyle\int_{z_1}^{z_2} f(z)\, dz$ and to ask whether, in some circumstances, the integral might be independent of the path from z_1 to z_2.

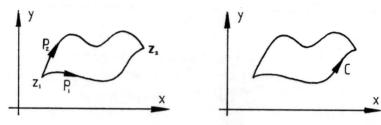

Figure 1.6

The implication of this proposition is that $\int f(z)dz$ is zero for any closed contour C passing through z_1 and z_2. For if the integral along P_1 is the same as the integral along P_2, the integral around the contour $C = P_1 - P_2$ must vanish, since $\int_{-P_2} f(z)dz = -\int_{P_2} f(z)dz = -\int_{P_1} f(z)dz$. The vanishing of an integral evaluated along a closed contour C seems at first to require very restrictive constraints upon the integrand f, but it transpires that the analyticity of $f(z)$ is sufficient to ensure this property. This theorem, which is surely one of the most appealing and powerful in the whole of mathematics, has many important applications.

Cauchy's Theorem is as follows. If $f(z)$ is analytic on and inside a simple closed contour C, then

$$\int_c f(z)dz = 0 \quad . \tag{1.47}$$

The result holds under weaker conditions for f at points *on* C, but this result is sufficient for many purposes.

Proof. An elementary proof is given here for the restrictive special case where the real and imaginary parts $u(x,y)$ and $v(x,y)$ of $f = u + iv$ are assumed to have continuous first derivatives on and inside C. [In fact any analytic function must have this property, but a proof of this usually needs Cauchy's theorem, and should not be assumed in its proof.]

Our starting point is the two–dimensional version of Stokes' theorem, that

$$\int_c (U\,dx + V\,dy) = \int_s \left(\frac{\partial V}{\partial x} - \frac{\partial U}{\partial y} \right) dS \quad ,$$

where S is the region enclosed by C, for any U and V with continuous derivatives on and inside C. Now from formula (1.42), the real part of $\int f(z)dz$ has the form

$$\int_c (u\,dx - v\,dy) = \int_s \left(-\frac{\partial v}{\partial x} - \frac{\partial u}{\partial y} \right) dS \quad ,$$

using Stokes' theorem with $(U, V) = (u, -v)$ and the latter integral vanishes on account of the Cauchy–Riemann equations (1.26). The imaginary part of $\int_c f(z)dz$ is similarly shown to be zero, using Stokes' theorem with $(U, V) = (v, u)$.

Several important corollaries follow immediately from Cauchy's theorem.

I. Independence of Path. If $f(z)$ is analytic inside a closed contour C and if z_1 and z_2 lie inside C, then the integral $\int_{z_1}^{z_2} f(z)dz$ has the same value for any path P, from z_1 to z_2, inside C. The proof is obtained by taking the difference of the integral evaluated along two different paths P_1 and P_2, whence

$$\int_{P_1} f(z)dz - \int_{P_2} f(z)dz = \int_{P_1 - P_2} f(z)dz = 0 \quad,$$

applying Cauchy's theorem to the path $P_1 - P_2$ which is a closed contour.

II. Deformation of Contour. If $f(z)$ is analytic on and between two nonintersecting contours C_1 and C_2, then

$$\int_{c_1} f(z)dz = \int_{c_2} f(z)dz \quad, \tag{1.48}$$

where here and henceforth the integral round a closed simple contour is taken in the positive (anti–clockwise) sense unless stated otherwise.

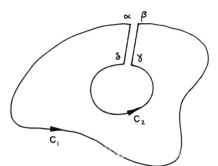

Figure 1.7

The proof follows by consideration of the integral $\int_c f(z)dz$ along the closed contour $C = \alpha\beta\gamma\delta\alpha$ shown in the diagram. Cauchy's theorem ensures that the integral round C is zero, whilst

$$\int_{\alpha\beta} = \int_{c_1} \quad, \quad \int_{\gamma\delta} = -\int_{c_2}$$

and

$$\int_{c_1} f dz - \int_{c_2} f dz + \int_{\beta\gamma} f dz + \int_{\delta\alpha} f dz = 0 \quad.$$

The last two integrals cancel and the desired result follows.

Corollaries I and II show that paths of integration may be deformed in any way whatsoever, provided they do not cross any singular points of $f(z)$. More generally, the same ideas can be extended to show that if C_1 encloses several distinct contours $C_2, C_3, \ldots, C_n$, with f analytic on each contour and in the region inside C_1 but outside $C_2, C_3, \ldots, C_n$, then

$$\int_{c_1} f(z)dz = \int_{c_2} f(z)dz + \int_{c_3} f(z)dz + \ldots + \int_{c_n} f(z)dz \quad . \tag{1.49}$$

III. **Indefinite Integrals.** Corollary I can be used to define functions (indefinite integrals) by integration. For with f analytic on and inside a closed contour C, the integral $\int_{z_1}^{z_2} f(z)dz$ can be thought of as a function of the upper limit of integration, inside C. Changing the notation slightly, we have

$$F(z) = \int_{z_1}^{z} f(\zeta)d\zeta \quad , \tag{1.50}$$

with z_1 and z inside C, and z_1 is considered to be fixed. The analyticity of f ensures that the integral is the same for any path within C from z_1 to z. Furthermore, as with real variables, F is differentiable with derivative $f(z)$. To prove this we have to show that

$$\lim_{h \to 0} \left\{ \frac{F(z+h) - F(z)}{h} - f(z) \right\} \equiv \lim_{h \to 0} R = 0 \quad .$$

Now the function R inside the curly bracket can be expressed, using (1.50) as

$$R = \frac{1}{h} \int_{z}^{z+h} \left\{ f(\zeta) - f(z) \right\} d\zeta \quad ,$$

hence $|R| \le |(1/h)| \times |h| \times \max |f(\zeta) - f(z)|$, where the maximum is taken over those values of ζ on the straight line from z to $z+h$. Thus $R \to 0$ as $h \to 0$ and

$$\frac{dF}{dz} = f(z) \quad . \tag{1.51}$$

Residues

Suppose that a function f has an isolated singularity at z_0, with a Laurent series expansion

$$f(z) = \ldots + \frac{b_2}{(z - z_0)^2} + \frac{b_1}{(z - z_0)} + a_0 + a_1(z - z_0) + a_2(z - z_0)^2 + \ldots,$$

$$\text{for} \quad 0 < |z - z_0| < R \quad . \tag{1.52}$$

The terms involving negative powers of $(z - z_0)$ reflect the singular nature of $f(z)$ at z_0. The coefficient b_1 of the term $(z - z_0)^{-1}$ plays a key role in the sequel and is identified by a special name: b_1 is called the *residue* of the function $f(z)$ at the point z_0. It is clear that the residue at an analytic point is always zero, and that the residue at an isolated singularity may or may not be zero.

The significance of this particular term of the full series (1.52) becomes apparent by considering the integral of each term of the series round a circular path C centred at z_0, with radius δ that is arbitrary except for the stipulation $\delta < R$ to ensure the validity of the series (1.52) on C. Thus C is given by

$$z - z_0 = \delta e^{i\theta}, \quad 0 \leq \theta \leq 2\pi$$

and a typical term $(z - z_0)^n$ of the series (with n positive or negative) can be integrated explicitly to give

$$\int_c (z - z_0)^n dz = \int_0^{2\pi} i\delta^{n+1} e^{i(n+1)\theta} d\theta = (n+1)^{-1} \delta^{n+1} \left[e^{i(n+1)\theta} \right]_0^{2\pi} = 0$$

for $n \neq -1$. If $n = -1$, then

$$\int_c (z - z_o)^{-1} dz = \int_0^{2\pi} i d\theta = 2\pi i \quad .$$

Thus

$$\int_c f(z) dz = 2\pi i b_1 \tag{1.53}$$

assuming for the moment that summation and integration can be interchanged; the result depends *only* on the coefficient b_1 and is independent of the other coefficients $a_0, a_1, a_2, \ldots, b_2, b_3, \ldots$ Confirmation of this result for any function $f(z)$ with an isolated singularity at z_0 will be given later. An important generalization of this remarkable result shows how to perform integrals round any closed contour enclosing isolated singular points of an otherwise analytic function.

Residue Theorem. If $f(z)$ is analytic on and inside a simple closed contour C, except at a finite number of isolated singularities, with residues $r_1, r_2, \ldots, r_n$ inside C, then

$$\int_c f(z)dz = 2\pi i \{r_1 + r_2 + \ldots + r_n\} \quad . \tag{1.54}$$

Proof. The contour of integration may be collapsed, according to (1.49), on to a series of circles $C_1, C_2, \ldots, C_n$, each of which encloses just one of the singularities $z_1, z_2, \ldots, z_n$. Each of these integrals may be evaluated separately, by means of (1.53), to get

$$\int_{C_n} f(z)dz = 2\pi i r_n \quad ,$$

and the formula (1.54) follows.

There are many applications of this theorem; the following typical example shows the astounding power of the residue theorem.

Example. Calculate the integral $I = \int_0^\infty (1 + x^2)^{-1} \cos x\, dx$.

Evidently $I = \frac{1}{2} \lim_{N \to \infty} I(N)$, where $I(N) = \int_{-N}^N (1 + x^2)^{-1} e^{ix} dx$, since the imaginary part of the latter integrand is odd, and integrates to zero. In order to utilize the residue theorem, the integral $I(N)$ is cast into the form of a complex contour integral. Thus

$$I(N) = \int_{-N}^N (1 + z^2)^{-1} e^{iz} dz = \int_{P_1 + P_2} (1 + z^2)^{-1} e^{iz} dz - \int_{P_2} (1 + z^2)^{-1} e^{iz} dz \quad ,$$

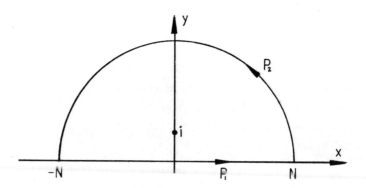

Figure 1.8

where P_1 and P_2 are the paths shown in the diagram.

The first integral, round the closed contour $(P_1 + P_2)$ is evaluated by means of the residue theorem, while the second integral is shown to tend to zero as $N \to \infty$. This latter result is verified by virtue of the estimates $|e^{iz}| = |e^{-y}| \leq 1$, and $|1 + z^2| \geq |z|^2 - 1 = N^2 - 1$, whence the integral round P_2 is absolutely bounded, according to (1.46), by $2\pi N/(N^2 - 1) \to 0$ as $N \to \infty$.

The integral round $(P_1 + P_2)$ is given by the residue theorem as $2\pi i r$, where r is the residue of the function $(1 + z^2)^{-1} e^{iz}$ at the simple pole $z = i$, since this is the only singularity within the contour. In order to find r, expand about the point $z = i$ to get

$$(1 + z^2)^{-1} e^{iz} = (z - i)^{-1} (z + i)^{-1} \left\{ e^{i(z-i)-1} \right\} = (2ie)^{-1}(z - i)^{-1} + \ldots \quad ,$$

whence $r = e^{-1}/2i$. Finally, then,

$$I = \frac{1}{2} \lim_{N \to \infty} 2\pi i e^{-1}/2i = \frac{1}{2}\pi/e \quad .$$

Two major steps involved in the above calculation are those of estimating an integral round an arc of large radius N, and of calculating the residue (or residues) of the integrand. Two brief sections are therefore devoted to techniques that are useful in accomplishing these steps.

Calculation of Residues

If $f(z)$ is known to have a *simple pole* at $z = z_0$, that is to say, it has a Laurent series expansion of the form

$$f(z) = b_1(z - z_0)^{-1} + a_0 + a_1(z - z_0) + a_z(z - z_0)^2 + \ldots \quad \text{for } 0 < |z - z_0| < R \quad ,$$

then the singularity can be removed by multiplying by $(z - z_0)$. Thus

$$(z - z_0)f(z) = b_1 + a_0(z - z_0) + a_1(z - z_0)^2 + \ldots, 0 < |z - z_0| < R \quad ,$$

and the residue at z_0 is given by

$$b_1 = \lim_{z \to z_0} (z - z_0)f(z) \quad . \tag{1.55}$$

A particular, and frequently occurring, case of a simple pole singularity arises when f takes the form

$$f(z) = p(z)/q(z) \quad ,$$

where p and q are both analytic at z_0, and $q(z)$ has a *zero of order one* at z_0. So (unless p has a zero there) f has a simple pole at z_0, with residue given by (1.55) or

$$b_1 = \lim_{z \to z_0} \frac{(z - z_0)}{q(z)} p(z) = \frac{p(z_0)}{q'(z_0)} \quad . \tag{1.56}$$

The result is still correct if p has a zero at z_0, whence $b_1 = 0$. An alternative way of arriving at formula (1.56) is to expand p and q by their Taylor series about z_0 (this is justified later), to get

$$f(z) = \frac{p(z_0) + (z - z_0)p'(z_0) + \dots}{0 + (z - z_0)q'(z_0) + \dots} = \frac{1}{(z - z_0)} \frac{p(z_0)}{q'(z_0)} + \dots \quad ,$$

and the result (1.56) follows by inspection.

A generalization is obtained for a pole of order m at z_0. Thus if

$$f(z) = b_m(z - z_0)^{-m} + \dots + b_1(z - z_0)^{-1} + a_0 + a_1(z - z_0) + \dots, \quad 0 < |z - z_0| < R,$$

then

$$(z - z_0)^m f(z) = b_m + \dots + b_1(z - z_0)^{m-1} + a_0(z - z_0)^m + \dots$$

and the residue b_1 is the coefficient of the $(z - z_0)^{m-1}$ in the Taylor Series for $(z - z_0)^m f(z)$, hence is given by

$$b_1 = \frac{1}{(m - 1)!} \left\{ \frac{d^{m-1}}{dz^{m-1}} [(z - z_0)^m f(z)] \right\}_{z=z_0} \quad . \tag{1.57}$$

Example. Find the residue of $(1 + z^2)^{-1} e^{iz}$ at $z = 1$. Thus formula (1.56) gives

$$b_1 = \left[e^{iz}/2z \right]_{z=i} = e^{-1}/2i \quad ,$$

as was found from first principles in the previous example.

Example. Find the residue of $f(z) = 1/\{(z - 1)(z - 2)(z - 3)\}$ at each of the singular points $z = 1, 2, 3$.

The residue at $z = 1$ is $\lim_{z \to 1} (z-1)f(z) = \frac{1}{2}$; residue at $z = 2$ is $\lim_{z \to 2} (z-2)f(z) = -1$; residue at $z = 3$ is $\lim_{z \to 3} (z - 3)f(z) = \frac{1}{2}$;

Example. Find the residue of $(1 + z^2)^{-2} e^{iz}$ at $z = i$. Since $z = i$ is a pole of order 2, formula (1.57) gives

$$\text{residue} = \frac{1}{1!} \frac{d}{dz} \left\{ \frac{e^{iz}}{(z + i)^2} \right\}_{z=i} = -\tfrac{1}{2} i e^{-1} \quad .$$

Turning now to the task of estimating an integral round an arc of large radius, the following general result is often useful when the integrand contains an exponential function: such a case occurred in our worked example ($f = (1+z^2)^{-1}\exp(iz)$), though the large denominator $(1 + z^2)$ was sufficient in that case to ensure the vanishing of the integral round a large contour, without any need to appeal to properties of the exponential factor. Jordan's Lemma deals with integrands of the form $g(z)e^{imz}$, $m > 0$, evaluated round arcs in the upper half plane, and exploits the fact that $|\exp(imz)| = e^{-my}$ is exponentially small on such a contour, except on a small section near the axis $y = 0$.

Jordan's Lemma. Suppose $g(z)$ is uniformly small as $|z| = N \to \infty$, $\Im z \geq 0$. (That is, given any small positive number ϵ there exists a number $N_0 = N_0(\epsilon)$, independent of θ, such that $|g(Ne^{i\theta})| < \epsilon$ for all $N > N_0$, $0 \leq \theta \leq \pi$.). Then

$$I = \int_c g(z)e^{imz}dz \to 0 \quad \text{as} \quad N \to \infty \quad , \tag{1.58}$$

where m is any positive number and C is the semi–circle of radius N in the upper half plane (or any arc of that semi–circle).

Proof. With the parametric representation $z = Ne^{i\theta}$, $0 \leq \theta \leq \pi$,

$$I = \int_0^\pi g(Ne^{i\theta})\exp\{im(N\cos\theta + iN\sin\theta)\}Nie^{i\theta}d\theta \quad ;$$

hence

$$|I| \leq \int_0^\pi |g(Ne^{i\theta})|e^{-mN\sin\theta}Nd\theta \leq 2N\epsilon\int_0^{\pi/2} e^{-mN\sin\theta}d\theta$$

$$\leq 2N\epsilon\int_0^{\pi/2} e^{-m\frac{N}{\pi}2\theta}d\theta, \quad \text{since} \quad \sin\theta \geq \frac{2\theta}{\pi} \quad \text{for} \quad 0 \leq \theta \leq \frac{2}{\pi}$$

$$= \frac{\epsilon\pi}{m}(1 - e^{-mN}) < \epsilon\pi/m \quad ,$$

and since ϵ is arbitrarily small, we have $I \to 0$ as $N \to \infty$.

A similar result holds for $g(z)e^{-imz}$ integrated round arcs in the lower half plane, $e^{-mz}g(z)$ round arcs in the right half plane, and so on.

Example.

$I(N) = \int (1 + z^2)^{-1}e^{iz}dz$, round the semi–circle of radius N in the upper half plane. With $g = (1 + z^2)^{-1}$, $|g| \leq (N^2 - 1)^{-1} \to 0$ uniformly as $N \to \infty$, hence $I(N) \to 0$ as $N \to \infty$.

Indentation of Contours

In using the residue theorem to calculate integrals, it is sometimes necessary to indent a contour of integration in order to circumnavigate a singular point that would otherwise lie on the path of integration. An evaluation of the integral round a *small arc* close to the singular point can be effected for the case of a *simple pole*, according to the following result.

Small Arc Lemma: If $f(z)$ has a *simple pole* at $z = z_0$ and the path of integration is the arc of a circle centred at z_0, of radius δ and subtending an angle θ_0 at z_0, taken in anticlockwise sense, then

$$I = \int_P f(z)dz \to i\theta_0 r \quad \text{as} \quad \delta \to 0 \quad , \tag{1.59}$$

where r is the residue of $f(z)$ at z_0.

Note that if $\theta_0 = 2\pi$, whence P is a closed contour enclosing z_0, then the integral has the value $2\pi i r$ in accordance with the formula (1.53): this is independent of δ and independent of the order of the pole singularity. The generalization(1.59) requires, however, the limit $\delta \to 0$ and the restrictive condition that the pole be simple.

Proof. Express $f(z)$ in the form

$$f(z) = \frac{r}{z - z_0} + g(z) \quad ,$$

where $g(z)$ is analytic on and inside a circle $|z - z_0| < \delta_0$ for sufficiently small δ_0. Thus $I = I_1 + I_2$ where

$$I_1 = r \int_P (z - z_0)^{-1} dz \quad \text{and} \quad I_2 = \int_P g(z)dz \quad ,$$

and we may take $\delta < \delta_0$. The integral I_1 can be evaluated directly, from the observation that P is given by $z - z_0 = \delta e^{i\theta}$ $(\alpha \le \theta < \alpha + \theta_0)$, whence

$$I_1 = r \int_\alpha^{\alpha+\theta_0} i d\theta = i\theta_0 r \quad .$$

The integral I_2 tends to zero as $\delta \to 0$, since

$$|I_2| = \left| \int_P g dz \right| \le \theta_0 \delta G \to 0 \quad ,$$

where G is the maximum value of $|g(z)|$ inside the circle $|z - z_0| = \delta$.

Example. Evaluate the integral $I = \int_0^\infty x^{-1} \sin x\, dx$.

In order to use the residue theorem, we are led to consider the integral $\int_c z^{-1} e^{iz} dz$ around the contour $C = P_1 + P_2 + P_3 + P_4$ shown;

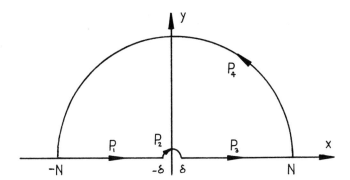

Figure 1.9

the path is indented at 0 in order to avoid the simple pole singularity there. Since there are no singularities inside C, we have

$$\int_c \frac{e^{iz}}{z}\, dz = 0 \quad.$$

The integral round P_4 tends to zero as $N \to \infty$, by Jordan's Lemma, and the integral round P_2 is given by the "small arc lemma" as $-\pi i r = -\pi i$, since $r = 1$ and the minus sign takes account of the clockwise sense of the path P_2. The integrals along P_1 and P_2 are found to be

$$\int_{P_1} \frac{e^{iz}}{z}\, dz = +\int_N^\delta \frac{e^{-ix}}{x}\, dx; \quad \int_{P_3} \frac{e^{iz}}{z}\, dz = \int_\delta^N \frac{e^{ix}}{x}\, dx \quad ;$$

hence $\displaystyle\int_0^\infty \frac{e^{ix} - e^{-ix}}{x}\, dx - \pi i = 0$, and $I = \pi/2$, on taking the limits $\delta \to 0$ and $N \to \infty$.

Example. Calculate $I = \displaystyle\int_0^\infty (1 + x^2)^{-2} \log x\, dz$.

Consider the contour integral $\displaystyle\int_c (1 + z^2)^{-2} \log z\, dz$ around the closed contour $C = P_1 + P_2 + P_3 + P_4$, shown in the previous example; $\log z$ is defined to be such that $\log 1 = 0$, with a branch cut along the negative *imaginary* axis.

On P_3, $z = t$, $\delta \le t \le N$, and $\log z = \log t$, so that

$$\int_{P_3} (1+z^2)^{-2} \log z \, dz = \int_{\delta}^{N} (1+t^2)^{-2} \log t \, dt = I(\delta, N), \quad \text{say} .$$

On P_1, $z = te^{i\pi}$, $N \ge t \ge \delta$, and $\log z = \log t + i\pi$, whence

$$\int_{P_1} (1+z^2)^{-2} \log z \, dz = I(\delta, N) + i\pi \int_{\delta}^{N} (1+t^2)^{-2} dt .$$

The integral along P_4 tends to zero as $N \to \infty$, since

$$\left| \frac{\log z}{(1+z^2)^2} \right| \le \frac{2 \log N}{N^4} \quad \text{for } N > N_0 ,$$

so the integral is absolutely bounded by $4\pi \log N / N^3 \to 0$ as $N \to \infty$. Similarly, the integral along P_2 tends to zero as $\delta \to 0$ since

$$\left| \frac{\log z}{(1+z^2)^2} \right| \le 2 \log \delta \quad \text{for } \delta < \delta_0 ,$$

and the integral is absolutely bounded by $4\pi\delta \log \delta \to 0$ as $\delta \to 0$. Thus letting $\delta \to 0$, $N \to \infty$, and noting that the only interior singularity is a pole of order two at $z = i$, we have $I(\delta, N) \to I$, and

$$2I + i\pi \int_0^\infty (1+t^2)^{-2} dt = 2\pi i \quad (\text{residue at } z = i)$$

$$= 2\pi i \frac{d}{dz} \left\{ \frac{\log z}{(z+i)^2} \right\}_{z=i} = -\frac{\pi}{2} + \frac{i\pi^2}{4} .$$

Thus $\int_0^\infty (1+t^2)^{-2} \log t \, dt = -\pi/4$, and also $\int_0^\infty (1+t^2)^{-2} dt = +\pi/4$.

Yet another important consequence of Cauchy's theorem is the following formula, which expresses an analytic function $f(z)$ in terms of its values round an enclosing boundary C, and which plays an important role in establishing some of the basic results of analytic functions.

1.7 THE CAUCHY INTEGRAL

If $f(z)$ is analytic on and within a simple closed contour C, then

$$f(z) = \frac{1}{2\pi i} \int_c \frac{f(\zeta)d\zeta}{\zeta - z} \qquad (1.60)$$

for any point z inside C; the integral is zero if z lies outside C.

The result would follow immediately if we allowed ourselves the luxury of the residue theorem, since the integrand is an analytic function of ζ except for a simple pole, with residue $f(z)$ at $\zeta = z$. But the residue theorem has made use of the Laurent series expansion, which will be established in the present development by means of the Cauchy formula (1.60). Thus a proof of (1.60) is needed from first principles, in order to avoid a circular argument.

Proof. The integral (1.60) can be written as

$$\frac{f(z)}{2\pi i} \int_c \frac{d\zeta}{\zeta - z} + \frac{1}{2\pi i} \int_c g(\zeta)d\zeta \qquad (1.61)$$

where

$$g(\zeta) = \{f(\zeta) - f(z)\}/(\zeta - z)$$

is analytic within C except at $\zeta = z$ where g is for the moment undefined. But f is analytic, therefore differentiable at $\zeta = z$ and the limit of $g(\zeta)$ equals $f'(z)$ as $\zeta \to z$. It follows that the function $g(\zeta)$ is analytic inside C for $\zeta \neq z$, and is bounded as $\zeta \to z$. Now Cauchy's theorem (corollary II) states that the integral (1.60) can be evaluated round a circle C_1 of small radius δ, centred at $\zeta = z$. Thus the first integral of (1.61) has the value $2\pi i$, as is seen by taking $\zeta - z = \delta e^{i\theta}$, $0 \leq \theta \leq 2\pi$. The second integral is absolutely bounded by $2\pi\delta$ times max $|g(\zeta)|$ and therefore tends to zero as $\delta \to 0$. The result (1.60) then follows by letting $\delta \to 0$. With z outside C_1, the Cauchy integral is immediately seen to be zero, using Cauchy's theorem.

Laurent Series

We are now in a position to prove the Laurent theorem referred to in formula (1.34). It is sufficient to prove the result for $z_0 = 0$: that is, if $f(z)$ is analytic within the annulus $R_1 < |z| < R_2$, then it can be expressed as

$$f(z) = \ldots + \frac{b_2}{z^2} + \frac{b_1}{z} + a_0 + a_1 z + a_2 z^2 + \ldots, R_1 < |z| < R_2 \quad . \qquad (1.62)$$

Note that the more general expansion (1.34) can be reduced to (1.62) by means of the substitution $z - z_0 = z'$.

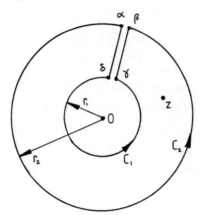

Figure 1.10

Let z be any point inside the annulus. As our starting point we use the Cauchy Integral (1.60) where C is the closed contour $\alpha\beta\gamma\delta\alpha$ shown in the diagram, with $R_1 < r_1 < |z| < r_2 < R_2$. Note that the contributions cancel from the two straight sections $\beta\gamma$ and $\delta\alpha$, and that $\alpha\beta = C_2$, $\gamma\delta = -C_1$. Thus

$$f(z) = \frac{1}{2\pi i}\int_{c_2}\frac{f(\zeta)d\zeta}{\zeta - z} - \frac{1}{2\pi i}\int_{c_1}\frac{f(\zeta)d\zeta}{\zeta - z}, \quad r_1 < |z| < r_2 \quad . \tag{1.63}$$

The radii r_1 and r_2 can be arbitrarily close to R_1 and R_2. The proof is almost complete since the first integral is, by inspection, analytic provided z does not cross the path C_2. We can certainly expand the function $1/(\zeta - z)$ (with ζ on C_2) in powers of z/ζ to get

$$\frac{1}{2\pi i}\int_{c_2}\frac{f(\zeta)}{\zeta}\sum_0^\infty\left(\frac{z}{\zeta}\right)^n d\zeta = \frac{1}{2\pi i}\sum_0^\infty z^n\int_{c_2}\frac{f(\zeta)}{\zeta^{n+1}}d\zeta, \quad |z| < r_2 \quad .$$

More precisely, we should expand $(\zeta - z)^{-1}$ up to $(z/\zeta)^N$, with an error term which is then shown to give negligible contribution to the integral as $N \to \infty$ with $|z| < |\zeta| = r_2$. Similarly, the second integral of (1.63) may be expanded in inverse powers of z to get

$$\frac{-1}{2\pi i}\int_{c_1}\frac{f(\zeta)}{-z}\sum_0^\infty\left(\frac{\zeta}{z}\right)^n d\zeta = \frac{1}{2\pi i}\sum_0^\infty z^{-n-1}\int_{c_1}\zeta^n f(\zeta)d\zeta, \quad |z| > r_1 \quad .$$

Thus the series (1.62) is established for $r_1 < |z| < r_2$, with coefficients given by

$$a_n = \frac{1}{2\pi i}\int_{c_2}\frac{f(\zeta)}{\zeta^{n+1}}d\zeta, \quad b_n = \frac{1}{2\pi i}\int_{c_1}\zeta^{n-1}f(\zeta)d\zeta \quad . \tag{1.64}$$

Since the series (1.62) converges for $R_1 < r_1 < |z| < r_2 < R_2$, it converges for $R_1 < |z| < R_2$. One important special case of (1.64) is

$$b_1 = \frac{1}{2\pi i} \int_{c_1} f(\zeta)d\zeta \quad,$$

which confirms (and generalizes) result (1.53).

1.8 ANALYTICITY

The Cauchy integral (1.60) can be used to establish the following result. If f is analytic within a simple closed contour C_0, it can be differentiated any number of times. That is, the nth derivative $f^{(n)}(z)$ is also analytic, for any n. [It is *not* true in real variable theory that if a function is differentiable once then it is differentiable any number of times.] As a starting point, it will be shown that the derivative $f'(z)$ of an analytic function is obtained from (1.60) by formally differentiating under the integral sign. That is

$$f'(z) = \frac{1}{2\pi i} \int_c \frac{f(\zeta)d\zeta}{(\zeta - z)^2}, \quad z \text{ inside } C \quad, \tag{1.65}$$

if f is analytic on and inside a simple closed contour C. To prove (1.65), use the definition $f' = \lim_{h \to 0} \{f(z + h) - f(z)\}/h$; so the difference between the left–hand side and the right–hand side of (1.65) can be expressed as

$$P = \lim_{h \to 0} \frac{h}{2\pi i} \int_c \frac{f(\zeta)d\zeta}{(\zeta - z)^2(\zeta - z + h)} \quad,$$

after using the Cauchy integral (1.60) Thus

$$|P| \leq \frac{h}{2\pi} \frac{M}{d^2(d - |h|)} \to 0 \quad \text{as} \quad h \to 0 \quad,$$

where $M = \max|f(\zeta)|$, ζ on C and $d = \min|\zeta - z|$, ζ on C. Similarly, the nth derivative of $f(z)$ exists and is given by

$$f^{(n)} = \frac{n!}{2\pi i} \int_c \frac{f(\zeta)d\zeta}{(\zeta - z)^{n+1}} \quad. \tag{1.66}$$

Analytic Continuation

This idea is perhaps best appreciated with reference to a specific example. It is well known, and easy to prove, that the function f_0 defined by the geometric series

$$f_0(z) = 1 + z + z^2 + z^3 + \dots , \qquad (1.67)$$

converges (and is analytic) for $|z| < 1$ but diverges for $|z| > 1$. For $|z| < 1$ it actually converges to the value $(1 - z)^{-1}$.

Similarly, the function f_1, defined by the series

$$f_1(z) = \frac{1}{1 - i} + \frac{z - i}{(1 - i)^2} + \frac{(z - i)^2}{(1 - i)^3} + \dots , \qquad (1.68)$$

converges to the value $(1 - z)^{-1}$ within the circle $|z - i| < \sqrt{2}$.

Evidently f_0 is analytic inside the unit circle R_0, whilst f_1 is analytic in a different region R_1 (the circle $|z - i| < \sqrt{2}$) which overlaps with R_0. We can now define a new function $f(z)$ as

$$f(z) = \begin{cases} f_0(z) & \text{if } z \text{ is in } R_0 , \\[2mm] f_1(z) & \text{if } z \text{ is in } R_1 . \end{cases} \qquad (1.69)$$

Note that the definition is consistent inside $R_0 \cap R_1$, the region common to both R_0 and R_1, since $f_0 = f_1$ there. Thus f can be thought of as extending the original function f_0 into a larger region $R_0 \cup R_1$. The function f_1 is called the analytic continuation of f_0 into $R_0 \cup R_1$. More generally, we have the following definition:

Definition. If f_0 is analytic in a region D_0 bounded by a simple closed contour C_0 and if f_1 is analytic in D_1 where D_0 and D_1 have a common region in which $f_0 = f_1$, then f_1 is called the analytic continuation of f_0 into $D_0 \cup D_1$.

Example. The function f_0 given by (1.67) inside the unit circle is continued by f_1 (1.68) into the region of two overlapping circles. It can be continued by a similar process, using overlapping circles to cover the whole complex plane, apart from the singular point $z = 1$. Thus f_0, together with all its analytic continuations, leads to the function

$$F(z) = (1 - z)^{-1}, \quad z \neq 1 \quad .$$

Example. It is possible for a function to have singularities that are so closely packed as to form a "natural boundary" across which the function cannot be continued.

Liouville's Theorem

If $f(z)$ is analytic in the whole complex plane, and if $f(z)$ is absolutely bounded for all z, i.e. $|f(z)| < A$, then $f(z)$ is a constant. In particular, if $f \to 0$ as $|z| \to \infty$, then $f \equiv 0$.

Proof. Since $f(z)$ is analytic within a circle C of *any* radius r, its Laurent series is a Taylor series:

$$f(z) = a_0 + a_1 z + a_2 z^2 + \dots$$

where

$$a_n = \frac{1}{2\pi i} \int_c \frac{f(\zeta)}{\zeta^{n+1}} d\zeta \tag{1.70}$$

evaluated along any circle C of radius r (formula (1.64)). Thus

$$|a_n| \leq \frac{1}{2\pi} \frac{A}{r^{n+1}} 2\pi r = A r^{-n}, \quad n = 0, 1, 2 \dots \quad .$$

If $n \geq 1$, letting r become increasingly large shows that $a_n = 0$. It follows that $f \equiv a_0 =$ constant. Further, $a_0 = 0$ if $f \to 0$ as $|z| \to \infty$.

A generalization of this result is as follows. If $f(z)$ is analytic in the whole plane and is bounded by a constant times $|z|^N$, i.e. $|f| < A|z|^N$ for all z, then $f(z)$ *is a polynomial* of degree $n \leq N$. For Equation (1.70) with $n > N$ shows that

$$|a_n| \leq \frac{1}{2\pi} \frac{A r^N}{r^{n+1}} 2\pi r = A r^{N-n}$$

for arbitrary radius r. Letting $r \to \infty$ ensures that $a_n = 0$ for $n > N$.

Fundamental Theorem of Algebra

If $f(z)$ is a polynomial of degree $n > 1$, with complex coefficients, then the equation $f(z) = 0$ has at least one solution.

Proof. This is a simple application of Liouville's theorem. For $f(z)$ is analytic everywhere. If there is no solution of the equation $f(z) = 0$, then $1/f(z)$ is also analytic everywhere and bounded. Thus $1/f(z) = A(\neq 0)$ and $f(z) = 1/A$ is of degree one, which contradicts the hypothesis that $f(z)$ has degree greater than one. It follows that the supposition of no solution of $f(z) = 0$ is incorrect.

Plemelj Formulae

Suppose the function $f(z)$ is analytic at all points on a smooth path P. Define the associated function $F(z)$ by the Cauchy integral

$$F(z) = \frac{1}{2\pi i} \int_P \frac{f(\zeta)d\zeta}{\zeta - z} \quad . \tag{1.71}$$

If P happens to be a closed path C, then (1.71) is the same as the Cauchy formula (1.60). The existence and analyticity does not actually need such a severe condition as that of f analytic, and this will be relaxed later. A good account of the use of the Plemelj formulae is given by Carrier, Krook & Pearson (1966).

The function $F(z)$ defined by (1.71) is analytic at all points z not on P — essentially because the integrand is an analytic function of $z(\zeta \neq z)$. Suppose z_0 is a point on P, other than an end point. Let us examine the behaviour of F as z approaches z_0 from either side. Looking along P in the direction of integration, the limiting value as $z \to z_0$ from the left is denoted by $F_+(z_0)$, while $F_-(z_0)$ is the limiting value as $z \to z_0$ from the right. To determine $F_+(z_0)$, deform the path P by indenting it with a small semi–circle P_1 of radius ϵ and centre at z_0, as shown in the diagram, where ϵ is small enough to ensure f is analytic on P_1. We may now let $z \to z_0$ from the left, without crossing the deformed path $P_0 + P_1$, and let $\epsilon \to 0$ to get

$$F_+(z_0) = \lim_{\epsilon \to 0} \frac{1}{2\pi i} \int_{P_1} \frac{f(\zeta)}{\zeta - z_0}d\zeta + \lim_{\epsilon \to 0} \frac{1}{2\pi i} \int_{P_0} \frac{f(\zeta)d\zeta}{\zeta - z_0} \quad .$$

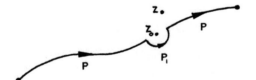

Figure 1.11

As the integrand has residue $f(z_0)$ at $\zeta = z_0$, the first integral has value $\frac{1}{2}f(z_0)$ in the limit $\epsilon \to 0$, by the small arc Lemma. Thus

$$F_+(z_0) = \tfrac{1}{2}f(z_0) + F_P(z_0) \quad , \tag{1.72}$$

where the "Cauchy Principal Value" Integral F_P is given by

$$F_P \equiv \frac{1}{2\pi i} \oint_P \frac{f(\zeta)d\zeta}{\zeta - z_0} = \lim_{\epsilon \to 0} \frac{1}{2\pi i} \int_{P_0(\epsilon)} \frac{f(\zeta)d\zeta}{\zeta - z_0} \quad . \tag{1.73}$$

Similarly, by indenting the contour with a semi–circle on the left side of P, we find that

$$F_-(z_0) = -\tfrac{1}{2}f(z_0) + F_P(z_0) \quad . \tag{1.74}$$

The first terms of (1.72) and (1.74) are different if P has corners.

Our derivation has assumed the analyticity of f at z_0, but the results can be shown to hold under weaker conditions. It is sufficient that f be continuous on P and satisfies a "Lipschitz condition"

$$|f(z) - f(z_0)| < A|z - z_0|^\alpha \tag{1.75}$$

for some constant A and $0 < \alpha \le 1$, at all points z on P in a neighbourhood of z_0.

Addition and subtraction of (1.72) and (1.74) give

$$f(z_0) = F_+(z_0) - F_-(z_0) \tag{1.76}$$

and

$$F_P(z_0) = \tfrac{1}{2}\{F_+(z_0) + F_-(z_0)\} \quad . \tag{1.77}$$

Equations (1.76), (1.77), or equivalently (1.72) and (1.74), are known as the Plemelj formulae.

Conversely, they may be used to solve the following problem. Given a continuous function $f(\zeta)$ prescribed along a smooth curve P, find a function $G(z)$ which is analytic at all points not on P and with the given jump discontinuity

$$G_+(z) - G_-(z) = f(z) \quad \text{on } P \quad . \tag{1.78}$$

By inspection, formulae (1.71) and (1.76) show that

$$G_0 = \frac{1}{2\pi i} \int_P \frac{f(\zeta)d\zeta}{\zeta - z} \tag{1.79}$$

is a possible solution. But we could add any analytic function to it unless there are subsidiary requirements. And we need end conditions on G to rule out the possibility of pole singularities there. Suppose then that we impose the subsidiary conditions that

(i) $G(z) = O(z^n)$ as $z \to \infty$, to limit the growth at infinity, and

(ii) $|G| < K|z - z_e|^\lambda$ for some A and $\lambda(\lambda > -1)$ at each end point z_e, to rule out pole singularities there.

It can then be shown that the general solution for G is

$$G(z) = G_0(z) + P_n(z) \qquad (1.80)$$

where G_0 is given by (1.79) and P_n is an arbitrary polynomial of degree n. A brief outline of the proof runs as follows. Consider $F(z) = G(z) - G_0(z)$, which satisfies the conditions (i) and (ii), with $F_+ - F_- = 0$ on P. It can then be shown that $F(z)$ is analytic even on P (since it now has no discontinuity there), and at the end points. The extension of Liouville's theorem then requires F to be a polynomial of degree n as asserted.

Example. Solve the "aerofoil" equation

$$\frac{1}{\pi} \fint_{-1}^{1} \frac{f(t)dt}{t-s} = g(s), \quad -1 < s < 1 \quad , \qquad (1.81)$$

where g is a given continuous function of s, $|s| < 1$. Define

$$F(z) = \frac{1}{2\pi i} \int_{-1}^{1} \frac{f(t)dt}{t-z} \quad .$$

Evidently $F(z) = 0(1/z)$ as $|z| \to \infty$, and the Plemelj formulae give, for $|x| < 1$

$$F_+(x) = \tfrac{1}{2}f(x) + F_P(x), \quad F_-(x) = -\tfrac{1}{2}f(x) + F_P(x) \quad .$$

Thus the integral equation (1.81) can be expressed as

$$F_+(x) + F_-(x) = -ig(x) \quad . \qquad (1.82)$$

If the $+$ sign on the left–hand side were a minus, we could solve as above. So we work instead with a related function $W(z)$ defined as

$$W(z) = (z^2 - 1)^{1/2}F(z)$$

where $(z^2-1)^{1/2}$ has a branch cut from -1 to $+1$ along the real axis, with $(z^2-1)^{1/2}$ real and positive when $z = x > 1$. In particular, $(z^2 - 1)^{1/2} \to \pm i(1 - x^2)^{1/2}$ as $z \to x \pm i0$, $-1 < x < 1$. Also $W = 0(1)$ as $|z| \to \infty$. Thus

$$W_+(x) = +i(1 - x^2)^{1/2}F_+(x)$$
$$W_-(x) = -i(1 - x^2)^{1/2}F_-(x) ,$$

and this change of sign is important since (1.82) becomes

$$W_+(x) - W_-(x) = (1 - x^2)^{1/2} g(x) \quad .$$

The problem for $W(z)$ is now of the type described above, with solution

$$W(z) = \frac{1}{2\pi i} \int_{-1}^{1} \frac{(1 - x^2)^{1/2} g(x) dx}{x - z} + A_0 \quad ,$$

the polynomial P_n being a constant in this case to ensure $W = 0(1)$ at ∞. Finally

$$f(x) = F_+ - F_-$$

$$= -i(1 - x^2)^{-1/2}(W_+ + W_-) = -2i(1 - x^2)^{-1/2} W_p(x) \quad ,$$

i.e. $\qquad f(x) = -\frac{1}{\pi}(1-x^2)^{-1/2} \fint_{-1}^{1} \frac{(1 - t^2)^{1/2} g(t) dt}{t - x} + A(1-x^2)^{-1/2} \quad , \quad (1.83)$

where A is an arbitrary constant.

REFERENCE

Carrier, G.R., Krook, M. & Pearson, C.E. (1966). Functions of a Complex Variable: Theory and Techniques. McGraw-Hill.

2. GENERALIZED FUNCTIONS

2.1 INTRODUCTION

It is an object of this Chapter to remove some of the mystique associated with generalized functions and to advocate their use in the analysis of physical problems. We will define their properties without proof and direct those interested in pursuing the fundamentals of generalized functions to Lighthill (1958), Gelfand & Shilov (1964) and Jones (1966) who give comprehensive accounts of the general theory with proofs of properties. We lean heavily on these works in recognizing that generalized functions can essentially be manipulated according to the usual rules of addition, differentiation and integration, though in general they may not be multiplied by other than ordinary functions. They provide a powerful extension of the normal mathematical equipment available to physicists.

As a general principle, functions that exist *only* in a generalized sense never feature in any definite description of a physical property but only at some intermediate point in an analysis. Generalized functions have integral properties that relate them uniquely to ordinary functions. The simplest of all generalized functions is the unit function, 1 say, which has the integral property that defines it,

$$\int 1G(g)dg = \int G(g)dg \ . \tag{2.1}$$

The "test function" $G(g)$ is arbitrary except for conditions requiring that G be sufficiently smooth for all g, and suitable small as $g \to \pm\infty$. To be definite, G is taken to be a "good function" in the descriptive notation due to Lighthill, and defined as follows.

A good function $G(g)$ is one that is everywhere differentiable any number of times, and is such that it and all its derivatives are $o(|g|^{-N})$ as $|g| \to \infty$, for *all* N. Thus a good function vanishes for large $|g|$ faster than any inverse power of $|g|$. A function having less stringent behaviour at infinity, namely one that does not grow too quickly for large $|g|$, is called "fairly good".

A fairly good function $\phi(g)$ is differentiable any number of times and is such that there exists a fixed number N with the property that ϕ and its derivatives are $0(|g|^N)$ as $|g| \to \infty$. For example, e^{-g^2} is a good function, e^{-g^4} is a good function; $e^{-|g|}$ is not a good function, since it is not differentiable at $g = 0$. Any polynomial is a fairly good function; e^g is not a fairly good function.

It is clear that the unit generalized function, defined by Equation (2.1), is sufficiently well described by the ordinary constant unity that we can unambiguously term it 1, though the only property required of it is that the integral definition (2.1) be satisfied.

The Heaviside unit function $H(g)$ is a similarly simple function in that it is again easily associated over most values of g with a simple ordinary function. Again this is defined by an integral property,

$$\int_{-\infty}^{+\infty} H(g - a)G(g)dg = \int_{a}^{\infty} G(g)dg . \qquad (2.2)$$

Once more, the "test" function $G(g)$ is any good function of g. Clearly we can associate the Heaviside function with ordinary constants over most of the range, i.e.

$$H(g - a) = 1(g > a)$$
$$= 0(g < a) .$$

To enquire into the value of $H(g - a)$ at $g = a$ is irrelevant. The generalized function is not defined at any particular point. The only property that we require of H is that the integral definition (2.2) be satisfied; for this the actual value at $g = 1$ (or for that matter any other particular point) is irrelevant.

Dirac's delta function $\delta(g)$ is the generalized function that we shall use most frequently and this is defined by the integral property,

$$\int_{-\infty}^{+\infty} \delta(g - a)G(g)dg = G(a) . \qquad (2.3)$$

Here again a is a parameter and $G(g)$ is any good ordinary function of g.

No ordinary function has either of the properties of definitions (2.2) and (2.3) so that H and δ exist only in a generalized sense. They can alternatively be defined by a regular sequence of good functions. For example, $e^{-ng^2}(n/\pi)^{1/2}$ is one of several equivalent sequences that can define the generalized function $\delta(g)$. It is easily shown, by, for example, a Taylor expansion of G about $g = 0$ and term–by–term integration, that

$$\lim_{n \to \infty} \int_{-\infty}^{+\infty} e^{-ng^2} (n/\pi)^{1/2} G(g) dg = g(0) . \tag{2.4}$$

This idea of generating a sequence of good functions can be extended to form a basic definition for generalized functions, by expressing integrals, such as (2.1) and (2.2), that involve generalized functions, in terms of ordinary integrals. A generalized function $f(g)$ is defined by a generating sequence $f_n(g)$ of good functions according to the definition,

$$\lim_{n \to \infty} \int_{-\infty}^{\infty} f_n(g) G(g) dg = \int_{-\infty}^{\infty} f(g) G(g) dg , \tag{2.5}$$

provided the limit exists, where G is any good function.

Two generalized functions f_1 and f_2 are said to be equal if and only if

$$\int_{-\infty}^{\infty} f_1(g) G(g) dg = \int_{-\infty}^{\infty} f_2(g) G(g) dg$$

for any good function G.

The definition (2.5) emphasizes a point already made that a generalized function $f(g)$ need not assume a particular value for each point g: the function is defined by the definite integral of f times G, over the whole range of values of g.

It may be the case that particular values can be assigned to a generalized function $f(g)$ for certain ranges of values of g. Indeed, a most important property that makes generalized function analyses meaningful to physical problems where properties are functions in the ordinary sense is that a very wide class of ordinary functions can be treated as generalized functions and vice–versa. Any ordinary function $f(g)$, such that $(1 + g^2)^N f(g)$ is absolutely integrable from $-\infty$ to $+\infty$, for some N, can be regarded as a generalized function. Also any generalized function expressible as an ordinary function in an interval is equal to the ordinary function in that interval. For example, the unit generalized function in (2.1) is equal to the ordinary unit integer in any interval from $-\infty$ to $+\infty$, while $H(g)$ is equal to unity in an interval from 0 to ∞ and equal to the ordinary zero in any interval in $-\infty$ to 0.

The definitions of generalized functions concern only the restricted class of good functions. However, their useful analytical properties concern their action on the very wide class of fairly good functions that are continuous with all derivatives non–singular, and diverging no worse than algebraically at infinity. Further, many of their properties are formally the same as those of ordinary functions. From now on we will deal solely with their use and check back occasionally with the definition (2.5) for proof of consistency.

The first important property of the delta function is that

$$\delta(g - a)\phi(g) = \delta(g - a)\phi(a) , \qquad (2.6)$$

where the function $\phi(g)$ is any fairly good function of g. The proof of this property follows immediately from the definition recognizing that the product ϕG is itself a good function of g. Thus

$$\int_{-\infty}^{+\infty} \delta(g - a)\phi(g)G(g)dg = \phi(a)G(a) = \int_{-\infty}^{+\infty} \delta(g - a)\phi(a)G(g)dg . \qquad (2.7)$$

The next important property is that $\delta(g - a)$ is the derivative of $H(g - a)$ in a formal sense. Again the proof of this follows directly from the definitions (2.2) and (2.3)

$$\int_{-\infty}^{+\infty} \frac{\partial H(g - a)}{\partial g}G(g)dg = -\int_{-\infty}^{+\infty} H(g - a)\frac{\partial G(g)}{\partial g}dg \qquad (2.8)$$

because G, being good, is vanishingly small at $\pm\infty$. The integral is then equal, by definition (2.2), to

$$-\int_{a}^{\infty} \frac{\partial G(g)}{\partial g}dg = G(a) = \int_{-\infty}^{+\infty} \delta(g - a)G(g)dg ; \qquad (2.9)$$

evidently

$$\delta(g - a) = \frac{\partial}{\partial g}H(g - a) = H'(g - a) , \qquad (2.10)$$

where the prime denotes differentiation with respect to the argument. As an obvious extension of this property, if we use Equation (2.6),

$$\frac{\partial}{\partial g}\{H(g - a)\phi(a)\} = \delta(g - a)\phi(a) = \delta(g - a)\phi(g) , \qquad (2.11)$$

we can define the indefinite integral to be the reverse of differentiation, so that

$$\int_{-\infty}^{g} \delta(g' - a)\phi(g')dg' = H(g - a)\phi(a) .$$

It follows immediately that

$$\int_{<a}^{>a} \delta(g-a)\phi(g)dg = \phi(a) , \qquad (2.12)$$

where the integral ranges from any number less than a to any number greater than a.

For example, these properties allow the direct integration of the equation

$$\frac{\partial^2 G}{\partial x^2}(x, x_0) = \delta(x - x_0) , \qquad (2.13)$$

to give $G(x, x_0) = (x - x_0)H(x - x_0) + AX + B$, which under the constraints that, $G = 0$ at $x = 0$ and $x = 1$, for

$$0 \le x \le 1; \ \ 0 < x_0 < 1 ,$$

becomes

$$G(x, x_0) = x(H(x - x_0) + x_0 - 1) - x_0 H(x - x_0) .$$

The delta function is an even function of its argument as is readily proved from the definition,

$$\int_{-\infty}^{+\infty} \delta(-g+a)G(g)dg = \int_{-\infty}^{+\infty} \delta(g_1-(-a))G(-g_1)dg_1 = G(a) = \int_{-\infty}^{+\infty} \delta(g-a)G(g)dg, \qquad (2.14)$$

i.e.

$$\delta(g - a) = \delta(a - g) , \qquad (2.15)$$

so that the half–range integral

$$\int_{<a}^{>a} \delta(g-a)\phi(g)dg = \frac{1}{2}\int_{<a}^{a} \delta(g-a)\phi(g)dg = \frac{1}{2}\phi(a) .$$

A variable change from x to $g(x)$ permits straightforward evaluation of the integral

$$\int_{<x_0}^{>x_0} \phi(x)\delta\{g(x)\}dx = \int \phi\delta(g)\frac{dg}{\frac{dg}{dx}} = \frac{\phi(x_0)}{\left|\frac{dg}{dx}(x_0)\right|}$$

where

$$g(x_0) = 0 ,$$

and x_0 is the only zero of g in the range of integration.

This allows the two generalized functions $\delta\{g(x)\}$ and $\dfrac{\delta(x - x_0)}{|g'(x_0)|}$ to be equated, $g(x_0)$ again being zero.

Other generalized functions that crop up frequently in what follows are $|g|$, $\mathrm{sgn}(g)$ and $\log|g|$. These generalized functions can be differentiated and manipulated according to the rules for ordinary functions with the proviso that negative integral powers of $|g|$ are admitted to contain a certain indeterminancy. These functions have the properties

$$
\begin{aligned}
|g| = {} & g, & g > 0 & \\
= {} & -g, & g < 0 & \qquad (2.16) \\
\mathrm{sgn}(g) = {} & 1, & g > 0 & \\
= {} & -1, & g < 0\,. & \qquad (2.17)
\end{aligned}
$$

Hence

$$
\mathrm{sgn}(g) = 2H(g) - 1 \qquad (2.18)
$$

$$
\frac{\partial}{\partial g}\mathrm{sgn}(g) = 2\delta(g) \qquad (2.19)
$$

$$
\frac{\partial|g|}{\partial g} = \mathrm{sgn}(g)\,. \qquad (2.20)
$$

There is a difficulty involved in assigning a suitable definition to the generalized functions g^{-1} and $|g|^{-1}$, the origin of the difficulty being associated with the ordinary algebra of these functions. The generalized function $f(g)$ satisfying the equation

$$
gf(g) = 1 \qquad (2.21)
$$

is reasonably defined as g^{-1}. However, a direct consequence of Equation (2.6),

$$
\delta(g - a)\phi(g) = \delta(g - a)\phi(a)
$$

is that

$$
g\,\delta(g) = 0\,, \qquad (2.22)
$$

from which it follows that $f = C\delta(g)$ is a solution of

$$
gf(g) = 0, \qquad (2.23)
$$

and that,

$$
f(g) = g^{-1} + C\delta(g) \qquad (2.24)
$$

is a solution of Equation (2.21) for an arbitrary constant C. Evidently g^{-1} is undetermined to within an arbitrary multiple of a delta function. This difficulty is overcome by formally defining g^{-1} to be the odd generalized function satisfying Equation (2.21).

In a similar way, $|g|^{-1}$ will be defined as an even generalized function satisfying the equation

$$gf(g) = \text{sgn}(g) . \tag{2.25}$$

Evidently, since this equation remains satisfied if any multiple of a delta function is added to $f(g)$, the generalized function $|g|^{-1}$ is undetermined to within an arbitrary constant times the delta function. This indeterminancy is important and is admitted in the definition; it persists when $|g|^{-1}$ is differentiated any number of times. With these definitions the functions g and $|g|$ can be given arbitrary inverse powers and they then behave like the ordinary functions as far as multiplication and differentiation is concerned.

The logarithm is also defined in a way that allows it to be manipulated via the ordinary rules of differentiation, so that for example

$$|g|^\alpha \log|g| = \frac{\partial}{\partial\alpha}|g|^\alpha . \tag{2.26}$$

Again a negative integral power of $|g|$ implies an indeterminancy, the function $|g|^{-1}\log|g|$ being defined as the solution of the equation

$$gf(g) = \log|g|\text{sgn}(g) , \tag{2.27}$$

so that it is clearly undetermined to within a constant multiple of the delta function.

Derivatives of the delta function are also defined but only inasmuch as they can be integrated to give the delta function defined by (2.3). The nth derivative of δ is denoted by $\delta^n(g)$, and the following important result follows immediately on integration by parts

$$\int_{-\infty}^{+\infty} \phi(g)\delta^n(g)dg = (-1)^n \int_{-\infty}^{+\infty} \phi^n(g)\delta(g)dg$$
$$= (-1)^n \int_{-\epsilon}^{+\epsilon} \phi^n(g)\delta(g)dg = (-1)^n\phi^n(0) = 2(-1)^n \int_0^\epsilon \phi^n(g)\delta(g)dg , \tag{2.28}$$

for any fairly good function ϕ.

These definitions and properties need no modification when g is made a scalar function of n independent variables, $x_1, x_2-, -, x_n$, or of the n–dimensional vector

x. The only proviso that must be made is that the function g, and its derivatives by continuous and that g must have a finite non–zero gradient everywhere where $g = 0$.

If we refer to the n–dimensional space as a volume, then $g(\mathbf{x}) = 0$ defines an $(n - 1)$ dimensional space, which may or may not be simply connected. If $n = 3$, $g(\mathbf{x}) = 0$ defines a surface (or surfaces); if $n = 2$, a line; and if $n = 1$, a point. The gradient of a function $g(\mathbf{x})$ is defined to be the vector $(\partial g/\partial n)\mathbf{n}$, where $\mathbf{n} = \mathbf{n}(\mathbf{x})$ is the unit normal to the "level surfaces" $g = $ constant, and directed in the sense of increasing g. In terms of a Cartesian coordinate system $\mathbf{x} = (x_1, x_2, x_3, \ldots, x_n)$, grad g is given by

$$\text{grad } g(\mathbf{x}) = \left(\frac{\partial g}{\partial x_1}, \frac{\partial g}{\partial x_2}, \frac{\partial g}{\partial x_3}, \cdots \frac{\partial g}{\partial x_n} \right) . \tag{2.29}$$

The direction cosines of the surface normal are

$$n_i = (\partial g/\partial x_i)/|\text{grad } g|, \qquad i = 1, 2, 3, \ldots, n , \tag{2.30}$$

where

$$|\text{grad } g| = \left\{ \frac{\partial g}{\partial x_i} \frac{\partial g}{\partial x_i} \right\}^{1/2} \tag{2.31}$$

in the double–suffix summation convention.

Consider now the line integral in the particular direction x_α,

$$I = \int_{x_\alpha} \phi(\mathbf{x})\delta(g(\mathbf{x}))dx_\alpha . \tag{2.32}$$

Contributions to the integral come only from the point (or points) where $g = 0$. To evaluate the integral we change variables from x_α to g, (x_i being held constant for $i \neq \alpha$). We have

$$dx_\alpha = \frac{dg}{\dfrac{\partial g}{\partial x_\alpha}} \tag{2.33}$$

and the integral then becomes

$$I = \int_{g \gtrless 0}^{g \lessgtr 0} \phi(\mathbf{x})\delta(g) \frac{dg}{\left(\dfrac{\partial g}{\partial x_\alpha} \right)} . \tag{2.34}$$

the inequalities being chosen according to the sign of $\frac{\partial g}{\partial x_\alpha}$. The integral can be made one over g increasing (i.e. $dg + ve$) by writing the integral in the form

$$I = \int_{g<0}^{g>0} \phi(\mathbf{x})\delta(g)\frac{dg}{\left|\dfrac{\partial g}{\partial x_\alpha}\right|} = \left\{\dfrac{\phi}{\left|\dfrac{\partial g}{\partial x_\alpha}\right|}\right\}(\mathbf{x}_\alpha^*) \ , \tag{2.35}$$

using (2.12), $g(\mathbf{x}_\alpha^*) = 0$ being the equation that selects the particular value of $\mathbf{x}_\alpha$ that makes $\mathbf{x} = \mathbf{x}_\alpha^*$.

Therefore, in general

$$\int \phi(\mathbf{x})\delta(g(\mathbf{x}))dx_\alpha = \Sigma \frac{\phi(\mathbf{x}_\alpha^*)}{|\partial g/\partial x_\alpha|} \ , \tag{2.36}$$

the sum being taken over all zeros of g included in the integration interval.

In particular, if $n = 1$ and $g = ax + b$, then

$$\int_{<(-b/a)}^{>(-b/a)} \phi(x)\delta(ax + b)dx = \frac{\phi(-b/a)}{|a|} \ , \tag{2.37}$$

$$\int_{-\infty}^{x} x^2\delta(x^2 - y^2)dx = \frac{1}{2}|y|H(|y| - |x|) + |y|H(x - |y|)$$

and

$$\int x_1\delta(\cos|\mathbf{x}|)dx_1 \text{ is } \frac{N\pi}{2} \ ,$$

where $|\mathbf{x}|^2 = x_i x_i$ and N is an integer determined by the precise range of integration.

Two results similar to (2.36) will sometimes prove useful. The first is the identity

$$\int_V \phi(\mathbf{x})\delta(g(\mathbf{x}))d\mathbf{x} = \int_{S(\mathbf{x}^*)} \phi(\mathbf{x}^*)\frac{d\mathbf{x}^*}{|\text{grad } g|(\mathbf{x}^*)} \ , \tag{2.38}$$

where S is the "hypersurface" defined by the equation $g(\mathbf{x}^*) = 0$. The proof follows that of Equation (2.36) with the volume split into the hypersurface S and the curvilinear coordinate normal to S.

The second useful integral property is the identity

$$\int_V \phi(\mathbf{x})\delta\{g(\mathbf{x})\}d\mathbf{x}$$
$$= \sum_{i=1}^{n}\int_S \left\{\frac{\phi\frac{\partial g}{\partial x_i}}{|\text{grad } g|^2}\right\}(\mathbf{x}^*)d\mathbf{x}_i \tag{2.39}$$

Again $g(\mathbf{x}^*) = 0$ defines a hypersurface on which the integrand is to be evaluated. $d\mathbf{x}_i$ is an element of the coordinate hyperplane with normal in the x_i direction. The result follows directly from (2.38) when the element of hypersurface area is expressed in terms of the coordinate hyperplanes through the expression

$$dx^* = ds = \sum_i n_i dx_i = \sum_i \frac{(\partial g/\partial x_i)dx_i}{|\text{grad } g|} . \tag{2.40}$$

As an example of how these results can be used to reduce the number of dimensions in an integral, consider the three–dimensional volume integral

$$\int_V \frac{\partial f}{\partial x_i}dV = \int_S \ell_i f \; ds$$

by Green's Lemma .

Here f could be an n–dimensional vector, an arbitrary tensor, or a scalar, but in each case a fairly good function of $\mathbf{x}$. V is a finite volume bounded by the closed surface S and ℓ_i the outward normal from V. This identity is simply proved by the use of (2.38) in three dimension as follows.

Write

$$\int_V \frac{\partial f}{\partial x_i}dV = \int_\infty H(g)\frac{\partial f}{\partial x_i}dV , \tag{2.41}$$

where the integral is now over the infinite volume and g is a function positive in V and negative elsewhere, with $g = 0$ defining the bounding surface S. The integral can then be manipulated as follows:

$$\int_\infty H(g)\frac{\partial f}{\partial x_i}dV = \int_\infty \frac{\partial}{\partial x_i}\{H(g)f\}dV - \int_\infty f\frac{\partial H(g)}{\partial x_i}dV$$

$$-- \int_\infty f\delta(g) \, \frac{\partial g}{\partial x_i} \, dV$$

$$= - \int_S f\frac{\frac{\partial g}{\partial x_i}ds}{|\text{grad } g|} = \int_S \ell_i f \; ds .$$

Here we have used the fact that $\{(\partial g/\partial x_i)/|\text{grad } g|\}$ is the direction cosine of S in the direction of increasing g: thus it is the negative of ℓ_i.

A special case of the identities (2.38) and (2.39) is the two–dimensional divergence theorem, or Green's theorem for a plane, which is, in the above notation

$$\int_{\text{area}} \nabla \cdot \mathbf{v} \; dA = \int_{\text{contour}} \mathbf{v} \cdot d\mathbf{c} = \int v_1 dx_1 - \int v_2 dx_2 .$$

Consider next the line integral in the particular direction

$$I = \int_{x_\alpha} f(\mathbf{x})\delta^n(g(\mathbf{x}))dx_\alpha \ , \tag{2.42}$$

where x_α is one of the coordinates in a space of one or more dimensions. This integral can be written as

$$I = \int f(\mathbf{x})\delta^n(g(\mathbf{x})) \ \frac{dg}{\left|\dfrac{\partial g}{\partial x_\alpha}\right|} \ , \tag{2.43}$$

where the integral is to be taken in the sense of g increasing and the integration range includes the neighbourhood of $g = 0$. Equation (2.28) shows this integral to be

$$(-1)^n \frac{\partial^n}{\partial g^n} \left[\frac{f(\mathbf{x})}{\left|\dfrac{\partial g}{\partial x_\alpha}\right|} \right] (x_\alpha{}^*) \ , \tag{2.44}$$

where $g(x_\alpha{}^*) = 0$, and, since x_α is the only component of $\mathbf{x}$ that is being varied in the integral, the operator

$$\frac{\partial}{\partial g} = \left(\frac{\partial}{\partial x_\alpha} \right)^{-1} \frac{\partial g}{\partial x_\alpha} \ . \tag{2.45}$$

As a particular case of (2.39), we can evaluate the one–dimensional integral

$$\int_{<(-b/a)}^{>(-b/a)} \phi(x)\delta^n(ax + b)dx$$

$$= \frac{(-1)^n}{|a|} \frac{1}{a^n} \phi^n(-b/a) \ . \tag{2.46}$$

2.2 THE TWO–DIMENSIONAL DELTA FUNCTION

We now define the delta function with vector argument. It has similar integral properties to the one–dimensional delta function, with respect to integration over the entire space. Consider first the delta function $\delta(\mathbf{x})$, where $\mathbf{x}$ is a two–dimensional vector. This function is defined to have the property

$$\int f(\mathbf{x})\delta(\mathbf{x}) = f(\mathbf{0}) \ , \tag{2.47}$$

where dx is an area element and the integration range includes the neighbourhood of the point $x = 0$.

This two–dimensional delta function can be written in terms of one–dimensional delta functions in two different ways, using the property (2.6).

$$\int f(x)\delta(x)dx = \int f(x)\delta(x)dx_1\,dx_2 = f(0)$$

$$= f(x_1 = 0, x_2 = 0) = \int f(x)\delta(x_1)\delta(x_2)dx_1\,dx_2 \ , \tag{2.48}$$

thus

$$\delta(x) = \delta(x_1)\delta(x_1) \ ,$$

irrespective of the orientation of the orthogonal coordinate axes.

This demonstrated isotropy of the delta function implies it to be a function of $r(= |x|)$ only, so that the area integral could be performed in polar coordinates. Thus

$$f(0) = \int_0^\epsilon f(x)\delta(x)2\pi r\ dr$$

$$= 2\int_0^\epsilon f(x)\delta(r)dr,$$

whence

$$r\delta(x) = \frac{\delta(r)}{\pi} \ . \tag{2.49}$$

Now an application of Equation (2.28) shows the equivalence of the two integrals

$$\left.\begin{array}{l} \displaystyle\int_0^\epsilon f(x)r\delta'(r)dr = -\frac{1}{2}f(0) \\[2mm] \displaystyle= -\int_0^\epsilon f(x)\delta(r)dr \ , \end{array}\right\} \tag{2.50}$$

so that

$$r\ \delta'(r) = -\delta(r) \ , \tag{2.51}$$

and from (2.49)

$$\delta(x) = -\ \frac{\delta'(r)}{\pi} \ . \tag{2.52}$$

The homogeneous equation

$$\nabla^2 G \equiv \left(\frac{\partial^2}{\partial r^2} + \frac{1}{r}\frac{\partial}{\partial r}\right)G = 0 \ ,$$

has the solution

$$G = A \log r + B \qquad \text{for } r \neq 0 .$$

Taken together with the divergence theorem

$$\int_v \nabla \cdot \mathbf{v} dV = \int_S \mathbf{n} \cdot \mathbf{v} dS$$

applied to $\mathbf{v} = \nabla G$, it is straightforward to show that

$$\nabla^2 G = \delta(\mathbf{x})$$

has the solution

$$G = \frac{1}{2\pi} \log r + B .$$

2.3 THE THREE–DIMENSIONAL DELTA FUNCTION

The delta function in three dimensions possesses the above integral properties when integrated over a volume. Accordingly

$$\int_V f(\mathbf{x})\delta(\mathbf{x})d\mathbf{x} = f(0) . \tag{2.53}$$

This time $d\mathbf{x}$ is a three–dimensional volume element and the integration range includes the neighbourhood of the point $\mathbf{x} = \mathbf{0}$.

Again it is readily show that (cf. (2.48))

$$\delta(\mathbf{x}) = \delta(x_1)\delta(x_2)\delta(x_3) , \tag{2.54}$$

and again the integration of this isotropic function can be carried out in polar coordinates. Thus

$$f(0) = \int_V f(\mathbf{x})\delta(\mathbf{x})d\mathbf{x} = 4\pi \int_0^\epsilon f(\mathbf{x})\delta(\mathbf{x})r^2 dr = 2 \int_0^\epsilon f(\mathbf{x})\delta(r)dr , \tag{2.55}$$

whence

$$2\pi r^2 \delta(\mathbf{x}) = \delta(r) . \tag{2.56}$$

From Equation (2.28) we have again an equality of two integrals

$$\int_0^\epsilon f(\mathbf{x})r^2\delta''(r)dr = f(0) = \int_0^\epsilon f(\mathbf{x})2\delta(r)dr ,$$

so that

$$r^2\delta''(r) = 2\delta(r)$$

and, from (2.56)

$$\delta(\mathbf{x}) = \frac{\delta''(r)}{4\pi r} . \tag{2.57}$$

2.4 CONVOLUTION ALGEBRA

The convolution product of two functions, f and g, of an n–dimensional vector argument $\mathbf{x}$ is written $f * g$, a symbol that represents the operation

$$f * g = \int f(\mathbf{y}) g(\mathbf{x} - \mathbf{y}) d\mathbf{y} \; ; \tag{2.58}$$

$d\mathbf{y}$ is a volume element in the n–dimensional space and the integration ranges over the infinite volume. Obviously, the product is defined only for the restricted class of functions for which the integral exists so that either f or g must vanish with sufficient rapidity as $\mathbf{x}$ tends to infinity in any direction. It is certainly sufficient that one of the functions be "fairly good" and the other a generalized function, since the latter is generated from a sequence of good functions, each of which is small at infinity.

It is easily proved from the definition (2.58) that the convolution product has the following important properties

$$f * (\lambda g + \mu h) = \lambda (f * g) + \mu (f * h) \; , \tag{2.59}$$

where λ and μ are constants

$$f * g = g * f; \tag{2.60}$$

$$f * (g * h) = (f * g) * h; \tag{2.61}$$

$$\frac{\partial}{\partial x_i}(f * g) = \frac{\partial f}{\partial x_i} * g = f * \frac{\partial g}{\partial x_i} \tag{2.62}$$

$$f * \delta = f \; . \tag{2.63}$$

The first of these properties is obvious, since

$$f * (\lambda g + uh) = \int f(\mathbf{y})\{\lambda g(\mathbf{x} - \mathbf{y}) + \mu h(\mathbf{x} - \mathbf{y})\} d\mathbf{y}$$

$$= \lambda \int f(\mathbf{y}) g(\mathbf{x} - \mathbf{y}) d\mathbf{y} + \mu \int f(\mathbf{y}) h(\mathbf{x} - \mathbf{y}) d\mathbf{y} = \lambda (f * g) + \mu (f * h) \; .$$

The second property (2.60), that the convolution product is commutative, is readily established by writing $\mathbf{z} = \mathbf{x} - \mathbf{y}$ as a new variable to replace $\mathbf{y}$ in the definition (2.58). The volume element $d\mathbf{y}$ is replaced by $d\mathbf{z}$, since the Jacobian of the transformation is unity, whence

$$f * g = \int f(\mathbf{y}) g(\mathbf{x} - \mathbf{y}) d\mathbf{y} = \int f(\mathbf{x} - \mathbf{z}) g(\mathbf{z}) d\mathbf{z} = g * f \; .$$

In order to establish (2.62), we have

$$\frac{\partial}{\partial x_i}(f*g) = \frac{\partial}{\partial x_i} \int f(\mathbf{y})g(\mathbf{x}-\mathbf{y})d\mathbf{y} = \int f(\mathbf{y})\frac{\partial g}{\partial x_i}(\mathbf{x}-\mathbf{y})d\mathbf{y} = f*\frac{\partial g}{\partial x_i} \ ,$$

and similarly

$$\frac{\partial}{\partial x_i}(f*g) = \frac{\partial}{\partial x_i}(g*f) = g\frac{\partial f}{\partial x_i} = \frac{\partial f}{\partial x_i}*g \ ,$$

where use has been made of the commutative law (2.60).

Evidently, the differential operator $(\partial/\partial x_i)$ acting on the convolution product $f*g$ can be performed on either f or g before forming the product. A repeated use of this law reveals the more general result that

$$\left(\frac{\partial}{\partial x_i}\frac{\partial}{\partial x_j}\frac{\partial}{\partial x_k}\cdots\right)(f*g) = \left(\frac{\partial}{\partial x_i}\frac{\partial}{\partial x_j}\frac{\partial}{\partial x_k}\cdots f\right)*g$$

$$= \left(\frac{\partial}{\partial x_i}f\right)*\left(\frac{\partial}{\partial x_j}\frac{\partial}{\partial x_k}\cdots g\right), \quad \text{etc.,} \tag{2.64}$$

and the differential operators $(\partial/\partial x_i)$, $(\partial/\partial x_j)$, ... may be applied to either f or g or in any combination.

Further, if this result is combined with formula (2.59), it is seen that any linear differential operator, with constant coefficients can be taken to act on either of the functions of the product, i.e.

$$L(f*g) = (Lf)*g = f*(Lg) \ . \tag{2.65}$$

Finally, the formula (2.63) is verified by writing

$$f*\delta = \int f(\mathbf{y})\delta(\mathbf{x}-\mathbf{y})d\mathbf{y} = \int f(\mathbf{x}-\mathbf{z})\delta(\mathbf{z})d\mathbf{z} = f(\mathbf{x})$$

using (2.47). This result shows that the delta function $\delta(\mathbf{x})$ acts as the "unit" element in the algebra of convolution products.

The one–dimensional convolution product of a good function with a polynomial is itself a polynomial. The proof follows by use of Equation (2.64). For if $f(\mathbf{x})$ is a polynomial of degree N,

$$\left(\frac{d}{dx}\right)^{N+1}(f*g) = \left(\frac{d^{N+1}f}{dx^{N+1}}\right)*g = 0 \ ,$$

which may be integrated $N+1$ times to show that $(f*g)$ is a polynomial (of degree $\leq N$).

The following integral is useful.

$$\int f(\mathbf{y}) \, \frac{\partial^p}{\partial y_i \partial y_j \partial y_k \dots} \delta(\mathbf{y} - \mathbf{x}) = (-1)^p \frac{\partial^p}{\partial x_i \partial x_j \partial x_k \dots} f(\mathbf{x}) \, , \qquad (2.66)$$

this being the generalization to n dimensions of the one–dimensional Equation (2.28).

The proof follows by use of formulae (2.63) and (2.65), whence

$$Lf = L(f*\delta) = f*(L\delta) \, .$$

In particular, if L is the operator $(-1)^p \partial^p / \partial x_i \partial x_j \partial x_k \dots$, then

$$(-1)^p \frac{\partial^p}{\partial x_i \partial x_j \partial x_k \dots} f(\mathbf{x}) = (-1)^p \int f(\mathbf{y}) \frac{\partial^p}{\partial x_i \partial x_j \partial x_k \dots} \delta(\mathbf{x} - \mathbf{y}) d\mathbf{y}$$

$$= \int f(\mathbf{y}) \frac{\partial^p}{\partial y_i \partial y_j \partial y_k \dots} \delta(\mathbf{y} - \mathbf{x}) d\mathbf{y} \, .$$

2.4.1 Solution of Non–homogeneous Linear Equations

An important application of convolution algebra is that of finding solutions to inhomogeneous linear differential equations in terms of elementary solutions. Consider the equations

$$L_1 \phi = L_2 Q \qquad (2.67)$$

and

$$L_1 G = \delta \, , \qquad (2.68)$$

where L_1 and L_2 are linear operators. Suppose that $L_2 Q$ is given and that we wish to solve for ϕ knowing also the solution of the elementary Equation (2.68), i.e. G is supposed known and is termed the *Green function* of the problem. First, we write

$$\phi = \phi * \delta \qquad \text{from (2.63)} \qquad (2.69)$$

$$= \phi * (L_1 G) \qquad \text{from (2.68)} \qquad (2.70)$$

$$= (L_1 \phi) * G \qquad \text{from (2.65)} \qquad (2.71)$$

$$= (L_2 \phi) * G \qquad \text{from (2.67)} \qquad (2.72)$$

thus $\phi = L_2(Q * G)$ from (2.65) is the solution to equation (2.67). $\qquad$ (2.73)

2.4.2 Green Functions

Field quantities in problems of sound and vibration are usually governed by a linear partial–differential equation with an inhomogeneity acting as the source term. Symbolically, the equation can be written

$$L_{\mathbf{x}}\phi(\mathbf{x}) = Q(\mathbf{x}), \qquad \text{for } \mathbf{x} \text{ in } V . \tag{2.74}$$

Here $L_{\mathbf{x}}$ denotes a linear combination of partial–differential operators with constant coefficients acting on the field quantity ϕ, which is to be found in terms of the known source field $Q(\mathbf{x})$. Equation (2.74) holds for $\mathbf{x}$ in V, which might consist of the whole n–dimensional space $\mathbf{x}$, or a given subregion of this space; the variables $\mathbf{x}$ denotes position, or one of them could refer to time. The corresponding homogeneous equation is

$$L_{\mathbf{x}}\phi(\mathbf{x}) = 0 ; \tag{2.75}$$

any two solutions of (2.74) differ by a function that satisfies (2.75).

A simpler version of (2.74) can usefully be considered first by assuming that the inhomogeneity is concentrated at a point, i.e. it is a delta function. The field generated by the delta–function inhomogeneity is called a "Green function" for the problem. If the differential operators are with respect to position $\mathbf{x}$, and the delta function acts at the point $\mathbf{y}$, then the Green function will be a function of both $\mathbf{x}$ and $\mathbf{y}$, and is written

$$G = G(\mathbf{x}; \mathbf{y}) . \tag{2.76}$$

This function satisfies the equation

$$L_{\mathbf{x}}G(\mathbf{x}; \mathbf{y}) = \delta(\mathbf{x} - \mathbf{y}), \quad \text{for } \mathbf{x} \text{ in } V , \tag{2.77}$$

and represents the response at $\mathbf{x}$ due to point forcing at $\mathbf{y}$. It is not uniquely defined by (2.77), since any solution of the homogeneous Equation (2.75) could obviously be added.

The importance of the Green function lies in the fact that solutions of the more general Equation (2.74) can easily be generated in terms of G. For if (2.77) is multiplied by $Q(\mathbf{y})$ and integrated over the entire n–dimensional volume V in which (2.74) holds, it is seen that the differential operations act through $\mathbf{x}$, independently of $\mathbf{y}$, whence the operator $L_{\mathbf{x}}$ may be taken outside the integral. Thus

$$\int_{V(\mathbf{y})} L_{\mathbf{x}}G(\mathbf{x}; \mathbf{y})Q(\mathbf{y})d\mathbf{y} = \int_{V(\mathbf{y})} \delta(\mathbf{x} - \mathbf{y})Q(\mathbf{y})d\mathbf{y}$$

implies that

$$L_\mathbf{x} \int_V G(\mathbf{x}; \mathbf{y})Q(\mathbf{y})d\mathbf{y} = Q(\mathbf{x}) , \qquad (2.78)$$

for $\mathbf{x}$ within the region V.

By comparison of (2.78) with (2.74), it is evident that

$$\phi(\mathbf{x}) = \int_V G(\mathbf{x}; \mathbf{y})Q(\mathbf{y})d\mathbf{y} \qquad (2.79)$$

is a solution of the inhomogeneous Equation (2.74), and is simply a superposition of the elementary solutions $G(\mathbf{x}; \mathbf{y})$ weighted by the "source" strength $Q(\mathbf{y})$. The *general solution* of (2.74) is

$$\phi(\mathbf{x}) = \int_V G(\mathbf{x}; \mathbf{y})Q(\mathbf{y})d\mathbf{y} + \phi_h(\mathbf{x}) , \qquad (2.80)$$

ϕ_h satisfying the homogeneous Equation (2.75). The particular solution for a given problem is that which satisfies further *auxiliary conditions* (boundary conditions, initial conditions and/or a radiation condition) appropriate to the physical situation under consideration. Typical examples of auxiliary conditions are as follows.

Causality condition. In problems where the field is induced by a forcing term that is zero for $t < t_0$ and starts at $t = t_0$, it is argued on physical grounds that the response cannot anticipate the source; thus we impose the "causality constraint" that

$$\phi = 0 \quad \text{for} \quad t < t_0 . \qquad (2.81)$$

Boundary condition. If ϕ represents the velocity potential of a fluid motion and the volume V is bounded by a fixed rigid surface S ("acoustically hard" surface), then it is required that the velocity has no component normal to the surface: thus

$$\frac{\partial \phi}{\partial n} = 0 \quad \text{for} \quad \mathbf{x} \text{ on } S , \qquad (2.82)$$

where n denotes the normal direction from the surface.

Alternatively, if the surface is such that it cannot support a force ("acoustically soft" surface), and if ϕ represents the pressure, then

$$\phi = 0 \quad \text{for} \quad \mathbf{x} \text{ on } S . \qquad (2.83)$$

Exact Green functions. If the auxiliary conditions on ϕ are homogeneous, as is the case with (2.81), (2.82) and (2.83), then it is often found convenient to choose the Green function G to satisfy the corresponding conditions, together with (2.77); the superposition of such solutions given by (2.79) will also satisfy the auxiliary condition, as is readily verified. Thus, the complementary function ϕ_h of (2.80) is zero and the solution for ϕ is complete. Such a function G is called the exact Green function of the problem. Its usefulness will be seen to extend to more general cases in which, for example, $\dfrac{\partial \phi}{\partial n}$ takes a prescribed value over part S_1 of the bounding surface and ϕ is given over the remaining surface S_2. The exact Green function for this problem is that which satisfies the Equation (2.77), with $\dfrac{\partial G}{\partial n} = 0$ on S_1 and $G = 0$ on S_2.

We will demonstrate the method of solution by the Green function technique by solving for the function $\phi(\mathbf{x})$ that is governed by the one–dimensional wave equation and two boundary conditions

$$\frac{\partial^2 \phi}{\partial x^2} + k^2 \phi = \alpha^2 H \left(\frac{\pi}{2\alpha} |x| \right) \cos \alpha x; \quad \phi(0) = \phi \left(\frac{\pi}{2\alpha} \right) = 0 \; .$$

The Green function $G(x; y)$ is a solution to

$$\frac{\partial^2 G}{\partial x^2} + k^2 G = \delta(x - y) \; ,$$

and it is easy to verify by direct differentiation that

$$G(x; y) = \frac{1}{2k} \; \sin \; k|x - y|$$

is such a solution

$$\frac{\partial G}{\partial x} = k \; \mathrm{sgn}(x - y) \frac{1}{2k} \; \cos \; k|x - y|$$

$$\frac{\partial^2 G}{\partial x^2} = 2k\delta(x - y)\frac{1}{2k} \; \cos \; k|x - y| - k^2 [\mathrm{sgn}(x - y)]^2 \frac{1}{2k} \; \sin \; k|x - y|$$

$$= \delta(x - y) - k^2 G \; ;$$

$\phi_h = A \, \cos \, kx + B \, \sin \, kx$, A and B being arbitrary constants, is the general solution to the homogeneous equation, so that

$$\phi(x) = \int\limits_{-\pi/2\alpha}^{\pi/2\alpha} \frac{1}{2k} \sin(k|x - y|)\alpha^2 \cos \alpha y \, dy + A \cos kx + B \sin kx,$$

$$\phi(x) = \frac{\alpha^2}{k^2 - \alpha^2} \cos \alpha x - \frac{\alpha^3}{k(k^2 - \alpha^2)} \sin \frac{k\pi}{2\alpha} \cos kx + A \cos kx + B \sin kx \; .$$

The boundary conditions require that

$$0 = \frac{\alpha^2}{k^2 - \alpha^2} - \frac{\alpha^3}{k(k^2 - \alpha^2)} \sin \frac{k\pi}{2\alpha} + A \ ,$$

$$0 = \frac{-\alpha^3}{k(k^2 - \alpha^2)} \sin \frac{k\pi}{2\alpha} \cos \frac{k\pi}{2\alpha} + A \cos \frac{k\pi}{2\alpha} + B \sin \frac{k\pi}{2\alpha} \ .$$

The complete solution is

$$\pi(x) = \frac{\alpha^2}{k^2 - \alpha^2} \left\{ \cos \alpha x - \cos kx + \cot \frac{k\pi}{2\alpha} \sin kx \right\} \ .$$

We could alternatively have derived this solution by using the 'exact" Green function G^e which itself satisfies the boundary conditions imposed on ϕ,

$$G^e(x; y) = \frac{1}{2k} \left\{ \sin k|x - y| - \sin k|y| \cos kx + \frac{[\sin k|y| \cos \frac{k\pi}{2\alpha} - \sin k|\frac{\pi}{2\alpha} - y|]}{\sin \frac{k\pi}{2\alpha}} \sin kx \right\}$$

in which case

$$\phi(x) = \int\limits_{-\pi/2\alpha}^{\pi/2\alpha} G^e(x; y) \alpha^2 \cos \alpha y \, dy$$

$$= \frac{\alpha^2}{k^2 - \alpha^2} \left\{ \cos \alpha x - \cos kx + \cot \frac{k\pi}{2\alpha} \sin kx \right\} \ .$$

Symmetry properties. If the Equation (2.77) for $G(\mathbf{x}; \mathbf{y})$ holds for all $n-$ dimensional $\mathbf{x}$–space, then its solution is a function of the combination $(\mathbf{x} - \mathbf{y})$. To establish this fact, note that the linear operator $L_{\mathbf{x}}$ is unchanged by an origin shift, i.e. $L_{\mathbf{x}} = L_{(\mathbf{x}-\mathbf{y})}$, whence (2.77) may be written as

$$L_{(\mathbf{x}-\mathbf{y})} G = \delta(\mathbf{x} - \mathbf{y}), \quad \text{for all } \mathbf{x} \ , \tag{2.84}$$

i.e.

$$L_{\mathbf{x}} G = \delta(\mathbf{x}'), \qquad \text{for all } \mathbf{x}' \ , \tag{2.85}$$

where $\mathbf{x}' = |\mathbf{x} - \mathbf{y}|$. Since the right side of (2.85) is independent of $\mathbf{y}$, the function G must be a function only of $\mathbf{x}' = (\mathbf{x} - \mathbf{y})$, i.e.

$$G(\mathbf{x}; \mathbf{y}) = G(\mathbf{x} - \mathbf{y}) \ . \tag{2.86}$$

This symmetry property allows derivatives of G with respect to $\mathbf{x}$ and $\mathbf{y}$ to be interchanged

$$\frac{\partial}{\partial x_i} G(\mathbf{x} - \mathbf{y}) = -\frac{\partial}{\partial y_i} G(\mathbf{x} - \mathbf{y}) \ , \tag{2.87}$$

$$\frac{\partial^n}{\partial x_i \partial x_j \dots} G(\mathbf{x} - \mathbf{y}) = (-1)^n \frac{\partial^n}{\partial y_i \partial y_j \dots} G(\mathbf{x} - \mathbf{y}) \ . \tag{2.88}$$

Note that Equation (2.86) allows (2.79), if valid over infinite space, to be written in the compact notion

$$\phi = G * Q \ ,$$

and that the derivation of this equation can be expressed as follows (cf. (2.69)–(2.73))

$$L\phi = Q$$

$$LG = \delta$$

$$\phi = \delta * \phi = LG * \phi \ = G * L\phi = G * Q \ .$$

2.5 DEVELOPMENT OF INTEGRAL EQUATIONS

When Equation (2.74) is supplemented by additional boundary conditions on some hypersurface S, then the field ϕ can usually be expressed as a sum of elements generated by the distributed source Q together with other elements arising from the boundary conditions imposed on S. These second terms are surface integrals of ϕ and are not know *a priori*, so that ϕ is then posed as the solution to an integral equation. The surface terms vanish exactly only when a very special choice of Green function has been made: G is then the *exact* Green function, but this cannot always be calculated explicitly. For any other G, an integral equation results.

There are several ways of generating integral equations from the governing differential equation, most methods being based on applications of Green's theorem in the appropriate number of dimensions for the problems. There is one particularly straightforward way that makes use of generalized functions, as follows.

Consider a field $\phi(\mathbf{x})$ described in a bounded n–dimensional volume V_0 by Equation (2.74). Thus

$$L_\mathbf{x}\phi(\mathbf{x}) = Q(\mathbf{x}) \ \text{in} \ V_0 \ . \tag{2.89}$$

A solution is required in V_0 with ϕ subject to certain boundary conditions on the boundary hypersurface S. The field outside V_0 is of no interest and can be arbitrary. Since (2.89) has no meaning outside V_0, we do not alter the sense of the equation in the region of interest if we multiply (2.89) by a function equal to unity inside V_0 and zero outside V_0. That is, we can multiply (2.89) by a Heaviside function $H(g\{\mathbf{x}\})$, where

$$
\begin{aligned}
g(\mathbf{x}) &> 0 \quad \text{inside} \quad V_0 \\
g(\mathbf{x}) &< 0 \quad \text{outside} \quad V_0 \\
g(\mathbf{x}) &= 0 \qquad \text{on} \quad S \ .
\end{aligned}
\tag{2.90}
$$

Inside V_0, ϕ satisfies the equation

$$
H(g)L_{\mathbf{x}}\phi(\mathbf{x}) = H(g)Q(\mathbf{x}) \ ,
\tag{2.91}
$$

or

$$
HL\phi = HQ \ .
\tag{2.92}
$$

The next step is to transfer H inside the operator L on the left side of this equation. If L is a constant, the operation is straightforward since $HC\phi = CH\phi$. But if L involves differential operators, as it invariably does, then additional terms are generated in this step. For example,

$$
H\frac{\partial}{\partial x_i}\phi = \frac{\partial}{\partial x_i}(H\phi) - \phi\frac{\partial H}{\partial x_i} = \frac{\partial}{\partial x_i}(H\phi) - \phi\delta(g)\frac{\partial g}{\partial x_i} \ .
\tag{2.93}
$$

The term $\phi\delta(g)(\partial g/\partial x_i)$ can be treated as an additional source term concentrated on the hypersurface $g = 0$ (i.e. on S). Each time H is taken inside any differential operator, then an extra term involving delta functions is introduced so that (2.92) can in general be written

$$
L(\phi H) = HQ + \Sigma f(\phi, \delta(g)) \ ,
\tag{2.94}
$$

the term $\Sigma f(\phi, \delta(g))$ signifying a collection of terms involving ϕ and its derivatives and the delta function and its derivatives.

Expressed in this way, Equation (2.94) is an equation valid throughout all space for the generalized function ϕH, equal to ϕ in the volume of interest V_0, but identically zero elsewhere. The problem has been transformed from a finite space problem with boundary conditions to an infinite space problem with additional sources concentrated on the surface S. That is,

$$
H\phi = \int_V G(\mathbf{x}; \mathbf{y})\{HQ + \Sigma f(\phi, \delta(\mathbf{g}))\}d\mathbf{y}
\tag{2.95}
$$

is an integral equation for the field ϕ. If G is chosen in such a way that the hypersurface concentrations vanish, then G has been chosen to be the exact Green function and (2.95) is then an explicit solution rather than an integral equation.

That there is an arbitrariness in the choice of G is easily demonstrated. We require a solution to (2.94), an equation defined over the entire infinite space, but such that outside V_0 the right–hand side is zero. The right–hand side is non–zero only within the n–dimensional volume V_0 and the bounding hypersurface S, where the delta functions are non–zero.

Now choose G to satisfy

$$L_{\mathbf{x}} G = \delta |\mathbf{x} - \mathbf{y}| \quad \text{for all } \mathbf{y} \text{ on } S \text{ and in } V_0 \tag{2.96}$$

but to satisfy *any* other condition outside V_0 and S. Now multiply (2.96) by the right–hand side of (2.94) and integrate over the finite volume V_0^+ that includes S and V_0 where (2.96) is valid. Thus

$$L_{\mathbf{x}} \int_{V_0^+} G\{HG + \sum f(\phi, \delta(g))\} dy = HQ + \sum f(\phi, \delta(g)) . \tag{2.97}$$

A comparison of (2.97) and (2.94) shows that (2.95) is indeed the required solution to (2.94) for any G whatsoever that satisfies (2.96) within and on S. It can be subject to arbitrary source fields outside S. The step implied by choosing the exact Green function is a selection of exterior source fields that generate within S the same effect as is produced by the boundary conditions on S.

We can make this procedure more specific by considering a particular form of $L_{\mathbf{x}}$. Consider the problem of n–dimensional non–dispersive wave generation in a finite volume V_0 bounded by a hypersurface S. Inside V_0 the field satisfies the equation

$$\frac{\partial^2 \phi}{\partial t^2}(\mathbf{x}, t) - c^2 \nabla^2 \phi(\mathbf{x}, t) = Q(\mathbf{x}, t) , \tag{2.98}$$

where c is the constant wave speed.

In three dimensions this problem is the well–known Kirchoff problem of specifying the field ϕ within and on S in terms of Q and boundary conditions on S.

We proceed as before by multiplying the equation by a Heaviside function $H(g)$ to generate a form valid over all space. Again $g(\mathbf{x})$ is positive within V_0. Thus

$$H(g) \left\{ \frac{\partial^2 \phi}{\partial t^2} - c^2 \nabla^2 \phi \right\} = QH , \tag{2.99}$$

or, equivalently, since

$$H\nabla^2\phi = \nabla^2(H\phi) - \frac{\partial}{\partial x_i}\left\{\phi\frac{\partial H}{\partial x_i}\right\} - \frac{\partial\phi}{\partial x_i}\frac{\partial H}{\partial x_i} , \qquad (2.100)$$

$$\frac{\partial^2}{\partial t^2}(\phi H) - c^2\nabla^2(\phi H) = QH - c^2\frac{\partial}{\partial x_i}\left\{\frac{\phi\partial H}{\partial x_i}\right\} - c^2\frac{\partial\phi}{\partial x_i}\frac{\partial H}{\partial x_i}$$

$$= Q^*(\mathbf{x}, t), \quad \text{say} . \qquad (2.101)$$

Now we select a Green function that satisfies the equation

$$\frac{\partial^2 G}{\partial t^2}(\mathbf{x}, t; \mathbf{y}, \tau) - c^2\nabla_{\mathbf{x}}^2 G(\mathbf{x}, t : \mathbf{y}, \tau) = \delta(\mathbf{x} - \mathbf{y})\delta(t - \tau) , \qquad (2.102)$$

for all points on and within the boundary hypersurface S. G has the arbitrariness that the right–hand side of this equation can be supplemented by any function that is zero on and within S. As before, we multiply this equation by the right–hand side of (2.101) at position $\mathbf{y}$ and time τ and integrate over all time τ and over the finite volume V_0^+ that includes S and its interior volume V_0

$$\left\{\frac{\partial^2}{\partial t^2} - c^2\nabla_{\mathbf{x}}^2\right\}\int_{V_0^+,\tau} G(\mathbf{x}, t; \mathbf{y}, \tau)Q^*(\mathbf{y}, t)dyd\tau = Q^*(\mathbf{x}, t) . \qquad (2.103)$$

On comparing this with (2.101), we see that

$$H\phi = \int_{V_0^+,\tau} G(\mathbf{x}, t : \mathbf{y}, t)Q^*(\mathbf{y}, \tau)dyd\tau$$

$$= -c^2\int_\infty G(\mathbf{x}, t; \mathbf{y}, \tau)\left\{\frac{\partial}{\partial y_i}\left[\phi(\mathbf{y}, \tau)\frac{\partial H}{\partial y_i}\right] + \frac{\partial\phi}{\partial y_i}\frac{\partial H}{\partial y_i}\right\}dyd\tau \quad (2.104)$$

$$+ \int_{V_0,\tau} G(\mathbf{x}, t; \mathbf{y}, \tau)Q(\mathbf{y}, \tau)dyd\tau . \qquad (2.105)$$

The first term on the right–hand side can be made a divergence by absorbing G inside the $\partial/\partial y_i$ operation to leave a correction term proportional to $\partial G/\partial y_i$. The divergence integrates to zero since the presence of the H factor in the integrand makes all hypersurface integrals vanish at large enough $|\mathbf{y}|$, where $g < 0$. Equation (2.105) can therefore be written

$$H\phi(\mathbf{x}, t) = \int_{V_0,\tau} G(\mathbf{x}, t; \mathbf{y}, \tau)Q(\mathbf{y}, \tau)dyd\tau$$

$$- c^2\int_\infty\left\{G\frac{\partial\phi}{\partial y_i} - \phi\frac{\partial G}{\partial y_i}\right\}\frac{\partial H}{\partial y_i}dyd\tau . \qquad (2.106)$$

$$\frac{\partial H(g)}{\partial y_i} = \frac{\partial g}{\partial y_i} \delta(g) \text{ and}$$

$$\int_\infty f(\mathbf{y}) \frac{\partial g}{\partial y_i} \delta(g) d\mathbf{y} = \int_{S(g=0)} f(\mathbf{y}) \frac{\frac{\partial g}{\partial y_i}}{|\nabla g|} = -\int_S f(\mathbf{y}) \ell_i dS, \tag{2.107}$$

where ℓ_i is the direction cosine of the outward normal from V_0 at the hypersurface S. Thus Equation (2.106) can be rewritten

$$H\phi(\mathbf{x}, t) = \int_{V_0, \tau} G(\mathbf{x}, \mathbf{t}; \mathbf{y}, \tau) Q(\mathbf{y}, \tau) d\mathbf{y} d\tau$$

$$- c^2 \int_{S, \tau} \ell_i \left\{ \phi \frac{\partial G}{\partial y_i} - G \frac{\partial \phi}{\partial y_i} \right\} d\mathbf{y} d\tau . \tag{2.108}$$

Inside V_0, $H\phi = \phi$, so that (2.108) is an integral equation for ϕ in terms of any Green function G satisfying (2.102) in V_0 but forced by any arbitrary source field outside V_0.

At this point the choice of the exact Green function is made clear. Suppose that the wave equation is to be solved subject to the condition $\phi = 0$ on S. Then if we can find the exact Green function, with $G = 0$ on S, the entire surface term vanishes and the volume integral then gives the exact solution. Further, if ϕ takes *any* prescribed values over the surface S, the solution is given explicitly by (2.108) in terms of the Green function, with $G = 0$ on S. Evidently the field is determined once the source distribution Q is given and ϕ is prescribed on S (the Dirichlet boundary condition).

Similarly, suppose the boundary condition is that $\partial\phi/\partial n \equiv \lambda_i \partial\phi/\partial y_i = 0$ on the boundary surface, i.e. the Neumann condition. Then the exact Green function of the problem has $\partial G/\partial n = 0$ on S, again the surface terms of (2.108) vanish to leave the solution as a volume integral of the source field. It is also clear that if $\partial\phi/\partial n$ takes *any* given values on S, then a selection of G such that $\partial G/\partial n = 0$ on S, makes (2.108) an explicit formula for ϕ.

Once the exact Green function is known, the problem is formally solved, so that the emphasis must now be placed on methods of obtaining Green functions. We shall deal first with the Green functions for an unbounded domain, these being obtained in a straightforward manner by the application of Fourier transform techniques.

2.6 THE SIMPLE HARMONIC OSCILLATOR

One of the simplest of the equations we shall deal with is that for the simple harmonic oscillator. This is a one—dimensional problem with

$$L = \frac{\partial^2}{\partial t^2} + \omega_0^2, \quad \omega_0 \text{ being a constant.} \tag{2.109}$$

The Green functions for the infinite space problem is then the function that satisfies (2.77) for all time, with L given by (2.109). Thus

$$\frac{\partial^2}{\partial t^2} G(t; \tau) + \omega_0^2 G(t; \tau) = \delta(t - \tau); \tag{2.110}$$

each side of this equation is a generalized function.

Both G and δ in terms of their Fourier transform are

$$G(t; \tau) = \frac{1}{2\pi} \int_{-\infty}^{\infty} G(\omega, \tau) e^{-i\omega t} d\omega \tag{2.111}$$

$$\delta(t - \tau) = \frac{1}{2\pi} \int_{-\infty}^{\infty} e^{i\omega(\tau - t)} d\omega . \tag{2.112}$$

Equation (2.110) then becomes

$$\int_{-\infty}^{\infty} (\omega_0^2 - \omega^2) G(\omega, \tau) e^{-i\omega t} d\omega = \int_{-\infty}^{\infty} e^{i\omega\tau} e^{-i\omega\tau} d\omega , \tag{2.113}$$

so that the Fourier transform of Equation (2.110) is

$$(\omega_0^2 - \omega^2) G(\omega, \tau) = e^{i\omega\tau} = (\omega_0 + \omega)(\omega_0 - \omega) G(\omega, \tau) . \tag{2.114}$$

The general solution of this algebraic equation for the generalized function G can be written down immediately, bearing in mind that (2.24) is the general solution of (2.21).

$$G(\omega, \tau) = A\delta(\omega - \omega_0) + B\delta(\omega + \omega_0) + \frac{e^{i\omega\tau}}{(\omega_0^2 - \omega^2)} , \tag{2.115}$$

where A and B are two arbitrary functions of τ. The required Green function is then obtained by inserting (2.115) into the transformation formula (2.111) to give

$$G(t; \tau) = \frac{A}{2\pi} e^{i\omega_0 t} + \frac{B}{2\pi} e^{-i\omega_0 t} + \frac{1}{2\pi} \int_{-\infty}^{\infty} \frac{e^{i\omega\tau} e^{-i\omega t}}{(\omega_0^2 - \omega^2)} d\omega . \tag{2.116}$$

The remaining integral can be accomplished by reference to the standard result for generalized functions

$$\int_{-\infty}^{\infty} \frac{e^{i\omega t}}{(\omega - \omega_0)^m} d\omega = \frac{(\pi i)(it)^{m-1}}{(m-1)!} e^{i\omega_0 t} \operatorname{sgn} t \,, \tag{2.117}$$

where ω_0 is wholly real. Using this result, we have

$$\frac{1}{2\pi} \int_{-\infty}^{\infty} \frac{e^{i\omega(\tau - t)}}{(\omega_0^2 - \omega^2)} d\omega = \frac{1}{4\pi\omega_0} \int_{-\infty}^{\infty} \left\{ \frac{e^{i\omega(\tau - t)}}{(\omega_0 - \omega)} + \frac{e^{i\omega(\tau - t)}}{(\omega_0 + \omega)} \right\} d\omega$$

$$= \frac{1}{4\pi\omega_0} \left\{ \pi i e^{i\omega_0((\tau - t))} \operatorname{sgn}(t - \tau) + \pi i e^{-i\omega_0((\tau - t))} \operatorname{sgn}((\tau - t)) \right\}$$

$$= \frac{1}{4\omega_0} \left\{ e^{i\omega_0(\tau - t)} - e^{-i\omega_0(\tau - t)} \right\} \operatorname{sgn}(t - \tau)$$

$$= \frac{\operatorname{sgn}(\tau - t)}{2\omega_0} \sin \omega_0(\tau - t) \,, \quad \text{so that} \,, \tag{2.118}$$

$$G(t; \tau) = \frac{A}{2\pi} e^{i\omega_0 t} + \frac{B}{2\pi} e^{-i\omega_0 t} + \frac{\sin \omega_0(t - \tau)}{2\omega_0} \operatorname{sgn}(\tau - t) \,. \tag{2.119}$$

The two functions A and B can be eliminated by an appear to the "causality" condition which requires that the field G should not anticipate the source, i.e. $G = 0$ for all $t < \tau$. This requires

$$\frac{A}{2\pi} e^{i\omega_0 t} + \frac{B}{2\pi} e^{-i\omega_0 t} = \frac{\sin \omega_0(t - \tau)}{2\omega_0} \tag{2.120}$$

so that

$$G(t; \tau) = \frac{\sin \omega_0(t - \tau)}{\omega_0} H(t - \tau) \,, \tag{2.121}$$

is the causal Green function for the simple harmonic oscillator.

This function would alternatively be obtained by first solving the homogeneous equation $(d^2/dt^2 + \omega_0^2)G = 0$ in the intervals $t < \tau$ and $t > \tau$, invoking the causality condition for $t < \tau$, continuity of G at $t = \tau$ and the appropriate discontinuity of dG/dt at $t = \tau$.

2.7   GREEN FUNCTIONS FOR THE WAVE EQUATION

For the wave equation we have

$$L = \frac{\partial^2}{\partial t^2} - c^2 \nabla^2 \ , \tag{2.122}$$

so that

$$\frac{\partial^2 G}{\partial t^2}(\mathbf{x}, t; \mathbf{y}, \tau) - c^2 \nabla_\mathbf{x}^2 G(\mathbf{x}, t; \mathbf{y}, \tau) = \delta(\tau - t)\delta(\mathbf{x} - \mathbf{y}) \ . \tag{2.123}$$

In terms of Fourier transforms, this is

$$\frac{\partial^2}{\partial t^2} G(\mathbf{k}, t; \mathbf{y}, \tau) + c^2 k^2 G(\mathbf{k}, t; \mathbf{y}, \tau) = e^{i\mathbf{k}\cdot\mathbf{y}}\delta(t - \tau) \ ,$$

i.e.

$$\left\{ \frac{\partial^2}{\partial t^2} + c^2 k^2 \right\} G(\mathbf{k}, t; \mathbf{y}, \tau)e^{-i\mathbf{k}\cdot\mathbf{y}} = \delta(t - \tau) \ . \tag{2.124}$$

This is the oscillator Equation (2.110), so that from (2.121) we have

$$G(\mathbf{k}, t; \mathbf{y}, \tau) = e^{i\mathbf{k}\cdot\mathbf{y}} \, \frac{\sin ck(t - \tau)}{ck} \, H(t - \tau) \ . \tag{2.125}$$

The required causal Green function is the Fourier inverse of this result; thus

$$G(\mathbf{x}, t; \mathbf{y}, \tau) = \frac{1}{(2\pi)^n} \int_\infty e^{i\mathbf{k}\cdot(\mathbf{y}-\mathbf{x})} \, \frac{\sin ck(t - \tau)}{ck} H(t - \tau) d\mathbf{k} \ , \tag{2.126}$$

where n is the number of space dimensions.

2.7.1   One–dimensional Wave Equation; $n = 1$.

The function $(ck)^{-1} \sin ck(t - \tau)$ is an even function in k, so that only the cosine terms need be retained in the exponential in Equation (2.126). Then

$$G(x, t; y, \tau) = \frac{1}{\pi} \int_0^\infty \cos k(y - x) \sin ck(t - \tau)\frac{dk}{ck} H(t - \tau) \tag{2.127}$$

$$= \frac{H(t - \tau)}{2\pi} \int_0^\infty \left\{ \sin k(y - x + c[t - \tau]) - \sin k(y - x - c[t - \tau]) \right\} \frac{dk}{ck} \ . \tag{2.128}$$

Now we use the standard result

$$\int_0^\infty \frac{\sin mx}{x} \, dx = \frac{\pi}{2} \, \mathrm{sgn}\, m \ , \tag{2.129}$$

so that

$$G(x,t;y,\tau) = \frac{H(t-\tau)}{4c} \left\{ \mathrm{sgn}(y - x + c[t-\tau]) - \mathrm{sgn}(y - x - c[t-\tau]) \right\}. \quad (2.130)$$

$$= \frac{1}{2c} \, H(c(t-\tau) - |y - x|) \qquad (2.131)$$

is the causal Green function for the one–dimensional wave equation.

It is straightforward to verify by direct differentiation that this is the required solution of (2.123).

2.7.2 Two–dimensional Wave Equation; $n = 2$

The integral over $\mathbf{k}$ in (2.126) is best conducted in cylindrical polar coordinates by writing

$$\mathbf{k} \cdot (\mathbf{y} - \mathbf{x}) = k|\mathbf{y} - \mathbf{x}|\cos\theta, \text{ and } \int d\mathbf{k} = \int_0^\infty \int_0^{2\pi} k \, dk \, d\theta . \qquad (2.132)$$

Thus

$$G(\mathbf{x},t;\mathbf{y},\tau) = \frac{1}{(2\pi)} \int_0^\infty \int_0^{2\pi} e^{ik|\mathbf{y}-\mathbf{x}|\cos\theta} d\theta \, \frac{\sin \, ck(t-\tau)}{ck} H(t-\tau) k \, dk$$

$$= \frac{1}{2\pi} \int_0^\infty J_0(k|\mathbf{x} - \mathbf{y}|) \, \frac{\sin \, ck(t-\tau)}{c} \, H(t-\tau) dk \qquad (2.133)$$

$$= \frac{1}{2\pi c} \left\{ c^2(t-\tau)^2 - |\mathbf{x} - \mathbf{y}|^2 \right\}^{-1/2} H(c(t-\tau) - |\mathbf{x} - \mathbf{y}|) \quad (2.134)$$

2.7.3 Three–dimensional Wave Equation; $n = 3$

We now evaluate the wavenumber integral in (2.126) in spherical polar coordinates, by writing

$$\mathbf{k} \cdot (\mathbf{y} - \mathbf{x}) = k|\mathbf{y} - \mathbf{x}|\cos\theta, \text{ and } \int d\mathbf{k} = \int_{k=0}^\infty \int_0^\pi 2\pi k^2 \sin\theta d\theta \, dk , \qquad (2.135)$$

$$G(\mathbf{x},t;\mathbf{y}\tau) = \frac{1}{(2\pi)^2} \int_{k=0}^\infty \int_0^\pi e^{ik|\mathbf{x}-\mathbf{y}|\cos\theta} H(t-\tau) \sin\theta \, d\theta \, 2\pi k^2 \frac{\sin \, ck(t-\tau)}{ck} \, dk$$

$$= \frac{1}{2\pi^2 c} \int_0^\infty \sin k|\mathbf{x} - \mathbf{y}| \sin ck(t-\tau) \frac{H(t-\tau)dk}{|\mathbf{x} - \mathbf{y}|} \qquad (2.136)$$

$$= \frac{H(t-\tau)}{(2\pi)^2 c|\mathbf{x} - \mathbf{y}|} \int_0^\infty \cos(k[|\mathbf{x} - \mathbf{y}| - c(t-\tau)]) - \cos(k[|\mathbf{x} - \mathbf{y}| + c(t-\tau)])dk.$$

$$(2.137)$$

The integral can be evaluated by use of the result that

$$\int_0^\infty k^m e^{-ikx} dk = \pi(i)^m \delta^m(x) = \frac{m!}{(ix)^{m+1}} \; ; \qquad (2.138)$$

in particular when $m = 0$

$$\int_0^\infty e^{ikx} dk = \int_0^\infty \cos kx \; dk - i \int_0^\infty \sin kx \; dk = \pi\delta(x) - \frac{1}{x} \; . \qquad (2.139)$$

Thus,

$$G(\mathbf{x}, t; \mathbf{y}, \tau) = \frac{H(t-\tau)}{4\pi c|\mathbf{x} - \mathbf{y}|} \{\delta(|\mathbf{x} - \mathbf{y}| - c(t-\tau)) - \delta(|\mathbf{x} - \mathbf{y}| + c(t-\tau))\} \quad (2.140)$$

The second of these terms is zero, so that the Green function for the three–dimensional wave equation under the causality restriction is the well–known retarded potential

$$G(\mathbf{x}, t; \mathbf{y}, \tau) = \frac{\delta(|\mathbf{x} - \mathbf{y}| - c(t-\tau))}{4\pi c|\mathbf{x} - \mathbf{y}|} \qquad (2.141)$$

$$= \frac{\delta((t-\tau) - |\mathbf{x} - \mathbf{y}|/c)}{4\pi c|\mathbf{x} - \mathbf{y}|} \; . \qquad (2.142)$$

2.8 THE BENDING WAVE EQUATION

Thin elastic plates have small–amplitude vibration governed by the time–dependent bi–harmonic equation,

$$L = \frac{\partial^2}{\partial t^2} + \gamma^2 \nabla^4 \; . \qquad (2.143)$$

The Green function for bending waves is therefore the solution of

$$\left\{ \frac{\partial^2}{\partial t^2} + \gamma^2 \nabla_\mathbf{x}^4 \right\} G(\mathbf{x}, t; \mathbf{y}, \tau) = \delta(|\mathbf{x} - \mathbf{y}|)\delta(t-\tau) \; . \qquad (2.144)$$

This equation has Fourier transforms in space given by

$$\left\{ \frac{\partial^2}{\partial t^2} + \gamma^2 k^4 \right\} g(\mathbf{k}, t; \mathbf{y}, \tau) = e^{i\mathbf{k} \cdot \mathbf{y}} \delta(t - \tau) , \qquad (2.145)$$

so that, from (2.121), we have

$$g(\mathbf{k}, t; \mathbf{y}, \tau) = e^{i\mathbf{k} \cdot \mathbf{y}} \frac{\sin \gamma k^2 (t - \tau)}{\gamma k^2} H(t - \tau) , \qquad (2.146)$$

and

$$G(\mathbf{x}, t; \mathbf{y}, \tau) = \frac{1}{(2\pi)^n} \int e^{i\mathbf{k} \cdot (\mathbf{y} - \mathbf{x})} \frac{\sin \gamma k^2 (t - \tau)}{\gamma k^2} d\mathbf{k} \, H(t - \tau) . \qquad (2.147)$$

For a two–dimensional plate, $n = 2$, the $\mathbf{k}$ integration can best be carried out in cylindrical polar coordinates, using Equation (2.132) to show that

$$\begin{aligned} G(\mathbf{x}, t; \mathbf{y}, \tau) &= \frac{1}{2\pi} \int_0^\infty J_0(k|\mathbf{x} - \mathbf{y}|) \frac{\sin \gamma k^2 (t - \tau)}{\gamma k} \, dk \, H(t - \tau) \\ &= \frac{H(t - \tau)}{4\pi\gamma} \int_0^\infty J_0 \left(z^{1/2} |\mathbf{x} - \mathbf{y}| \right) \sin\{\gamma(t - \tau)\} \, \frac{dz}{z} . \end{aligned} \qquad (2.148)$$

This integral can be expressed alternatively in terms of the well–tabulated si function. Thus

$$G(\mathbf{x}, t; \mathbf{y}, \tau) = -\frac{H(t - \tau)}{4\pi\gamma} \, si \left\{ \frac{|\mathbf{x} - \mathbf{y}|^2}{4\gamma(t - \tau)} \right\} , \qquad (2.149)$$

where

$$si(\alpha) = -\int_\alpha^\infty \frac{\sin t}{t} \, dt . \qquad (2.150)$$

2.9 THE REDUCED WAVE EQUATION

The Green function for the wave equation satisfies the equation

$$\left(\frac{\partial^2}{\partial t^2} - c^2 \nabla^2 \right) G(\mathbf{x}, \mathbf{y}; t) = \delta|\mathbf{x} - \mathbf{y}| . \qquad (2.151)$$

Now we define $G(\mathbf{x}, \mathbf{y}; \omega)$ to be the transform of $G(\mathbf{x}, \mathbf{y}; t)$ with respect to time,

$$G(\mathbf{x}, \mathbf{y}; \omega) = \int_{-\infty}^\infty G(\mathbf{x}, \mathbf{y}; t) e^{i\omega t} dt . \qquad (2.152)$$

The governing equation (2.151) then takes the form

$$(-\omega^2 - c^2\nabla^2)G(\mathbf{x}, \mathbf{y}; \omega) = \delta(\mathbf{x} - \mathbf{y}) ,$$

i.e.

$$(\nabla^2 + k_0^2)G(\mathbf{x}, \mathbf{y}; \omega) = -\frac{1}{c^2} \delta(\mathbf{x} - \mathbf{y}) , \qquad (2.153)$$

where $k_0 = \omega/c$. It follows that the solution of the "reduced wave equation" or Helmholtz equation

$$(\nabla^2 + k_0^2)g|\mathbf{x} - \mathbf{y}| = \delta(\mathbf{x} - \mathbf{y}) \qquad (2.154)$$

is

$$g(\mathbf{x} - \mathbf{y}) = -c^2 \int_0^\infty G(\mathbf{x}, \mathbf{y}; t)e^{i\omega t}dt . \qquad (2.155)$$

Using this formula in conjunction with the solution (2.131) for the one–dimensional wave equation, it is seen that

$$g(\mathbf{x}, \mathbf{y}) = -\frac{1}{2}c \int_{-\infty}^\infty H(ct - |\mathbf{x} - \mathbf{y}|)e^{i\omega t}dt$$

$$= -\frac{1}{2}c \left[\frac{1}{i\omega}e^{i\omega t}H(ct - |\mathbf{x} - \mathbf{y}|)\right]_{-\infty}^\infty$$

$$+ \frac{c}{2i\omega} \int_{-\infty}^\infty e^{i\omega t}c\delta(ct - |\mathbf{x} - \mathbf{y}|)dt$$

$$= 0 + \frac{c}{2i\omega}e^{i\omega|\mathbf{x}-\mathbf{y}|/c} ,$$

i.e.

$$y(x, y) = \frac{1}{2ik_0} e^{ik_0|x-y|} \qquad (2.156)$$

is the Green function for the one–dimensional form of the Helmholtz Equation (2.154).

It is straightforward to verify by direct differentiation that

$$g(x, y) = \frac{1}{2ik_0} e^{ik_0|x-y|}$$

satisfies

$$\left(\frac{\partial^2}{\partial x^2} + k_0^2\right) g(x, y) = \delta(x - y) .$$

Similarly, in two dimensions, using (2.115) together with Equation (2.134) we have

$$g(\mathbf{x} - \mathbf{y}) = -\frac{c^2}{2\pi c} \int_{-\infty}^{\infty} \frac{H(ct-r)}{(c^2t^2-r^2)^{1/2}} e^{i\omega t} dt, \quad \text{where} \quad r = |\mathbf{x} - \mathbf{y}|$$

$$= -\frac{c}{2\pi} \int_{r/c}^{\infty} \frac{e^{i\omega t}}{(c^2t^2-r^2)^{1/2}} dt$$

$$= -\frac{1}{2\pi} \int_{0}^{\infty} e^{ik_0 r \cosh u} du \; ;$$

which is obtained by putting $t = \frac{r}{c} \cosh u$, $k_0 = \omega/c$.

Evidently then,

$$g(\mathbf{x} - \mathbf{y}) = \frac{1}{4i} H_0^{(1)}(k_0 r) \;, \tag{2.157}$$

where $H_0^{(1)}(z) = \frac{1}{\pi i} \int_{-\infty}^{\infty} e^{iz \cosh u} du$ is the Hankel function of the first kind and zero order, is the two–dimensional Green function for the Helmholtz Equation (2.154).

Finally, the solution of Equation (2.154) in three dimensions is obtained from (2.155) and (2.141). Then

$$g(\mathbf{x} - \mathbf{y}) = -\frac{c^2}{4\pi cr} \int_{-\infty}^{\infty} \delta(r - ct) e^{i\omega t} dt, \quad r = |\mathbf{x} - \mathbf{y}| \;,$$

$$= -\frac{1}{4\pi r} e^{ik_0 t} \;, \tag{2.158}$$

is the Green function for the three–dimensional form of the Helmholtz equation.

2.10 SOUND WAVES WITH DAMPING

Propagating sound waves in a slightly viscous fluid are described by the operator

$$\frac{\partial^2}{\partial t^2} - 2\beta\nabla^2 \frac{\partial}{\partial t} - c^2\nabla^2 \;,$$

β being a small constant proportional to the kinematic viscosity of the fluid; squares of β are neglected. The Green function for this operator is

$$G(\mathbf{x}, t; \mathbf{y}, \tau) = \frac{1}{(2\pi)^n} \int e^{-i\mathbf{k}\cdot(\mathbf{x}-\mathbf{y})-\beta k^2(t-\tau)} \frac{\sin ck(t-\tau)}{ck} H(t - \tau) d\mathbf{k} \;.$$

By evaluating the integral for three space dimensions, we can show that

$$G(\mathbf{x},t;\mathbf{y},\tau) = \frac{1}{8\pi^{3/2}cr(\beta t')^{1/2}} \left[\exp\left\{\frac{(r-ct')^2}{4\beta t'}\right\} - \exp\left\{\frac{(r+ct')^2}{4\beta t'}\right\} \right] H(t') \,,$$

where $r = |\mathbf{x} - \mathbf{y}|$ and $t' = t - \tau$.

2.11 INTERNAL WAVES

Small oscillations of a fluid with uniform density gradient in the vertical (x_N) direction are described by

$$L = \frac{\partial^2}{\partial t^2} \nabla^2 + N^2 \nabla_h^2 \,,$$

where ∇_h^2 denotes the Laplacian operator in the horizontal coordinates, and N is constant. The Green function for this internal wave equation,

$$\left(\frac{\partial^2}{\partial t^2}\nabla^2 + N^2 \nabla_h^2\right) G = \delta(\mathbf{x}_h)\delta(x_N)\delta(t)$$

which satisfies causality is

$$G(\mathbf{x}_h, x_N, t) =$$
$$-\frac{H(t)}{(2\pi)^3} \int \frac{1}{N k_h k} \sin\left(\frac{N k_h t}{k}\right) e^{-i\mathbf{k}_h \mathbf{x}_h - i k_N x_N} \, d\mathbf{k}_h dk_N \,,$$

where $k = \sqrt{k_h^2 + k_N + 2}$. By writing $k_h = k \cos \alpha$, $k_N = k \sin \alpha$ and evaluating the integral over k, one can show that

$$G(\mathbf{x}_h, x_N, t) = -\frac{H(t)}{2\pi N r} \int_0^\theta \frac{\sin(N t \cos \alpha)d\alpha}{\sqrt{\cos^2 \alpha - \cos^2 \theta}} \,,$$

where $\mathbf{x}_h = r \sin \theta$, $x_N = r \cos \theta$

$$G_{t\to 0} \sim -\frac{tH(t)}{4\pi r} \,.$$

REFERENCES

Lighthill, M.J. (1958) Fourier Analysis and Generalised Functions. Cambridge University Press.

Gelfand, I.M. & Shilov, G.E. (1964) Generalised Functions. Academic Press.

Jones, D.S. (1066) Generalised Functions. McGraw–Hill.

3. FOURIER TRANSFORMS, RANDOM PROCESSES, DIGITAL SAMPLING AND WAVELETS

3.1 DEFINITIONS AND FORMAL PROPERTIES OF FOURIER TRANSFORMS

Consider ordinary functions $f(t)$ of a real variable t, $-\infty < t < \infty$, with at most a finite number of simple jump discontinuities and integrable singularities (i.e. no worse than $|t|^{-\nu}$ for some $\nu < 1$, if the singularity is at $t = 0$). Suppose also that $|f(t)| < A_+ \exp(-\epsilon_+ t)$ as $t \to +\infty$ for some A_+ and $\epsilon_+ > 0$. Then the *half–range Fourier transform*

$$\tilde{f}_+(\omega) = \int_0^\infty f(t) \exp(i\omega t)dt \tag{3.1}$$

exists as an analytic function of the complex "frequency" ω in the upper half–plane $R_+ : \Im\omega > -\epsilon_+$. In the same way, if $|f(t)| < A_- \exp(\epsilon_- t)$ as $t \to -\infty$, the half–range Fourier transform

$$\tilde{f}_-(\omega) = \int_{-\infty}^0 f(t) \exp(i\omega t)dt \tag{3.2}$$

exists as an analytic function in the lower half–plane $R_- : \Im\omega < +\epsilon_-$. If both conditions on $f(t)$ are satisfied together, the *full–range Fourier transform*

$$\tilde{f}(\omega) = \tilde{f}_+(\omega) + \tilde{f}_-(\omega) = \int_{-\infty}^\infty f(t) \exp(i\omega t)dt \tag{3.3}$$

exists as an analytic function in the strip D of overlap between R_+ and R_-,

$$D : -\epsilon_+ < \Im\omega < \epsilon_- . \tag{3.4}$$

In some applications one deals with the Fourier transform over a finite interval of t, corresponding to functions $f(t)$ of a compact support (i.e. vanishing outside some interval that can usually be taken as $(0, \ell)$). Then

$$\widetilde{f_\ell}(\omega) = \int_0^\ell f(t) \exp(i\omega t) dt \tag{3.5}$$

is an entire function of ω, analytic in the whole ω- plane.

These various Fourier transforms have distinct and important behaviour as $|\omega| \to \infty$. Let $|\omega| \to \infty$ along the positive imaginary axis. Then

$$\widetilde{f}_+(iu) = \int_0^\infty f(t) \exp(-ut) dt$$

is the one–sided *Laplace transform* of $f(t)$. As $u \to +\infty$, only values of t close to zero make any significant contribution to the integral. Indeed Watson's Lemma (see Chapter 4) asserts that to within an exponentially small error, the asymptotic expansion of $\widetilde{f}_+(iu)$ as $u \to +\infty$ is obtained by inserting for $f(t)$ its asymptotic expansion as $t \to 0+$ and evaluating the Laplace transform term by term. For example, if $f(0)$ is finite and not zero,

$$\widetilde{f}_+(iu) \sim \int_0^\infty f(0) \exp(-ut) dt \quad \text{as} \quad u \to +\infty$$

$$= f(0)/u.$$

Alternatively, if f is singular at $t = 0$, for example $f(t) \sim At^{-1/2}$ as $t \to 0+$, then

$$\widetilde{f}_+(iu) \sim A \int_0^\infty \frac{\exp(-ut)}{t^{1/2}} dt \quad \text{as} \quad u \to \infty$$

$$= A \left(\frac{\pi}{u}\right)^{1/2} .$$

If the asymptotic expansion of $f(t)$ contains $\ln(t)$ terms, the Laplace transform contains corresponding $\ln(u)$ terms; for example, if $f(t) \sim At^{-1/2} \ln(t)$ for small positive t,

$$\widetilde{f}_+(iu) \sim A \int_0^\infty \frac{\ln(t)}{t^{1/2}} \exp(-ut) dt \quad \text{as} \quad u \to \infty .$$

A change of variable $x = ut$ shows the integral to be equal to

$$\frac{-\ln(u)}{u^{1/2}} \int_0^\infty \frac{e^{-x}}{x^{1/2}} dx + \frac{1}{u^{1/2}} \int_0^\infty \frac{\ln(x)}{x^{1/2}} e^{-x} dx .$$

These integrals are standard (Gradshteyn & Ryzhik 1980) and lead to

$$\widetilde{f}_+(iu) \sim -A \left(\frac{\pi}{u}\right)^{1/2} (\ln(4u) + \gamma) ,$$

where $\gamma = 0.57722...$ is Euler's constant.

Similar arguments give the behaviour of $\tilde{f}_+(\omega)$ elsewhere at infinity in R_+. Thus $\tilde{f}_+(\omega)$ is analytic in R_+ and has algebraic (or at worst algebraic–logarithmic) decay at infinity there, and correspondingly for $\tilde{f}_-(\omega)$ in R_-. Moreover, the asymptotic behaviour of $f(t)$ as $t \to 0+$ can be inferred from that of $\tilde{f}_+\omega$ as $|\omega| \to \infty$ in R_+, while that of $f(t)$ as $t \to 0-$ can be inferred from that of $\tilde{f}_-(\omega)$ as $|\omega| \to \infty$ in R_-.

The transforms $\tilde{f}_\ell(\omega)$ of functions of compact support do not have the algebraic decay at infinity of the half–range transforms (for otherwise, since they are entire, they would be identically zero). As $|\omega| \to \infty$ in R_+, the integral defining $\tilde{f}_\ell(\omega)$ is dominated by the contributions from $t = 0+$ and $\tilde{f}_\ell(\omega)$ therefore has algebraic decay in the upper half–plane. As $|\omega| \to \infty$ in R_-, however, the dominant contribution comes from the vicinity of $t = \ell-$; in fact, with $t = \ell - \alpha$,

$$\tilde{f}_\ell(-iv) = \exp(v\ell) \int_0^\ell f(\ell - \alpha) \exp(-v\alpha) d\alpha$$

$$\sim \frac{\exp(v\ell)}{v} f(\ell-), \text{ if } f(\ell-) \text{ is finite, for example .}$$

More generally, if we assume the behaviour of $f(t)$ to be described by algebraic–logarithmic functions of t near $t = 0$ and $t = \ell$, the integral contributes only algebraic–logarithmic terms in v. It is therefore clear that $\tilde{f}_\ell(\omega)$ is exponentially large, like $\exp(i\omega\ell)$, in the lower half–plane.

3.1.1 Analytic Continuation

In the first instance a full–range Fourier transform $\tilde{f}(\omega)$ is defined only in a strip D, a half–range Fourier transform in an upper or lower half–plane; the integrals defining $\tilde{f}(\omega), \tilde{f}_+(\omega)$ and $\tilde{f}_-(\omega)$ diverge elsewhere. Nonetheless, the functions $\tilde{f}(\omega)$, $\tilde{f}_+(\omega)$ and $\tilde{f}_-(\omega)$ can usually be extended to have meaning in a larger domain. Indeed this is essential if one is to take advantage of the contour deformation possibilities offered by Cauchy's theorem in performing integrals of the transforms in the ω–plane (as is necessary in the inversion process giving $f(t)$ from $\tilde{f}(\omega)$). Generally one has specific functions $\tilde{f}(\omega)$, $\tilde{f}_+(\omega)$ or $\tilde{f}_-(\omega)$. Then it is obvious, or almost obvious, how they should be analytically continued, and obvious that there is, perhaps, a branch cut or singularity in one half–plane that prevents analytic continuation everywhere. Applications in later chapters will make this clear. As one example here, suppose simply that $f(t) = \exp(-at)$, where a is a real, positive constant. Then, for $\Im \omega > -a$, $\tilde{f}_+(\omega) = \int_0^\infty \exp[-(a - i\omega)t]dt = i(\omega + ia)^{-1}$, which constant. Then, for $\Im\omega > -a$, $\tilde{f}_+(\omega) = \int_0^\infty \exp[-(a - i\omega)t]dt = i(\omega + ia)^{-1}$, which is indeed analytic for $\Im\omega > -a$ and has $\tilde{f}_+(\omega) \sim i\omega^{-1}$ as expected, as $|\omega| \to \infty$ in R_+. But $i(\omega + ia)^{-1}$ can immediately be analytically continued to the whole of the ω–plane, as an analytic function except for a non–removable pole singularity at $\omega = -ia$.

3.1.2 Fourier Transform of a Derivative

One important property of the full–range Fourier transform is that it converts differentiation in t–space into multiplication in ω–space. The Fourier transform of $f'(t)$ is

$$\int_{-\infty}^{\infty} f'(t)e^{i\omega t}d\omega = \left[f(t)e^{i\omega t}\right]_{-\infty}^{\infty} - i\omega \int_{-\infty}^{\infty} f(t)e^{i\omega t}dt$$

$$= -i\omega\tilde{f}(\omega) , \qquad (3.6)$$

where it has been assumed that ω is restricted to some strip D in which the Fourier transforms of $f'(t)$ and $f(t)$ both exist, and that $|f(t)e^{i\omega t}| \to 0$ as $t \to \pm\infty$ for all ω in D. (The usual practice is to assume such conditions to be met, to execute formal manoeuvres and then having constructed a formal solution to verify that it satisfies all the required conditions.) Since

$$\text{the Fourier transform of } f'(t) \text{ is } -i\omega\tilde{f}(\omega) , \qquad (3.7)$$

the operation of taking Fourier transforms reduces an ordinary differential equation with constant coefficients to an algebraic equation. By extension, the order of a constant–coefficient partial differential equation is reduced by one for each variable in which the Fourier transform is made.

3.1.3 Linearity of Fourier Transforms

The Fourier transform operation is linear, i.e.

$$\text{the Fourier transform of } \alpha f(t) + \beta g(t) \text{ is } \alpha\tilde{f}(\omega) + \beta\tilde{g}(\omega) , \qquad (3.8)$$

for any complex constants α, β with appropriate restrictions on ω. The half–range Fourier transform is also a linear operation.

Examples.

(i) The Fourier transform of $\exp(-a|t|)$, $a > 0$.

Straightforward integration shows that

$$\tilde{f}(\omega) = \frac{2a}{a^2 + \omega^2}, \quad \tilde{f}_+(\omega) = (a - i\omega)^{-1} \text{ and } \tilde{f}_-(\omega) = (a + i\omega)^{-1} \ .$$

These have the expected properties. Note that $\tilde{f}(\omega)$ is real and even, as it always is when $f(t)$ is real and even.

(ii) The Fourier transform of $\exp(-a^2 t^2)$, $a > 0$.

The integral for $\tilde{f}(\omega)$ converges in every strip $-\epsilon_1 < \Im\omega < \epsilon_2$ and so $\tilde{f}(\omega)$ will be an entire function. We have

$$\tilde{f}(\omega) = \int_{-\infty}^{\infty} \exp\left(-\left(at - \frac{i\omega}{2a}\right)^2 - \frac{\omega^2}{4a^2}\right) dt \ .$$

Deform the contour of integration from the real t-axis to the line $\Im t = i\omega/2a^2$. Cauchy's theorem shows that this is legitimate. On this path we can put $at - i\omega/2a = u$ where u is real, $-\infty < u < \infty$. Thus the original integral over t is equal to one over u on the shifted path and is

$$\tilde{f}(\omega) = \frac{e^{-\omega^2/4a^2}}{a} \int_{-\infty}^{\infty} e^{-u^2} du$$

$$= \pi^{1/2} a^{-1} \exp(-\omega^2/4a^2) \ .$$

(iii) The Fourier transform of 1 **for** $0 < t < \ell$.

We have

$$\tilde{f}(\omega) = \tilde{f}_\ell(\omega) = \int_0^\ell e^{i\omega t} dt$$

$$= \frac{e^{i\omega\ell} - 1}{i\omega}$$

which is analytic everywhere. $\tilde{f}(\omega)$ behaves like $i\omega^{-1}$ as $|\omega| \to \infty$ in R_+, where $e^{i\omega\ell} \to 0$, and like $e^{i\omega\ell}/i\omega$ as $|\omega| \to \infty$ in R_-.

(iv) The Fourier transform of $|t|^{-1/2}$.

$$\tilde{f}_+(\omega) = \int_0^{\infty} \frac{e^{i\omega t}}{t^{1/2}} dt$$

actually converges for real ω even though the integrand only tends to zero slowly as $t \to \infty$. For complex ω with $\Im\omega > 0$, $\widetilde{f}_+(\omega)$ can be evaluated by using Cauchy's theorem to replace the integration along the positive real axis of t by one along a line with arg $t = \pi/2 - \arg \omega$. On this line $i\omega t$ is real and negative and the substitution $i\omega t = -u^2$ converts the integral into standard form to show

$$\widetilde{f}_+(\omega) = \frac{2e^{i\pi/4}}{\omega^{1/2}} \int_0^\infty e^{-u^2} du = e^{i\pi/4} \left(\frac{\pi}{\omega}\right)^{1/2} \quad,$$

where the cut for $\omega^{1/2}$ must lie in $\Im\omega < 0$.

For $\widetilde{f}_-(\omega)$ we have similarly

$$\widetilde{f}_-(\omega) = e^{-\pi i/4} \left(\frac{\pi}{\omega}\right)^{1/2} \quad,$$

where now, however, the cut for $\omega^{1/2}$ must now lie in $\Im(\omega) > 0$.

The integrals for $\widetilde{f}_+(\omega)$ and $\widetilde{f}_-(\omega)$ only converge simultaneously, when ω is real. Then if $\omega > 0$, $\omega^{1/2}$ has its ordinary meaning for both $\widetilde{f}_+(\omega)$, and $\widetilde{f}_-(\omega)$, while when $\omega < 0$, $\omega^{1/2}$ means $|\omega|^{1/2}e^{i\pi/2}$ for $\widetilde{f}_+(\omega)$ and $|\omega|^{1/2}e^{-i\pi/2}$ for $\widetilde{f}_-(\omega)$. Thus for ω real,

$$\widetilde{f}(\omega) = 2\cos\left(\frac{\pi}{4}\right) \frac{\pi^{1/2}}{|\omega|^{1/2}} \quad,$$

whether ω is positive or negative. Similar arguments apply to the half–range and full–range Fourier transforms of all functions $|t|^{-\gamma}$, $0 < \gamma < 1$. For $\gamma > 1$ the defining integrals do not converge for any complex ω. Nevertheless the Fourier transforms of these functions exist as generalized functions. The way in which some of these generalized Fourier transforms can be calculated will be discussed in Section 3.2.

3.1.4 Fourier Inversion Theorem

The Fourier inversion theorem will just be stated here (see Jeffreys & Jeffreys (1949) for a formal proof). If

$$\widetilde{f}(\omega) = \int_{-\infty}^\infty f(t')e^{i\omega t'} dt' \quad,$$

is analytic in a strip $D : -\epsilon_+ < \Im\omega < \epsilon_-$, then for any path C from $-\infty$ to $+\infty$ in that strip

$$f(t) = \frac{1}{2\pi} \int_C e^{-i\omega t} \widetilde{f}(\omega) d\omega \quad. \tag{3.9}$$

Let us decompose $\tilde{f}(\omega)$ into

$$\tilde{f}(\omega) = \tilde{f}_+(\omega) + \tilde{f}_-(\omega) \ ,$$

where $\tilde{f}_+(\omega)$ is analytic in $\Im\omega > -\epsilon_+$ and vanishes algebraically at infinity therein. $\tilde{f}_-(\omega)$ is to be analytic in $\Im\omega < \epsilon_-$ and similarly decay algebraically at infinity in that region. Then we have

$$f(t) = \frac{1}{2\pi} \int_C \tilde{f}_+(\omega) e^{-i\omega t} d\omega + \frac{1}{2\pi} \int_C \tilde{f}_-(\omega) e^{-i\omega t} d\omega \ ,$$

and for the first term we deform C to infinity in the upper-half ω-plane. The integrand is small at infinity in $\Im\omega > 0$ for negative t, and no singularities are crossed, so the first integral vanishes for all $t < 0$, and

$$f(t) = \frac{1}{2\pi} \int_C \tilde{f}_-(\omega) e^{-i\omega t} d\omega \ \text{ for } \ t < 0 \ .$$

We see that $f(t)$ for negative t is determined solely by the values of $\tilde{f}_-(\omega)$, just as $\tilde{f}_-(\omega)$ was defined solely by the values of $f(t)$ for negative t. In a similar way, $f(t)$ for positive t is determined by $\tilde{f}_+(\omega)$ alone.

3.1.5 Convolution Theorem

The convolution product notation was introduced in Chapter 2; it defines

$$f * g(t) = \int_{-\infty}^{\infty} f(t') g(t - t') dt' \ . \tag{3.10}$$

The convolution theorem states that the Fourier transform of a convolution is the product of the Fourier transform of the factors,

$$\widetilde{(f * g)}(\omega) = \tilde{f}(\omega)\tilde{g}(\omega) \ . \tag{3.11}$$

Proof.

$$\widetilde{(f * g)}(\omega) = \int_{-\infty}^{\infty} e^{i\omega t} \left\{ \int_{-\infty}^{\infty} f(t') g(t - t') dt' \right\} dt \ .$$

After reversing the orders of integration, this is

$$\int_{-\infty}^{\infty} f(t') \left\{ \int_{-\infty}^{\infty} e^{i\omega t} g(t - t') dt \right\} dt'$$

$$= \int_{-\infty}^{\infty} f(t') e^{i\omega t'} \left\{ \int_{-\infty}^{\infty} e^{i\omega(t - t')} g(t - t') d(t - t') \right\} dt'$$

$$= \int_{-\infty}^{\infty} f(t') e^{i\omega t'} dt' \int_{-\infty}^{\infty} g(\tau) e^{i\omega\tau} d\tau$$

$$= \tilde{f}(\omega)\tilde{g}(\omega) \ .$$

In an analogous way the Fourier transform of a product $f(t)g(t)$ is related to the convolution of their transform,

$$\widetilde{f(t)g(t)} = \frac{1}{2\pi} \, \widetilde{f} * \widetilde{g}(\omega) \; . \tag{3.12}$$

This can be proved by considering the inverse Fourier transform of $(2\pi)^{-1} \widetilde{f} * \widetilde{g}(\omega)$:

$$\frac{1}{(2\pi)^2} \int_{-\infty}^{\infty} e^{-i\omega t} \widetilde{f} * \widetilde{g}(\omega) d\omega = \frac{1}{(2\pi)^2} \int_{-\infty}^{\infty} e^{-i\omega t} \left\{ \int_{-\infty}^{\infty} \widetilde{f}(\lambda) \widetilde{g}(\omega - \lambda) \right\} d\omega \; .$$

After reversing the orders of integration, this is

$$= \left\{ \frac{1}{2\pi} \int_{-\infty}^{\infty} e^{-i\lambda t} \widetilde{f}(\lambda) d\lambda \right\} \left\{ \frac{1}{2\pi} \int_{-\infty}^{\infty} e^{-i(\omega - \lambda)t} \widetilde{g}(\omega - \lambda) d(\omega - \lambda) \right\}$$

$$= f(t)g(t) \; .$$

The product $f(t)g(t)$ is equal to the inverse Fourier transform of $(2\pi)^{-1} \widetilde{f} * \widetilde{g}(\omega)$. Equation (3.12) then follows.

3.1.6 Parseval's Theorem.

As a particular case of the convolution theorem, take $t = 0$. Then

$$\int_{-\infty}^{\infty} f(t')g(-t')dt' = \frac{1}{2\pi} \int_{-\infty}^{\infty} \widetilde{f}(\omega) \widetilde{g}(\omega) d\omega \; . \tag{3.13}$$

Putting, further, $g(-t') = h(t') = f^*(t')$, where the star denotes a complex conjugate, the left–hand side becomes

$$\int_{-\infty}^{\infty} |f(t)|^2 dt \; ,$$

a measure of the "total energy" of the process $f(t)$. It follows directly from the definition of $g(t)$ that $\widetilde{g}(\omega) = \widetilde{f}^*(\omega)$, and hence the right–hand side of (3.13) becomes

$$\frac{1}{2\pi} \int_{-\infty}^{\infty} |\widetilde{f}(\omega)|^2 d\omega \; ,$$

in which $|\widetilde{f}(\omega)|^2$ is a measure of the "energy" (per unit ω) at the frequency ω and the integral represents the total energy expressed in terms of its distribution in frequency rather than in time

$$\int_{-\infty}^{\infty} |f(t)|^2 dt = \int_{-\infty}^{\infty} |\widetilde{f}(\omega)|^2 \frac{d\omega}{2\pi} \; . \tag{3.14}$$

3.2 TRANSFORMS OF GENERALIZED FUNCTIONS

We recall the definition introduced in Chapter 2 of a good function of the real variable t, $G(t)$: $G(t)$ is to be everywhere differentiable and such that it and all its derivatives are $O(|t|^{-N})$ as $|t| \to \infty$ for all N. It can be shown directly from the definition of the Fourier transform that the Fourier transform of any good $G(t)$ is itself a good function of ω. Generalized functions were defined in Chapter 2 in terms of the way in which they operate on good functions. The Fourier transform of a generalized function can be defined in a similar way.

We will denote the Fourier transform of the generalized function $f(t)$ by $\tilde{f}(\omega)$ and define it by

$$\int_{-\infty}^{\infty} \tilde{f}(\omega)G(\omega)d\omega = \int_{-\infty}^{\infty} f(t)\tilde{G}(t)dt \ . \tag{3.15}$$

The integrals in this expression are to be interpreted in the way indicated in Equation (2.5). $G(\omega)$ is an arbitrary good function of ω and its Fourier transform is some good function $\tilde{G}(t)$,

$$\tilde{G}(t) = \int_{-\infty}^{\infty} G(\omega)e^{i\omega t}d\omega \ . \tag{3.16}$$

Suppose a particular $f(t)$ is sufficiently well–behaved for its Fourier transform to exist in the classical way outlined in Section 3.1. We can show that the definition in (3.15) coincides with the classical Fourier transform of $f(t)$, $\tilde{f}_c(\omega)$, say. For such a function $f(t)$, the integral on the right–hand side of (3.15) exists in the ordinary sense. We can write

$$\int_{-\infty}^{\infty} f(t)\tilde{G}(t)dt = \int_{-\infty}^{\infty} f(t)\left(\int_{-\infty}^{\infty} G(\omega)e^{i\omega t}d\omega\right)dt \ .$$

After exchanging the order of integration, this becomes

$$\int_{-\infty}^{\infty} f(t)\tilde{G}(t)dt = \int_{-\infty}^{\infty} G(\omega)\left(\int_{-\infty}^{\infty} f(t)e^{i\omega t}dt\right)d\omega$$

$$= \int_{-\infty}^{\infty} G(\omega)\tilde{f}_c(\omega)d\omega \ , \tag{3.17}$$

which is identical to (3.15). Two generalized functions f_1 and f_2 are said to be equal (see Chapter 2), when

$$\int_{-\infty}^{\infty} f_1(t)G(t)dt = \int_{-\infty}^{\infty} f_2(t)G(t)dt$$

for all good functions $G(t)$. Hence our definition of the Fourier transform of a generalized function is identical to the classical Fourier transform whenever the latter exists. However, Equation (3.15) enables a Fourier transform to be defined for all generalized functions, although, of course, the transform may itself be a generalized function. This is a powerful extension.

3.2.1 Fourier Inversion Theorem

Since $G(t)$ is a good function, we know from Section 3.1 that Fourier inversion applies to it. In our notation this can be expressed in the form

$$\widetilde{\widetilde{G}}(-t) = 2\pi G(t). \tag{3.18}$$

The inversion theorem for $f(t)$ then follows from the definition (3.15). First we note that a simple change of variable shows that

$$\int_{-\infty}^{\infty} \widetilde{\widetilde{f}}(-t)G(t)dt = \int_{-\infty}^{\infty} \widetilde{\widetilde{f}}(t)G(-t)dt \ . \tag{3.19}$$

Then, after two applications of (3.15), we can write

$$\int_{-\infty}^{\infty} \widetilde{\widetilde{f}}(-t)G(t)dt = \int_{-\infty}^{\infty} f(t)\widetilde{\widetilde{G}}(-t)dt \ . \tag{3.20}$$

Finally, using the inversion theorem for $G(t)$ expressed in (3.18), we note that

$$\int_{-\infty}^{\infty} \widetilde{\widetilde{f}}(-t)G(t)dt = \int_{-\infty}^{\infty} 2\pi f(t)G(t)dt \ . \tag{3.21}$$

Equality occurs for all good functions, $G(t)$, and hence

$$\widetilde{\widetilde{f}}(-t) = 2\pi f(t) \ , \tag{3.22}$$

which proves the inversion theorem.

3.2.2 Fourier Transform of a Derivative

The derivative of a generalized function is defined by

$$\int_{-\infty}^{\infty} f'(t)G(t)dt = -\int_{-\infty}^{\infty} f(t)G'(t)dt \ , \tag{3.23}$$

for any good function $G(t)$. We can find the relationship between the Fourier transform of $f(t)$ and its derivative $f'(t)(= g(t)$, say) by working directly from the definitions.

The definition (3.15) for the Fourier transform shows that

$$\int_{-\infty}^{\infty} \widetilde{g}(\omega)G(\omega)d\omega = \int_{-\infty}^{\infty} f'(t)\widetilde{G}(t)dt$$

$$= -\int_{-\infty}^{\infty} f(t)\widetilde{G}'(t) \tag{3.24}$$

from the definition of the derivative. Now $\widetilde{G}(t)$ is the Fourier transform of the good function $G(\omega)$, and straightforward differentiation demonstrates that $\widetilde{G}'(t)$ is the Fourier transform of $i\omega G(\omega)$. Hence application of the definition (3.15) to the right–hand side of (3.24) shows that

$$\int_{-\infty}^{\infty} \widetilde{g}(\omega)G(\omega)d\omega = -\int_{-\infty}^{\infty} \widetilde{f}(\omega)i\omega G(\omega)d\omega \ . \tag{3.25}$$

Since this applies for all good functions, $G(\omega)$, we have shown that

$$\text{the Fourier transform of } f'(t) \text{ is } -i\omega\widetilde{f}(\omega) \ . \tag{3.26}$$

This extends the result in (3.6) to generalized functions.

3.2.3 Evaluation of Fourier Transforms of Generalized Functions

The Fourier transforms of generalized functions can be evaluated by

(a) working directly from the definition in (3.15);

(b) using the fact that the Fourier transform of $f'(t)$ is $-i\omega\widetilde{f}(\omega)$;

(c) noting the legality, in generalized function theory, of inserting "convergence factors" and letting a parameter $\epsilon \to 0$ at a suitable stage.

Examples.

(i) **The Fourier transform of $\delta(t)$.**

We see from the definition of the Fourier transform in (3.15) that

$$\int_{-\infty}^{\infty} \widetilde{\delta}(\omega)G(\omega)d\omega = \int_{-\infty}^{\infty} \delta(t)\widetilde{G}(t)dt \tag{3.27}$$

$$= \widetilde{G}(0) \ , \tag{3.28}$$

from the integral property of the δ–function. Now $\widetilde{G}(0) = \int_{-\infty}^{\infty} G(\omega)d\omega$. Hence Equation (3.28) is equivalent to

$$\int_{-\infty}^{\infty} \widetilde{\delta}(\omega)G(\omega)d\omega = \int_{-\infty}^{\infty} G(\omega)d\omega \ ,$$

i.e.

$$\widetilde{\delta}(\omega) = 1 \quad \text{or the unit function } I \ . \tag{3.29}$$

(ii) **The Fourier transform of the unit function $I(t)$.**

The inversion of the result in (3.29) is

$$\widetilde{I}(\omega) = 2\pi\delta(\omega) \ . \tag{3.30}$$

(iii) **The Fourier transform of sgn t.**

Take $f(t) = \text{sgn } t$. Then $f'(t) = 2\delta(t)$ and

$$-i\omega\widetilde{f}(\omega) = 2 \tag{3.31}$$

from (3.26) and (3.29). We will call the solution to $y\,g(y) = 1$, y^{-1}. One particular solution is $g(y) = d(\ln|y|)/dy$ and the general solution can be written

$$g(y) = \frac{d}{dy}\ln|y| + C\delta(y) \ ,$$

where C is a constant; we must choose $C = 0$ because we want the function called y^{-1} to be odd. Therefore we have

$$\text{the Fourier transform of sgn } t \text{ is } 2i\omega^{-1} \ . \tag{3.32}$$

(iv) **The Fourier transform of $H(t)$.**

Since sgn $t = 2H(t) - I(t)$, it follows at once from (3.30) and (3.32) that

$$\text{the Fourier transform of } H(t) \text{ is } i\omega^{-1} + \pi\delta(\omega) \ . \tag{3.33}$$

(v) **The Fourier transform of $t^\alpha H(t)$, $\alpha > -1$.**

The integral $\int_0^\infty t^\alpha e^{i\omega t}dt$ does not converge at infinity except when $-1 < \alpha < 0$. We therefore use method (c) and consider

$$\int_0^\infty e^{-\epsilon t + i\omega t}t^\alpha dt = \alpha!/(\epsilon - i\omega)^{\alpha+1} \ ,$$

where we use

$$\int_0^\infty t^\alpha e^{-zt}dt = \alpha!/z^{\alpha+1} \quad \text{for } Re \ \alpha > -1 \text{ and } Re \ z > 0 \ .$$

$z^{\alpha+1}$ must be regarded as having a cut from 0 to ∞ in $Re\ z < 0$. Letting $\epsilon \to 0+$, we have

$$\text{the Fourier transform of } t^\alpha H(t) = \frac{\alpha!}{|\omega|^{\alpha+1}} \exp\left[\frac{\pi i}{2}(\alpha+1)\text{sgn }\omega\right] . \qquad (3.34)$$

(vi) **The Fourier transform of $t^\alpha H(t)$, for nonintegral α less than -1.**

We remark that $t^\beta H(t)$ is a generalized function for $\beta > -1$ with derivative $\beta t^{\beta-1} H(t)$. Repeated differentiation and the rule that the Fourier transform of the nth derivative of a function is equal to $(-i\omega)^n$ times the Fourier transform of the function can be used to show that the expression in (3.34) for the Fourier transform of $t^\alpha H(t)$ for $\alpha > -1$ is, in fact, valid for all nonintegral α.

(vii) **The Fourier transform of $t^n, n \geqslant 1$.**

Consider $f(t) = \delta^{(n)}(t)$. Then $\widetilde{f}(\omega) = (-i\omega)^n \ \widetilde{\delta}(\omega) = (-i\omega)^n$. The inverse of this relationship (see Equation (3.22)) is that the Fourier transform of $(-i\omega)^n$ is $2\pi\delta^{(n)}(-t)$, or with a change of variable

$$\text{the Fourier transform of } t^n \text{ is } 2\pi(-i)^n \delta^{(n)}(\omega) . \qquad (3.35)$$

(viii) **The Fourier transform of $t^n, n \leq -1$.**

We already have that the Fourier transform of sgn t is $2i\omega^{-1}$, for which the inversion theorem and a change of notation gives

$$\text{the Fourier transform of } t^{-1} \text{ is } \pi i \text{sgn } \omega . \qquad (3.36)$$

Higher inverse powers are defined by differentiation of t^{-1} and their Fourier transforms are obtained by repeated multiplication of the Fourier transform of t^{-1} by $(-i\omega)$. Observe that only in a few cases do the Fourier transforms exist in any ordinary sense, yet in the majority of cases the generalized Fourier transforms exist as ordinary functions.

(ix) **The Fourier transform of $|t|^{-1}$.**

Let $f(t) = \ln|t|$, $f'(t) = t^{-1}$. Then from (3.36) the Fourier transform of $f'(t)$ is πi sgn ω, and is also $-i\omega \widetilde{f}(\omega)$. Therefore

$$\widetilde{f}(\omega) = -\pi|\omega|^{-1} + D\delta(\omega) \qquad (3.37)$$

for some constant D. Use of the inversion theorem then gives

$$\text{the Fourier transform of } |t|^{-1} \text{ is } -2\ln|\omega| + C \text{ for some } C . \qquad (3.38)$$

The Fourier transform of higher inverse powers can be determined from repeated differentiation.

Lighthill (1978) gives a table of the Fourier transforms of all fractional powers, $|t|^\alpha$, $|t|^\alpha \text{sgn } t$, $t^\alpha H(t)$ and positive integral powers t^n and negative integral powers t^{-m} and $|t|^{-m}$. Functions like $t^\alpha H(t) \ln t$ are defined as

$$\frac{\partial}{\partial \alpha} [t^\alpha H(t)],$$

and their Fourier transforms obtained by taking $\partial/\partial \alpha$ of the Fourier transform of $t^\alpha H(t)$.

The importance of these particular results is that, according to a technique developed rigorously in detail by Lighthill, the asymptotic behaviour (for real ω) of a Fourier transform can be used to infer the behaviour of $f(t)$ in the vicinity of its singularities. Often the asymptotic behaviour of $\tilde{f}(\omega)$ can be obtained in terms of algebraic–logarithmic functions, and then Lighthill's method and the table of Fourier transforms can be used to infer a rather simple form for $f(t)$ in the neighbourhood of points at which $f(t)$ (or perhaps some derivative of $f(t)$) is singular (see Chapter 4 of Lighthill 1978).

3.2.4 Analytic Continuation

Although the Fourier transforms of generalized functions are defined in the first place only for real ω, they may in some cases be analytically continued. The Fourier transform of a generalized function which may be interpreted as zero for $t < 0$ ($t^\alpha H(t)$, for example) is an analytic function in an upper half ω–plane (which includes the positive real ω–axis, but not the negative) and any branch cut (for example, for the function $\omega^{\alpha+1}$) must lie in the lower half ω–plane (which includes the negative real axis). Correspondingly, for generalized functions which can be regarded as zero for $t > 0$ are analytic in the lower half ω–plane. Functions which can be regarded as vanishing outside a finite range (a, b) in t–space have generalized transforms which are entire functions of ω, the simplest such functions are $\delta(t)$ and its derivatives, whose Fourier transforms are 1 and polynomials in ω. Some functions do not have Fourier transforms which can be continued off the real axis of the ω–plane; for example, the Fourier transform of t^n is proportional to $\delta^{(n)}(\omega)$, and delta functions of complex argument are an example of the *ultra distributions*, which do not fall within the scope of elementary generalized function theory.

3.3 RANDOM PROCESSES

Many problems of practical importance concern noise and vibration fields driven by unsteady fluid flows. In almost all flows occurring in nature and technology, the Reynolds number is large and the flow is turbulent. Then the flow parameters vary in a random way. This means that if an experiment were to be repeated a number of times, the values of the pressure or the velocity at the same position and the same time after the start of the experiment would be different. This lack of reproducibility means that the actual values of the flow parameters are relatively unimportant and it is more useful to consider their mean values.

We will begin by defining a method of taking averages and evaluating correlations, which is suitable for theoretical work. Then we define some useful spectral functions and in Section 3.4 we compare these theoretical measures with those which can be determined experimentally.

3.3.1 Correlation Functions

Consider a random process characterized by a single scalar function $p(t)$ of the continuous variable t (e.g. with t representing time, $p(t)$ might be the pressure at a fixed point on a surface over which a turbulent boundary layer is formed). The basic idea for defining mean values is that of an *ensemble of realizations* of $p(t)$. That is an endless sequence of experiments performed under identical conditions. Each realization, identified by the suffix m which is *not* a tensor suffix, gives a measured value $p_m(t)$ for each value of t. We define an *ensemble average*, denoted by an overbar, of any function of p, as

$$\overline{f(p)} = \lim_{N \to \infty} \frac{1}{N} \sum_{m=1}^{N} f(p_m) \qquad (3.39)$$

The ensemble average is just the limit of the arithmetic average over the infinite ensemble.

Let us suppose that the mean value of $p(t)$ is subtracted, so that

$$\overline{p(t)} = 0 .$$

Then the simplest and most important statistical parameter associated with p is $P(t, t')$, where

$$P(t, t') = \overline{p(t)p(t')} . \qquad (3.40)$$

$P(t, t')$ is usually called the *autocovariance* or the *autocorrelation*. We can go on to define higher–order averages, e.g. a triple covariance for two times, as $\overline{p(t)p(t)p(t')}$.

The random function $p(t)$ is said to be *stationary* or *statistically steady* if the averages $\overline{p(t)p(t')p(t'')}$.... for an arbitrary number of products p are unchanged by the addition of an arbitrary constant to each of $t, t', t'', \dots$.

$$\overline{p(t)p(t')p(t'')}... = \overline{p(t+\tau)p(t'+\tau)p(t''+\tau)}... \qquad (3.41)$$

for any τ. Then all one–point averages are independent of t, while many point averages depend only on the time differences $t - t', t - t'', \dots$. Stationarity in fluid mechanics is only possible when there is an external source of energy to maintain the flow against viscous decay, though in many cases the timescale of viscous decay is much longer than the timescales of interest in the flow, and then the flow is approximately stationary.

When p is a function of a space coordinate x, the random function $p(x)$ is said to be *homogeneous* in x if

$$\overline{p(x)p(x')p(x'')}... = \overline{p(x+\Delta)p(x'+\Delta)p(x''+\Delta)}... \qquad (3.42)$$

for any Δ and an arbitrary number of products of p. A process $p(\mathbf{x})$ may be homogeneous in any one, two or three of x_1, x_2, x_3, and inhomogeneous in the others. For example, in turbulent flow down a long pipe, the motion is homogeneous in the coordinate parallel to the pipe axis, but not in the coordinates normal to the axis. In the boundary–layer flow over a flat plate, the motion is nearly homogeneous in the two coordinates in the plane of the plate, but inhomogeneous in the normal coordinate. Turbulence generated in the wake of a grid in a wind tunnel is nearly homogeneous in all three coordinates.

It should be noted that no process can in reality be completely homogeneous. Any real fluid flow, for instance, takes place in a bounded region, and components of the motion on lengthscales comparable with the scale of the region cannot be homogeneous. In pipe flow, components of the motion with lengthscale comparable with pipe length will not be homogeneous. One can only expect to find homogeneity in a local sense in general, i.e. only components on lengthscales much smaller than the characteristic lengthscale of the inhomogeneity can be homogeneous.

Consider now a function which is stationary in t. If the statistical connection between the values of $p(t)$ occurring at different times drops to zero rapidly enough as the separation between the times increases, we can imagine time as divided up in many intervals which have no statistical connection with each other. In each interval we have a realization of $p(t)$ taking place, and stationarity implies that these realizations are taking place under similar conditions. Therefore, instead of performing a sequence of realizations to obtain an ensemble, we have the ensemble of realizations taking place in the different time intervals.

We can then obtain the ensemble average by averaging a single realization over all time (which itself constitutes an ensemble of realizations), so that

$$\overline{f(p)} = \lim_{T \to \infty} \frac{1}{2T} \int_{-T}^{T} f(p(t)) dt \quad . \tag{3.43}$$

Functions for which the ensemble average is equal to the space or time averages are called *ergodic* random processes. Birkhoff's famous Ergodic Theorem proves the equality for homogeneous functions under certain conditions on the rapidity with which the statistical connections decrease with increasing separations.

When the random function $p(t)$ is stationary, $P(t, t')$ is a function only of the difference $t - t'$ and we will write it as $P(\tau)$, where

$$P(\tau) = \overline{p(t)p(t + \tau)} \quad . \tag{3.44}$$

It then follows from the definition that $P(\tau)$ is an even function of τ,

$$P(\tau) = P(-\tau) \quad . \tag{3.45}$$

Also

$$P(0) = \overline{p^2} \quad , \tag{3.46}$$

and the Schwarz inequality shows that this is the maximum value of $P(\tau)$ (although there may be other local maxima or minima at which $|P(\tau)| < \overline{p^2}$).

It is usual to assume that events at different times ultimately become statistically independent for large separations, so that

$$P(\tau) \to 0 \quad \text{as} \quad \tau \to \infty \quad .$$

When the autocovariance is normalized on its value at $\tau = 0$, the result is called an *autocorrelation function*,

$$\widehat{P}(\tau) = P(\tau)/\overline{p^2} \quad . \tag{3.47}$$

Even though $p(t)$ is a randomly varying function, there are nevertheless various ways in which scales that characterize the signal can be defined. For example, the *integral timescale L_τ* is defined by

$$L_\tau = \frac{1}{\overline{p^2}} \int_{0}^{\infty} P(\tau) d\tau \quad , \tag{3.48}$$

and is a measure of the time beyond which the autocovariance is effectively zero. In a crude way, one can regard a random process $p(t)$ as built up from perfectly correlated events with $\hat{P}(\tau) = 1$ for $\tau < L_\tau$ and from perfectly uncorrelated events with $\hat{P}(\tau) = 0$ for $\tau > L_\tau$. The existence of L_τ implies that the process p has a finite "memory". It is useful, for estimation purposes, to consider the meaning of a non–zero value of L_τ

$$\overline{p^2} L_\tau = \int_0^\infty P(\tau) d\tau$$

$$= \int_0^\infty \overline{p(t)p(t')} dt' \quad .$$

On replacing the overbar by an average over t, we obtain

$$\overline{p^2} L_\tau = \lim_{T \to \infty} \frac{1}{T} \int_0^T \left\{ \int_0^\infty p(t)p(t') dt' \right\} dt$$

$$= \lim_{T \to \infty} \left(\frac{1}{T^{1/2}} \int_0^T p(t) dt \right) \left(\frac{1}{T^{1/2}} \int_0^T p(t') dt' \right) \quad .$$

Thus if L_τ is finite and non–zero, the integral of $p(t)$ over a large interval T is of order $T^{1/2}$;

$$\left| \int_0^T p(t) dt \right| = 0(T)^{1/2} \quad \text{as} \quad T \to \infty \quad . \tag{3.49}$$

The *differential scale* ℓ_τ is defined as the scale for variations in the derivatives of p, by

$$\overline{\left(\frac{\partial p}{\partial t} \right)^2} = \frac{\overline{p^2}}{\ell_\tau^2} \quad . \tag{3.50}$$

It is related to the autocovariance as follows:

Since $\tau = t' - t$, we have

$$\frac{\partial}{\partial \tau} = \frac{\partial}{\partial t'} \bigg|_t \quad \text{and} \quad \frac{\partial}{\partial \tau} = -\frac{\partial}{\partial t} \bigg|_{t'} \quad .$$

Hence

$$\frac{\partial^2 P(\tau)}{\partial \tau^2} = -\frac{\partial^2}{\partial t \partial t'} \overline{p(t)p(t')} = -\overline{\frac{\partial p(t)}{\partial t} \frac{\partial p(t')}{\partial t'}} \quad ,$$

and so

$$\overline{\left(\frac{\partial p}{\partial t} \right)^2} = -\frac{\partial^2 P}{\partial \tau^2}(0) \quad , \tag{3.51}$$

i.e.

$$\ell_\tau^2 = -\frac{P}{\partial^2 P/\partial \tau^2}\bigg|_{\tau=0} \ .$$

If p is a function of two variables, $p = p(x,t)$, say, then we may want to consider mean values like $\overline{p(x,t)p(x+\Delta,t+\tau)}$. Averages of this kind are called *cross–covariances*. More general cross–covariances are obtained from mean values formed from different functions, e.g. $\overline{\phi(t)\theta(t+\tau)}$, the previous cross–covariance being a special case of this with x, serving to distinguish the $\phi(t) = p(x,t)$ from the $\theta(t+\tau) = p(x+\Delta,t+\tau)$.

Slightly different generalizations of these one–dimensional formulae are needed for turbulence and noise problems. Then the Eulerian velocity vector u_i is a function of time t and the position vector $\mathbf{x} = (x_1, x_2, x_3)$, and cross–covariance of the form $\overline{u_i(\mathbf{x},t)u_j(\mathbf{x}+\Delta,t+\tau)}$ arise. These can sometimes be simplified by Taylor's (1938) hypothesis. Taylor asserted that the appearance of a time variation of the turbulent field at a fixed point is due mainly to the convection past that point of an almost rigid spatial pattern of turbulence by the mean flow $(U,0,0)$. Time variation in a frame convected with the mean flow is substantially slower than that in a fixed frame, so that

$$u_i(x_1, x_2, x_3, t) \simeq u_i(x_1 - Ut, x_2, x_3, 0) \ .$$

Then, for example, the cross–covariance $\overline{u_i(\mathbf{x},t)u_j(\mathbf{x}+\Delta,t+\tau)}$ can be expressed as a spatial covariance $\overline{u_i(\mathbf{x},t)u_j(\mathbf{x}+\Delta - \mathbf{U}\tau, t)}$.

In noise and vibration problems, it is customary to assume that the entire space–time statistics of the source field is known and to determine the corresponding statistics of the noise and vibration field.

Example.

A rigid circular cylinder of diameter D and length $2H(D << H)$ is placed with its ends at $(0,0,-H)$ and $(0,0,H)$, normal to a steady uniform stream, $(U,0,0)$. This stream induces unsteady vortex shedding producing a fluctuating lift on the cylinder, $(0, f(z,t), 0)$ per unit length. The distant pressure perturbation at a position $\mathbf{x} = (x_1, x_2, 0)$ is given by

$$p'(\mathbf{x},t) = -\frac{\cos\theta}{4\pi|\mathbf{x}|c}\frac{\partial}{\partial t}\int\limits_{-H}^{H} f(z, t - |\mathbf{x}|/c)dz \ , \tag{3.52}$$

where $\cos\theta = x_2/|\mathbf{x}|$. Determine the distant mean–square pressure perturbation in terms of the mean–square value of f and its differential and integral scales.

The mean–square pressure is obtained by multiplying together two terms similar to that in Equation (3.52) and evaluating the mean value.

$$\overline{p^2}(\mathbf{x}) = \left(\frac{\cos\theta}{4\pi|\mathbf{x}|c}\right)^2 \int_{-H}^{H}\int_{-H}^{H} \overline{\frac{\partial f}{\partial t}(z, t - |\mathbf{x}|/c)\frac{\partial f}{\partial t}(z', t - |\mathbf{x}|/c)}dz\ dz' \qquad (3.53)$$

An *integral lengthscale* L_w for any function $w(t)$ can be defined in an analogous way to the integral timescale in (3.48):

$$L_w = \frac{1}{\overline{w^2}}\int_0^\infty \overline{w(z,t)w(z+\Delta, t)}\, d\Delta \quad .$$

With z' rewritten as $z + \Delta$, Equation (3.53) becomes

$$\overline{p^2}(\mathbf{x}) = \left(\frac{\cos\theta}{4\pi|\mathbf{x}|c}\right)^2 \int_{-H}^{H}\int_{-H-z}^{H-z} \overline{\frac{\partial f}{\partial t}(z,t)\frac{\partial f}{\partial t}(z+\Delta, t)}\, d\Delta dz \quad .$$

If L_{ft}, the integral lengthscale of $\partial f/\partial t$, is much smaller than H, the limits of the Δ–integration are effectively $\mp\infty$. It then follows from the definition of the integral lengthscale that

$$\overline{p^2}(\mathbf{x}) = \left(\frac{\cos\theta}{4\pi|\mathbf{x}|c}\right)^2 \int_{-H}^{H} 2L_{ft}\overline{\left(\frac{\partial f}{\partial t}\right)^2}(z,t) \quad .$$

Equation (3.50) can then be used to replace the time derivatives by the differential timescale τ_f leading to

$$\overline{p^2}(\mathbf{x}) = \left(\frac{\cos\theta}{4\pi|\mathbf{x}|c}\right)^2 \frac{1}{\tau_f^2}2L_{ft}\int_{H}^{H} \overline{f^2}dz$$

$$= \left(\frac{\cos\theta}{4\pi|\mathbf{x}|c}\right)^2 \frac{4HL_{ft}}{\tau_f^2}\overline{f^2} \quad . \qquad (3.53)$$

In general it is reasonable to assume that L_{ft}, which is of the order of an eddy size, is much less than the cylinder length. But if the cylinder vibrates in response to the eddy shedding, the flow can become correlated along the whole length of the cylinder. Then all the forces are in phase, and

$$\int_{-H}^{H}\int_{-H}^{H} \overline{\frac{\partial f}{\partial t}(z,t)\frac{\partial f}{\partial t}(z', t)}\, dz\ dz' = 4H^2\overline{\left(\frac{\partial f}{\partial t}\right)^2} \quad ,$$

giving the much larger mean–square pressure

$$\overline{p^2}(\mathbf{x}) = \left(\frac{\cos\theta}{4\pi|\mathbf{x}|c} \right)^2 \frac{4H^2}{\tau_f^2} f^2 .$$

3.3.2 Power Spectral Density

It is convenient first to consider the transforms of mean values of a random process, and later to relate these to the transforms of the fluctuating quantities themselves. Consider, as the simplest case, the autocovariance

$$P(\tau) = \overline{p(t)p(t+\tau)}$$

of the stationary random variable $p(t)$ with zero mean. If $P(\tau) \to 0$ sufficiently rapidly as $|\tau| \to \infty$, then $\int_{-\infty}^{\infty} |P(\tau)|^2 d\tau$ will exist, and the Fourier transform $P(\tau)$ will exist as an ordinary function. The transform is defined by

$$\widetilde{P}(\omega) = \int_{-\infty}^{\infty} P(\tau)e^{i\omega\tau}\,d\tau \ , \tag{3.54}$$

and the Fourier inversion theorem gives

$$P(\tau) = \frac{1}{2\pi} \int_{-\infty}^{\infty} \widetilde{P}(\omega)e^{-i\omega\tau}\,d\omega \ .$$

Since $P(\tau)$ is real and even, the same is true of $\widetilde{P}(\omega)$, and (3.54) can be written as

$$\widetilde{P}(\omega) = 2 \int_{0}^{\infty} P(\tau)\cos\omega\tau \ d\tau \ .$$

Similarly,

$$P(\tau) = \frac{1}{\pi} \int_{0}^{\infty} \widetilde{P}(\omega)\cos\omega\tau \ d\omega \ .$$

When $\tau = 0$, we have

$$P(0) = \overline{p^2} = \frac{1}{2\pi} \int_{-\infty}^{\infty} \widetilde{P}(\omega)d\omega \ ,$$

so that $\widetilde{P}(\omega)d\omega/2\pi$ is the contribution to the "power" $\overline{p^2}$ from the frequency range $(\omega, \omega + d\omega)$. $\widetilde{P}(\omega)$ is called the *power spectral density*.

We can generalize these results to n space dimensions. Thus the covariance of the homogeneous random variable, $p(\mathbf{x})$, is defined by

$$P(\boldsymbol{\Delta}) = \overline{p(\mathbf{x})p(\mathbf{x}+\boldsymbol{\Delta})} \ . \tag{3.55}$$

Then, with the *power spectral density* of $P(\Delta)$ defined by

$$\widetilde{P}(\mathbf{k}) = \int P(\Delta)e^{i\mathbf{k}\cdot\Delta}d\Delta \ , \tag{3.56}$$

the Fourier inversion theorem gives

$$P(\Delta) = \frac{1}{(2\pi)^n} \int \widetilde{P}(\mathbf{k})e^{-i\mathbf{k}\cdot\Delta}d\mathbf{k} \ . \tag{3.57}$$

Setting $\Delta = 0$ in (3.57), we see that $\widetilde{P}(\mathbf{k})d\mathbf{k}/(2\pi)^n$ is the contribution to $\overline{p^2}$ from the volume element $d\mathbf{k}$ in $\mathbf{k}$–space.

The interpretation of $\widetilde{P}(\mathbf{k})$ in terms of the power in a small wavenumber element fails near $\mathbf{k} = 0$. In many cases it is quite misleading to regard $\widetilde{P}(0)$ as a measure of the power in the modes of infinite wavelength, though it is usual to speak of the form of the spectrum near $\mathbf{k} = 0$ as the "large eddy structure " in turbulence theory. The only safe interpretation of $\widetilde{P}(0)$ appears to come from the relationship

$$\widetilde{P}(0) = \int P(\Delta)d\Delta$$
$$= \overline{p^2}V_p \ ,$$

where V_p is the integral correlation volume. Thus $\widetilde{P}(0)$ is related to the total integrated covariance, and may or may not vanish depending upon the value of V_p.

In the case when $p(\mathbf{x},t)$ is homogeneous in $\mathbf{x}$ and stationary in t, we define a power spectrum from the covariance $P(\Delta,\tau) = \overline{p(\mathbf{x},t)p(\mathbf{x}+\Delta,t+\tau)}$ by

$$\ddot{P}(\mathbf{k},\omega) = \int P(\Delta,\tau)e^{i\mathbf{k}\cdot\Delta+i\omega\tau}d\Delta \ d\tau \ . \tag{3.58}$$

Then

$$P(\Delta,\tau) = \frac{1}{(2\pi)^4} \int \widetilde{P}(\mathbf{k},\omega)e^{-i\mathbf{k}\cdot\Delta-i\omega\tau}d\mathbf{k} \ d\omega \ ,$$

so that $\widetilde{P}(\mathbf{k},\omega)/(2\pi)^4$ is the spectral density of $\overline{p^2}$ in $(\mathbf{k},\omega)$ space.

3.3.3 Fourier Analysis of a Random Variable

If a function $p(t)$ is periodic in t, it can be expressed as a Fourier series; if $\int |p(t)|^2 dt$ over all time exists, $p(t)$ can be expressed as an ordinary Fourier integral. For a stationary random variable $p(t)$, neither of these properties holds, for as we have seen in Equation (3.49)

$$\left| \int_0^T p(t)dt \right| = 0(T^{1/2}) \text{ as } T \to \infty \ .$$

We can avoid these difficulties by simply regarding $p(t)$ as a generalized function over the whole range $(-\infty, +\infty)$ of t, because as we have seen, every generalized function has a Fourier transform (which may of course also be a generalized function).

Although the generalized function $\tilde{p}(\omega)$, the Fourier transform of $p(t)$, is defined formally through the techniques introduced in Section 3.2 (see Equation (3.15)), we will write it symbolically in the form

$$\tilde{p}(\omega) = \int p(t)e^{i\omega t}dt \ . \tag{3.59}$$

We will show that, if $p(t)$ is a stationary random function of t, the Fourier coefficients $\tilde{p}(\omega)$ are "uncorrelated" for distinct values of ω, or statistically orthogonal.

It follows from (3.59) that

$$\overline{\tilde{p}^*(\omega)\tilde{p}(\omega')} = \int e^{-i\omega t + i\omega' t'} \overline{p(t)p(t')}dt\, dt'$$

$$= \int e^{-i\omega t + i\omega'(t+\tau)}\overline{p(t)p(t+\tau)}\, dt\, d\tau \ . \tag{3.60}$$

Now in an homogeneous field, $\overline{p(t)p(t+\tau)} = P(\tau)$ is independent of t. The τ–integral in (3.60) exists in the ordinary sense and, after using the definition in (3.54), we obtain

$$\overline{\tilde{p}^*(\omega)\tilde{p}(\omega')} = \tilde{P}(\omega') \int e^{i(\omega'-\omega)t}dt \ .$$

The t–integral only exists as a generalized function and is to be interpreted in the way outlined in Section 3.2. It represents the Fourier transform of the unit function evaluated at $\omega' - \omega$. Equation (3.30) shows that this is equal to $2\pi\delta(\omega' - \omega)$. Hence

$$\overline{\tilde{p}^*(\omega)\tilde{p}(\omega')} = 2\pi\delta(\omega - \omega')\tilde{P}(\omega) \ . \tag{3.61}$$

Thus distinct frequency Fourier elements of a statistically stationary random function are uncorrelated. When $\omega = \omega'$, the mean–squared amplitude in unit frequency band, normalized on the infinite constant $2\pi\delta(\omega - \omega')$, is equal to the power spectral density $\widetilde{P}(\omega)$. The infinite constant appears here as a consequence of stationarity, which implies a constant mean energy in each interval of the infinite time considered. The spectral density emerges naturally as an appropriately normalized measure of the mean energy in each Fourier mode of zero bandwidth.

The result in (3.61) can be readily extended to many dimensions. In three–dimensional space, with $\widetilde{p}(\mathbf{k}) = \int p(\mathbf{x})e^{i\mathbf{k}\cdot\mathbf{\Delta}}d\mathbf{\Delta}$, we obtain

$$\overline{\widetilde{p}^*(\mathbf{k})\widetilde{p}(\mathbf{k}')} = (2\pi)^3 \delta(\mathbf{k} - \mathbf{k}')\widetilde{P}(\mathbf{k}) \quad, \tag{3.62}$$

where $\widetilde{P}(\mathbf{k})$ is defined in (3.56). When $p(\mathbf{x}, t)$ is homogeneous in $\mathbf{x}$ and stationary in t, its Fourier transform $\widetilde{p}(\mathbf{k}, \omega)$ will only exist as a generalized function. Then the result analogous to (3.61) and (3.62) is

$$\overline{\widetilde{p}^*(\mathbf{k}, \omega)\widetilde{p}(\mathbf{k}', \omega')} = (2\pi)^4 \delta(\mathbf{k} - \mathbf{k}')\delta(\omega - \omega')\widetilde{P}(\mathbf{k}, \omega) \quad. \tag{3.63}$$

$\widetilde{P}(\mathbf{k}, \omega)$, the power spectral density of $p(\mathbf{x}, t)$, was defined in (3.58).

Example. The wavenumber response of a transducer.

The interpretation of measured values of a continuous random variable has certain difficulties associated with the scale of the measuring device relative to the scales of the random process itself. The degree to which one has the confidence in measured values depends upon a proper appreciation of the properties of the device, and upon an independent knowledge of the features of the random process.

As an example, take the case of the pressure $p(\mathbf{x})$ as a function of position $\mathbf{x}$ over a plane surface. Assume that $p(\mathbf{x})$ is statistically homogeneous in $\mathbf{x}$ and is measured by a circular transducer with a rigid face of radius a, centred at the point $\mathbf{x}$. The transducer is obviously insensitive to lengthscales much finer than a. Its action in determining an effective pressure by an average over the area πa^2 averages out the fine–scale structure of the true pressure field. Consequently, we expect little spectral response for $k = |\mathbf{k}| >> a^{-1}$. We have, in fact, the measured pressure of $q(\mathbf{x})$ given by

$$q(\mathbf{x}) = \frac{1}{\pi a^2} \int p(\mathbf{x}')H(a - |\mathbf{x} - \mathbf{x}'|)d\mathbf{x}' \quad,$$

where $H(x)$ is the Heaviside function. Taking the transform we find

$$\widetilde{q}(\mathbf{k}) = \widetilde{p}(\mathbf{k})\widetilde{g}(\mathbf{k}) \quad,$$

where $\tilde{g}(\mathbf{k})$ is the two–dimensional transform of $H(a - |\mathbf{x}|)/\pi a^2$;

$$\tilde{g}(\mathbf{k}) = \int_0^a r\, dr \int_0^{2\pi} \frac{e^{ikr\cos\theta}}{\pi a^2} d\theta$$

$$= \frac{2J_1(ka)}{ka} .$$

Hence the measured and actual power spectra are related by

$$\tilde{Q}(\mathbf{k}) = \frac{4J_1^2(ka)}{(ka)^2} \tilde{P}(\mathbf{k}) .$$

For $0 < x << 1$, we have $J_1(x) \sim x/2$, so that wavenumbers less than about a^{-1} are passed essentially unchanged. For $ka > 4$, approximately, the large–argument asymptotic form of the Bessel function may be used to give

$$\tilde{Q}(\mathbf{k}) \sim \frac{8}{\pi(ka)^3} \cos^2\left(ka - \frac{3\pi}{4}\right) \tilde{P}(\mathbf{k}) .$$

The $(ka)^{-3}$ decay here is fairly rapid, and is helped by the oscillations of the cosine. The oscillations are the inevitable consequence of an abrupt cut off at at finite distance $|\mathbf{x}| = a$ in physical space. Apart from the oscillations, the box of width a in physical space transforms essentially into a box of width a^{-1} in Fourier space. The filter passes all scales greater than a virtually unchanged, but suppresses all scales less than a by averaging them out to zero.

3.4 DIGITAL SAMPLING

In Section 3.3 we defined and discussed correlation and spectral functions which are suitable for theoretical work. In this section we deal with measures that can be obtained experimentally, and compare them with the theoretical functions.

Consider the pressure at one position measured as a function of time. Although $p(t)$ might be measured as a continuous signal, only some of these values can be stored for subsequent analysis on a computer. It is customary to "sample" and store values of the signal $p(t)$ at equal time intervals. Suppose $p(t)$ is measured for a finite time interval T, and sampled at N different times with the time intervals Δ, i.e. at $t = n\Delta$, where $n = 0, 1, ..., N - 1$, and $T = N\Delta$. Let us denote these values by p_n, $n = 0, 1, ..., N - 1$. We can define a *discrete Fourier transform*, $\tilde{p}_m$, based on the data collected at the N sampling times, by

$$\tilde{p}_m = \frac{1}{N} \sum_{n=0}^{N-1} p_n e^{2\pi imn/N} \tag{3.64}$$

for $m = 0, ..., N - 1$.

Knowledge of the $\widetilde{p}_m$'s enables us to reconstruct p_n from

$$p_n = \sum_{m=0}^{N-1} \widetilde{p}_m e^{-2\pi imn/N} \quad . \tag{3.65}$$

This can be proved quite simply by substituting the expression for $\widetilde{p}_m$ into (3.65) and using

$$\sum_{m=0}^{N-1} e^{2\pi im(n-\ell)/N} = 0 \quad \text{if} \quad n \neq \ell + MN$$

$$= N \quad \text{if} \quad n = \ell + MN \quad \text{for} \quad M \quad \text{integer.}$$

The *fast Fourier transform* (FFT), introduced by Cooley & Tukey (1965), has had a dramatic effect on spectral analysis. It reduces the number of steps required to evaluate the $\widetilde{p}_m$ as in Equation (3.64) from order N^2 to $N \log_2 N$, thus enabling the discrete Fourier transform to be calculated quickly and conveniently (see Newland 1984 for a description of the FFT algorithm).

We will now investigate the relationship between this Fourier transform and the transform of the continuous variable $\widetilde{p}(\omega)$, where $\widetilde{p}(\omega) = \int_{-\infty}^{\infty} p(t)e^{i\omega t}dt$. We do this directly from the definition in (3.64):

$$\widetilde{p}_m = \frac{1}{N} \sum_{n=0}^{N-1} p_n e^{2\pi imn/N} \quad .$$

p_n in this equation can be written as

$$p_n = p(n\Delta) = \frac{1}{2\pi} \int_{-\infty}^{\infty} \widetilde{p}(\omega)e^{-i\omega n\Delta}d\omega \quad , \tag{3.66}$$

leading to

$$\widetilde{p}_m - \frac{1}{2\pi N} \int_{-\infty}^{\infty} \widetilde{p}(\omega) \sum_{n=0}^{N-1} e^{in(2\pi m/N - \omega\Delta)}d\omega \quad . \tag{3.67}$$

The series in this expression is a geometric progression and can be summed in a straightforward way to give

$$\widetilde{p}_m = \int_{-\infty}^{\infty} \widetilde{p}(\omega)\widetilde{f}(\omega - 2\pi m/T)d\omega \quad , \tag{3.68}$$

where

$$\widetilde{f}(\omega) = \frac{\sin(N\Delta\omega/2)}{2\pi N \sin(\Delta\omega/2)} e^{-i(N-1)\Delta\omega/2} \quad .$$

A sketch of $|\widetilde{f}(\omega)|$ is given in Figure 3.1.

For large N, $|\widetilde{f}(\omega)|$ is small

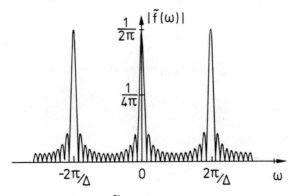

Figure 3.1 The variation of $|\widetilde{f}(\omega)|$ with frequency ω.

(of order N^{-1}) except near $\sin(\Delta\omega/2) = 0$ where it increases to a value of $(2\pi)^{-1}$, i.e. for ω near $2\pi M/\Delta$, M integer, $|\widetilde{f}(\omega)| \sim (2\pi)^{-1}$. Hence, from Equation (3.68), we have $\widetilde{p}_m$ not just related to the frequency component $2\pi m/T$, but to all components $2\pi m/T \pm 2\pi M/\Delta$. This contamination is called *aliasing*. It arises because it is impossible to distinguish between disturbances at frequency ω and $2\pi M/\Delta \mp \omega$ by sampling every Δ seconds. Suppose, for example, that we sample the function $\cos(2\pi M/\Delta \mp \omega)t$ at $t = n\Delta$, $n = 0, 1, ... N - 1$. Then, since

$$\cos\left(\frac{2\pi M}{\Delta} \mp \omega\right)n\Delta = \cos \omega n\Delta \ ,$$

disturbances at these different frequencies are indistinguishable. This is illustrated in Figure 3.2, where $\cos \omega_0 t$, $\cos 2\omega_0 t$ and $\cos 4\omega_0 t$ are sampled every $2\pi/3\omega_0$ seconds, i.e. $\Delta = 2\pi/3\omega_0$. We see that $\cos \omega_0 t, \cos(2\pi/\Delta - \omega_0)t$ ($= \cos 2\omega_0 t$) and $\cos(2\pi/\Delta + \omega_0)t$ ($= \cos 4\omega_0 t$) are equal at the sampling points.

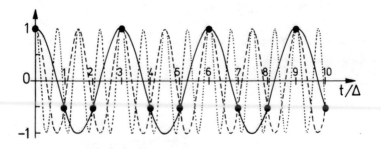

Figure 3.2 An example of aliasing: $\cos \omega_0 t$ (——), $\cos 2\omega_0 t$ (- - -) and $\cos 4\omega_0 t$ (······) are indistinguishable when sampled every $2\pi/3\omega_0$ seconds.

Aliasing was often observed in the motion of wagon–wheels in old westerns. There the frame speed of the film was too slow to reproduce the rapid motion of wheels, the motion was aliased and the wheels appeared to be rotating at a much slower speed, or to be going backwards.

Aliasing makes it impossible to interpret the information contained in $\tilde{p}_m$, and the only way of obtaining useful data is to ensure that the sampling rate is sufficiently fast that the Nyquist frequency, π/Δ, is larger than the maximum frequency in the signal. Since in practice the signal is contaminated by noise, the highest frequency in the signal is not known. The usual procedure is to decide the highest frequency of interest and then to pass the signal through a low–pass filter which cuts off higher frequencies. The sampling rate is then chosen to ensure that the Nyquist frequency is at least as high as this cut–off frequency. This ensures that $\tilde{p}(\omega)$ is zero at all values of ω greater than π/Δ or less than $-\pi/\Delta$, and hence $\tilde{p}(\omega)$ is zero at all values of $\omega = 2\pi m/T \pm 2\pi M/\Delta$, except when $M = 0$. The values of the discrete Fourier coefficients $\tilde{p}_m$, $0 \le m \le N/2$ are then only influenced by disturbances with frequencies nearly equal to $2\pi m/T$.

We can relate the power spectral density to the products of $\tilde{p}_m$. It follows from (3.68) that

$$\tilde{p}_m \tilde{p}_m^* = \int \tilde{p}^*(\omega)\tilde{p}(\alpha)\tilde{f}^*\left(\omega - \frac{2\pi m}{T}\right)\tilde{f}\left(\alpha - \frac{2\pi m}{T}\right) d\omega \, d\alpha \ . \tag{3.69}$$

If we average this over different ensembles for a statistically stationary process, or over different time sequences if the process is ergodic, we can use (3.61)

$$\overline{\tilde{p}^*(\omega)\tilde{p}(\alpha)} = 2\pi\delta(\omega - \alpha)\tilde{P}(\omega) \ , \tag{3.70}$$

to obtain from (3.69)

$$\overline{\tilde{p}_m \tilde{p}_m^*} = 2\pi \int \tilde{P}(\omega)\left|\tilde{f}\left(\omega - \frac{2\pi m}{T}\right)\right|^2 d\omega \ . \tag{3.71}$$

The measured data points enable us to evaluate a weighted average of the true power spectral density over the frequency band where $\tilde{f}$ is non–zero. The weighting function $|\tilde{f}|^2$ is called a *window function*. The width of the peak of $\tilde{f}(\omega)$ near the origin is of order $4\pi/T$, as shown in Figure 3.3. This is a measure of the frequency resolution that can be obtained with this window function. $|\tilde{f}(\omega)|$ only decays slowly with increasing ω (like $|\omega|^{-1}$ for large ω), and so the estimated spectrum $\overline{\tilde{p}_m \tilde{p}_m^*}$ is contaminated by values of $\tilde{P}(\omega)$ in many side lobes. This is referred to as "window leakage", and can be reduced by weighting the data in the time domain, so–called indirect methods, or equivalently by smoothing the predicted spectral density (direct methods). This is accompanied by a reduction in resolution.

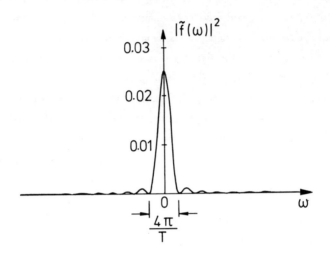

Figure 3.3 The variation of $|\widetilde{f}(\omega)|^2$ with ω for very large N.

The field of spectral analysis is extensive, and we will only give a brief introduction to some of the more common techniques. For a more complete description, see Bendat & Piersol (1974), Childers (1978), Glover (1984) or the IEEE special issue on Spectral Estimation (IEEE Volume 70, Number 9, 1982).

3.4.1 Indirect Methods

We can define an *autocorrelation* based on N samples of data, by

$$P_D(n) = \frac{1}{N-n} \sum_{m=0}^{N-n-1} p_m p_{m+n} \quad \text{for } 0 \leq n \leq N-1 \;, \tag{3.72}$$

with

$$P_D(n) = P_D(-n) \quad \text{for } -N+1 \leq n \leq 0 \;.$$

This is an unbiased estimate of the autocovariance $P(\tau)$ used in Section 3.3, because an ensemble average of (3.72) gives

$$\overline{P_D(n)} = \overline{p_m p_{m+n}} = \overline{p(m\Delta)p(m\Delta + n\Delta)} = P(n\Delta) \;. \tag{3.73}$$

Sometimes a biased lag sequence is used;

$$\widehat{P}_D(n) = \frac{1}{N} \sum_{m=0}^{N-n-1} p_m p_{m+n} \quad \text{for } 0 \leq n \leq N-1 \ . \tag{3.74}$$

This has a mean value of $P(n\Delta)(N - |n|)/N$.

We will weight the unbiased autocorrelation in Equation (3.72) by a function w_n. w_n is called a *lag-window*, and we put no constraints on it other than it be even, $w_{-n} = w_n$, that it vanish for large enough $|n|$, $w_n = 0$ for $|n| \geq M$, for some value of $M(\leq N)$.

Let $\widetilde{P}_m$ be the discrete Fourier transform of the weighted autocorrelation function,

$$\widetilde{P}_m = \sum_{n=-\infty}^{\infty} w_n P_D(n) e^{2\pi imn/N} \ . \tag{3.75}$$

We will investigate the relationship between this estimate and $\widetilde{P}(\omega)$, the power spectral density of the continuous autocovariance. An average of equation (3.75) gives, after using (3.73),

$$\overline{\widetilde{P}_m} = \sum_{n=-\infty}^{\infty} w_n P(n\Delta) e^{2\pi imn/N} \ . \tag{3.76}$$

We can rewrite the autocovariance $P(n\Delta)$ in terms of an integral of its Fourier transform, the power spectral density,

$$P(n\Delta) = \frac{1}{2\pi} \int \widetilde{P}(\omega) e^{-i\omega n\Delta} d\omega \ . $$

Then (3.76) becomes

$$\overline{\widetilde{P}_m} = \int W\left(\omega - \frac{2\pi m}{T}\right) \widetilde{P}(\omega) d\omega \ , \tag{3.77}$$

$$\text{where} \quad W(\omega) = \frac{1}{2\pi} \sum_{n=-\infty}^{\infty} w_n e^{-i\omega n\Delta} \ . \tag{3.78}$$

This is often called a Blackman–Tukey estimate because of their pioneering work in indirect methods (1959). The measured spectral density is again related to the integral of the true spectral density weighted by a *spectral window function*, $W(\omega)$, which depends on the lag–window.

If, for example, the lag–window is rectangular, i.e.

$$w_n = 1 \qquad \text{for } |n| \leq M$$
$$= 0 \qquad \text{for } |n| > M.$$

The series in (3.78) can be summed in a straightforward way, to give

$$W_R(\omega) = \frac{\sin(\omega(2M+1)\Delta/2)}{2\pi \sin(\omega\Delta/2)} . \qquad (3.79)$$

This window function has the disadvantage that it is not positive definite. For the Bartlett window

$$w_n = 1 - |n|/M \qquad \text{for } |n| \leq M$$
$$= 0 \qquad \text{for } |n| \geq M ,$$

the series in (3.78) sums to give

$$W_B(\omega) = \frac{1}{2\pi M} \frac{\sin^2(\omega M\Delta/2)}{\sin^2(\omega\Delta/2)} . \qquad (3.80)$$

[Note that this lag–window arises naturally if we use the biased estimate (3.74) for the autocovariance.]

The Hann window

$$w_n = \frac{1}{2} [1 + \cos(\pi n/M)] \quad \text{for } |n| \leq M$$
$$= 0 \qquad \text{for } |n| > M$$

leads to the window function

$$W_H(\omega) = \frac{1}{2} W_R(\omega) + \frac{1}{4} W_R(\omega - \frac{\pi}{M\Delta}) + \frac{1}{4} W_R(\omega + \frac{\pi}{M\Delta}) . \qquad (3.81)$$

These three window functions are plotted on a logarithmic scale in Figure 3.4. From the curves it is clear that the rectangular window has the best resolution, but that the window function only decays slowly (as $|\omega|^{-1}$ for large $|\omega|$), resulting in considerable leakage. The Hann window function decays more rapidly (like $|\omega|^{-3}$ as $|\omega|$ increases). The side lobes therefore have far less effect, but this is gained at the expense of a loss in resolution.

The relative performances of the rectangular and Hann window functions are illustrated in Figure 3.5 (from Ables 1974). Figure 3.5a shows the true spectral density used in a numerical experiment. The autocovariance was calculated from this, and then truncated, with just the first 15 terms being retained for subsequent analysis. The spectrum predicted from these 15 data points, by using a rectangular window function, is shown in Figure 3.5b. The resolution is good, with some evidence of the existence of the second largest peak. But away from the peak the predicted spectrum is swamped by ringing produced by window leakage. Figure 3.5c shows the effect of a Hann window. The ringing has been suppressed but the resolution is reduced.

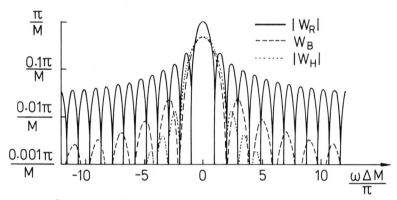

Figure 3.4 Spectral windows

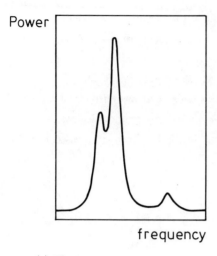

(a) The true spectrum

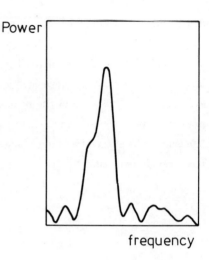

(b) Fourier transform of truncated and sampled autocorrelation function (rectangular lag–window).

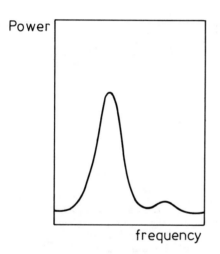

(c) Fourier transform of truncated and sampled autocorrelation function. Before transformation the function was multiplied by Hann lag–window.

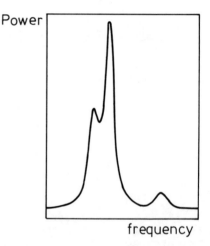

(d) The maximum entropy spectrum corresponding to the truncated and sampled autocorrelation function.

Figure 3.5 The resolution obtained from rectangular and Hann lag windows compared with the maximum entropy spectrum (from Ables 1974).

3.4.2 Direct Methods

The convenience of FFT routines means that it is often preferable to calculate the discrete Fourier transform of the data $\tilde{p}_m$ in Equation (3.64), and to use this to estimate the spectral density, rather than to calculate the autocovariance. This is the essence of the so–called *direct methods*. A reduction in window leakage can then be obtained by smoothing operations in the frequency domain.

Let $\tilde{p}_m$ be the discrete Fourier transform of the N data points

$$\tilde{p}_m = \frac{1}{N} \sum_{n=0}^{N-1} p_n e^{2\pi i m n / N} \quad .$$

We can *smooth* this spectrum by considering (see Bingham et al. 1967)

$$\tilde{q}_m = -\frac{1}{4}\tilde{p}_{m-1} + \frac{1}{2}\tilde{p}_m - \frac{1}{4}\tilde{p}_{m+1} \quad . \tag{3.82}$$

Expressing $\tilde{p}_m$ in terms of $\tilde{p}(\omega)$, the transform of the continuous variable from equation (3.68), we find

$$\tilde{q}_m = \int \tilde{p}(\omega)\tilde{g}\left(\omega - \frac{2\pi m}{T}\right) d\omega \quad , \tag{3.83}$$

where

$$\tilde{g}(\omega) = -\frac{1}{4}\tilde{f}\left(\omega + \frac{2\pi}{T}\right) + \frac{1}{2}\tilde{f}(\omega) - \frac{1}{4}\tilde{f}\left(\omega - \frac{2\pi}{T}\right) \quad ,$$

and

$$\tilde{f}(\omega) = \frac{\sin(N\Delta\omega/2)}{2\pi N \sin(\Delta\omega/2)} e^{-i(N-1)\Delta\omega/2} \quad .$$

We can go on to construct an estimate of the power spectral density from $\tilde{q}_m$, and a repetition of the steps leading from (3.68) to (3.71), with $\tilde{q}_m$ replacing $\tilde{p}_m$ and $\tilde{g}$ in place of $\tilde{f}$, shows that

$$\overline{\tilde{q}_m \tilde{q}_m^*} = 2\pi \int \tilde{P}(\omega)\left|\tilde{g}\left(\omega - \frac{2\pi m}{T}\right)\right|^2 d\omega \quad . \tag{3.84}$$

The window function $|\tilde{g}(\omega)|^2$ decays far more rapidly with ω than the window function $|\tilde{f}(\omega)|^2$ that occurs when no smoothing is applied. Problems of window leakage are therefore reduced by the smoothing, but so is the resolution.

Note that we could have obtained the functions $\widetilde{q}_m$ by multiplying the raw data by the weighting function $\frac{1}{2}(1 - \cos(2\pi t/T))$ before taking the discrete Fourier transform, because

$$\frac{1}{N} \sum_{n=0}^{N-1} p(n\Delta) \frac{1}{2} \left(1 - \cos \frac{2\pi n\Delta}{T} \right) e^{2\pi i m n/N}$$

$$= \frac{1}{2}\widetilde{p}_m - \frac{1}{4}\widetilde{p}_{m-1} - \frac{1}{4}\widetilde{p}_{m+1} \qquad (3.85)$$

$$= \widetilde{q}_m \ .$$

Smoothing in the frequency domain is therefore entirely equivalent to weighting in the time domain.

There are other ways of smoothing the predicted spectra and Welch (1967) recommends *sequential averaging*. This is a process whereby N measured data points are divided up into K (possibly overlapping) sets of data each of length L. The procedure is to evaluate the estimated spectral density for each set of data from $\widetilde{q}_m \widetilde{q}_m^*$ and then to average this over the K sets. The resolution is only of order $2\pi/(L\Delta)$, but the variance is also reduced.

There are other operations in the frequency domain which can be used to extract information about different parts of the spectrum. For example, if information is only required about the power spectral density in a frequency band with centre–frequency ω_c, *complex demodulation* can be used to shift the centre–frequency of interest to zero. The result is then passed through a low–pass filter so that all the analyzing power can be concentrated into the frequency band of interest. This can be achieved computationally by multiplying the data p_m by $e^{i\omega_c m\Delta}$, filtering the result and then working out its discrete Fourier transform. Suppose $p(t)$ has a component $\cos \omega t$ with ω near the centre–frequency. Then multiplication by $e^{i\omega_c t}$ gives

$$p(t)e^{i\omega_c t} = \frac{1}{2} \left(e^{i(\omega_c - \omega)t} + e^{i(\omega + \omega_c)t} \right) \ .$$

If we pass this signal through a low–pass filter, with a small cut–off frequency ω_f, the second term will be eliminated, and the first retained provided ω is within ω_f of the centre frequency ω_c. We can now sample the filtered signal with quite a long time delay Δ between successive data samples. Δ is restricted only by the requirement that ω_f be less than the Nyquist frequency π/Δ. This allows much larger Δ than if the signal $p(t)$ were sampled directly without complex modulation. Then to analyse frequencies near ω_c we require a sampling period Δ' for which $\pi/\Delta' > \omega_c$. With N

data points the resolution is of order $2\pi/(N\Delta')$ and so complex demodulation gives a dramatic increase in the resolution for fixed N. The time of computation can be reduced by using slightly different numerical schemes (Bingham et al 1967).

3.4.3 Maximum Entropy and Other Models

The methods we have described so far have involved rather *ad hoc* manipulation of the data (differently weighting different data points, for example) and have made implicit assumptions about the value $p(t)$ at times other than when it is measured. The discrete Fourier transform (3.64) assumes, for example, that p_m is a periodic function with $p_m = p_{N+m}$. Maximum entropy methods are an attempt to do something better with the available data.

Let us assume that the autocovariance function $P(n\Delta)$ for $n = 0, ..., N-1$ has been determined from the measurements. If $\widetilde{P}(\omega)$ is the power spectral density,

$$P(n\Delta) = \frac{1}{2\pi} \int_{-\infty}^{\infty} \widetilde{P}(\omega)e^{-i\omega n\Delta} d\omega, \quad n = 0, ..., N-1 \ . \tag{3.86}$$

The aim of *maximum entropy methods* (Burg 1967) is to maximize

$$\int_{-\infty}^{\infty} \log \widetilde{P}(\omega) d\omega \ ,$$

which is a measure of the "randomness" in the process, subject to the constraints (3.86). This problem of constrained optimization can be solved by the introduction of a Lagrange multiplier for each constraint. Figure 3.5d shows the spectrum obtained by a maximum entropy method from the same 15 values of the autocorrelation function as Figures 3.5b and c. We see that the maximum entropy method gives a much closer approximation to the true spectral density (Figure 3.5a) than the indirect Blackman–Tukey approach (Figures 3.5b,c). The maximum entropy method works particularly well in this case, because there is no noise contaminating the sampled autocorrelation function.

Maximum entropy methods make no *a priori* assumptions about the unmeasured data and maximizes the entropy of the process. Another optimizing method is the *maximum likelihood method* which aims to find a minimum variance, unbiased estimator of the spectral components (Lacoss 1971). The methods of maximum entropy and maximum likelihood are often used when the spectrum contains one or more narrow peaks.

An alternative method involves the use of parametric models for the power spectral density. In the so–called *rational models*, the power spectral density is expressed as a magnitude of the squared ratio of polynomials in the variable $e^{i\omega}$, i.e.

$$\widetilde{P}(\omega) = \left| \frac{b_0 + b_1 e^{i\omega} + ... + b_j e^{ij\omega}}{1 + a_1 e^{i\omega} + ... + a_k e^{ik\omega}} \right|^2 . \tag{3.87}$$

Now

$$P(n\Delta) = \frac{1}{2\pi} \int_{-\infty}^{\infty} \widetilde{P}(\omega) e^{-i\omega n\Delta} d\omega, \quad n = 0, ..., N-1 , \tag{3.88}$$

and so if the number of data points N exceeds the number of unknowns, $j + k + 1$, in the expression (3.87), we can use (3.88) to determine the a's and b's that give the best fit to the observed data (see Cadzow 1982 for a summary of the methods of performing this optimization).

If $a_m = 0$ for $m \geq 1$,

$$\widetilde{P}(\omega) = \left| b_0 + b_1 e^{i\omega} + ... + b_j e^{ij\omega} \right|^2 . \tag{3.89}$$

The model then is said to be a *moving average*. The Blackman–Turkey estimation in (3.77) is one special case of such a model.

Alternatively, a model in which $b_m = 0$ $m \geq 1$ and

$$\widetilde{P}(\omega) = \frac{b_0}{1 + a_1 e^{i\omega} + ... + a_k e^{ik\omega}} , \tag{3.90}$$

is said to be *autoregressive*. Burg (1967) shows that if the first M lags of the autocovariance function are known exactly, the maximum entropy method reduces to that of finding an autoregressive model of order M.

3.5 WAVELETS

While Fourier decomposition is ideal for nearly harmonic signals, it is often not convenient for signals of short–time duration. We saw in (3.29) that the Fourier transform of the pulse, $\delta(t)$, is equal to unity for all frequencies ω. The reconstruction of the pulse from its Fourier components involves extensive cancellation between different frequency elements and is prone to instability in numerical calculations. Another practical difficulty is that the Fourier transform of a short–duration signal, embedded in statistically stationary noise, tends to be swamped by the Fourier transform of the noise. These problems arise because the basis functions of Fourier analysis, $e^{-i\omega t}$, extend over an infinite time interval and are poorly matched to a transient signal.

A classical method for dealing with short–time or non–stationary signals is Fourier analysis with a variable time window. This leads to a time–frequency transform of the form

$$\widehat{p}_F(\omega, b) = \int_{-\infty}^{\infty} p(t)\psi(t - b)e^{i\omega t}dt \quad ; \tag{3.91}$$

ω denotes the frequency and b the centre of the time window. The weighting function $\psi(t)$ is real and following Gabor (1946) it is common to take it to be Gaussian. A disadvantage of the time–frequency transform in (3.91) is that the window function is fixed and independent of ω. The number of cycles within the window is therefore a function of the frequency. In particular at high frequencies many cycles are included. The reconstruction of the short–time signal $p(t)$ from this transform involves extensive cancellation between these high–frequency elements and like the infinite–range Fourier transform is prone to numerical instability.

The wavelet transform was introduced by Goupillaud et al. (1984) to avoid this difficulty. Wavelet transforms are again functions of two parameters. One corresponds, as for the short–time Fourier transform in (3.91), to a time shift. The other parameter is not a frequency, but rather a *dilation, a*. The *wavelet transform* $\widehat{p}_w(a, b)$ is defined by

$$\widehat{p}_w(a, b) = \frac{1}{|a|^{1/2}} \int_{-\infty}^{\infty} p(t)w^* \left(\frac{t - b}{a} \right) dt \quad . \tag{3.92}$$

The function $w(t)$ is called the *analyzing wavelet* and the family of functions $w((t - b)/a)$ are referred to as *wavelets*.

We make no assumptions about $w(t)$ other than to require that it be square integrable and satisfy an admissibility condition

$$\int_{-\infty}^{\infty} |\widetilde{w}(\omega)|^2 \frac{d\omega}{|\omega|} \quad is \quad finite \quad , \tag{3.93}$$

where $\widetilde{w}(\omega)$ denotes the Fourier transform of $w(t)$. Provided $\widetilde{w}(\omega)$ is continuous near the origin, Equation (3.93) implies that

$$\widetilde{w}(0) = 0 \quad , \tag{3.94}$$

i.e. $\int_{-\infty}^{\infty} w(t)dt$ vanishes. It follows then that $w(t)$ must possess at least some oscillations. Hence the name "wavelet".

Many forms of analyzing wavelet have been used. We will see that it is particularly advantageous if the analyzing wavelet is localized in time and frequency space. Derivatives of Gaussians readily satisfy the condition (3.93) and have the advantage of decaying rapidly for large $|t|$ and $|\omega|$ (t and ω real). Hence

$$w(t) = t\, e^{-t^2/2} \quad \text{or} \quad w(t) = (1 - t^2)e^{-t^2/2}$$

are appropriate analyzing wavelets. In their pioneering paper, Goupillaud et al. (1984) introduced the analyzing wavelet

$$w(t) = \exp(-ict - t^2/2) - 2^{1/2}\exp(-ict - t^2 - c^2/4)$$

to investigate seismic signals. This has the Fourier transform

$$\widetilde{w}(\omega) = (2\pi)^{1/2}e^{-(c-\omega)^2/2} - (2\pi)^{1/2}e^{-(c-\omega)^2/4 - c^2/4}$$

and is therefore localized near $t = 0$ and near $\omega = c$. Since $\widetilde{w}(0)$ vanishes, it satisfies the admissibility condition (3.93).

Insight into the filtering effect of a wavelet transform can be obtained by expressing the right–hand side of (3.92) in terms of the Fourier transforms of $p(t)$ and $w(t)$:

$$\widehat{p}_w(a, b) = \frac{1}{(2\pi)^2|a|^{1/2}} \int_{-\infty}^{\infty}\int_{-\infty}^{\infty} \widetilde{p}(\omega)e^{-i\omega t}d\omega \int_{-\infty}^{\infty} \widetilde{w}^*(\omega')e^{i\omega'(t-b)/a}d\omega' \quad .$$

After exchanging the order of integration and using the relationship $\int e^{-it(\omega - \omega'/a)}\, dt = 2\pi\delta(\omega - \omega'/a)$, we obtain

$$\widehat{p}_w(a, b) = \frac{|a|^{1/2}}{2\pi} \int_{-\infty}^{\infty} \widetilde{p}(\omega)\widetilde{w}^*(a\omega)e^{-i\omega b}d\omega \quad . \tag{3.95}$$

If $w(t)$ is localized in time so that it is only significant when $t_{\min} \le t \le t_{\max}$, Equation (3.92) shows that $\widehat{p}_w(a, b)$ is only influenced by $p(t)$ for t in the range

$$at_{\min} + b \le t \le at_{\max} + b \quad . \tag{3.96}$$

Similarly, if $\widetilde{w}(\omega)$ is localized so that it is only significant for $\omega_{\min} \le \omega \le \omega_{\max}$, it is evident from (3.95) that $\widehat{p}(a, b)$ is only influenced by the frequency components of $\widetilde{p}(\omega)$ in the range

$$\omega_{\min}/a \le \omega \le \omega_{\max}/a \quad . \tag{3.97}$$

For large a, $\widehat{p}(a, b)$ contains the large–scale features of the signal: the signal is analyzed over a large–time interval and low–frequency components of $\widehat{p}(\omega)$ are investigated. Small a means that high–frequency components of $\widehat{p}(\omega)$ are explored by analyzing a short–time interval.

Frequency bands with different centre–frequencies are investigated when the scale parameter a is varied. However we see from (3.97) that bandwidth also increases in proportion to a^{-1}, so that the ratio of centre–frequency to bandwidth is kept constant. This is sometimes referred to as *constant-Q filtering* (Flandrin 1990), and wavelet transforms provide a way of implementing constant–Q filtering in the time domain.

The conditions (3.96) and (3.97) ensure that the time interval investigated is inversely proportional to the centre–frequency. In contrast to the short–time Fourier transform, the number of cycles analyzed remains constant as the frequency varies.

3.5.1 Inversion of a Wavelet Transform

If $p(t)$ is square integrable, it can be reconstructed from its wavelet transform (Grossmann & Morlet 1984; Daubechies 1989)

$$p(t) = \frac{1}{c_w} \int_{-\infty}^{\infty} \int_{-\infty}^{\infty} \widehat{p}_w(a, b) \frac{1}{|a|^{1/2}} w \left(\frac{t - b}{a} \right) \frac{da \, db}{a^2} \quad , \qquad (3.98)$$

where

$$c_w = \int_{-\infty}^{\infty} |\widetilde{w}(\omega)|^2 \frac{d\omega}{|\omega|} \quad .$$

A similar reconstruction can be made from the short–time Fourier transform defined in Equation (3.91)

$$p(t) = \frac{1}{c_F} \int_{-\infty}^{\infty} \int_{-\infty}^{\infty} \widehat{p}_F(\omega, b) \psi(t - b) e^{-i\omega t} d\omega \, db \qquad (3.99)$$

where

$$c_F = 2\pi \int_{-\infty}^{\infty} \psi^2(t) dt \quad .$$

We have seen that wavelet transforms provide a means of analyzing and reconstructing short–time signals. The time–scale decomposition they provide may be particularly useful if there is some *a priori* reason to suspect fractal behaviour or self–similarity of different scales in the signal. They may, for example, prove to be useful in the analysis of experimental data of turbulent flows. Another application has been in the analysis of speech and music, possibly because some researchers believe that the ear provides constant–Q filtering to an incoming sound signal. Wavelet transforms may therefore analyze a signal in a way which is directly analogous to the ear's response. If there is no reason to expect a dependence on scale rather than frequency, there is little to be gained from using the decomposition into continuous wavelets (3.98) rather than the short–time Fourier elements (3.99). However, wavelets have some advantage when discrete transforms are considered (Flandrin 1990).

3.5.2 Discrete Wavelet Transform

A greater advantage of the wavelet transform is that analyzing wavelets $w(t)$ have been found, such that both $w(t)$ and its Fourier transform have fast decay (e.g. faster than any inverse polynomial power), for which the family of wavelets $2^{j/2}w(2^j t - k)$, j, k integer, form a complete orthonormal basis for square–integrable functions. In contrast, the Balian–Low theorem (Daubechies 1989) shows that there is no orthonormal basis for the short–time Fourier transform with reasonable localization properties. Any square–integrable function, $p(t)$, can therefore be written as an expansion over the wavelet basis functions:

$$p(t) = \sum_{j,k} p_{jk} \, 2^{j/2} w^*(2^j t - k) \quad . \tag{3.100}$$

In this expansion the scale factor a takes discrete values $a = 2^{-j}$. An increase in j by 1, halves the value of a and results in an octave increase in frequency (see Equation (3.97)).

The coefficients p_{jk} follow directly from the expansion (3.100) and the orthogonality condition which can be expressed in the form

$$2^{(j+j')/2} \int_{-\infty}^{\infty} w^*(2^j t - k)w(2^{j'} t - k')dt = \delta_{jj'} \, \delta_{kk'} \quad , \tag{3.101}$$

where $\delta_{jj'}$ is the Kronecker delta function

$$\delta_{jj'} = 1 \quad \text{for} \quad j = j'$$
$$= 0 \quad \text{for} \quad j \neq j' \; .$$

After multiplying both sides of Equation (3.100) by $2^{j'/2}w(2^{j'} t - k')$, integration shows that

$$p_{jk} = 2^{j/2} \int_{-\infty}^{\infty} p(t)w(2^j t - k) \, dt \quad . \tag{3.102}$$

The first orthogonal bases with fast decay were reported by Lemarié & Meyer (1986). Since then emphasis has been placed on analyzing wavelets of compact support, i.e. such that $w(t)$ vanishes outside a finite interval.

The Haar analyzing wavelet is an elementary function of compact support which generates an orthonormal basis. The Haar wavelet is defined by

$$w(t) = \quad 1 \quad \text{for} \quad 0 < t < 1/2$$
$$= -1 \quad \text{for} \quad 1/2 < t < 1$$
$$= \quad 0 \quad \text{outside this range .}$$

It is evident from the definition that the Haar wavelet satisfies the orthogonality condition (3.101). However, it is not ideal because its discontinuities mean that its Fourier transform only decays slowly for large $|\omega|$.

Daubechies (1988) gives a prescription for calculating a sequence of analyzing wavelets of compact support, which generate orthonormal bases. These wavelets, $w_N(t)$, are indexed by an integer N, the smoothness of the analyzing wavelet increasing with increasing N. For $N = 1$, the Daubechies wavelet is just the Haar wavelet and is discontinuous. The wavelet for $N = 2$ is shown in Figure 3.6. It is continuous but not differentiable. Large values of N lead to more regular analyzing wavelets and therefore to functions for which $\widetilde{w}(\omega)$ decays more rapidly for large $|\omega|$. Strang (1989) gives a review of Daubechies's work and describes an efficient algorithm to calculate her wavelets.

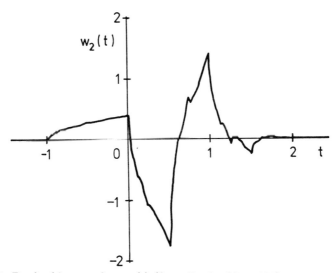

Figure 3.6 Daubechies wavelet, $w_2(t)$ (from Daubechies 1989).

In summary, wavelet transforms provide information about the scale–time decomposition of a signal. Wavelet theory is developing quickly and proving to be a useful tool to analyze a range of physical signals. Some recent applications are described in the conference proceedings edited by Combes et al. (1989).

REFERENCES

Ables, J.G. (1974) Maximum entropy spectral analysis. Astron. Astrophys. Suppl. Series 15:383–393 (reprinted in Childers 1978).

Bendat, J.S. & Piersol, A.G. (1974) Random Data: Analysis and Measurement Procedures. Wiley.

Bingham, C., Godfrey, M.D. & Tukey, J.W. (1967) Modern techniques of power spectrum estimation. IEEE Trans. Audio & Electroacoust. AU–15:56–66 (reprinted in Childers 1978).

Blackman, R.B. & Tukey, J.W. (1959) The Measurement of Power Spectrum from the Point of View of Communications Engineering. Dover, New York.

Burg, J.P. (1967) Maximum entropy spectral analysis, Proc. 37th Meeting of Society of Exploration Geophysicists (reprinted in Childers 1978).

Cadzow, J.A. (1982) The spectral estimation: an overdetermined rational model equation approach. Proc. IEEE 70:907–939.

Childers, D.G.(ed) (1978) Modern Spectrum Analysis. IEEE Press.

Combes, J.M., Grossmann, A. & Tchamitchian, Ph. (1989) Wavelets: Time–Frequency Methods and Phase Space. Springer–Verlag.

Cooley, J.W. & Tukey, J.W. (1965) An algorithm for the machine calculation of complex Fourier Series. Math. Comput. 19:297–301.

Daubechies, I. (1988) Orthonormal bases of compactly supported wavelets. Comm. Pure & Appl. Math.49:909–996.

Daubechies, I. (1989) Orthonormal bases of wavelets with finite support–connection with discrete filters. Wavelets: Time–Frequency Methods and Phase Space. Eds. Combes, J.M., Grossmann, A. & Tchamitchian, Ph. Springer–Verlag, pp.38–66.

Flandrin, P. (1990) Wavelets and related time–scale transforms. Advanced Signal Processing Algorithms, Architectures and Implementations. Ed. F.T. Luk, SPIE Proceedings, 1348.

Gabor, D (1946) Theory of communication, J.I.E.E. 93 (III):429–441.

Glover, K. (1984) Frequency Domain Identification Methods, Encyclopaedia of Systems and Control (ed. M. Singh). Pergamon Press.

Goupillaud, P., Grossmann, A. & Morlet, J. (1984) Cycle–octave and related transforms in seismic signal analysis. Geoexploration 23:85–102.

Gradshteyn, I. S. & Ryzhik, I.M. (1980) Table of Integrals, Series and Products. Academic Press.

Grossmann, A. & Morlet, J. (1984) Decomposition of Hardy functions into square integrable wavelets of constant shape. SIAM J. Math. Anal 15:723–736.

Jeffreys, H. & Jeffreys, B.S. (1949) Mathematical Physics. Cambridge University Press.

Lacoss, R.T. (1971) Data adaptive spectral analysis methods. Geophys. 36:661–675 (reprinted in Childers 1978).

Lemarié, P.G. & Meyer, Y. (1986) Ondelettes et bases hilbertiennes. Rev. Mat. Iberoamericana 2:1–18.

Lighthill, M.J. (1978) Fourier Analysis and Generalised Functions. Cambridge Univ. Press.

Newland, D.E. (1984) An Introduction to Random Vibrations and Spectral Analysis. Longman.

Strang, G. (1989) Wavelets and dilation equations: a brief introduction. SIAM Review 31:614–627.

Taylor, G.I. (1938) The spectrum of turbulence. Proc. Roy. Soc. London A164:476–490.

Welch, P.D. (1967) The use of Fast Fourier Transform for the estimation of power spectra: a method based on time averaging over short modified periodograms. IEEE Trans. Audio and Electroacoust. AU–15:70–73 (reprinted in Childers 1978).

4. ASYMPTOTIC EVALUATION OF INTEGRALS

4.1 INTRODUCTION

This Chapter concerns methods of estimating certain integrals that depend on a large (or small) parameter. The essential idea in each case is to compare the given integral with a simpler one that can be evaluated exactly yet approximates closely to the original. In the most elementary case the integrand can be expressed as a Taylor series in increasing powers of the small parameter; if the approximating series is a uniformly good approximation to the integrand (for all values of the dummy variable of integration) it can be integrated term by term.

Example. The elliptic integral

$$K(m) = \int_0^{\frac{\pi}{2}} (1 - m \cos^2 \theta)^{-1/2} d\theta, \qquad m \ll 1 . \tag{4.1}$$

If m is small compared with unity, then $m \cos^2 \theta \le m$ is also small for all values of θ. Thus we may expand the integrand as a Taylor series

$$(1 - m \cos^2 \theta)^{-1/2} = 1 + \frac{m}{2} \cos^2 \theta + \frac{3m^2}{8} \cos^4\theta + R_3 , \tag{4.2}$$

where the error after the first three terms (say) is bounded by a constant times m^3, i.e.

$$|R_3| < C_3 m^3 ,$$

where C_3 is some constant. (Here and henceforth $C_1, C_2, \ldots$ denote constants.) On integrating term by term

$$K(m) = \frac{\pi}{2} (1 + \frac{m}{4} + \frac{9}{64} m^2) + 0(m^3) . \tag{4.3}$$

Higher–order terms can of course be obtained by expanding $(1 - m \cos^2)^{-1/2}$ up to terms higher than those given by (4.2).

The snag is that, in most cases of interest, a single expansion of the integrand is not valid for all values of the integration variable. Thus two or more approximations to the integrand may be called for to deal with different ranges of θ.

Example. Estimate the elliptic integral for small positive values of $\epsilon = 1 - m$. Here

$$K = \int_0^{\frac{\pi}{2}} (1 - (1 - \epsilon) \cos^2 \theta)^{-1/2} d\theta = \int_0^{\frac{\pi}{2}} (\sin^2 \theta + \epsilon \cos^2 \theta)^{-1/2} d\theta .$$

For fixed $\theta > 0$, the integrand can be expanded as a power series in ϵ. In particular the leading term is

$$(\sin^2 \theta + \epsilon \cos^2 \theta)^{-1/2} \sim \operatorname{cosec}\theta, \quad \epsilon \to 0, \qquad \theta > 0 . \tag{4.4}$$

But $\operatorname{cosec} \theta$ is infinite at $\theta = 0$ and the approximation (4.4) clearly fails near $\theta = 0$. The approximation is said to be uniformly valid for $\theta \geq \theta_0 > 0$, but is not uniformly valid for $\theta > 0$. This is, of course, because the $\epsilon \cos^2 \theta$ term is important when θ is also small. But in this vicinity (θ small) we can simplify the integrand in a different way, by making the comparisons $\sin \theta \sim \theta$ and $\cos^2 \theta \sim 1$. Thus we divide the range of integration as follows:

$$I = \int_0^{\delta} (\quad)^{-1/2} d\theta + \int_{\delta}^{\frac{\pi}{2}} (\quad)^{-1/2} d\theta \equiv I_1 + I_2 , \tag{4.5}$$

where δ is chosen to be such that

$$\epsilon \lll \delta^2 \lll 1 . \tag{4.6}$$

In integral I_1, $\theta \leq \delta << 1$ so that $(\sin^2 \theta + \epsilon \cos^2 \theta)^{-1/2} \sim (\theta^2 + \epsilon)^{-1/2}$, and

$$I_1 \sim \int_0^{\delta} (\theta^2 + \epsilon)^{-1/2} d\theta$$
$$= \sinh^{-1}(\delta/\epsilon^{1/2}) = \log(2\delta/\epsilon^{1/2}) + o(1) , \tag{4.7}$$

since δ has been chosen (see (4.6)) to be such that $\delta/\epsilon^{1/2} >> 1$. As for I_2, (4.4) shows

$$I_2 \sim \int_{\delta}^{\frac{\pi}{2}} \operatorname{cosec}\theta \, d\theta = \log \frac{1 + \cos \delta}{\sin \delta} \sim \log \frac{2}{\delta} , \tag{4.8}$$

since δ is small. Adding (4.7) and (4.8), we have

$$K = I_1 + I_2 \sim \frac{1}{2} \log \frac{16}{1-m} \quad \text{for } m \sim 1 . \tag{4.9}$$

Note that the parameter δ cancels when (4.7) and (4.8) are added: indeed, it must as this parameter is arbitrary except for the constraint (4.6).

Asymptotic Expansion

Suppose we want an estimate for a function $f(z)$ as $z \to \infty$, with z real and positive for the moment. An asymptotic expansion is defined as follows. The symbolism

$$f(z) \sim \sum_0^\infty a_n z^{-n} \quad \text{as} \quad |z| \to \infty \tag{4.10}$$

means that for any positive integer m

$$f(z) = \sum_0^m a_n z^{-n} + R_m(z) , \tag{4.11}$$

where

$$|z^m R_m(z)| \to 0 \quad \text{as} \quad |z| \to \infty, \quad m \quad \text{fixed} . \tag{4.12}$$

The condition (4.12), sometimes written equivalently as $R_m(z) = o(z^{-m})$ as $z \to \infty$, ensures that the error R_m is of smaller order than the last term retained in the expansion. The definition may be extended to deal with complex z; in this case the expansion (4.10) will typically hold for a given range of values for arg z.

Note that there is no requirement that the sum (4.10) be convergent. A convergent series is a different concept from an asymptotic series. With a convergent series one can think of improving the estimate by taking more and more terms (i.e. let $m \to \infty$ with z fixed); an asymptotic estimate is improved by letting $|z| \to \infty$ with m fixed.

Example.

$$I(N) = \int_0^\infty (1+t)^{-1} \exp(-Nt)dt, \qquad N \to +\infty . \tag{4.13}$$

The integrand is exponentially small as $N \to \infty$ except near $t = 0$. Thus if we divide the range of integration as

$$\int = \int_0^a + \int_a^\infty \quad \text{with} \quad \frac{1}{N} \ll a \ll 1 ,$$

the contribution from the second range is exponentially small, and negligible compared with *any* inverse power of N. In the small interval $0 \le t \le a$, $(1+t)^{-1}$ is little different from unity, so we expect the leading–order estimate

$$I \sim \int_0^a e^{-Nt} dt \sim \int_0^\infty e^{-Nt} dt = \frac{1}{N} \quad \text{as} \quad N \to \infty .$$

To go to higher order, note that

$$(1+t)^{-1} = 1 - t + t^2 - + \ldots + (-1)^{m-1} t^{m-1} + (-1)^m (1+t)^{-1} t^m ;$$

so

$$I = \int_0^\infty e^{-Nt} dt - \int_0^\infty t e^{-Nt} dt + \ldots + (-1)^{m-1} \int_0^\infty t^{m-1} e^{-Nt} dt + R_m ,$$

where

$$|R_m| \le \int_0^\infty t^m e^{-Nt} = m!/N^{m+1} .$$

Thus

$$I = \frac{1}{N} - \frac{1}{N^2} + \frac{2!}{N^3} - \ldots + \frac{(-1)^{m-1}(m-1)!}{N^m} + R_m ,$$

with

$$N^m R_m \to 0 \quad \text{as} \quad N \to \infty, \quad m \text{ fixed} .$$

Thus the complete asymptotic expansion for I is

$$I \sim \frac{1}{N} - \frac{1}{N^2} + \frac{2!}{N^3} - \frac{3!}{N^4} \ldots , \tag{4.14}$$

which is not convergent. The result (4.14) is that which would have been obtained by expanding

$$(1+t)^{-1} = 1 - t + t^2 - t^3 \ldots \tag{4.15}$$

and integrating term–by–term in the integral (4.13), from 0 to ∞, even though the series (4.15) converges only for $t < 1$. The point is that the result comes from the region very close to $t = 0$, where (4.15) does converge; for $t > t_0$ the integrand is dominated by the exponential factor e^{-Nt}. A generalisation of this particular example is easy to establish and is of crucial important in what follows.

4.2 WATSON'S LEMMA:

Consider the integral

$$I = \int_0^a t^\lambda f(t) e^{-Nt} dt, \quad N \to \infty , \qquad (4.16)$$

where $a(> 0)$ may be finite or infinite, and with $\lambda > -1$. Suppose that $f(t)$ is analytic (i.e. differentiable any number of times) for $t > 0$ (this condition can be relaxed later), and that it has a Taylor series with non–vanishing radius of convergence t_1 about $t = 0$. Suppose also that f has the bound

$$|f(t)| < A e^{\alpha t}$$

for some constant α and positive constant A.

Then the asymptotic expansion for $I(N)$ is obtained by using the local Taylor series expansion

$$f(t) = a_0 + a_1 t + a_2 t^2 + a_3 t^3 + \ldots (t < t_1) \qquad (4.17)$$

for f, substituting into (4.16) and integrating term by term from $t = 0$ to $t = \infty$, to get

$$I \sim \frac{a_0 \lambda!}{N^{\lambda+1}} + \frac{a_1(\lambda+1)!}{N^{\lambda+2}} + \frac{a_2(\lambda+2)!}{N^{\lambda+3}} + \cdots .$$

In terms of our original definition of an asymptotic expansion, this means that

$$I = N^{-(\lambda+1)} J ,$$

with $J \sim a_0 \lambda! + a_1(\lambda+1)! N^{-1} + \ldots .$

Proof. Choose a fixed number t_0 less than the radius of convergence t_1 of the Taylor series (4.17). Thus for $t < t_0$

$$f(t) = a_0 + a_1 t + \ldots + a_{m-1} t^{m-1} + r_m(t)$$

where

$$|r_m(t)| < A_m t^m \qquad (4.18)$$

for some positive constant A_m. The integral I can now be written as

$$I = \int_0^\infty t^\lambda (a_0 + \ldots + a_{m-1} t^{m-1}) e^{-Nt} dt \quad \int_{t_0}^\infty t^\lambda (a_0 + \ldots + a_{m-1} t^{m-1}) e^{-Nt} dt$$

$$+ \int_0^{t_0} t^\lambda r_m(t) e^{-Nt} dt + \int_{t_0}^a t^\lambda f(t) e^{-Nt} dt$$

$$= I_1 + I_2 + I_3 + I_4 , \text{ say} .$$

Now

$$I_1 = \frac{a_0\lambda!}{N^{\lambda+1}} + \ldots + \frac{a_{m-1}(\lambda+m-1)!}{N^{\lambda+m}}$$

and it remains to show that I_2, I_3 and I_4 are negligible (i.e. $o(N^{-\lambda-m})$) for fixed m as $N \to \infty$. To estimate I_2 note that $|a_0 + \ldots + a_{m-1}t^{m-1}| < B_m t^{m-1}$ (with $t \geqslant t_0$) where B_m is a constant, so

$$|I_2| < B_m \int_{t_0}^{\infty} t^{\lambda+m-1}e^{-Nt}dt = B_m e^{-Nt_0} \int_0^{\infty} (t_0+s)^{\lambda+m-1}e^{-Ns}ds < C_m e^{-Nt_0}$$

which is $o(N^{-\lambda-m})$ as $N \to \infty$, as required. Also, (4.18) ensures that

$$|I_3| < A_m \int_0^{\infty} t^{\lambda+m}e^{-Nt}dt = A_m(\lambda+m)!N^{-\lambda-m-1} = o(N^{-\lambda-m}),$$

and

$$|I_4| < A \int_{t_0}^{\infty} t^{\lambda}e^{\alpha t - Nt}dt < B \int_{t_0}^{\infty} e^{\alpha_1 t - Nt}dt = \frac{B}{N-\alpha_1}e^{-(N-\alpha_1)t_0},$$

where $\alpha < \alpha_1 < N$. The required result then follows. Note that the exponential function $\exp(-Nt)$ dominates the integrand of (4.16), and ensures that contributions to the asymptotic expansion (to any order $N^{-\mu}$) come from the immediate vicinity of the origin. Changing the upper limit changes the result by an exponentially small amount. So does a change in the function f away from the origin. So the same expansion is obtained if f is a piecewise continuous function, but with the same behaviour (4.17) in a small neighbourhood of the origin, and if $|f| < Ae^{\alpha t}$ at infinity.

Example.

$$I(N) = \int_0^{\infty} |\sin t|e^{-Nt}dt.$$

Here $f(t) = \sin t$ for $0 < t < \pi$, $f(t) = \pm\sin t$ otherwise. Hence

$$I(N) \sim \int_0^{\infty} \left(t - \frac{t^3}{3!} + \frac{t^5}{5!} - \frac{t^7}{7!}\cdots\right)e^{-Nt}dt,$$

i.e.

$$I(N) \sim \frac{1}{N^2} - \frac{1}{N^4} + \frac{1}{N^6} - \frac{1}{N^8}\cdots \text{ as } N \to \infty.$$

4.3 RAPIDLY OSCILLATORY INTEGRALS

In the integrals estimated by Watson's Lemma, the vicinity of the origin gives the main contribution; elsewhere the integrand is exponentially small and therefore negligible. A related type of integral has a purely imaginary exponent. Hence the integrand is rapidly *oscillatory* (rather than rapidly decaying as above); the argument that identifies the most important part of the integration range is now more subtle since it relies on destructive interference (i.e. cancellations) rather than smallness of the integrand away from the critical points. The main contributions come from the regions where the destructive interference is least effective. The integrals under consideration are of the type

$$I(N) = \int_a^b A(t) \exp\{i\,Nu(t)\}dt \,, \qquad (4.19)$$

where $u(t)$, a and b are real. The easiest case is as follows.

No Stationary Points. Suppose that A and u are smooth, in the sense that A', u' and u'' are continuous in the integration range $a \leq t \leq b$, and suppose that the phase function $u(t)$ has no stationary points, i.e. $u'(t) \neq 0$ for $a \leq t \leq b$.

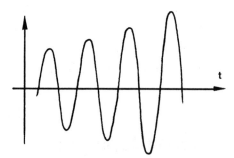

Figure 4.1

Now although $A(t)$ is of order unity (with respect to N), the integral I will be much less because of destructive interference. The integrand of (4.19) has real and imaginary parts that are shown symbolically in the diagram, where it is seen that neighbouring lobes nearly cancel. As N increases the oscillations become more closely packed and the cancellation becomes more effective. Analytically, the result is obtained by partial integration.

Thus

$$I = \int_a^b \frac{A}{i\,Nu'} \frac{d}{dt} \left(e^{i\,Nu} \right) dt = \left[\frac{A}{i\,Nu'} e^{i\,Nu} \right]_a^b - \int_a^b \left\{ \frac{d}{dt} \left(\frac{A}{i\,Nu'} \right) \right\} e^{i\,Nu} dt \,, \quad (4.20)$$

which is of order N^{-1}. Indeed, if A'' and u''' are continuous, the leading contribution comes from the end points since the integral is found to be $0(N^{-2})$ by a further partial integration. The contributions from near a and b arise because the interference effect is not quite complete there — the end lobes do not quite cancel. This analysis is not valid, however, if u' vanishes at some intermediate point $t = t_0$. In terms of the destructive interference phenomenon, such a stationary point t_0 has crucial importance since the phase function Nu varies relatively slowly in its vicinity.

We expect that a stationary point t_0 will lead to a relatively large contribution to $I(N)$, and the following analysis due to Kelvin shows that it gives a term of order $N^{-1/2}$ (or more), and is bigger than the order N^{-1} term that has been obtained by (4.20) in the absence of a stationary point.

Method of Stationary Phase (Kelvin)

If

$$I(N) = \int_a^b A(t) \exp\{i\,Nu(t)\} dt$$

where A', A'', u' and u'' are continuous in the interval $a \le t \le b$ with just one stationary point t_0 inside the interval $a < t_0 < b$, with $u'(t_0) = 0$ and $u''(t_0) \ne 0$, then

$$I(N) \sim A(t_0) \left\{ \frac{2\pi}{N|u''(t_0)|} \right\}^{1/2} \exp\left\{ i\,Nu(t_0) \pm i\,\frac{\pi}{4} \right\} \text{ as } N \to \infty \quad (4.21)$$

with the $+$ or $-$ sign according as $u''(t_0) > 0$ or $u''(t_0) < 0$.

Rough Idea. We anticipate that the main contribution comes from the vicinity of $t = t_0$, so we make the approximations

$$A(t) \sim A(t_0) \text{ and } u(t) \sim u(t_0) + \frac{1}{2}(t - t_0)^2 u''(t_0)$$

in which the quadratic term in $u(t)$ is the first non–constant term in the Taylor series expansion for $u(t)$. Furthermore, since the region away from t_0 is of secondary significance, we can now replace the range of integration by the whole t–axis. Thus the actual integral is replaced by the comparison integral

$$I \sim A(t_0)e^{i\,Nu(t_0)} \int_{-\infty}^{\infty} \exp\left\{\frac{1}{2}\,iN(t-t_0)^2\,u''(t_0)\right\}dt$$

which can be evaluated exactly. For on making the substitution

$$(t-t_0)^2 = \pm\left\{\frac{2}{Nu''(t_0)}\right\}x^2\ ,$$

$$\int_{-\infty}^{\infty} \exp\left\{\ \right\}\,dt = 2\left\{\frac{2}{N|u''(t_0)|}\right\}^{1/2} \int_{0}^{\infty} e^{\pm ix^2}\,dx = \left\{\frac{2\pi}{N|u''(t_0)|}\right\}^{1/2} e^{\pm i\pi/4}$$

after deforming the integration path on to the line $x = e^{\pm i\pi/4}p$ and using the result $\int_0^\infty e^{-p^2}\,dp = \frac{1}{2}\sqrt{\pi}$. This gives the stated result (4.21) and provides a method that is easy to remember (and to generalize) in practice. But it is reasonable rather than rigorous and can be put on a firmer footing as follows.

Proof.   Near $t = t_0$, $u(t) = u(t_0) + \frac{1}{2}\,u''(t_0)(t-t_0)^2 + \ldots\ $.

Make the substitution $x = x(t)$, defined by $u(t) = u(t_0) + \frac{1}{2}\,u''(t_0)x^2$.

It is a one–to–one mapping for $a \le t \le b$, (with $x \sim t - t_0$ as $t \to t_0$) since $u' = 0$ only at t_0. Thus

$$I = \exp(i\,Nu(t_0)) \int_{-x_1}^{x_2} A_1(x)\ \exp\left\{\frac{1}{2}\,i\,Nu''(t_0)x^2\right\}dx\ ,$$

where $-x_1$ and $+x_2$ are the values of x corresponding to $t = a$ and $t = b$ (x_1 and x_2 both positive) and where

$$A_1(x) = A(t(x))dt/dx$$

is such that $A_1(0) = A(t_0)$, since $dx/dt = 1$ at $t = t_0$. Now write

$$I = \exp(i\,Nu(t_0)) \int_{-x_1}^{x_2} (A_1(x) - A_1(0))e^{\frac{1}{2}\,i\,Nu''(t_0)x^2}\,dx$$

$$+ A(t_0)e^{i\,Nu(t_0)} \int_{-x_1}^{x_2} e^{\frac{1}{2}\,i\,Nu''(t_0)x^2}\,dx\ .$$

The first of these integrals is of order N^{-1} since it can be expressed as

$$\int \frac{A_1(x) - A_1(0)}{i\,Nu''(t_0)x}\,\frac{d}{dx}\left(e^{\frac{1}{2}\,i\,Nu''(t_0)x^2}\right)dx$$

and integrated by parts, since $A_1(x) - A_1(0) = 0(x)$ and $\{A_1(x) - A_1(0)\}/x$ is differentiable at $x = 0$. As for the second integral, its limits can be replaced by $-\infty$ and ∞ since, for example, $\int_{x_2}^{\infty} \exp\{\frac{1}{2} i N u''(t_0)x^2\}dx$ is $0(1/N)$ by partial integration. Finally

$$\int_{-\infty}^{\infty} \exp\left\{\frac{1}{2} i N u''(t_0)x^2\right\} dx = \left\{\frac{2\pi}{N|u''(t_0)|}\right\}^{1/2} \exp(\pm i\pi/4)$$

and the result follows.

Extension of Results

If there are m distinct stationary points in the interval, divide the range of integration into m suitable intervals, with one stationary point in each interval and their contributions simply add together. A similar subdivision is appropriate if one or more of the functions A', A'', u', u'' is piecewise continuous, provided they are smooth within small neighbourhoods of the stationary points. If a stationary point is at an end point, say at $t = a$, then its contribution is halved.

Higher–Order Stationary Points

If t_0 is a stationary point (i.e. $u(t_0) = 0$) but $u''(t_0) = 0$, $u'''(t_0) = 0, \ldots u^{(n-1)}(t_0) = 0$, $u^{(n)}(t_0) \neq 0$, then the phase function $u(t)$ varies even more slowly and we expect an estimate of larger magnitude than the $N^{-1/2}$ term above. In the spirit of Kelvin's method, we have

$$I \sim A(t_0) \exp\left\{i N u(t_0)\right\} \int_{-\infty}^{\infty} \exp\left\{iN \frac{(t-t_0)^n}{n!} u^{(n)}(t_0)\right\} dt .$$

On substituting $\dfrac{N(t-t_0)^n}{n!} u^{(n)}(t_0) = \pm x^n$ with $+$ or $-$ according as $u^{(n)}(t_0) > 0$ or < 0, we find

$$I \sim A(t_0) \left\{\frac{n!}{|u^{(n)}(t_0)|}\right\}^{1/n} N^{-1/n} J_n \exp(i N u(t_0))$$

where

$$J_n = \int_{-\infty}^{\infty} e^{\pm i x^n} dx .$$

The magnitude of I is proportional to $N^{-1/n}$, which is bigger as n increases.

Example. The Bessel function $J_0(N)$ has the integral representation

$$J_0(N) = \frac{2}{\pi} \int_1^\infty \frac{\sin(Nx)dx}{(x^2-1)^{1/2}} = \frac{2}{\pi} \Im \int_0^\infty e^{iN\cosh t} dt \, ,$$

with the substitution $x = \cosh t$. The stationary point is at $t = 0$, near which

$$\cosh t = 1 + \frac{t^2}{2} + \dots \, .$$

Thus

$$J_0(N) \sim \frac{2}{\pi} \Im\, e^{iN} \int_0^\infty e^{\frac{i}{2}Nt^2} dt = \left(\frac{2}{\pi N}\right)^{1/2} \sin(N + \pi/4) \, . \qquad (4.22)$$

4.4 INTEGRALS WITH A LARGE EXPONENT

A much more general class of complex integrals can now be analyzed and a good account is given by Carrier et al. (1966), whose approach is followed here. Consider

$$I(N) = \int_A^B g(z) \exp\{Nf(z)\}dz, \qquad N \to \infty \qquad (4.23)$$

along some path from A to B in the complex z–plane, where $f(z)$ and $g(z)$ are taken to be analytic in the region of interest. We can consider N to be real; for if not, we write $N = N_0 e^{i\alpha}$ and the phase factor $e^{i\alpha}$ can be absorbed into f. Our previous results have been for f purely real or purely imaginary.

The magnitude of the integral depends crucially on the real part of $f(z)$; if it is written as $f = u + iv$ (u and v real), then

$$|\exp(Nf)| = \exp(Nu)$$

and a very crude bound for the integral is given by

$$|I| \leq LG e^{NU} \qquad (4.24)$$

where L is the path length, G the maximum value of $|g|$ on the path and U is the maximum value of u on the path. Clearly (4.24) might be a vast over–estimate since a path deformation might produce a lesser value of U. The best bound is obtained by taking a path such that U is as small as possible. As a preliminary we need an analysis to determine the behaviour of $u(x, y)$ as a function of x and y.

Contour Map of $u(x, y)$

Imagine $u(x, y)$ to be the height of a surface above a reference plane $u = 0$ at "sea level". It will be found expedient to deform our path of integration (picking up the residue contributions where appropriate) so that it joins the end points A and B and passes along the "low ground" of the surface, where u is as small as possible. An overall picture of $u(x, y)$ can be obtained by considering the contours $u = $ constant; it is useful to visualize also the family of contours $v = $ constant which are the lines of constant phase $\Im(f) = $ constant. The two families of curves are easily seen to be mutually orthogonal, as

$$\nabla u \cdot \nabla v = \frac{\partial u}{\partial x}\frac{\partial v}{\partial x} + \frac{\partial u}{\partial y}\frac{\partial v}{\partial y} = -\frac{\partial u}{\partial x}\frac{\partial u}{\partial y} + \frac{\partial u}{\partial y}\frac{\partial u}{\partial x} = 0$$

using the Cauchy–Riemann equations for the real and imaginary parts of an analytic function $f(z) = u + iv$. It is also easy to see that u (and also v) cannot have a maximum. For at a stationary point (x_0, y_0) where $\partial u/\partial x = \partial u/\partial y = 0$ (hence $df/dz = 0$), the discriminant

$$u_{xx}u_{yy} - u_{xy}^2 = -u_{xx}^2 - u_{xy}^2 \le 0 ,$$

since $u_{xx} + u_{yy} = 0$ from the Cauchy–Riemann equations. If $f''(z_0) \ne 0$, the discriminant is strictly negative and the point (x_0, y_0) is a saddle point. (More generally, the value of any harmonic function u at a point (x_0, y_0) equals the average of u over a circle centred at (x_0, y_0): so it can not have a maximum or minimum at (x_0, y_0).)

Example. $f = z^2$ so $u = x^2 - y^2$ and $v = 2xy$.

The contours $u = $ constant are shown by solid lines and the lines $v = $ constant are shown by broken lines. The paths of steepest descent (most rapid decrease of u) are orthogonal to the level surfaces $u = $ constant and are therefore given by the broken lines. Arrows signify the direction of descent. Note that there is a saddle point at $z = 0$ (where $f'(z) = 0$). Moving from $(x, y) = (0, 0)$ along the y–axis in either direction corresponds to a decrease in u, whilst the x–axis corresponds to an increase in u. The "valleys" $u(x, y) < u(0, 0)$ are the regions $y^2 > x^2$ whilst the "ridges" $u(x, y) > u(0, 0)$ are given by $x^2 > y^2$.

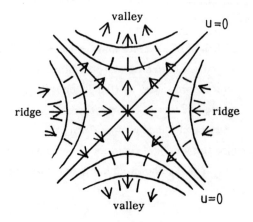

Figure 4.2

We are now in a position to analyze the integral (4.23). There are two possible cases which have to be dealt with separately.

Case I. End Points in the Same Valley

Deform the integration contour onto the path AA_1B_1B so that v is constant on AA_1 and B_1B, u is constant on A_1B_1 with $u(A) > u(A_1)$ and $u(B) > u(B_1)$; this possibility is meant by the statement that the end points A and B are in the same valley. Suppose first that $u(A) > u(B)$; then the main contribution to the integral is from the path AA_1, in the vicinity of A as the integrand decays rapidly along AA_1.

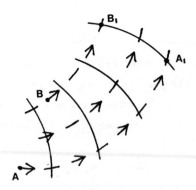

Figure 4.3

On the path AA_1,

$$f(z) - f(A) = u(z) - u(A) < 0$$

$$= -t \text{ say}, \quad 0 < t < t_1 .$$

Thus

$$f(z) = f(A) - t, \quad 0 < t < t_1 , \tag{4.25}$$

and

$$\int_{AA_1} = - \exp(Nf(A)) \int_0^{t_1} e^{-Nt}(g/f')dt ,$$

where g/f' is a function of z that can be expressed (in principle) as a function of t, on inverting the relation (4.25) to get z as a function of t. Near $t = 0$

$$g/f' = a_0 + a_1 t + a_2 t^2 + \ldots , \quad 0 < t < t_2 \quad \text{say} .$$

Watson's Lemma ensures that

$$I_{AA_1} \sim - \exp(Nf(A)) \left\{ \frac{a_0}{N} + \frac{a_1}{N^2} + \frac{2!a_2}{N^3} + \ldots \right\} . \tag{4.26}$$

The leading term is easy to obtain since $t = 0$ corresponds to $z = z_A$; hence $a_0 = g(A)/f'(A)$, in the obvious notation. The contribution from $B_1 B$ is calculated similarly, is of order $N^{-1} \exp(Nu(B))$ and is negligible compared with (4.26) if $u(A) > u(B)$.

The contribution from $A_1 B_1$ is negligible since

$$|I_{A_1 B_1}| \leq GL \exp(Nu(A_1))$$

where L is the length of the path $A_1 B_1$ and G the maximum value of $|g|$ on the path $A_1 B_1$. Thus $I_{A_1 B_1}$ is asymptotically negligible compared with each term of the series (4.26) since $u(A) > u(A_1)$. In particular, the leading term is

$$I \sim - \left\{ g(A)/f'(A) \right\} N^{-1} \exp \left\{ Nf(A) \right\} \quad \text{as} \quad N \to \infty . \tag{4.27}$$

Generalization

If $u(B) > u(A)$, then the main contribution is from near B, whence

$$I \sim +\left\{g(B)/f'(B)\right\}N^{-1} \, \exp\left\{Nf(B)\right\} \quad \text{as} \quad N \to \infty \, .$$

If both end points A and B are on the same level (i.e. $u(A) = u(B)$), then there are similar contributions from AA_1 and B_1B, whence

$$I \sim -\frac{g(A)}{f'(A)} \frac{e^{Nf(A)}}{N} + \frac{g(B)}{f'(B)} \frac{e^{Nf(B)}}{N}$$

to leading order.

Example. $I(N) = \displaystyle\int_1^3 \exp(iNz^2)d$. Here $g(z) = 1$ and $f(z) = i\,z^2 = i(x^2 - y^2) - 2xy$.

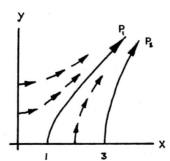

Figure 4.4

The broken lines show the paths of steepest descent from the two end points. So we put

$$I = \int_{P_1} - \int_{P_2}$$

where the paths P_1 and P_2 are as shown, and are linked at great distance by a joining arc on which the integrand is exponentially small.

On path P_1, $iz^2 = i - t$, i.e. $z^2 = 1 + it$, $0 < t < \infty$. Since $2z\,dz = i\,dt$,

$$I_1 = e^{iN} \int_0^\infty (i/2z)e^{-Nt}dt = \frac{1}{2}ie^{iN} \int_0^\infty (1+it)^{-1/2}e^{-Nt}dt$$

$$\sim \frac{1}{2}ie^{iN} \int_0^\infty e^{-Nt}(1 - \frac{1}{2}\,it\ldots)dt = \frac{1}{2}ie^{iN}\left\{\frac{1}{N} - \frac{i}{2N^2} + \ldots\right\}$$

Similarly, P_2 is given by $z^2 = 9 + it$; hence

$$I_2 = \frac{1}{6} i e^{9iN} \int_0^\infty (1 + \frac{it}{9})^{-1/2} e^{-Nt} dt \sim \frac{1}{6} i e^{9iN} \left\{ \frac{1}{N} - \frac{i}{18N^2} \cdots \right\} ,$$

and finally

$$I \sim \frac{1}{2} i e^{iN} \left\{ \frac{1}{N} - \frac{i}{2N^2} + \cdots \right\} - \frac{1}{6} i e^{9iN} \left\{ \frac{1}{N} - \frac{i}{18N^2} + \cdots \right\} .$$

Case II. End Points in Different Valleys

The elementary treatment given above fails if the two end points A and B are in different valleys. Then we have to go over a ridge (or several ridges) to join the end points. Our aim is to do this by staying on as low ground as possible, which means taking a path through an appropriate saddle point (or saddle points). It is not enough to locate the various saddle points of $f(z)$; we have to determine, from the global topography of the function, which of these saddle points are relevant in our journey from A to B.

Saddle Point Analysis

At a saddle point, $\partial u/\partial x = \partial u/\partial y = 0$, i.e.

$$f'(z_0) = 0 \quad \text{at} \quad (x, y) = (x_0, y_0) .$$

Near such a point, if we restrict attention to the simplest case with $f''(z_0) \neq 0$,

$$f(z) = f(z_0) + \frac{1}{2}(z - z_0)^2 f''(z_0) + \cdots ,$$

Using local polar coordinates $r e^{i\theta} = (z - z_0)$ and writing $f''(z_0) = A e^{i\alpha}$, we have

$$u - u_0 = \frac{1}{2} A r^2 \cos(2\theta + \alpha) \quad \text{and} \quad v - v_0 = \frac{1}{2} A r^2 \sin(2\theta + \alpha)$$

together with terms of order r^3.

The level surfaces $u = u_0$ are given by

$$2\theta + \alpha = \pm\pi/2 \quad \text{or} \quad \pm 3\pi/2 .$$

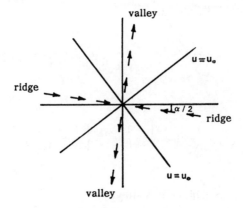

Figure 4.5

The paths of steepest descent from z_0 are given by

$$\theta = \frac{\pi}{2} - \frac{\alpha}{2} \quad \text{and} \quad \theta = -\frac{\pi}{2} - \frac{\alpha}{2} \,,$$

and the paths of steepest ascent are given by

$$\theta = -\frac{\alpha}{2} \quad \text{and} \quad \theta = \pi - \frac{\alpha}{2} \,.$$

The paths of ascent and descent are shown in the broken lines in the diagram. The valleys are given locally by

$$-\frac{\pi}{4} - \frac{\alpha}{2} < \theta < \frac{3\pi}{4} - \frac{\alpha}{2} \quad \text{and} \quad -\frac{3\pi}{4} - \frac{\alpha}{2} < \theta < -\frac{\pi}{4} - \frac{\alpha}{2} \,.$$

The contribution of an integral passing through a saddle point is obtained as follows. Near $z = z_0$, $f(z) = f(z_0) + \frac{1}{2}(z - z_0)^2 A e^{i\alpha} + \dots$, so write

$$f(z) = f(z_0) - t^2 \tag{4.28}$$

which changes from the complex variable z to the real variable t, and describes the path from valley to valley along a path that descends most rapidly on either side of z_0. The transformation (4.28) is analytic near $z = z_0$ since for small t,

$$t^2 \sim -\frac{1}{2} A e^{i\alpha}(z - z_0)^2 \,;$$

hence

$$t \sim (A/2)^{1/2}(z - z_0)e^{i\alpha/2}e^{-i\pi/2} \tag{4.29}$$

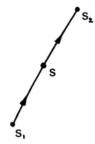

Figure 4.6

on choosing the branch, such that $t < 0$ on $S_1 S$ and $t > 0$ on $S S_2$. Thus

$$\int_{S_1}^{S_2} e^{Nf(z)} g(z)dz = e^{Nf(z_0)} \int_{-t_1}^{t_2} e^{-Nt^2} \left(g \frac{dz}{dt} \right) dt \quad .$$

In principle one can express $g \, dz/dt$ as a function of t (on inverting the Equation (4.28)), whence

$$g \, dz/dt = a_0 + a_1 t + a_2 t^2 + \dots \, (-T < t < T \text{ say}) \, .$$

In particular, the leading term is

$$a_0 = g(z_0)(dz/dt)_0 = g_0 \left(\frac{2}{A} \right)^{1/2} e^{i\pi/2} e^{-i\alpha/2}$$

and the first approximation to $I_{S_1 S_2}$ is

$$I \sim e^{Nf(z_0)} g(z_0) \left(\frac{2}{A} \right)^{1/2} e^{i\pi/2 - i\alpha/2} \int_{-\infty}^{\infty} e^{-Nt^2} dt$$

$$= e^{Nf(z_0)} g(z_0) \left(\frac{2\pi}{AN} \right)^{1/2} i e^{-i\alpha/2} \quad .$$

$$(4.30)$$

Higher-Order Approximation

Taking the expansion further, we have

$$I \sim e^{Nf(z_0)} \int_{-\infty}^{\infty} e^{-Nt^2} (a_0 + a_1 t + a_2 t^2 + \dots) dt \, .$$

The odd powers of t give zero contribution, since the integral of an odd function is zero, from $-\infty$ to ∞. Put $|t| = x^{1/2}/N^{1/2}$ to get

$$I \sim 2e^{Nf(z_0)} \frac{1}{2N^{1/2}} \int_0^{\infty} e^{-x} \left(a_0 + \frac{a_2 x}{N} + \frac{a_4 x^2}{N^2} \right) \frac{dx}{x^{1/2}} \, ,$$

i.e.

$$I \sim N^{-1/2} e^{Nf(z_0)} \left\{ (-\tfrac{1}{2})! a_0 + (\tfrac{1}{2})! a_2 N^{-1} + (\tfrac{3}{2})! a_4 N^{-2} + \ldots \right\} .$$

Example. Find the leading term for the integral

$$I(N) = \int_A^B z^2 \exp \left\{ iN \left(z^3 + 3z \right) \right\} dz \quad \text{as} \quad N \to \infty$$

with end points $A = -1 + i$ and $B = 1 + i$.

Here

$$f(z) = iz^3 + 3iz = u + iv$$

with

$$u = -3x^2 y + y^3 - 3y, \quad v = x^3 - 3xy^2 + 3x .$$

The level surface $u = 0$ is given by $y = 0$ or by the two branches of the hyperbola $3x^2 - y^2 = -3$. The level surface $u = \text{constant} = \kappa$ has asymptotes $y \sim 0$ or $y \sim \pm x\sqrt{3}$ and the family of level surfaces is shown in the diagram with solid lines.

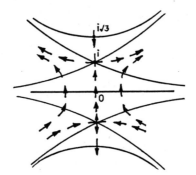

Figure 4.7

The level surfaces $v = \text{constant}$ are shown by broken lines, and the arrows show the direction of descent (u decreasing). Note that $u(A) = u(-1, 1) = -5$, $u(B) = -5$ and $u(S) = u(i) = -2$. An appropriate integration path is shown below, with v constant on $A_1 B_1$, and the main contribution comes from near the saddle point at $S = i$.

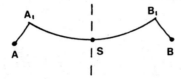

Figure 4.8

The path $A_1 B_1$ is given by

$$f(z) = -2 - t^2, \qquad (-t_1 \le t \le t_2)$$

$$= -2 + \frac{(z-i)^2}{2} (-6) + \dots$$

so $t \sim \pm 3^{1/2}(z-i)$ near S. To get the correct direction, we need the positive sign, so $z = i + 3^{-1/2}t$ near S; hence

$$I \sim \int_{-\infty}^{\infty} (i)^2 \exp(-2N - Nt^2) \frac{1}{\sqrt{3}} dt = -(\pi/3N)^{\frac{1}{2}} e^{-2N} .$$

Inversion of Analytic Function

Higher–order approximations in the expansions given above, in both cases I and II, usually require the inversion of a functional equation $w = f(z)$, to get z in terms of w (in the form of a power series for our purpose).

Inverse Function Theorem: Suppose $w = f(z)$ is analytic at z_0, with $w_0 = f(z_0)$ and with $f'(z_0) \ne 0$. Then in the vicinity of z_0 the transformation $w = f(z)$ is one–to–one, with $z = g(w)$ analytic at w_0, $z_0 = g(w_0)$ and $f'(z) = 1/g'(w)$. That is, there is a region R (including z_0) in the z–plane and a corresponding region R_1 in the w–plane such that $w = f(z)$ is a one–to–one mapping from R on to R_1 and with the corresponding inverse formula $z = g(w)$.

Lagrange's Formula

Given $w = f(z)$, the inverse formula has a power–series representation of the form

$$z - z_0 = b_1(w - w_0) + b_2(w - w_0)^2 + b_3(w - w_0)^3 + \dots \qquad (4.31)$$

for w sufficiently close to w_0, where the coefficients b_n can be determined as follows. Given a point z (near z_0) in the region R, consider the integral $\dfrac{1}{2\pi i} \displaystyle\int_P \dfrac{dz'}{f(z') - w}$ evaluated round a contour P that encloses z and z_0, but lies within R.

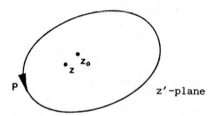

Figure 4.9

Since $f(z') = w$ has just one solution ($z' = z$), the integrand has just one singularity at $z' = z$ with residue $1/f'(z) = dg/dw$. Hence

$$\frac{dg}{dw} = \frac{1}{2\pi i} \int_P \frac{dz'}{\{f(z') - w_0 - (w - w_0)\}}$$

$$= \frac{1}{2\pi i} \int_P \frac{dz'}{f(z') - w_0} \left\{ 1 + \frac{w - w_0}{f - w_0} + \left(\frac{w - w_0}{f - w_0}\right)^2 + \dots \right\}$$

for w sufficiently close to w_0 (i.e. $|w - w_0| < \max |f(z') - w_0|$, z' on P). But from (4.31), $dg/dw = b_1 + 2b_2(w - w_0) + 3b_3(w - w_0)^2 + \dots$, so that a comparison of the last two expressions for dg/dw shows that

$$nb_n = \frac{1}{2\pi i} \int_P \frac{dz}{(f(z) - w_0)^n} .$$

Now the integrand

$$F(z) = (f(z) - w_0)^{-n} = \frac{B_n}{(z - z_0)^n} + \dots + \frac{B_1}{(z - z_0)^1} + \dots$$

has a pole of order n at $z = z_0$; hence

$$nb_n = B_1 = \frac{1}{(n-1)!} \left[\frac{d^{n-1}}{dz^{n-1}} \frac{(z - z_0)^n}{(f(z) - f(z_0))^n} \right]_{z=z_0} ,$$

which gives an explicit formula for b_n in terms of $f(z)$.

Example. $w = 2z + z^2$, $z = 0$ when $w = 0$; we find the nth term in the expansion $z = b_1 w + b_2 w^2 + b_3 w^3 + \dots$.

Lagrange's formula gives

$$nb_n = \frac{1}{(n-1)!} \left[\frac{d^{n-1}}{dz^{n-1}} (z + 2)^{-n} \right]_{z=0} .$$

Hence

$$b_1 = \frac{1}{2}, \ b_2 = -\frac{1}{8}, \ b_3 = \frac{1}{16}, \dots b_n = (-1)^{n-1} \frac{(2n-2)!}{n!(n-1)!2^{2n-1}}$$

4.5 DIFFRACTION INTEGRALS

The use of Fourier transforms in acoustic diffraction problems, with wavenumber k, often leads to solutions in the integral form

$$I(x,y) = \int_B f(s)\exp(-isx - \gamma y)ds ,$$

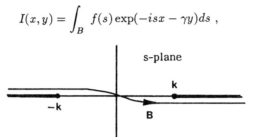

Figure 4.10

where $\gamma = (s^2 - k^2)^{1/2}$, $\gamma(0) = -ik$, has branch cuts from $s = \pm k$ to $\pm\infty$. The path B is below the cut on the positive real axis and above the cut on the negative real axis. It is required to find the "far–field" form of the integral I when $r \to \infty$, where r is the polar coordinate defined by $x = r\cos\theta$, $y = r\sin\theta$. The following procedure is given by Noble (1958).

The exponent $(-isx - \gamma y)$ is complicated, with variable real and imaginary parts. It is useful to deform the path B on to either (i) a steepest descents path (where the exponent has constant imaginary part), or (ii) a path where the real part of the exponent is constant, with a view to using a stationary phase argument. The latter option will be followed here.

It is easy to verify that a suitable path is that given by

$$s = s_1 + is_2 = -k\cos(\theta + it), \quad \infty > t > -\infty ,$$

i.e.

$$s_1 = -k\cos\theta\cosh t, \quad s_2 = k\sin\theta\sinh t ,$$

so the path is one branch of the hyperbola

$$(s_1/k\cos\theta)^2 - (s_2/k\sin\theta)^2 = 1$$

being the left or right branch according as $0 < \theta < \pi/2$ or $\pi/2 < \theta < \pi$. The value of γ is seen to be $\gamma = -ik\sin(\theta + it)$, so the exponent is $-isx - \gamma y = ikr\cosh t$, which is purely imaginary as required.

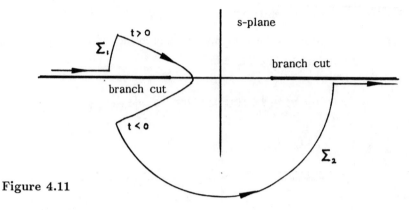

Figure 4.11

Now we deform B on to the path shown in the diagram for $0 < \theta < \pi/2$; there may be residue contributions from poles of $f(s)$ that are crossed in the deformation. It is easy to verify that the integrand is exponentially *small* on the circular linking arcs Σ_1 and Σ_2, and give vanishingly small contributions as their radii tend to infinity, if $|f| \to 0$ at infinity. Noting that $ds = ik \sin(\theta + it)dt$, we have

$$I = -\int_{-\infty}^{\infty} ik \sin(\theta + it) \, f(-k \cos(\theta + it)) \, e^{ikr \cosh t} dt$$

together with any residue contributions. This integral can now be estimated by the method of stationary phase as $kr \to \infty$. The stationary point is at $t = 0$; hence

$$I \sim -ik \sin \theta f(-k \cos \theta) \int_{-\infty}^{\infty} \exp\{ikr(1 + \frac{1}{2} t^2)dt$$

$$= -ik \sin \theta f(-k \cos \theta)(2\pi/kr)^{1/2} \exp(ikr + i\pi/4) \quad ,$$

together with any residue contributions.

Example. Compressible fluid, of wave speed c, occupies the half–space $y > 0$. The plane $y = 0$ is fixed except for the motion of a strip ($|x| < a$) which has normal velocity $v_0 \cos \omega t$. Find the distant field as $kr \to \infty$.

Solution. The velocity potential $\Re\phi(x, y) \exp(-i\omega t)$ satisfies the equations

$$(\partial^2/\partial x^2 + \partial^2/\partial y^2 + k^2) \, \phi = 0, \qquad k = \omega/c \quad ,$$

$$\frac{\partial\phi}{\partial y} = \begin{cases} V_0, & |x| < a \\ 0, & |x| > a \end{cases}$$

and an outgoing wave condition at infinity. Fourier transformation leads to the solution

$$\phi = -\frac{V_0}{\pi} \int_B \frac{\sin as}{s\gamma} \exp(-isx - \gamma y)ds \ .$$

Hence, since this is one of the above type with $f = \dfrac{-V_0}{\pi} \dfrac{\sin as}{s\gamma}$,

$$\phi \sim -\frac{V_0}{\pi} \frac{\sin(ka \cos \theta)}{k \cos \theta} \sqrt{\frac{2\pi}{kr}} \exp(ikr + i\pi/4) \quad \text{as} \quad kr \to \infty \quad .$$

This method of evaluating the distant field of a linear acoustic field has many extensions and has been put to considerable use in the literature.

REFERENCES

Carrier, G.F., Krook, M. & Pearson, C.E. (1966). Functions of a Complex Variable. McGraw–Hill.

Noble, B. (1958). Methods Based on the Wiener–Hopf Technique. Pergamon.

5. WIENER-HOPF TECHNIQUE

5.1 INTRODUCTION

This is a method for solving certain linear partial–differential equations subject to
mixed boundary conditions on semi–infinite geometries. Equivalently, it is applicable to integral equations of convolution type, such as

$$\int_0^\infty k(x-t)\phi(t)dt + \phi(x) = f(x), \qquad x > 0 \ . \tag{5.1}$$

The method hinges on the use of complex Fourier transforms and exploits their
known analyticity properties.

A complete exposition, with an excellent bibliography up to its date of publication, is given by Noble (1958). A briefer, more elementary, treatment is given by
Carrier, Krook & Pearson (1966).

Generalized Transforms

For functions $f(x)$ that belong to a suitable class (e.g. (i) piecewise continuously
differentiable and absolutely integrable, or (ii) generalized functions), the ordinary
Fourier transform is defined in Chapter 3 as

$$F(s) = \int_{-\infty}^\infty f(x)\exp(isx)dx \ , \tag{5.2}$$

with s real for the moment; the inversion formula is

$$f(x) = \frac{1}{2\pi}\int_{-\infty}^\infty F(s)\exp(-isx)ds \ . \tag{5.3}$$

Analyticity Properties. Regarding s as a complex variable, $s = s_1 + is_2$, the following properties are crucial to the Wiener–Hopf procedure, and are given without
proof.

(1) If $f(x)$ vanishes outside the finite interval (a,b), then

$$F(s) = \int_a^b f(x)\exp(isx)dx \tag{5.4}$$

is an *analytic function* of the complex variable s. The result is obvious if f is continuous since the integrand and its s–derivative are uniformly continuous and we can differentiate under the integral sign to get $F'(s) = \int_a^b ixf\exp(isx)dx$. If f has an integrable singularity, at $x = a$, say, then the contribution from the vicinity of $x = a$ has to be bounded separately and leads to the same conclusion.

(2) If $f(x)$ vanishes for $x < 0$, and if $|f(x)| < Ae^{\alpha x}$ as $x \to \infty$ for some constants $A(>0)$ and α, then we can define its *generalized Fourier transform*

$$F_+(s) = \int_0^\infty f(x)e^{isx}dx = \int_0^\infty f(x)e^{-s_2 x}e^{is_1 x}dx , \tag{5.5}$$

which exists and is analytic when the imaginary part s_2 of s satisfies

$$s_2 > \alpha . \tag{5.6}$$

Note that $F_+(s)$ is just the ordinary transform of the function $f(x)\exp(-s_2 x)$ with s_2 chosen by (5.6) to ensure $f\exp(-s_2 x)$ is exponentially small as $x \to +\infty$. The inversion formula is therefore

$$f(x)e^{-s_2 x} = \frac{1}{2\pi}\int_{-\infty}^\infty F_+ e^{-is_1 x}ds_1 \quad .$$

$$\text{i.e. } f(x) = \frac{1}{2\pi}\int_P F_+(s)e^{-isx}ds , \tag{5.7}$$

which is just like the ordinary inversion integral (5.3) but evaluated along the path P such that $s_2 > \alpha$.

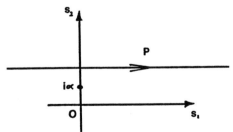

Figure 5.1

(3) Similarly, if f vanishes for $x > 0$ and $|f| < Be^{\beta x}$ as $x \to -\infty$, then

$$F_-(s) = \int_{-\infty}^0 f(x)e^{isx}dx \tag{5.8}$$

exists and is analytic for $s_2 < \beta$; its inversion formula is

$$f(x) = \frac{1}{2\pi} \int_P F_-(s) e^{-isx} ds , \qquad (5.9)$$

where P is the path $s = s_1 + is_2$, $-\infty < s_1 < \infty$, with s_2 fixed and less than β.

Combining the results of F_+ and F_-, suppose that $f(x)$ is defined for all x, with $|f| < Ae^{\alpha x}$ as $x \to \infty$, $|f| < Be^{\beta x}$ as $x \to -\infty$, $(\alpha < \beta)$, then its full range transform

$$F(s) = F_+(s) + F_-(s) = \int_{-\infty}^{\infty} f(x) e^{isx} dx \qquad (5.10)$$

is analytic for all complex s within the strip

$$\alpha < s_2 < \beta . \qquad (5.11)$$

Its inversion formula is of the form (5.3), evaluated along the path $P(s = s_1 + is_2, -\infty < s_1 < \infty)$ with s_2 in the range (5.11).

Example. Solve $f'' - \alpha^2 f = 0$ $(\alpha > 0)$, for $x > 0$, with f bounded as $x \to \infty$ and $f(0) = 1$. The answer $f = e^{-\alpha x}$ is obvious since the general solution of the ordinary differential equation is elementary. It is nevertheless instructive to reach the solution using transforms. To do this we need f to be defined over the whole range $(-\infty < x < \infty)$; hence define

$$f_+(x) = \left\{ \begin{array}{ll} f, & x > 0 \\ 0, & x < 0 . \end{array} \right.$$

The transform of f_+'' is

$$\int_0^{\infty} f_+'' e^{isx} dx = -s^2 F_+ + is - A, \quad s_2 > 0 , \qquad (5.12)$$

on integrating by parts, where $A = f'(0)$ is unknown. Thus the differential equation has the transform

$$-(s^2 + \alpha^2) F_+ + is - A = 0 ,$$

i.e.

$$F_+(s) = -\frac{A - is}{s^2 + \alpha^2} .$$

Now F_+ must be analytic in the whole half–plane $s_2 > 0$, so the apparent singularity at $s = +i\alpha$, where the denominator vanishes, must be cancelled by a corresponding zero in the numerator. Thus $A - is$ must vanish at $s = i\alpha$; hence $A = \alpha$ and $F_+ = i(s + i\alpha)^{-1}$, and the inversion formula gives

$$f_+ = \frac{1}{2\pi} \int_P \frac{i}{(s + i\alpha)} e^{-isx} ds \ ,$$

evaluated along a path from $s_1 = -\infty$ to $s_1 = +\infty$ above the pole at $s = -i\alpha$; when $x > 0$ a deformation into the lower half–plane gives the result $f = e^{-\alpha x}$.

Abelian Theorem (see Noble 1958).

Suppose $f = 0$ for $x < 0$, $|f| < Ae^{\alpha x}$ as $x \to \infty$, f is infinitely differentiable for $x > 0$ and $f \sim x^\lambda$ as $x \to +0$ with $\lambda > -1$. Then

$$F_+ \sim \int_0^\infty x^\lambda e^{isx} dx = \frac{\lambda!}{(-is)^{\lambda+1}} \ , \tag{5.13}$$

as $|s| \to \infty$ in the half–plane of analyticity $s_2 > \alpha$. That is, the form of $F_+(s)$ at large $|s|$ corresponds to the form of $f(x)$ at small (positive) x. Similarly if $f = 0$ for $x > 0$, $|f| < Be^{\beta x}$ as $x \to -\infty$ and $f \sim (-x)^\mu$ as $x \to -0$, $(\mu > -1)$ then

$$F_-(s) \sim \int_{-\infty}^0 (-x)^\mu e^{isx} dx = \frac{\mu!}{(is)^{\mu+1}} \ , \tag{5.14}$$

as $|s| \to \infty$, $s_2 < \beta$. Note: these results are obvious, by Watson's Lemma, when $s_2 \to \infty$ (or $s_2 \to -\infty$), and the derivatives of f do not even have to be continuous. When $|s| \to \infty$ with s_2 fixed, on the other hand, a more delicate argument is needed (cf. Lighthill 1959) and the continuity of $f^{(n)}$ is required.

We are now in a position to use the Wiener–Hopf method. It will first be illustrated with a very elementary example that shows the essential stages with little algebra.

5.2 WIENER–HOPF PROCEDURE FOR VIBRATING STRING PROBLEM

The method is illustrated with reference to the elementary problem of waves on a string with a discontinuity of density. The equilibrium position is at $y = 0$, with strings of different densities at $x < 0$ and $x > 0$. An incident wave from $x = -\infty$ has displacement

$$y_{inc} = \exp(ik_1 x - i\omega t)$$

and the time factor $\exp(-i\omega t)$ will henceforth be suppressed. Writing the deflections as

$$y = \begin{cases} e^{ik_1 x} + u_-(x), & x < 0 \\ \\ u_+(x), & x > 0 \end{cases}$$

the governing equations are

$$\frac{d^2 u_-}{dx^2} + k_1^2 u_- = 0, \qquad x < 0$$
$$\frac{d^2 u_+}{dx^2} + k_2^2 u_+ = 0, \qquad x > 0 \qquad ,$$

where $k_i^2 = \omega^2 \rho_i / T$, ρ_i = densities, T = tension. The boundary conditions are

$$u_- \sim R e^{-ik_1 x} \quad \text{as} \quad x \to -\infty$$
$$u_+ \sim T e^{+ik_1 x} \quad \text{as} \quad x \to +\infty$$
$$(u_+ - u_-) = 1, \quad (u_+{}' - u_-{}') = ik_1 \quad \text{at} \quad x = 0 .$$

It is mathematically convenient to assign small imaginary parts to k_1 and k_2, in order to improve the convergence of subsequent integrals. Thus

$$k_1 \to k_1 + \epsilon, \quad k_2 \to k_2 + i\epsilon \quad (\epsilon > 0) ,$$

so that $u_\pm$ are exponentially small as x a $\pm \infty$; we finally let $\epsilon \to +0$ at the end of the calculation. The transformed versions of the differential equations are readily found to be

$$(k_1^2 - s^2)U_-(s) + u_-{}'(0) - is\, u_-(0) = 0, \quad s_2 < \epsilon \qquad (5.16)$$
$$(k_2^2 - s^2)U_+(s) + u_+{}'(0) + is\, u_+(0) = 0, \quad s_1 > \epsilon . \qquad (5.17)$$

Adding, and using the boundary conditions at $x = 0$ to eliminate $u_\pm(0)$, $u_\pm'(0)$, one finds

$$\left(\frac{s^2 - k_2^2}{s^2 - k_1^2}\right) U_+(s) + U_-(s) = \frac{i}{s + k_1}, \quad -\epsilon < s_2 < \epsilon , \qquad (5.18)$$

which is a typical Wiener–Hopf equation for the two unknown functions U_+ and U_-. [In this elementary problem, one could have solved (5.16) and (5.17) independently since they are essentially uncoupled: this is not typical but a property of the simple governing equations in this problem.]

Wiener–Hopf Procedure.

The main features of the technique can be illustrated through example (5.18) which is of the general form

$$K(s)U_+(s) + U_-(s) = P(s), \quad \alpha < s_2 < \beta , \tag{5.19}$$

where P and K are known, and are analytic in the strip $\alpha < s_2 < \beta$. The function K is called the *kernel* of the Wiener–Hopf equation: the difficulty of a given problem is characterized by the complexity of this function. The kernel K has no zeros in the strip $\alpha < s_2 < \beta$ and K is of algebraic growth as $|s| \to \infty$ as in example (5.18), where $K = (s^2 - k_2^2)/(s^2 - k_1^2)$. These properties will be assumed in the general analysis and procedure that follows.

Stage 1: The first and crucial step is to factorize the kernel $K(s)$ as a product of the form

$$K(s) = K_+(s)K_-(s), \quad \alpha < s_2 < \beta , \tag{5.20}$$

where K_+ is analytic and free from zeros in the half–plane $s_2 > \alpha$, and K_- is analytic and free from zeros in the half–plane $s_2 < \beta$. Each of the functions K_+ and K_-^{-1} is to be of algebraic growth at most, as $|s| \to \infty$ in the respective half–plane (that is, $K_+ = 0(s^n)$ for some constant n, and similarly for K_-^{-1}). Although $K(s)$ is originally defined only in the strip $\alpha < s_2 < \beta$, the constituent functions K_+ and K_- are defined in overlapping half–planes. Explicit formulae for K_+ and K_-, for a wide class of kernels $K(s)$, will be constructed later (see (5.39), (5.40)). In the present problem, the factorization can be done by inspection; thus

$$K(s) = \left(\frac{s^2 - k_2^2}{s^2 - k_1^2} \right), \quad K_+ = \frac{s + k_2}{s + k_1}, \quad K_- = \frac{s - k_2}{s - k_1} . \tag{5.21}$$

Quite generally the Wiener–Hopf equation (5.19) can then be recast as

$$K_+(s)U_+(s) + U_-(s)/K_-(s) = P(s)/K_-(s), \quad \alpha < s_2 < \beta , \tag{5.22}$$

and our aim is to separate the "plus" functions and the "minus" functions. To achieve this, we have to perform a sum decomposition of the known function $R(s) = P(s)/K_-(s)$ on the right–hand side of (5.22). Thus

Stage 2: Given a function $R(s) = P(s)/K_-(s)$, analytic and of algebraic growth in the strip $\alpha < s_2 < \beta$, express it as a sum of the form

$$R(s) = R_+(s) + R_-(s) ,$$

where R_+ and R_- are analytic and of algebraic growth, at most, in the respective half–planes $s_2 > \alpha$ and $s_2 < \beta$.

Completion of solution: Equation (5.22) can be written as

$$K_+(s)U_+(s) - R_+(s) = -U_-(s)/K_-(s) + R_-(s), \quad \alpha < s_2 < \beta, \tag{5.23}$$

where K_+, K_-, R_+ and R_- are known; the two functions U_+ and U_- have to be found. The key fact is that the left–hand side of (5.23) is analytic in the whole half–plane $s_2 > \alpha$, and the right–hand side is analytic in the half–plane $s_2 < \beta$. Thus each side analytically continues the other to define a function $E(s)$ *that is analytic in the whole s–plane.* Although (5.19) and hence (5.23) were defined originally only in the strip, the analyticity properties of U_+ and U_- ensure that each side of (5.23) (together with their analytic continuations) must be analytic for all s. To find $E(s)$, hence U_+ and U_-, we appeal to edge conditions on $u_+(x)$ and $u_-(x)$ as $x \to \pm 0$. According to the Abelian theorems discussed above, this provides information about U_+ and U_- as $|s| \to \infty$ in their respective half–planes, hence gives information about the left–hand side and the right–hand side of (5.23), hence about $E(s)$. Thus if

$$\text{LHS} = 0(s^{n_1}) \quad \text{and} \quad \text{RHS} = 0(s^{m_1})$$

as $|s| \to \infty$ in the respective half–planes, the extension of Liouville's theorem ensures that $E(s) =$ polynomial of degree n, ($n =$ lesser of n_1 and m_1), and our solution is complete. In particular, a frequent case is that in which LHS $\to 0$ and RHS $\to 0$; hence $E(s) = 0$.

String problem. Let us complete the analysis pertaining to the string problem, for which the Wiener–Hopf equation is (5.18). Using the factorization (5.21), the equation can be written as

$$\left(\frac{s+k_2}{s+k_1}\right)U_+(s) + \left(\frac{s-k_1}{s-k_2}\right)U_-(s) = \frac{i(s-k_1)}{(s+k_1)(s-k_2)}, \quad -\epsilon < s_2 < \epsilon, \tag{5.24}$$

on completing the stage 1 factorization by inspection. Note that the right–hand side is neither a "plus" nor a "minus" function, since it has poles at $s = -k_1$ and at $s = k_2$. Stage 2 of the procedure is to express it as a *sum* of "plus" and "minus" functions, and this is readily achieved using partial fractions, thus

$$\frac{i(s-k_1)}{(s+k_1)(s-k_2)} = \frac{2ik_1}{(k_1+k_2)}\frac{1}{(s+k_1)} + \frac{i(k_2-k_1)}{(k_2+k_1)}\frac{1}{(s-k_2)} \equiv R_+(s) + R_-(s),$$

and Equation (5.24) can be written as

$$\left(\frac{s + k_2}{s + k_1}\right)U_+ - \frac{2ik_1}{(k_1 + k_2)}\frac{1}{(s + k_1)} = -\left(\frac{s - k_1}{s - k_2}\right)U_- + i\left(\frac{k_2 - k_1}{k_2 + k_1}\right)\frac{1}{(s - k_2)} = E(s),$$

(5.25)

where $E(s)$ is *analytic for all* s. Since u_+ and u_- are bounded at $x = 0$, it follows ((5.13) and (5.14)) that $U_+ = 0(1/s)$ and $U_- = 0(1/s)$; hence that the LHS and RHS of (5.25) tend to zero as $|s| \to \infty$ in the respective half–planes. Thus $E(s) \to 0$ as $|s| \to \infty$ and $E(s)$ is analytic for all s; therefore $E(s) = 0$ by Liouville's theorem. It follows that

$$U_+(s) = \left(\frac{2ik_1}{k_1 + k_2}\right)\frac{1}{(s + k_2)} \quad \text{and} \quad U_-(s) = i\left(\frac{k_2 - k_1}{k_2 + k_1}\right)\frac{1}{(s - k_1)} \,.$$

Finally, u_+ and u_- are determined by Fourier inversion. For example,

$$u_+(x) = \frac{ik_1}{\pi(k_1 + k_2)}\int_{-\infty}^{\infty}\frac{e^{-isx}}{s + k_2}ds = \frac{2k_1}{k_1 + k_2}e^{ik_2 x}, \quad x > 0 \,,$$

on collapsing the contour in the lower half–plane for positive x.

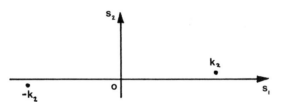

Figure 5.2

5.3 WIENER–HOPF PROCEDURE FOR INTEGRAL EQUATION

The method is illustrated with reference to the particular integral equation

$$f(x) = \frac{1}{2}\int_0^{\infty} f(t)\exp\{-|x - t|\}\,dt, \quad 0 < x < \infty \,,$$

(5.26)

and it is required to find the eigensolution of this homogeneous equation. Since the integral term is

$$e^{-x}\int_0^x f(t)e^t dt + e^x\int_x^{\infty} f(t)e^{-t}dt \,,$$

and the latter integral exists, we are led to seek solutions such that

$$|f| < Ae^{t(1-\delta)} \quad \text{for large } t, \text{ with } \delta > 0 \,.$$

(5.27)

Now convolution integrals of the type

$$\int_{-\infty}^{\infty} k(x-t)f(t)dt \,, \tag{5.28}$$

where the integration runs from $-\infty$ to $+\infty$, have Fourier transform $K(s)F(s)$, in which $K(s)$ and $F(s)$ denote the transforms of $k(x)$ and $f(x)$. It is therefore expedient to recast the integral equation (5.26) so as to have an integral from $-\infty$ and $+\infty$. Thus we redefine the unknown function

$$f_+(x) = \begin{cases} f(x), & x > 0 \\ 0, & x < 0 \end{cases}$$

and the equation (5.26) can be written as

$$\frac{1}{2}\int_{-\infty}^{\infty} f_+(t)e^{-|x-t|}dt = \begin{cases} f_+(x), & x > 0 \\ g_-(x), & x < 0 \end{cases}. \tag{5.29}$$

Here g_- is unknown and is in fact defined for negative x by Equation (5.29). We now need a little information about the analyticity properties of the transforms $F_+(s)$ and $G_-(s)$ of $f_+(x)$ and $g_-(x)$. From (5.27), it is seen that

$$F_+(s) = \int_0^{\infty} f_+(x)e^{isx}dx$$

exists and is analytic for $s_2 > 1 - \delta$. As $x \to +0$, $f \to \frac{1}{2}\int_0^{\infty} f(t)e^{-t}dt$ which exists; hence $F_+ = 0(1/s)$ as $|s| \to \infty$. Similarly, $G_-(s)$ is analytic for $s_2 < 1$, and $G_- = 0(1/s)$ as $|s| \to \infty$.

Now take the Fourier transform of Equation (5.29); that is, multiply by $\exp(isx)$ $= \exp(is_1x - s_2x)$, with $1 - \delta < s_2 < 1$, and integrate from $x = -\infty$ to $x = +\infty$. The result is

$$\frac{1}{2}F_+(s)\frac{2}{s^2+1} = F_+(s) + G_-(s), \quad 1 - \delta < s_2 < 1\,,$$

since the Fourier transform of $e^{-|x|}$ is $2(s^2+1)^{-1}$. Thus

$$\frac{s^2}{s^2+1} F_+(s) + G_-(s) = 0, \quad 1 - \delta < s_2 < 1\,,$$

which is of the standard type (5.19), with $P(s) = 0$ and with kernel

$$K(s) = \frac{s^2}{s^2+1}, \quad 1 - \delta < s_2 < 1\,.$$

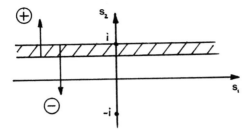

Figure 5.3

There are two poles at $s = \pm i$ and a zero at $s = 0$, which is in the "minus" region $s_2 < 1$. By inspection

$$K = \left(\frac{s^2}{s+i}\right) \cdot \left(\frac{1}{s-i}\right) = K_+ K_- ,$$

so that

$$\frac{-s^2 F_+(s)}{s+i} = (s-i)G_-(s) = E(s) ,$$

where $E(s)$ is analytic for all s, by analytic continuation. Since F_+ and G_- are $0(1/s)$ as $|s| \to \infty$, $E(s) = 0(1)$; hence $E(s) = A$ is constant by the extension of Liouville's theorem. In particular,

$$F_+(s) = -A\frac{s+i}{s^2}$$

and

$$f_+(x) = \frac{-A}{2\pi} \int_{is_2-\infty}^{is_2+\infty} \frac{s+i}{s^2} e^{-isx} ds, \quad 1-\delta < s_2 < 1 .$$

For positive x, $f_+ = iA$ (times residue at $s = 0$), i.e.

$$f(x) = B(1+x), \quad x > 0 ,$$

where $B = iA$ is an arbitrary constant. One can readily check that this eigenfunction satisfies the integral Equation (5.26).

5.4 PROPERTIES OF WIENER–HOPF DECOMPOSITIONS

Uniqueness of Decomposition

Product factorization: Given $K(\neq 0)$ in a strip $\alpha < s_2 < \beta$, with K_+ and K_- required to be free from zeros in their respective half–planes, and $K_\pm, K_\pm^{-1}$ of algebraic growth at infinity (i.e. $K_\pm, K_\pm^{-1} = 0(s^u)$ for some finite u) the factorization

$$K(s) = K_+(s)K_-(s), \quad \alpha < s_2 < \beta ,$$

is unique apart from multiplying by constants $(K = (aK_+)(\frac{1}{a}K_-))$.

Proof. Suppose that $K(s)$ has two possible factorizations:

$$K(s) = K_+ K_- = Y_+ Y_-, \quad \alpha < s_2 < \beta .$$

Then we can define $P_+ = Y_+/K_+$ and $P_- = Y_-/K_-$ since $K_\pm \neq 0$ in their respective half-planes. Thus

$$K_+ K_- = K_+ P_+ K_- P_-, \quad \alpha < s_2 < \beta ,$$

i.e.

$$P_+ P_- = 1 \quad \text{and} \quad P_+ = 1/P_- = E(s)$$

is analytic for all s, by analytic continuation. Now P_+ has no zeros in the half-plane $s_2 > \alpha$, so neither has E, and P_- has no poles in the half-plane $s_2 < \beta$; hence E has no zeros there. Thus E has no zeros in the whole plane and is of algebraic growth, i.e. $E = 0(s^N)$ for some N. Liouville's theorem ensures that E is a polynomial of degree N, but E has no zeros and hence is a constant. Thus $P_+ = a$, $P_- = 1/a$.

Sum Decomposition:

$$R(s) = R_+(s) + R_-(s) .$$

This decomposition is *not* unique, since we could add and subtract any *polynomial* to R_+ and from R_-. Apart from this arbitrariness, the decomposition is unique if we again insist on at most algebraic growth. For if

$$R = R_+ + R_- = Q_+ + Q_-$$

are two possible decompositions, then

$$R_+ - Q_+ = Q_- - R_- = F(s) ,$$

which is analytic for all s, by continuation. If $R_\pm$ and $Q_\pm$ are of algebraic growth, then $F(s)$ is a polynomial, by Liouville's theorem.

Decomposition Formulae.

Explicit decomposition formulae can now be given for a wide class of functions. Given $K(s)$ and $R(s)$ in a strip $\alpha < s_2 < \beta$, we have to find

$$K(s) = K_+ K_- \tag{5.30}$$

and

$$R(s) = R_+ + R_- , \tag{5.31}$$

where $K_\pm$, $K_\pm^{-1}$, $R_\pm$ are to have only algebraic growth at infinity. All the "plus" functions are analytic for $s_2 > \alpha$, and the "minus" functions are analytic for $s_2 < \beta$; K_+ and K_- are to be free from zeros in their respective half–planes. Since $\log K = \log K_+ + \log K_-$, the product factorization (5.30) reduces to a sum decomposition for $\log K$.

Basic Sum Decomposition

Suppose R is analytic in the strip $\alpha < s_2 < \beta$ and suppose $R = 0(s^{-\delta})$ as $|s| \to \infty$ in the strip, for some $\delta > 0$. Then for any point s within the strip, Cauchy's integral formula may be used to give $R(s) = \frac{1}{2\pi i} \int_c \frac{R(\zeta)d\zeta}{\zeta - s}$ evaluated round the closed contour $C = L_1 + P_1 + L_2 + P_2$ shown in the diagram, lying within the strip.

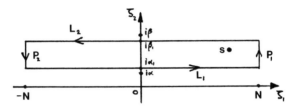

Figure 5.4

As $N \to \infty$ the contribution from the paths P_1 and P_2 tend to zero, since the integrand has absolute value of order $\zeta^{-1-\delta}$. Thus

$$R(s) = \frac{1}{2\pi i} \int_{i\alpha_1 - \infty}^{i\alpha_1 + \infty} \frac{R(\zeta)d\zeta}{\zeta - s} - \frac{1}{2\pi i} \int_{i\beta_1 - \infty}^{i\beta_1 + \infty} \frac{R(\zeta)d\zeta}{\zeta - s} \tag{5.32}$$

$$\equiv R_+(s) + R_-(s) , \tag{5.33}$$

where

$$\alpha < \alpha_1 < s_2 < \beta_1 < \beta .$$

The first integral R_+ exists and is analytic for all s such that $s_2 > \alpha_1$, and the path of integration can be shifted so that α_1 is arbitrarily close to α. Similarly, the second integral R_- exists and is analytic for all $s_2 < \beta_1$ arbitrarily close to β.

This basic decomposition result is at the heart of the Wiener–Hopf method.

Example. Find a sum decomposition for $R(s) = (s^2 + 1)^{-1}$, $-1 < s_2 < 1$. From (5.32) and (5.33)

$$R_+(s) = \frac{1}{2\pi i} \int_{i\alpha-\infty}^{i\alpha+\infty} \frac{d\zeta}{(\zeta^2 + 1)(\zeta - s)}, \quad -1 < \alpha < s_2 .$$

Collapsing the contour in the lower half–plane gives $R_+ = -\{$residue at $\zeta = -i\}$
$= \frac{1}{2}i(s + i)^{-1}$.

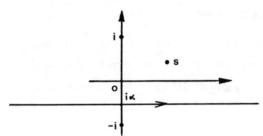

Figure 5.5

Similarly, $R_- = -\frac{1}{2}i(s - i)^{-1}$ (which could also be found from the identity $R_- = R - R_+$). These results could also have been found (more simply) using partial fractions. One could add and subtract arbitrary polynomials to and from R_+ and R_-: in most applications it is expedient to choose $|R_\pm|$ to be as small as possible as $|s| \to \infty$, in order to minimize the number of arbitrary constants in the polynomial $E(s)$ such as appears in (5.25). The integral expressions (5.32) usually lead to the most appropriate values for R_+ and R_-.

The above example was elementary. The next one is not.

Example. Find the sum decomposition for the function

$$R(s) = (s^2 - k^2)^{-1/2} = 1/\gamma , \tag{5.34}$$

where $k = k_1 + i\epsilon$ has a small positive imaginary part ϵ. The function $\gamma = (s^2 - k^2)^{1/2}$ is defined so that $\gamma = -ik$ at $s = 0$, with branch cuts from $\pm k$ in the first and third quadrants.

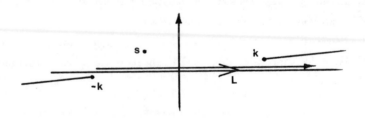

Figure 5.6

Thus R is analytic in the strip $-\epsilon < s_2 < \epsilon$ and is of order s^{-1} as $|s| \to \infty$ in the strip. Formula (5.32) gives

$$R_+ = \frac{1}{2\pi i} \int_{L_1} \frac{d\zeta}{(\zeta^2 - k^2)^{1/2}(\zeta - s)} \, ,$$

where s can now be anywhere above L_1. On collapsing the contour in the lower half–plane, and choosing (as a matter of convenience) the branch cut to be a straight line from $-k$ and pointing away from the origin, it is found that

$$R_+ = -\frac{1}{\pi i} \int_1^\infty \frac{dp}{(p^2 - 1)^{1/2}(kp + s)} = -\frac{1}{\pi i} \int_0^{\pi/2} \frac{d\theta}{k + s\cos\theta}$$

after setting $p = \sec\theta$. The integral can be evaluated by the usual substitution $t = \tan\frac{1}{2}\theta$, whence

$$R_+ = \frac{2}{\pi\gamma} \tan^{-1} \left(\frac{k - s}{k + s} \right)^{1/2} , \tag{5.35}$$

and similarly

$$R_- = \frac{2}{\pi\gamma} \tan^{-1} \left(\frac{k + s}{k - s} \right)^{1/2} . \tag{5.36}$$

Here $\tan^{-1} z$ is such that $\tan^{-1} 0 = 0$, and with branch cuts from $z = \pm i$ to $z = \pm i\infty$. To verify that R_+ is indeed a "plus" function, note that near $s = k$ (where there seems to be a branch–point singularity),

$$R_+ \sim \frac{2}{\pi(s^2 - k^2)^{1/2}} \left(\frac{k - s}{k + s} \right)^{1/2} \sim \frac{-1}{i\pi k} \, ,$$

and the singularity is removable.

Generalization of the Basic Sum Decomposition

If $R(s)$ does not tend to zero as $|s| \to \infty$ in the strip (or does not tend to zero sufficiently quickly), one can sometimes obtain R_+ and R_- by taking the limit $N \to \infty$ in the Cauchy integral described above. As an alternative it is sometimes simpler to factorise by a polynomial.

Example. $R = (s^2 - k^2)^{1/2}$ is analytic in the strip $-\epsilon < s_2 < \epsilon$, when $(s^2 - k^2)^{1/2}$ is the square root function that appeared in the last example. But here $|R| = 0(s)$ as $|s| \to \infty$ and the integrals (5.32) are divergent as they stand. To deal with this, note that

$$R(s) = (s^2 - k^2) \frac{1}{(s^2 - k^2)^{1/2}}$$

$$= (s^2 - k^2) \left\{ \frac{2}{\pi\gamma} \tan^{-1} \left(\frac{k-s}{k+s}\right)^{1/2} + \frac{2}{\pi\gamma} \tan^{-1} \left(\frac{k+s}{k-s}\right)^{1/2} \right\}$$

and

$$R_{\pm} = \frac{2}{\pi} (s^2 - k^2)^{1/2} \tan^{-1} \left(\frac{k \mp s}{k \pm s}\right)^{1/2} . \tag{5.37}$$

Basic Product Factorization

If $K(s)$ has no zeros in the strip $\alpha < s_2 < \beta$, and if

$$K(s) = 1 + 0(s^{-\delta}), \text{ with } \delta > 0, \text{ as } |s| \to \infty \text{ in the strip }, \tag{5.38}$$

then $\log K$ is analytic in the strip and $\log K = 0(s^{-\delta})$ as $|s| \to \infty$. Thus the problem reduces to the basic sum decomposition for the function $\log K = \log K_+ + \log K_-$. Hence

$$K_+ = \exp\left\{ \frac{1}{2\pi i} \int_{i\alpha_1 - \infty}^{i\alpha_1 + \infty} \frac{\log K(\zeta) d\zeta}{\zeta - s} \right\}, \quad s_2 > \alpha_1 , \tag{5.39}$$

$$K_- = \exp\left\{ \frac{-1}{2\pi i} \int_{i\beta_1 - \infty}^{i\beta_1 + \infty} \frac{\log K(\zeta) d\zeta}{\zeta - s} \right\}, \quad s_2 < \beta_1 . \tag{5.40}$$

Generalization. The stringent requirement (5.38) can be relaxed somewhat. Suppose then that, within the strip

$$K = Ae^{i\mu} |s|^\sigma (1 + 0(s^{-\delta_1})) \quad \text{as } s_1 \to \infty ,$$

and $\tag{5.41}$

$$K = Ae^{i\nu} |s|^\sigma (1 + 0(s^{-\delta_2})) \quad \text{as } s_1 \to -\infty ,$$

Then one writes (see Noble 1958)

$$K = Ae^{i\mu} (s-a)^p (s-b)^q K_1(s) , \tag{5.42}$$

where a is above the strip and b is below it. The functions $(s-a)^p$ and $(s-b)^q$ have branch cuts that do not cross the strip and are such that $\arg(s-a)^p$ and $\arg(s-b)^q \to 0$ as $s_1 \to +\infty$.

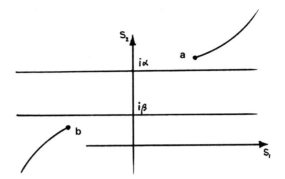

Figure 5.7

As $s_1 \to \infty$ within the strip

$$K \sim Ae^{i\mu}|s|^{p+q}K_1 \ . \tag{5.43}$$

As $s_1 \to -\infty$ within the strip, $\arg(s - a) \to -\pi$ and $\arg(s - b) \to +\pi$, so

$$K \sim Ae^{i\mu}|s|^{p+q}e^{i\pi q - i\pi p}K_1 \ . \tag{5.44}$$

If we choose $p + q = \sigma$ and $\mu + \pi q - \pi p = \nu$, i.e. $p = \frac{1}{2}\sigma + \frac{1}{2\pi}(\mu - \nu)$ and $q = \frac{1}{2}\sigma - \frac{1}{2\pi}(\mu - \nu)$, then (5.41)–(5.44) show that $K_1 = 1 + 0(s^{-\delta_1}) + 0(s^{-\delta_2}) + O(s^{-1})$ as $|s_1| \to \infty$; thus the basic factorization integrals (5.39) and (5.40) can be used to factorise $K_1(s) = K_{1+}(s)K_{1-}(s)$. Finally

$$K_+(s) = Ae^{i\mu}(s - b)^q K_{1+}(s)$$

and

$$K_-(s) = (s - a)^p K_{1-}(s) \ .$$

5.5 DIFFRACTION BY SEMI–INFINITE RIGID PLATE: SOMMERFELD PROBLEM

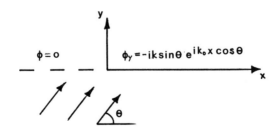

Figure 5.8

A plane sound wave of potential $\phi_{inc}e^{-i\omega t} = e^{i(k_0 x \cos\theta + k_0 y \sin\theta - \omega t)}$ is scattered by a rigid screen at $y = 0$, $x > 0$. Working with the scattered potential ϕ and dropping the time factor $\exp(-i\omega t)$, it is easy to see that ϕ is an odd function of y; hence $\phi(x,0) = 0$ for $x < 0$: we may confine attention to the half space $y > 0$, since for $y < 0$, $\phi(x,y) = -\phi(x,-y)$. The governing wave equation is the Helmholtz equation

$$(\partial^2/\partial x^2 + \partial^2/\partial y^2 + k_0^2)\ \phi = 0, \quad y \geq 0 , \tag{5.45}$$

where $k_0 = \omega/c$, and c is the sound speed. Boundary conditions on the plane $y = 0$ are

$$\phi = 0, \qquad\qquad\qquad x < 0, \quad y = 0 , \tag{5.46}$$

$$\phi_y = -ik_0 \sin\theta\ \exp(ik_0 x \cos\theta), \qquad x > 0, \quad y = 0 , \tag{5.47}$$

where the suffix denotes partial differentiation. Conditions at infinity require cylindrically spreading waves as $x \to -\infty$, and cancellation of the incident wave plus cylindrically spreading waves as $x \to +\infty$. Thus

$$\phi \sim -\exp(ik_0 x\ \cos\theta + ik_0 y\ \sin\theta) + f_1(x)\ \exp(ik_0 x) \text{ as } x \to +\infty , \tag{5.48}$$

$$\phi \sim f_2(x)\ \exp(-ik_0 x) \text{ as } x \to -\infty \tag{5.49}$$

for any fixed $y > 0$, where f_1 and f_2 are algebraic functions. Finally, a condition to ensure no energy creation at the edge is specified by

$$\phi \text{ bounded, } |\nabla\phi| = 0(x^\alpha), \text{ for some } \alpha > -1, \text{ as } x \to \pm 0 . \tag{5.50}$$

It is mathematically convenient to assign a small positive imaginary part to $k_0 = k + i\epsilon$, to improve convergence of subsequent Fourier integrals, and finally we let $\epsilon \to +0$. Define the full and half-range transforms as follows:

$$\Phi(s,y) = \int_{-\infty}^{\infty} \phi(x,y)e^{isx}dx = \Phi_-(s,y) + \Phi_+(s,y) , \tag{5.51}$$

with

$$\Phi_+ = \int_0^{\infty} \phi e^{isx}dx, \text{ analytic for } s_2 > -\epsilon \cos\theta , \tag{5.52}$$

$$\Phi_- = \int_{-\infty}^{0} \phi e^{isx}dx, \text{ analytic for } s_2 < \epsilon . \tag{5.53}$$

The half-planes of analyticity (5.52) and (5.53) follow at once from the boundary conditions (5.48) and (5.49). Evidently the full-range transform $\Phi(s,y)$ is analytic in the strip

$$S: \ -\epsilon \cos\theta < s_2 < \epsilon . \tag{5.54}$$

The Fourier transform of Equation (5.45) is

$$(\partial^2/\partial y^2 - s^2 + k_0^2)\, \Phi(s,y) = 0, \ s \text{ in the strip } S \ ,$$

with solutions

$$\Phi = \exp(\pm\gamma y) \ ,$$

where

$$\gamma = (s^2 - k_0^2)^{1/2}, \quad \gamma(0) = -ik_0 \tag{5.55}$$

has branch cuts from $\pm k_0$ to ∞ in the first and third quadrants. With this choice of cuts $\Re\,\gamma \geq 0$ for all s in S. The solution that remains bounded as $y \to \infty$ is

$$\Phi(s,y) = A(s)e^{-\gamma y} \ , \tag{5.56}$$

and $A(s)$ has to be determined from the (mixed) boundary conditions (5.46) and (5.47). Thus (5.46) gives

$$\Phi_-(s,0) = 0 \ , \tag{5.57}$$

and (5.47) gives

$$\Phi'_+(s,0) = k_0 \ \sin\theta/(s + k_0\cos\theta) \ , \tag{5.58}$$

where Φ'_+ denotes $\partial\Phi_+/\partial y$. Henceforth the arguments $(s,0)$ will be suppressed from (5.57) and (5.58). Now setting $y = 0$ in (5.56) gives

$$\Phi_+ + \Phi_- = A \ , \tag{5.59}$$

and from the y–derivative (5.56), at $y = 0$, we have

$$\Phi'_+ + \Phi' = -\gamma A \ . \tag{5.60}$$

On eliminating $A(s)$ between these two equations, and on using the two boundary conditions (5.57) and (5.58), one is led to the Wiener–Hopf equation

$$\gamma\Phi_+ + \Phi'_- = -k_0 \ \sin\theta/(s + k_0 \ \cos\theta), \quad s \text{ in } S \ . \tag{5.61}$$

This is of the standard form (5.19), with kernel

$$K(s) = \gamma(s) = \left(s^2 - k_0^2\right)^{1/2} \ . \tag{5.62}$$

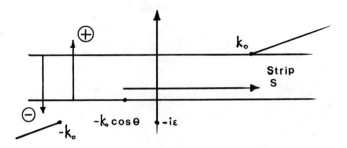

Figure 5.9

The factorization of K can be done by inspection; thus $K_+K_- = (s + k_0)^{1/2}(s - k_0)^{1/2}$ where $K_\pm = (s \pm k_0)^{1/2}$. For example, $K_+ = (s + k_0)^{1/2}$ has a branch cut in the third quadrant only, and is analytic in the "plus" region $s_2 > -\epsilon \cos \theta$. Dividing Equation (5.61) by $K_- = (s - k_0)^{1/2}$, we have

$$(s + k_0)^{1/2}\Phi_+ + (s - k_0)^{-1/2}\Phi'_- = \frac{-k_0 \sin \theta}{(s + k_0 \cos \theta)(s - k_0)^{1/2}} \text{ , for } s \text{ in } S . \quad (5.63)$$

Stage 2 requires a sum split of the function on the right–hand side. There is a branch cut in the "plus" region and a simple pole in the "minus" region. This pole is removable by the rearrangement

$$\frac{1}{(s + k_0 \cos \theta)(s - k_0)^{1/2}} = \frac{1}{(s + k_0 \cos \theta)} \left\{ \frac{1}{(s - k_0)^{1/2}} - \frac{1}{(-k_0 \cos \theta - k_0)^{1/2}} \right\}$$

$$+ \frac{1}{(-k_0 \cos \theta - k_0)^{1/2}(s + k_0 \cos \theta)} \text{ ,}$$

in which the first term is a "minus" function (with no pole at $-k_0 \cos \theta$) and the final term is a "plus" function. Thus (5.63) can be written as

$$(s + k_0)^{1/2}\Phi_+ + \frac{k_0 \sin \theta}{(-k_0 \cos \theta - k_0)^{1/2}(s + k_0 \cos \theta)}$$

$$- (s - k_0)^{-1/2}\Phi'_- - \frac{k_0 \sin \theta}{s + k_0 \cos \theta} \left\{ \frac{1}{(s - k_0)^{1/2}} - \frac{1}{(-k_0 \cos \theta - k_0)^{1/2}} \right\} \equiv E(s) .$$

$$(5.64)$$

The usual argument now follows: though (5.64) holds at first only in the strip S, each side analytically continues the other to define a function $E(s)$ that is analytic in the whole s–plane, and $E(s)$ is determined from the behaviour of ϕ near the discontinuity at $x = 0$. According to (5.50), $\phi(x,0)$ is bounded as $x \to +0$, from which the Abelian theorem ensures that $\Phi_+ = 0(1/s)$ as $|s| \to \infty$ in the half–plane $s_2 > -\epsilon \cos \theta$. It follows that the left–hand side of (5.64) tends to zero as $|s| \to \infty$ in the upper half–plane. Similarly, (5.50) requires $\phi_y(x,0) = 0(x^\alpha)$ as $x \to -0$, with $\alpha > -1$; hence $\Phi'_- = 0(s^{-1-\alpha})$ and the right–hand side of (5.64) tends to zero as $|s| \to \infty$ in the lower half–plane. It follows that $E(s) \to 0$ as $|s| \to \infty$; hence that $E(s) \equiv 0$ by Liouville's theorem, and the solution is uniquely determined. Note that Equations (5.57) and (5.59) imply that $\Phi_+(s) = A(s)$; hence from (5.64) with $E = 0$ we have

$$A(s) = \Phi_+ = \frac{-i \sin \theta}{(1 + \cos \theta)^{1/2}} \frac{k_0^{1/2}}{(s + k_0 \cos \theta)(s + k_0)^{1/2}} .$$

From (5.56) together with the inversion integral, we have

$$\phi(x,y) = \frac{-i}{2\pi} \frac{k_0^{1/2} \sin \theta}{(1 + \cos \theta)^{1/2}} \int \frac{\exp(-\gamma y - isx)}{(s + k_0)^{1/2}(s + k_0 \cos \theta)} ds , \qquad (5.65)$$

along a path (such as the real axis) that is within the strip S. Finally, we let $\epsilon \to +0$, so that the appropriate integration path is shown in the diagram.

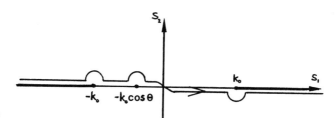

Figure 5.10

The integral (5.65) can be written in terms of the Fresnel function, which is well tabulated.

REFERENCES

Carrier, G.F., Krook, M. & Pearson, C. (1966). Functions of a Complex Variable. McGraw–Hill.

Lighthill, M.J. (1959). Introduction to Fourier Analysis and Generalised Functions. Cambridge University Press.

Noble, B. (1958). Methods Based on the Wiener–Hopf Technique. Pergamon.

6. MATCHED ASYMPTOTIC EXPANSIONS APPLIED TO ACOUSTICS

6.1 INTRODUCTION

Many problems in acoustics contain a small dimensionless parameter ϵ, and it is useful, both conceptually and from the point of view of numerical computation, to seek a solution in the form of a perturbation series. In the simplest case the series would proceed by integral powers of ϵ,

$$\phi = \phi_0 + \epsilon\phi_1 + \epsilon^2\phi_2 + \ldots$$

where, in a typical application, ϕ might be an acoustic potential or pressure and ϵ a frequency parameter, or Helmholtz number. In most cases, however, the problem of determining the functions ϕ_i is a *singular perturbation problem*, that is, one in which no single series like the one quoted will be valid both in the near field (where boundary data are specified and where surface loading may be of interest) and in the far field (where the signal directivity and level are required). Separate series must be developed describing the near and far fields, but neither can be completely constructed independently of the other because each series lacks sufficient boundary data for its unique determination.

This chapter describes how the method of Matched Asymptotic Expansions (MAE) can be used safely and systematically (1) to indicate the appropriate form taken by the inner (near–field) and outer (wave–field) series and (2) to determine all unknown functions and constants appearing in both series by "matching" the series according to a clear–cut rule. These points are illustrated by detailed study of several very simple low–frequency acoustic scattering problems which serve to demonstrate that physical arguments are unreliable in these problems and that they are no substitute for the unambiguous matching rule. Two–dimensional scattering problems are used to introduce logarithmic gauge functions; it is shown that the matching rule can easily accommodate these functions and moreover, that insistence upon satisfaction of the matching rule can in some cases be used to greatly improve the rapidity of convergence of series involving logarithmic functions.

6.2 HEURISTIC APPROACH TO MATCHING AND ITS PITFALLS

A heuristic approach to matching has been widely used in the past. Our purpose here is to show how simple acoustic problems demonstrate that *anything short of a proper mathematical rule for matching is likely to lead to erroneous results*, even if the physical basis for heuristic matching may seem clear. The correct solutions will be found later with the aid of an Asymptotic Matching Principle.

We start by looking at the solution given by Landau & Lifschitz (1959) for the sound field radiated into still fluid by the forced oscillation of a body about a fixed mean position. If the velocity of the (rigid) body is $\mathbf{V} \exp(-i\omega t)$, we want a solution of

$$(\nabla^2 + k_0^2)\phi = 0 \quad,$$

with

$$\phi \sim r^{-1} \exp(ik_0 r) f(\theta, \psi) \text{ as } r \to \infty \quad, \tag{6.1}$$

and

$$\mathbf{n} \cdot \nabla \phi = \mathbf{n} \cdot \mathbf{V} \text{ on the body,}$$

$k_0 = \omega/c_0$ being the acoustic wavenumber at frequency ω. If the body, of typical dimension L say, is *compact* — of small extent compared with the wavelength $2\pi k_0^{-1}$ — then we argue that, with axes fixed on an origin somewhere in the body, the motion is incompressible within a wavelength of the body. Thus for $r \ll k_0^{-1}$ we have

$$\nabla^2 \phi = 0 \quad,$$

and

$$\mathbf{n} \cdot \nabla \phi = \mathbf{n} \cdot \mathbf{V} \quad, \tag{6.2}$$

the radiation condition being inapplicable for this range of r. It is argued, however, that ϕ cannot grow when $r \gg L$ and therefore that no eigensolution, which would necessarily grow algebraically for $r \gg L$, can be present. (The term *eigensolution* is used for any solution of $\nabla^2 \phi = 0$ with $\mathbf{n} \cdot \nabla \phi = 0$ on the body. For a sphere of radius L, for example, the general axisymmetric eigensolution is

$$\phi = \sum_{n=1}^{\infty} A_n P_n(\cos\theta)\left\{r^n + \frac{nL^{2n+1}}{(n+1)r^{n+1}}\right\} \quad ,$$

where P_n denotes the nth Legendre polynomial.) Then for $r \gg L$ this inner potential ϕ has an harmonic multipole representation, in which the monopole must be absent,

$$\phi \sim a_i \frac{\partial}{\partial x_i}\left(\frac{1}{r}\right) + a_{ij}\frac{\partial^2}{\partial x_i \partial x_j}(\frac{1}{r}) + \dots \quad , \tag{6.3}$$

the leading term of which is a potential dipole,

$$\phi \sim a_i \frac{\partial}{\partial x_i}\left(\frac{1}{r}\right) \quad . \tag{6.4}$$

For example, in the case of a sphere of radius L the boundary condition on ϕ is

$$\frac{\partial \phi}{\partial r} = V \cos\theta \text{ on } r = L$$

if the velocity $\mathbf{V}$ is in direction $\theta = 0$, and the potential is

$$\phi = -\frac{VL^3}{2r^2}\cos\theta \quad ,$$

which is precisely Equation (6.3) if $\mathbf{a} = (VL^3/2, 0, 0)$ and $a_{ij} = \dots = 0$.

Now we argue in a complementary fashion that if $r \gg L$ then the body appears, to leading order, as a singularity at the origin, so that for $r \gg L$ we seek a solution of the full Helmholtz equation, satisfying the radiation condition, with singular behaviour at $r = 0$. The most general such solution can be written as a series of wave multipoles,

$$\phi = b\frac{e^{ik_0 r}}{r} + b_i \frac{\partial}{\partial x_i}\left(\frac{e^{ik_0 r}}{r}\right) + b_{ij}\frac{\partial^2}{\partial x_i \partial x_j}\left(\frac{e^{ik_0 r}}{r}\right) + \dots \tag{6.5}$$

$$\sim \frac{b}{r} + b_i \frac{\partial}{\partial x_i}(\frac{1}{r}) + \dots$$

when $r \ll k_0^{-1}$. Thus in an overlap domain

$$L \ll r \ll k_0^{-1}$$

appropriate terms from

$$\frac{b}{r} + b_i \frac{\partial}{\partial x_i}\left(\frac{1}{r}\right) + \ldots$$

and

$$a_i \frac{\partial}{\partial x_i}\left(\frac{1}{r}\right) + \ldots$$

should be identical. The obvious choice is

$$b = 0 \ , \ a_i = b_i \tag{6.6}$$

and thus, since $\mathbf{a}$ is known in terms of $\mathbf{V}$ for a body of given shape, we arrive at the leading–order wavefield in the form of an *acoustic dipole*,

$$\phi \sim a_i \frac{\partial}{\partial x_i}\left(\frac{e^{ik_0 r}}{r}\right) \ \sim ik_0 \frac{(\mathbf{a} \cdot \mathbf{x})}{r^2}\exp(ik_0 r) \quad . \tag{6.7}$$

This result is in fact correct for a compact body of any shape. The argument is presented with such authority by Landau & Lifschitz (1959) that one is only inclined to examine its weak points when one uses the same argument only to arrive at a plainly wrong result. Although this simple kind of argument was used a lot by Lord Rayleigh (1945) and by Lamb (1932) (to examine sound transmission through rows of parallel slits in a screen, for example), it is not difficult to find simple problems in which a strictly comparable argument goes badly wrong, and we now give two such examples.

Take, for instance, the case of plane wave scattering by the said compact body. If the incident potential is

$$\phi^i = \exp ik_0 x \quad ,$$

then the scattered field satisfies

$$(\nabla^2 + k_0^2)\phi = 0 \ ,$$

$$\phi \sim r^{-1}\exp(ik_0 r)f(\theta) \qquad \text{as } r \to \infty \quad , \tag{6.8}$$

$$\mathbf{n} \cdot \nabla\phi = -\mathbf{n} \cdot \nabla\phi^i \text{ on the body,}$$

which is almost the problem (6.1) again. For $r \ll k_0^{-1}$ the potential is harmonic and the incident stream is effectively a uniform stream, $\phi^i \sim (ik_0)x$, and therefore in the case of a sphere the leading–order inner field is precisely as before if we make

172

the substitution $V = -ik_0$. Whether or not the scatterer is a sphere, the inner potential still has the form (6.3), dominated for $r \gg L$ by an harmonic dipole. Again, we can evidently "match" this dipole to an outer wave–field acoustic dipole

$$\phi \sim b_i \frac{\partial}{\partial x_i}\left(\frac{e^{ik_0 r}}{r}\right)$$

by choosing the strength b_i of the acoustic dipole to agree with the a_i of the harmonic dipole in (6.4), and taking the acoustic monopole strength b to vanish.

This time, however, the casual approach is quite inadequate. In fact the directivity function for the leading–order wavefield scattered from a compact sphere is

$$f(\theta) = 2 - 3\cos\theta \quad,$$

i.e. the sum of equally important dipole *and monopole* terms. The correct monopole strength b cannot be found from first–order matching alone, for the dipole is more singular than the monopole as $r \to 0$, and so dominates at leading order as we come in towards the scatterer. The monopole strength is in fact determined by compressibility effects in the small scattering region, expressed analytically by the *second* term in the expansion near the body. We shall comment on this again after looking at this scattering problem in detail with MAE.

The Landau–Lifschitz problem of sound emission by an oscillating body also goes wrong when there is a uniform mean flow past the body. Then the answer one gets depends on whether pressures, or potentials — or some other field variables — are matched. For again, the outer behaviour of the potential perturbation near the body is a dipole

$$\phi \sim a_i(t)\frac{\partial}{\partial x_i}\left(\frac{1}{r}\right) \quad, \tag{6.9}$$

and the leading term for the pressure $(p = -\rho(\partial/\partial t + U\partial/\partial x)\phi)$ is

$$p \sim -\rho\dot{a}_i(t)\frac{\partial}{\partial x_i}\left(\frac{1}{r}\right) \quad. \tag{6.10}$$

It is argued that this should be matched to the inner limit of a solution to the *convected* wave equation, for when $r \gg L$ the steady perturbation due to the body is small and only the uniform flow at speed U remains. For simplicity suppose that the free stream Mach number $M = U/c_0$ is small, so that we can neglect M^2

compared with unity. Then a wave field ϕ which agrees with (6.9) in $L \ll r \ll k_0^{-1}$ is

$$\phi = \frac{\partial}{\partial x_i} \left(\frac{a_i(t - \frac{r - Mx}{c_0})}{r} \right) \quad . \tag{6.11}$$

The associated pressure in the wavefield is then

$$p = -\rho \left(\frac{\partial}{\partial t} + Mc_0 \frac{\partial}{\partial x_i} \right) \frac{\partial}{\partial x_i} \left(\frac{a_i(t - \frac{r - Mx}{c_0})}{r} \right) \quad ,$$

and for $r \ll k_0^{-1}$ that has the inner limit

$$-\rho M c_0 a_i(t) \frac{\partial^2}{\partial x_i \partial x} \left(\frac{1}{r} \right) \quad . \tag{6.12}$$

where $x = r \cos \theta$ and the stream is in the positive x direction. Equation (6.12) does not match (6.10); similar differences arise if one matches the pressures and then considers whether the velocities agree.

The message of this section is then simply this: that even to leading order some care is needed in ordering terms correctly, and a mathematical rule for matching must be devised and observed without the use of physical reasoning to shortcut any steps.

6.3   FORMAL APPROACH TO MATCHING

It is a fairly general criterion that a perturbation problem (a problem containing a parameter ϵ, where we are interested in the approximate behaviour of the solution as ϵ tends to some value, usually zero or infinity) is singular and singular in a way calling for the use of MAE, if the problem involves two dynamically significant lengthscales whose ratio becomes either large or small as ϵ approaches its limiting value.

Low–frequency acoustic generation and scattering problems exemplify perfectly this kind of singular perturbation problem and its treatment by MAE. (In one problem (Crighton & Leppington 1973), to which we shall return later, an acoustic problem served to highlight an extremely subtle, though probably common, failure of the Asymptotic Matching Rule as widely practiced in the literature, and to show how a modified rule could be devised to rectify matters.) The perturbation parameter is the Helmholtz number $\epsilon = k_0 L$ which tends to zero, and the two obviously significant lengthscales, L characterizing the body geometry and k_0^{-1} characterizing the wave propagation, become asymptotically disparate as $\epsilon \to 0$.

Consider for definiteness a scattering problem, in which r' denotes the dimensional position variable and $\phi'(r', k_0, L)$ is the scattered potential for an incident potential $\exp(ik_0 x')$. Throughout this section the additional dependence on angular variables (θ, ψ) will be suppressed. To assess the smallness of various terms as $\epsilon \to 0$ we need dimensionless variables. The potential is already dimensionless, and either L or k_0^{-1} can be used to normalize lengths. We shall use the following notation:

$r = k_0 r'$ denotes the *outer* position coordinate,

$\phi'(r' = \frac{r}{k_0}, k_0, L) \equiv \phi(r, \epsilon)$ denotes the outer potential,

$R = \frac{r'}{L} = \frac{r}{\epsilon}$ denotes the *inner* position coordinate,

$\phi'(r' = LR, k_0, L) \equiv \Phi(R, \epsilon)$ denotes the inner potential.

We want to know how $\phi(r, \epsilon)$ behaves as $\epsilon \to 0$ for *all* values of r from ϵ to ∞, or equivalently how $\Phi(R, \epsilon)$ behaves as $\epsilon \to 0$ for *all* R from 1 to ∞, because in general one needs to know the directivity of the scattered field at infinity and also the pressure on the scattering body. It is in the nature of a singular perturbation problem that it is necessarily impossible to find a single Poincaré asymptotic expansion, which in the simplest case would be a power series in ϵ,

$$\phi(r, \epsilon) \sim \phi_0(r) + \epsilon \phi_1(r) + \epsilon^2 \phi_2(r) + \cdots \quad , \qquad (6.13)$$

which holds over the whole range of values of r of interest. This is because when such a series is inserted into differential equations and boundary conditions and like powers of ϵ are equated one has to assume that r is fixed and $O(1)$. Usually in these problems r can be allowed to increase indefinitely without affecting the process of equating powers of ϵ, and so boundary conditions at infinity — the radiation condition, for example — can be imposed on the functions $\phi_0, \phi_1, \ldots$. If r is allowed to become small, however, the whole basis for the expansion (6.13) is undermined as the terms then have different orders of magnitude from those postulated in the expansion. In particular this means that the terms $\phi_0, \phi_1, \ldots$ cannot usually be made to satisfy the required boundary condition on the body, where $r = O(\epsilon)$. Consequently

(a) we cannot find the pressure on the scattering body from the outer expansion

$$\phi(r, \epsilon) \sim \phi_0(r) + \epsilon \phi_1(r) + \epsilon^2 \phi_2(r) + \ldots$$

which is unlikely to hold when r is as small as ϵ;

(b) we cannot at the moment find even the far–field directivity from the outer expansion, because that expansion contains undetermined constants, and even functions, arising from the fact that only one boundary condition (at infinity) can be applied to the outer series.

Similar considerations apply to the inner potential which also, in the simplest case, might have a power series expansion

$$\Phi(R, \epsilon) \sim \Phi_0(R) + \epsilon \Phi_1(r) + \epsilon^2 \Phi_2(R) + \dots \quad . \tag{6.14}$$

Formal processes (substitution in equations, equating powers of ϵ, etc.) will lead to a partial determination of Φ_0, Φ_1, etc., only for values of R of order unity. One expects therefore to be able to apply the boundary condition on the body, but not the condition at infinity, so that we can certainly not find the directivity function from the inner expansion nor can we at the moment find the pressure on the scatterer from that inner expansion because of the presence of undetermined constants and functions in $\Phi_0, \Phi_1, \dots$.

The inner and outer expansions are not *uniformly valid*. The outer holds in the wavefield $r = O(1)$ and usually right out to infinity. It may also hold for some smaller values of r, down to $r = O(\epsilon^{1/2})$ perhaps, for example, but it will not usually hold uniformly down to values as small as $r = O(\epsilon)$. The inner holds around the body, $R = O(1)$, and perhaps also for larger values of R, up to $R = O(\epsilon^{-1/4})$ say, but is unlikely to hold for values as large as $O(\epsilon^{-1})$ which are in the wave zone where r is the natural variable.

Now the inner and outer series are just different approximate representations for the *same* function $\phi'(r', k_0, L)$, so the question is "Can that fact be utilized to pin down the indeterminacies which exist in the inner or outer expansions separately, and so to find approximate solutions covering the whole range of interest?"

An affirmative answer to that question can be given for the case when the inner and outer expansions *overlap strongly*, that is, when both are simultaneously valid for a range of intermediate values of $r, r = O(\epsilon^\alpha)$ say, where

$$0 \leq \alpha_1 < \alpha < \alpha_2 \leq 1 \quad . \tag{6.15}$$

The number α_2 is connected with the smallest value of r for which the outer expansion holds, while α_1 is connected with the largest value of r for which the inner expansion holds. In the example just quoted, the outer expansion held down to $r = O(\epsilon^{1/2})$, so that $\alpha_2 = 1/2$, the inner up to $R = O(\epsilon^{-1/4})$, this corresponding to $r = O(\epsilon^{3/4})$ and giving $\alpha_1 = 3/4$. Since $\alpha_1 > \alpha_2$ there is no overlap in this case, there is no known way of "matching" the inner and outer expansions, and indeed no general method may exist at all. If the inner holds up to $R = O(\epsilon^{-1/2})$, i.e. to $r = O(\epsilon^{1/2})$, then $\alpha_1 = \alpha_2 = 1/2$ and there is only *marginal overlap* along the line $\alpha = 1/2$ rather than a domain $\alpha_1 < \alpha < \alpha_2$. In this case it may or may not be possible to match the expansions. No general rule is known and examples can be given of both possibilities (see Crighton & Leppington 1973; Fraenkel 1969). If, on the other hand, the outer holds down to $r = O(\epsilon^{5/8})$ say, the inner up to $R = O(\epsilon^{-1/2})$, then there is strong overlap in the domain $r = O(\epsilon^\alpha)$, where

$$1/2 < \alpha < 5/8 \quad .$$

Then there is a matching rule, which, in the simpler problems at any rate, is sufficient to uniquely determine the various unknowns in the inner and outer expansions.

One version of the rule runs as follows (the *Intermediate Matching Principle*). Suppose one knows in advance that an overlap domain $r = O(\epsilon^\alpha)$ exists for α given by (6.15). (We should note that in general the values α_1 and α_2 will depend upon the orders to which the inner and outer expansions, respectively, are carried; α_1 will be a non–decreasing function of the number of terms retained in the inner expansion, α_2 will be a non–increasing function of the number retained in the outer series, so that the width of the overlap domain will, if anything, reduce as the expansions are taken to higher order.) Then one introduces an intermediate variable r_* by $r = \epsilon^\alpha r_*$, $R = r_*/\epsilon^{1-\alpha}$, so that $r \to 0$ and $R \to \infty$ as $\epsilon \to 0$ with r_* fixed. Inner and outer series are then expanded as $\epsilon \to 0$ for fixed values of r_* and corresponding terms, functions of r_* and ϵ, are made to agree. If one does not know the range of permissible values of α in advance one has to proceed tentatively, introducing the intermediate variable and trying to see whether there is a range of values for α which will allow the intermediate expansions of the inner and outer series to be matched term by term.

This is, at present, the most fundamental of the available matching procedures, though probably not the most common nor the simplest to use in practice. Difficulties arise, for instance, when the inner and outer expansions are re–expanded in terms of the intermediate variable r_*, it not being clear how many terms in the expansion are to be retained unless information is to hand on the domain of validity of the inner and outer series. The book by Cole (1968) gives an extensive illustration of the use of the intermediate matching idea, although it avoids the difficulties inherent in this approach when carried to high order.

The most common procedure, and perhaps the most straightforward to use in a routine fashion, is a version of the principle put forward by Van Dyke (1975) as the *Asymptotic Matching Principle*. In its original version it is ambiguous in some cases, so we propose here, as in Crighton & Leppington (1973), to use the following notation and principles:

we write

$$\phi^{(m)}(r, \epsilon) = \phi_0(r) + \epsilon\phi_1(r) + .. + \epsilon^m\phi_m(r) \tag{6.16}$$

for the outer expansion truncated beyond $O(\epsilon^m)$. If the outer series really is asymptotic, then

$$|\phi(r, \epsilon) - \phi^{(m)}(r, \epsilon)| = o(\epsilon^m) \tag{6.17}$$

for appropriate values of r. Now we write $r = \epsilon R$ in $\phi^{(m)}$, hold R fixed and expand through terms $O(\epsilon^n)$ say. This gives a perfectly definite and readily calculated set of terms which will be denoted by

$$\phi^{(m,n)} \quad . \tag{6.18}$$

Next we take the inner expansion and truncate it beyond $O(\epsilon^n)$, giving

$$\Phi^{(n)}(R, \epsilon) = \Phi_0(R) + ... + \epsilon^n\Phi_n(R) \quad . \tag{6.19}$$

In this we make the transformation $R = r/\epsilon$, keep r fixed and expand as $\epsilon \to 0$ through terms $O(\epsilon^m)$, giving another set of terms denoted by

$$\Phi^{(n,m)} \quad . \tag{6.20}$$

As things now stand $\phi^{(m,n)}$ is a function of (R, ϵ), $\Phi^{(n,m)}$ a function of (r, ϵ), but we shall understand that both are expressed *without any further expansion or approximation* in terms of, say, r, ϵ.

Then the Asymptotic Matching Principle states that

$$\phi^{(m,n)} \equiv \Phi^{(n,m)} \quad . \tag{6.21}$$

It is easy to believe the truth of the principle on the basis that each side of (6.21) is an asymptotic representation for the potential in the overlap domain, and asymptotic representation with respect to a given set of gauge functions like integral powers of ϵ is unique, so that the two sides of (6.21) must be identical. That reasoning is generally quite false, though the conclusion as expressed in (6.21) is correct. Although $\phi^{(m)}$ and $\Phi^{(n)}$ *do* have asymptotic significance with respect to ϕ', the functions $\phi^{(m,n)}$ and $\Phi^{(n,m)}$ in general have *no asymptotic significance* whatever.

This point is perhaps made clearer if we consider unequal values of m and n, say $m > n$. Then it appears that genuine information has been thrown away in forming $\phi^{(m,n)}$ from $\phi^{(m)}$, while information of no significance has been retained in forming $\Phi^{(n,m)}$ from $\Phi^{(n)}$. The principle (6.21) must not, however, be regarded as saying anything about the common asymptotics of the two expansions; it merely defines two functions associated with the expansions in a clearly prescribed way, and asserts their equality.

As far as justification of the MAE procedures goes, one can say the following. It is necessary first to prove that the formal substitution of an assumed series like (6.13), (6.14) into equations and boundary conditions will produce an asymptotic series for any values of r, R. This is a difficulty common to all perturbation methods, and to many other processes (solution of differential equations by series and contour integrals, for example) as well, and is not a difficulty peculiar to MAE. Under this heading one needs to establish the range of values of r, R for which the outer and inner series are asymptotically valid, and thus to find the overlap domain, if it exists. Proofs of these points can be given in some simple cases, but the proofs fall far short of covering the kinds of problems of serious research interest, especially nonlinear problems. Granted the resolution of these difficulties, however, the two overlapping expansions *must* match according to (6.21). This has been proven (Crighton & Leppington 1973; Fraenkel 1969; Eckhaus 1973) under various different assumptions about the overlap, and with greater generality than we have so far required. The proof given in Crighton & Leppington (1973) involves showing that there exist numbers (m_1, n_1) such that all terms inside the (m, n) "block" (the one of interest) plus some other terms outside the (m, n) block but inside the (m_1, n_1) block together constitute an asymptotic representation for ϕ' with an error smaller than the smallest term in the (m, n) block and for some intermediate range of values of r. This is done for both ϕ and Φ and the two asymptotic representations are equated term by term. The rule (6.21) is just a statement of this term–by–term equality for a particularly efficiently calculated block of terms.

The Asymptotic Matching Principle has been proven also under the following more general conditions, which are sufficient to cover most applications. Suppose that the inner and outer expansions proceed in fractional or integral powers of ϵ multiplied by integral powers of $\ln \epsilon$,

$$\phi(r, \epsilon) \sim \phi_0(r) + \epsilon \ln \epsilon \phi_1(r) + \epsilon \phi_2(r) + \epsilon^2 \ln^2 \epsilon \phi_3(r)$$

$$+ \epsilon^2 \ln \epsilon \phi_4(r) + \epsilon^2 \phi_5(r) + \epsilon^3 \ln^3 \epsilon \phi_6(r) + ... \tag{6.22}$$

$$\Phi(R, \epsilon) \sim \epsilon^{1/2} \Phi_0(R) + \epsilon \Phi_1(R) + \epsilon^{3/2} \ln \epsilon \Phi_2(R) + \epsilon^{3/2} \Phi_3(R)$$

$$+ \epsilon^2 \ln \epsilon \Phi_4(R) + \epsilon^2 \Phi_5(R) + \epsilon^{5/2} \ln^2 \epsilon \Phi_6(R) + ... \qquad (6.23)$$

being a case that arises in diffraction by a thick rigid plate (Crighton & Leppington 1973) and is interesting in several respects. Then a function like $\phi^{(m)}$ or $\Phi^{(n)}$ will be constructed as before, by terminating the expansion beyond $O(\epsilon^m)$ or $O(\epsilon^n)$ as the case may be, where now m and n can be any rational numbers which may or may not appear in either of the series. In truncating the series, *all logarithmic terms are to be grouped together according to the order of their algebraic multiplier*, and are never to be separated out. Thus, with the above forms for example, we have

$$\phi^{(1)} = \phi_0 + \epsilon \ln \epsilon \phi_1 + \epsilon \phi_2 \quad ,$$

$$\phi^{(1/2)} = \phi_0 \quad ,$$

$$\Phi^{(1)} = \epsilon^{1/2} \Phi_0 + \epsilon \Phi_1 \quad ,$$

$$\Phi^{(3/2)} = \epsilon^{1/2} \Phi_0 + \epsilon \Phi_1 + \epsilon^{3/2} \ln \epsilon \Phi_2 + \epsilon^{3/2} \Phi_3 \quad .$$

Having taken any two rational numbers m and n, and formed $\phi^{(m)}$ and $\Phi^{(n)}$, one then writes $r = \epsilon R$ in $\phi^{(m)}$ and expands up to and including all terms which are not smaller than ϵ^n, writes $R = r/\epsilon$ in $\Phi^{(n)}$ and expands keeping all terms not smaller than ϵ^m, giving the quantities $\phi^{(m,n)}$ and $\Phi^{(n,m)}$. *Again, algebraic order only is* taken into account, $\epsilon^\lambda \ln^\mu \epsilon$ being regarded as $O(\epsilon^\lambda)$ whatever the value of μ. Then the matching principle

$$\phi^{(m,n)} \equiv \Phi^{(n,m)} \qquad (6.24)$$

has been established also (Crighton & Leppington 1973; Fraenkel 1969) for expansions of this kind.

Occasionally it is necessary to generalize this rule further, as for example in the case of scattering by a soft body (Lesser & Crighton 1975). But that example deserves separate study for other reasons, and for most problems the principle (6.24) is adequate.

6.4 SOUND GENERATION BY FORCED OSCILLATIONS

We return now to the problem of sound generation by the forced oscillation of a rigid compact body. For definiteness we take a sphere of radius L whose velocity in the positive x' direction ($\theta = 0$) is $V \exp(-i\omega t)$, so that we need an axisymmetric solution of

$$(\nabla'^2 + k_0^2)\phi' = 0 \ ,$$

$$\phi' \sim r'^{-1} \exp(ik_0 r')f(\theta) \text{ as } r' \to \infty \ , \tag{6.25}$$

$$\frac{\partial\phi'}{\partial r'} = V \cos\theta \text{ on } r' = L \ .$$

In terms of the outer variable $r = k_0 r'$ and with an appropriate normalization for ϕ', say $\phi' = VL\phi$, the complete problem is

$$(\nabla_r^2 + 1)\phi = 0 \ ,$$

$$\phi \sim r^{-1} \exp(ir)g(\theta) \text{ as } r \to \infty \ , \tag{6.26}$$

$$\frac{\partial\phi}{\partial r} = \frac{1}{\epsilon}\cos\theta \text{ on } r = \epsilon \ ,$$

with $\epsilon = k_0 L$. Letting $\epsilon \to 0$ makes the body boundary condition tell us no more than that ϕ must be singular in some fashion as $r \to 0$. Whatever the form taken by the outer expansion (and that cannot usually be assumed in advance, for it is dictated to some extent by information from the inner region) all terms in the outer expansion must therefore be radiating axisymmetric solutions of the Helmholtz equation, the general solution of which is a multipole series

$$a\left(\frac{e^{ir}}{r}\right) + a_i \frac{\partial}{\partial x_i}\left(\frac{e^{ir}}{r}\right) + a_{ij}\frac{\partial^2}{\partial x_i \partial x_j}\left(\frac{e^{ir}}{r}\right) + \dots \ . \tag{6.27}$$

A little thought shows that the most general axisymmetric solution must be

$$a\left(\frac{e^{ir}}{r}\right) + a_1 \frac{\partial}{\partial x}\left(\frac{e^{ir}}{r}\right) + a_{11}\frac{\partial^2}{\partial x^2}\left(\frac{e^{ir}}{r}\right) + a_{22}\nabla_\perp^2\left(\frac{e^{ir}}{r}\right) + a_{111}\frac{\partial^3}{\partial x^3}\left(\frac{e^{ir}}{r}\right)$$

$$+ a_{122}\frac{\partial}{\partial x}\nabla_\perp^2\left(\frac{e^{ir}}{r}\right) + (\text{higher multipoles than octupoles}) \ . \tag{6.28}$$

Let us assume therefore that the outer series starts off with a term of algebraic order, $\epsilon^\lambda \phi_0$ say, where ϕ_0 has the above form, in which $\nabla_\perp^2$ denotes the Laplace operator in the two transverse coordinates,

$$\nabla_\perp^2 \equiv \frac{\partial^2}{\partial y^2} + \frac{\partial^2}{\partial z^2} \ .$$

The coefficients a, a_1, a_{11}, etc. will then be given a superscript $^{(0)}$ to indicate that they refer to ϕ_0. We can also simplify (6.28) somewhat, using the fact that for $r \neq 0$

$$\left(\nabla_\perp^2 + \frac{\partial^2}{\partial x^2} + 1\right)\left(\frac{e^{ir}}{r}\right) = 0 \quad,$$

and therefore the terms with coefficients a_{22} and a_{122} can be assumed to be already included elsewhere.

Turn now to the inner problem, obtained from (6.26) by writing $\Phi(R, \epsilon)$ for $\phi(r, \epsilon)$ and $r = \epsilon R$, so that

$$\left(\nabla_R^2 + \epsilon^2\right)\Phi = 0 \quad,$$

$$\Phi \sim R^{-1}\exp(i\epsilon R)h(\theta) \text{ as } \epsilon R \to \infty \quad, \tag{6.29}$$

$$\frac{\partial \Phi}{\partial R} = \cos\theta \text{ on } R = 1 \quad,$$

the radiation condition obviously being unenforceable in the limit $\epsilon \to 0$.

The body boundary condition indicates (with the chosen normalization, $\phi' = VL\Phi$) that Φ is of order unity in the inner region, so that the inner series should take the form

$$\Phi \sim \Phi_0 + (\text{terms which vanish with } \epsilon) \quad.$$

(The only other possibility would be that Φ starts with a term which becomes infinite as $\epsilon \to 0$, $\Phi \sim \epsilon^{-1}\Phi_{-1}$ say. Then Φ_{-1} would have to be an *inner eigensolution*, the general axisymmetric form of which was given following Equation (6.2). If one persists with this and attempts to match with the outer solution, one finds of course that $\Phi_{-1} \equiv 0$. The reader should verify this after seeing how the first nontrivial matching is carried out.) The problem for Φ_0 is then

$$\nabla_R^2 \Phi_0 = 0 \quad,$$
$$\frac{\partial \Phi_0}{\partial R} = \cos\theta \text{ on } R = 1 \quad, \tag{6.30}$$

whose general solution is

$$\Phi_0 = -\frac{1}{2R^2}\cos\theta + \sum_{\nu=1}^{\infty} A_\nu^{(0)} P_\nu(\cos\theta)\left\{R^\nu + \frac{\nu}{\nu+1}\frac{1}{R^{\nu+1}}\right\} \quad, \tag{6.31}$$

the sum of a particular solution which decays at infinity, plus a general axisymmetric eigensolution satisfying

$$\nabla_R^2 \Phi = 0, \quad \frac{\partial \Phi}{\partial R} = 0 \text{ on } R = 1 \quad.$$

For matching, note that when $\phi^{(\lambda)} \equiv \epsilon^{\lambda}\phi_0$ is expanded with $r = \epsilon R$, the leading term will involve only inverse powers of R, while when $\Phi^{(0)} \equiv \Phi_0$ is expanded with $R = r/\epsilon$, the leading term will involve the positive power r^{ν} unless we choose $A_{\nu}^{(0)} \equiv 0$. Thus no inner eigensolutions are needed to leading order (though they are needed at higher order to match the ascending powers of R which arise from the expansion of the phase factors $\exp(i\epsilon R)$ in (6.28)).

We then have

$$\Phi^{(0)} = \Phi_0 = -\frac{1}{2R^2}\cos\theta$$

$$\Phi^{(0)}\left(\frac{r}{\epsilon}\right) = -\frac{\epsilon^2}{2r^2}\cos\theta \quad,$$

and so

$$\Phi^{(0,0)} = \Phi^{(0,1)} = 0 \quad,$$

$$\Phi^{(0,2)} = -\frac{\epsilon^2}{2r^2}\cos\theta = \Phi^{(0,3)} = \Phi^{(0,4)} = \ldots\ldots \quad.$$

On the other hand, for the leading outer solution we have

$$\phi^{(\lambda)} = \epsilon^{\lambda}\phi_0 \quad,$$

and $\phi^{(\lambda)}(\epsilon R) = O(\epsilon^{\lambda-N})$ where N is the highest multipole order present in (6.28) ($N = 1$ monopole, $N = 2$ dipole, $N = 3$ quadrupole, $N = 4$ octupole, etc.). Suppose then that λ were equal to 0. Then $\phi^{(0)}(\epsilon R) = O(\epsilon^{-1})$ at the very least, and because that term must match with $\Phi^{(0,0)} = 0$ the only possibility is that $\phi_0 = 0$. The same conclusion is reached if we take $\lambda = 1$, because $\Phi^{(0,1)} = 0$. We therefore try taking $\lambda = 2$ — i.e. *the order of the leading outer term is that of the first nontrivial term in the outer expansion of the leading inner term* — so that we have

$$\phi^{(2)} = \epsilon^2\phi_0 \quad.$$

Since the expansion of $\Phi^{(0)}(r/\epsilon)$ contains no inverse power of ϵ it follows that only the values $N = 1$ and $N = 2$ are permitted in ϕ_0, and hence

$$\phi_0 = u^{(0)}\left(\frac{e^{ir}}{r}\right) + u_1^{(0)}\frac{\partial}{\partial x}\left(\frac{e^{ir}}{r}\right) \quad, \tag{6.32}$$

$$\phi^{(2)}(\epsilon R) = \epsilon a^{(0)}\left(\frac{e^{i\epsilon R}}{R}\right) + a_1^{(0)}\frac{\partial}{\partial X}\left(\frac{e^{i\epsilon R}}{R}\right) \quad,$$

$$\phi^{(2,0)} = a_1^{(0)}\frac{\partial}{\partial X}\left(\frac{1}{R}\right) = -a_1^{(0)}\frac{1}{R^2}\cos\theta \quad,$$

$$\phi^{(2,1)} = \epsilon a^{(0)} \left(\frac{1}{R} \right) + \phi^{(2,0)} \quad , \tag{6.33}$$

$$\phi^{(2,2)} = -\epsilon^2 a_1^{(0)} \frac{\partial}{\partial X} \left(\frac{R}{2} \right) + \phi^{(2,1)} = -\epsilon^2 \frac{a_1^{(0)}}{2} \cos \theta + \phi^{(2,1)} \quad .$$

From the matching

$$\phi^{(2,0)} \equiv \Phi^{(0,2)}$$

we thus get

$$a_1^{(0)} = \tfrac{1}{2} \quad , \tag{6.34}$$

which determines the dipole strength, but not the monopole $a^{(0)}$. We see, however, that this could be found from $\phi^{(2,1)}$ in (6.33) by matching to $\Phi^{(1,2)}$, and therefore, prompted by the fact that the inner expansion of $\phi^{(2)}(\epsilon R)$ contains terms $O(1)$ and $O(\epsilon)$, we try to continue the inner expansion with

$$\Phi \sim \Phi_0 + \epsilon \Phi_1 + o(\epsilon) \quad . \tag{6.35}$$

Φ_1 is an inner eigensolution

$$\Phi_1 = \sum_{\nu=1}^{\infty} A_\nu^{(1)} P_\nu(\cos \theta) \left\{ R^\nu + \left(\frac{\nu}{\nu+1} \right) \frac{1}{R^{\nu+1}} \right\} \quad , \tag{6.36}$$

and the expansion of

$$\Phi^{(1)} \left(\frac{r}{\epsilon} \right) = \Phi_0 \left(\frac{r}{\epsilon} \right) + \epsilon \Phi_1 \left(\frac{r}{\epsilon} \right)$$

must start with terms $O(1)$ and $O(\epsilon)$ in order that $\Phi^{(1,2)}$ can be matched to $\phi^{(2,1)}$ in (6.33). It follows that only the term with $\nu = 1$ can be present in (6.36) and that then

$$\Phi^{(1,2)} = -\frac{\epsilon^2}{2r^2} \cos \theta + A_1^{(1)} r \cos \theta \quad . \tag{6.37}$$

This is identical with

$$\phi^{(2,1)} = \epsilon a^{(0)} \frac{1}{R} - \frac{1}{2R^2} \cos \theta$$

if and only if

$$A_1^{(1)} = a^{(0)} = 0 \quad . \tag{6.38}$$

Thus matching of the second–order inner solution to the first–order outer solution shows that the second–order inner solution is identically zero in this case, and that the previously undetermined leading–order outer monopole strength is also zero.

We have recovered the Landau–Lifschitz result (6.7) and the "obvious" choice (6.6) is seen to be correct — but for a deeper reason.

At this stage we will break off and make a similar examination of the plane wave scattering problem, where we shall see that the second–order inner problem is not trivial, as it was above, and this has a profound influence on the leading–order outer field, making the "obvious" choice (6.6) of zero monopole strength not only not obvious, but actually incorrect.

6.5 PLANE WAVE SCATTERING

For the scattering by a fixed rigid sphere of radius L of a plane incident wave of potential $\exp(ik_0 x')$, the problem for the scattered field ϕ' is

$$(\nabla'^2 + k_0^2)\phi' = 0 \; ,$$

$$\phi' \sim r'^{-1} \exp(ik_0 r')f(\theta) \quad \text{as } r' \to \infty \; , \tag{6.39}$$

$$\frac{\partial \phi'}{\partial r'} = -ik_0 \cos\theta \exp(i\epsilon\cos\theta) \quad \text{on } r' = L \; .$$

The potentials are already dimensionless and hence the problem in outer variables takes the form

$$(\nabla_r^2 + 1)\phi = 0 \; ,$$

$$\phi \sim r^{-1} \exp(ir)g(\theta) \quad \text{as } r \to \infty \; , \tag{6.40}$$

$$\frac{\partial \phi}{\partial r} = -i\cos\theta \exp(i\epsilon\cos\theta) \quad \text{on } r = \epsilon \; ,$$

while in inner variables we have

$$(\nabla_R^2 + \epsilon^2)\Phi = 0 \; ,$$

$$\Phi = R^{-1} \exp(i\epsilon R)h(\theta) \quad \text{as } \epsilon R \to \infty \; , \tag{6.41}$$

$$\frac{\partial \Phi}{\partial R} = -i\epsilon\cos\theta \exp(i\epsilon\cos\theta) \quad \text{on } R = 1 \; .$$

As in Section 6.4 all outer solutions, to any order in ϵ, are outer *eigensolutions*, with the general form (6.28), but the order λ of the leading outer potential, $\phi \sim \epsilon^\lambda \phi_0$, is not yet known. It can be argued, as before, that the order of the leading inner

potential is that *distinguished order* (a term used by Cole (1968)) for which the
solution is not just an inner eigensolution. Thus

$$\Phi \sim \epsilon \Phi_0 + \cdots \quad ,$$

where

$$\Phi_0 = \frac{i}{2R^2} \cos\theta + \sum_{\nu=1}^{\infty} A_\nu^{(0)} P_\nu(\cos\theta) \left\{ R^\nu + \left(\frac{\nu}{\nu+1}\right) \frac{1}{R^{\nu+1}} \right\} \quad , \qquad (6.42)$$

and again, as in Section 6.4, all the $A_\nu^{(0)} \equiv 0$. We then find that λ has the value 3
this time, with monopoles and dipoles permitted in ϕ_0, so that

$$\phi \sim \epsilon^3 \phi_0 + \cdots \quad ,$$

$$\phi_0 = a^{(0)} \left(\frac{e^{ir}}{r}\right) + a_1^{(0)} \frac{\partial}{\partial x} \left(\frac{e^{ir}}{r}\right) \quad . \qquad (6.43)$$

It might be thought that, since λ is 3 here rather than 2, quadrupoles should
also be permitted in ϕ_0 this time. That cannot be, however, for then $\epsilon^3 \phi_0(\epsilon R)$
would contain an $O(1)$ term from the quadrupole, which could not be matched to
the inner solution which is $O(\epsilon)$.

The matching rule

$$\phi^{(3,1)} \equiv \Phi^{(1,3)}$$

determines the dipole strength as

$$a_1^{(0)} = -\frac{i}{2} \quad , \qquad (6.44)$$

but again fails to determine the monopole strength. We therefore move on to the
second inner problem, the inner expansion continuing with

$$\Phi \sim \epsilon \Phi_0 + \epsilon^2 \Phi_1 + \cdots \quad .$$

The difference between the generation and scattering problems is apparent now, in
that Φ_1 is not an eigensolution but a solution of

$$\nabla_R^2 \Phi_1 = 0 \quad ,$$

$$\frac{\partial \Phi_1}{\partial R} = \cos^2\theta \quad \text{on } R = 1 \quad . \qquad (6.45)$$

Writing the boundary condition as

$$\frac{\partial \Phi_1}{\partial R} = \frac{2}{3} P_2(\cos\theta) + \frac{1}{3} \quad,$$

we easily find a particular solution in the form

$$\Phi_1 = -\frac{2}{9} \frac{P_2(\cos\theta)}{R^3} - \frac{1}{3R} \quad,$$

to which is to be added the general eigensolution

$$\sum_{\nu=1}^{\infty} A_\nu^{(1)} P_\nu(\cos\theta) \left\{ R^\nu + \left(\frac{\nu}{\nu+1}\right) \frac{1}{R^{\nu+1}} \right\} \quad.$$

Application of the matching rule

$$\phi^{(3,2)} \equiv \Phi^{(2,3)}$$

now determines the monopole strength as

$$a^{(0)} = -\frac{1}{3} \quad, \tag{6.46}$$

and shows that $A_\nu^{(1)} \equiv 0$, so that the inner eigensolutions play no part in the second term also.

The leading–order wave field is then

$$\phi \sim \epsilon^3 \left\{ -\frac{1}{3} \frac{e^{ir}}{r} - \frac{i}{2} \frac{\partial}{\partial x} \left(\frac{e^{ir}}{r}\right) \right\} \quad,$$

from which the directivity function, as $r \to \infty$, then follows at once as proportional to $2 - 3\cos\theta$, as quoted in Section 6.2. Note that the monopole term in the far–field is directly comparable with the dipole term, yet it arises from a weak monopole $-\epsilon^2/3R$ in the near–field as compared with the dipole $i\epsilon\cos\theta/2R^2$. Such subtleties are the essence of acoustics, even in the low–frequency limit.

The above procedure may seem rather laborious but we have gone into it in some detail because a routine framework of this kind is indispensible in more difficult problems. Moreover, with practice the way in which the expansions are likely to proceed can often be anticipated (though as Van Dyke (1975) emphasizes, many workers have obtained incorrect results by failing to suspect the presence of certain kinds of terms, usually logarithmic terms, whereas adherence to the procedure used here will greatly reduce that risk) and it is not always necessary to include the most general eigensolution if one keeps in mind the forms of the terms which need to be matched (though again, failure to find the most general eigensolution, particularly in nonlinear problems, has led to many incorrect results in the past, and general eigensolutions should be included if there is the slightest suspicion of difficulty).

6.6 HIGHER APPROXIMATIONS

We continue here with the scattering problem to obtain one more term of each series. The inner series so far is

$$\Phi \sim \epsilon \left\{ \frac{i}{2R^2} \cos\theta \right\} - \epsilon^2 \left\{ \frac{2}{9} \frac{P_2(\cos\theta)}{R^3} + \frac{1}{3R} \right\} + o(\epsilon^2) \quad . \tag{6.47}$$

If $\Phi^{(2)}(r/\epsilon)$ is expanded for $\epsilon \to 0$, the first terms are $O(\epsilon^3)$ and $O(\epsilon^5)$, so we attempt to continue the outer expansion in the form

$$\phi \sim \epsilon^3 \phi_0 + \epsilon^5 \phi_1 + o(\epsilon^5) \tag{6.48}$$

using the matching rule

$$\Phi^{(2,5)} \equiv \phi^{(5,2)} \quad . \tag{6.49}$$

Since $\Phi = O(\epsilon)$, the expansion of $\phi(\epsilon R)$ cannot contain any terms larger than $O(\epsilon)$, and so ϕ_1 can contain monopole, dipole, quadrupole and octupole elements, with the general form

$$\phi_1 = a^{(1)} \left(\frac{e^{ir}}{r} \right) + a_1^{(1)} \frac{\partial}{\partial x} \left(\frac{e^{ir}}{r} \right) + a_{11}^{(1)} \frac{\partial^2}{\partial x^2} \left(\frac{e^{ir}}{r} \right) + a_{111}^{(1)} \frac{\partial^3}{\partial x^3} \left(\frac{e^{ir}}{r} \right) \quad . \tag{6.50}$$

Then

$$\phi^{(5)}(\epsilon R) = \epsilon^3 \phi_0(\epsilon R) + \epsilon^5 \phi_1(\epsilon R) \quad ,$$

and expanding gives

$$\phi^{(5)}(\epsilon R) = \epsilon^3 \left\{ \frac{(1 + i\epsilon R - \frac{\epsilon^2 R^2}{2} + ...)}{\epsilon R} - \frac{i}{2\epsilon^2} \frac{\partial}{\partial X} \frac{(1 + i\epsilon R - \frac{\epsilon^2 R^2}{2} + ...)}{R} \right\}$$

$$+ \epsilon^5 \left\{ a^{(1)} \frac{(1 + i\epsilon R + ...)}{\epsilon R} + \frac{a_1^{(1)}}{\epsilon^2} \frac{\partial}{\partial X} \frac{(1 + i\epsilon R + ...)}{R} \right. \tag{6.51}$$

$$\left. + \frac{a_{11}^{(1)}}{\epsilon^3} \frac{\partial^2}{\partial X^2} \frac{(1 + i\epsilon R + ...)}{R} + \frac{a_{111}^{(1)}}{\epsilon^4} \frac{\partial^3}{\partial X^3} \frac{(1 + i\epsilon R + ...)}{R} \right\} \quad .$$

Truncating this expansion beyond $O(\epsilon^2)$ gives

$$\phi^{(5,2)} = -\frac{1}{3}\epsilon^2 \frac{1}{R} - \frac{i\epsilon}{2}\frac{\partial}{\partial X}\left(\frac{1}{R}\right) + \epsilon^2 a_{11}^{(1)}\frac{\partial^2}{\partial X^2}\left(\frac{1}{R}\right) + \epsilon a_{111}^{(1)}\frac{\partial^3}{\partial X^3}\left(\frac{1}{R}\right) , \qquad (6.52)$$

while from (6.47)

$$\Phi^{(2,5)} = \frac{i\epsilon^3}{2r^2}\cos\theta - \frac{2}{9}\epsilon^5\frac{P_2(\cos\theta)}{r^3} - \frac{1}{3}\epsilon^3\frac{1}{r} . \qquad (6.53)$$

Express (6.52) and (6.53) in terms of the same variable and use the fact that $P_2(\cos\theta) = (1/2)(3\cos^2\theta - 1)$. Then the matching rule (6.49) requires

$$\frac{i\epsilon}{2R^2}\cos\theta - \frac{1}{9}\epsilon^2\frac{(3\cos^2\theta - 1)}{R^3} - \frac{1}{3}\epsilon^2\frac{1}{R} \equiv -\frac{1}{3}\epsilon^2\frac{1}{R} + \frac{i\epsilon}{2R^2}\cos\theta$$

$$+\epsilon^2 a_{11}^{(1)}\frac{(3\cos^2\theta - 1)}{R^3} + \epsilon a_{111}^{(1)}\left[\frac{9\cos\theta}{R^4} - \frac{15\cos^3\theta}{R^4}\right] ,$$

where the differentiations in (6.52) have been performed using $\partial R/\partial X = X/R = \cos\theta$. Thus matching gives

$$a_{11}^{(1)} = -\frac{1}{9} , \quad a_{111}^{(1)} = 0 , \qquad (6.54)$$

and determines the most singular (octupole and quadrupole) terms in ϕ_1, but fails to determine the dipole and monopole terms.

To obtain these it is necessary to go further with the inner expansion. Consideration of (6.51) shows that the dipole term in (6.50), with coefficient $a_1^{(1)}$, makes a contribution to $\phi^{(5)}(\epsilon R)$ of order $O(\epsilon^3)$, while the monopole, with coefficient $a^{(1)}$, makes a contribution of $O(\epsilon^4)$. Therefore $a_1^{(1)}$ will be determined if we use the matching rule

$$\phi^{(5,3)} = \Phi^{(3,5)} , \qquad (6.55)$$

which needs the third term of the inner expansion, while the monopole can only be determined from the fourth term of the inner series and use of the rule

$$\phi^{(5,4)} = \Phi^{(4,5)} . \qquad (6.56)$$

Accordingly, we assume an inner expansion

$$\Phi \sim \epsilon\Phi_0 + \epsilon^2\Phi_1 + \epsilon^3\Phi_2 + \epsilon^4\Phi_3 + \cdots , \qquad (6.57)$$

which gives the problem

$$\nabla_R^2 \Phi_2 = -\Phi_0 = -\frac{i}{2R^2} \cos\theta \ ,$$

$$\frac{\partial \Phi_2}{\partial R} = \frac{i}{2} \cos^3\theta \ \text{ on } R = 1 \ ,$$

(6.58)

for the third term. This is typical of the higher–order inner problems; not only are the boundary conditions inhomogeneous (as they were for Φ_0 and Φ_1), but the Laplace equation itself now has a forcing function related to the lower–order potentials. Here the form of the forcing suggests looking for a particular integral of the form $f(R)\cos\theta$, and then one quickly finds that $f(R) = (i/4)$. We therefore write

$$\Phi_2 = \frac{i}{4}\cos\theta + \psi \ , \tag{6.59}$$

and then

$$\nabla_R^2 \psi = 0 \ ,$$

$$\frac{\partial \psi}{\partial R} = \frac{i}{2}\cos^3\theta$$

$$= \frac{i}{5} P_3(\cos\theta) + \frac{3i}{10} P_1(\cos\theta) \ , \tag{6.60}$$

where $P_1(\cos\theta) = \cos\theta$, $P_3(\cos\theta) = (1/2)(5\cos^3\theta - 3\cos\theta)$ and the transformation to Legendre functions is made so that we can most readily use the facts that $R^n P_n(\cos\theta)$ and $R^{-(n+1)} P_n(\cos\theta)$ are axisymmetric solutions of Laplace's equation (and are the only ones which are finite on the axis, where $\cos\theta = \pm 1$). We now try to eliminate the forcing terms from the boundary condition (6.60) by looking for a particular solution

$$\alpha \frac{P_3(\cos\theta)}{R^4} + \beta \frac{P_1(\cos\theta)}{R^2} \ ,$$

which requires $\alpha = -i/20$, $\beta = -3i/20$.

To this we have to add the general inner eigensolution, as given in (6.31) with coefficients $A_\nu^{(2)}$, so that the general solution for Φ_2 is

$$\Phi_2 = \frac{i}{4}\cos\theta - \frac{i}{20}\frac{P_3(\cos\theta)}{R^4} - \frac{3i}{20}\frac{P_1(\cos\theta)}{R^2}$$

$$+ \sum_{\nu=1}^{\infty} A_\nu^{(2)} P_\nu(\cos\theta)\left\{ R^\nu + \left(\frac{\nu}{\nu+1}\right)\frac{1}{R^{\nu+1}}\right\} \ . \tag{6.61}$$

Note here that, whatever the values taken by the $A_\nu^{(2)}$, Φ_2 does not tend to zero as $R \to \infty$ and there is no way of making it do so. At still higher orders the inner potentials will not merely remain finite as $R \to \infty$, but will actually tend to infinity.

The function $\Phi^{(3)}$ is now given by

$$\Phi^{(3)}\left(\frac{r}{\epsilon}\right) = \frac{i\epsilon^3}{2r^2}\cos\theta - \frac{2}{9}\epsilon^5\frac{P_2(\cos\theta)}{r^3} - \frac{\epsilon^3}{3r} + \frac{i\epsilon^3}{3r} + \frac{i\epsilon^3}{4}\cos\theta - \frac{3i}{20}\epsilon^5\frac{P_1(\cos\theta)}{r^2}$$

$$-\frac{i}{20}\epsilon^7\frac{P_3(\cos\theta)}{r^4} + \epsilon^3 \sum_{\nu=1}^{\infty} A_\nu^{(2)} P_\nu(\cos\theta)\left\{\left(\frac{r}{\epsilon}\right)^\nu + \left(\frac{\nu}{\nu+1}\right)\left(\frac{\epsilon}{r}\right)^{\nu+1}\right\} \quad , \quad (6.62)$$

in which the leading–order terms must be at most $O(\epsilon^3)$ in order for this to be matched to the outer series which starts with $\epsilon^3\phi_0$. Therefore

$$A_\nu^{(2)} = 0 \tag{6.63}$$

and again it seems that the inner eigensolutions are "too singular" as $R \to \infty$ to be tolerated. We shall return to this in a moment. For matching according to (6.55) we have

$$\Phi^{(3,5)} = \epsilon\frac{i}{2R^2}\cos\theta - \frac{2}{9}\epsilon^2\frac{P_2(\cos\theta)}{R^3} - \frac{\epsilon^2}{3R} + \frac{i\epsilon^3}{4}\cos\theta - \frac{3i}{20}\epsilon^3\frac{P_1(\cos\theta)}{R^2} \quad , \quad (6.64)$$

after truncating (6.62) at $O(\epsilon^5)$ and then returning to the variable R, while from (6.51) we have

$$\phi^{(5,3)} = -\frac{1}{3}\frac{\epsilon^2}{R} - \frac{i\epsilon^2}{3} - \frac{i\epsilon}{2}\left[-\frac{\cos\theta}{R^2} - \frac{1}{2}\epsilon^2\cos\theta\right]$$

$$-\epsilon^3 a_1^{(1)}\frac{1}{R^2}\cos\theta + \frac{1}{9}\epsilon^2\left[\frac{1}{R^3} - \frac{3\cos^2\theta}{R^3}\right] \tag{6.65}$$

(use having been made of the results (6.54)).

The equality of (6.64) and (6.65) determines the dipole coefficient as

$$a_{12}^{(1)} = \frac{3i}{20} \quad , \tag{6.66}$$

and all other terms in (6.65) have identical counterparts in (6.64) except for the term $-i\epsilon^3/3$. This outer term is unmatched because of a *failure to take the most general inner eigensolution* in (6.61). To that eigensolution should be added the *constant* term

$$A_0^{(2)} \quad ,$$

which is all too easily discarded as irrelevant. It has a serious effect at higher order, however, because later the function Φ_2 will act as a forcing function for the inner term $\epsilon^5 \Phi_4$, and the constant part $A_0^{(2)}$ of Φ_2 will induce a variable part of Φ_4 and hence, through the matchings, of the wavefield too. If we add $A_0^{(2)}$ to (6.61), (6.64) acquires a term

$$\epsilon^3 A_0^{(2)} \quad ,$$

which can be matched to the corresponding term in (6.65) provided

$$A_0^{(2)} = -\frac{i}{3} \quad . \tag{6.67}$$

Thus the lowest–order (least rapidly growing) inner eigensolution is called into play in the third–order inner solution.

To find Φ_3 in (6.57) we have the problem

$$\nabla_R^2 \Phi_3 = -\Phi_1 = \frac{2}{9} \frac{P_2(\cos\theta)}{R^3} + \frac{1}{3R} \quad ,$$

$$\frac{\partial \Phi_3}{\partial R} = -\frac{1}{6} \cos^4 \theta \quad \text{on } R = 1 \quad . \tag{6.68}$$

This is a complicated but straightforward problem which is tackled as before. First we find a particular integral to annihilate the forcing in the differential equation. Recall that

$$\nabla_R^2 \equiv \frac{1}{R^2} \frac{\partial}{\partial R} \left(R^2 \frac{\partial}{\partial R} \right) + \frac{1}{R^2} \frac{1}{\sin\theta} \frac{\partial}{\partial \theta} \left(\sin\theta \frac{\partial}{\partial \theta} \right)$$

and that

$$\frac{1}{\sin\theta} \frac{\partial}{\partial \theta} \left(\sin\theta \frac{\partial}{\partial \theta} P_n(\cos\theta) \right) + n(n+1) P_n(\cos\theta) = 0 \quad ,$$

and then it is not hard to construct a particular integral

$$\frac{1}{6} R - \frac{1}{27} \frac{P_2(\cos\theta)}{R} \quad ,$$

so that we write

$$\Phi_3 = \frac{1}{6} R - \frac{1}{27} \frac{P_2(\cos\theta)}{R} + \psi \quad . \tag{6.69}$$

Then

$$\nabla_R^2 \psi = 0 \quad ,$$

$$\frac{\partial \psi}{\partial R} = -\frac{1}{6} \cos^4 \theta - \frac{1}{6} - \frac{1}{27} P_2(\cos\theta) \quad \text{on } R = 1 \quad . \tag{6.70}$$

Rewriting $\cos^4 \theta$ as

$$\frac{8}{35} P_4(\cos \theta) + \frac{4}{7} P_2(\cos \theta) + \frac{1}{5} \quad,$$

we find

$$\psi = \frac{4}{525} \frac{P_4(\cos \theta)}{R^5} + \frac{5}{567} \frac{P_2(\cos \theta)}{R^3} + \frac{1}{5R} \quad,$$

plus an inner eigensolution. Since the outer expansion begins with $O(\epsilon^3)$, $\epsilon^4 \Phi_3(r/\epsilon)$ cannot contain any terms larger than ϵ^3 and therefore the most general eigensolution possible at this stage is

$$A_0^{(3)} + A_1^{(3)} P_1(\cos \theta) \{R + \tfrac{1}{2}\tfrac{1}{R^2}\} \quad.$$

Hence

$$\Phi_3 = \frac{1}{6} R - \frac{1}{27} \frac{P_2(\cos \theta)}{R} + \frac{4}{525} \frac{P_4(\cos \theta)}{R^5} + \frac{5}{567} \frac{P_2(\cos \theta)}{R^3} + \frac{1}{5R}$$

$$+ A_0^{(3)} + A_1^{(3)} P_1(\cos \theta) \left\{ R + \frac{1}{2} \frac{1}{R^2} \right\} \quad, \tag{6.71}$$

and we now have $\Phi^{(4)}$, that is, the inner expansion through $O(\epsilon^4)$, determined apart from $A_0^{(3)}$ and $A_1^{(3)}$.

Next we calculate $\Phi^{(4,5)}$, and transform it back to the variable R, with the result

$$\Phi^{(4,5)} = \Phi^{(3,5)} \text{ (Eqn. 6.64)} + \epsilon^4 \left\{ \frac{1}{6} R - \frac{1}{27} \frac{P_2(\cos \theta)}{R} + \frac{1}{5R} \right.$$

$$\left. + A_0^{(3)} + A_1^{(3)} R \cos \theta \right\} \quad, \tag{6.72}$$

while

$$\phi^{(5,4)} = \phi^{(5,3)} \text{ (Eqn. 6.65)} + \epsilon^4 \frac{R}{6} - \frac{\epsilon^4}{12} \frac{\partial}{\partial X}(R^2) + \epsilon^4 \frac{a^{(1)}}{R} - \frac{\epsilon^4}{2} a_{11}^{(1)} \frac{\partial^2}{\partial X^2}(R) \quad. \tag{6.73}$$

Performing the differentiations, using (6.54) for $u_{11}^{(1)}$ and matching (6.72) to (6.73) gives

$$\frac{1}{6} R - \frac{1}{54} \frac{(3\cos^2 \theta - 1)}{R} + \frac{1}{5R} + A_0^{(3)} + A_1^{(3)} R \cos \theta$$

$$\equiv \frac{1}{6} R - \frac{1}{6} R \cos \theta + \frac{a^{(1)}}{R} + \frac{1}{18} \left(\frac{1}{R} - \frac{\cos^2 \theta}{R} \right) \quad,$$

so that

$$A_0^{(3)} = 0 \;, \quad A_1^{(3)} = -\frac{1}{6} \;, \quad a^{(1)} = \frac{22}{135} \;. \tag{6.74}$$

Thus when four inner terms and two outer terms are matched these terms are all uniquely determined. Note that in the fourth inner term an eigensolution is present which grows linearly as $R \to \infty$. In still higher approximations the further eigensolutions are also needed in the inner series in order to match the positive powers of R which arise from the expansion of the phase terms $\exp(i\epsilon R)$ in the outer series.

There is no particular merit, nor any great difficulty, in carrying the inner and outer expansions further, and so one can determine the near and far fields to any order in ϵ, though the algebraic complexity increases rapidly with the order. The point we want to demonstrate here is that adherence to the formal procedures provides a straightforward and routine way of carrying the approximations through a number of terms, and that the matching rule can be used to indicate how the expansions should proceed and what form of eigensolution is sufficient at each stage.

6.7 TWO–DIMENSIONAL PROBLEMS: LOGARITHMIC GAUGE FUNCTIONS

Logarithmic gauge functions arise in many problems, and the aim of this section is to show how they arise and how they can be handled by MAE in the context of two–dimensional acoustic scattering problems.

We start by looking at the scattering of a plane wave $\phi^i = \exp(ik_0 x')$ by the rigid fixed cylinder $r' = L$, so that the problem for the scattered field ϕ' is

$$(\nabla'^2 + k_0^2)\phi' = 0 \;,$$

$$\frac{\partial \phi'}{\partial r'} = -ik_0 \cos\theta \exp(ik_0 L \cos\theta) \quad \text{on } r' = L \;, \tag{6.75}$$

$$\phi' \sim (r')^{-1/2} \exp(ik_0 r')f(\theta) \quad \text{as } r' \to \infty \;,$$

the last of these being the two–dimensional radiation condition. In terms of outer and inner coordinates, we have

$$(\nabla_r^2 + 1)\phi = 0 \;,$$

$$\frac{\partial \phi}{\partial r} = -i\cos\theta \exp(i\epsilon \cos\theta) \quad \text{on } r = \epsilon \;, \tag{6.76}$$

$$\phi \sim r^{-1/2} \exp(ir)g(\theta) \ \text{ as } r \to \infty \ ,$$

and

$$(\nabla_R^2 + \epsilon^2)\Phi = 0 \ ,$$

$$\frac{\partial \Phi}{\partial R} = -i\epsilon \cos\theta \exp(i\epsilon \cos\theta) \ \text{ on } R = 1 \ , \qquad (6.77)$$

$$\Phi \sim R^{-1/2} \exp(i\epsilon R)h(\theta) \qquad \text{as } \epsilon R \to \infty \ ,$$

the scattered field being, of course, symmetric about the line $\theta = 0$.

An inner eigensolution $\Lambda_\nu(R)$ of integral order ν is defined by

$$\Lambda_\nu(R) = (R^\nu + R^{-\nu})\cos\nu\theta \qquad (6.78)$$

for $\nu = 0, \ 1, 2...$, and satisfies

$$\nabla_R^2 \Lambda_\nu(R) = 0 \ , \quad \frac{\partial \Lambda_\nu}{\partial R} = 0 \ \text{ on } R = 1 \quad .$$

The general outer eigensolution can be represented in various forms. The commonest representation would be in terms of a sum of Hankel functions of the first kind and of all integral orders,

$$\sum_{\nu=0}^{\infty} a_\nu H_\nu^{(1)}(r)\cos\nu\theta \ ,$$

the typical term of which varies with r like $r^{-1/2}\exp(ir)$ for large r and so satisfies the radiation condition. Here we stick to our previous type of representation in terms of multipole derivatives of the fundamental solution $H_0^{(1)}(r)$ of the Helmholtz equation in two dimensions. Thus we write the general outer eigensolution as

$$aH_0^{(1)}(r) + a_i \frac{\partial}{\partial x_i} H_0^{(1)}(r) + a_{ij} \frac{\partial^2}{\partial x_i \partial x_j} H_0^{(1)}(r) + \ ... \quad , \qquad (6.79)$$

(x_1, x_2) being the position coordinates in a plane perpendicular to the axis of the cylinder. As before, the solution symmetric about the x_1-axis, $\theta = 0$, is of the form

$$aH_0^{(1)}(r) + a_1 \frac{\partial}{\partial x} H_0^{(1)}(r) + a_{11} \frac{\partial^2}{\partial x^2} H_0^{(1)}(r) + \ ... \quad . \qquad (6.80)$$

An advantage of using the form (6.80) is that we need only to know the behaviour of the zeroth–order Hankel function as $r \to 0$, which is

$$H_0^{(1)}(r) \sim \left[1 - \frac{1}{4}r^2 + O(r^4) \right]$$

$$+ \frac{2i}{\pi}(\ln r + \gamma_E - \ln 2)\left(1 - \frac{1}{4}r^2 + O(r^4) \right) + \frac{2i}{\pi}\frac{1}{4}r^2 + O(r^4) \qquad (6.81)$$

where $\gamma_E = 0.5772...$ is the Euler constant.

As in the previous sections, the first inner problem determines how the expansions start off. Motivated by the boundary condition in (6.77) we assume

$$\Phi \sim \epsilon \Phi_0 + ... \quad ,$$

and then

$$\nabla_R^2 \Phi_0 = 0 \quad , \quad \frac{\partial \Phi_0}{\partial R} = -i \cos \theta \text{ on } R = 1 \quad ,$$

so that

$$\Phi_0 = \frac{i \cos \theta}{R} + \sum_{\nu=0}^{\infty} A_\nu^{(0)} \Lambda_\nu(R) \quad . \qquad (6.82)$$

Then

$$\Phi^{(1)}\left(\frac{r}{\epsilon}\right) = \epsilon \Phi_0\left(\frac{r}{\epsilon}\right) = \frac{i\epsilon^2 \cos \theta}{r} + \sum_{\nu=0}^{\infty} \epsilon A_\nu^{(1)} \left\{ \left(\frac{r}{\epsilon}\right)^\nu + \left(\frac{\epsilon}{r}\right)^\nu \right\} \cos \nu \theta \quad ,$$

and unless all the $A_\nu^{(0)}$ are zero, the leading term of the expansion of $\Phi^{(1)}(r/\epsilon)$ will contain terms which are either constant or which grow with r. Such terms cannot be matched to the leading–order solution, because that is an eigensolution like (6.80) whose *leading-order* terms all contain either $\ln r$ or inverse powers of r. Therefore we take

$$\Phi^{(1)}\left(\frac{r}{\epsilon}\right) = \frac{i\epsilon^2 \cos \theta}{r} \quad , \qquad (6.83)$$

which shows that the leading order term in the outer expansion must be $O(\epsilon^2)$, so that

$$\phi \sim \epsilon^2 \phi_0 + ... \quad , \qquad (6.84)$$

where ϕ_0 is given by (6.80) with superscript $^{(0)}$ on the coefficients.

Because the leading inner term is $O(\epsilon)$, $\epsilon^2 \phi_0(\epsilon R)$ cannot be larger than $O(\epsilon)$, and so

$$\phi_0 = a^{(0)} H_0^{(1)}(r) + a_1^{(0)} \frac{\partial}{\partial x} H_0^{(1)}(r) \qquad (6.85)$$

is the most general possibility, consisting of an acoustic monopole and an acoustic dipole respectively. Then

$$\epsilon^2 \phi_0(\epsilon R) = a^{(0)} \epsilon^2 \frac{2i}{\pi} \left(\ln R + \ln \epsilon + \gamma_E - \ln 2 - \frac{\pi i}{2} \right)$$

$$+ O(\epsilon^4 \ln \epsilon, \epsilon^4) + \epsilon a_1^{(0)} \frac{2i}{\pi} \frac{\cos \theta}{R} + O(\epsilon^3 \ln \epsilon, \epsilon^3) \quad , \tag{6.86},$$

which gives

$$\phi^{(2,1)} = \epsilon a_1^{(0)} \frac{2i}{\pi} \frac{\cos \theta}{R} \quad . \tag{6.87}$$

Applying the matching rule

$$\phi^{(2,1)} = \Phi^{(1,2)}$$

gives

$$a_1^{(0)} = \frac{\pi}{2} \quad . \tag{6.88}$$

The situation is precisely as in the spherical problem. First–order matching determines the dipole coefficient, but does not determine the monopole part of the first–order outer field. To find that we have to take careful account of small phase variations in the scattering region which are responsible for an acoustically efficient source mechanism. Analytically, we expand (6.86) to next order to see how the inner series should progress and find that the next terms are $O(\epsilon^2 \ln \epsilon)$ and $O(\epsilon^2)$, just as indicated in (6.86). We therefore assume

$$\Phi \sim \epsilon \Phi_0 + \epsilon^2 \ln \epsilon \Phi_1 + \epsilon^2 \Phi_2 \ldots \quad , \tag{6.89}$$

and we shall assume the matching rule

$$\phi^{(2,2)} = \epsilon^2 a^{(0)} \frac{2i}{\pi} \left(\ln \epsilon + \ln R + \gamma_E - \ln 2 - \frac{\pi i}{2} \right) + \frac{i\epsilon}{\pi R} \cos \theta = \Phi^{(2,2)} \quad , \tag{6.90}$$

where we understand that $\epsilon^2 \ln \epsilon \Phi_1$ and $\epsilon^2 \Phi_2$ are to be taken as a single term $O(\epsilon^2)$.

Φ_1 is of course an inner eigensolution, and $\epsilon^2 \Phi_1(r/\epsilon)$ can be no larger than $O(\epsilon^2)$, since that is the order of the outer potential. Thus

$$\Phi_1 = A_0^{(1)} \tag{6.91}$$

is the only possibility. For Φ_2 we have the problem

$$\nabla_R^2 \Phi_2 = 0 \quad , \quad \frac{\partial \Phi_2}{\partial R} = \cos^2 \theta \quad \text{on } R = 1 \quad . \tag{6.92}$$

This problem introduces the two–dimensional source potential $\ln R$. Writing the boundary condition as

$$\frac{\partial \Phi_2}{\partial R} = \tfrac{1}{2}(1 + \cos 2\theta)$$

we look for a solution

$$\Phi_2 = \alpha \ln R + \beta \frac{\cos 2\theta}{R^2} \quad,$$

and we find $\alpha = 1/2$, $\beta = -1/4$, so that

$$\Phi_2 = \tfrac{1}{2} \ln R - \tfrac{1}{4} \tfrac{1}{R^2} \cos 2\theta + A_0^{(2)} \quad, \tag{6.93}$$

the form of the eigensolution being determined by the same argument as for (6.91). Thus we have

$$\Phi^{(2)}(R) = \frac{i\epsilon \cos \theta}{R} + \epsilon^2 \ln \epsilon A_0^{(1)} + \epsilon^2 \left\{ \tfrac{1}{2} \ln R - \frac{1}{4} \frac{\cos 2\theta}{R^2} + A_0^{(2)} \right\} \quad,$$

$$\Phi^{(2)}\left(\frac{r}{\epsilon}\right) = \frac{i\epsilon^2 \cos \theta}{r} + \epsilon^2 \ln \epsilon A_0^{(1)} + \epsilon^2 \left\{ \tfrac{1}{2} \ln r - \tfrac{1}{2} \ln \epsilon \right\} \tag{6.94}$$

$$- \frac{1}{4} \epsilon^4 \frac{\cos 2\theta}{r^2} + \epsilon^2 A_0^{(2)} \quad,$$

which gives

$$\Phi^{(2,2)} = \frac{i\epsilon^2 \cos \theta}{r} + \epsilon^2 \ln \epsilon \left(A_0^{(1)} - \tfrac{1}{2} \right) + \epsilon^2 \tfrac{1}{2} \ln r + \epsilon^2 A_0^{(2)} \quad. \tag{6.95}$$

Matching (6.95) to (6.90) gives

$$A_0^{(1)} = \tfrac{1}{2} \quad, \quad a^{(0)} = \frac{\pi}{4i} \quad, \quad A_0^{(2)} = \tfrac{1}{2}\left(\gamma_E - \ln 2 - \frac{\pi i}{2} \right) \quad, \tag{6.96}$$

which yields the monopole strength $a^{(0)}$ and *non–zero* eigenfunctions in the second and third inner terms Φ_1 and Φ_2.

If we now go back to $\Phi^{(2)}(R/\epsilon)$, as given by (6.94), we see that since $A_0^{(1)} = 1/2$ it contains terms $O(\epsilon^2)$ which are matched to the outer solution, and then continues with an $O(\epsilon^4)$ quadrupole term $\cos 2\theta/r^2$. This suggests that we continue the outer series in the form

$$\phi \sim \epsilon^2 \phi_0 + \epsilon^4 \phi_2 + \ldots \quad. \tag{6.97}$$

The choice of ϕ_2 here, rather than ϕ_1, is deliberate. ϕ_2 is again an outer eigensolution, and as in the sphere problem we can allow up to three derivatives of $H_0^{(1)}(r)$ in ϕ_2, but not more, in order that $\epsilon^4 \phi_2(\epsilon R)$ be no larger, as $\epsilon \to 0$, than the leading inner term, which is $O(\epsilon)$. Thus

$$\phi_2 = a^{(2)} H_0^{(1)}(r) + a_1^{(2)} \frac{\partial}{\partial x} H_0^{(1)}(r) + a_{11}^{(2)} \frac{\partial^2}{\partial x^2} H_0^{(1)}(r) + a_{111}^{(2)} \frac{\partial^3}{\partial x^3} H_0^{(1)}(r) \quad , \quad (6.98)$$

and without taking the inner solution further, the only matching rule we can use is that

$$\Phi^{(2,4)} \text{ (Eqn. 6.94)} = \phi^{(4,2)} \quad .$$

The right–hand side of this equation can be calculated directly, with some effort, from the assumed form (6.98) and the expansion (6.81). One finds then that

$$a_{111}^{(2)} = 0 \quad , \quad -\frac{2i}{\pi} a_{11}^{(2)} = -\frac{1}{4} \quad , \quad (6.99)$$

and the monopole and dipole coefficients remain undetermined.

Expansion of $\phi^{(4)} \equiv (\epsilon^2 \phi_0 + \epsilon^4 \phi_2)(\epsilon R)$ beyond $O(\epsilon^2)$ produces terms $O(\epsilon^3 \ln \epsilon)$ and $O(\epsilon^3)$, so that the inner expansion must take the form

$$\Phi \sim \epsilon \Phi_0 + \epsilon^2 \ln \epsilon \Phi_1 + \epsilon^2 \Phi_2 + \epsilon^3 \ln \epsilon \Phi_3 + \epsilon^3 \Phi_4 \cdots \quad . \quad (6.100)$$

The solutions for Φ_3 and Φ_4 are

$$\Phi_3 = A_0^{(3)} + A_1^{(3)}(R + R^{-1}) \cos\theta \quad ,$$

$$\Phi_4 = A_0^{(4)} + A_1^{(4)}(R + R^{-1}) \cos\theta$$

$$- \frac{i}{2} R \ln R \cos\theta - \frac{i}{24} \frac{\cos 3\theta}{R^3} - \frac{7i}{8} \frac{\cos\theta}{R} \quad , \quad (6.101)$$

and the matching rule to be applied next is

$$\Phi^{(3,4)} = \phi^{(4,3)} \quad . \quad (6.102)$$

Most of the details of this matching are unimportant. One aspect serves, however, as a vital warning. We see from (6.101) that in $\Phi^{(3,4)}$ there will be a term

$$- \frac{i}{2} \epsilon^4 r (\ln r - \ln \epsilon) \cos\theta \quad (6.103)$$

in terms of outer variables, and to this order there are no other inner terms of this form. In the outer expansion (6.97) there are also no terms of the required form; certainly there are terms which contain $\ln \epsilon$ *when expressed in inner variables,* but when these terms are re–expressed in outer variables the $\ln \epsilon$ disappears from them. Therefore the outer expansion as it stands contains no term which can possibly match (6.103) according to the rule (6.102); and (6.103) has a definite non–zero coefficient, because the term in (6.101) which gives rise to it, namely $(-i/2)R \ln R \cos \theta$, is a particular integral for the equation

$$\nabla_R^2 \Phi_4 = -\Phi_1 = -\frac{i \cos \theta}{R} \quad .$$

The fault lies with the assumed form (6.97) for the outer expansion. Although the expansion of $\Phi^{(2)}(r/\epsilon)$ produced only algebraic terms $O(\epsilon^2, \epsilon^4)$, indicating the form (6.97), *the presence of* $\ln \epsilon$ *terms in the inner expansion at one stage must be taken as a warning that they will probably arise at the next outer stage.* Thus in place of (6.97) we should have anticipated

$$\phi \sim \epsilon^2 \phi_0 + \epsilon^4 \ln \epsilon \phi_1 + \epsilon^4 \phi_2 + \dots \quad , \tag{6.104}$$

and this choice enables all the matchings to be effected without difficulty — though the algebraic details are horrendous. A further point to be noted is that because ϕ_1 necessarily contains Hankel functions which have logarithmic singularities, the expansion of $\epsilon^4 \ln \epsilon \phi_1$ will involve a term $\epsilon^4 \ln^2 \epsilon$ — and therefore the inner expansion presumably takes the form

$$\Phi \sim \epsilon \Phi_0 + \epsilon^2 \ln \epsilon \Phi_1 + \epsilon^2 \Phi_2 + \epsilon^3 \ln \epsilon \Phi_3 + \epsilon^3 \Phi_4$$

$$+\epsilon^4 \ln^2 \epsilon \Phi_5 + \epsilon^4 \ln \epsilon \Phi_6 + \epsilon^4 \Phi_7 + \dots \quad . \tag{6.105}$$

It is necessary in fact to determine all the eight functions in (6.105) in order to uniquely determine the three functions in the outer wave–field (6.104).

Once again the low–frequency acoustic problems bring out subtleties in the matching procedure. Although we have not proved it here, it can be shown that failure to regard $\epsilon^3 \ln \epsilon \Phi_3$ and $\epsilon^3 \Phi_4$ in (6.100) as a single "block" leads to an incorrect determination of some of the coefficients in the outer terms ϕ_1 and ϕ_2 (see Lesser & Crighton 1975; Crighton & Leppington 1973). We have also seen how the presence of logarithmic terms may be indicated by systematic use of the matching principle, and that once they have arisen their presence must be suspected in all subsequent terms of both expansions.

6.8 PURELY LOGARITHMIC GAUGE FUNCTIONS:
SCATTERING BY SOFT BODIES

The problem of plane wave scattering by a soft cylinder — on which the total potential is zero — provides an excellent illustration of the way in which purely logarithmic gauge functions may arise, leading to series which are useless for practical purposes because of the slowness of their convergence. We show how naive application of the matching principle fails, but then show how modification of the gauge functions together with an *insistence* on all matchings enables the slowly convergent series to be "renormalized" into a rapidly convergent series.

If the cylinder is circular, of radius L, then the inner and outer problems for the scattered field ϕ corresponding to an incident potential $\exp(ik_0 x')$ are, respectively,

$$(\nabla_R^2 + \epsilon^2)\Phi = 0 \ , \quad \Phi = -\exp(i\epsilon\cos\theta) \ \text{on} \ R = 1 \ , \tag{6.106}$$

$$(\nabla_r^2 + 1)\phi = 0 \ , \quad \phi \sim r^{-1/2}\exp(ir)f(\theta) \ \text{as} \ r \to \infty \ . \tag{6.107}$$

The inner eigensolutions this time satisfy

$$\nabla_R^2\Phi = 0, \quad \Phi = 0 \ \text{on} \ R = 1, \ \text{and are}$$

$$\Lambda_0 = \ln R \tag{6.108}$$

$$\Lambda_\nu = (R^\nu - R^{-\nu})\cos\nu\theta \ \ (\nu = 1, 2, ...) \ ,$$

while the outer eigensolutions are, as before, of the general form

$$aH_0^{(1)}(r) + a_1\frac{\partial}{\partial x}H_0^{(1)}(r) + a_{11}\frac{\partial^2}{\partial x^2}H_0^{(1)}(r) + ... \quad . \tag{6.109}$$

The boundary condition in (6.106) indicates that $\Phi = O(1)$ as $\epsilon \to 0$, so that if

$$\Phi \sim \Phi_0 + ... \ ,$$

then

$$\Phi_0 = -1 + \sum_{\nu=0}^{\infty} A_\nu^{(0)}\Lambda_\nu(R) \ , \tag{6.110}$$

and we cannot decide which eigensolutions may be permitted without looking next at the outer field. Suppose that ϕ starts off with an algebraic gauge function $\phi \sim \epsilon^\alpha\phi_0$ with $\alpha > 0$ (on physical grounds). Then the leading term of $\epsilon^\alpha\phi_0(\epsilon R)$

must be $O(1)$ as $\epsilon \to 0$ in order to match Φ_0. The only possibility is that α is an integer and that the first term permitted in (6.109) for ϕ_0 contains α derivatives. But then that term would contain a factor $R^{-\alpha}$, whereas the leading term in the expansion of $\Phi_0(r/\epsilon)$ contains only r^β or $\ln r$ where $\beta \geq 0$. Matching to leading order is therefore impossible if ϕ starts off with an algebraic gauge function. A little trial and error, based on the fact that a monopole $a^{(0)} H_0^{(1)}(r)$ is certain to be present in ϕ_0, suggests that instead we should try

$$\phi \sim \frac{1}{\ln \epsilon} \phi_0 + \dots \quad . \tag{6.111}$$

Then *only* the monopole can be present in ϕ_0, for the presence of a dipole $a_1(\partial/\partial x)$ $H_0^{(1)}(r)$ would make $(1/\ln \epsilon)\phi_0(\epsilon R)$ infinite like $(\epsilon \ln \epsilon)^{-1}$, rather than $O(1)$ which it needs to be. Thus

$$\phi_0 = a^{(0)} H_0^{(1)}(r) \quad , \tag{6.112}$$

and with an obvious extension of our previous notation,

$$\phi^{\left(\frac{1}{\ln \epsilon}\right)}(\epsilon R) = \frac{1}{\ln \epsilon} a^{(0)} H_0^{(1)}(\epsilon R) \sim \frac{a^{(0)}}{\ln \epsilon} \frac{2i}{\pi} \ln \epsilon \quad \text{to leading order} \quad . \tag{6.113}$$

Hoping that we can extend the matching principle to cover this sort of case we write

$$\phi^{\left(\frac{1}{\ln \epsilon}, 0\right)} = \frac{2i}{\pi} a^{(0)} \quad ,$$

and try matching this to the terms up to $O(1/\ln \epsilon)$ in

$$\Phi^{(0)}\left(\frac{r}{\epsilon}\right) = -1 + A_0^{(0)}(\ln r - \ln \epsilon) + \sum_{\nu=1}^{\infty} A_\nu^{(0)} \left\{ \left(\frac{r}{\epsilon}\right)^\nu - \left(\frac{\epsilon}{r}\right)^\nu \right\} \cos \nu\theta \quad .$$

This will only be possible if all the $A_\nu^{(0)} \equiv 0$, and then

$$\Phi^{\left(0, \frac{1}{\ln \epsilon}\right)} = -1 \quad ,$$

which gives us

$$a^{(0)} = \frac{\pi i}{2} \quad . \tag{6.114}$$

This determines the leading–order solutions as

$$\Phi \sim -1, \quad \phi \sim \frac{1}{\ln \epsilon} \frac{\pi i}{2} H_0^{(1)}(r) \quad . \tag{6.115}$$

To improve upon these approximations we have to expand $(1/\ln\epsilon)\phi_0(\epsilon R)$ up beyond the $O(1)$ term, for expansion of $\Phi_0 = -1$ is not helpful. This gives

$$\phi^{\left(\frac{1}{\ln\epsilon}\right)}(\epsilon R) = -1 - \frac{1}{\ln\epsilon}\left\{\ln R + \gamma_E - \ln 2 - \frac{\pi i}{2}\right\} + O\left(\frac{\epsilon^2}{\ln\epsilon}\right) \quad,$$

and suggests

$$\Phi \sim -1 + \frac{1}{\ln\epsilon}\Phi_1 + \cdots \quad. \tag{6.116}$$

Φ_1 is an inner eigensolution, and some thought shows that the algebraic type of eigensolution cannot yet enter into the solutions. (Note that our previous arguments as to how each expansion determines the form of the other have to some extent broken down with purely logarithmic gauge functions; for example, the outer expansion of $\Phi_0 = -1$ is -1, but this does not mean here that the outer series starts with an $O(1)$ term.) We try

$$\Phi_1 = A_0^{(1)}\ln R \quad, \tag{6.117}$$

and then we have

$$\Phi^{\left(\frac{1}{\ln\epsilon}\right)} = -1 + \frac{1}{\ln\epsilon}A_0^{(1)}\ln R \quad, \quad \Phi^{\left(\frac{1}{\ln\epsilon},\frac{1}{\ln\epsilon}\right)} = -1 - A_0^{(1)} + \frac{A_0^{(1)}}{\ln\epsilon}\ln r \quad. \tag{6.118}$$

On the other hand,

$$\phi^{\left(\frac{1}{\ln\epsilon},\frac{1}{\ln\epsilon}\right)} = -1 - \frac{1}{\ln\epsilon}\left(\ln R + \gamma_E - \ln 2 - \frac{\pi i}{2}\right) = -\frac{\ln r}{\ln\epsilon} - \frac{1}{\ln\epsilon}\left(\gamma_E - \ln 2 - \frac{\pi i}{2}\right) \quad.$$

$$\tag{6.119}$$

Clearly, (6.118) and (6.119) do not match, though the choice $A_0^{(1)} = -1$ does at least match the term in $(\ln r)/\ln\epsilon$. We can either be satisfied with the matching that does work, and which determines $A_0^{(1)}$ uniquely, and merely ignore the failure of other terms to match, or we can think about trying to modify the expansions a little in order that matching of all terms can take place satisfactorily.

This difficulty often arises in the case of purely logarithmic gauge functions. Fraenkel (1969) has discussed it at some length, showing how the matching failure can be traced to at best "marginal overlap" of the inner and outer expansions when inverse powers of $\ln\epsilon$ are used as gauge functions. He also shows how, in principle, the terms which cannot be expected to match can be traced at each stage. It should, therefore, be possible to persist with our expansions in inverse powers of

$\ln \epsilon$, and to correctly determine all terms, regardless of the matching failure, by careful attention to the points made by Fraenkel. But the result would be useless for practical purposes, for the coefficients would all be of order one and an ϵ of 10^{-10} would still only make consecutive terms smaller by a factor of 10 or so.

We recommend the unrelenting insistence on the matching rule, combined with a slight change to the gauge functions, as a systematic and effective way of redeeming the situation on all counts. As a first step, we try taking $(\ln \epsilon + K)^{-1}$ as the basic gauge function, in place of $(\ln \epsilon)^{-1}$, K being a constant. Assume

$$\phi \sim \frac{1}{(\ln \epsilon + K)}\bar{\phi}_0 + ... \quad , \quad \Phi \sim -1 + \frac{1}{(\ln \epsilon + K)}\bar{\Phi}_1 + ... \quad , \tag{6.120}$$

and as before, appropriate choices for the eigenfunctions are

$$\bar{\phi}_0 = \bar{a}^{(0)} H_0^{(1)}(r) \quad , \quad \bar{\Phi}_1 = \bar{A}_0^{(1)} \ln R \quad . \tag{6.121}$$

Now,

$$\bar{\phi}^{\left(\frac{1}{\ln \epsilon + K}\right)}(\epsilon R) = \frac{\bar{a}^{(0)}}{(\ln \epsilon + K)}\left\{1 + \frac{2i}{\pi}(\ln R + \ln \epsilon + K + \gamma_E - \ln 2 - K) + ...\right\} \quad ,$$

so that

$$\bar{\phi}\left(\frac{1}{\ln \epsilon + K}, \frac{1}{\ln \epsilon + K}\right) = \frac{\bar{a}^{(0)}}{(\ln \epsilon + K)}\left[1 + \frac{2i}{\pi}(\ln r + \gamma_E - \ln 2)\right] \quad , \tag{6.122}$$

while

$$\bar{\Phi}\left(\frac{1}{\ln \epsilon + K}, \frac{1}{\ln \epsilon + K}\right) = -1 - \bar{A}_0^{(1)} + \frac{(K + \ln r)\bar{A}_0^{(1)}}{(\ln \epsilon + K)} \quad . \tag{6.123}$$

Equations (6.122) and (6.123) can now be matched, and give

$$\bar{a}^{(0)} = \frac{\pi i}{2} \quad , \quad \bar{A}_0^{(1)} = -1 \quad , \tag{6.124}$$

as before, while the remaining terms can be matched if we choose

$$K = \gamma_E - \ln 2 - \frac{\pi i}{2} \quad . \tag{6.125}$$

Now we have the solutions

$$\phi \sim \frac{1}{(\ln \epsilon + K)} \frac{\pi i}{2} H_0^{(1)}(r) \ , \quad \Phi \sim -1 - \frac{\ln R}{(\ln \epsilon + K)} \ , \tag{6.126}$$

and to take the expansions further we expand $\phi(\epsilon R)$ beyond $(\ln \epsilon + K)^{-1}$. We find

$$\phi(\epsilon R) \sim -1 - \frac{\ln R}{(\ln \epsilon + K)} + O\left(\frac{\epsilon^2}{\ln \epsilon + K}, \epsilon^2\right) \ ,$$

which indicates an inner expansion

$$\Phi \sim -1 - \frac{\ln R}{(\ln \epsilon + K)} + \frac{\epsilon^2}{(\ln \epsilon + K)} \Phi_2 + \epsilon^2 \Phi_3 + \dots \quad . \tag{6.127}$$

This shows the value of enforcing the matching rules in terms of slightly strained gauge functions. Instead of having the slow series in inverse powers of $\ln \epsilon$, we have effectively now summed that series, or at any rate a subset of that series, into the term $(\ln \epsilon + K)^{-1}$, and the next term, $O(\epsilon^2)$, is now very much smaller than $1/(\ln \epsilon)^2$, as it would otherwise have been. Note, of course, that now we are out of the infinite sequence of purely logarithmic functions we must expect to have to treat

$$\frac{\epsilon^2}{(\ln \epsilon + K)} \Phi_2 \quad \text{and} \quad \epsilon^2 \Phi_3 \ ,$$

together as effectively a single $O(\epsilon^2)$ term in the matching.

Failure of matching at higher orders in this and similar problems indicates that a further straining of the gauge functions is necessary; the overlap has become marginal, or perhaps has disappeared altogether, at higher order in terms of powers of ϵ and inverse powers of $(\ln \epsilon + K)$. The remedy may be to try

$$\ln \epsilon + K \rightarrow \ln \epsilon + K + K_1 \epsilon \ ,$$

or more generally $\ln \epsilon + K \rightarrow \ln \epsilon + K + K_1(\epsilon)$ where $K_1(\epsilon) \rightarrow 0$ as $\epsilon \rightarrow 0$ and is to be determined by the enforcement of all matchings.

6.9 COMPOSITE EXPANSIONS

Occasionally it is useful to be able to combine the inner and outer expansions into a single smooth approximation, valid everywhere from the scattering body to infinity in the scattering type of problem, for example. (More often, however, one needs just to know the directivity pattern of the field at infinity, or the pressure on the scatterer, and these can be obtained directly from the outer expansion alone, or the inner expansion alone, respectively.) There are two common ways of combining the inner and outer series — the methods of *additive* and *multiplicative composition*. In the additive method it is customary to add $\phi^{(m)}(r)$ say to $\Phi^{(n)}(R)$ and take off the part $\phi^{(m,n)}$ (or $\Phi^{(n,m)}$, they are identical) which expansions $\phi^{(m)}$ and $\Phi^{(n)}$ have in common so that it is not counted twice. This gives a composite

$$\phi_{mn}^{AC} \equiv \phi^{(m)} + \Phi^{(n)} - \phi^{(m,n)} \quad , \qquad (6.128)$$

which is often claimed to be as good as the inner or outer series in their respective domains, and so, if the expansions overlap, to provide a single smooth function equivalent in accuracy to the two separate expansions. In the method of multiplicative composition one defines

$$\phi_{mn}^{MC} \equiv \frac{\phi^{(m)} \cdot \Phi^{(n)}}{\phi^{(m,n)}} \quad , \qquad (6.129)$$

which again is claimed to provide a different, though asymptotically equivalent, single smooth function, good through $O(\epsilon^m)$ in the outer region and through $O(\epsilon^n)$ in the inner.

As Van Dyke (1975) points out in the Notes to the revised edition of his book, these two composites are particular instances of a very general class of composites. Let F be a general function, F^{-1} its inverse. Then a general composite formed from the outer expansion $\phi^{(m)}$ through $O(\epsilon^m)$ and the inner expansion $\Phi^{(n)}$ through $O(\epsilon^n)$ is defined by

$$\phi_{mn}^{F} = F^{-1}\left\{ F(\phi^{(m)}) + F(\Phi^{(n)}) - F(\phi^{(m,n)}) \right\} \quad . \qquad (6.130)$$

The function $F(\phi) \equiv \phi$ gives the additive composition, the function $F(\phi) \equiv \ln\phi$ gives the multiplicative rule. Other functions F give different composites, as when $F(\phi) = \exp\phi$ for example, which gives

$$\phi_{mn} = \ln\left\{ \exp\phi^{(m)} + \exp\Phi^{(n)} - \exp\phi^{(m,n)} \right\} \quad . \qquad (6.131)$$

Now in general none of these composites have the accuracy often claimed for them. This follows from the fact that in general neither $\phi^{(m,n)}$ nor $\Phi^{(n,m)}$ has any asymptotic significance with respect to ϕ or Φ, as stated already in connection with the Asymptotic Matching Principle. The only thing that can be said about the additive or multiplicative composites is that if, for example, $\phi_{mn}^{AC}(r, R = r/\epsilon, \epsilon)$ is expanded for fixed r and $\epsilon \to 0$, and if that expansion is truncated beyond $O(\epsilon^m)$ the result will be precisely $\phi^{(m)}$, the outer expansion through $O(\epsilon^m)$. Likewise if $\phi_{mn}^{AC}(r = \epsilon R, R, \epsilon)$ is expanded through $O(\epsilon^n)$ for fixed R, the result will be the inner expansion $\Phi^{(n)}$ through $O(\epsilon^n)$ — and the same is true of the multiplicative composite.

Therefore a composite is at best only as good as $\phi^{(m)}$ or $\Phi^{(n)}$ when it has been expanded and truncated, and otherwise may not have the accuracy of $\phi^{(m)}$ in the outer region or of $\Phi^{(n)}$ in the inner region, while its accuracy or otherwise in between — say for $r = O(\epsilon^{1/2})$ — is in general quite unknown. This last point is emphatically made by Schneider (1973) who gives an example in which the multiplicative composite is hopeless in the intermediate range of r because of a zero in the denominator, $\phi^{(m,n)}$. Van Dyke (1975) gives examples in which different functions F can be used to produce widely differing accuracies of the composites formed according to (6.0).

It seems then that there is no known rule which invariably produces a composite as good as the inner or outer expansions taken separately, let alone in between, and therefore that composites are to be avoided wherever possible. In simple cases, and often to leading order only, it may be possible to form a composite by careful inspection of the behaviour of the inner and outer expansions in their own domains and in the overlap domain. If that can be done, the additive rule should be used, as the most fundamental, though the multiplicative rule usually gives much neater results.

Just to show how these composites are formed, according to the usual, though generally unsafe prescription, we go back to the field scattered by a compact rigid sphere, for which

$$\phi \sim \epsilon^3 \left\{ -\frac{1}{3}\frac{e^{ir}}{r} - \frac{i}{2}\frac{\partial}{\partial x}\frac{e^{ir}}{r} \right\} = \phi^{(3)} \quad , \quad \Phi \sim \epsilon \frac{i}{2R^2}\cos\theta = \Phi^{(1)} \quad , \qquad (6.132)$$

to leading order. We have

$$\phi^{(3,1)} = \Phi^{(1,3)} = \epsilon \frac{i}{2R^2}\cos\theta = \Phi^{(1)} \quad ,$$

so that additive and multiplicative composition both produce the same composite which is actually $\phi^{(3)}$ itself. In other words

$$\epsilon^3 \left\{ -\frac{1}{3}\frac{e^{ir}}{r} - \frac{i}{2}\frac{\partial}{\partial x}\frac{e^{ir}}{r} \right\}$$

is a uniformly valid description of the whole field, to leading order. However, the accuracy of this leading–order approximation is *not* uniform, as is only to be expected in a singular perturbation problem. In the outer field the next term is $O(\epsilon^5)$ while $\phi^{(3)}$ is $O(\epsilon^3)$, whereas in the inner field $\phi^{(3)}$ is $O(\epsilon)$ and the next term is $O(\epsilon^2)$.

To improve accuracy in the near field we can try using

$$\phi \sim \epsilon^3 \left\{ -\frac{1}{3}\frac{e^{ir}}{r} - \frac{i}{2}\frac{\partial}{\partial x}\frac{e^{ir}}{r} \right\} = \phi^{(3)}$$

with

$$\Phi \sim \epsilon \frac{i}{2R^2}\cos\theta - \epsilon^2 \left\{ \frac{2}{9}\frac{P_2(\cos\theta)}{R^3} + \frac{1}{3R} \right\} = \Phi^{(2)}$$

and

$$\phi^{(3,2)} = \Phi^{(2,3)} = -\epsilon^2 \frac{1}{3R} + \epsilon \frac{i}{2R^2}\cos\theta \quad .$$

Then the additive $(3,2)$ composite is

$$\epsilon^3 \left\{ -\frac{1}{3}\frac{e^{ir}}{r} - \frac{i}{2}\frac{\partial}{\partial x}\frac{e^{ir}}{r} \right\} - \epsilon^2 \left\{ \frac{2}{9}\frac{P_2(\cos\theta)}{R^3} \right\} ,$$

and for once this is neater than the multiplicative composite. No significance is to be attached to the last term here in the wave zone $r = O(1)$, for there it is $O(\epsilon^5)$ and the outer expansion itself is only good to $O(\epsilon^3)$ so far. The inclusion of the last term when $R = O(1)$ may make a distinct difference to the value of ϕ for moderately small values of ϵ, and may apparently improve upon the asymptotic representation $\phi^{(3)}$. In general, however, such an improvement is coincidental, and the ϕ_{32} composite has no asymptotic correctness until it is expanded for fixed r and terms smaller than $O(\epsilon^3)$ thrown away, which of course takes us back to the outer expansion $\phi^{(3)}$ alone.

6.10 CONCLUSIONS

This chapter has attempted to show MAE at work on some simple, but hopefully representative, problems of classical linear acoustics. Despite their apparent physical and mathematical simplicity, these low–frequency sound generation and scattering problems illustrate very effectively a number of subtleties which have often gone unnoticed in more complicated, but perhaps less delicate, problems. They show how careful application of the method leads in a reasonably straightforward way to solutions in the near and far fields whose accuracy is limited only by the length of the algebraic manipulations that have to be carried out at high order.

Modern acoustics, and especially underwater acoustics, deals with many problems involving one or more small parameters. For example, the fluid is generally only weakly compressible, so that often the Helmholtz numbers $k_0 L$ may be *small*; a vibrating surface coupled to an acoustic fluid may be subject to *heavy* or *light* fluid loading; the acoustic field may be in bulk motion at a *low* Mach number; the phase speed of an elastic surface wave may be only *slightly less* than the sonic speed, or the frequency may be *very much less* than the coincidence frequency, etc. All such problems can be regarded as governed by two distinct lengthscales, typically an acoustic lengthscale and the other scale determined by hydrodynamic or elastic effects regardless of compressibility. As such, we must expect them to be singular perturbation problems, and must look to MAE as a promising method of attack. A variety of problems in both linear and nonlinear acoustics is discussed from this viewpoint in Lesser & Crighton (1975), while more recent work deals with MAE in problems involving coupled wave–bearing media.

REFERENCES

Cole, J.D. (1968). Perturbation Methods in Applied Mathematics. Ginn (Blaisdell), Waltham, Massachusetts.

Crighton, D.G. & Leppington, F.G. (1973). Proc. R. Soc. Lond. A 335:313.

Eckhaus, W. (1973). Matched Asymptotic Expansions and Singular Perturbations. North Holland, Amsterdam.

Fraenkel, L.E. (1969). Proc. Camb. Phil. Soc. 65:209 (3 papers).

Lamb, H. (1932). Hydrodynamics. Dover, New York

Landau, L.D. & Lifschitz, E.M. (1959). Fluid Mechanics (transl. J.B. Sykes & W.H. Reid). Pergamon Press, London.

Lesser, M.B. & Crighton, D.G. (1975). Physical Acoustics and the Method of Matched Asymptotic Expansions. In: Physical Acoustics Vol.11 (eds. W.P. Mason & R.N. Thurston), Academic Press, New York.

Rayleigh, Lord (1015). Theory of Sound (2 volumes), Dover, New York.

Schneider, W. (1973). J. Fluid Mech. 50:785.

Van Dyke, M.D. (1975). Perturbation Methods in Fluid Mechanics. Revised annotated edition published by Parabolic Press, Stanford, California.

7. MULTIPLE SCALES

Sometimes there are two different scales in a physical problem as, for example, when high–frequency sound travels through a region in which the sound speed varies slowly. Then the short wavelength and the long length characterizing the sound–speed variation are two distinct disparate scales. The method of multiple scales exploits this difference. The method works because the short and very long lengthscales (or a high–frequency disturbance and slow modulation) are only weakly coupled. The slowly varying quantity is approximately constant over one cycle of variation of the fast parameter, while the longer lengths, over which the slow parameter variation is appreciable, involve many cycles of the fast parameter and its net effect is almost cancelled. We will illustrate the method by considering a damped harmonic oscillator. A clear description of the method of multiple scales is given by Nayfeh (1973).

7.1 THE DAMPED HARMONIC OSCILLATOR

The damped harmonic oscillator is convenient as an introductory example because it can be solved exactly giving an insight into the method and a useful check on the accuracy of the approximate solution. Consider

$$\ddot{u} + 2\epsilon\dot{u} + u = 0 \ , \tag{7.1}$$

for $t \geq 0$, where the dot denotes differentiation with respect to t. When Equation (7.1) is solved with the initial conditions

$$\left.\begin{array}{c} u(0) = 0 \\ \dot{u}(0) = 1 \end{array}\right\} \ , \tag{7.2}$$

the exact solution is

$$u(t) = e^{-\epsilon t}\frac{\sin(\sqrt{1 - \epsilon^2}t)}{\sqrt{1 - \epsilon^2}} \ . \tag{7.3}$$

We will now go on to determine an asymptotic solution of Equation (7.1) for small values of ϵ which could be useful for more complicated equations whose exact solution is not so easily found. A simple–minded approach would be to rewrite $u(t)$ as an asymptotic series in powers of ϵ (called a *Poincaré expansion*),

$$u(t) = u_0(t) + \epsilon u_1(t) + \epsilon^2 u_2(t) + \dots \ . \tag{7.4}$$

By such a series we mean that

$$u(t) \sim \sum_{n=0}^{p} \epsilon^n u_n(t) \ ,$$

where the symbol $\sim$ denotes that

$$u(t) - \sum_{n=0}^{p} \epsilon^n u_n(t) = o(\epsilon^p) \ ,$$

i.e. the sum of a given number of terms in the series p, say, tends to $u(t)$ as ϵ tends to zero. This is quite a different condition from, and does not imply that, the infinite sum converges for fixed ϵ and t.

When the form for $u(t)$ in Equation (7.4) is substituted into (7.1) and all powers of ϵ equated to zero, it leads to a sequence of equations:

$$\ddot{u}_0 + u_0 = 0 \tag{7.5a}$$

$$\ddot{u}_1 + u_1 = -2\dot{u}_0 \tag{7.5b}$$

$$\ddot{u}_2 + u_2 = -2\dot{u}_1 \ , \tag{7.5c}$$

and so on. The initial conditions (7.2) show that $u_n(0) = 0$ for $n \geq 0$, $\dot{u}_0(0) = 1$ and $\dot{u}_n(0) = 0$ for $n > 0$. The equations for the $u_n(t)$ can be solved subject to the initial conditions in a straightforward way, and give

$$u_0 = \sin t \tag{7.6a}$$

$$u_1 = -t \sin t \tag{7.6b}$$

$$u_2 = \tfrac{1}{2}(1 + t^2) \sin t - \tfrac{1}{2} t \cos t \ . \tag{7.6c}$$

Hence the asymptotic expansion obtained is

$$u(t) = \sin t - \epsilon t \sin t + \tfrac{1}{2} \epsilon^2 ((1 + t^2) \sin t - t \cos t) + 0(\epsilon^3) \ . \tag{7.7}$$

An inspection of this form shows it only to be a good approximation to $u(t)$ when ϵt is small. For t as large as ϵ^{-1} this expansion breaks down, the second and third terms in (7.7) being just as large as the "leading" term. Investigation of the exact solution (7.3) shows why this straightforward expansion in powers of ϵ is not useful for large times. The solution contains a factor $\exp(-\epsilon t)$, which can be expanded as an infinite sum,

$$\exp(-\epsilon t) = 1 - \epsilon t + \tfrac{1}{2}\epsilon^2 t^2 + \cdots .$$

Truncation of this series after any finite number of terms will be a poor approximation to the exponential for large enough values of t. A better approach would be not to expand the exponential but to treat the product ϵt as one parameter, a slow time. That is the idea behind the method of multiple scales.

We begin by deciding somehow that a slow time

$$s = \epsilon t \tag{7.8}$$

and a fast time

$$\theta = t(1 + \epsilon^2 \alpha_2 + \epsilon^2 \alpha_3 ...) \tag{7.9}$$

are appropriate. The constants $\alpha_2, \alpha_3, \ldots$ are to be found in the course of the calculation. There is no need for an ϵt term in θ because ϵt is the slow timescale s. We seek an expansion for $u(t)$ of the form

$$u(t, \epsilon) \sim \sum_{n=0}^{p} \epsilon^n w_n(\theta, s) \quad , \tag{7.10}$$

where w_n and α_n are yet to be determined.

Initially we had a single independent variable t and $u(t)$ satisfied the ordinary differential equation (7.1). We have now replaced t by the two variables θ and s, and the functions w_n will satisfy partial differential equations. The extra freedom this introduces can be exploited to increase the range of validity of the asymptotic expansion. The Poincaré expansion in Equation (7.7) broke down for large t because higher–order terms increased more rapidly with t than the lower–order terms. We can ensure that this does not happen in our multiple scale expansion by insisting that "$w_n(\theta, s)$ should be no worse than $w_{n-1}(\theta, s)$" for $n = 1, ..., p$, i.e.

$$\frac{w_n}{w_{n-1}} < \infty \quad \text{for all} \quad t \quad . \tag{7.11}$$

This rule is sufficient to determine w_n and α_n.

It follows the definitions of the two timescales in Equations (7.8) and (7.9) that

$$\frac{d}{dt} w_n(\theta, s) = (1 + \epsilon^2 \alpha_2 + ...) \frac{\partial w_n}{\partial \theta} + \epsilon \frac{\partial w_n}{\partial s}$$

and

$$\frac{d^2}{dt^2} w_n(\theta_s) = (1 + \epsilon^2 2\alpha_2 ..) \frac{\partial^2 w_n}{\partial \theta^2} + 2\epsilon(1 + \epsilon^2 \alpha_2 + ..) \frac{\partial^2 w_n}{\partial \theta \partial s} + \epsilon^2 \frac{\partial^2 w_n}{\partial s^2} .$$

Substituting the series (7.10) into Equation (7.1) gives

$$\frac{\partial^2 w_0}{\partial \theta^2} + w_0 = 0 , \tag{7.12a}$$

$$\frac{\partial^2 w_1}{\partial \theta^2} + w_1 = -2 \frac{\partial^2 w_0}{\partial \theta \partial s} - 2 \frac{\partial w_0}{\partial \theta}. \tag{7.12b}$$

$$\frac{\partial^2 w_2}{\partial \theta^2} + w_2 = -2 \frac{\partial^2 w_1}{\partial \theta \partial s} - 2 \frac{\partial w_1}{\partial \theta} - 2\alpha_2 \frac{\partial^2 w_0}{\partial \theta^2} - \frac{\partial^2 w_0}{\partial s^2} - 2 \frac{\partial w_0}{\partial s} . \tag{7.12c}$$

The solution of Equation (7.12a) is

$$w_0 = A_0(s) \cos \theta + B_0(s) \sin \theta , \tag{7.13}$$

the initial conditions giving

$$\left.\begin{array}{l} A_0(0) = 0 \\ B_0(0) = 1 \end{array}\right\} \tag{7.14}$$

Evidently, the complete expression for w_0 cannot be determined from Equation (7.12a) alone. With the form for w_0 in (7.13), Equation (7.12b) becomes

$$\frac{\partial^2 w_1}{\partial \theta^2} + w_1 = -2(\dot{B}_0(s) + B_0(s)) \cos \theta + 2(\dot{A}_0(s) + A_0(s)) \sin \theta . \tag{7.15}$$

The forcing terms, $\cos \theta$ and $\sin \theta$, on the right–hand side of (7.15) are resonant and would produce secular terms in w_1 of the form $\theta \cos \theta$ and $\theta \sin \theta$. Such terms have worse behaviour as $t \to \infty$ than the $\cos \theta$ and $\sin \theta$ terms in w_0. They therefore violate the rule (7.11) and must be annihilated. This requires that

$$\left.\begin{array}{l} \dot{A}_0(s) + A_0(s) = 0 \\ \dot{B}_0(s) + B_0(s) = 0 \end{array}\right\} \tag{7.16}$$

It then follows from the initial conditions (7.14) that

$$A_0(s) = 0, \quad B_0(s) = e^{-s}$$

and so

$$w_0(\theta, s) = e^{-s} \sin \theta . \tag{7.17}$$

Continuing on to the next order, the w_1 Equation (7.15) can then be solved to give

$$w_1(\theta, s) = A_1(s)\cos\theta + B_1(s)\sin\theta \ . \tag{7.18}$$

It follows immediately from the initial conditions that

$$\left.\begin{matrix} A_1(0) = 0 \\ B_1(0) = 0 \ . \end{matrix}\right\} \tag{7.19}$$

The full form for $A_1(s)$ and $B_1(s)$ can be found by considering Equation (7.12c) and ensuring that no secular terms appear in w_2. Equation (7.12c) becomes

$$\frac{\partial^2 w_2}{\partial\theta^2} + w_2 = -2(\dot{B}_1(s) + B_1(s))\cos\theta + 2(\dot{A}_1(s) + A_1(s) + \alpha_2 e^{-s} + \tfrac{1}{2}e^{-s})\sin\theta \ . \tag{7.20}$$

Annihilating the forcing terms gives

$$\dot{B}_1(s) + B(s) = 0 \tag{7.21}$$

and

$$\dot{A}_1(s) + A_1(s) = -(\alpha_2 + \tfrac{1}{2})e^{-s} \ . \tag{7.22}$$

We see immediately from Equation (7.21) and the initial conditions (7.19) that

$$B_1(s) \equiv 0 \ . \tag{7.23}$$

Equation (7.22) leads to secular terms of the form se^{-s} in $A_1(s)$ (violating the rule (7.11)), unless

$$\alpha_2 = -\tfrac{1}{2} \ . \tag{7.24}$$

We therefore insist on this value of α_2 and then the initial condition and equation give

$$A_1(s) \equiv 0 \ . \tag{7.25}$$

Collecting these terms together, we have the result

$$w_1(\theta, s) = 0 \ , \tag{7.26}$$

and so our asymptotic solution is

$$u(t) \sim e^{-\epsilon t}\sin((1 - \tfrac{1}{2}\epsilon^2)t) \ . \tag{7.27}$$

If we compare this with the exact solution

$$u(t) = e^{-\epsilon t} \frac{\sin(\sqrt{1-\epsilon^2} t)}{\sqrt{1-\epsilon^2}} \quad , \tag{7.28}$$

we see that it is a good approximation when $\epsilon^4 t$ is small. For larger t higher–order terms in the fast–time expansion (7.9) become important. A comparison of (7.27) and (7.7) shows that the two–term multiple scales expansion extends over a considerably larger time range than the two–term Poincaré expansion which requires ϵt small.

In this example the fast and slow–time variables had a particularly simple form. The slow time was $s = \epsilon t$ and the fast time was slightly stretched from t to $\theta = t(1 + \epsilon^2 \alpha_2 + ...)$. Sometimes it is more convenient to stretch the slow time variable and not the fast, i.e. to take

$$\theta = t, \quad s = \epsilon t(1 + \epsilon \alpha_1 + ...) \ .$$

The method can be generalized to more complicated time variables in which θ (or s) is a nonlinear function of time, perhaps with

$$\theta = f(t), \quad s = \epsilon t(1 + \epsilon \alpha_1 + ...) \ ,$$

for some function $f(t)$, or even with a nonlinear stretching of θ;

$$\theta = f_1(t) + \epsilon f_2(t) + \epsilon^2 f_3(t) + ..., \quad s = \epsilon t \ .$$

Sometimes it is more convenient to write the expansion in terms of an asymptotic sequence of $\epsilon, \delta_n(\epsilon)$, say, rather than in powers of ϵ. The next example is one in which more thought must be given to the appropriate choice of variables.

7.2 THE EFFECT OF A GRADUAL SOUND–SPEED VARIATION ON PLANE WAVES

A one–dimensional acoustic pressure perturbation, $p(x,t)$, propagating in a region with varying sound speed satisfies the wave equation

$$\frac{\partial^2 p}{\partial x^2} - \frac{1}{c^2} \frac{\partial^2 p}{\partial t^2} = 0 \ , \tag{7.29}$$

where $c(x)$ is a function of position and varies over a lengthscale ℓ. We will consider waves of frequency ω travelling in the positive x–direction with the pressure prescribed on $x = 0$. Then the acoustic pressure can be written as $p(x,t) = \widetilde{p}(x)e^{-i\omega t}$

(the real part of the right–hand side is implied), and $\widetilde{p}$ satisfies

$$\frac{d^2\widetilde{p}}{dx^2} + \frac{\omega^2}{c^2}\widetilde{p} = 0 \ , \tag{7.30}$$

with $\widetilde{p}$ given on $x = 0$,

$$\widetilde{p}(0) = I, \quad \text{say.} \tag{7.31}$$

There are two distinct lengthscales in the problem for $\widetilde{p}(x)$, the wavelength $2\pi c/\omega$ and the lengthscale ℓ, over which the sound speed c varies. For high–frequency waves, the wavelength is small in comparison with ℓ. We will bring out this difference by introducing a nondimensional lengthscale $X = x\omega/c_0$, where c_0 is the sound speed at $x = 0$. The sound–speed variation $c(x)$ can be expressed as

$$\frac{c_0}{c(x)} = f\left(\frac{x}{\ell}\right) \ ,$$

where $f(0) = 1$ and the function f and its derivatives are of order unity. In terms of the nondimensional parameter X,

$$\frac{c_0}{c(x)} = f\left(\frac{c_0 X}{\omega\ell}\right) = f(\epsilon X) \ , \tag{7.32}$$

where $\epsilon = c_0/(\omega\ell)$ is a small parameter. The wave equation (7.30) then becomes

$$\frac{d^2 P}{dX^2} + f^2(\epsilon X)P = 0 \ , \tag{7.33}$$

where $P(X) = \widetilde{p}(x)$. The boundary condition (7.31) shows that

$$P(0) = I \ . \tag{7.34}$$

We wish to solve Equation (7.33) asymptotically, subject to the boundary condition (7.34), for small ϵ.

A straightforward expansion in powers of ϵ, with

$$P(X) \sim \sum_{n=0}^{p} \epsilon^n P_n(X) \ ,$$

gives

$$P_0(X) = Ie^{iX} \tag{7.35a}$$

$$P_1(X) = \tfrac{1}{2}\dot{f}(0)(iX^2 - X)Ie^{iX} \tag{7.35b}$$

and so on. Because of the appearance of the secular terms, such an expansion is only valid for small ϵX^2.

The Poincaré expansion rapidly breaks down because the phase speed of the disturbance is wrongly predicted to be unity rather than $(f(\epsilon X))^{-1}$. Over reasonable distances this leads to an appreciable error in the predicted form for the pressure perturbation. As an attempt to do better by using multiple scales, we introduce a short lengthscale which will account for the rapid variations in phase. We write

$$\theta = \int_0^X f(\epsilon \chi)d\chi \ , \tag{7.36}$$

$$s = \epsilon X \ , \tag{7.37}$$

and

$$P(X) \sim \sum_{n=0}^p \epsilon^n w_n(\theta, s) \ . \tag{7.38}$$

Then

$$\frac{dw_n}{dX} = f(\epsilon X)\frac{\partial w_n}{\partial \theta} + \epsilon \frac{\partial w_n}{\partial s}$$

and

$$\frac{d^2 w_n}{dX^2} = f^2(\epsilon X)\frac{\partial^2 w_n}{\partial \theta^2} + \epsilon \dot{f}(\epsilon X)\frac{\partial w_n}{\partial \theta} + 2\epsilon f(\epsilon X)\frac{\partial^2 w_n}{\partial \theta \partial s} + \epsilon^2 \frac{\partial^2 w_n}{\partial s^2}$$

$$= f^2(s)\frac{\partial^2 w_n}{\partial \theta^2} + \epsilon \dot{f}(s)\frac{\partial w_n}{\partial \theta} + 2\epsilon f(s)\frac{\partial^2 w_n}{\partial \theta \partial s} + \epsilon^2 \frac{\partial^2 w_n}{\partial s^2} \ .$$

Substituting for P in Equation (7.33) gives

$$\frac{\partial^2 w_0}{\partial \theta^2} + w_0 = 0 \tag{7.39a}$$

$$f^2(s)\left(\frac{\partial^2 w_1}{\partial \theta^2} + w_1\right) = -\dot{f}(s)\frac{\partial w_0}{\partial \theta} - 2f(s)\frac{\partial^2 w_0}{\partial \theta \partial s} \ , \tag{7.39b}$$

etc.

Equation (7.39a), the radiation condition and (7.34), the boundary condition at $x = 0$, immediately give

$$w_0(\theta, s) = A_0(s)e^{i\theta} \ , \tag{7.40}$$

with $A_0(0) = I$ from the boundary condition. The full form A_0 is found by investigating the w_1 equation. When (7.40) is used to replace w_0 in Equation (7.39b), it leads to

$$f^2(s)\left(\frac{\partial^2 w_1}{\partial \theta^2} + w_1\right) = -i(\dot{f}(s)A_0(s) + 2f(s)\dot{A}_0(s))e^{i\theta} \ . \tag{7.41}$$

The terms on the right–hand side of Equation (7.41) are again resonant and would produce a term of the form $\theta e^{i\theta}$ in w_1. If w_1 is to obey the rule and be no worse that w_0, the forcing terms in Equation (7.41) must be identically equal to zero, i.e.

$$\frac{\dot{A_0}(s)}{A_0(s)} = -\frac{1}{2}\frac{\dot{f}(s)}{f(s)} \quad , \tag{7.42}$$

which, together with the boundary condition, give

$$A_0(s) = I\,[f(s)]^{-1/2}\,. \tag{7.43}$$

The expansion therefore has the form

$$P(X) = I\left(\frac{1}{[f(s)]^{1/2}} + \epsilon A_1(s) + ...\right)e^{i\theta} \quad , \tag{7.44}$$

where $s = \epsilon X$ and the higher–order terms A_n for $n \geq 1$ can be found from repeated applications of the rule. Returning to dimensional parameters this gives

$$p(x,t) = I\left(\left[\frac{c(x)}{c_0}\right]^{1/2} + \frac{c_0}{\omega\ell}B_1\left(\frac{x}{\ell}\right) + ...\right)\exp i\omega\left[\int_0^x \frac{dx}{c(\chi)} - t\right] \,. \tag{7.45}$$

The form of the leading–order term in the pressure perturbation could in fact have been deduced from simple physical arguments. Just as we anticipated, the phase of the disturbance is consistent with a mode travelling with phase speed $c(x)$. Over a cycle in the rapidly varying phase the sound speed is approximately constant and the root–mean–square pressure amplitude is $I(c(x)/2c_0)^{1/2}$. Hence $\overline{p^2}/\rho_0 c(x)$ is constant, which is just what we would expect for sound waves travelling through a region in which the sound speed varies so slowly on a wavelength scale that $\overline{p^2}/\rho_0 c(x)$ is the acoustic energy flux and negligible energy is reflected.

In determining the multiple–scales expansion we used physical arguments to decide on a suitable short lengthscale. This simplified the calculation. We will now go on to consider a somewhat more general lengthscale and show that in fact the form of the short length variable can be deduced directly from the differential equation and the rule "w_n should be no worse than w_{n-1}". Suppose we take

$$\theta = \tau(X) \tag{7.46}$$

$$s = \epsilon X \tag{7.47}$$

where $\tau(X)$ is at the moment an arbitrary function of X, except that since θ, the short length scale, is to be of order X, $\dot{\tau}(X)$ is of order unity. In a region with uniform sound speed, for which

$$f(\epsilon X) \equiv 1 \ ,$$

$\tau(X)$ would just be proportional to X. Hence $\ddot{\tau}$ is nonzero only because of variations in the sound speed and so $\dot{\tau}$ varies over the long lengthscale ϵX. We will therefore write $\dot{\tau}(X) = g(\epsilon X)$. Then differentiation of $w_n(\theta, s)$ gives

$$\frac{dw_n}{dX} = g(\epsilon X)\frac{\partial w_n}{\partial \theta} + \epsilon\frac{\partial w_n}{\partial s},$$

and

$$\frac{d^2 w_n}{dX^2} = g^2(\epsilon X)\frac{\partial^2 w_n}{\partial \theta^2} + \epsilon \dot{g}(\epsilon X)\frac{\partial w_n}{\partial \theta} + 2\epsilon g(\epsilon X)\frac{\partial^2 w_n}{\partial \theta \partial s} + \epsilon^2 \frac{\partial^2 w_n}{\partial s^2} \ .$$

The equation for w_0 is

$$g^2(s)\frac{\partial^2 w_0}{\partial \theta^2} + f^2(s)w_0 = 0 \ , \tag{7.48}$$

with solution

$$w_0(\theta, s) = A_0(s)\exp\left[\frac{f(s)}{g(s)}i\theta\right] \ . \tag{7.49}$$

Now w_1 satisfies

$$g^2(s)\frac{\partial^2 w_1}{\partial \theta^2} + f^2(s)w_1 = -\dot{g}(s)\frac{\partial w_0}{\partial \theta} - 2g(s)\frac{\partial^2 w_0}{\partial \theta \partial s} \ . \tag{7.50}$$

The forcing term $-2g(s)\partial^2 w_0/\partial\theta\partial s$ is equal to

$$-2ig(s)\left[\frac{d}{ds}\left\{\frac{f(s)A_0(s)}{g(s)}\right\} + \frac{f(s)A_0(s)}{g(s)}\frac{d}{ds}\left\{\frac{f(s)}{g(s)}\right\}i\theta\right]\exp\left[\frac{f(s)}{g(s)}i\theta\right] \ . \tag{7.51}$$

We see that unless

$$\frac{d}{ds}\left(\frac{f(s)}{g(s)}\right) = 0 \ ,$$

the right–hand side of Equation (7.50) involves one resonant term proportional to θ. Such a forcing term would inevitably lead to w_1 having worse behaviour as X tends to infinity than w_0 and so we must insist that

$$\frac{d}{ds}\left(\frac{f(s)}{g(s)}\right) = 0 \ , \tag{7.52}$$

i.e.

$$g(s) = \lambda f(s) \tag{7.53}$$

where λ is an arbitrary constant. Integration of this result gives

$$\tau(X) = \int_0^X \dot{\tau}(\chi)d\chi = \lambda \int_0^X f(\epsilon\chi)d\chi \ , \tag{7.54}$$

and $\tau(X)$ is found to be identical to our previous short lengthscale (defined in Equation (7.36)) apart from an arbitrary multiplicative constant. The form of a suitable short lengthscale can therefore be determined by applying the rule "w_n should be no worse than w_{n-1}", although less algebra is involved if the scale is chosen from physical arguments and then shown to work.

7.3 COMPARISON WITH THE WKB METHOD

The *WKB (WENTZEL–KRAMERS–BRILLOUIN) method* is a way of determining approximate solutions to equations of the form

$$\frac{d^2w}{dz^2} + q(z)w = 0 \ , \tag{7.55}$$

where q is a slowly varying function of z. Although the method predates the method of multiple scales, we will see that it can be considered to be a particular application of multiple scales.

The idea behind the WKB method is that if $q(z)$ were constant, Equation (7.55) would have solutions of the form

$$w(z) = ae^{i\theta(z)} \ , \tag{7.56}$$

with a constant and $\theta(z) = \mp q^{1/2}z$. If q varies slowly and the solution is written in the form

$$w(z) = a(z)e^{i\theta(z)} \ , \tag{7.57}$$

it is reasonable therefore to expect $\theta(z)$ to vary rapidly and $a(z)$ slowly. Differentiation of (7.57) gives

$$\frac{d^2w}{dz^2} = (\ddot{a} + 2i\dot{a}\dot{\theta} + ia\ddot{\theta} - a\dot{\theta}^2)e^{i\theta} \ , \tag{7.58}$$

and Equation (7.55) becomes

$$\ddot{a} + 2i\dot{a}\dot{\theta} + ia\ddot{\theta} - a\dot{\theta}^2 + aq = 0 \ . \tag{7.59}$$

We then choose θ such that $\dot{\theta}^2 - q = 0$, i.e.

$$\theta(z) = \mp \int^z q^{1/2}(z')dz' \ . \tag{7.60}$$

So far no approximation has been made, but the essence of the WKB approximation is to neglect $\ddot{a}$ in comparison with $a\ddot{\theta}$. Equation (7.59) then reduces to

$$2\frac{\dot{a}}{a} + \frac{\ddot{\theta}}{\dot{\theta}} = 0 \ , \tag{7.61}$$

which can be integrated immediately to give

$$a(z) \propto \dot{\theta}^{-1/2} = \mp q^{-1/4} \ . \tag{7.62}$$

Hence the form of the approximate solution to Equation (7.55) is found to be

$$w(z) = q^{-1/4}\left\{ A\exp\left(i\int_0^z q^{1/2}(z')dz'\right) + B\exp\left(-i\int_0^z q^{1/2}(z')dz'\right)\right\} \ , \tag{7.63}$$

where A and B are arbitrary constants.

Equation (7.55) is exactly the equation we solved by multiple scales in Section 7.2 (with f^2 rewritten as q). That method showed that the leading term in an asymptotic expansion had the form given in Equation (7.63) and also leads to a prescription for obtaining higher–order terms.

As an illustration of the WKB method, we will use it to investigate the asymptotic form of two standard equations as z tends to infinity. We will first determine the asymptotic form required of $q(z)$ for the method to be valid, by checking when $\ddot{a}$ is small in comparison with $a\ddot{\theta}$. Suppose that for large $|z|$, $q(z) = 0(|z|^n)$ for some n. Then Equations (7.60) and (7.62) show that $\dot{\theta} = 0(|z|^{n/2})$, and $a(z) = 0(|z|^{-n/4})$, and hence

$$\frac{\ddot{a}}{a\ddot{\theta}} = 0(|z|^{-(n+2)/2}) \ . \tag{7.64}$$

This is negligible as z tends to infinity, provided

$$n \geq -2 \ . \tag{7.65}$$

Let us consider asymptotic solutions of Bessel's equation as a first example.

$$\ddot{w} + \frac{\dot{w}}{z} + \left(1 - \frac{\nu^2}{z^2}\right)w = 0 \ . \tag{7.66}$$

This can be converted into the form in Equation (7.55) by introducing a new variable $W(z)$, defined by

$$W(z) = z^{1/2} w(z) \ . \tag{7.67}$$

The equation for $W(z)$ is

$$\ddot{W} + \left\{ 1 + \frac{\frac{1}{4} - \nu^2}{z^2} \right\} W = 0 \ , \tag{7.68}$$

which satisfies the criterion (7.65). Hence for large $|z|$ Equation (7.68) has solutions of the form

$$W(z) \sim A \left[1 + \frac{\frac{1}{4} - \nu^2}{z^2} \right]^{-1/4} e^{\mp i\theta(z)} \ , \tag{7.69}$$

where A is an arbitrary constant, and

$$\theta(z) = \int^z \left[1 + \frac{\frac{1}{4} - \nu^2}{z'^2} \right]^{1/2} dz' \ . \tag{7.70}$$

This leads to

$$w(z) \sim \frac{A}{z^{1/2}} \left[1 + \frac{\frac{1}{4} - \nu^2}{z^2} \right]^{-1/4} e^{\mp i\theta(z)} \ . \tag{7.71}$$

For very large z this reduces to the usual form $Az^{-1/2} e^{\mp iz}$, but instead of a power-series expansion in z^{-1}, the next order terms are presented as a change in phase and amplitude. This is sometimes particularly convenient, for example when finding the positions of zeros.

The asymptotic solutions of Airy's equation

$$\ddot{w} - zw = 0 \ , \tag{7.72}$$

can also be investigated by this method. Here $a(z) = z^{-1/4}$ and $\theta(z) = i\frac{2}{3} z^{3/2}$. Hence for z real and positive,

$$w \sim Az^{-1/4} \exp \left[\mp \frac{2}{3} z^{3/2} \right] \ , \tag{7.73}$$

while for z real and negative

$$w \sim Az^{-1/4} \exp \left[\mp \frac{2}{3} i |z|^{3/2} \right] \ , \tag{7.74}$$

demonstrating that Airy's equation has exponential behaviour for z positive and oscillatory behaviour for negative real z. These asymptotic forms can also be obtained by taking Laplace transforms and then evaluating the inversion integral by the method of steepest descents. But that involves considerably more work!

7.4 RAY THEORY

The high frequency, or ray theory, description of sound waves travelling through a fluid in which the mean density and sound speed are functions of position can be developed from the equations of motion by the method of multiple scales. The treatment is very similar to that described in Section 7.2 for one–dimensional waves. We will denote the mean density and sound speed by $\rho_0(\mathbf{x})$ and $c(\mathbf{x})$ respectively, and then the linearized equation of mass conservation states that

$$\frac{D\rho}{Dt} = -\rho_0 \frac{\partial u_i}{\partial x_i} \; , \tag{7.75}$$

where $\mathbf{u}$ is the particle velocity and D/Dt the material derivative $\partial/\partial t + u_i \, \partial/\partial x_i$. From the definition of the sound speed, we have

$$\frac{D\rho}{Dt} = \frac{1}{c^2} \frac{Dp}{Dt} \; , \tag{7.76}$$

and after linearization, this simplifies to

$$\frac{D\rho}{Dt} = \frac{1}{c^2} \frac{\partial p}{\partial t} \; . \tag{7.77}$$

We can eliminate $D\rho/Dt$ from Equations (7.75) and (7.77) to obtain

$$\frac{1}{c^2} \frac{\partial p}{\partial t} = -\rho_0 \frac{\partial u_i}{\partial x_i} \; . \tag{7.78}$$

Combining this with the linearized momentum equation, $\rho_0 \partial \mathbf{u}/\partial t = -\nabla p$, gives the wave equation

$$\frac{1}{c^2} \frac{\partial^2 p}{\partial t^2} = \rho_0 \frac{\partial}{\partial x_i} \left(\frac{1}{\rho_0} \frac{\partial p}{\partial x_i} \right) \; . \tag{7.79}$$

We will investigate harmonic solutions of this equation by writing

$$p(\mathbf{x}, t) = \widetilde{p}(\mathbf{x}) e^{-i\omega t} \; . \tag{7.80}$$

$\widetilde{p}(\mathbf{x})$ therefore satisfies

$$\rho_0 \frac{\partial}{\partial x_i} \left(\frac{1}{\rho_0} \frac{\partial p}{\partial x_i} \right) + \frac{\omega^2}{c^2} \widetilde{p} = 0 \; . \tag{7.81}$$

There are again two different lengthscales in the problem, a typical wavelength of order c/ω and the lengthscale ℓ over which the mean density and sound speed vary.

For high-frequency waves, the wavelength is very much smaller than ℓ and the method of multiple scales is appropriate.

As in Section 7.2, we will introduce a nondimensional variable $\mathbf{X} = \mathbf{x}\omega/c_0$, where c_0 is the sound speed at, say, the origin. Then with $P(\mathbf{X}) = \widetilde{p}(\mathbf{x})$, $P(\mathbf{X})$ satisfies

$$\rho_0 \frac{\partial}{\partial X_i} \left(\frac{1}{\rho_0} \frac{\partial P}{\partial X_i} \right) + \frac{c_0^2}{c^2} P = 0 \ . \tag{7.82}$$

ρ_0 and c vary over length ℓ and so it is convenient to introduce nondimensional density and sound–speed functions $\widehat{\rho}_0$ and $\widehat{c}$, defined by

$$\frac{\rho_0(\mathbf{x})}{\rho_0(0)} = \widehat{\rho}\left(\frac{\mathbf{x}}{\ell}\right) = \widehat{\rho}(\epsilon\mathbf{X}) \ ,$$

$$\frac{c(\mathbf{x})}{c_0} = \widehat{c}\left(\frac{\mathbf{x}}{\ell}\right) = \widehat{c}(\epsilon\mathbf{X}) \ ,$$

where $\epsilon = c_0/\omega\ell$ is a small parameter. $\widehat{\rho}$ and $\widehat{c}$ are of order unity and so are their derivatives with respect to their arguments. Suitably nondimensionalized, then, the wave equation (7.82) becomes

$$\frac{\partial^2 P}{\partial X_i \partial X_i} + \frac{\partial P}{\partial X_i} \widehat{\rho}_0 \frac{\partial}{\partial X_i} \left(\frac{1}{\widehat{\rho}_0} \right) + \frac{P}{\widehat{c}^2} = 0 \ . \tag{7.83}$$

It is then appropriate to introduce a short lengthscale to describe the phase variations,

$$\theta = \tau(\mathbf{X}) \ , \tag{7.84}$$

say, where the function τ has yet to be determined, and a slowly varying spatial variable

$$\eta = \epsilon\mathbf{X} \ , \tag{7.85}$$

to describe the gradual variations in amplitude. If c and ρ_0 were uniform, $\widehat{c}$ and $\widehat{\rho}_0$ would be unity everywhere and the phase of a plane wave travelling in the direction of the unit vector $\mathbf{k}$ would be $\mathbf{k}.\mathbf{X}$, i.e. the phase variation would be described by $\theta = \tau(\mathbf{X}) = \mathbf{k}.\mathbf{X}$. It follows then that $\nabla\tau$ only changes because of the gradual variations in sound speed and so alters over a lengthscale ϵ^{-1}. It is therefore appropriate to write

$$\nabla\tau = \mathbf{f}(\epsilon\mathbf{X}) = \mathbf{f}(\boldsymbol{\eta}) \tag{7.86}$$

for some vector $\mathbf{f}$ with magnitude of order unity.

We expand $P(\mathbf{X})$ as an asymptotic series

$$P(\mathbf{X}) = \sum_{n=0} \epsilon^n w_n(\theta, \boldsymbol{\eta}) \ . \tag{7.87}$$

Then

$$\frac{\partial P}{\partial X_i} = \sum_{n=0} \epsilon^n \left[\frac{\partial w_n}{\partial \theta} \frac{\partial \tau}{\partial X_i} + \frac{\partial w_n}{\partial \eta_j} \frac{\partial \eta_j}{\partial X_i} \right]$$

$$= \sum_{n=0} \epsilon^n \left[\frac{\partial w_n}{\partial \theta} f_i(\boldsymbol{\eta}) + \frac{\partial w_n}{\partial \eta_i} \epsilon \right] \ ,$$

and differentiating again, gives

$$\frac{\partial^2 P}{\partial X_i \partial X_i} = \sum_{n=0} \epsilon^n \left[\frac{\partial^2 w_n}{\partial \theta^2} f_i f_i + \frac{\partial w_n}{\partial \theta} \frac{\partial f_i}{\partial \eta_i} \epsilon + 2 \frac{\partial^2 w_n}{\partial \theta \partial \eta_i} \epsilon f_i + \frac{\partial^2 w_n}{\partial \eta_i \partial \eta_i} \epsilon^2 \right] \ .$$

Also we note

$$\frac{\partial}{\partial X_i} \left(\frac{1}{\hat{\rho}_0} \right) = \epsilon \frac{\partial}{\partial \eta_i} \left(\frac{1}{\hat{\rho}_0(\boldsymbol{\eta})} \right) \ .$$

Substituting these derivatives into the differential Equation (7.83) and equating the coefficients of each power of ϵ to zero, gives a sequence of equations for the unknown functions w_n. We can determine these and $\tau(\mathbf{X})$ from the rule that "w_n should be no worse than w_{n-1}".

For convenience we will suppress the circumflex denoting the nondimensionalized quantities in the following analysis. Then, equating the coefficient of ϵ^0 to zero, gives

$$\frac{\partial^2 w_0}{\partial \theta^2} |\mathbf{f}(\boldsymbol{\eta})|^2 + \frac{w_0}{c^2(\boldsymbol{\eta})} = 0 \ . \tag{7.88}$$

This can be integrated immediately to give

$$w_0(\theta, \boldsymbol{\eta}) = A_0(\boldsymbol{\eta}) e^{i\theta/|\mathbf{f}|c} \ . \tag{7.89}$$

The equation for $w_1(\theta, \boldsymbol{\eta})$ is

$$\frac{\partial^2 w_1}{\partial \theta^2} |\mathbf{f}(\boldsymbol{\eta})|^2 + \frac{w_1}{c^2(\boldsymbol{\eta})} = -2 \frac{\partial^2 w_0}{\partial \theta \partial \eta_i} f_i - \frac{\partial w_0}{\partial \theta} \frac{\partial f_i}{\partial \eta_i} - \frac{\partial w_0}{\partial \theta} f_i \rho_0 \frac{\partial}{\partial \eta_i} \left(\frac{1}{\rho_0} \right) \ .$$

Substituting for w_0 from Equation (7.89) gives

$$\frac{\partial^2 w_1}{\partial \theta^2} |\mathbf{f}(\boldsymbol{\eta})|^2 + \frac{w_1}{c^2(\boldsymbol{\eta})} = \frac{-i}{|\mathbf{f}|c} \left\{ 2 f_i \frac{\partial A_0}{\partial \eta_i} + A_0 \left(\frac{\partial f_i}{\partial \eta_i} + f_i \rho_0 \frac{\partial}{\partial \eta_i} \left(\frac{1}{\rho_0} \right) \right) \right\} e^{i\theta/|\mathbf{f}|c}$$

$$- 2 i A_0 f_i \frac{\partial}{\partial \eta_i} \left\{ \frac{1}{|\mathbf{f}|c} \exp \left[\frac{i\theta}{|\mathbf{f}(\boldsymbol{\eta})|c(\boldsymbol{\eta})} \right] \right\} \ .$$

$$\tag{7.90}$$

It is immediately apparent that the last term on the right–hand side of Equation (7.90) will produce a resonant forcing term proportional to θ unless the product $|\mathbf{f}(\boldsymbol{\eta})|c(\boldsymbol{\eta})$ is constant. This defines the short lengthscale apart from an arbitrary constant, which without loss of generality we may take to be unity. Since $\mathbf{f}(\boldsymbol{\eta})$ is equal to $\nabla\tau$, this condition gives

$$|\nabla\tau| = \frac{1}{c} \ . \tag{7.91}$$

Returning to (7.90) we see that w_1 has worse behaviour than w_0 unless the right–hand side of Equation (7.90) vanishes. Hence, in order to satisfy the rule, we must have

$$2f_i\frac{\partial A_0(\boldsymbol{\eta})}{\partial \eta_i} + A_0(\boldsymbol{\eta})\left(\frac{\partial f_i}{\partial \eta_i} + f_i\rho_0\frac{\partial}{\partial \eta_i}\left(\frac{1}{\rho_0}\right)\right) = 0 \ . \tag{7.92}$$

When (7.86) is used to rewrite $\mathbf{f}$ as $\nabla\tau$, Equation (7.92) becomes

$$2(\nabla\tau.\nabla)A_0 + A_0(\boldsymbol{\eta})(\nabla^2\tau + \rho_0\nabla\tau\cdot\nabla(\rho_0{}^{-1})) = 0 \ . \tag{7.93}$$

The first term in the asymptotic series is therefore given by

$$w_0(\theta, \boldsymbol{\eta}) = A_0(\boldsymbol{\eta})e^{i\tau(\mathbf{X})} \ , \tag{7.94}$$

where the phase $\tau(\mathbf{X})$ is given by Equation (7.91) and the amplitude $A_0(\boldsymbol{\eta})$ is to be determined by solving Equation (7.93).

Higher–order terms in the asymptotic series also have phase $\tau(\mathbf{X})$ and the asymptotic series for the pressure perturbation has the form

$$P(\mathbf{X}) = e^{i\tau(\mathbf{X})}\{A_0(\boldsymbol{\eta}) + \epsilon A_1(\boldsymbol{\eta}) + ...\} \ . \tag{7.95}$$

Such a representation is called a *ray series*.

Curves of constant phase are *wavefronts*, and we define *rays* to be curves that are everywhere normal to the wavefronts, as shown in Figure 7.1. Equation (7.91) implies that

$$|\nabla\tau|^2 = \frac{1}{c^2} \ . \tag{7.96}$$

This is called the *eikonal equation*; it determines the phase and hence also the shape of the rays. Let a ray be denoted by the parametric curve $\mathbf{X} = \mathbf{X}(s)$, where s denotes arc length along the ray. The tangent to the ray $d\mathbf{X}/ds$ is always perpendicular to surfaces of constant phase, i.e. it is perpendicular to the surfaces $\tau(\mathbf{X}) = $ constant. It is therefore parallel to $\nabla\tau$, and we can write

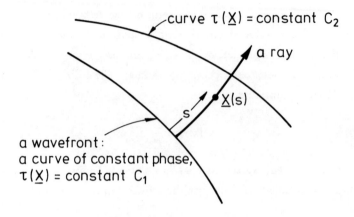

Figure 7.1 The geometry of the wavefronts and a ray.

$$\frac{d\mathbf{X}}{ds} = \alpha \nabla \tau \ ,$$

where α is a scalar. Now $|d\mathbf{X}/ds|$ is unity because $dX^2 + dY^2 + dZ^2 = ds^2$, but Equation (7.96) gives $|\nabla \tau|$ as c^{-1} and so α must be equal to c, i.e.

$$\frac{1}{c}\frac{d\mathbf{X}}{ds} = \nabla \tau \ . \tag{7.97}$$

Differentiating with respect to s gives

$$\frac{d}{ds}\left(\frac{1}{c}\frac{d\mathbf{X}}{ds}\right) = \frac{d}{ds}(\nabla \tau) \ . \tag{7.98}$$

We can evaluate $d(\nabla \tau)/ds$ from Equation (7.96). The gradient of (7.96) shows that

$$(\nabla \tau . \nabla)\nabla \tau = \frac{1}{c}\nabla \left(\frac{1}{c}\right) \ . \tag{7.99}$$

The operator $\nabla \tau . \nabla$ involves differentiation in the ray direction and is equal to $c^{-1}d/ds$. Hence (7.99) reduces to

$$\frac{d}{ds}\nabla \tau = \nabla \left(\frac{1}{c}\right) \ ,$$

which we can use to replace the right–hand side of Equation (7.98) to obtain

$$\frac{d}{ds}\left(\frac{1}{c}\frac{d\mathbf{X}}{ds}\right) = \nabla\left(\frac{1}{c}\right) \quad . \tag{7.100}$$

This represents three differential equations for the three components of $\mathbf{X}(s)$. Given suitable initial conditions and the sound speed as a function of position, these equations can be integrated to determine the ray path.

An alternative representation is to express the ray path by $X = X(Z)$ and $Y = Y(Z)$. Then $ds = (X'^2 + Y'^2 + 1)^{1/2}dZ$ where $X' = dX/dZ$, $Y' = dY/dZ$ and Equation (7.100) gives

$$\frac{d}{dZ}\left(\frac{X'}{c(X'^2 + Y'^2 + 1)^{1/2}}\right) = (X'^2 + Y'^2 + 1)^{1/2}\frac{\partial}{\partial X}\left(\frac{1}{c}\right) \tag{7.101a}$$

and

$$\frac{d}{dZ}\left(\frac{Y'}{c(X'^2 + Y'^2 + 1)^{1/2}}\right) = (X'^2 + Y'^2 + 1)^{1/2}\frac{\partial}{\partial Y}\left(\frac{1}{c}\right) \quad , \tag{7.101b}$$

which again are differential equations describing the path of the rays. These equations are precisely the Euler equations for the integral

$$\int ds/c = \int (X'^2 + Y'^2 + 1)^{1/2}dZ/c$$

to be stationary (see Margenau & Murphy 1956 for a derivation of the Euler equations). The ray path between two fixed points is the path along which the time of travel is stationary (Fermat's principle).

If c is a function of Z only , and the coordinate system is chosen so that Y' is initially zero, Equations (7.101) reduce to

$$Y'(Z) = 0 \tag{7.102}$$

and

$$\frac{\sin\theta}{c} = \text{constant} \quad , \tag{7.103}$$

where $\sin\theta = X'/(X'^2 + 1)^{1/2}$, i.e. θ is the angle between the Z–axis and the tangent to the ray, as shown in Figure 7.2. This is the familiar *Snell's law*.

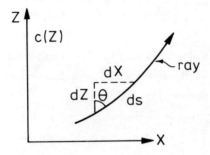

Figure 7.2 Snell's law $\sin\theta/c = $ constant along a ray, when the speed of sound is a function of Z only .

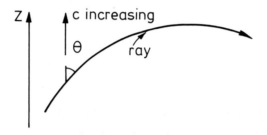

Figure 7.3 A ray propagating into a region of *increasing* sound speed is bent *away* from the direction of stratification.

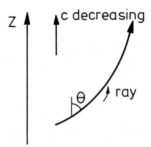

Figure 7.4 A ray propagating into a region of *decreasing* sound speed is bent *towards* the direction of stratification.

Snell's law states that a ray travelling in a direction in which the sound speed increases is bent away from the Z–direction (see Figure 7.3), while one travelling into a region of decreasing sound speed is bent towards the direction of stratification (as shown in Figure 7.4).

This refraction can lead to some interesting effects in the ocean. For example, near as minimum of the sound speed any ray propagating downwards will eventually come into a region where the sound speed increases with depth, and will be bent back. Similarly, a ray propagating upwards will also find itself travelling into a region where the sound speed increases, and will also be refracted back. Figure 7.5 illustrates the ray pattern. We see that a ray can thus become "trapped" or confined within certain levels. These regions are called *sound channels*. In them the sound can only spread out in a circle rather than over a sphere as it would in a homogeneous medium. As a result, the sound decays more slowly and can be detected at far greater distances. One such minimum occurs at a depth of about 1000 m in sea water and enables sound to be heard over large distances in the upper regions of the ocean.

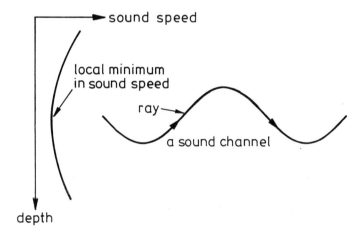

Figure 7.5 A sound channel.

A maximum in the sound speed leads to the formation of *shadow zones*, as shown in Figure 7.6. Rays propagating upwards move into regions where the sound speed decreases and are refracted upwards. Similar rays moving downwards are refracted downwards more steeply. It is evident from Figure 7.6 that there is a region into which no rays can penetrate.

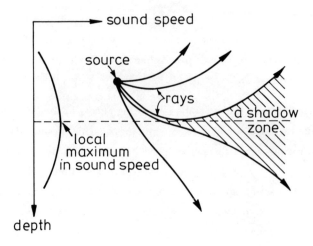

Figure 7.6 The formation of a shadow zone.

Equation (7.93) gives information about the way in which the amplitude of the sound varies along a ray. First we note that since $\nabla\tau.\nabla$ is equal to $c^{-1}d/ds$, Equation (7.93) simplifies to

$$2\frac{\partial A_0}{\partial s} + A_0\left(c\nabla^2\tau + \rho_0\frac{\partial}{\partial s}\left(\frac{1}{\rho_0}\right)\right) = 0 \ . \tag{7.104}$$

This can be solved immediately to give

$$A_0 = A_0(s_0)\exp\left[-\frac{1}{2}\int_{s_0}^{s}\left\{c\nabla^2\tau + \rho_0\frac{\partial}{\partial s}\left(\frac{1}{\rho_0}\right)\right\}ds\right] \ , \tag{7.105}$$

where s_0 is some reference point on the ray. Now straightforward integration shows

$$\exp\left[-\frac{1}{2}\int_{s_0}^{s}\rho_0\frac{\partial}{\partial s}\left(\frac{1}{\rho_0}\right)ds\right] = \exp\left[-\frac{1}{2}\ln\left(\frac{\rho_0(s_0)}{\rho_0(s)}\right)\right] = \left(\frac{\rho_0(s)}{\rho_0(s_0)}\right)^{1/2} \ . \tag{7.106}$$

We can also simplify the terms $\exp[-\frac{1}{2}\int_{s_0}^{s}c\nabla^2\tau ds]$. Consider a short length δs of a ray tube of cross–sectional area $\Delta\Sigma(s)$. Let S be the surface of the ray tube and V its volume, as in Figure 7.7. By the divergence theorem

$$\int_V \nabla^2\tau dV = \int_S \mathbf{n}.\nabla\tau dS \ . \tag{7.107}$$

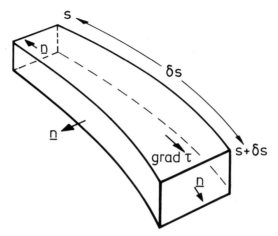

Figure 7.7 A ray tube.

Now on the sides of the ray tube $\mathbf{n}$ is perpendicular to $\nabla\tau$:

$$\mathbf{n}.\nabla\tau = 0 \ .$$

At the ends of the tube $\mathbf{n} = \pm c\nabla\tau$ and $\mathbf{n}.\nabla\tau = \pm 1/c$, from the eikonal equation. Therefore Equation (7.107) states that

$$\nabla^2\tau\delta s\Delta\Sigma = \frac{\Delta\Sigma(s+\delta s)}{c(s+\delta s)} - \frac{\Delta\Sigma(s)}{c(s)} = \frac{\partial}{\partial s}\left(\frac{\Delta\Sigma}{c}\right)\delta s \ .$$

Thus

$$c\nabla^2\tau = \frac{c}{\Delta\Sigma}\frac{\partial}{\partial s}\left(\frac{\Delta\Sigma}{c}\right) \ , \tag{7.108}$$

and

$$\exp\left[-\tfrac{1}{2}\int_{s_0}^{s}c\nabla'\tau ds\right] = \exp\left[-\tfrac{1}{2}\ln\frac{\Delta\Sigma(s)c(s_0)}{\Delta\Sigma(s_0)c(s)}\right]$$

$$= \left(\frac{c(s)\Delta\Sigma(s_0)}{\Delta\Sigma(s)c(s_0)}\right)^{1/2} \ . \tag{7.109}$$

Returning to Equation (7.105), we see that the solution can be expressed

$$A_0(s) = A_0(s_0)\left(\frac{\rho_0(s)c(s)}{\Delta\Sigma(s)}\frac{\Delta\Sigma(s_0)}{\rho_0(s_0)c(s_0)}\right)^{1/2} \ , \tag{7.110}$$

i.e. $A_0^2(s)\Delta\Sigma(s)/(\rho_0(s)c(s))$ is constant along a ray tube. This is just what one obtains heuristically by using the plane–wave result that the mean acoustic intensity is $\overline{p^2}/\rho_0 c$ and applying conservation of energy flow along a ray tube. The change in pressure amplitude can therefore by found by tracing a bundle of rays and investigating the change in area of the ray tube.

We have just determined the leading term for the pressure in an expansion in powers of ϵ (i.e. in inverse powers of ω). For high frequencies it is sufficient to consider just this term but the method of multiple scales can, if need be, be extended to include further terms in the asymptotic series.

REFERENCES

Margenau, H. & Murphy, G.M. (1956). The Mathematics of Physics and Chemistry. Van Nostrand.

Nayfeh, A. (1973). Perturbation Methods. Wiley.

8. STATISTICAL ENERGY ANALYSIS

8.1 INTRODUCTION

An exact analysis of large structures or rooms becomes very cumbersome and time consuming because many modes have to be considered and because the frequency response changes rapidly for each frequency step, so that many points have to be calculated in order to derive the average response to a certain frequency band. But even if one were willing to do all this work (which in principle is possible) and to spend a large amount of money for computer time, there is still the fundamental difficulty that complex structures, especially their boundary conditions, cannot be specified accurately enough to allow calculations that agree with measured data in all details (especially when more than ten modes are involved).

Having these facts in mind it is not surprising that a method was developed, which does not have the drawbacks just mentioned and which permits the description of sound and vibration fields in complicated structures in a rather simple way. This method is the statistical energy analysis (SEA) which deals with the average behaviour of complex structures and rooms when they are excited by frequency bands, especially by random noise, as is often the case when the sound is a byproduct of turbulence.

Obviously, SEA is not a method of solving all vibration problems. As a matter of fact, it has some serious disadvantages which pose rather stringent limits to its application (see Section 8.6). But it has proved to be very helpful in many practical applications.

8.2 POWER FLOW BETWEEN TWO RESONATORS

Important aspects of SEA can be understood already by studying the power flow in two coupled resonators. We investigate the system shown in Figure 8.1.

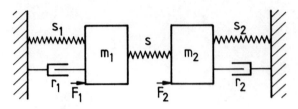

Figure 8.1 Two coupled resonators.

It consists of masses m_1, m_2, springs s_1, s_1, dashpots r_1, r_2, and the loss–free coupling spring s. (If the coupling were of another type, the calculations would be similar (Maidanik & Lyon 1962; Scharton & Lyon 1968)). The equations of motion are obtained by balancing the forces, i.e.

$$m_1\ddot{\xi}_1 + r_1\dot{\xi}_1 + (s_1 + s)\xi_1 - s\xi_2 = F_1$$
$$-s\xi_1 + m_2\ddot{\xi}_2 + r_2\dot{\xi}_2 + (s_2 + s)\xi_2 = F_2 \quad . \tag{8.1}$$

Solving these equations for purely harmonic motions and omitting the common time factor $e^{-i\omega t}$, we get with $v = -i\omega\xi$

$$v_1 = \frac{i\omega}{m_1 m_2 D(\omega)}\left[m_2(\omega^2 + 2i\omega\delta_2 - {\omega_2}^2)F_1 + sF_2\right]$$
$$v_2 = \frac{-i\omega}{m_1 m_2 D(\omega)}\left[sF_1 + m_1(\omega^2 + 2i\omega\delta_1 - {\omega_1}^2)\right] \quad , \tag{8.2}$$

where

$${\omega_1}^2 = (s_1 + s)/m_1; \quad {\omega_2}^2(s_2 + s)/m_2; \quad \delta_1 = r_1/2m_1;$$
$$\delta_2 = r_2/2m_2;$$
$$D(\omega) = \omega^4 + 2i\omega^3(\delta_1 + \delta_2) - \omega^2({\omega_1}^2 + {\omega_2}^2 + 4\delta_1\delta_2) - 2i\omega(\delta_1{\omega_2}^2 + \delta_2{\omega_1}^2)$$
$$+ {\omega_1^2}{\omega_2}^2 - s^2/m_1 m_2 \quad .$$

The force acting on the coupling spring is

$$F = \frac{s}{-i\omega}(v_1 - v_2) \quad ; \tag{8.3}$$

therefore the net power flow through the coupling element is

$$P_{12} = \frac{1}{2} Re\{Fv_2{}^*\} = \frac{s}{2} Re\left\{\frac{v_1 v_2{}^*}{-i\omega} - \frac{|v_2{}^2|}{-i\omega}\right\} = \frac{s}{2} Re\left\{\frac{v_1 v_2{}^*}{-i\omega}\right\}$$

$$P_{21} = \frac{1}{2} Re\{-Fv_1{}^*\} = \frac{-s}{2} Re\left\{\frac{|v_1{}^2|}{-i\omega} + \frac{v_1{}^* v_2}{-i\omega}\right\} = \frac{s}{2} Re\left\{\frac{v_1{}^* v_2}{-i\omega}\right\} = -P_{12} \quad .$$

$$(8.4)$$

Here we used the fact that $Re\{i|v|^2\} = 0$ and

$$Re\left\{\frac{v_1 v_2{}^*}{-i}\right\} = |v_1 v_2| Re\left\{\frac{1}{-i} e^{i(\phi_1 - \phi_2)}\right\} = -|v_1 v_2| \sin(\phi_1 - \phi_2)$$

$$Re\left\{\frac{v_1{}^* v_2}{-i}\right\} = |v_1 v_2| Re\left\{\frac{1}{-i} e^{-i(\phi_1 - \phi_2)}\right\} = |v_1 v_2| \sin(\phi_1 - \phi_2)$$

$$(8.5)$$

The relation $P_{21} = -P_{12}$ was to be expected because no power is lost in the coupling element.

Inserting (8.2) into (8.4) gives

$$P_{12} = \frac{\omega^2 s^2}{m_1{}^2 m_2{}^2 |D(\omega)|^2}\left[m_2 \delta_2 |F_1|^2 - m_1 \delta_1 |F_2|^2 + \alpha F_1 F_2\right] \quad . \tag{8.6}$$

Here α denotes the factor in front of the cross terms $F_1 F_2$. Next we assume that F_1 and F_2 are of white noise type with bandwidth $\Delta\omega$ and that they are uncorrelated, so that $\overline{F_1 F_2}$ vanishes. Averaging over frequency we get

$$\overline{P_{12}} = \frac{\pi}{2\Delta\omega}\left[\frac{|F_1{}^2|}{m_1 \delta_1} - \frac{|F_2{}^2|}{m_2 \delta_2}\right]\frac{(\delta_1 + \delta_2)s^2}{m_1 m_2 Q} \quad , \tag{8.7}$$

where

$$Q = (\omega_1{}^2 - \omega_2{}^2)^2 + 4(\delta_1 + \delta_2)(\delta_1 \omega_2{}^2 + \delta_2 \omega_1{}^2) + (\delta_1 + \delta_2)s^2/m_1 m_2 \delta_1 \delta_2 \quad ,$$

($\Delta\omega$ is assumed to contain all resonances).

Next we calculate the kinetic energies averaged over frequency. As can be seen from (8.2), here again we have to integrate $1/|D(\omega)|^2$; therefore the calculations are very similar and the result after some rearrangement is

$$\frac{1}{2}\left[m_1|v_1|^2 - m_2|v_2|^2\right] = \frac{\pi}{2\Delta\omega}\left[\frac{|F_1|^2}{m_1 \delta_1} - \frac{|F_2|^2}{m_2 \delta_2}\right] A$$

with

$$A = \frac{(\omega_1{}^2 - \omega_2{}^2)^2 + (\delta_1 + \delta_2)(\delta_1 \omega_2{}^2 + \delta_2 \omega_1{}^2)}{2Q} \quad . \tag{8.8}$$

This allows us to write finally

$$\overline{P_{12}} = \beta \left(\frac{m_1 |\overline{v_1}|^2}{2} - \frac{m_2 |\overline{v_2^2}|}{2} \right) \quad . \tag{8.9}$$

The importance of this formula lies in the fact that the frequency average of the power flowing from system 1 to system 2 is proportional to the difference in the mean kinetic energies. The proportionality factor is obtained by comparing (8.7) with (8.8). Thus

$$\beta = \frac{2}{m_1 m_2} \frac{s^2(\delta_1 + \delta_2)}{({\omega_1}^2 - {\omega_2}^2)^2 + (\delta_1 + \delta_2)(\delta_1 {\omega_2}^2 + \delta_2 {\omega_1}^2)} \quad . \tag{8.10}$$

It obviously depends on the coupling spring s, the distance between the resonance frequencies and the damping.

It is of interest to note, that (8.9) is analogous to the heat flow equation. To see this we have to view, as in thermodynamics, the energy per resonator as its temperature and the power flow as the heat flow.

8.3   POWER FLOW IN MULTI–MODAL SYSTEMS

8.3.1 Modal Density

Before we discuss SEA in its most general form, we have to introduce the concept of modal density, i.e. the number of modes within a certain frequency band.

In one–dimensional systems, the resonance frequencies f_N are given by

$$f_N = \frac{Nc}{2(\ell + \Delta \ell)} \quad . \tag{8.11}$$

Here $N = 1, 2, 3, \ldots$, c = phase speed of sound on the system, ℓ = length, $\Delta \ell$ = "endcorrection", which depends on the boundary conditions and the frequency. Since $\Delta \ell$ is always smaller than half a wavelength, we get in the high–frequency limit, where $\ell >> \Delta \ell$,

$$N \approx 2\ell f_N / c \quad . \tag{8.12}$$

If f_N is taken as an upper frequency limit, (8.12) gives the number of modes up to this limit. The number of modes within a frequency band $\Delta \omega = 2\pi \Delta f$ is obtained by differentiating (8.12). The result is

$$\frac{\Delta N}{\Delta \omega} = \frac{\ell}{\pi} \frac{d\omega/c}{d\omega} = \frac{\ell}{\pi} \left[\frac{1}{c} + \omega \frac{d1/c}{d\omega} \right] \quad . \tag{8.13}$$

Here it has been taken into account that the medium may be dispersive, i.e. c may be a function of frequency. For non–dispersive media, i.e. air in pipe, longitudinal or torsional waves on rods, (8.13) becomes

$$\Delta N / \Delta \omega = \ell / c \pi \quad . \tag{8.14}$$

For the most important dispersive waves, the bending waves, where

$$c = \sqrt{\omega} \left(\frac{D}{m} \right)^{1/4} \quad , \tag{8.15}$$

we find

$$\frac{\Delta N}{\Delta \omega} = \frac{1}{2 \pi \sqrt{\omega}} \left(\frac{m}{D} \right)^{1/4} \quad , \tag{8.16}$$

(D = bending stiffness, m = mass per unit length.)

Equation (8.12), which is the basis of our calculations, states that the number of modes is approximately equal to the dimension of the system divided by half the wavelength. If we apply the same argument to a two–dimensional system of area S, we would find the number of modes to be proportional to $\omega^2 S / c^2$. The exact formula is obtained by assuming again that the boundary conditions do not have a strong influence on N; therefore we can work with the most simple ones. In such a case, the resonances are given by

$$\omega_{n_1 n_2}^2 \approx \pi^2 c^2 \left[\left(\frac{n_1}{\ell_1} \right)^2 + \left(\frac{n_2}{\ell_2} \right)^2 \right] \quad . \tag{8.17}$$

Now we have to find out how many combinations of n_1 and n_2 are possible so that all resonances are below a certain value of ω. To solve this problem, we use the simple geometric representation of (8.17) which is shown in Figure 8.2. Obviously, each intersection represents a possible resonance frequency which has a distance $\omega_{n_1 n_2} / \pi c$ from the origin. The number of modes with frequencies below ω are given approximately by dividing the quarter circle with radius $\omega \pi c$ through the area of one area element, which is $1/\ell_1 \ell_2$. Thus we have

$$N \approx \frac{\pi}{4} \left(\frac{\omega}{\pi c} \right)^2 \left(\frac{\ell_2 \ell_2}{1} \right) = \frac{\omega^2 S}{4 \pi c^2} \quad . \tag{8.18}$$

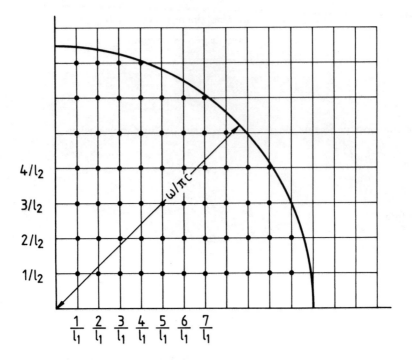

Figure 8.2 Geometric representation of Equation (8.17).

For a non–dispersive medium, we get from (8.18)

$$\frac{\Delta N}{\Delta \omega} = \frac{\omega S}{2\pi c^2} \quad ,$$

(8.19)

and for a medium with the dispersion law given by (8.15), we find

$$\frac{\Delta N}{\Delta \omega} = \frac{S}{4\pi} \left(\frac{m}{D}\right)^{1/2} \quad .$$

(8.20)

For three dimensions, the arguments are similar. The results are

$$N \approx \frac{\omega^2 V}{6\pi^2 c^3}$$

(8.21)

$$\frac{\Delta N}{\Delta \omega} = \frac{\omega^2 V}{2\pi^2 c^3} \quad .$$

Here V is the volume. The second formula holds when there is no dispersion.

8.3.2 Mean–Square Response of a Point–Driven Structure

According to Equation (20.70) the sound pressure in a room can be expressed as a sum of its modes. The same can be done for the velocity of a finite structure, provided the damping is not too high and no energy is propagating through the boundaries of the structure. Thus we can write

$$v(x) = \sum_{n=1}^{\infty} \frac{\phi_n(x) F_n}{\underline{\omega}_n^2 - \omega^2} \quad . \tag{8.22}$$

Here $\underline{\omega}_n^2 = \omega_n^2(1 - i\eta)$ are the complex eigenfrequencies, η the loss factor, F_n the modal excitation parameter. ϕ_n are the modal functions; they have the important orthogonality property

$$\int \phi_n(x) m \phi_\nu(x) dx = \delta_{n\nu} \quad . \tag{8.23}$$

(see (20.64)).

In the last equation, where the integral has to be taken over the whole volume or area, m is the mass per unit volume or area.

For the total kinetic energy, we find

$$E_k = \frac{1}{2} \int m|v(x)|^2 dx = \frac{1}{2} \sum_n \sum_\nu \frac{|F_n F_\nu|}{|(\underline{\omega}_n^2 - \omega^2)(\underline{\omega}_\nu^2 - \omega^2)|} \int \phi_n(x) m \phi_\nu(x) dx$$

$$= \frac{1}{2} \sum_n \frac{|F_n|^2}{|\underline{\omega}_n^2 - \omega^2|^2} \quad . \tag{8.24}$$

Taking the frequency average over a band $\Delta\omega$ gives

$$\overline{E}_k = \frac{1}{2} \sum_n \frac{1}{\Delta\omega} \int \frac{|F_n|^2}{|\underline{\omega}_n^2 - \omega^2|^2} d\omega \quad . \tag{8.25}$$

To proceed further we distinguish between resonating and non–resonating modes. Resonating modes (they shall be numbered from n_1 to n_2) are those that have their resonance frequency in the range of integration. The remainder are non–resonating modes. Thus

$$2\overline{E}_k = \frac{1}{\Delta\omega} \sum_{n_1}^{n_2} \int \frac{|F_n|^2}{|\underline{\omega}_n^2 - \omega^2|^2} d\omega + \frac{1}{\Delta\omega} \sum \int \frac{|F_n|^2}{|\underline{\omega}_n^2 - \omega^2|^2} d\omega \quad . \tag{8.26}$$

The integrands in the first sum have a pronounced peak when $\omega_n \approx \omega$. This allows us to make the following approximations:

$$\int \frac{d\omega}{|\underline{\omega}_n^2 - \omega^2|^2} = \int \frac{d\omega}{(\omega_n^2 - \omega^2)^2 + \eta^2 \omega_n^4} \approx \int \frac{d\omega}{4\omega_n^2(\omega_n - \omega)^2 + \eta^2 \omega_n^4}$$

$$= \frac{1}{2\omega_n^3 \eta} \tan^{-1}\left(\frac{2(\omega_n - \omega)}{\eta \omega_n}\right) \approx \frac{\pi}{2\eta\omega_n^3} \quad . \tag{8.27}$$

Here we have assumed that $\eta \omega_n$ is small compared with the bandwidth $\Delta \omega$, so that the argument of the $\tan^{-1}$–function is always large. The integration is taken from ω_1 to ω_2 so that $\Delta \omega = \omega_2 - \omega_1$. For further discussion, we assume that $|F_n|^2$ is rather independent of frequency. This gives

$$2\overline{E}_k \approx \frac{\pi}{2\eta \omega_n^3 \Delta \omega} \sum_{n_1}^{n_2} |F_n|^2 + \sum_{n<n_1} \frac{|F_n|^2}{\Delta \omega} \int \frac{d\omega}{\omega^4} + \sum_{n>n_2} \frac{|Fn|^2}{\Delta \omega} \int \frac{d\omega}{\omega_n^4}$$

$$\approx \frac{\pi}{2\eta \omega_M^3 \Delta \omega} \sum_{n_1}^{n_2} |F_n|^2 + \sum_{n<n_1} \frac{|F_n|^2}{\omega_M^4} + \sum_{n>n_2} \frac{|F_n|^2}{\omega_n^4} \quad . \tag{8.28}$$

Here ω_M denotes the centre frequency of the band $\omega_1 < \omega_M < \omega_2$. The last sum in (8.28) usually is unimportant because for $n > n_2$ we have $\omega_n > \omega_M$; furthermore, high modes seldom are excited very much. The situation is somewhat more complicated with respect to the second sum. If all values of $|F_n|^2$ are of the same order of magnitude, it can be neglected because in most cases of practical interest $\eta \Delta \omega << \omega_M$, which makes the first sum dominant. There are, however, situations where F_n depends very strongly on mode number and where the following calculations may lead to incorrect results. We shall discuss this point later. For the time being, we assume that the second sum in (8.28) can also be neglected, and that $|\overline{F_n^2}|$ is an average value of $|F_n^2|$. This way we get

$$\overline{E}_k \approx \frac{\pi \overline{|F_n|}^2}{4\eta \omega^3} \frac{\Delta N}{\Delta \omega} \quad . \tag{8.29}$$

Here we replaced the centre frequency ω_M by ω and took into account that $n_2 - n_1 = \Delta N$ is equal to the number of modes in the frequency band $\Delta \omega$.

This is a remarkably simple result. It states that the average kinetic energy of the system depends only on the excitation parameter, the frequency, the loss factor and, most notably, on the number of modes in the excited frequency band. This result is correct provided the motion of the system is governed solely by those modes which are in resonance.

A special but rather common example is excitation by a point force or a combination of uncorrelated point forces (rain on the roof). In this case, the source term in (20.62) is

$$F(x) = -i\omega F_0 \delta(x - x_0) \quad . \tag{8.30}$$

F_0 is the amplitude of the force acting at $x = x_0$. The factor $-i\omega$ appears, because in going from (20.70) to (8.22), we replaced the general field quantity $p(x)$ by the velocity. If we do this, the basic relation (20.62) has to be an equation balancing forces. This is only the case when we multiply it by $-i\omega$., A combination of (8.30) with (20.66) gives

$$F_n = -i\omega \int F_0 \delta(x - x_0)\phi_n(x)dx = -i\omega F_0 \phi_n(x_0) \quad ,$$

$$|F_n|^2 = \omega^2 F_0^2 \phi_n^2(x_0) \quad .$$

(8.31)

This result is not yet very helpful, because it contains the eigenfunction ϕ_n, which we usually do not know explicitly. Therefore, we calculate an ensemble average, i.e. we assume that there are many structures with the same eigenfunctions and each of them is driven by a point force but at different position x_0. To average over the whole ensemble, we take

$$|\overline{F_n^2}| = \omega^2 |F_0^2| \sum \phi_n^2(x_0)\Delta m / \sum \Delta m$$

$$= \omega^2 |F_0^2| \int \phi_n^2(x_0)m dx_0 / \int m dx_0 = \omega^2 |F_0^2|/M \quad .$$

(8.32)

Here M is the total mass of the system. In taking the average, we "weighted" each value with the mass density at this position. This allows us to apply the orthogonality relation (8.23). Whether this method is realistic is another question, but for slowly varying mass density (as is usually found in real life), this does not make any difference.

If we now insert (8.32) into (8.29), we get for the total kinetic energy of a point-driven structure in the average

$$\overline{E}_k \approx \frac{\pi F_0^2}{4\eta\omega M} \frac{\Delta N}{\Delta\omega} \quad .$$

(8.33a)

The mean–square velocity is consequently

$$|\overline{v^2}| \approx \frac{\pi F_0^2}{2\eta\omega M^2} \frac{\Delta N}{\Delta\omega} \quad .$$

(8.33b)

In deriving this result we made may assumptions which, taken at their face value, may make Equation (8.33) appear rather unrealistic. Although there are limits of the applicability of (8.29), (8.33), which will be discussed in Section 8.6, the equations are surprisingly "robust" provided one works with third octave or octave frequency bands and is satisfied with an accuracy of a few dB (which, however, may be more than a 50% error in amplitude).

We obtain some further insight into (8.33) by comparing this result with the behaviour of a simple system of mass M, resonance frequency ω_0, and a loss factor η. The equation of motion for such a system is

$$\left[-\omega^2 + \omega_0^2(1 - i\eta)\right] \; v = -i\omega F_0/M \quad .$$

For broad–band excitation we make the same calculations that led us to (8.27), and obtain

$$|\overline{v^2}| = \frac{1}{\Delta\omega} \int |v^2|d\omega = \frac{\pi F_0^2}{2\eta\omega_0 M^2} \; \frac{1}{\Delta\omega} \quad . \tag{8.34}$$

The only difference between (8.34) and (8.33b) lies in the factor ΔN. Or, in other words, a system with ΔN modes within the frequency band $\Delta\omega$ has the same average response as ΔN simple mass–spring systems with the same resonances.

8.3.3 Comparison of Finite and Infinite Systems

Beams, plates, shells, etc. of infinite extent are much easier to handle mathematically than finite structures, because there are only waves away from the source and boundary conditions do not have to be considered. Therefore, it would be advantageous to recast the previous calculations in such a way that the properties which are known for the corresponding infinite structure can be used for the finite structure. The quantity best suited for this purpose is power, because it is well known from many experiments that the sound power transmitted by a given source into a large room, when averaged over frequency and a few positions, is equal to the power transmission in free space. The same is true for power transmission into plates, beams, etc.

In the following we restrict ourselves to beams, plates and shells, where the power transmitted by a driving pressure $p(x)$ is given by

$$P = \frac{1}{2}Re\left\{\int_S p(x)v^*(x)dx\right\} = \frac{1}{2}Re\left\{\sum \frac{F_n^*}{(\omega_n^2 - \omega^2)^*} \int_S p(x)\phi_n(x)dx\right\}$$

$$= \frac{1}{2\omega}Re\left\{\sum \frac{i|F_n|^2}{(\omega_n^2 - \omega^2)^2}\right\} = \frac{1}{2\omega}\sum \frac{\eta\omega_n^2|F_n|^2}{(\omega_n^2 - \omega^2)^2 + \eta^2\omega_n^4} \quad . \tag{8.35a}$$

Here we introduced (8.22) and made use of the fact that, similar to (8.31), the excitation parameter for a distributed source is given by

$$F_n = -i\omega \int p(x)\phi_n(x)dx \quad .$$

Integration over frequency is done similar to (8.24)—(8.27). If we again consider only resonating modes, we get, in the frequency average

$$\overline{P} = \frac{|\overline{F_n{}^2}|}{2} \frac{\pi}{2\omega^2} \frac{\Delta N}{\Delta\omega} \quad . \tag{8.35b}$$

Comparing this with (8.29) gives

$$\overline{E}_k = \frac{\overline{P}}{\eta\omega}; \quad \eta = \frac{\overline{P}}{\omega \overline{E}_k} \quad . \tag{8.36}$$

The second version of (8.36) is just the definition of the loss factor, which fortunately is recovered. For point–driven structures, we have already found (8.32); furthermore, we know from (8.20) the number of modes in a plate. Combining these result, we have, because of $M = mS$

$$\overline{P} = \frac{1}{2} \frac{F_0^2 \omega^2}{M} \frac{\pi}{2\omega^2} \frac{S}{4\pi} \left(\frac{m}{D}\right)^{1/2} = \frac{F_0^2}{2} \frac{1}{8\sqrt{Dm}} \quad . \tag{8.37}$$

This is a remarkable result because it shows that in the frequency average the power transmission into a large plate is independent of size, shape and boundary conditions. Furthermore, $1/8\sqrt{Dm}$ is just the real part of the point input admittance (or mobility) of an infinite plate with bending stiffness D and mass per unit area m. This quantity can be derived quite independently (see Cremer, Heckl & Ungar 1988).

If we accept the equality of power transfer for finite, and the corresponding infinite structure as a fact, and there is no reason for doubt, we can derive an interesting general relation between the point admittance A of a structure and its asymptotic number of modes. To this end, we write for an infinite structure

$$P_\omega = \frac{1}{2} Re\left\{F_0 v^*(x_0)\right\} = \frac{|F_0|^2}{2} Re\{A\} \quad . \tag{8.38}$$

(see also Cremer, Heckl & Ungar 1988).

On the other hand, if we insert (8.32) into (8.35b), we get

$$\overline{P} = \frac{|F_0|^2}{2} \frac{\pi}{2M} \frac{\Delta N}{\Delta\omega} \quad ; \tag{8.39}$$

thus

$$\frac{\Delta N}{\Delta\omega} = \frac{2M}{\pi} Re\{A\} \quad . \tag{8.40}$$

8.3.4 Energy Balance Equations

The results of this paragraph rely on some rather sweeping statements but still they provide the most useful relations of SEA. The easiest way to derive them is to do some energy bookkeeping.

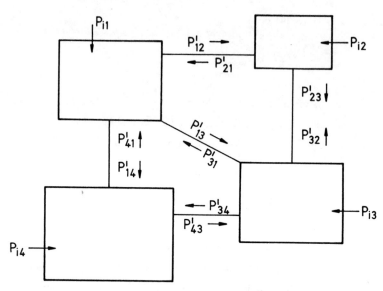

Figure 8.3 Power in coupled systems.

In Figure 8.3 several coupled elements are shown. Conservation of energy applied to the first of them requires

$$P_{\ell 1} + \sum_{n \neq 1} P'_{1n} = \sum_{n \neq 1} P'_{n1} + P_{i1} \quad . \tag{8.41}$$

Here $P_{\ell 1}$ is the power transformed into heat in system 1. P_{i1} is the power supplied by an outside source. P'_{1n} is the power flowing from element 1 to element n, and P'_{n1} is the power flow in the opposite direction. Note that P'_{mn} is not the net power flow in the coupling element but the flow in the direction from n to m. The net power flow, the quantity we used in Section 8.2, would be $P'_{nm} - P'_{mn}$. The coupling elements are assumed not to absorb energy. According to the definition of the (average) loss factor, the power lost in the nth system is given by

$$P_{\ell n} = \omega \eta_n E_{kn} = \frac{1}{2} M_n \omega \eta_n |v_n|^2 \tag{8.42}$$

(see also (8.36)).

To describe the power flow from n to m, we introduce a similar quantity, the so-called transmission loss factor η_{nm}, which gives the power that is "lost" from element n because it propagated to m:

$$P'_{nm} = \frac{1}{2} M_n \omega \eta_{nm} |v_n|^2 \quad . \tag{8.43}$$

With this definition we tacitly assume that the power flow is proportional to the energy density in the "source" element.

Combining (8.41) with (8.43), we get, for the mth element,

$$M_m |v_m^2| (\eta_m + \sum_{n \neq 1} \eta_{mn}) - \sum_{n \neq 1} M_n \eta_{nm} |v_n^2| = \frac{2 P_{im}}{\omega} \tag{8.44}$$

$$\text{for } m = 1 \dots N \; .$$

As a simple check we can add all the N equations of (8.44) and get

$$\frac{\omega}{2} \sum_{m=1}^{N} M_m \eta_m |v_m|^2 = \sum_{n=1}^{N} P_{im} \quad , \tag{8.45}$$

which just states that the total incoming power must be equal to the total power absorbed by the losses.

Before proceeding to the important question of how to find the coupling loss factors η_{nm}, we prove a very helpful reciprocal property of this quantity. To this end we consider just three elements. Then (8.44) becomes

$$a_{11} v_1^2 + a_{12} v_2^2 + a_{13} v_3^2 = Q_1$$
$$a_{21} v_1^2 + a_{22} v_2^2 + a_{23} v_3^2 = Q_2 \quad . \tag{8.46}$$
$$a_{31} v_1^2 + a_{32} v_2^2 + a_{33} v_3^2 = Q_3$$

Here

$$a_{12} = -\frac{1}{2} \omega M_2 \eta_{21}; \; a_{13} = -\frac{1}{2} \omega M_3 \eta_{31}; \; a_{21} = -\frac{1}{2} \omega M_1 \eta_{12}, \text{ etc.} \tag{8.47}$$

If we first assume that only the first system is excited by a point force, then we have, because of (8.38)

$$Q_1 = \frac{F_0^2|}{2} Re\{A_1\} = \frac{|F_0|^2}{2} A_{R1} \tag{8.48}$$

$$Q_2 = 0; \quad Q_3 = 0 \quad .$$

Solving (8.46) gives

$$v_2^2 = -\frac{|F_0|^2}{2}A_{R1}(a_{21}a_{33} - a_{31}a_{23})/\text{Det} \quad .$$

In a second thought experiment we excite element 2 by the point force F_0. This gives us

$$Q_1 = 0; \quad Q_2 = \frac{|F_0^2|}{2}A_{R2}, \quad Q_3 = 0 \quad ,$$

and

$$v_1^2 = -\frac{F_0^2}{2}A_{R2}(a_{12}a_{33} - a_{13}a_{32})/\text{Det} \quad . \tag{8.49}$$

Because of reciprocity, $v_1^2 = v_2^2$, or

$$A_{R1}(a_{21}a_{33} - a_{31}a_{23}) = A_{R2}(a_{12}a_{33} - a_{13}a_{32}) \quad .$$

We obtain similar relations by applying reciprocity to elements 1 and 3, and then to 2 and 3. The results are

$$A_{R1}(a_{21}a_{32} - a_{22}a_{31}) = A_{R3}(a_{12}a_{23} - a_{22}a_{13})$$
$$A_{R3}(a_{11}a_{23} - a_{13}a_{21}) = A_{R2}(a_{11}a_{32} - a_{12}a_{31}) \quad . \tag{8.50}$$

These equations can be fulfilled only if

$$a_{12} = a_{21}A_{R1}/A_{R2}; \quad a_{13} = a_{31}A_{R1}/A_{R3}; \quad a_{23} = a_{32}A_{R2}/A_{R3} \quad .$$

Using (8.47) gives

$$M_2\eta_{21} = M_1\eta_{12}A_{R1}/A_{R2} \text{ etc.} \tag{8.51}$$

The quantity A_R which we introduced here is the real part of the point admittance. Since it is connected to the mode number via (8.40), we finally get the general relation

$$\eta_{nm}\frac{\Delta N_n}{\Delta\omega} = \eta_{mn}\frac{\Delta N_m}{\Delta\omega} \quad . \tag{8.52}$$

So we need to know the coupling loss factor only for one direction; the other one follows from (8.52).

8.3.5 Coupling Loss Factors

The most difficult problem in SEA is finding the coupling loss factors. Sometimes they can be obtained from independent measurements, but generally they have to be approximated by treating the power flow in corresponding infinite systems.

Example 1. Radiation efficiency

We assume the SEA system (see Figure 8.4) to consist of a vibrating structure (element 1) with mean–square velocity $|v_1|^2$ and a reasonably large room (element 2) filled with a gas or liquid with density ρ_2 and speed of sound c_2. If the room is large enough, the radiated power is independent of room shape and size and we can write for the radiated power, i.e. the power related from 1 to 2

$$P'_{12} = \frac{1}{2}\rho_2 c_2 S_1 |v_1|^2 \sigma \quad . \tag{8.53}$$

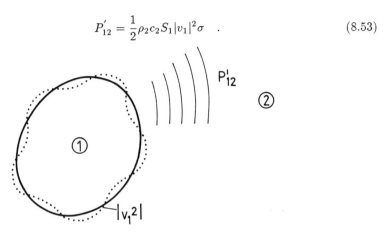

Figure 8.4 Radiation from a vibrating structure.

Here S_1 is the radiating area and σ the so–called radiation efficiency. Comparing (8.43) with (8.53), we find for the coupling loss factor of this solid–fluid coupling

$$\eta_{12} = \frac{\rho_2 c_2}{\omega M_1/S_1}\sigma = \frac{\rho_2 c_2}{\omega m_1}\sigma \quad . \tag{8.54}$$

The radiation efficiency σ can be calculated, found in the literature, measured, or roughly estimated. The estimation rules are:

- when the average distance between nodal lines on the vibrating structure is larger than half the wavelength in the surrounding medium, it turns out that $\sigma \approx 1$; typically, this is the case when the centre frequency of excitation ω is so high that the relation holds

$$\omega > \sqrt{12}\frac{c_2^2}{c_{L1}h_1} \tag{8.55}$$

(see also Equation (17.77)); c_{L1} = speed of longitudinal waves in the outer wall material of the structure, h_1 = thickness of the wall structure;

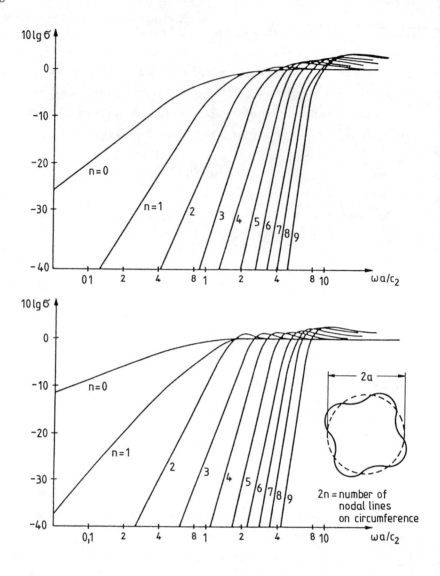

Figure 8.5a Radiation efficiencies of spheres (top) and cylinders (bottom).

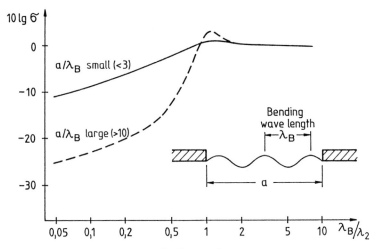

Figure 8.5b Radiation efficiencies of a finite plate.

- when the "coincidence" condition (8.55) is not fulfilled, then $\sigma < 1$. How small σ becomes depends on the mechanical properties of the structure, the frequency, the boundary conditions, and also the type of excitation. Figure 8.5 gives examples of radiation efficiencies. An upper limit is given approximately by

$$\sigma < \frac{\omega^2 D_1^2/4c_2^2}{1 + \omega^2 D_1^2/4c_2^2} \quad, \tag{8.56}$$

where D_1 is a typical dimension of the structure.

Example 2. Transmission coefficients

For some combinations of structural components, transmission coefficients τ_{12} are known. They determine how much power propagates from structure 1 to structure 2 under idealized conditions and without any power flowing backwards from 2 to 1. As seen from structure 1, the transmission coefficient is the same as an absorption coefficient α, because the definitions of the two quantities are closely related:

$$\alpha = \frac{P_a}{P_{i1}}; \quad \tau_{12} = \frac{P'_{12}}{P_{i1}} \quad. \tag{8.57}$$

Here P_{i1} is the power that impinges on the boundary. P_a and P'_{12} are the powers that are not reflected (and therefore appear as losses as far as system 1 is concerned).

A relation between α or τ_{12} and the reverberation time T_1 of system 1 is obtained from the general considerations on reverberation in Chapter 22. There it is found that the reverberation time depends only on the mean free path ℓ_m (see Equations (22.20)—(22.22)), the wave speed c_{gr1}, which is the group velocity in the present case where the energy propagation is concerned, and the average absorption coefficient α_m:

$$T_1 = 13.815 \, \ell_m/\alpha_m c_{gr1} \quad . \tag{8.58}$$

Next we restrict ourselves to planar systems with area S_1 and perimeter P_1. Usually only part of the perimeter is connected to the adjacent. Provided this part has length ℓ_{12}, and absorption coefficient α or transmission coefficient τ_{12}, then the average absorption coefficient is

$$\alpha_m = \frac{\ell_{12}}{P_1}\alpha \quad \text{or} \quad \frac{\ell_{12}}{P_1}\tau_{12} \quad .$$

Using now the statistical formula for the mean free path in two dimensions (Equation (22.21)), we find

$$T_1 = 13.815\pi S_1/\tau_{12}\ell_{12}c_{gr1} \quad . \tag{8.59}$$

The transition to the coupling loss factor is given by the general relation between T and η as shown in Table 1 of Chapter 22. So we finally arrive at

$$\eta_{12} = \frac{13.815}{\omega T_1} \approx \frac{\tau_{12}\ell_{12}}{\omega\pi S_1}c_{gr1} \tag{8.60}$$

(see Wöhle 1984).

Figure 8.6 gives examples of calculated transmission coefficients for one–dimensional structure–borne sound problems (see Cremer, Heckl & Ungar 1988). The results presented there can also be used as approximations when plates are connected, provided that the group velocity in both elements is almost the same. The situation is more complicated when plates or sheets with very different values of c_{gr} are connected; no simple results can be given for such cases.

8.3.6 Strong Coupling

The equations of SEA become very simple and very useful when the coupling between the elements is strong, i.e. when the material loss factor is smaller than the coupling loss factor. This can be seen most easily when two elements are coupled, because in this case, the energy balance equations are

$$\omega(\eta_1 + \eta_{12})M_1|v_1^2| - \omega\eta_{21}M_2|v_2^2| = 2P_{i1}$$
$$-\omega\eta_{12}M_1|v_1^2| + \omega(\eta_2 + \eta_{21})M_2|v_2^2| = 2P_{i2} \quad . \tag{8.61}$$

From this we can find

$$\frac{|v_1|^2}{|v_2|^2} = \frac{M_2}{M_1} \frac{\eta_2 + \eta_{21} + \eta_{21} P_{i2}/P_{i1}}{\eta_{12} + (\eta_1 + \eta_{12})P_{i2}/P_{i1}} \quad . \tag{8.62}$$

For strong coupling, i.e. $\eta_2 << \eta_{21}$ and $\eta_1 << \eta_{12}$, this becomes

$$\frac{|v_1^2|}{|v_2^2|} \approx \frac{M_2}{M_1} \frac{\eta_{21}}{\eta_{12}} = \frac{M_2}{M_1} \frac{\Delta N_1}{\Delta N_2} \tag{8.63}$$

or

$$\frac{M_1|v_1|^2}{M_2|v_2|^2} = \frac{E_{1kin}}{E_{2kin}} \approx \frac{\Delta N_1}{\Delta N_2} \quad . \tag{8.64}$$

Here we have used the reciprocity relation (8.52).

It is easy to extend (8.64) to systems with several elements. The results are

$$E_{1kin}/E_{2kin}/E_{3kin}\ldots \approx \Delta N_1/\Delta N_2/\Delta N_3\ldots \quad . \tag{8.65}$$

Equation (8.65) is very useful because it shows that in the case of a strongly coupled system the kinetic energies, and consequently the r.m.s. velocities, are simply related by the number of modes. The type of coupling, the excitation, etc., do not play a role.

If only one of two systems in excited, say $P_{i2} = 0$, the condition that *all* material loss factors have to be smaller than the transmission loss factors can be relieved somewhat. As (8.62) shows, it is enough in this case if $\eta_2 << \eta_{21}$; there is no condition on η_1.

Loss factors of real structures, including friction losses at interfaces, are typically of the order of 10^{-2} to 10^{-3}. Coupling loss factors are, according to (8.60), of the order of $\tau_{nm}\ell_n/\lambda_n$, where ℓ_n is a typical dimension and λ_n a wavelength. Thus the condition of strong coupling is fulfilled when the transmission coefficient is 10^{-2} or higher. As Figure 8.6 shows, this may well be the case for plate and beam combinations, but it is rather unlikely when two air spaces are coupled via a wall. Similarly, a brief look at (8.54) reveals that transmission loss factors between air or gas spaces and solid structures are usually very small (because $\rho c << \omega m$), giving no strong coupling, whereas in the case of liquid–filled spaces coupled to solid structures, strong coupling may very well occur.

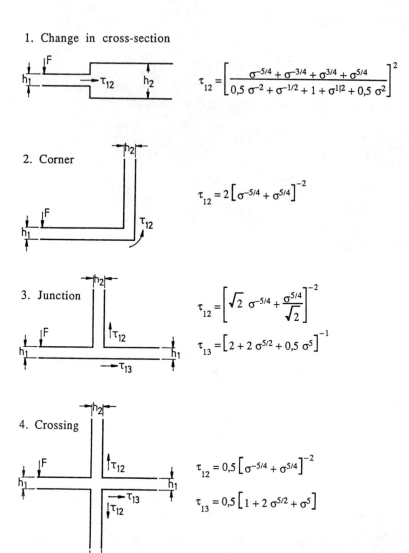

1. Change in cross-section

$$\tau_{12} = \left[\frac{\sigma^{-5/4} + \sigma^{-3/4} + \sigma^{3/4} + \sigma^{5/4}}{0,5\,\sigma^{-2} + \sigma^{-1/2} + 1 + \sigma^{1/2} + 0,5\,\sigma^{2}} \right]^{2}$$

2. Corner

$$\tau_{12} = 2 \left[\sigma^{-5/4} + \sigma^{5/4} \right]^{-2}$$

3. Junction

$$\tau_{12} = \left[\sqrt{2}\,\sigma^{-5/4} + \frac{\sigma^{5/4}}{\sqrt{2}} \right]^{-2}$$

$$\tau_{13} = \left[2 + 2\,\sigma^{5/2} + 0,5\,\sigma^{5} \right]^{-1}$$

4. Crossing

$$\tau_{12} = 0,5 \left[\sigma^{-5/4} + \sigma^{5/4} \right]^{-2}$$

$$\tau_{13} = 0,5 \left[1 + 2\,\sigma^{5/2} + \sigma^{5} \right]$$

Figure 8.6 Examples of transmission coefficients for bending waves. All elements consist of the same material; $\sigma = h_1/h_2$. Excitation of other wave types (longitudinal, torsional) is not considered.

In passing, it is method that the weak coupling case is a good possibility for measuring a transmission loss factor. To this end, and element 1, which is large enough to have the required high modal density, is connected via the coupling element, which is to be tested, to an element 2. Element 2 should again be large enough and it should have a loss factor $\eta_2 >> \eta_{21}$).* If only element 1 is excited, Equation (8.62) becomes

$$\frac{|v_1^2|}{|v_2^2|} = \frac{M_2}{M_1} \frac{\eta_2}{\eta_{12}} \quad . \tag{8.66}$$

With the exception of η_{12}, all quantities in (8.66) can be measured directly (η_2 via a reverberation time measurement). In principle, (8.66) is the equation underlying the standard transmission loss measurement of walls and floors in architectural acoustics. All measurement problems that arise there also appear in η_{12} measurements (and vice versa).

8.4 THERMODYNAMIC ANALOGY

In thermodynamics it is quite common to consider heat energy of a solid body as the total mechanical energy of all oscillating atoms and molecules. Temperature is then understood as the average energy ΔE of one oscillator and the total energy is equal to

$$E = \Delta N.\Delta E \sim \Delta N.T_T \quad . \tag{8.67}$$

Here ΔN is the number of oscillators and T_T the thermodynamic temperature.

In thermodynamics with its vast number of oscillators, it can usually be assumed that in the steady–state case, in an isolated system, there is equipartition of energy, i.e. in the average, all modes are excited equally well.

Another fact in thermodynamics is that the net energy flow P_{12} between two bodies is given by the temperature difference and the heat conductivity, k_{12}

$$P_{12} = k_{12}(T_{T1} - T_{T2}) \quad . \tag{8.68}$$

Finally, we can learn from thermodynamics that, according to the second law, the entropy (which is related to the number of oscillators) has a tendency to increase, which usually means that coupled elements (if they are isolated from outside) settle at a common temperature sooner or later.

It is easily seen that SEA has many features in common with thermodynamics.

(a) According to (8.9), the energy flow between two coupled oscillators is proportional to the "conductivity" β and the difference of the two "temperatures" (energies per mode).

---

* But the loss factor also must not be too high, because then the conditions for SEA are violated.

(b) the same argument as in (a) can be applied to the energy balance equation (8.44), which holds for multi–modal systems. If we consider only two elements, and write (8.44) only for $m = 1$, it reads

$$P_{i1} = \frac{1}{2}\omega M_1\eta_1|v_1^2| + \frac{1}{2}\omega M_1\eta_{12}|v_1^2| - \frac{1}{2}\omega M_2\eta_{21}|v_2^2| \quad . \tag{8.69}$$

Using the reciprocity relation (8.52), this can be written as

$$P_{i1} - \frac{1}{2}\omega M_1\eta_1|v_1^2| = \omega\eta_{12}\Delta N_1\left(\frac{1}{2}\frac{M_1|v_1^2|}{\Delta N_1} - \frac{1}{2}\frac{M_2|v_2^2|}{\Delta N_2}\right) \quad ,$$

or

$$P_{12} = \omega\eta_{12}\Delta N_1(T_{1\text{SEA}} - T_{2\text{SEA}}) \sim \tau_{12}(T_{1\text{SEA}} - T_{2\text{SEA}}) \quad . \tag{8.70}$$

Here we have introduced the net power flow which is incoming power minus dissipated power, and the "SEA–temperature"

$$T_{\text{SEA}} = \frac{1}{2}M|v^2|/\Delta N \quad . \tag{8.71}$$

(c) The equipartition of energy, which holds under certain conditions in SEA, was discussed when (8.34) and (8.33b) were compared.

(d) Strong coupling just means that all elements have the same temperature, because (8.65) could also be written as

$$T_{1\text{SEA}} = T_{2\text{SEA}} = T_{3\text{SEA}}\ldots \quad . \tag{8.72}$$

(e) Even the concept of SEA–entropy (i.e. the number of excited modes) can give some insight into noise control problems in general.

• Sound energy has a tendency to spread over all available modes, i.e. "entropy" increases; the opposite never happens.

• One might judge the "difficulty" of a noise control job by setting it equal to the "SEA–entropy" (not the energy) of the sound source. The higher the "SEA–entropy", the more space is needed for additional secondary noise control devices, and the more transducers are necessary when active control (anti sound) is used.

8.5   APPLICATIONS

8.5.1 Vibration Excitation by a Sound Field

Element 1 shall be a reverberant room of volume V_1, density ρ_1, speed of sound c_1. It has, when measured without power flow from element 1 to 2, the reverberation time T_1. It is excited by a sound source of power P_{i1}. Element 2 is a plane or curved plate of mass per unit area m_2, area S_1 and radiation efficiency σ_2. Its reverberation time, without power flow from 2 to 1, shall be T_1. Element 2 is not excited by an external source.

Under these conditions, (8.44) or (8.61) become

$$\omega \left(\frac{13.8}{\omega T_1} + \eta_{12} \right) \frac{\rho_1 V_1}{2} |v_1^2| - \frac{1}{2}\omega \eta_{21} m_2 S_2 |v_2^2| = P_{i1}$$

(8.73)

$$-\frac{1}{2}\omega \eta_{12} \rho_1 V_1 |v_1^2| + \frac{1}{2}\omega \left(\frac{13.8}{\omega T_2} + \eta_{21} \right) m_2 S_2 |v_2^2| = 0 \quad .$$

Using the relations (8.52) and (8.54), this can be rewritten as

$$\frac{1}{2} \left(\frac{13.8}{T_1} + \sigma_2 \frac{\rho_1 c_1}{m_2} \frac{\Delta N_2}{\Delta N_1} \right) \frac{V_1}{\rho_1 c_1^2} |p_1|^2 - \frac{1}{2}\rho_1 c_1 \sigma_2 S_2 |v_2^2| = P_{i1}$$

(8.74)

$$-\frac{1}{2}\sigma_2 \frac{\rho_1 c_1}{m_2} \frac{\Delta N_2}{\Delta N_1} \frac{V_1}{\rho_1 c_1^2} |p_1|^2 + \frac{1}{2} \left(\frac{13.8}{T_2} + \frac{\rho_1 c_1}{m_2}\sigma_2 \right) m_2 S_2 |v_2^2| = 0 \quad .$$

Here we have replaced the total kinetic energy $\frac{1}{2}\rho_1 V_1 |v_1^2|$ by the total potential energy $\frac{1}{2}V \frac{|p_1|^2}{\rho_1 c_1^2}$, which is of equal magnitude, when p_1 is the r.m.s. sound pressure in the reverberant room. Equation (8.74) contains only quantities which are known; therefore $|v_2^2|/P_{i1}$, $|p_1^2|/P_{i1}$; and $|v_2^2|/|p_1^2|$ can be calculated.

8.5.2 Coupling of a Beam and a Plate

When a beam is rigidly connected to a plate and when the damping of the two systems is small, the conditions of "strong coupling" hold. Therefore we can apply Equation (8.63), and find

$$\frac{|v_1^2|}{|v_2^2|} \approx \frac{M_2}{M_1} \frac{\Delta N_1}{\Delta N_2} \quad .$$

(8.75)

This very simple result agrees fairly well with experimental results, provided there are enough resonating modes in each frequency band.

8.5.3 Multiple Wall with Many Rigid Sound Bridges

Double walls and multiple walls are generally assumed to have better sound isolation at high frequencies than single walls of the same weight because *the sound has to overcome many obstacles*. This belief can be wrong when the multiple wall consists of several lightly damped plates which are connected by many rigid sound bridges. Structures of this type are fairly common in aircraft and spacecraft because they can be made very light and very stiff.

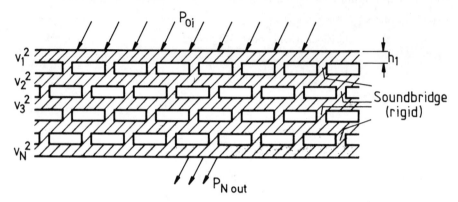

Figure 8.7 Application of S.E.A.   Multiple wall with many sound bridges.

We consider a construction shown in Figure 8.7. It consists of N equal plates of mass per unit area m_1. The plates are connected via rigid sound bridges. We assume that the plate sections between the sound bridges are large enough for the conditions of SEA to hold, i.e. the distance between sound bridges shall be at least a wavelength. This restricts the applicability of the following calculations to the high–frequency range. In the low–frequency range, the construction can be considered as a homogeneous plate with a certain mass and rigidity. It can then be treated by standard methods. Between these two limiting cases there remains a gap within which FEM or a similar more method has to be used.

The energy balances equations for the plates at high frequencies are

$$\frac{\omega}{2}(\eta_1 + \eta_{12})m_1 S_1 |v_1^2| - \frac{\omega}{2}\eta_{21} m_2 S_2 |v_1^2| = P_{0i}$$

$$-\frac{\omega}{2}\eta_{12}m_1 S_1 |v_1^2| + \frac{\omega}{2}(\eta_2 + \eta_{21} + \eta_{23})m_2 S_2 |v_2^2| - \frac{\omega}{2}\eta_{32} m_3 S_3 |v_3^2| = 0$$

$$-\frac{\omega}{2}\eta_{23}m_2 S_2 |v_2^2| + \frac{\omega}{2}(\eta_3 + \eta_{32} + \eta_{34})m_3 S_3 |v_3^2| - \frac{\omega}{2}\eta_{43} m_4 S_4 |v_4^2| = 0$$

$$-\frac{\omega}{2}\eta_{N-1,N}m_{N-1} S_{N-1} |v_{N-1}^2| + \frac{\omega}{2}(\eta_N + \eta_{N,N-1})m_N S_N |v_N^2| + P_{N\ \text{out}} = 0 \quad .$$

Next we add all these equations. For the sake of simplicity, we also assume that all areas are equal to S_1, all masses equal to m_1, and all loss factors equal to η_1. This gives

$$\frac{\omega}{2}m_1 S_1 \eta_1 \left(|v_1|^2 + |v_2|^2 + |v_3^2| \ldots |v_N^2| \right) + P_{N \text{ out}} = P_{0i} \quad .$$

If the elements are strongly coupled (as they usually are) we can use (8.65). Furthermore, we can, in air, safely neglect $P_{N \text{ out}}$ and find

$$\frac{\omega}{2}m_1 S_1 \eta_1 |v_N^2| \left(\frac{\Delta N_1}{\Delta N_N} + \frac{\Delta N_2}{\Delta N_N} + \ldots 1 \right) = N\frac{\omega}{2}m_1 S_1 \eta_1 |v_n^2| = P_{01} \quad .$$

Here we set $\Delta N_n = \Delta N_N = \Delta N_1$.

In the literature we find

$$P_{0i} = \frac{\pi}{4} \frac{S_1 \sigma_1}{\sqrt{m_1 D_1}} \frac{c_0^2}{\omega^2}|p_0^2|$$

(Cremer, Heckl & Ungar 1988, Section VI.8.d). This gives

$$|v_N^2| = \frac{\pi}{2} \frac{\sigma_1}{\eta_1} \frac{c_0^2}{\omega^3} \frac{|p_0^2|}{m_1\sqrt{m_1 D_1} N} = \pi\sqrt{3}\frac{1}{\eta_1} \frac{c_0^2}{\omega^3} \frac{|p_0^2|}{\rho_1^2 c_{L1}^2} \frac{\sigma_1}{N h_1^3} \quad . \tag{8.76}$$

Here p_0 is the r.m.s. sound pressure in the room before the wall, ρ_1 the density of the wall material, c_{L1} the speed of longitudinal waves in the wall material, and h_1 the wall thickness.

If we want to compare the velocity v_N^2 of this multiple wall with the vibration of a compact wall made of the same material but with thickness $N h_1$, then we see that the velocity ratio would be

$$\frac{v_{\text{mult}}^2}{v_{\text{comp}}^2} \approx \frac{\sigma_{\text{mult}}}{N h_1^3} \frac{(N h_1)^3}{\sigma_{\text{comp}}} > 1 \quad , \tag{8.77}$$

i.e. the strongly coupled multiple wall vibrates more (and at least at the high frequencies has poorer sound isolation) than the compact wall. In experiments, this effect does show up rather clearly, with Equation (8.77) giving an upper limit for the velocity ratio.

8.5.4 Calculation of Transmission Loss Coefficients Using the Energy Balance Equations

Equation (8.44), which forms the fundamental equation of SEA, can also be applied "in reverse" to measure the coupling loss factors, etc. To this end the structure of interest, which may consist of N elements, is excited at position 1 and all the velocities v_n^2 are measured; then the structure is excited at another position and again all velocities are measured, etc. This way a system of linear equations can be established, where v_n^2 are the known coefficients and η the unknowns. In principle, it is easy to get as many equations (or more) as there are unknowns and therefore the η–values can be calculated. As was pointed out by Woodhouse (1981), this method is not very accurate, but it is still worthwhile making such experiments because much can be learned from this "inverse problem".

8.6   LIMITS OF APPLICABILITY

8.6.1   Number of Modes

SEA is based on the assumption that the vibration of a structure or the sound field in a room is determined by the resonant vibration of many modes. This requirement obviously is fulfilled only if the excitation is of broad–band type containing all frequencies, and if the structure or room has enough resonating modes within the frequency band of interest. If one is willing to accept errors of 2—3 dB, three or more modes within the frequency band (octave or third octave) seem to be sufficient. Since the number of modes generally increases with frequency, SEA is better suited for high frequencies than for low ones. A case where SEA would definitely be the wrong method would be a structure with only a few modes when it is excited by a narrow band source (i.e. gear box, hydraulic pump, blade passage noise, etc.).

8.6.2   Relative Mode Spacing

When a structure is excited by a broad–band source, it acts as a sort of filter emphasizing the resonance frequencies. Therefore the sound which is transmitted to the second (coupled) structure is already heavily filtered. If the resonance frequencies of the second structure were the same ("modal line–up") as those of the first, the transmission would be rather strong; if, however, the resonances would always be "in between" ("modal anti–line–up"), the transmission would be weak. SEA does not give reliable results in these two extreme cases, because it assumes that the resonance frequencies are spaced "randomly". What the phrase "randomly–spaced resonances" actually means is hard to quantify. It certainly is a more stringent condition than having a large mode number; generally it is equivalent to the requirement that the shapes of the coupled structure should be somewhat different (a difference in the linear dimensions by 1/6 of a wavelength seems to be sufficient).

If the average distance between neighbouring resonances is $\delta\omega$, and if this distance is larger than the half–width $\eta\omega$ of a resonance, then it can be shown (Hodges & Woodhouse 1986) that "modal line–up" leads to a coupling loss factor which is by a factor 2 $\delta\omega/\pi\omega(\eta_1 + \eta_2)$ too high. In the other extreme case of "modal anti–line–up", it is too small by the same factor. In case $\delta\omega$ is smaller than the half–width $\eta\omega$, modal line–up or anti–line–up does not have a pronounced effect.

8.6.3 Forced Vibrations

In the basic relations of SEA (e.g. (8.44)), all modes that do not have their resonance frequency in the frequency band of interest are neglected. This can lead to serious errors when the damping is so high that the resonating modes are no longer dominant. it can be shown that this is the case when $\eta\ell_m/\lambda > 1$ (ℓ_m = mean free path, λ = wavelength) (Lesueur 1988).

The neglect of the non–resonating modes can also lead to errors when the spatial distribution of the exciting field is such that non–resonating modes are well coupled to the excitation, whereas the resonating modes are not well coupled, i.e. when in (8.28) the "generalized forces F_n" are small for $n_1 < n < n_2$ and large for $n < n_1$ or $n > n_2$. This phenomenon hardly every occurs when the excitation is localized to a small region, but when the spatial distribution of the excitation matches well with the low–order modes, the sum over $n < n_1$ in (8.28) may not be negligible. A typical example where this happens is the excitation of a thin plate by sound waves at low frequencies.

REFERENCES

Cremer, L., Heckl, M. & Ungar, E.E. (1988) Structure–Borne Sound. Springer, Berlin, Chapter IV,V.

Hodges, C.H. & Woodhouse, J. (1986) Rep. Prog. Phys. 49:107.

Lesueur, C. (1988) Rayonnement acoustique des structures. Editions Eyrolles, Paris, Chapter 7.

Lyon, R.H. & Maidanik, G. (1962) J. Acoust. Soc. Amer. 34:623.

Scharton, T. & Lyon, R.H. (1968) J. Acoust. Soc. Amer. 43:1332.

Wöhle, W. (1984) Statistische Energienanalyse der Schalltransmission.

Chapter 1.10 in Taschenbuch Akustik, VEB–Verlag Technik, Berlin.

Woodhouse, J. (1981) Appl. Acoustics 14:455.

9. MEAN ENERGY AND MOMENTUM EFFECTS IN WAVES

9.1 INTRODUCTION

Although energy is a second–order quantity in acoustics (and therefore very small), it is important because there is a fundamental conservation law for energy and also because energy considerations are most useful for deriving equations of motion. An important application of the conservation law in acoustics is the Statistical Energy Analysis (SEA) (see Chapter 8). The most common example of the use of energy methods for deriving equations of motion is the application of Hamilton's principle (or some related energy principle). This is equivalent to the statement that, subject to the initial and boundary conditions, the motion of a system is such that the mean difference between kinetic and potential energy is as small as possible (Morse & Feshbach 1953) (see also Rayleigh's principle in Chapter 20).

The importance of energy principles can also be seen by the fact that one of the most widely used numerical methods, the Finite Element Method (FEM) (see Chapter 10) can be considered as an application of Hamilton's principle.

Energy is not only a mathematical tool. Together with intensity and power, it has during the last decade (Fahy 1989) become a rather popular measured quantity in acoustics. These quantities help determine radiated and transmitted sound power, in order to distinguish near fields and far fields, and to follow the acoustic energy flow in gases, fluids and on the surface of solids.

Apart from energy, this Chapter also deals with the time–averaged momentum associated with sound waves. This is also a second–order quantity which, in its mathematical form, is very similar to intensity. The mean momentum is of importance in explaining stationary effects in acoustics, such as radiation pressure and acoustic streaming. It also helps to calculate the so–called wave drag.

9.2 HAMILTON'S PRINCIPLE AND THE WAVE EQUATION

9.2.1 Derivation of the Sound Wave Equation

Hamilton's principle is usually written as

$$\delta \int_{t_0}^{t_1} (E_{\text{Kin}} - E_{\text{Pot}})\, dt = 0 \,. \tag{9.1}$$

Here E_{Kin} and E_{Pot} are the total kinetic and potential energy of the systems, δ the symbol of variation and t is time. In a homogeneous medium of density ρ the total kinetic energy is

$$E_{\text{Kin}} = \frac{1}{2} \rho \int_V \left(\dot\xi_1^2 + \dot\xi_2^2 + \dot\xi_3^2 \right) dx_1 dx_2 dx_3 \quad . \tag{9.2}$$

ξ_i are the components of the displacements and $\dot\xi_i$ the components of the velocities. The integral has to be taken over the volume of interest V. If K is the compressibility of the medium, the potential energy per unit volume is

$$\frac{1}{2} K \left(\frac{\Delta V}{V} \right)^2 = \frac{1}{2} K (\xi_1' + \xi_2' + \xi_3')^2 \quad .$$

In the last expression the divergence

$$\xi_1' + \xi_2' + \xi_3' = \frac{\partial \xi_1}{\partial x_1} + \frac{\partial \xi_2}{\partial x_2} + \frac{\partial \xi_3}{\partial x_3} = \text{div}\ \xi$$

appears. If we integrate over the whole volume and combine the result with (9.1) and (9.2), we find

$$\delta \int_{t_0}^{t_1} \int_V \left[\frac{\rho}{2} \left(\dot\xi_1^2 + \dot\xi_2^2 + \dot\xi_3^2 \right) - \frac{K}{2} (\xi_1' + \xi_2' + \xi_3')^2 \right] dV dt = 0 \quad . \tag{9.3}$$

Applying the rules of the calculus of variations (which loosely speaking mean to replace δ by the corresponding derivative) gives

$$\int_{t_0}^{t_1} \int_V \left[\rho \left(\dot\xi_1 \delta\dot\xi_1 + \dot\xi_2 \delta\dot\xi_2 + \dot\xi_3 \delta\dot\xi_3 \right) \right.$$

$$\left. -K (\xi_1' + \xi_2' + \xi_3') (\delta\xi_1' + \delta\xi_2' + \delta\xi_3') \right] dV dt = 0 \quad . \tag{9.4}$$

Here we have to deal with two types of integrals. The first can be transformed by partial integration to give

$$\int_{t_0}^{t_1} \int_V \dot{\xi}_1 \delta \dot{\xi}_1 \, dV \, dt = \int_{t_0}^{t_1} \int_V \dot{\xi}_1 \delta \left(\frac{d\xi_1}{dt} \right) \, dV \, dt$$

$$= \int_V \dot{\xi}_1 \delta \xi_1 \, dV \Big|_{t_0}^{t_1} - \int_{t_0}^{t_1} \int_V \ddot{\xi}_1 \delta \xi_1 \, dV \, dt \quad .$$

(9.5)

For a representative term of the second type, we make a similar transformation

$$\int_{t_0}^{t_1} \int_V \xi_1' \delta \xi_2' \, dV \, dt = \int_{t_0}^{t_1} \int_V \xi_1' \delta \left(\frac{\partial \xi_2}{\partial x_2} \right) \, dx_2 \, dx_1 \, dx_3 \, dt$$

$$= \int_{t_0}^{t_1} \int \xi_1' \delta \xi_2 \, dx_1 \, dx_3 \, dt \Big|_{x_{2a}}^{x_{2b}} - \int_{t_0}^{t_1} \int_V \frac{\partial \xi_1'}{\partial x_2} \delta \xi_2 \, dx_1 \, dx_2 \, dx_3 \, dt .$$

(9.6)

In the following we are interested in the derivation of a wave equation in an unbounded medium, i.e. the limits x_{2a} and x_{2b} (similar values for the other coordinates) are so far away that all field quantities are zero. We also assume that at time t_0, the field has not yet started, and that it ended before t_1. Thus in (9.5) and (9.6) we can neglect the terms involving the integration limits, and we get from (9.4)

$$\int_{t_0} \int_V \left[-\rho \ddot{\xi}_i \delta \xi_i + K \frac{\partial \text{div} \xi}{\partial x_i} \delta \xi_i \right] \, dV \, dt = 0 \quad .$$

(9.7)

Here we apply the summation convention. Since (9.7) has to hold for all values of $\delta \xi_i$, it follows that

$$K \frac{\partial \text{div} \xi}{\partial x_i} - \rho \ddot{\xi}_i = 0 \qquad i = 1, 2, 3 \ .$$

(9.8)

The well–known equation for sound waves is obtained by differentiating each equation in (9.8) with respect to x_i, adding all equations, and replacing $K \, \text{div} \, \xi$ by the pressure fluctuation (sound pressure). This way we finally get

$$\frac{\partial^2 p}{\partial x_1^2} + \frac{\partial^2 p}{\partial x_2^2} + \frac{\partial^2 p}{\partial x_3^2} - \frac{\rho}{K} \frac{\partial^2 p}{\partial t^2} = 0 \quad \text{or} \quad \Delta p - \frac{1}{c^2} \frac{\partial^2 p}{\partial t^2} = 0 \quad .$$

(9.9)

Here we approximated the total time derivative indicated by a dot, by the partial time derivative. As long as we do linear acoustics, this does not cause problems. Furthermore, we introduced the speed of sound

$$c = \sqrt{K/\rho} \quad .$$

(9.10)

9.2.2 Bending, shear, and longitudinal waves in a three–layer beam

The full beauty of energy calculations combined with Hamilton's principle can be seen when more complicated systems are investigated. In such systems the energy can very often be found in a rather straightforward way, whereas the alternative method — the force and moment balancing — is more complicated and prone to error. As an example, we treat a three–layer beam (sandwich) which is allowed to have shear and longitudinal motion in its plane and bending motion perpendicular to the plane (but the y–component of the motion is the same over the entire thickness).

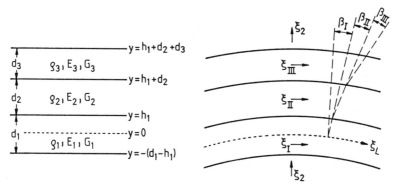

Figure 9.1 Three-layer beam with densities ρ_1, ρ_2, ρ_3, Young's modulus E_1, E_2, E_3, shear modulus G_1, G_2, G_3.

As can be seen from Figure 9.1, the in–plane displacement in the three layers, is given by

$$\xi_I = \xi_L + y\beta_I; \qquad \xi_{II} = \xi_L + h_1\beta_I + (y - h_1)\beta_{II}$$

$$(9.11)$$

$$\xi_{III} = \xi_L + h_1\beta_I + d_2\beta_{II} + (y - h_1 - d_2)\beta_{III} \quad .$$

The displacement in the normal direction is ξ_2 which is assumed to be independent of y. β_I, β_{II}, β_{III} are the angles of the deflection in the three layers. The reference plane $y = 0$ is arbitrary. In Figure 9.1 it is placed inside part I, but it might also be somewhere else. Its position has no effect on the final results, although it does show up in the coefficients of Equation (9.16).

For the calculation of the potential energy we need the strains, which are related, in our beam structure, to the displacements by

(a) longitudinal strains:

$$\epsilon_I = \frac{\partial \xi_I}{\partial x} = \xi'_L + y\beta'_I$$

$$\epsilon_{II} = \frac{\partial \xi_{II}}{\partial x} = \xi'_L + h_1\beta'_I + (y - h_1)\beta'_{II} \tag{9.12}$$

$$\epsilon_{III} = \frac{\partial \xi_{III}}{\partial x} = \xi'_L + h_1\beta'_I + d_2\beta'_{II} + (y - h_1 - d_2)\beta'_{III} \quad ;$$

(b) shear strains:

$$\gamma_I = \frac{\partial \xi_I}{\partial y} + \frac{\partial \xi_2}{\partial x} = \beta_I + \xi'_2$$

$$\gamma_{II} = \frac{\partial \xi_{II}}{\partial y} + \frac{\partial \xi_2}{\partial x} = \beta_{II} + \xi'_2 \tag{9.13}$$

$$\gamma_{III} = \frac{\partial \xi_{III}}{\partial y} + \frac{\partial \xi_2}{\partial x} = \beta_{III} + \xi'_2 \quad .$$

The prime denotes a derivative with respect to x. With these expressions we get, for the kinetic energy

$$E_{\mathrm{Kin}} = \frac{1}{2} \int\limits_{-\ell}^{\ell} \Big[\int\limits_{-(d_1-h_1)}^{h_1} \rho_1 \left(\dot{\xi}_I^2 + \dot{\xi}_2^2 \right) dy + \int\limits_{h_1}^{h_1+d_2} \rho_2 \left(\dot{\xi}_{II}^2 + \dot{\xi}_2^2 \right) dy$$

$$+ \int\limits_{h_1+d_2}^{h_1+d_2+d_3} \rho_3 \left(\dot{\xi}_{III}^2 + \dot{\xi}_2^2 \right) dy \Big] dx \quad , \tag{9.14}$$

and for the potential energy

$$E_{\mathrm{Pot}} = \frac{1}{2} \int\limits_{-\ell}^{\ell} \Big[\int\limits_{-(d_1-h_1)}^{h_1} \left(E_1\epsilon_I^2 + G_1\gamma_I^2 \right) dy + \int\limits_{h_1}^{h_1+d_2} \left(E_2\epsilon_{II}^2 + G_2\gamma_{II}^2 \right) dy$$

$$+ \int\limits_{h_1+d_2}^{h_1+d_2+d_3} \left(E_3\epsilon_{III}^2 + G_3\gamma_{III}^2 \right) dy \Big] dx \quad . \tag{9.15}$$

Using the same procedure and arguments as in the last paragraph, we obtain a set of partial differential equations for $\xi_L, \xi_2, \beta_I, \beta_{II}, \beta_{III}$. If we next assume that all motions consist of plane waves of the type $e^{jkx} e^{-i\omega t}$, each time derivative $(\cdot)$

corresponds to a multiplication with $-i\omega$ and each space derivative (') corresponds to a multiplication with ik. This way we find the following "dispersion relation"

$$
\begin{pmatrix}
\alpha_{11} & 0 & \alpha_{13} & \alpha_{14} & \alpha_{15} \\
0 & \alpha_{22} & \alpha_{23} & \alpha_{24} & \alpha_{25} \\
\alpha_{13} & \alpha_{23} & \alpha_{33} & \alpha_{34} & \alpha_{35} \\
\alpha_{14} & \alpha_{24} & \alpha_{34} & \alpha_{44} & \alpha_{45} \\
\alpha_{15} & \alpha_{25} & \alpha_{35} & \alpha_{45} & \alpha_{55}
\end{pmatrix}
\begin{pmatrix}
\xi_L \\
\xi_2 \\
\beta_I \\
\beta_{II} \\
\beta_{III}
\end{pmatrix}
=
\begin{pmatrix}
0 \\
p_A \\
0 \\
0 \\
0
\end{pmatrix} .
\tag{9.16}
$$

The coefficients are

$$
\alpha_{11} = k^2 E_E - \omega^2 m_E; \quad \alpha_{13} = \left(E_E h_1 - \frac{1}{2} E_1 d_1^2 \right) k^2 - \left(m_E h_1 - \frac{1}{2} \rho_1 d_1^2 \right) \omega^2
$$

$$
\alpha_{14} = \left(\frac{1}{2} E_2 d_2^2 + E_3 d_2 d_3 \right) k^2 - \left(\frac{1}{2} \rho_2 d_2^2 + \rho_3 d_2 d_3 \right) \omega^2 ;
$$

$$
\alpha_{15} = \frac{1}{2} E_3 d_3^2 k^2 - \frac{1}{2} \rho_3 d_3^2 \omega^2 ; \quad \alpha_{22} = -G_E k^2 + \omega^2 m_E ;
$$

$$
\alpha_{23} = i k\, G_1 d_1 ; \quad \alpha_{24} = i k\, G_2 d_2 ;
$$

$$
\alpha_{25} = i k\, G_3 d_3 ; \quad \alpha_{34} = \alpha_{14} h_1 ; \quad \alpha_{35} = \alpha_{15} h_1 ;
$$

$$
\alpha_{33} = \left[E_E h_1^2 - E_1 d_1^2 \left(h_1 - \frac{1}{3} d_1 \right) \right] k^2 - \left[m_E h_1^2 - \rho_1 d_1^2 \left(h_1 - \frac{1}{3} d_1 \right) \right] \omega^2 + G_1 d_1
$$

$$
\alpha_{44} = \left(\frac{1}{3} E_2 d_2^3 + E_3 d_2^2 d_3 \right) k^2 - \left(\frac{1}{3} \rho_2 d_2^3 + \rho_3 d_2^2 d_3 \right) \omega^2 + G_2 d_2 ;
$$

$$
\alpha_{45} = \alpha_{15} d_2 ; \quad \alpha_{55} = \frac{1}{3} E_3 d_3^3 k^2 - \frac{1}{3} \rho_3 d_3^3 \omega^2 + G_3 d_3 ;
$$

$$
E_E = E_1 d_1 + E_2 d_2 + E_3 d_3 ; \quad m_E = \rho_1 d_1 + \rho_2 d_2 + \rho_3 d_3; \quad G_E = G_1 d_1 + G_2 d_2 + G_3 d_3 .
$$

In (9.16) we also introduced a driving pressure p_A which acts from the outside in a direction perpendicular to the plate. Although (0.16) looks rather complicated, a few general results can be derived from it.

- Hamilton's principle gives in this, and in similar cases when no flow is present, a symmetric system of equations. (Sometimes in the literature asymmetric equations for such problems can be found but they contain small errors or inconsistent approximations.) As a consequence of the symmetry, the validity of the principle of reciprocity is assured.

- The three types of motion — longitudinal, shear, bending — are coupled with each other. If the layer configuration does not have a plane of symmetry there is no way (e.g. by another choice of the $y = 0$ plane) of decoupling the motions for all frequencies. Only for static problems can a so-called neutral plane be found. It can be calculated by omitting all terms that contain the frequency and by adjusting the line $y = 0$, i.e. the value of h_1 in such a way that α_{13} vanishes.

- The zeros of the determinant of (9.16) give the free–wave numbers and (when $E_1, G_1 \ldots$ are complex) also give the decay constants or loss factors. It turns out that there are five different free–wave numbers for each frequency, but in the frequency range where (9.11) describes correctly the motion, only two of them correspond to travelling waves. The wave numbers are independent of the choice of the reference line $y = 0$.

- Equation (9.16) gives the equations for longitudinal waves and for a Timoshenko beam as limiting cases. The quasi–static equations for the damping of a three–layer plate (Ungar 1971) can also be recovered.

- A considerable simplification of (9.16) can be achieved if the assumption is made underlying the Bernoulli beam theory. In this case we set $\gamma_I = \gamma_{III} = 0$, which is the same as setting $\beta_I = \beta_{III} = -ik\xi_2$, multiply the third and fifth equation with $-ik$, and then add them to the second one.

As was previously mentioned, there are many other applications of Hamilton's principle in vibration and acoustics, e.g. the derivation of shell equations (see Junger & Feit 1972), or the treatment of complex practical situations such as band/wheel systems (see Wang & Mote 1986).

9.3   SOUND INTENSITY IN GASES AND LIQUIDS

9.3.1 Definition

The intensity vector $\vec{I}$, which describes the energy flow in a sound field, is defined as the time average of the product of sound pressure p and particle velocity $\vec{v}$

$$\vec{I} = \overline{p\vec{v}} \quad .$$
(9.17)

Obviously the intensity vector has the same direction as the velocity vector (see e.g. Figure 9.2). Integrated over a certain area S with surface elements $\vec{dS}$ the intensity gives the power flow P through this area

$$P = \int_S \overline{\vec{I}\,\vec{dS}} \quad .$$
(9.18)

If the surface S totally encloses a sound source (or sink), (9.18) gives the sound power generated (or absorbed) by whatever is inside S. Here we have assumed that sound is not dissipated or otherwise absorbed during propagation. For audio frequencies and reasonable distances, this assumption agrees with practical situations.

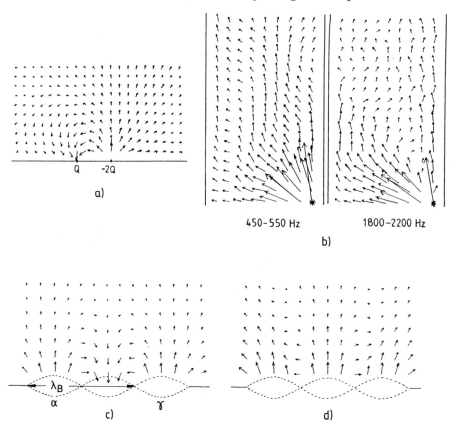

$$450-550 \text{ Hz} \qquad 1800-2200 \text{ Hz}$$

Figure 9.2 Examples of time–averaged intensities after Fahy (1989).

(a) monopoles of strength Q and $-2Q$;

(b) monopole in a duct;

(c) plate: $\lambda_B = 0.1\,\lambda_0$;

(d) plate: $\lambda_B = \lambda_0$ (λ_0 = wavelength in the medium).

When pressure and velocity consist of pure tones, e.g. the result of an FTT–analysis, we have

$$p = \Re\left\{\Sigma \mathbf{p}_n e^{-i\omega_n t}\right\} \; ; \quad \vec{\mathbf{v}} = \Re\left\{\Sigma \vec{\mathbf{v}}_n e^{-i\omega_n t}\right\} \quad . \tag{9.19}$$

Here

$$\mathbf{p}_n = p'_n + i p''_n \; ; \quad \vec{\mathbf{v}}_n = \vec{v}'_n + i\vec{v}''_n; \quad \vec{\mathbf{v}}^*_n = \vec{v}'_n - i\vec{v}''_n \tag{9.20}$$

are the complex amplitudes and ω_n the frequencies. p'_n, p''_n, v'_n, v''_n are real.

Introducing (9.19) into (9.17) gives

$$\vec{I} = \frac{1}{2} \Sigma \, \Re \{\mathbf{p}_n \, \vec{\mathbf{v}}_n^*\} \quad . \tag{9.21}$$

For some purposes it is also useful to distinguish between an "active intensity", which is given by (9.21), and a "reactive intensity", which is given by

$$\vec{Q} = \frac{1}{2} \Sigma \, \Im \{\mathbf{p}_n \vec{\mathbf{v}}_n^*\} \quad . \tag{9.22}$$

(see Fahy 1989).

With modern FFT–analyzers, the measurement of $\vec{Q}$ makes very little additional work when $\vec{I}$ can be measured.

Some insight into the active and reactive intensity can be gained if we write in one dimension

$$p = g(x)e^{-i(\omega t + \phi(x))}$$

$$\rho \frac{\partial v}{\partial t} = -\frac{\partial p}{\partial x} \quad \text{or} \quad v = \frac{+1}{i\omega\rho} \left(\frac{\partial g}{\partial x} - ig \frac{\partial \phi}{\partial x} \right) e^{-i(\omega t + \phi(x))} \quad .$$

This gives

$$I = \frac{-1}{2\omega\rho} \, g^2(x) \, \frac{\partial \phi(x)}{\partial x} \tag{9.23}$$

$$Q = \frac{1}{4\omega\rho} \frac{\partial g^2(x)}{\partial x} \quad .$$

We see that the active component is proportional to the phase gradient and the reactive part to the gradient of the mean square pressure.

9.3.2 Measurement Principles

Apart from a few exceptions where the particle velocity is measured directly, e.g. by a Doppler shift method, the measurement of $\vec{v}$ in the intensity equation is replaced by a pressure gradient measurement using two closely–spaced microphones (see Figure 9.3.). Thus the general linearized momentum equation for sound fields

$$\rho \frac{\partial v_n}{\partial t} = -\frac{\partial p}{\partial n} \tag{9.24}$$

is replaced by

$$v_n = \frac{-1}{\rho} \int_{-\infty}^{t} \left(\frac{\partial p}{\partial n}\right) dt \simeq \frac{1}{\rho d} \int_{-\infty}^{t} \left(p_1(\tau) - p_2(\tau)\right) d\tau \quad .$$

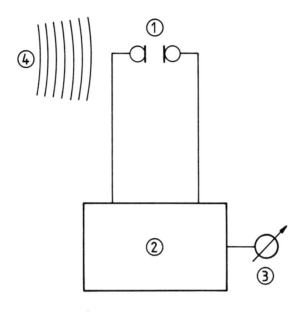

Figure 9.3 Two microphone intensity meter.

    (1) two identical pressure microphones separated by a distance $d \ll \lambda$

       (λ = wavelength);

    (2) analyzer;

    (3) data output unit;

    (4) sound field.

Here d is the microphone spacing which should be very small compared with the wavelength and p_1, p_2 are the pressures measured by the two microphones. Taking the average of p_1 and p_2 as the relevant pressure, we get for the intensity component in the direction of the microphone separation

$$I_d = \overline{\frac{1}{2\rho d} \left[p_1(t) + p_2(t)\right] \int_{-\infty}^{t} \left[p_1(\tau) - p_2(\tau)\right] d\tau} \quad . \tag{9.25}$$

If we introduce the abbreviation

$$\int_{-\infty}^{t} p_1(\tau) d\tau = g_1(t); \quad \text{i.e.} \quad p_1(t) = \frac{\partial g_1}{\partial t} \quad ,$$

and a similar one for p_2, then (9.25) becomes

$$I_d = \frac{1}{2\rho d} \left[\overline{\frac{\partial g_1}{\partial t} g_1} - \overline{\frac{\partial g_1}{\partial t} g_2} + \overline{\frac{\partial g_2}{\partial t} g_1} - \overline{\frac{\partial g_2}{\partial t} g_2} \right] \quad,$$

$$= \frac{1}{2\rho d} \left[\frac{1}{2} \overline{\frac{\partial g_1^2}{\partial t}} - \frac{1}{2} \overline{\frac{\partial g_2^2}{\partial t}} + \overline{\frac{\partial g_1 g_2}{\partial t}} - 2 \overline{\frac{\partial g_1}{\partial t} g_2} \right] \quad.$$

If the signals are stationary in time, the first three terms vanish because the average of a time derivative is zero for long averaging times. Thus we end up with

$$I_d = \frac{1}{\rho d} \overline{p_1(t) \int_{-\infty}^{t} p_2(\tau)\, d\tau} \quad. \tag{9.26}$$

So the intensity can be obtained by integrating one sound pressure, multiplying it with the second one and averaging the product.

Another rather elegant way to measure intensities with two closely–spaced microphones (Fahy 1989) employs the Fourier transforms of the pressure. If we write

$$p(t) = \frac{1}{2\pi} \int \breve{p}(\omega) e^{-i\omega t} d\omega, \quad v(t) = \frac{1}{2\pi} \int \breve{v}(\omega) e^{-i\omega t} d\omega \quad, \tag{9.27}$$

the pressure spectra $\breve{p}_1(\omega)$ and $\breve{p}_2(\omega)$ yield, when (9.23) and (9.24) are used

$$p(\omega) = \frac{\breve{p}_1(\omega) + \breve{p}_2(\omega)}{2} \quad; \quad v(\omega) = \frac{\breve{p}_1(\omega) - \breve{p}_2(\omega)}{-i\omega\rho d} \quad. \tag{9.28}$$

This way we obtain

$$\overline{p\, v} = \frac{1}{T} \int p v\, dt$$

$$= \frac{1}{4\pi^2 T} \int \frac{\breve{p}_1(\omega) + \breve{p}_2(\omega)}{2} \frac{\breve{p}_1(\omega') - \breve{p}_2(\omega')}{-i\omega'\rho d} e^{-i(\omega'+\omega)t} d\omega\, d\omega'\, dt$$

$$= \frac{1}{2\pi T} \int \frac{|\breve{p}_1(\omega)|^2 - |\breve{p}_2(\omega)|^2}{2i\omega\rho d} d\omega + \frac{1}{2\pi T} \int \frac{\breve{p}_2(\omega)\breve{p}_1^*(\omega) - \breve{p}_2^*(\omega)\breve{p}_1(\omega)}{2i\omega\rho d} \quad.$$

$$\tag{9.29}$$

Here we applied the orthogonality relation of the exponential function (see Chapter 3.2, Fourier Transform of the delta function) and the fact that for real-valued functions $\breve{p}(-\omega) = \breve{p}^*(\omega)$. The first integral in (9.29) is purely imaginary; it therefore represents the reactive intensity

$$Q_d = \frac{-1}{2\omega\rho d} \frac{1}{2\pi T} \int \left\{ |\breve{p}_1(\omega)|^2 - |\breve{p}_2(\omega)|^2 \right\} d\omega = \frac{-1}{2\omega\rho d} \left[G_{p_1 p_1}(\omega) - G_{p_2 p_2}(\omega) \right] .$$

(9.30)

In the second integral the numerator can be written as

$$|\breve{p}_2(\omega)\breve{p}_1(\omega)| \left(e^{i\phi_2 - i\phi_1} - e^{-i\phi_2 + i\phi_1} \right) = 2i |\breve{p}_2(\omega)\breve{p}_1(\omega)| \sin(\phi_2 - \phi_1) .$$

It is purely imaginary; therefore the second integral represents the (more important) active intensity

$$I_d = \frac{-1}{\omega\rho d} \frac{1}{2\pi T} \int |p_1(\omega)p_2(\omega)| \sin(\phi_2 - \phi_1) d\omega = \frac{-1}{\omega\rho d} \Im \left\{ G_{p_1 p_2}(\omega) \right\} . \quad (9.31)$$

Here G is the cross–power spectral density of quantities indicated by the indices.

Using (9.31) the intensity measurement is the same as the cross–power spectral density measurement of two sound pressures. With conventional two–channel F.F.T. analyzers, this is currently a routine operation. As a consequence, the difficulties of intensity measurements are not a problem of signal processing; the real problems lie in getting identical microphones (with minimum phase mismatch), in the errors caused by the finite microphone spacing, the signal–to–noise ratio, the relative magnitude of the active and reactive field, and, most importantly, the correct interpretation of the results.

9.3.3 Applications

9.3.3.1 Separation of incoherent noise sources

Intensity measurements are very useful when the total sound power of a noise source has to be measured in the presence of other noise generators. Such situations arise quite often in noise control problems where one usually wants to find out the most powerful noise sources in a complicated arrangement of sources. The method is sketched in Figure 9.4. The equation underlying the measurement is (9.18). In principle the shape of the surface S is not important as long as it completely encloses the source. In practical terms, it may be important because it influences the measurement errors. Such errors may be due to the non–radiating near field of the source or to the influence of strong outside sources. The near–field effects are especially strong when the source is a multipole of high order (e.g. rotating propeller) or when the sound radiating surfaces vibrate with a short wavelength, e.g. thin plates. The near–field effect is greatly reduced when the distance between the sound source and the measuring microphone is more than half a wavelength. The influence of strong outside sources is due to the fact that intensity measurements rely on determining the difference of sound pressure (see Equation (9.25)). Such measurements are inherently inaccurate when the measured quantity is the difference of two almost equal quantities.

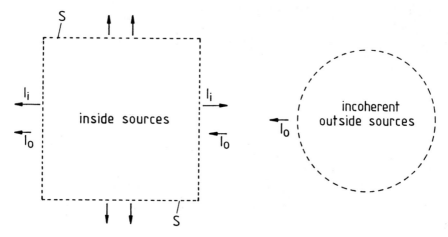

Figure 9.4 Measurement of sound power through a closed surface in the presence of incoherent outside sources. (If there is negligible absorption inside S, the sound power from outside sources is subtracted on one side and added at another one.)

A good indication for error in sound–power measurements is the difference between the sound pressure level and the sound intensity level (re $10^{-12}\mathrm{W/m^2}$) at the measuring points. According to ISO Standard 3740–3748 this difference should be less than 15 dB for the so–called "Survey grade".

9.3.3.2 Measurement of sound absorption and transmission

Since sound intensity can be positive or negative, power absorption can be measured as well as power generation in a closed region. Quite often part of the enclosing surface S consists of hard walls; such walls can be omitted because they have zero velocity and therefore there is no power flow through them.

The method is used, for example, to determine the sound absorption of acoustic terminations in tubes. Basically one uses the fact that by properly processing the signals of two microphones, one can distinguish the amplitudes of two waves travelling in different directions.

Measurements of the sound transmission through partitions, etc., can be accomplished in two ways. The first is to consider the transmitted sound as being lost on the source side; in this case the measurement is identical to an absorption measurement. For high transmission losses, this type of measurement is inaccurate because of the "difference problem". The second way is to treat the partition as a sound source and to measure the sound power it radiates into the receiving region.

9.3.3.3 Sound power of coherent sources

When intensity measurements are used for sound source identification or localization difficulties arise when the sources are coherent (or partly) coherent. The reason for this is obvious. The superposition principle of linear acoustics holds for field quantities such as pressure and velocities but not for powers and intensities. It is only for incoherent sound fields ("where all cross–terms vanish") that energy and power happen to be additive.

Figure 9.5 gives an example showing that a combination of coherent sources can lead to results that, at first glance, are unexpected. In the Figure two sources are assumed. For the sake of simplicity, they are in a baffle and only two dimensions are considered. With the symbols shown in Figure 9.5, the powers transmitted through the surfaces S_1 and S_2 are

$$P_1 = \int_{S_1} \overline{[p_{11}(t) + p_{12}(t)]\, v_{1n}(t)}\, dS$$
$$P_2 = \int_{S_2} \overline{[p_{21}(t) + p_{22}(t)]\, v_{2n}(t)}\, dS$$

(9.32)

Here p_{ij} is the pressure generated by the j-th source in front of source i. Since only the normal components of the velocity are important, and since S_1 and S_2 are very close to the baffle, no contributions from the velocity of the "other" source have to be considered.

If all quantities are expressed as wavenumber spectra, (9.32) becomes, after a few manipulations (see also Section 9.5.2.1)

$$P_1 = \frac{\omega \rho}{4\pi} \int_{-k_0}^{k_0} \frac{|\breve{v}_1(k)|^2}{\sqrt{k_0^2 - k^2}}\, dk + \frac{\omega \rho}{4\pi} \int_{-\infty}^{\infty} \Re \left\{ \frac{\breve{v}_2(k)\breve{v}_1^*(k)}{k_y} \right\}\, dk$$
$$P_2 = \frac{\omega \rho}{4\pi} \int_{-k_0}^{k_0} \frac{|\breve{v}_2(k)|^2}{\sqrt{k_0^2 - k^2}}\, dk + \frac{\omega \rho}{4\pi} \int_{-\infty}^{\infty} \Re \left\{ \frac{\breve{v}_1(k)\breve{v}_2^*(k)}{k_y} \right\}\, dk \quad .$$

(9.33)

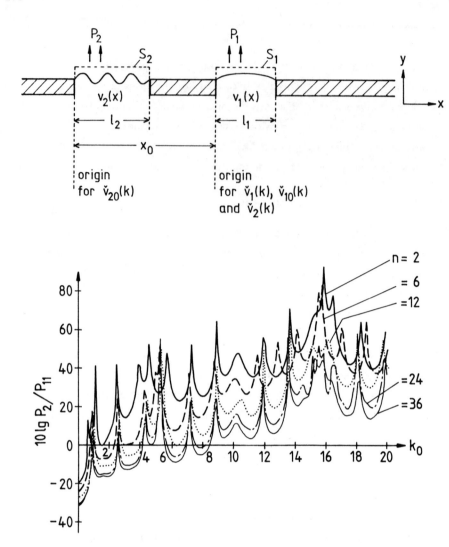

Figure 9.5 Power radiated from source 2 when its amplitude and phase is adjusted in such a way that no power is transmitted through S_1. P_{11} = power from source 1 when source 2 is not present. Source 1 is a piston of size $\ell_1 = 0.4$m; source 2 is of type $v_2(x) = A \sin n\pi x/\ell_2$ with $\ell_2 = 1.2$m; x_0 is 2.4 m. Combination of coherent radiators in a baffle.

Here

$$k_y = \begin{cases} \sqrt{k_0^2 - k^2} & \text{for} \quad k_0^2 > k^2 \\ i\,\sqrt{k^2 - k_0^2} & \text{for} \quad k_0^2 < k^2 \end{cases}$$

$$\check{v}(k) = \int_{-\infty}^{+\infty} v(x)e^{ikx}\,dx \quad . \tag{9.34}$$

$k_0 = \omega/c_0$ is the wavenumber in the surrounding medium and $\check{v}(k)$ are the wavenumber spectra of the source velocity distributions $v(x)$ at frequency ω. (Both use the same origin.)

In (9.33) the integrals ranging from $-k_0$ to k_0 correspond to the powers P_{11} and P_{22} that would be radiated from each source if it were alone The addition of the two equations in (9.33) gives

$$P_1 + P_2 = P_{11} + P_{22} + \frac{2\omega\rho}{4\pi} \int_{-k_0}^{k_0} \Re\left\{\check{v}_1(k)\check{v}_2^*(k)\right\} \frac{dk}{\sqrt{k_0^2 - k^2}} \neq P_{11} + P_{22} \; . \tag{9.35}$$

Thus in general the powers do not add.

More insight can be gained if we take into account that the sources are separated by a distance x_0. Thus if we use the index 0 to characterize the two wavenumber spectra when they are taken using their "own" origin, we get, because of the "shift formula" for Fourier transforms,

$$\check{v}_1(k) = \check{v}_{10}(k) = |\check{v}_{10}|\,e^{i\phi_{10}}$$
$$\check{v}_2(k) = \check{v}_{20}(k)\,e^{-ikx_0} = |v_{20}|\,e^{i\phi_{20}}\,e^{-ikx_0} \quad .$$

Here the wavenumber spectra are expressed by their absolute values and their phases ϕ_{10} and ϕ_{20} (which depend on k).

Introduction into (9.33), together with the substitution $k = k_0 \sin\gamma$, give

$$P_1 + P_2 = \frac{\omega\rho}{4\pi} \int_{-k_0}^{k_0} \left[\,|\check{v}_{10}^2| + |\check{v}_{20}^2| + 2|\check{v}_{10}|\,|\check{v}_{20}|\cos(\phi_{10} - \phi_{20} + kx_0)\right] \frac{dk}{\sqrt{k_0^2 - k^2}}$$

$$= \frac{\omega\rho}{4\pi} \int_{-\pi/2}^{\pi/2} \left[\,|\check{v}_{10}^2| + |\check{v}_{20}^2| + 2|\check{v}_{10}^2|\,|\check{v}_{20}^2|\cos(\phi_{10} - \phi_{20} + k_0x_0\sin\gamma)\right] d\gamma.$$

$$\tag{9.36}$$

One can easily see that for small values of k_0x_0 the cross term is of equal magnitude as the other two. Therefore it can have a substantial influence on the total result. Only when $k_0x_0 > 1$ and when frequency averages are taken (provided the wavenumber spectra are reasonably frequency independent) the cos–term averages to zero and the total power is $P_{11} + P_{22}$, i.e. only in this case can the powers from coherent sources be added.

In Figure 9.5 the amplitude and phase of source 2 are adjusted in such a way that P_1, i.e. the power through surface S_1 becomes zero (this is always possible). The quantity that is plotted is the power generated by source 2 which is necessary to "quieten" source 1. As one might expect, this power is small when x_0 is small but grows to very large values when $k_0 x_0 >> 1$.

The conclusions of these calculations dealing with the power of coherent sources hold even more for the intensity vectors. In principle the intensity vectors in the vicinity of a source can be brought to almost any position by a neighbouring coherent source; only when frequency averages are taken and when sources are separated by more than a wavelength do they not influence each other very much.

As an example of an incorrect conclusion drawn from arrow pictures, one may look at Figure 9.2c, where the sound appears to come from region α and γ. If one would shield these two regions, leaving the plate vibration unchanged, the sound power would not go down; it would instead be radiated from the centre region which in Figure 9.2c acts as a "sink".

9.4 INTENSITY OF WAVES IN SOLIDS

Analogous to Equation (9.17), one may also define an intensity for waves in structures. The difference is that in a solid, the forces acting on a volume element can no longer be expressed by a scalar quantity, pressure. We have to use the stress tensor, σ_{ij} instead. Therefore the components of the intensity vector in rectilinear coordinates are defined as

$$I_i = - \overline{\sigma_{ij} v_j} \tag{9.37}$$

(summation convection).

Here σ_{ij} is the stress tensor which in a homogeneous medium with Young's modulus E and Poisson's ratio ν, is given by

$$\sigma_{ij} = \frac{E\nu}{1-\nu}\,\epsilon_K\,\delta_{ij} + \frac{E}{1+\nu}\,\epsilon_{ij}$$

$$\epsilon_K = \epsilon_{11} + \epsilon_{22} + \epsilon_{33}; \quad \epsilon_{ij} = \frac{1}{2}\left(\frac{\partial \xi_i}{\partial x_j} + \frac{\partial \xi_j}{\partial x_i}\right) \quad . \tag{9.38}$$

ξ_i is the component of the displacement of a volume element.

For three–dimensional solids, the intensity is not a very useful concept because there is no way of making measurements inside a solid body. But for one– and two–dimensional bodies, the first successful attempts to work with intensities have already been made (Noiseux 1970; Pavic 1986; Rasmussen 1987). Without proof the final formula for the intensity in a plate, including bending and in–plane waves, is given (see Figure 9.6)

$$
I_1 = \frac{P_1'}{h} = \frac{E}{1-\nu^2} \overline{\left(\frac{\partial \xi_1}{\partial x_1} + \nu \frac{\partial \xi_3}{\partial x_3}\right) \dot{\xi}_1} + G \overline{\left(\frac{\partial \xi_1}{\partial x_3} + \frac{\partial \xi_3}{\partial x_1}\right) \dot{\xi}_3}
$$

$$
+ \frac{D}{h} \left[\overline{\Delta \xi_2 \frac{\partial \dot{\xi}_2}{\partial x_1}} - \overline{\frac{\partial \Delta \xi_2}{\partial x_1} \dot{\xi}_2} + (1-\nu) \left(\overline{\frac{\partial^2 \xi_2}{\partial x_1 \partial x_3} \frac{\partial \dot{\xi}_2}{\partial x_3}} - \overline{\frac{\partial^2 \xi_2}{\partial x_3^2} \frac{\partial \dot{\xi}_2}{\partial x_1}} \right) \right] .
$$

$$(9.39)$$

G = shear modulus, D = bending stiffness, Δ = Laplace operator, P_1' = power per unit width in the x_1 direction, $\xi_1 \xi_2, \xi_3$ are the displacements of the centre line; the dot ˙ indicates the time derivative. For I_3 only the indices 1 and 3 have to be exchanged. For beams we have to set $\xi_3 = 0$ and $\partial/\partial x_3 = 0$. the formula for intensity of shells is also known. In can be found in a paper by Pavic (1986).

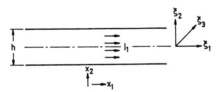

Figure 9.6 Coordinate system for Equation (9.39).

As Equation (9.39) shows, the exact measurement of the intensity on plates involves several steps, some of which are rather delicate:

- to get ξ_1 and ξ_3 in the centre line, one has to place two pick–ups on both sides of the plate and average the signal (otherwise the rotation associated with bending motion causes difficulties);
- the space derivatives have to be replaced by difference measurements of closely–spaced pick–ups; in one case a third derivative has to be known; it would involve a third–order difference measurement with four absolutely identical pick–ups.

At present the difficulties associated with a measurement according to (9.39) are not yet resolved. Therefore the so–called structure–borne sound intensity meters which are available at present rely on rather stringent assumptions. These are that the in–plane motions ξ_1 and ξ_3 are negligibly small, the bending wave near fields are unimportant (which means that $\Delta \xi_2 \approx -k_B^2 \xi_2 = -\omega \sqrt{\rho h/D} \, \xi_2$), and the derivative in one direction is dominant (so that terms involving derivatives in both directions can be omitted). When all these restrictions are accepted, an intensity meter for structure–borne sound is very similar to a two–microphone intensity measurement unit; the only difference is that the two microphones are replaced by two accelerometers. It cannot be expected that such a device will give the correct intensity, but it can give some results about the direction of energy flow in plates.

9.5 MEAN MOMENTUM AND WAVE DRAG

9.5.1 Mean Momentum

High–amplitude sound waves generate stationary effects such as radiation pressure and acoustic streaming. These effects cause phenomena such as "quartz wind" , the flow patterns in a Kundt's tube, the deflection of a Rayleigh disc in an intense sound field, etc. (see e.g. Beyer 1974). It is not the purpose of this Chapter to explain fully all these effects and go into the subtleties of the radiation pressure or the acoustic streaming around obstacles or near strong sound sources. Instead, only the main principle underling these phenomena is explained briefly.

In order to get the mean momentum in a sound field, we start from the conservation of momentum which, for homogeneous gases and fluids (see Chapters 11 and 12), is

$$\frac{\partial \rho v_i}{\partial t} + \frac{\partial p}{\partial x_i} = \frac{\partial T_{ij}}{\partial x_j} \approx \frac{\partial \rho v_i v_j}{\partial x_j} \quad . \tag{9.40}$$

Here T_{ij} is Lighthill's tensor. Its most important term is retained in the last expression of (9.40). v_i and v_j are the velocity components, which in the absence of any flow are the sound particle velocities.

If we take the time average of (9.40), the first term disappears because it is a time derivative, and we are left with

$$\frac{\partial \overline{p}}{\partial x_i} = \frac{\partial \overline{T_{ij}}}{\partial x_j} \approx \frac{\partial \overline{\rho v_i v_j}}{\partial x_i} \quad . \tag{9.41}$$

The overbar denotes time averages.

In a linear approximation the right–hand side would be considered to be of second order and would be neglected leaving no constant term for the sound–induced pressure. The situation is different when we go to the next approximation, i.e when we introduce for v_i and v_j the usual linear values. This way, when we consider only one component (i.e. plane waves), we get

$$\frac{\partial \overline{p}}{\partial x} \approx \frac{\partial \overline{\rho v^2}}{\partial x} \sim \frac{\partial E}{\partial x} \quad . \tag{9.42}$$

Here E is the mean energy density of the sound field. As a result of these simple considerations, we find that there are constant pressures (e.g. radiation pressure) in the sound field which are related to the energy density. We also see that there can be pressure gradients, which are the driving forces for streaming processes, provided there is also a spatial change in the energy density (e.g. caused by the attenuation of the sound waves).

9.5.2 Wave Drag

The last paragraph gave a brief description of the mean forces inside a gas or fluid. Now we turn to the forces acting on a vibrating surface.

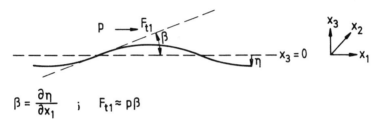

$$\beta = \frac{\partial \eta}{\partial x_1} \quad ; \quad F_{t1} \approx p\beta$$

Figure 9.7 Forces acting on a deformed surface.

As Figure 9.7 shows, there is a tangential force. It is given simply by the sound pressure p in front of the vibrator multiplied with the slope of the vibrating area. Thus for the mean tangential force per unit area, we get

$$F_{t1} = \overline{p \frac{\partial \eta}{\partial x_1}} \quad \text{and} \quad F_{t2} = \overline{p \frac{\partial \eta}{\partial x_2}} \quad , \tag{9.43}$$

Integrated over the whole area, Equation (9.43) gives the total drag (or propulsive force) caused by the wavy motion of the surface.

The similarity of (9.43) to the definition of the acoustic intensity is obvious, because if we replace the space derivative by the time derivative, we arrive at Equation (9.17).

9.5.2.1 Wave drag of a vibrating plate in a baffle

As a simple example we consider a one–dimensional problem, which consists of a plate vibrating with displacement $\eta(x_1)$ in an infinite baffle. All motions are assumed to have the same frequency ω; the common time factor is omitted. We apply spatial Fourier transforms, i.e. the displacement is expressed as

$$\boldsymbol{\eta}(x_1) = \frac{1}{2\pi} \int \check{\eta}(k_1) e^{+ik_1 x_1} dk_1 \quad . \tag{9.44}$$

$\check{\eta}(k)$ is the wavenumber spectrum of the motion. The sound pressure generated by this motion must be a solution of the sound wave equation; therefore it must be of the type (see also Chapter 17)

$$\mathbf{p}(x_1, x_2) = \frac{1}{2\pi} \int \check{p}(k_1) e^{+ik_1 x_1} e^{+ik_2 x_2} dk_1 \quad ,$$

with

$$k_2 = \begin{cases} \left[(\omega^2/c_0^2) - k_1^2 \right]^{1/2} & \text{for} \quad k_1^2 < \omega^2/c_0^2 \\ i \left[k_1^2 - (\omega^2/c_0^2) \right]^{1/2} & \text{for} \quad k_1^2 > \omega^2/c_0^2 \quad . \end{cases} \tag{9.45}$$

The speed of sound and the density in the medium in front of the plate shall be c_0 and ρ_0.

Continuity of motion at the interface requires

$$-\frac{\partial \mathbf{p}(x_1, x_2)}{\partial x_2} = -\omega^2 \rho_0 \boldsymbol{\eta}(x_1) \quad \text{for} \quad x_2 = 0 \quad .$$

Applying this to Equations (9.44) and (9.45) gives

$$\check{p}(k_1) = \frac{\omega^2 \rho_0}{i k_2} \check{\eta}(k_1) \quad . \tag{9.46}$$

The next quantity we need is

$$\frac{\partial \eta}{\partial x_1} = \frac{1}{2\pi} \int ik_1 \, \check{\eta}(k_1) e^{ik_1 x_1} dk_1 \quad . \tag{9.47}$$

If we first look for the sound power which is radiated from the surface, we have to calculate

$$P - \int I \, d\tau_1 = \frac{1}{2} \Re \left\{ \int \mathbf{p}(x_1, 0) \left[i\omega \boldsymbol{\eta}^*(x_1) \right] dx_1 \right\}$$

$$= \frac{\omega^3 \rho_0}{8\pi^2} \Re \left\{ \int \frac{1}{k_2} \check{\eta}(k_1) \check{\eta}^*(-k_1') \int e^{i(k_1 - k_1') x_1} dx_1 dk_1 dk_1' \right\} \tag{9.48}$$

$$= \frac{\omega^3 \rho_0}{4\pi} \int_{-\omega/c_0}^{\omega/c_0} \frac{|\check{\eta}(k_1)|^2}{\left[(\omega^2/c_0^2) - k_1^2 \right]^{1/2}} dk_1 \quad .$$

Here we have used the facts that the integral over x_1, which is written separately, is equal to $2\pi\delta(k_1 - k_1')$ and that k_2 is real only in the range $-\omega/c_0 < k_1 < \omega/c_0$. Equation (9.48) shows that the radiated power is determined by the wavenumber spectrum of the displacement in the low wavenumber range.

Next we calculate the total drag $F_{1\,tot}$ generated by the vibrating surface

$$F_{1\,tot} = \int F_{1\,t}\, dx_1 = \frac{1}{2}\,\Re\left\{\int \mathbf{p}(x_1, 0)\,\frac{\partial \eta^*(x_1)}{\partial x_1}\, dx_1\right\} \quad . \tag{9.49}$$

Introducing Equations (9.46) and (9.47) and making the same type of calculations that led to (9.49) results in

$$F_{1\,tot} = \frac{-\omega^2 \rho_0}{4\pi} \int_{-\omega/c_0}^{\omega/c_0} k_1 \frac{|\breve{\eta}(k_1)|^2}{\left[(\omega^2/c_0^2) - k_1^2\right]^{1/2}}\, dk_1 \quad . \tag{9.50}$$

Equation (9.50) is very similar to (9.48) and again only the small wavenumbers are important. There is, however, a decisive difference. When the wavenumber spectrum is symmetrical, i.e. $|\breve{\eta}(k_1)| = \breve{\eta}(-k_1)|$, as is always the case for standing wave patterns, there would be a power radiation but there would not be any wave drag because of the multiplication with k_1. Thus if we think of the displacement $\eta(x_1)$ as consisting of waves running in both directions, we get the wave drag only if the waves have different amplitudes, i.e. when the wavenumber spectrum is asymmetric.

Another way of proving the importance of the asymmetry is to expand the wavenumber spectrum in a Taylor series

$$|\breve{\eta}(k_1)|^2 = |\breve{\eta}(0)|^2 + k_1 \frac{\partial|\breve{\eta}(0)|^2}{\partial k_1} + \frac{k_1^2}{2}\frac{\partial^2|\breve{\eta}(0)|^2}{\partial k_1^2} + \dots \quad . \tag{9.51}$$

If we introduce this into (9.48) and (9.50), we see that the radiated power depends on $|\breve{\eta}(0)|^2$, the second derivative, the fourth derivative, etc. The drag force, however, is given by the first, third ... derivative. So again we see that in order to generate large drag we need a highly asymmetric wavenumber spectrum.

For the special case of a wave with wavenumber k_A running in a positive direction of a plate of length l, we would have

$$\eta(x_1) = \begin{cases} Ae^{ik_A x_1} & \text{for} -\ell/2 < x_1 < \ell/2 \\ \\ 0 & \text{otherwise} \end{cases} \quad . \tag{9.52}$$

The wavenumber spectrum is known to be

$$\breve{\eta}(k_1) = Al\, \frac{\sin(k_A + k_1)\ell/2}{(k_A + k_1)2} \quad . \tag{9.53}$$

If this expression is inserted into (9.48) or (9.50), the power and the drag can be found. Results for this problem and others are reported by Ffowcs Williams & Hill (1987). There it is also shown how the radiated sound is distributed over different directions and that the time average of the momentum in the distant sound field is equal to the surface drag.

The results derived here hold not only for surfaces which are forced to vibrate in a certain way. They also hold for compliant surfaces which adjust themselves to the fluctuating forces in a fluid. Thus one would expect that in this case there is also a strong relation between energy flow and wave drag and that compliant surfaces must react differently to upstream and downstream waves in order to get the required asymmetry.

The skewness required for large wave drag certainly is not a natural feature of continuous mechanical surfaces of conventional construction so that compliant surfaces that should reduce drag by harnessing the available "form drag" will be quite unusual, possible more like fish scales than a bending plate (Ffowcs Williams 1988).

REFERENCES

Beyer, R.T. (1974). Nonlinear Acoustic. Naval Ship Systems Command Department of the Navy, Washington, D.C.

Fahy, F. (1989). Sound Intensity. Elsevier, London.

Ffowcs Williams, J.E. & Hill, D.C. (1987). J. Fluid Mech. 184:101.

Ffowcs Williams, J.E. (1988). Private communication.

Junger, M.C. & Feit, D. (1972). Sound Structures and their Interaction. Chapter 9, MIT Press, Cambridge, USA.

Morse, P.M. & Feshbach, H. (1953). Methods of Theoretical Physics. Chapter 3. McGraw–Hill.

Noiseux, D.U. (1970). J. Acoustic Soc. Amer. 47:238.

Pavic, G. (1986). The influence of curvature on structure-borne acoustical power propagation in a cylindrical circular shell. Paper D 6–6, ICA, Toronto.

Rasmussen, G. (1987). Inter–Noise 87, Peking, 1183.

Ungar, E.E. (1971). Damping of Panels. In: Noise and Vibration Control. (Ed. L.L. Beranek). McGraw–Hill.

Wang, K.W. & Mote, C.D. (1986). J. Sound Vib. 109:237.

10. NUMERICAL METHODS

10.1 INTRODUCTION

In the days of Bernoulli, Euler, Rayleigh and other great scientists, acoustics was always in the forefront of new mathematical developments. New mathematical techniques were sometimes initiated and most often applied in acoustics for the first time. When the widespread application of high–speed computers opened a new area of numerical mathematics, sound and vibration problems were among the first ones where the new methods were used. At present there is a large number of numerical techniques and program packages which are applied in acoustics. It would be far beyond the scope of this text — also beyond the knowledge of the author — to give even a brief description of all of them. Therefore only a small number of numerical methods is treated here. The selection is made solely by the author's taste.

10.2 FINITE–ELEMENT METHODS (FEM) IN ACOUSTICS

With the exception of very high–amplitude noise or strong ultrasound fields the principle of superposition holds in acoustics. Therefore the majority of acoustic problems can be described by linear differential or linear integral equations. Thus if they are discretized a (large) number of linear algebraic equations is generated and all the many techniques which have been developed for this branch of mathematics can be applied.

In this respect, a typical example is the method of finite elements when applied to sound and vibration problems (in other fields of physics FEM may also include nonlinearities).

When FEM is explained, one usually starts with a variation principle for energies (see also Chapter 9, Hamilton's principle). If we restrict ourselves for the moment to gases and liquids inside a volume V with no energy transport through the surface, the function which has to be a minimum is

$$L = U - T = \frac{1}{2} \int_V \rho c^2 (\text{div } \vec{\xi})^2 \, dV - \frac{1}{2} \int_V \rho \, (\dot{\vec{\xi}})^2 \, DV$$

$$= \frac{1}{2} \int_V \left[\frac{p^2}{\rho c^2} - \frac{(\text{grad } p)^2}{\rho \omega^2} \right] dV \quad .$$

(10.1)

Here $\vec{\xi}$ is the displacement vector, $\dot{\vec{\xi}}$ the velocity, p the sound pressure, U the total potential and T the total kinetic energy. The density ρ and speed of sound c need not be constant; they may change as a function of position. To make the calculations short, we assume that there are no losses and that there is only a single frequency ω. this allows us to use the linearized momentum equation

$$\rho \, \ddot{\vec{\xi}} = -\text{grad } p; \text{ or } \dot{\xi}^2 = \frac{(\text{grad } p)^2}{\rho^2 \omega^2} \quad .$$

The next step is to divide the volume V into small subvolumes V_n, the so-called elements. Thus we have to find the minimum of the function

$$L = \sum_{n=1}^{N} \int_{V_n} \left[\frac{p_n^2}{2\rho c^2} - \frac{(\text{grad } p_n)^2}{2\rho \omega^2} \right] dV \quad ;$$

(10.2)

(p_n = pressure in the n-th element).

There are many ways to divide a given volume. Rectangular elements and tetrahedra are most often used, but wedges or curved elements may also be advantageous for certain shapes of V. For more information on that topic and FEM in general, see e.g. Zienkiewicz (1971); Gladwell (1965). When the volume is divided into elements there are many intersection points, the so-called nodes. In Figure 10.1 they are indicated by dots.

The essential feature of FEM is to express the field quantities in each element, in our case the sound pressure $p_n(x, y, z)$, in terms of the values at the nodes. These nodal values shall be called P_ν. A connection between the continuous functions $p_n(x, y, z)$ and the discrete values P_ν is established by the shape functions or interpolation functions $a_{n\nu}(x, y, z)$. They are chosen in such a way that the pressure in each element can be expressed as

$$p_n(x, y, z) = \sum_\nu a_{n\nu}(x, y, z) P_\nu \quad .$$

(10.3)

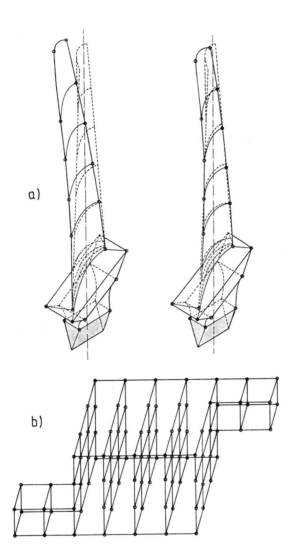

Figure 10.1 Examples of FEM nets.

    (a) Turbine blade;

    (b) expansion chamber in a muffler.

The summation is taken over all values of ν, i.e. over all nodal points. But since $a_{n\nu}(x, y, z)$ is zero outside the n-th element, only a few terms are nonzero in Equation (10.3). These are those terms for which the nodal points are inside or at the boundary of the n-th element. Obviously the shape functions have to be such that at a nodal point the pressure $p_n(x_\nu, y_\nu, z_\nu)$ is equal to P_ν.

There is a large variety of shape functions, which is described in the literature or used in standard computer codes. The simplest ones are linear combinations of the coordinates. In this case, which is the most common, the volume elements V_n have to have dimensions which are smaller than at least a sixth of a wavelength and smaller than any other length which is representative of the problem (e.g. the source dimension). Larger elements can be used when higher–order shape functions are available which give a better representation of the field inside an element. When bending is involved the shape functions have to be at least of third order to assure continuity at the boundaries of each element.

Sometimes an exact solution for the interior field of a large element is known, e.g. a combination of free waves with known wavenumbers. Such elements are called *substructures* and the FEM procedure is used to combine these fairly large substructures.

If (10.3) is inserted into (10.2), the result is

$$L = \frac{1}{2\rho} \sum_n \left\{ \frac{1}{c^2} \int_{V_n} \left[\sum_\nu P_\nu a_{n\nu} \right]^2 dV - \frac{1}{\omega^2} \int_{V_n} \left[\sum_\nu P_\nu \, \mathrm{grad} \, a_{n\nu} \right]^2 dV \right\} .$$

$$(10.4)$$

Here and in the following, the argument of the shape function is omitted.

Now the Lagrangian is expressed in terms of the finite number of values for P_ν. The next step is to minimize (10.4), i.e. to find

$$\frac{\partial L}{\partial P_\mu} = 0$$

for each value of μ for which the nodal pressure is not known. This way we obtain

$$\sum_n \sum_{\nu'} P_{\nu'} \left\{ \frac{1}{c^2} \int_{V_n} a_{n\mu} a_{n\nu'} \, dV - \frac{1}{\omega^2} \int_{V_n} \mathrm{grad} \, a_{n\mu} \, \mathrm{grad} \, a_{n\nu'} \, dV \right\} = 0$$

$$\text{for} \quad \mu = 1, 2 \dots . \qquad (10.5)$$

This is a set of linear equations for the unknown nodal pressure P_ν. The matrix of (10.5) is sparsely occupied; therefore the solution is reasonably straightforward even if there are several thousand nodal points, i.e. unknowns.

The scheme described here can also be extended to include sources acting from outside or sinks drawing energy from the structure. All that has to be done is to add the external work function W to (10.1). This function, which in our case would be

$$W = - \int_s p\xi dS$$

describes the energy flow in and out of the system through the area S when the displacement ξ on this surface is known.

Boundary conditions, which have not yet been mentioned, can also be included. To this end some values of P_ν, or differences of neighbouring values representing derivatives, have to be introduced as constraints.

When FEM is applied to elasto–acoustic problems the calculations are a little more complicated because one has to work with the displacement vector instead of the scalar pressure, but the general scheme of the calculation is the same:

- calculation of the Lagrangian, which in addition to (10.1) may include a work function that represents the work done by external sources;
- representation of the boundaries that are not connected to an outside source by constraints unless the boundaries are free;
- division of the whole system into small elements and selection of the shape functions;
- expression of the strains, stresses and energies as functions of the displacement vectors at the nodes (using the shape functions);
- minimization of the Lagrangian to obtain a set of linear equations for the field quantities at the nodes;
- solution of the linear equations.

For the calculation of resonance frequencies and mode shapes, FEM is now a well–established technique which has found widespread applications. The situation is more difficult when the frequency response of the amplitudes or the r.m.s values within a certain frequency band have to be found for a specific excitation. The reason for the difficulties lies not in the numerical method or in the computational effort, but in the lack of necessary data. A notorious problem in this respect is damping, which on one hand determines the magnitude of the resonance amplitude, but on the other hand is very hard to quantify, especially when it is due to friction at the interfaces of many individual parts that make up a complicated structure.

10.3 BOUNDARY ELEMENT METHOD (BEM)

The most general "solution" of the radiation and scattering problem in acoustics is the Kirchoff–Helmholtz equation, which represents the sound field by the strength of the sound sources Q which may be present, and by the values of the sound pressure and the sound velocity on a closed surface. If we restrict ourselves to single–frequency sounds, the Kirchoff–Helmholtz equation is

$$
\left.\begin{array}{c} p(x,y,z) \\ p(x,y,z)/2 \\ 0 \end{array}\right\} = \frac{1}{4\pi} \int_V Q \, \frac{e^{ikr_Q}}{r_Q} \, dV - \frac{i\omega\rho}{4\pi} \int_S v_S \, \frac{e^{ikr}}{r} \, dS
$$

$$
+ \frac{1}{4\pi} \int_S p_S \, \frac{\partial}{\partial n} \left(\frac{e^{ikr}}{r} \right) dS \quad ;
$$

(10.6)

Q = source strength of any (transparent) source inside the volume V which is enclosed by the surface S,

r_Q = distance between the source point and the receiver point,

v_S = normal component of the velocity on the surface S,

p_S = sound pressure on the surface S,

r = distance between the surface element and the receiver point,

$\partial/\partial n$ = derivative in a direction normal to the surface,

$p(x,y,z)$= sound pressure at the receiving point if it is inside the closed surface S (which may extend to infinity); zero sound pressure is obtained, when the receiving point is outside the surface S (see Figure 10.2); the value $p(x,y,z)$ /2 holds for points exactly on the surface S.

If a radiation problem has to be solved, the velocity v_S on the surface S of the radiator is known; in addition, there may be other sources of strength Q. The unknown quantities are the sound pressures $p(x,y,z)$ and p_S at the receiving point and on the surface, respectively. If a scattering problem has to be solved, the source strength Q (or the incoming sound which has to be specified in some other way) and the boundary conditions on the scatterer of surface S are given. The unknown quantities again are $p(x,y,z)$ and p_S. The surface velocity v_S is either zero (when the scatterer is rigid and fixed at its position) or it is related to the surface pressure p_S by an impedance condition

$$
v_S = p_S/Z_S \quad .
$$

(10.7)

Here Z_S is the impedance of the scatterer at a specific surface element. Z_S has to be a local impedance, i.e. it depends only on the p_S/v_S ratio at a certain element and is not influenced by the field in the neighbourhood; thus scatterers that consist of plates or shells that are excited to flexural motion are not covered by (10.7). More complicated methods have to be applied in such cases.

Whether Equation (10.6) represents a radiation or a scattering problem, it is always an integral equation which contains the unknown pressure on the left–hand side as well as under the integral.

For a numerical solution the surface is divided into many boundary elements S_n. This way the integrals over S become

$$\frac{-i\omega\rho}{4\pi} \int_S v_S \frac{e^{ikr}}{r} dS \approx \frac{-i\omega\rho}{4\pi} \sum_n v_{Sn} \int_{S_n} \frac{e^{ikr}}{r} dS$$

$$\frac{1}{4\pi} \int_S p_S \frac{\partial}{\partial n} \left(\frac{e^{ikr}}{r} \right) dS \approx \frac{1}{4\pi} \sum_n p_{Sn} \int_{S_n} \frac{\partial}{\partial n} \left(\frac{e^{ikr}}{r} \right) dS \quad .$$

$$(10.8)$$

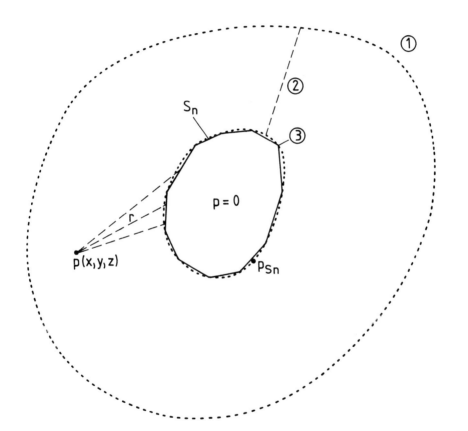

Figure 10.2 Discretization of the Kirchoff–Helmholtz Equation.
The total surface S consists of the infinite surface (1), which does not contribute to the integral, the negligibly small connection area (2), and the radiating or scattering surface (3). For the sake of simplicity (3) has no sharp corners. The surface (3) is divided into the elements S_n.

The boundary elements are assumed to be so small that the surface velocity v_{Sn} and the pressure p_{Sn} can be taken as constants in front of the integrals over S_n (see Figure 10.2).

The next task is to calculate the integrals. Whenever r is the distance to a receiving point outside the boundary element S_n this poses no problem, because r is always greater than a certain minimum distance. A minor difficulty arises, when the pressure on one of the surface elements is calculated. In such a case the distance r can become very small and even zero. Fortunately this does not generate any problems because for small elements, we can approximate $e^{ikr} \approx 1$, which leaves integrals of the type

$$\int_{S_n} \frac{1}{r}\, dS \quad \text{and} \quad \int_{S_n} \frac{1}{r^2} \frac{\partial r}{\partial n}\, dS \ .$$

Both these integrals remain finite, because with decreasing distance the surface elements become smaller, too.

Having calculated the integrals in (10.8), Equation (10.6) becomes a set of linear equations

$$\left.\begin{array}{c} p_\nu \\ p_\nu/2 \to \\ 0 \end{array}\right\} = \frac{1}{4\pi} \int Q \frac{e^{ikr_Q}}{r_Q} - \frac{i\omega\rho}{4\pi} \sum_n v_{Sn} a_{\nu n} + \frac{1}{4\pi} \sum p_{Sn}\, b_{\nu n} \ . \qquad (10.9)$$

Here p_ν is the sound pressure at position ν (which may be in the middle of one of the boundary elements); $a_{\nu n}$ and $b_{\nu n}$ are the integrals in (10.8) with $r = r_{n\nu}$ being the distance from a surface element dS on the boundary element S_n to the position ν.

The necessary steps to solve a radiation or scattering problem, therefore, are:

(a) Take as position ν the centre of the boundary elements consecutively; this way as many linear equations as there are boundary elements are obtained;

(b) replace p_{Sn} by the appropriate value of p_ν;

(c) solve the set of linear equations;

(d) take as ν any point of interest and calculate the pressure by using (10.9) with the known values of p_{Sn}.

This procedure sounds rather straightforward, but there is a difficulty which makes it necessary to modify the calculation scheme, when the radiating or scattering surface is much larger than the wavelength. The reason is that the set of linear equations (10.9) becomes singular when the frequency coincides with one of the resonance frequencies (for Dirichlet boundary conditions) of the volume enclosed by the surface of the scatterer or radiator.

To circumvent this deficiency, a Combined Helmholtz Integral Equation Formulation (CHIEF) was proposed by Schenk (1968). The basic idea is to form an overdetermined system of equations by adding a few more points ν to step (a) of the above–mentioned scheme; these points are inside the scatterer or radiator (i.e. "outside" the total surface S). For such points the pressure, according to (10.6), is zero. A least–square technique is then used to solve the system of $N + N_1$ equations (N = number of boundary elements, N_1 = number of additional "inner" points). This method yields unique results even at the characteristic frequencies, provided the "inner" points do not lie on a nodal line of the interior standing wave (which is rather unlikely). There are several program packages already which use this method successfully (Seybert et al. 1985).

Another method for avoiding the difficulties associated with the "inner resonances" is proposed by Burton & Miller (1971). In this case a linear combination is formed consisting of (10.6) and the same equation when the normal derivative of the pressure is taken. The method is known as CONDOR (Composite Outward Normal Derivative Overlap Relation) (Reut 1985). By this method the matrices never become singular.

For the calculation of the sound field inside a closed surface (e.g. the passenger cabin of a car) one can compare FEM and BEM calculations, because both can be used in this case. It turns out that FEM needs many more equations (because there are more volume elements than surface elements), but the matrices are sparsely occupied and can be brought to a *band structure*. In the BEM calculations the matrices are much smaller but practically all coefficients in the matrix are nonzero. Thus it is more a matter of taste and experience which method is used.

10.4 SOURCE SUBSTITUTION METHODS

10.4.1 Multipole Synthesis

The standard way to solve the scattering problem for spherical (spheroidal) or cylindrical bodies is to replace the scatterer by a series of multipole sources and to adjust the strength of each multipole in such a way that the boundary conditions are fulfilled (see, e.g. Morse 1948). In the textbook examples, the calculations are rather easy because functions with the orthogonality property are used. For other, less simple geometries, as well as for sound radiators, the same general idea can be applied but the calculations are more tedious because the advantages of orthogonality can no longer be employed.

If a sound–radiating body is considered, the first step is to take it out of the field and instead replace it with a combination of multipoles at a position where the original "center" of the body was. The "center" in this context is not defined. In principle it can be any point, but it should be somewhere in the middle to keep the computational effort and the accuracy at a reasonable level. The sound pressure p_A generated by these auxiliary multipoles in an homogeneous medium at point $\mathbf{r}$ is given by

$$p_A(\mathbf{r}) = \sum_{\nu=1}^{N} Q_\nu\, \phi_\nu(\mathbf{r}) \quad . \tag{10.10}$$

Here Q_ν is the strength of the ν–th multipole. The functions ϕ_ν are

- in three dimensions, spherical coordinates $\mathbf{r} = (r, \theta, \phi)$

$$\phi_\nu(\mathbf{r}) = P_n^m(\cos\theta)\, h_n(k_0 r)_{\sin n\phi}^{\cos n\phi}$$

- in two dimensions, cylindrical coordinates $\mathbf{r} = (r, \phi)$

$$\phi_\nu(\mathbf{r}) = H_n^{(1)}(k_0 r)_{\sin n\phi}^{\cos n\phi} \quad .$$

The abbreviations are

$P_n^m(\cos\theta)$ = associated Legendre function;

$h_n(kr)$ = spherical Hankel function for outgoing waves;

$H_n^{(1)}(kr)$ = cylindrical Hankel function.

The origin of the coordinate system is the position of the multipoles. $k_0 = \omega/c$ is the wavenumber. The index ν stands for n or a combination of n and m.

The normal component of the velocity which the multipoles generate on a contour S that is the same as the surface of the original body is calculated from (10.10) by using conservation of momentum. the result is

$$v_A(\mathbf{r}_s) = \frac{1}{i\omega\rho} \sum_{\nu=1}^{N} Q_\nu\, \frac{\partial \phi_\nu(\mathbf{r}_s)}{\partial n} \quad . \tag{10.11}$$

$\mathbf{r}_S$ is the coordinate of a point on the contour S; $\partial/\partial n$ the derivative in the normal direction. Figure 10.3 shows in a qualitative way the shape of $\partial \phi_\nu / \partial n$ for a monopole, dipole, quadrupole, etc. on a rectangular contour.

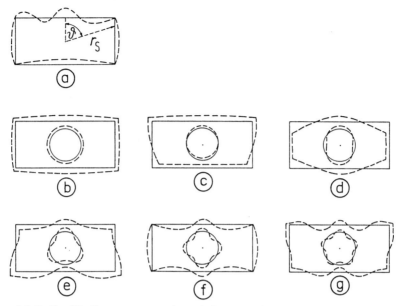

Figure 10.3 Qualitative representation of the normal component of the velocity
of a vibrating box. (a) surface velocity v_S; (b) – (g) surface velocity
generated by multipoles of order 0 – 5.

One possible way to find the unknown values Q_ν (see Williams et al. 1964) is
to minimize on the contour S the difference between the velocity $v_A(\mathbf{r}_s)$ generated
by the multipoles and the velocity $v_S(\mathbf{r}_s)$ of the radiating body, i.e.

$$\int_s \left| v_s(\mathbf{r}_s) - v_A(\mathbf{r}_s) \right|^2 dS = \int_s \left| v_s(r_s) + \frac{i}{\omega\rho} \sum_\nu Q_\nu \frac{\partial\phi_\nu(\mathbf{r}_s)}{\partial n} \right|^2 dS = \text{Min} . \quad (10.12)$$

After differentiating with respect to Q_μ for $0 \leqq \mu \leqq N$, the result is

$$\sum_\nu Q_\nu \, \beta_{\mu\nu} = -i\omega\rho \int_S v_S \frac{\partial}{\partial n} \phi_\mu^* (\mathbf{r}_S) \, dS$$

with $\qquad\qquad\qquad\qquad\qquad\qquad\qquad\qquad\qquad\qquad\qquad (10.13)$

$$\beta_{\mu\nu} = \int_S \frac{\partial}{\partial n} \phi_\nu (\mathbf{r}_S) \frac{\partial}{\partial n} \phi_\mu^* (\mathbf{r}_S) \, dS \quad .$$

Equation (10.13) is a set of linear equations for the unknown source strengths Q_ν.
It can be solved by the usual methods and then the pressure can be calculated
according to (10.10). If v_S is a reasonably smooth function, it seems to be sufficient
to make $N \approx 10\ell/\lambda$, where ℓ is the greatest length of the radiator.

An alternative procedure (Cremer 1984) is to apply a so–called moment method, i.e. to multiply (10.11) with a proper set of functions and equate the result. Since intensity is a very important quantity for radiation, it seems to make sense to multiply (10.11) with $\phi_\mu^*(\mathbf{r}_S)$ and integrate over the surface. (ϕ_μ is the pressure generated by the μ–th multipole.) In this way by setting $v_s = v_{SA}$, we get

$$\sum_\nu Q_\nu \, \gamma_{\mu\nu} = -i\omega\rho \int_S v_S \, \phi_\mu^* \, (\mathbf{r}_S) \, dS$$

with

$$\gamma_{\mu\nu} = \int_S \phi_\mu^* \, (\mathbf{r}_S) \, \frac{\partial}{\partial n} \, \phi_\nu \, (\mathbf{r}_S) \, dS \quad .$$

(10.14)

Equation (10.14) is again a set of linear equations, which can be used to find Q_ν. It can be shown (Heckl 1989) that the moment method is a consequence of (10.6).

The procedure described here can also be applied to scattering problems. The only difference is that the surface velocity $v_s(\mathbf{r}_s)$ has to be modified by the incoming sound and the boundary conditions.

10.4.2 Substitution by Many Monopoles and Dipoles

Instead of replacing the radiator or scatterer by a combination of multipoles which are all at the same place in the *centre* of the body, one may also replace the original by several monopoles and dipoles which are located at different positions $\mathbf{r}_\nu$ inside the contour S. Thus instead of (10.10) and (10.11), we have to write

$$p_A(\mathbf{r}) = \sum_{\nu=1}^{N} \left[Q_{M\nu} \phi_M \, (\mathbf{r} - \mathbf{r}_\nu) + Q_{D\nu} \phi_D \, (\mathbf{r} - \mathbf{r}_\nu) \right]$$

$$v_A(\mathbf{r}_s) = \frac{1}{i\omega\rho} \sum_{\nu=1}^{N} \left[Q_{M\nu} \frac{\partial \phi_M(\mathbf{r}_s - \mathbf{r}_\nu)}{\partial n} + Q_{D\nu} \frac{\partial \phi(\mathbf{r}_s - \mathbf{r}_\nu)}{\partial n} \right] \quad .$$

(10.15)

The functions ϕ_M and ϕ_D which represent the sound fields generated by a monopole or dipole are given by the functions ϕ_ν of the previous paragraph for $\nu = 0$ and $\nu = 1$, respectively. The same procedures that have been applied already (minimum error or moment method) can now be used to find the unknowns $Q_{M\nu}$ and $Q_{D\nu}$.

Whether the method described in Section 10.4.1 or in Section 10.4.2 is more suitable for a specific problem is hard to decide. There is some indication that for bodies that do not deviate too much from the spherical or cylindrical shape, the multipole method is more effective, whereas for long bodies or for those that have a shape similar to L, T, U the monopole and dipole representation seems to be better. Little can be said about the ambiguity with respect to the position $\mathbf{r}_\nu$ of the auxiliary sources. It usually takes a few trial runs before "good" positions are found.

If substitution methods are compared with the Kirchoff–Helmholtz equation and the corresponding BEM procedure (see Section 10.3), the following can be said:

- BEM calculations need many more elements and therefore more linear equations than the substitution methods (Ochmann 1990);
- the eigenvalue problem which makes the matrices singular and which can be overcome in the BEM only by additional calculations does not arise in the substitution model;
- BEM gives a zero field inside the radiator or scatterer; substitution methods do not have this property.

10.4.3 Other Methods

The T–matrix method or nullfield method (Watermann 1969) that has been applied successfully to acoustic, electromagnetic and elastomechanic problems (Varadan 1980) can also be considered a substitution method. The basic idea is to express the Green's function e^{ikr}/r and its derivative as a sum of orthogonal functions, and to use the nullfield part of (10.6) to determine the unknown quantities. For details, see the relevant literature where it is also shown that the method does not break down at the eigenfrequencies (Martin 1982).

The source–substitution method has also been proposed (Tomilina 1989) for the solution of elastoacoustic problems with fluid loading. In this case one starts with the equation of motion for the elastic system and takes into account fluid loading by a surface pressure p_{RAD}. In general terms, the equation to be solved is

$$L\{v\} = p_E - p_{RAD} \quad . \tag{10.16a}$$

$L\{v\}$ is the appropriate linear differential operator representing the equation of motion of the system and p_E the exciting pressure which constitutes the outside source (see Figure 10.4). For the simple case of a plate with bending stiffness B and bending wavenumber k_B, (10.16a) would be

$$\frac{-1}{i\omega} B(\Delta\Delta v - k_B^4\, v) = p_E - p_{RAD} \quad . \tag{10.16b}$$

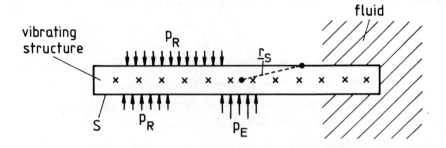

Figure 10.4 Source substitution method for elasto– acoustic problems.
× × location of auxiliary sources. S = surface of structure, $\mathbf{r}_S$ = coordinate of a point on S.

The next step is to express (10.16a) in terms of the auxiliary sources which are assumed to be inside the structure. Thus the normal component of the surface velocity and the fluid pressure on the surface S are approximated by

$$v_A(\mathbf{r}_s) = \frac{1}{i\omega\rho} \sum_{\nu=1}^{N} Q_\nu \, \frac{\partial\phi_\nu(\mathbf{r}_s)}{\partial n}$$

$$p_{RAD}(\mathbf{r}_s) = \sum_{\nu=1}^{N} Q_\nu \, \phi_\nu(\mathbf{r}_s) \quad .$$

(10.17)

If these approximations are inserted into (10.16a), we obtain

$$\sum_{\nu=1}^{N} Q_\nu \left[\frac{1}{i\omega\rho} L \left\{ \frac{\partial\phi_\nu(\mathbf{r}_s)}{\partial n} \right\} + \phi_\nu(\mathbf{r}_s) \right] = \widetilde{p}_E \quad . \tag{10.18}$$

Since we want to find a good approximation for (10.16) the difference between the known excitation outside pressure p_E and the auxiliary approximate quantity $\widetilde{p}_E$ should be as small as possible. Thus we demand, similar to (10.12)

$$\int_S \left| p_E - \widetilde{p}_E \right|^2 dS = \text{Min} \quad . \tag{10.19}$$

This is a quadratic expression in Q_ν; therefore the usual minimization procedure gives a set of linear equations that can be solved by standard techniques.

10.5 APPLICATIONS OF FOURIER TRANSFORMS

10.5.1 Plane Radiators

Fourier transforms (see Chapter 3) are in widespread use in acoustics and since the time Fast Fourier Transform algorithms have been available they have been a common tool for processing time signals of sound pressures and other measured quantities. Apart from this almost standard application, Fourier Transforms can also be used in the space domain to find the so–called wavenumber spectra which are nothing but a decomposition of a complicated field into a series of plane waves (see also Chapter 17). This is especially useful when one is working in plane geometrics. A typical example of this type is the far–field directivity of a plane sound radiator with surface velocity $v(x, y)$ which is imbedded in a large baffle (see Figure 10.5).

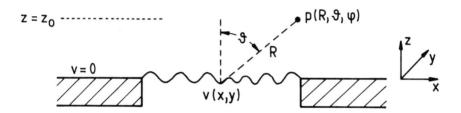

Figure 10.5 Radiation from a plane radiator.

The wavenumber spectrum in this case is

$$\check{v}(k_x, k_y) = \int v(x, y)e^{-ik_x x}\, e^{-ik_y y}\, dx\, dy \quad . \tag{10.20}$$

From that the far–field sound pressure can be found to be

$$p(r, \theta, \phi) = \frac{-i\omega\mu}{2\pi r}\, e^{ik_0 r}\, \check{v}(k_x = k_0 \sin\theta \cos\phi, \quad k_y = k_0 \sin\theta \sin\phi) \quad . \tag{10.21}$$

The total radiated sound power is

$$P = \frac{k_0^2 \rho c}{8\pi^2} \int_0^{2\pi} \int_0^{\pi/2} \left|\check{v}(k_x = k_0 \sin\theta \cos\phi, k_y = k_0 \sin\theta \sin\phi)\right|^2 \sin\theta\, d\theta\, d\phi \quad . \tag{10.22}$$

Here $k_0 = \omega/c$ in the wavenumber in the surrounding medium.

The equations show that the directivity is directly proportional to the lower part of the wavenumber spectrum and the radiated power is given by a simple integration.

10.5.2 Acoustic Near–field Holography

When we see a source of light or an illuminated object (i.e. a scatterer of light waves), we get a fairly good impression of its size, shape and source strength. Since acoustics also deals with waves, we can ask the question: can we reconstruct size and shape of a sound source (or scatterer) by measuring the sound it emits? For the very special case of planar objects, or more precisely if we are interested in the field quantities in a certain plane, the answer is given by acoustic holography (see, e.g. Maynard et al. 1985).

In the most simple case we assume that in the plane $z = 0$ there is a certain velocity distribution $v(x, y)$ which we want to reconstruct by measuring the sound field it generates.

If the wavenumber spectrum of the velocity we are looking for is $\breve{v}(k_x.k_y)$, the wavenumber spectrum of the sound pressure it generates in the plane $z = z_0 > 0$ is

$$\breve{p}(k_x, k_y, z_0) = \frac{\omega\rho}{k_z} \, \breve{v}(k_x, k_y) \, e^{ik_z z_0} \tag{10.23}$$

with

$$k_z = \begin{cases} (k_o^2 - k_x^2 - k_y^2)^{1/2} & \text{for} \quad k_x^2 + k_y^2 \le k_o^2 \\ i \, (k_x^2 + k_y^2 - k_o \,)^{1/2} & \text{for} \quad k_x^2 + k_y^2 > k_o^2 \end{cases}$$

and $k_o = \omega/c$.

To derive this formula we require that $p(x, y, z)$ as well as $\breve{p}(k_x, k_y, z_0)$ have to be solutions of the acoustic wave equation and that because of conservation of momentum, the relation $i\omega\rho\, v(x, y) = \partial p(x, y, 0)/\partial z$ or $i\omega\rho\, \breve{v}(k_x, k_y) = ik_z \, \breve{p}(k_x, k_y, z_0 = 0)$ must hold. Therefore, the steps necessary to deduce the radiator velocity $v(x, y)$ from the measured sound pressure are:

- measure the sound pressures in a plane $z = z_0$;
- calculate from these sound pressures the Fourier Transform $\breve{p}(k_x, k_y, z_0)$;
- apply (10.23) to find $\breve{v}(kx, k_y)$;
- use the "back–transformation"

$$v(x, y) = \frac{1}{4\pi} \int \breve{v}(k_x, k_y) \, e^{ik_x x} \, e^{ik_y y} \, dk_x \, dk_y \tag{10.24}$$

to get the desired result.

Obviously there are some difficulties in this procedure. Some of them are due to the number of sampling points and to the finite size of the sampling area. These are problems which appear in all digital sampling and FFT applications. But there is also one difficulty which is typical for acoustic holography and which stems from the $e^{ik_z z_0}$ term in (10.23). This term is just a phase factor when $k_1^2 + k_2^2 < \omega^2/c^2$, but it is an amplitude factor which may be very small when $(k_x^2 + k_y^2)^{1/2} z_0$ becomes large. As a consequence any small error in $\check{p}(k_x, k_y, z_0)$ becomes amplified in the reconstruction process and can easily lead to completely wrong results for $v(x, y)$. The solution for this problem is either to make $z_0^2 < |k_x^2 + k_y^2|^{-1}$, i.e. to make near–field holography, or to be content with the wavenumber spectrum for $\sqrt{k_x^2 + k_y^2} < \omega/c$, which means that the overall behaviour is reconstructed, but the fine details with dimensions smaller than the wavelength in the surrounding medium cannot be recovered. Thus we can conclude that the resolution of acoustic holography is determined either by the wavelength in the medium or by the measuring distance z_0. This is sometimes called the nonuniqueness property because a full reconstruction is impossible from far–field measurements.

If the sound field radiated by the radiator $v(x, y)$, which shall be reconstructed, is disturbed by other sources or by reflectors, then in principle acoustic holography can still be used, provided the disturbances are generated outside the radiator and the measuring plane, i.e. if they are in the region $z > z_0$. Since in this case sound waves are travelling in positive and negative z–direction (10.23) has to be replaced by

$$\check{p}(k_x, k_y\, z_0) = \frac{\omega\rho}{k_z}\, \check{v}(k_x, k_y)\, e^{ik_z z_0} + \check{A}(k_x, k_y)\, e^{-ik_z z_0} \quad . \tag{10.25a}$$

$\check{A}(k_y, k_y)$ is the wavenumber spectrum of the disturbing sources. At the moment it is unknown. The usual technique to find $\check{A}$ is to make a separate set of pressure measurements in a plane $z = z_1$. The Fourier transform of these measurements yields

$$\check{p}(k_x, k_y\, z_1) = \frac{\omega\rho}{k_z}\, \check{v}(k_x, k_y)\, e^{ik_z z_1} + \check{A}(k_x, k_y)\, e^{-ik_z z_1} \quad . \tag{10.25b}$$

Equation (10.15a,b) are a set of linear equations for each value of k_x, k_y. They can easily be solved to give the unknown quantities $\check{v}$ and $\check{A}$. From $\check{v}$ the desired surface distribution $v(x, y)$ can be found, while A is no longer needed.

This method has two drawbacks. It relies on the measurement of differences and is therefore less accurate; it does not work when the determinant of the matrix vanishes. This happens for $\sin(z_0 - z_1) = 0$. The second difficulty can be overcome by making measurements in a third plane and solving the overdetermined linear system which is obtained in this case.

If the geometry of interest is not plane but cylindrical, the method is very similar. The main difference is that

$$\int \ldots e^{ik_x x}\, e^{ik_y y}\, e^{ik_z z}\, dk_x, dk_y$$

has to be replaced by the transform

$$\sum_{n=-\infty}^{\infty} \int \ldots e^{in\phi}\, e^{ik_y y}\, H_n^{(1)}\left(k_r r\right) dk_y \quad,$$

where r, ϕ, y are cylindrical coordinates and $H_n^{(1)}$ the Hankel function.

10.5.3 Sound in Layered Media

In a layered medium the material parameters depend only on one coordinate. Let this coordinate in a Cartesian system be z. If we assume that the layers have a finite thickness (which can also be used as an approximation for continuously varying parameters) the sound field can be represented (see Figure 10.6) in terms of the wavenumber spectrum in each layer

$$v_n(x,y) = \int \quad \left[\breve{v}_{n+} e^{ik_{zn} z} + \breve{v}_{n-} e^{ik_{zn} z}\right] e^{ik_x x}\, e^{ik_y y}\, dk_x\, dk_y$$

$$p_n(x,y) = \int \frac{\omega\rho}{k_{zn}} \left[\breve{v}_{n+} e^{ik_{zn} z} - \breve{v}_{n-} e^{-ik_{zn} z}\right] e^{ik_x x}\, e^{ik_y y}\, dk_x\, dk_y \quad.$$

$$(10.26)$$

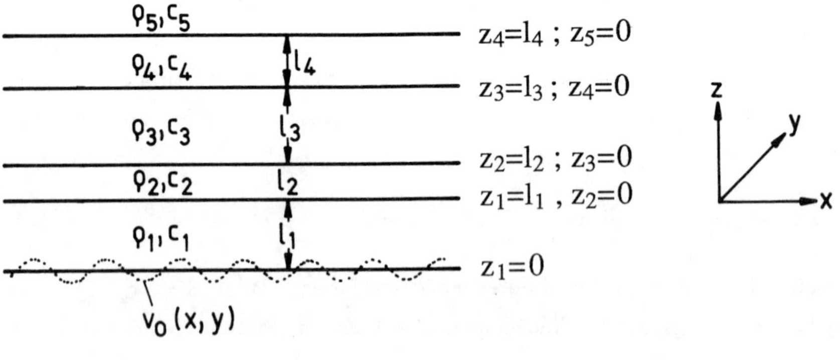

Figure 10.6 Coordinates in a layered medium

Here

$$k_{zn} = \begin{cases} \left(-k_x{}^2 - k_y{}^2 + \omega^2/c_n^2\right)^{1/2} & \text{if } \quad k_x{}^2 + k_y{}^2 < \omega^2/c_n^2 \\[2mm] i\left(k_x{}^2 + k_y{}^2 - \omega^2/c_n^2\right)^{1/2} & \text{if } \quad k_x{}^2 + k_y{}^2 \geq \omega^2/c_n^2 \quad. \end{cases}$$

If, instead of the wavenumber spectra $\breve{v}_{n+}$ and $\breve{v}_{n-}$ which describe the velocities, the wavenumber spectra $\breve{p}_{n+}$ and $\breve{p}_{n-}$ for the pressures would have been used, we would have had to replace

$$\breve{v}_{n+} = k_{zn}\breve{p}_{n+}/\omega\rho_n; \qquad \breve{v}_{n-} = -k_{zn}\breve{p}_{n-}/\omega\rho_n \quad . \tag{10.27}$$

For the following discussion we can omit the factor $e^{ik_x x}e^{ik_y y}$ as well as the integration, because for plane layers there is no conversion from one wavenumber to another.

As an example of the application (10.26), we assume that at $z_1 = 0$ a certain velocity $v_0(x, y)$ is given, and that layer number five extends to infinity (see Figure 10.6). The calculation scheme is then as follows:

(a) make a Fourier transform of $v_0(x, y)$ to get the wavenumber spectrum $\breve{v}_0$;

(b) equate $\breve{v}_0$ with the wavenumber spectrum given by (10.26) for $z_1 = 0$, i.e.

$$\breve{v}_0 = \breve{v}_{1+} + \breve{v}_{1-} \quad ; \tag{10.28}$$

(c) apply the continuity condition for pressure and velocity at the interfaces, i.e.

$$\breve{v}_{1+}e^{ik_{z1}\ell_1} + \breve{v}_{1-}e^{-ik_z\ell_1} = \breve{v}_{2+} + \breve{v}_{2-}$$

$$\tag{10.29}$$

$$\frac{\rho_1}{k_{z1}}\left[\breve{v}_{1+}e^{ik_{z1}\ell_1} - \breve{v}_{1-}e^{-ik_{z1}\ell_1}\right] = \frac{\rho_2}{k_{z2}}\left(\breve{v}_{2+} - \breve{v}_{2-}\right) \quad ,$$

etc. for $z_3 = 0$, $z_4 = 0$, $z_5 = 0$;

(d) set $\breve{v}_{5-} = 0$, because no sound is coming from infinity;

(e) solve the eleven linear equations (in general $2N + 1$) consisting of (10.28) and (10.29) to get the eleven unknowns (in general $2N + 1$) $\breve{v}_{1+}$, $\breve{v}_{1-} \dots \breve{v}_{4+}$, $\breve{v}_{4-}$, $\breve{v}_{5+}$;

(f) insert these quantities in (10.26) to get the whole field.

If at $z_1 = 0$ the pressure is given instead of the velocity and/or the last boundary condition is represented by an impedance, the calculations are more or less the same.

In many applications the square brackets in (10.26) are replaced by

$$\breve{v}_{no} \cos k_{zn}z_n + i\,\frac{k_{zn}}{\omega\rho_n}\,\breve{p}_{no}\,\sin k_{zn}z_n$$

$$\tag{10.30}$$

$$i\,\frac{\omega\rho_n}{k_{zn}}\,\breve{v}_{no}\,\sin k_{zn}z_n + \breve{p}_{no}\,\cos k_{zn}z_n \quad .$$

This is the so–called *transfer matrix method*, where $\breve{v}_{no}$ and $\breve{p}_{no}$ are the values at $z_n = 0$. If this approach is used, the continuity conditions at the interfaces lead to matrix multiplication, which is sometimes easier to handle.

The method described above for ordinary sound waves in a layered gas or liquid can also be applied to layered solids (Brekhovskikh 1960) and to viscous media (Heckl & Jackson 1967), the difference being that the square brackets in (10.26) contain more terms because there are two types of waves travelling backwards and forwards. Furthermore, there are more variables (tangential velocities and tangential stresses) but at the same time there are more conditions at the interfaces, so that again a full set of linear equations is obtained.

Finally, it should be mentioned that with some modifications, the method can also be used to investigate the transmission through a medium which has a flow speed that varies along the coordinate z. In this case, the wavenumber k_z depends also on the flow speed and most importantly the boundary condition at an interface is continuity of pressure and displacement (not velocity because of the convective term (see Miles 1957).

10.6 RAY TRACING

10.6.1 Rays in an Homogeneous Medium with Boundaries

Rays are the solution of the wave equation for the limiting case of vanishing wavelength. The ray paths can therefore be found if the wave equation is solved for $\lambda \to 0$, i.e. $k^2 \to \infty$. If we take a medium with speed of sound $c(x_j)$, i.e. the speed of sound is not constant in space but is constant in time, the wave equation is $(x_j = x_1, x_2, x_3)$

$$\nabla^2 p + \frac{\omega^2}{c^2(x_j)}\, p = 0 \quad . \tag{10.31}$$

As a solution, we try

$$p = p_0(k_j)\, e^{ik_0 S(x_j)} \quad . \tag{10.32}$$

Here $k_o = \omega/c_0$ is the constant wavenumber in a medium with speed of sound c_0. Usually for c_0 the average of $c(x_j)$ is taken. $S(x_j)$ is the equation of the wavefronts, i.e. the surfaces of constant phase. Inserting (10.32) into (10.31) and retaining only the highest power in k_o gives the so–called *Eikonal* equation (see e.g. Dowling & Ffowcs Williams 1983)

$$\left(\frac{\partial S}{\partial x_1}\right)^2 + \left(\frac{\partial S}{\partial x_2}\right)^2 + \left(\frac{\partial S}{\partial x_3}\right)^2 = |\mathrm{grad}\, S(x_j)|^2 = c_0{}^2/c^2(x_j) \quad . \tag{10.33}$$

The rays $X_i(s)$ are lines that are normal on the wavefronts; therefore, in parameter form, they are given by the equations

$$\frac{dX_i(s)}{ds} = \frac{1}{|\text{grad } S|} \frac{\partial S(x_j)}{\partial x_i}, \qquad i = 1, 2, 3 \quad . \tag{10.34}$$

(see Figure 10.7).

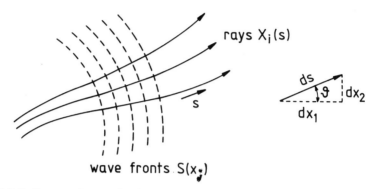

Figure 10.7 Rays and wave fronts.

Here s is the arc length along a ray, i.e. $dX_1^2 + dX_2^2 + dX_3^2 = ds^2$. In an homogeneous medium $c(x_j) = c_0$ is a constant. This gives

$$S(x_j) = \alpha_1 x_1 + \alpha_2 x_2 + \alpha_3 x_3 \quad \text{with} \quad \alpha_1^2 + \alpha_2^2 + \alpha_3^2 = 1$$

as a solution of (10.33). As one might have expected in this special case, the wavefronts are planes and the rays are straight lines normal to the planes.

Instead of deriving the Eikonal equation, one might also have used Fermat's principle, which states that sound always takes the path of minimum "flight time". From this principle one can also deduce that in homogeneous media the reflection at boundaries follows the laws of geometrical optics. (Angles of incidence and reflection are equal.)

These facts are often applied for the calculation of sound fields in large rooms when the modal approach becomes too complicated. The basic idea (Krokstadt et al. 1968) is to follow the paths of "sound particles" (i.e. rays) that are emitted from a source in different directions and to include a certain number of reflections. If the reflection is not perfect, this can be taken into account by introducing a reflection factor $r < 1$.

An alternative method is to calculate all the images and to add the sound pressures coming from them (by this method even phases can be taken into account to a certain degree). For simple rectangular geometries, this alternative method is very useful, but for complicated room shapes consisting of N plane walls, it becomes rather tedious because there are approximately $(N-1)^K$ images, where K is the order of the reflection.

10.6.2 Ray Tracing in Inhomogeneous Media

The atmosphere, as well as the ocean, are inhomogeneous media; therefore the variations of the speed of sound have to be taken into account (for many examples, see Lee et al. 1990). As a first approximation and for long–range propagation, this can be done by ray calculations because the speed of sound is a slowly varying function compared with the wavelength. The equations that have to be solved for ray calculations are (10.34) and (10.33). To eliminate the wavefronts $S(x_j)$, we first make use of the fact that because of (10.33), we can write for (10.34)

$$\frac{c_0}{c(x_j)}\frac{dX_i(s)(}{ds} = \frac{\partial S(x_j)}{\partial x_i} ; \qquad i = 1,2,3 \quad . \tag{10.35}$$

Differentiation with respect to s gives

$$\frac{d}{ds}\left[\frac{c_0}{c(x_j)}\frac{dX_i(s)}{ds}\right] = \frac{d}{ds}\left[\frac{\partial S(x_j)}{\partial x_i}\right] = \frac{\partial^2 S}{\partial x_i \partial x_1}\frac{\partial X_1}{ds} + \frac{\partial^2 S}{\partial x_i \partial x_2}\frac{dX_2}{ds} + \frac{\partial^2 S}{\partial x_i \partial x_3}\frac{dX_3}{ds}$$

$$= \frac{c(x_j)}{c_0}\left[\frac{\partial^2 S}{\partial x_i \partial x_1}\frac{\partial S}{\partial x_1} + \frac{\partial^2 S}{\partial x_i \partial x_2}\frac{\partial S}{\partial x_2} + \frac{\partial^2 S}{\partial x_i \partial x_3}\frac{\partial S}{\partial x_3}\right]$$

$$= \frac{c(x_j)}{2c_0}\frac{\partial}{\partial x_i}\left[\left(\frac{\partial S}{\partial x_1}\right)^2 + \left(\frac{\partial S}{\partial x_2}\right)^2 + \left(\frac{\partial S}{\partial x_3}\right)^2\right] = \frac{c(x_j)}{2c_0}\frac{\partial}{\partial x_i}\left[\frac{c_0{}^2}{c^2(x_j)}\right]$$

$$= -\frac{c(x_j)}{c_0}\frac{c_0{}^2}{c^3(x_j)}\frac{\partial c(x_j)}{\partial x_i} = -\frac{c_0}{c^2(x_j)}\frac{\partial c(x_j)}{\partial x_i} \quad .$$

$$\tag{10.36}$$

In this sequence of equations the following transformations have been used: chain rule and $x_i = X_i$; Equation (10.35); differentiation of squared functions; Equation (10.33); differentiation of $1/c^2(x_j)$. The final result of (10.36) giving the rays $X_i(s)$ in terms of speed distribution, is

$$\frac{d}{ds}\left[\frac{1}{c(x_j)}\frac{dX_i(s)}{ds}\right] = \frac{-1}{c^2(x_j)}\frac{\partial c(x_j)}{\partial x_i} \quad . \tag{10.37}$$

For computational purposes it is convenient to transform (10.37) into a set of first–order differential equations by using the expressions in square brackets as separate functions. Thus the system that has to be solved (usually numerically) is

$$\frac{dX_i(s)}{ds} = c(x_j)\alpha_i; \qquad \frac{d\alpha_i}{ds} = \frac{-1}{c^2(x_j)} \frac{\partial c(x_j)}{\partial x_i} \qquad . \qquad (10.38)$$

If cylindrical coordinates r, θ, z are used, and if the wave speed is independent of the angle θ, one has to replace x_1 by r and x_2 by z and omit the equation with $i = 3$.

It may be mentioned here that for layered media, i.e. when $c(x_j)$ depends on one coordinate only, say x_1, (10.38) gives $\alpha_2 = $ constant and therefore

$$\frac{dX_2(s)}{ds} = c(x_1) \text{ constant} \quad \text{or} \quad \sin\theta/c(x_1) = \text{constant}$$

which is just Snell's law. (Since $dX_1^2 + dX_2^2 = ds^2$, one can write $dX_2 = ds\sin\theta$; see Figure 10.7.)

Figure 10.8 shows the results (taken from Porter & Bucker 1987) of ray calculations for a so–called Munk profile in very deep water. In this profile, the sound speed is 1550m/s at the sea surfaces, decreases to a minimum of 1500m/s at a depth of 1300m and then increases monotonically with depth.

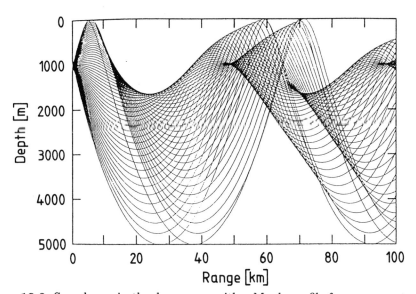

Figure 10.8 Sound rays in the deep ocean with a Munk–profile for a source at a depth of 1000 m (After Porter & Bucker 1987).

10.6.3 Gaussian Beam Method

The principle of the Gaussian beam method and the rather similar parabolic equation method can be understood fairly easily when a simple example is considered.

We start with a cylindrical coordinate system r, z and assume all quantities to be independent of circumferential angle. Thus the wave equation for single frequencies is

$$\frac{1}{r} \frac{\partial}{\partial r} \left[r \frac{\partial p}{\partial r} \right] + \frac{\partial^2 p}{\partial z^2} + k_o{}^2 p = 0 \quad . \tag{10.39}$$

As a solution we assume

$$p = u(r, z) \, e^{ik_o z} \quad . \tag{10.40}$$

The important point here is that we have "factored out" the rapidly varying term $e^{ik_o z}$. Combining (10.39) and (10.40) gives

$$\frac{1}{r} \frac{\partial}{\partial r} \left[r \frac{\partial u}{\partial r} \right] + \frac{\partial^2 u}{\partial z^2} + 2ik_o \frac{\partial u}{\partial z} = 0 \quad . \tag{10.41}$$

The essential step now is to assume that the amplitude function $u(r, z)$ depends only slowly on z so that the second derivative $\partial^2 u / \partial z^2$ can be neglected. This way the parabolic equation

$$\frac{1}{r} \frac{\partial}{\partial r} \left[r \frac{\partial u}{\partial r} \right] + 2ik_o \frac{\partial u}{\partial z} = 0 \tag{10.42}$$

is obtained. It can be solved by the function

$$u(r, z) = g(z) \, e^{f(z) r^2} \quad ; \tag{10.43}$$

(see e.g. Wen & Breazeale 1988). Introducing (10.43) into (10.42) gives

$$\left[4fg(1 + fr^2) + 2ik_0 \left(\frac{\partial g}{\partial z} + r^2 g \frac{\partial f}{\partial z} \right) \right] e^{f(z) r^2} = 0 \quad . \tag{10.44}$$

This expression must hold for each value of r, i.e. separately for the coefficients of r^0 and r^2. (Note that f and g do not depend on r.) Thus (10.44) can be split up into the following

$$2\,fg + ik_o\,\frac{\partial g}{\partial z} = 0 \qquad\qquad (10.45a)$$

$$2\,f^2 + ik_o\,\frac{\partial f}{\partial z} = 0 \quad . \qquad\qquad (10.45b)$$

Integration of (10.45b) gives

$$f = \frac{-1}{B + 2iz/k_o} \quad .$$

Introducing this into (10.45a) results in

$$g = \frac{A}{B + 2izk_o} \quad .$$

A and B are arbitrary, complex constants. The general expression for the sound pressure in this approximation is finally

$$p(r, z) = \sum_n \frac{A_n}{B_n + 2iz/k_o}\, e^{ik_o z}\, e^{-r^2/(B_n + 2iz/k_o)} \quad . \qquad (10.46)$$

The summation has been taken because, for each value of n, (10.46) is a special solution of the parabolic equation (10.42). The general solution is obtained by summing up special solutions. The actual values of A_n, B_n are determined by adjusting the solution in such a way that the boundary conditions at the surface of the source are fulfilled (see e.g. Wen & Breazeale 1988). The important aspect is that (10.46) consists of a series of Gaussian curves which are centered at $r = 0$ and have a width that increases with distance z.

A more useful application of the Gaussian beam method is to calculate the sound field in a medium with slowly varying speed of sound (e.g. the atmosphere or ocean). To this end the field in the vicinity of each ray is expressed in terms of a Gaussian beam which has its centre on the ray. It would require a rather lengthy derivation of the Gaussian beam method for media with slowly varying parameters; therefore only the main steps and the final results are given here (for details, see e.g. Cerveny et al. 1982; Porter & Bucker 1987; Schutz & Saied 1990).

The first step is to calculate the rays by using (10.38). The next step is to express the wave equation in an orthogonal coordinate system that is made up of the arc length s along the ray and the normal n to this line (Figure 10.9).

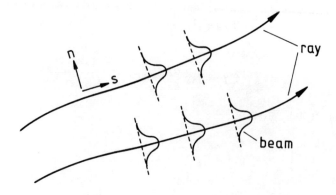

Figure 10.9 Coordinate system for the Gaussian beam method.

In the circumferential direction around the ray all quantities are equal. The third — and most tricky — step is to neglect some terms which are slowly varying with s. In this way a parabolic equation similar to (10.42) is obtained. The solution of this equation is

$$p(s,n) = A \sqrt{\frac{c}{rq_2}} \; e^{-i\omega(\tau + n^2\alpha/2)} \quad . \tag{10.47}$$

Here

$$\tau = \int \frac{ds}{c}$$

is the travel time along a ray with $c = c(s)$ being the sound speed along the ray. A is an amplitude that depends on the initial conditions, α an abbreviation for $\alpha = -q_1/q_2$, and r the distance to the original sound source. In some versions of the Gaussian beam method r is a constant which is incorporated in A. The functions q_1 and q_2 depend only on the coordinate s. They are given by the following differential equations

$$\frac{dq_1}{ds} = -\frac{c_{nn}}{c^2} q_2$$

$$\frac{dq_2}{ds} = cq_1 \quad .$$

Again c is the sound speed along the ray. c_{nn} is defined by

$$c_{nn} = \frac{\partial^2 c(s,n)}{\partial n^2} \quad ;$$

it is thus related to the curvature of the sound speed profile in the vicinity of the ray in a normal direction.

The Gaussian beam method is applied in ocean acoustics and for calculations of the sound propagation in the atmosphere. In this case wind effects can also be included. One advantage of the Gaussian beam method is that caustics and sharp boundaries at shadow zones do not cause problems.

10.7 CONCLUDING REMARKS

In this text only a few numerical methods are listed. Many others that are also very important are not even mentioned, e.g. computer tomography. which is essential in ultrasonic diagnostics, or echo ranging, which finds wide application in oil exploration, underwater sonar, and non–destructive testing.

Another wide application of numerical methods which is just at its early state is nonlinear acoustics. In this field analytical methods are very scarce (see Chapters 23—25), especially when one wants not only to deal with nearly linear but also with fully nonlinear problems. In such cases one has to work with very time–consuming computational methods or programs which try to simulate the essential features of nonlinear effects. It is certain that in this field many important and interesting developments will emerge in the future.

REFERENCES

Brekkhovskikh, L.M. (1960) Waves in Layered Media. Academic Press, New York.

Burton, A.J. & Miller, G.F. (1971) Proc. Roy. Soc. A323:201.

Cerveny, V., Popov, M.M. & Psencik, I. (1982) Geophys. J. R. Astron. Soc. 70:109.

Cremer, L. (1984) Acustica 55:44.

Dowling, A.P. & Ffowcs Williams, J.E. (1983) Sound and Sources of Sound. Chapter 5. Ellis Horwood, Chichester.

Gladwell, G.M.L. (1965) A finite element method for acoustics. Proc. 5th ICA (Liège), Paper L33.

Heckl, M. (1989) Acustica 68:251.

Heckl, M. & Jackson, F.J. (1967) Acustica 18:61.

Krokstadt, A., Strom, S. & Sorsdal, S. (1968) J. Sound. Vib. 8:118.

Lee, D., Cakmak, A. & Vichnevetsky, R. (ed.) (1990) Computational Acoustics, esp. Vol.1. Ocean–acoustic Models and Supercomputing. North–Holland, Amsterdam.

Martin, P.A. (1982) Wave Motion 4:391.

Maynard, J.D., Williams, E.G. & Lee, Y. (1985) J. Acoust. Soc. Amer. 78:1395.

Miles, J.A. (1957) J. Acoust. Soc. Amer. 29:226.

Morse, P.M. (1948) Vibration and Sound. Chapter VII. McGraw–Hill, New York.

Ochmann, M. (1990) Acustica 72:233.

Porter, M.B. & Bucker, H.P. (1987) J. Acoust. Soc. Amer. 82:1349.

Reut, Z. (1985) J. Sound Vib. 103:297.

Schenk, H.A. (1968) J. Acoust. Soc. Amer. 44:41.

Schultz, M. & Saied, F. (1990) Solving the wave equation on Hypercube architecture. In: Computational Acoustics (ed. Lee et al.). Vol.I, p.267. North–Holland, Amsterdam.

Seybert, A.F., Soenarko, B., Rizzo, F.J. & Sippy, D.J. (1985) J. Acoust. Soc. Amer. 77:362.

Tomilina, T.M. (1989) Fast algorithm for sound field analysis based on fictitious sources method. 19. ICA, Belgrade, p.433.

Varadan, V.K. & Varadan, V.V. (ed.) (1980) Acoustic, Electromagnetic and Elastic Wave Scattering. Focus on the T–Matrix approach. Pergamon Press.

Waterman, P.C. (1969) J. Acoust. Soc. Amer. 45:1417.

Wen, J.J. & Breazeale, M.A. (1988) J. Acoust. Soc. Amer. 83:1752.

Williams, W., Parke, N.G., Moran, D.A. & Sherman, C.H. (1964) J. Acoust. Soc. Amer. 36:2316.

Zienkiewicz, O.C. (1971) The Finite Element Method in Engineering Sciences. McGraw–Hill, London.

PART II
The Generation of Unsteady Fields

11. NOISE SOURCE MECHANISMS

11.1 THE EQUATIONS OF FLUID MOTION

Each property of a fluid flow is a function of position $\mathbf{x}$ and time t. The density ρ is the mass per unit volume. The specific volume, or volume per unit mass, is ρ^{-1}. The velocity at a point is the speed and direction at which the fluid particle currently at that point is moving. The Eulerian approach considers conditions at a fixed point $\mathbf{x}$ as time progresses. The Lagrangian approach considers a particular fluid particle which occupies successively different locations $\mathbf{x}$, as time progresses. Rates of change in a Lagrangian frame are rates of change as seen by an observer moving with the fluid velocity. This rate of change is written D/Dt. Thus

$$\frac{D}{Dt} f(\mathbf{x}, t) = \frac{\partial f}{\partial t} + \frac{\partial x_i}{\partial t}\bigg|_{\text{particular particle}} \frac{\partial f}{\partial x_i} \quad , \tag{11.1}$$

where the double suffix summation convention is applied. But

$$\frac{\partial x_i}{\partial t}\bigg|_{\text{particular particle}} = v_i \quad ; \tag{11.2}$$

therefore

$$\frac{Df}{Dt} = \left\{ \frac{\partial}{\partial t} + v_i \frac{\partial}{\partial x_i} \right\} f \quad . \tag{11.3}$$

The operator D/Dt is often called the substantive or total derivative.

A fluid particle of constant mass m occupies a volume m/ρ. The rate of increase of its volume is m/ρ times $\nabla \cdot \mathbf{v}$, $\nabla \cdot \mathbf{v}$ being the rate of volume increase per unit volume

$$\frac{D}{Dt}\left\{\frac{m}{\rho}\right\} = \frac{m}{\rho} \nabla \cdot \mathbf{v} \tag{11.4}$$

or, equivalently

$$\frac{1}{\rho}\frac{D\rho}{Dt} + \nabla \cdot \mathbf{v} = 0 \quad , \tag{11.5}$$

$$\frac{\partial \rho}{\partial t} + \frac{\partial}{\partial x_i}(\rho v_i) = 0 \quad . \tag{11.6}$$

If a stress p_{ij} is defined over any space, there is a force on unit volume in that space equal in magnitude and direction to $-\partial p_{ij}/\partial x_j$. The stress tensor p_{ij} is the force exerted in the i direction or a surface with outward normal in the j direction. Fluid flows support a stress field in which p_{ij} has the form

$$p_{ij} = p\delta_{ij} + \mu\left(-\frac{\partial v_i}{\partial x_j} - \frac{\partial v_j}{\partial x_i} + \frac{2}{3}\nabla \cdot \mathbf{v}\delta_{ij}\right) \quad , \tag{11.7}$$

μ being the coefficient of viscosity and p the pressure.

Any particular fluid particle of constant mass m, occupying a volume m/ρ will therefore be subjected to a force equal in magnitude and direction to $-m/\rho\,\partial p_{ij}/\partial x_j$, so that it will have an acceleration Dv_i/Dt given by Newton's laws of motion

$$m\frac{Dv_i}{Dt} = -\frac{m}{\rho}\frac{\partial p_{ij}}{\partial x_j} \quad ,$$

i.e.

$$\rho\frac{Dv_i}{Dt} + \frac{\partial p_{ij}}{\partial x_j} = 0 \quad ; \tag{11.8}$$

this is the Navier–Stokes equation of fluid motion, which can alternatively be written as

$$\frac{\partial}{\partial t}(\rho v_i) + \frac{\partial}{\partial x_j}(\rho v_i v_j + p_{ij}) = 0 \quad . \tag{11.9}$$

At this stage we dispense with the viscosity term on the grounds that the coefficient of viscosity in the fluid which interest us most is small. Viscous stresses can therefore be important only in regions where the velocity gradient, which in Equation (11.8) multiples the coefficient of viscosity μ, is very large. Such regions exist near boundaries where "inviscid" theory predicts a velocity discontinuity. In such regions, known as *boundary layers*, the viscous term will obviously be important; away from these boundary layers the viscous stress is small. The inviscid form of the stress tensor is

$$p_{ij} = p\delta_{ij} \quad , \tag{11.10}$$

and (11.8) then becomes

$$\rho \frac{D\mathbf{v}}{Dt} + \nabla p = 0 \quad , \tag{11.11}$$

or

$$\frac{\partial \rho v_i}{\partial t} + \frac{\partial p}{\partial x_i} + \frac{\partial}{\partial x_j}(\rho v_i v_j) = 0 \quad . \tag{11.12}$$

Vorticity

The detailed motion of a fluid particle can be considered as a superposition of three basic terms. For apart from its translational motion, such a particle will in general undergo a rotation and a distortion of shape, i.e. strain. The velocity of fluid instantaneously at $x_j + \delta x_j$, relative to that at x_j, is

$$\delta v_i = \frac{\partial v_i}{\partial x_j} \delta x_j \quad . \tag{11.13}$$

The rate–of–strain tensor, $\partial v_i/\partial x_j$, may be written as the sum of symmetric and anti–symmetric parts,

$$\frac{\partial v_i}{\partial x_j} = \frac{1}{2}\left(\frac{\partial v_i}{\partial x_j} + \frac{\partial v_j}{\partial x_i}\right) + \frac{1}{2}\left(\frac{\partial v_i}{\partial x_j} - \frac{\partial v_j}{\partial x_i}\right) \quad . \tag{11.14}$$

The symmetric part of (11.14)

$$e_{ij} = \frac{1}{2}\left(\frac{\partial v_i}{\partial x_j} + \frac{\partial v_j}{\partial x_i}\right) \tag{11.15}$$

is called *the* rate–of–strain tensor; this part of (11.14) corresponds to a stretching of the fluid particle along three mutually perpendicular directions, the "principal axes".

We now define

$$\omega_{ij} = \frac{1}{2}\left(\frac{\partial v_i}{\partial x_j} - \frac{\partial v_j}{\partial x_i}\right) \tag{11.16}$$

which has only three independent terms, since $\omega_{11} = \omega_{22} = \omega_{33} = 0$ and $\omega_{ij} = -\omega_{ji}$. The nonzero terms can conveniently be represented in the form of a vector; $\boldsymbol{\omega} = \operatorname{curl} \mathbf{v} = \nabla \wedge \mathbf{v}$ is the *vorticity* and gives a measure of the angular rotation of fluid particles; the anti–symmetric rate of strain $\boldsymbol{\omega}_{ij}$ represents an angular velocity $\frac{1}{2}\boldsymbol{\omega}$ of a particle about its centroid. A rigid body with angular velocity $\boldsymbol{\Omega}$ about the origin (i.e. the velocity $\mathbf{v}$ of any point with position vector $\mathbf{r}$ is $\mathbf{v} = \boldsymbol{\Omega} \wedge \mathbf{r}$), has vorticity

$$\boldsymbol{\omega} = \operatorname{curl} \mathbf{v} = 2\boldsymbol{\Omega} \quad .$$

This rotation terms matters a great deal because it is often identically zero. Most of the motions we shall treat can be considered as being induced in an otherwise still fluid by boundary motions. Since rotation of any particular particle can be brought about only by the application of a torque, and since in an inviscid fluid there are very rarely means of inducing such torques, the induced motion must be one in which each particle does not rotate about its centre. Such a motion is called *irrotational* and has zero vorticity, $\boldsymbol{\omega} = 0$. The equation governing vorticity is obtained by applying the curl operator to Equation (11.11) making use of the identities

$$\nabla \wedge \nabla \phi = 0 \quad ,$$

$$\nabla \wedge \left[\rho \frac{D\mathbf{v}}{Dt} \right] = \nabla \rho \wedge \frac{D\mathbf{v}}{Dt} + \rho \nabla \wedge \frac{D\mathbf{v}}{Dt} \quad ,$$

$$\nabla \wedge \left[(\mathbf{v} \cdot \nabla)\mathbf{v} \right] = \boldsymbol{\omega}(\nabla \cdot \mathbf{v}) - (\boldsymbol{\omega} \wedge \nabla)\mathbf{v} - (\mathbf{v} \wedge \nabla)\boldsymbol{\omega} \quad .$$

On making use of the continuity Equation (11.5), the curl of Equation (11.11) reduces to the vorticity equation

$$\frac{D}{Dt}\left(\frac{\boldsymbol{\omega}}{\rho} \right) = \left(\frac{\boldsymbol{\omega}}{\rho} \cdot \nabla \right)\mathbf{v} - \frac{1}{\rho^2}\nabla\rho \wedge \frac{D\mathbf{v}}{Dt} \tag{11.17}$$

We will now make the additional restriction and assume that the density is a function of pressure only. Then

$$\rho = f(p) \tag{11.18}$$

and

$$\nabla\rho = f'\nabla p \quad . \tag{11.19}$$

In this case ∇p is parallel to $\nabla\rho$, which, according to Equation (11.11), is parallel to $D\mathbf{v}/Dt$, so that

$$\nabla\rho \wedge \frac{D\mathbf{v}}{Dt} \equiv 0 \quad . \tag{11.20}$$

The vorticity equation is then

$$\frac{D}{Dt}\left(\frac{\boldsymbol{\omega}}{\rho} \right) = \left(\frac{\boldsymbol{\omega}}{\rho} \cdot \nabla \right)\mathbf{v} \quad . \tag{11.21}$$

The statement that only irrotational motion can be induced in a fluid originally at rest ($\mathbf{v} = \nabla \wedge \mathbf{v} = \boldsymbol{\omega} = 0$) is now proved, for when $\boldsymbol{\omega} = 0$ there is no rate of increase in $\boldsymbol{\omega}$ so that it must always remain zero.

Thus we have $\boldsymbol{\omega} = 0$ for all t if $\boldsymbol{\omega} = 0$ at $t = 0$, provided that all the derivatives $\partial v_i / \partial x_j$ are uniformly bounded in t. The irrotational state is permanent whenever $\rho = f(p)$.

The condition $\nabla \wedge \mathbf{v} = \boldsymbol{\omega} = 0$ is most easily stated by writing

$$\mathbf{v} = \nabla \phi \quad . \tag{11.22}$$

The scalar function ϕ is called the velocity potential and is known to exist whenever $\omega = 0$.

The Navier–Stokes equation for an incompressible fluid is

$$\frac{\partial v}{\partial t} + (\mathbf{v} \cdot \nabla)\mathbf{v} = -\frac{1}{\rho} \nabla p + \nu \nabla^2 \mathbf{v} \quad ,$$

where $\nu = \mu / \rho$. By taking the curl of this equation and using div $\mathbf{v} = 0$, the vorticity $\omega = \operatorname{curl} \mathbf{v}$ is easily shown to satisfy

$$\frac{D\boldsymbol{\omega}}{Dt} = (\boldsymbol{\omega} \cdot \nabla)\mathbf{v} + \nu \nabla^2 \boldsymbol{\omega} \quad .$$

In two–dimensional motion, this reduces to

$$\frac{D\boldsymbol{\omega}}{Dt} = \nu \nabla^2 \boldsymbol{\omega} \quad ,$$

i.e. $\boldsymbol{\omega}$ then satisfies a convected diffusion equation.

Equation (11.11) can be rearranged as Bernoulli's equation

$$\frac{\partial \phi}{\partial t} + \frac{1}{2} v^2 + \int \frac{dp}{\rho} = \text{ a function of time only} \quad . \tag{11.23}$$

The equations of motion (11.5) and (11.11) are extremely complicated and very few nontrivial solutions are known; however, the equations simplify considerably in certain important situations. For example, if the compressibility of the fluid can be neglected, then $D\rho/Dt$ is zero and Equation (11.5), with $\mathbf{v} = \nabla \phi$, reduces to Laplace's equation

$$\nabla^2 \phi = 0 \quad . \tag{11.24}$$

The potential in incompressible flow therefore satisfies an extremely simple linear differential equation. The velocity is known once the potential is known and the pressure follows immediately from Bernoulli's equation (11.23).

11.2 WAVE EQUATION FOR COMPRESSIBLE FLUIDS

Wherever compressibility is important, then (11.24) is not true and the equations remain extremely complex and nonlinear. The acoustic equations are their approximate form for vanishingly small–amplitude motion and are obtained by linearizing (11.5) and (11.12). When density varies a little about its mean value ρ_0 and the fluid is at rest apart from small perturbation velocities, $\mathbf{v} = \nabla\phi$, then we have, from the continuity equation,

$$\frac{\partial \rho}{\partial t} + \rho_0 \nabla^2 \phi = 0 \quad , \tag{11.25}$$

and from the momentum equation

$$\rho_0 \frac{\partial \phi}{\partial t} + p = 0 \quad , \tag{11.26}$$

or

$$\rho_0 \frac{\partial^2 \phi}{\partial t^2} + \frac{\partial p}{\partial t} = 0 \quad . \tag{11.27}$$

We can combine (11.25) and (11.27) in the form

$$\frac{\partial^2 \phi}{\partial t^2} - c^2 \nabla^2 \phi = \frac{1}{\rho_0} \frac{\partial}{\partial t} \left(p - c^2 \rho \right) \quad . \tag{11.28}$$

The left–hand side of this equation is the wave operator acting on ϕ, and since we know that unsteady small-amplitude motions form the sound waves that travel with the speed of sound c in accordance with the *wave equation*

$$\frac{\partial^2 \phi}{\partial t^2} - c^2 \nabla^2 \phi = 0 \quad . \tag{11.29}$$

The speed of sound c is set by the vanishing of the right–hand side of (11.28). Thus

$$\frac{\partial}{\partial t} \left(p - c^2 \rho \right) = 0; \quad \text{i.e.} \quad c^2 = \frac{dp}{d\rho} \quad . \tag{11.30}$$

The linear relationships (11.26) and (11.30) between p, ρ and ϕ imply that, to this order of approximation, the density and pressure fluctuations satisfy the same wave equation,

$$\left(\frac{\partial^2}{\partial t^2} - c^2 \nabla^2 \right) (\rho - \rho_0) = 0 = \left(\frac{\partial^2}{\partial t^2} - c^2 \nabla^2 \right) (p - p_0) \quad . \tag{11.31}$$

11.3 INHOMOGENEOUS WAVE EQUATION

The source effect of "creating" fluid or momentum introduces unphysical modifications to the conservation equation and causes the addition of a function on the
right–hand side of the wave equation

$$\left(\frac{\partial^2}{\partial t^2} - c^2\nabla^2\right)(\rho - \rho_0) = Q(\mathbf{x}, t) \quad . \tag{11.32}$$

Q the acoustic source density, is zero except in the source region where fluid or
momentum is being created. If an external force acts on the fluid, or part of the
fluid, Equation (11.8) is modified by the addition of an extra term F_i on its right-
hand side, and the wave equation takes the inhomogeneous form (11.32), where
$Q(\mathbf{x}, t)$ then has the form of a divergence of a vector force field. In the following
discussion of the inhomogeneous equation, it will be seen that the wave field has a
different character depending on whether Q has the form of a divergence $(\partial F_i/\partial x_i)$
or a double divergence $(\partial^2 T_{ij}/\partial x_i \partial x_j)$. In these circumstances $Q(\mathbf{x}, t)$ is described
as being respectively a dipole, or a quadrupole distribution of sources.

11.4 MONOPOLE AND MULTIPOLE SOURCES

A linear nonhomogeneous equation can be solved in terms of the Green function of
the problem. This function is the solution of the equation with the inhomogeneity
concentrated at a point, in both space and time. Thus it is defined by

$$\frac{\partial^2 G}{\partial t^2} - c^2\nabla^2 G = \delta(\mathbf{x})\delta(t) \quad , \tag{11.33}$$

together with the casunlity condition that $G = 0$ for $t < 0$. In three space dimensions

$$G(\mathbf{x}, t) = \frac{1}{4\pi c^2}\frac{\delta(t - x/c)}{x} \quad , \tag{11.34}$$

and the solution of the more general equation $\{\partial^2/\partial t^2 - c^2\nabla^2\}(\rho - \rho_0) = Q$, is

$$(\rho - \rho_0)(\mathbf{x}, t) = \int G(\mathbf{x} - \mathbf{y}, t - \tau)Q(\mathbf{y}, \tau)d\mathbf{y}\, d\tau \tag{11.35}$$

$$= \frac{1}{4\pi c^2}\int \frac{Q(\mathbf{y}, t - |\mathbf{x} - \mathbf{y}|/c)}{|\mathbf{x} - \mathbf{y}|}d\mathbf{y} \quad . \tag{11.36}$$

Were the source of strength $q(t)$ concentrated at the origin, then $Q(\mathbf{x}, t) = q(t)\delta(\mathbf{x})$ and

$$(\rho - \rho_0)(\mathbf{x}, t) = \frac{1}{4\pi c^2} \frac{q(t - r/c)}{r} \quad .$$

The function Q is called the monopole *source strength* density. An observer at distance r from the source feels the influence of the source that occurred at the earlier time, i.e. the *retarded time* $t - r/c$; this is of course due to the fact that information propagating from the origin takes time r/c to travel the distance r from source to observer.

11.5 DIPOLE SOURCES

An interesting situation arises when two monopoles, of equal and opposite strengths, are placed close together. The density fluctuations induced by a point source of strength $-q(t)$ at y_i together with a point source $q(t)$ at $y_i + \ell_i$, is

$$4\pi c^2 (\rho - \rho_0) = \frac{q(t - (R + \delta R)/c)}{R + \delta R} - \frac{q(t - R/c)}{R} \quad ,$$

where $R + \delta R = |y_i + \ell_i - x_i|$ and $R = |y_i \stackrel{.}{-} x_i|$,

$$4\pi c^2 (\rho - \rho_0) \sim \ell_i \frac{\partial}{\partial y_i} \left\{ \frac{q(t - R/c)}{R} \right\} \quad \text{for} \quad |\ell_i| << |x_i - y_i| \quad ,$$

$$= -\ell_i \frac{\partial R}{\partial y_i} \left\{ \frac{q(t - R/c)}{R^2} + \frac{q'(t - R/c)}{cR} \right\}$$

$$= \ell \cos\theta \left[\frac{q}{R^2} + \frac{q'}{cR} \right] \quad , \tag{11.37}$$

where $\ell = |\ell_i|$, θ is the angle between ℓ_i and $(x_i - y_i)$, and the square brackets indicate that the function inside is to be evaluated at the retarded time $(t - R/c)$. The product $\mu_i = \ell_i q(t)$ is called the *dipole strength*, or *dipole moment*, and gives both the magnitude and direction of the compact double–source combination.

Were the source function Q of Equation (11.32) of the form of a derivative, $Q(\mathbf{x}, t) = \partial F_i(\mathbf{x}, t) / \partial x_i$, this could be interpreted as a distribution of dipoles of strength $F_i(x, t)$ per unit volume.

The "far field" induced by a dipole source is seen from (11.37) to take the form

$$4\pi c^2(\rho - \rho_0) \underset{R \to \infty}{\sim} \frac{\ell \cos \theta}{cR} q'(t - R/c) \quad . \tag{11.38}$$

It is instructive to contrast the far fields generated by monopole and dipole sources. If the source $q(t)$ is periodic with angular frequency ω, then the density fluctuation induced by the dipole as compared with that induced by the monopole is smaller in magnitude by the factor

$$\frac{\omega \ell \cos \theta}{c} = \frac{2\pi \ell}{\lambda} \cos \theta \quad , \tag{11.39}$$

where $\lambda = 2\pi c/\omega$ is the wavelength of sound. The dipole with scale ℓ small in comparison with the acoustic wavelength is much less efficient than the monopole as a radiator of sound; the intensity is less by the factor $(2\pi \ell/\lambda)^2$. The relative inefficiency of the dipole source is a consequence of the destructive interference of the wavefields induced by its constituent–opposing monopoles, this cancellation being effective, of course, only if the separation is small on a wavelength scale, i.e. $\ell << \lambda$; in this situation the dipole is called *compact*. It is seen from Equation (11.37) that there are two terms that arise from the small difference δR between the source $-q$ at distance R and the source $+q$ at distance $R + \delta R$ from the observer. The first term arises directly from the fact that the nearer source is heard more loudly but by an amount decreasing as the square of distance; the second term, which dominates the far radiation field, depends on the retarded time term $(t - R/c)$ and reflects the fact that the signals received by an observer were launched at different times by the constituent monopoles. If this were not so they would interfere much more destructively.

11.6 QUADRUPOLE SOURCES

Multipoles of higher order can be generated by similarly combining dipoles. A *quadrupole* is obtained by placing a pair of equal and opposite dipoles close together. A dipole of strength $-\mu_i = \ell_i q$ at y_i together with a dipole of moment μ_i at $y_j + h_j$ forms a quadrupole of strength

$$t_{ij} = h_j \mu_i = \ell_i h_j q(t) \quad , \tag{11.40}$$

which induces the density fluctuation

$$4\pi c^2(\rho - \rho_0)(\mathbf{x}, t) \sim \ell_i h_j \frac{\partial^2}{\partial x_i \partial x_j} \left\{ q \frac{(t - R/c)}{R} \right\} \quad \text{as} \quad |\ell_j| \quad \text{and} \quad |h_j| \to 0 \quad ,$$

i.e.

$$4\pi c^2(\rho - \rho_0)(\mathbf{x}, t) = \frac{\partial^2}{\partial x_i \partial x_j}\left\{ \frac{t_{ij}(t - R/c)}{R} \right\} \quad . \tag{11.41}$$

The far field is readily shown to be less than that due to a monopole by a factor of order $(2\pi\ell/\lambda)^2$, and the intensity is less by a factor of order $(2\pi\ell/\lambda)^4$.

Quadrupoles with $i = j$ are called *longitudinal* being formed from four sources in a straight line; if $i \neq j$, then the quadrupole t_{ij} can be considered as being formed from four sources on the corner of a rectangle, and this arrangement is called a *lateral* quadrupole.

A volume distribution of quadrupoles, with strength T_{ij} per unit volume, induces in unbounded space a density fluctuation

$$(\rho - \rho_0)(\mathbf{x}, t) = \frac{1}{4\pi c^2} \frac{\partial^2}{\partial x_i \partial x_j} \int_V \left[\frac{T_{ij}}{r} \right] d\mathbf{y} \quad , \tag{11.42}$$

where $R = |\mathbf{x} - \mathbf{y}|$, the square brackets again denote that the function is evaluated at the retarded time $(t - R/c)$, and the integral ranges over the entire volume containing the quadrupoles.

11.7 THE FLOW NOISE EQUATIONS

Flow noise is the term used to describe the pressure fluctuations associated with unsteady flow, particularly turbulent flow. Turbulence relies on vorticity for its existence and may be thought of as chaotic vorticity. Turbulence results from an instability of smooth flow, the initially weak unsteady vorticity being amplified according to (11.17) by vortex stretching in the velocity gradients. Since noise is a pressure fluctuation, we first consider the pressure field induced in an incompressible flow by turbulence. The density ρ is constant and $\nabla \cdot \mathbf{v} = 0$. The divergence of Equation (11.12) shows p to satisfy a Poisson, or inhomogeneous Laplace, equation

$$\nabla^2 p = -\rho_0 \frac{\partial^2 v_i v_j}{\partial x_i \partial x_j} = \nabla \cdot \left[\rho_0 \mathbf{v} \wedge \boldsymbol{\omega} - \nabla \frac{1}{2}\rho_0 v^2 \right] \quad , \tag{11.44}$$

or

$$\nabla^2 \left(p + \frac{1}{2}\rho_0 v^2 \right) = \rho_0 \nabla \cdot (\mathbf{v} \wedge \boldsymbol{\omega}) \quad . \tag{11.45}$$

The *compressible* flow noise equations are essentially the equations describing the generation and propagation of sound by unsteady flow or turbulence. The rearrangement of the equations into a quadrupole–driven wave equation was Lighthill's pioneering contribution to flow–noise theory. Take the divergence of the momentum equation (11.12) and subtract it from the time derivative of the continuity equation as expressed in Equation (11.6). Then subtract $c_0^2 \nabla^2 \rho$ from both sides to yield Lighthill's equation

$$\frac{\partial^2 \rho}{\partial t^2} - c_0^2 \nabla^2 \rho = \frac{\partial^2 T_{ij}}{\partial x_i \partial x_j} \quad , \tag{11.46}$$

where

$$T_{ij} = \rho v_i v_j + (p - c_0^2 \rho) \delta_{ij} \quad . \tag{11.47}$$

The parameter c_0 is the speed of sound in the radiation field, and may be quite different from that in the turbulence, particularly if the turbulent fluid is aereated water where the speed of sound varies rapidly with air/water concentration. If the speed of sound is constant at c_0 and the fluids of uniform mean density, then T_{ij} is simple $\rho v_i v_j$. In this case, it is the unsteady Reynolds stresses that generate sound in exactly the same way as would a distribution of quadrupoles.

11.8 SOUND AND PSEUDO–SOUND

The Reynolds number is a measure of the inertial stresses in a fluid relative to the viscous stresses. In almost all flows of engineering interest, the Reynolds number is large so that the viscous stresses are negligible, except in singular regions of abnormally high shear, i.e. boundary layers. Viscosity is not a significant parameter in flow noise so that there is very little loss of generality in assuming the motions to be essentially inviscid on the length scales of interest. Unsteady flows are driven by pressure gradients. A random velocity field, or turbulence, is thus inevitably associated with a random unsteady pressure which can be sensed by the ear, or a microphone, as if it were sound. Sound, however, is essentially a small–amplitude, irrotational, compressible motion with velocity potential ϕ satisfying the wave equation

$$\frac{\partial^2 \phi}{\partial t^2} - c^2 \nabla^2 \phi = 0 \tag{11.48}$$

Sound waves propagate through space with the wave speed c. Since the motion is linear, all quantities in the wave are in direct proportion so that they, too, satisfy the same homogeneous wave equation. In particular, the pressure p is a solution of

$$\frac{\partial^2 p}{\partial t^2} - c^2 \nabla^2 p = 0 \quad , \tag{11.49}$$

under the appropriate boundary conditions. It is only when the pressure satisfies this equation that it is properly regarded as sound. Other pressure fluctuations, indistinguishable by a single microphone from proper sound, have been termed *pseudo sound*, only pseudo because they lack some essential characteristics of sound. They do not propagate through the fluid but are convected with the eddy structures in the flow, often in a chaotic path and usually with a speed very much smaller than the sonic velocity. Pseudo sound is related to the velocity field by the nonlinear terms in the Bernoulli equation so that we can say, quite generally, that the pseudo sound–pressure level, p, is of the same order as the fluctuation in dynamic pressure, ρU^2, ρ being the fluid density and U the characteristic velocity of the flow. It is in fact very difficult, indeed, to say anything more definite about the pressure level, the reason being simply that the pseudo sound–pressure field has very little individual character. It merely provides the local pressure gradients necessary to balance the inertial forces in fluid constrained to move without volumetric change.

The sound field proper contains distinct features associated with the fact that the waves propagate away from disturbance centres. Because fluid compressibility is an essential feature of sound waves, they are immediately identified by density changes, the density satisfying the homogeneous wave equation in the sound field proper. The full equations, however, provide an essential inhomogeneity to this wave equation, the inhomogeneity being equivalent to a space distribution of acoustic sources. This was the viewpoint taken by Lighthill in writing the exact equations of fluid motion as a quadrupole–driven wave equation

$$\frac{\partial^2 \rho}{\partial t^2} - c_0^2 \nabla^2 \rho = \frac{\partial^2 T_{ij}}{\partial x_i \partial x_j} \quad , \tag{11.50}$$

or equivalently,

$$\frac{\partial^2}{\partial t^2}(\rho - \rho_0) - c_0^2 \nabla^2 (\rho - \rho_0) = \frac{\partial^2 T_{ij}}{\partial x_i \partial x_j} \quad . \tag{11.51}$$

We approximate that equation here to inviscid flows by setting

$$T_{ij} = \rho u_i u_j + (p - c_0^2 \rho)\delta_{ij} \quad . \tag{11.52}$$

The flow–noise issue is then posed as a problem of finding the solution to Equation (11.51) under the appropriate boundary conditions with T_{ij}, the quadrupole strength density, assumed known.

11.9 SOUND INDUCED BY CONVECTED TURBULENCE

We shall now consider the solution of Lighthill's equation for the distant sound field induced by a finite volume of turbulence that is statistically stationary in time. Outside, the turbulence T_{ij} will be assumed zero. We consider first the case where there are no boundaries to the flow, so that Equation (11.51) is valid everywhere. The density perturbation field $(\rho - \rho_0)$ is then simply expressed by using the three-dimensional Green function satisfying the casuality condition. For convenience, we drop the zero suffice on c_0:

$$(\rho - \rho_0)(\mathbf{x}, t) = \frac{1}{4\pi c} \frac{\delta(x - ct)}{x} \star \frac{\partial^2 T_{ij}}{\partial x_i \partial x_j} \tag{11.53}$$

$$= \frac{1}{4\pi c} \frac{\partial^2}{\partial x_i \partial x_j} \left\{ \frac{\delta(x - ct)}{x} \right\} \star T_{ij} \quad . \tag{11.54}$$

This last expression may be simplified in the distant radiation field, $x \to \infty$, by noting that derivatives of x^{-1} are then negligible in comparison with x^{-1}. Thus x^{-1} may be regarded as a constant in Equation (11.54) to obtain the asymptotic form of the density field, as

$$(\rho - \rho_0)(\mathbf{x}, t) = \left\{ \frac{x_i x_j}{x^3} \frac{1}{4\pi c^3} \frac{\partial^2}{\partial t^2} \delta(x - ct) \right\} \star T_{ij} \tag{11.55}$$

$$= \left\{ \frac{x_i x_j}{x^3} \frac{1}{4\pi c^3} \delta(x - ct) \right\} \star \frac{\partial^2 T_{ij}}{\partial t^2} \quad . \tag{11.56}$$

Here we have used the fact that

$$\frac{\partial}{\partial x_i} x = \frac{x_i}{x}, \quad \text{and} \quad A' \star B = A \star B' \quad . \tag{11.57}$$

When Equation (11.56) is written out in long–hand form, and the delta function eliminated by integrating out the time variable, we have the Lighthill expression for the distant sound field

$$(\rho - \rho_0)(\mathbf{x}, t) = \frac{1}{4\pi c^4} \frac{x_i x_j}{x^3} \int_{V(\mathbf{y})} \frac{\partial^2 T_{ij}}{\partial t^2} \left(\mathbf{y}, t - \frac{|\mathbf{x} - \mathbf{y}|}{c} \right) d\mathbf{y} \quad , \tag{11.58}$$

$$= \frac{1}{4\pi c^4} \frac{x_i x_j}{x^3} \int_{V(\mathbf{y})} \frac{\partial^2 T_{ij}}{\partial t^2} \left(\mathbf{y}, t - \frac{\mathbf{x} \cdot \mathbf{y}}{xc} - \frac{x}{c} \right) d\mathbf{y} \quad . \tag{11.59}$$

A convenient statistical measure of the radiation is the mean square density fluctuation but a measure with more physical significance is the magnitude of the intensity vector, $\mathbf{I}$, which is proportional to the mean square density fluctuation

$$\mathbf{I} = \frac{c^3 \overline{(\rho - \rho_0)^2}}{\rho_0} \frac{\mathbf{x}}{x} \quad . \tag{11.60}$$

Equation (11.59) can be squared by multiplying it with itself expressed in a new position variable $\mathbf{z}$ with origin at $\mathbf{y}$

$$(\rho - \rho_0)(\mathbf{x}, t) = \frac{1}{4\pi c^4} \frac{x_k x_\ell}{x^3} \int_{V(\mathbf{z})} \frac{\partial^2 T_{k\ell}}{\partial t^2} \left(\mathbf{y} + \mathbf{z}, t + \frac{\mathbf{x} \cdot \mathbf{y}}{cx} + \frac{\mathbf{x} \cdot \mathbf{z}}{cx} - \frac{x}{c} \right) d\mathbf{z} \ , \tag{11.61}$$
$$x \to \infty$$

$$I(\mathbf{x}) = \frac{x_i x_j x_k x_\ell}{16\pi^2 \rho_0 c^5 x^6} \int_{V(\mathbf{y})} \int_{V(\mathbf{z})} \frac{\partial^4 W_{ijk\ell}}{\partial \tau^4} \left(\mathbf{y}, \mathbf{z}, \frac{\mathbf{x} \cdot \mathbf{z}}{cx} \right) d\mathbf{z}\, d\mathbf{y} \ . \tag{11.62}$$
$$x \to \infty$$

Here $W_{ijk\ell}$ is the stress tensor correlation function

$$\overline{T_{ij}(\mathbf{y}, t) T_{k\ell}(\mathbf{y} + \mathbf{z}, t + \tau)} = W_{ijk\ell}(\mathbf{y}, \mathbf{z}, \tau) \quad , \tag{11.63}$$

and

$$\overline{\frac{\partial^2 T_{ij}}{\partial t^2} \cdot \frac{\partial^2 T_{k\ell}}{\partial t^2}} = \frac{\partial^4 W_{ijk\ell}}{\partial \tau^4} \quad , \tag{11.64}$$

when T_{ij} is statistically stationary in time.

Equation (11.62) gives the magnitude of the intensity at $(\mathbf{x})$ as a volume integral over $\mathbf{y}$. The integrand can then be interpreted as the contribution to the intensity produced by unit volume of turbulence at $\mathbf{y}$, though it involves integration of the correlation function over *all* space. This interpretation is by no means unique but is satisfactory in being independent of the shape of the volume of turbulence considered at $\mathbf{y}$. The interpretation would be very different and highly misleading if we were to consider the mean square value of the integrand in Equation (11.58) as indicating the sound produced by unit volume of turbulence. That result would be sensitive to the detailed shape of the volume considered. We denote the contribution to the magnitude of the intensity from unit volume of turbulence at $\mathbf{y}$ by the symbol $i(\mathbf{x})$

$$I(\mathbf{x}) = \int i(\mathbf{x}) d^3 \mathbf{y} \quad . \tag{11.65}$$

Also, for compactness, we write

$$W_{(\mathbf{x})} = \frac{x_i x_j x_k x_\ell}{x^4} W_{ijk\ell} \tag{11.66}$$

$$i(\mathbf{x}) = \frac{1}{16\pi^2 \rho_0 c^5 x^2} \int \frac{\partial^4}{\partial \tau^4} W_{(\mathbf{x})} \left(\mathbf{y}, \mathbf{z}, \frac{\mathbf{x \cdot z}}{cx} \right) d\mathbf{z} \quad . \tag{11.67}$$

This result can be derived in another way that illustrates the role played by the different spectral components of the source field. We do this by taking the generalized Fourier transform of Equation (11.59) in time

$$(\rho - \rho_0)(\mathbf{x}, \omega) = \int (\rho - \rho_0)(\mathbf{x}, t) e^{i\omega t} dt \quad ,$$

$$= \frac{-1}{4\pi c^4} \frac{x_i x_j}{x^3} \int_{V(\mathbf{y})} \omega^2 T_{ij}(\mathbf{y}, \omega) e^{i\mathbf{x \cdot y}\omega/xc} \, d\mathbf{y} \; e^{ix\omega/c} \quad , \tag{11.68}$$

$$(\rho - \rho_0)(\mathbf{x}, \omega) = - \frac{1}{4\pi c^4} \frac{x_i x_j}{x^3} \omega^2 e^{i\omega x/c} T_{ij}\left(-\frac{\mathbf{x}\omega}{xc}, \omega \right) \quad , \tag{11.69}$$
$$x \to \infty$$

where

$$T_{ij}(\mathbf{k}, \omega) = \int \int T_{ij}(\mathbf{y}, t) e^{i\mathbf{k \cdot y}} e^{i\omega t} d\mathbf{y} \, dt \quad . \tag{11.70}$$

The statistical measure, most useful at any one particular frequency, is the power spectral density of the density fluctuation, $\pi(\mathbf{x}, w)$, that is related to the generalized Fourier transform. Since $(\rho - \rho_0)(\mathbf{x}, \omega) = (\rho - \rho_0)^*(\mathbf{x}, -\omega)$

$$\overline{(\rho - \rho_0)(\mathbf{x}, \omega)(\rho - \rho_0)(\mathbf{x}, \omega')} = \delta(\omega + \omega')\pi(\mathbf{x}, \omega) \quad . \tag{11.71}$$

Again the spectral density of the intensity at x, $I(\mathbf{x}, w)$, is simply $\rho_0^{-1} c^3 \pi(\mathbf{x}, \omega)$, so that by making a product of (11.69), averaging and integrating over ω', we have

$$\int \frac{\rho_0}{c^3} I(\mathbf{x}, \omega) \delta(\omega + \omega') d\omega' = \frac{1}{16\pi^2 c^8} \int \frac{x_i x_j x_k x_\ell}{x^6} \omega^2 \omega'^2 \, e^{i(\omega + \omega')x/c}$$

$$\times \quad T_{ij}\left(-\frac{\mathbf{x}\omega}{xc}, \omega \right) T_{k\ell}\left(-\frac{\mathbf{x}\omega'}{xc}, \omega' \right) d\omega' \quad . \tag{11.72}$$

Because

$$\frac{x_i x_j x_k x_\ell}{x^4} \overline{T_{ij}(\mathbf{k}, \omega) T_{k\ell}(-\mathbf{k}, \omega')} = 2\pi \int_{V(\mathbf{y})} \overline{W}_{(\mathbf{x})}(\mathbf{y}, \mathbf{k}, \omega) d\mathbf{y} \, \delta(\omega + \omega') \quad ,$$

where

$$\overline{W}_{(\mathbf{x})}(\mathbf{y}, \mathbf{k}, \omega) = \int_t \int_{V(\mathbf{z})} W_{(\mathbf{x})}(\mathbf{y}, \mathbf{z}, t) e^{i\mathbf{k} \cdot \mathbf{z}} e^{i\omega t} d\mathbf{z} \, dt \quad,$$

$W_{(\mathbf{x})}$, being defined in Equation (11.66), (11.72) can be expressed as

$$I(\mathbf{x}, \omega) = \frac{\omega^4}{8\pi c^5 \rho_0 x^2} \int_{V(\mathbf{y})} \overline{W}_{(\mathbf{x})} \left(\mathbf{y}, -\frac{\mathbf{x}\omega}{xc}, \omega \right) d\mathbf{y} \quad. \qquad (11.73)$$
$$x \to \infty$$

Again, we can regard the integrand of Equation (11.73) to be that contribution to I that comes from unit volume of turbulence at $\mathbf{y}$

$$I(\mathbf{x}, \omega) = \int i(\mathbf{x}, \omega) d^3 \mathbf{y}$$

$$i(\mathbf{x}, \omega) = \frac{\omega^4}{8\pi c^5 \rho_0 x^2} \overline{W}_{(\mathbf{x})} \left(\mathbf{y}, -\frac{\omega \mathbf{x}}{xc}, \omega \right) \quad. \qquad (11.74)$$
$$x \to \infty$$

The total intensity at $\mathbf{x}$ is $\frac{1}{2\pi} \int i(\mathbf{x}, \omega) d\omega$, so that

$$i(\mathbf{x}) = \frac{1}{2\pi} \int \frac{\omega^4}{8\pi c^5 \rho_0 x^2} \overline{W}_{(\mathbf{x})} \left(\mathbf{y}, -\frac{\omega \mathbf{x}}{xc}, \omega \right) d\omega \quad, \qquad (11.75)$$

is another form of Equation (11.67).

It is Equation (11.67) that indicates the magnitude of the acoustic field most easily in terms of readily measurable real space–time features of the source flow, but it is probably Equation (11.74) that provides the more intricate detail of the radiation field. In (11.74) we see that it is only those spectral elements of the source field that match *exactly* the direction of propagation, speed of propagation, wavelength and frequency of a particular sound wave that can generate that wave. Since turbulence has very little of these characteristics in common with sound, we see immediately that only a minute fraction of the turbulence field can, in fact, achieve the required degree of matching. Most of the turbulence is evidently silent and we can at this stage appreciate the extreme inefficiency of turbulence as a source of sound. We can in fact give a general measure of the acoustic inefficiency by examining Equation (11.67) in conjunction with the hypothesis that the generating flow is characterized by one length scale, L, and one velocity level, U. Two cases can be distinguished, depending on whether or not the retarded time variation in (11.67), $\mathbf{x} \cdot \mathbf{z}/cx$, is negligible or not in comparison with the timescale of the flow, which by our hypothesis must be LU^{-1}. The ratio of these terms, when $|z|$ is at its maximum value, L, is $(L/c)(U/L) = M$, the flow Mach number. This is also the inverse ratio of the radiated acoustic wavelength, $\lambda = cL/U$, to eddy length L, and is for that reason termed the *compactness ratio*. At low Mach number, eddies are compact on a wavelength scale and retarded time changes are negligible. Then the magnitude of the intensity field is easily assessed as follows:

$$W \sim (T_{ij})^2 \sim \rho_0 U^4 \quad , \tag{11.76}$$

$$\frac{\partial}{\partial \tau} \sim \frac{U}{L}; \quad d\mathbf{z} \sim L^3 \quad , \tag{11.77}$$

$$I(\mathbf{x}) \sim L^3 i(\mathbf{x}) \sim \frac{1}{\rho_0 c^5 x^2} \left(\frac{U}{L}\right)^4 \rho_0{}^2 U^4 L^6 \quad , $$

$$I(\mathbf{x}) \sim \rho_0 U^3 \frac{L^2}{x^2} M^5 \quad . \tag{11.78}$$

The total acoustic power, P, radiated from the flow is the integral of the intensity over a large spherical surface,

$$P \sim x^2 I \sim \rho_0 u^3 L^2 M^5 \quad . \tag{11.79}$$

The mechanical power in the flow is $\rho_0 U^3 L^2$, so that we see that only M^5 of this is converted into sound. At low Mach number this is an exceedingly inefficient process. The prediction (11.79) that the sound power increases with the eight power of velocity is the celebrated Lighthill velocity index.

The radiation from non–compact, or high Mach number, flows is quite different and is easily predicted from a formula derived from (11.67) by a simple rescaling. We rewrite the volume element $d\mathbf{z}$ as the product of an area element $d\mathbf{z}_n$, with normal in the $\mathbf{x}$–direction, with the length element $d z_{\mathbf{x}}$. Then we let

$$c\tau = z_{\mathbf{x}} = \frac{\mathbf{x} \cdot \mathbf{z}}{x} \quad , \tag{11.80}$$

so that Equation (11.67) can be rewritten exactly as

$$i(\mathbf{x}) - \frac{1}{16\pi^2 \rho_0 x^2} \int_\tau \int_{S(z_{\mathbf{x}})} \frac{\partial^4}{\partial z_{\mathbf{x}}^4} W_{(\mathbf{x})} \left(\mathbf{y}, \{\mathbf{z} = \mathbf{z}_{(\mathbf{x})}, c\tau\}, \tau\right) dz_{(\mathbf{x})} d\tau \quad . \tag{11.81}$$

Now, when $M \gg 1$, $c\tau$ varies a negligible fraction of an eddy scale L in the maximum eddy timescale $\tau = LU^{-1}$, so that $c\tau$ can be set equal to zero. The magnitude of the intensity field in the non–compact limit is consequently assessed as follows:

$$\frac{\partial}{\partial z_{(\mathbf{x})}} \sim L^{-1}$$

$$W \sim \rho_0^2\, U^4 \tag{11.82}$$

$$d\mathbf{z}_{(\mathbf{x})} \sim L^2, \quad \tau \sim LU^{-1}$$

$$P \sim x^2 i(\mathbf{x})L^3 \sim \frac{{\rho_0}^2 U^4}{\rho_0}\,\frac{L^2 L}{U}\, L^{-4}L^3 \quad ,$$

$$P \sim \rho_0 U^3 L^2; \quad M >> 1 \quad . \tag{11.83}$$

Evidently, at high Mach number, the acoustic power increases in direct proportion to the mechanical power in the flow. This is an experimentally verified fact, experiments giving the constant of proportionality to be of the order of 1%.

The foregoing argument is based on the idea that there is only one speed characterizing the turbulence, and that is the typical magnitude of the turbulence fluctuation level. There are good, though by no means conclusive, arguments that this level cannot exceed the speed of sound, so that we do not expect turbulent eddies to become non–compact. However, in turbulent flows, there are two velocity scales; the mean velocity and the turbulent velocity. The turbulence is convected with the mean flow which can be at very high Mach numbers. Based on this speed, the eddies can easily be considered non–compact. This is the mechanism by which high Mach number jet flows radiate and, to demonstrate this effect, we will rework the theory starting from Equation (11.67), written here in the form

$$i(\mathbf{x}) = \int_\tau \int_{\mathbf{z}} \frac{\delta\left(\tau - \frac{\mathbf{x}\cdot\mathbf{z}}{xc}\right)}{16\pi^2 \rho_0 c^5 x^2}\, \frac{\partial^4}{\partial\tau^4}\, W_{(\mathbf{x})}(\mathbf{y},\mathbf{z},\tau) d\mathbf{x}\, d\tau \quad , \tag{11.84}$$

or, equivalently

$$i(\mathbf{x}) = \int_\tau \int_{\mathbf{z}} \frac{W_{(\mathbf{x})}(\mathbf{y},\mathbf{z},\tau)}{16\pi^2 \rho_0 c^5 x^2}\, \delta^{(iv)}\left(\tau - \frac{\mathbf{x}\cdot\mathbf{z}}{xc}\right) d\mathbf{z}\, d\tau \quad . \tag{11.85}$$

If the turbulence is convected with a constant speed $\mathbf{U}$, then the correlation function can be recognized to be of the functional form

$$W_{(\mathbf{x})}(\mathbf{y},\mathbf{z},\tau) = R(\mathbf{y},\mathbf{z} - \mathbf{U}\tau,\lambda) \quad , \tag{11.86}$$

where R is the correlation function as seen by an observer moving with the convection speed (i.e. attached to the eddies). We now introduce the moving coordinate, or Lagrangian space variable, η,

$$\eta = z - U\tau = z - \frac{U\lambda}{\epsilon U} \quad , \tag{11.87}$$

and a scaled time variable

$$\lambda = \epsilon U \tau \quad ; \tag{11.88}$$

λ is simply the timescale τ normalized to an equivalent length scale by multiplying τ by the speed characteristic of the turbulence. This is the typical turbulence velocity fluctuation level which we can write as ϵU, ϵ being a number smaller than unity. Volume elements are left unchanged by this transformation so that Equation (11.85) can be written as an integral over the η space

$$i(\mathbf{x}) = \int_\tau \int_y \frac{R(\mathbf{y}, \boldsymbol{\eta}, \lambda)}{16\pi^2 \rho_0 c^5 x^2} \delta^{(iv)} \left\{ \frac{\lambda}{\epsilon U} \left[1 - \frac{\mathbf{M} \cdot \mathbf{x}}{x} \right] - \frac{\mathbf{x} \cdot \boldsymbol{\eta}}{xc} \right\} d\boldsymbol{\eta} \, \frac{d\lambda}{\epsilon U} \quad . \tag{11.89}$$

Here we have written $\mathbf{M} = \mathbf{U}/c$ as the Mach number (vector) at which the eddies are being convected through a medium of uniform sound speed c.

The delta function can be integrated directly by making use of the formula (2.44). In the notation of that formula, we choose x_α to be parallel to ∇g

$$g = \frac{\lambda}{\epsilon U} \left[1 - \frac{\mathbf{M} \cdot \mathbf{x}}{x} \right] - \frac{\mathbf{x} \cdot \boldsymbol{\eta}}{xc} \quad , \tag{11.90}$$

$$|\nabla g| = \frac{\partial g}{\partial x_\alpha} = \left\{ \left(\frac{\partial g}{\partial \eta_i} \right)^2 + \left(\frac{\partial g}{\partial \lambda} \right)^2 \right\}^{1/2} \quad .$$

$$\frac{\partial g}{\partial x_\alpha} = \frac{1}{\epsilon U} \left\{ \epsilon^2 M^2 + \left(1 - \frac{\mathbf{M} \cdot \mathbf{x}}{x} \right)^2 \right\}^{1/2} \tag{11.91}$$

$$i(\mathbf{x}) = \int_S \frac{(\epsilon U)^4 \frac{\partial^4}{\partial x_\alpha^4} R(\mathbf{y}, [\boldsymbol{\eta}, \lambda]^*), d(\boldsymbol{\eta}, \lambda)^*}{16\pi^2 \rho_0 c^5 x^2 \left\{ \left(1 - \frac{\mathbf{M} \cdot \mathbf{x}}{x} \right)^2 + \epsilon^2 M^2 \right\}^{5/2}} \quad , \tag{11.92}$$

where S is the plane hypersurface in the $(\boldsymbol{\eta}, \lambda)$ space along which $g = 0$. The asterisks indicate that the variables are to be evaluated on that hypersurface.

Progress must cease at this result unless we can say something more about R. It is usual in aerodynamic noise theory to suppose R to be isotropic in space–time. Then

$$R(\mathbf{y}, \boldsymbol{\eta}, \lambda) \text{ is a function only of } \mathbf{y} \text{ and } |\boldsymbol{\eta}|^2 + \lambda^2 \quad , \tag{11.93}$$

and

$$(\epsilon U)^4 \frac{\partial^4}{\partial x_\alpha^4} R(\mathbf{y}, [\boldsymbol{\eta}, \lambda]^*) = (\epsilon U)^4 \frac{\partial^4}{\partial \lambda^4} R(\mathbf{y}, \boldsymbol{\eta}, 0) \quad ,$$

$$= \frac{\partial^4}{\partial \tau^4} R(\mathbf{y}, \boldsymbol{\eta}, \tau = 0) \quad . \tag{11.94}$$

R, being isotropic in $[\boldsymbol{\eta}, \lambda]$, has allowed x_α (in the notation of Equation (2.44)) to be set equal to λ without loss of generality, so that the element of hypersurface S, $d[\boldsymbol{\eta}, \lambda]^*$, is then simply the instantaneous volume element,

$$d[\boldsymbol{\eta}, \lambda]^* = d\boldsymbol{\eta} \quad . \tag{11.95}$$

This leads to the special result, valid over the entire Mach number range,

$$i(\mathbf{x}) = \frac{1}{16\pi^2 \rho_0 c^5 x^2} \left\{ \left(1 - \frac{\mathbf{M} \cdot \mathbf{x}}{x} \right)^2 + \epsilon^2 M^2 \right\}^{-5/2} \int \frac{\partial^4}{\partial \tau^4} R(\mathbf{y}, \boldsymbol{\eta}, 0) \, d\boldsymbol{\eta} \quad . \tag{11.96}$$

We can recover the non–convected result by setting the convection speed to zero, so that we see the effect of eddy convection at Mach number $\mathbf{M}$ is simply to multiply the non–convected intensity field of the isotropic source by the factor

$$\left\{ \left(1 - \frac{\mathbf{M} \cdot \mathbf{x}}{x} \right)^2 + \epsilon^2 M^2 \right\}^{-5/2} \quad . \tag{11.97}$$

In particular, the intensity field is given by Equation (11.78) in conjunction with this rule

$$I(\mathbf{x}) \sim \frac{\rho_0 U^3 M^5 L^2 / x^2}{\left\{ \left(1 - \frac{\mathbf{M} \cdot \mathbf{x}}{x} \right)^2 + \epsilon^2 M^2 \right\}^{5/2}} \quad . \tag{11.98}$$

At the Mach angle, the eddies move towards the observer at precisely the speed of sound so that they are inevitably non–compact. Then

$$\left(1 - \frac{\mathbf{M} \cdot \mathbf{x}}{x} \right) = 0 \quad , \tag{11.99}$$

and Equation (11.98) simply reduces to the non–compact case described by Equation (11.83).

The general results of steady convection source motion are seen to be that sound is beamed preferentially forward. At supersonic convection speeds, there is a sharp maximum in the radiation field at the Mach angle. The mechanism of eddy Mach–wave generation is almost identical to that of Cherenkov light emission by ultra relativistic charged particles.

The effect of eddy convection on the radiation spectrum is easily determined by inserting the moving axis correlation function (11.86) into Equation (11.93). We can anticipate that the effect of eddy motion is to Döppler shift the radiation spectrum, but, in detail, the Döppler factor has to be generalized to cope with the case of Mach–wave radiation

$$
W_{(\mathbf{x})}\left(\mathbf{y}, -\frac{\omega \mathbf{x}}{xc}, \omega\right) = \int_t \int_{\mathbf{z}} R(\mathbf{y}, \mathbf{z} - \mathbf{U}t.\epsilon Ut)\, e^{-i\omega(\mathbf{x}\cdot\mathbf{z}/xc)} e^{i\omega t} d\mathbf{z}\, dt\ ,
$$

(11.100)

$$
= \frac{\overline{R}}{\epsilon U}\left\{\mathbf{y}, -\frac{\omega \mathbf{x}}{xc}, \frac{\omega}{\epsilon U}\left(1 - \frac{\mathbf{M}\cdot\mathbf{x}}{x}\right)\right\}\ ,
$$

where $\overline{R}(\mathbf{y}, \mathbf{k}, \alpha)$ is the space–time Fourier transform of $R(\mathbf{y}, \mathbf{z}, \lambda)$.

With this result, Equation (11.74) becomes a formula relating the radiation frequency spectrum to the moving–axis wavenumber frequency spectrum of the quadrupole source field,

$$
i(\mathbf{x}, \omega) = \frac{\omega^4}{8\pi c^5 \rho_0 x^2}\, \frac{1}{\epsilon U}\, R\left\{\mathbf{y}, -\frac{\omega \mathbf{x}}{xc}, \frac{\omega}{\epsilon U}\left(1 - \frac{\mathbf{M}\cdot\mathbf{x}}{x}\right)\right\}\ .
$$

(11.101)

The only components of the wavenumber spectrum that radiate sound are those that precisely match the wavenumber of the sound. This feature was already established in Equation (11.75) and the result remains unmodified by eddy convection. However, the source and field frequencies are different by a factor $(1 - \mathbf{M}\cdot\mathbf{x}/x)$, which is usually known as the Döppler factor, and the effect as the Döppler effect. A moving source element at frequency ω radiates sound at frequency $\omega(1 - \mathbf{M}\cdot\mathbf{x}/x)^{-1}$.

Now if we again assume that the moving axis correlation function is scaled to be isotropic in space time, then the wavenumber frequency spectrum is also isotropic. Then

$$
R(\mathbf{y}, \boldsymbol{\alpha}, \beta) = R(\mathbf{y}, 0, \sqrt{\alpha^2 + \beta^2})\ ,
$$

and Equation (11.101) becomes

$$i(\mathbf{x}, \omega) = \frac{\omega^4}{8\pi c^5 \rho_0 x^2} \frac{1}{\epsilon U} R\left(\mathbf{y}, 0, \frac{\omega}{\epsilon U} \sqrt{\left[1 - \frac{\mathbf{M}\cdot\mathbf{x}}{x}\right]^2 + \epsilon^2 M^2}\right) \quad, \qquad (11.102)$$

and, by a frequency scaling,

$$i\left(\mathbf{x}, \frac{\omega}{\left\{\left[1 - \frac{\mathbf{M}\cdot\mathbf{x}}{x}\right]^2 + \epsilon^2 M^2\right\}^{1/2}}\right) = \frac{\omega^4 R\left(\mathbf{y}, 0, \frac{\omega}{\epsilon U}\right)}{8\pi c^5 \rho_0 x^2 \epsilon U \left\{\left[1 - \frac{\mathbf{M}\cdot\mathbf{x}}{x}\right]^2 + \epsilon^2 M^2\right\}^2} \, .$$

$$(11.103)$$

The main change due to source motion alters the frequency of the sound from that of the source by the factor

$$\left\{\left[1 - \frac{\mathbf{M}\cdot\mathbf{x}}{x}\right]^2 + \epsilon^2 M^2\right\}^{-1/2} \, .$$

This "Döppler" factor does not tend to zero at the Mach angle as it does for moving *point* sources. Its non–vanishing at the Mach angle is a generalization of the Döppler effect to a distribution of sources, which is never a singular condition

11.10 BOUNDARY EFFECTS ON FLOW NOISE

There is an acoustic by–product of surfaces subjected to unsteady hydrodynamic forces. Those surfaces occupy some of the space previously occupied by fluid in Lighthill's original formulation of the problem. The equations of motion conse-quently apply to only the limited domain occupied by the fluid. There are several schemes for deriving the equations for this problem. Here we will describe one of the ways that that has proved powerful in developing new results.

Let us define the surface, S, which may be in one or several closed parts, by the equation

$$f(\mathbf{x}, t) = 0 \quad \text{on} \quad S \, . \qquad (11.104)$$

Let f be such that it is positive in every part of space occupied by the fluid and negative in every part of space enclosed by the surface S,

$$f > 0 \quad \text{in fluid}$$

$$(11.105)$$

$$f < 0 \quad \text{within } S \quad .$$

Further, we let the surface be in motion with velocity $\mathbf{v}$. Therefore, since the surface is always defined by $f = 0$,

$$\frac{\partial f(\mathbf{x}, t)}{\partial t} + v_i \frac{\partial f(\mathbf{x}, t)}{\partial x_i} = 0 = \frac{\partial f(\mathbf{y}, t)}{\partial \tau} + v_i \frac{\partial f(\mathbf{y}, \tau)}{\partial y_i} \quad . \tag{11.106}$$

The Heaviside function $H(f)$ is by definition equal to unity in the entire region occupied by fluid and is zero in all the volume enclosed by the surface S. The function $(\rho - \rho_0) H(f)$ is defined throughout all space, being equal to the density fluctuation in that region occupied by the fluid and zero elsewhere. By deriving an equation for $(\rho - \rho_0) H(f)$, we therefore derive an equation valid for *all* space and determining $(\rho - \rho_0)$ in the only region of space that is of interest; that actually occupied by fluid. We do this as follows.

The continuity equation is

$$\frac{\partial}{\partial t}(\rho - \rho_0) + \frac{\partial}{\partial x_i}(\rho u_i) = 0 \quad . \tag{11.107}$$

Now multiply this equation by $H(f)$ and take $H(f)$ inside the differential operators,

$$\frac{\partial}{\partial t}\{(\rho - \rho_0)H(f)\} + \frac{\partial}{\partial x_i}\{\rho u_i H(f)\} = \rho_0 v_i \frac{\partial H}{\partial x_i} + \rho(u_i - v_i)\frac{\partial H}{\partial x_i} \quad . \tag{11.108}$$

Similarly, for the momentum equation,

$$\frac{\partial}{\partial t}(\rho u_i) + \frac{\partial}{\partial x_i}(p\delta_{ij} + \rho u_i u_j) = 0 \quad ,$$

$$\frac{\partial}{\partial t}\{\rho u_i H(f)\} + \frac{\partial}{\partial x_j}\{(p\delta_{ij} + \rho u_i u_j)H(f)\} = \{p\delta_{ij} + \rho u_i(u_j - v_j)\}\frac{\partial H}{\partial x_j} \quad . \tag{11.109}$$

Now we form the wave equation for $(\rho - \rho_0)H$ in the same way as Lighthill formed his equation by cross differentiation,

$$\left\{\frac{\partial^2}{\partial t^2} - c^2 \nabla^2\right\} [(\rho - \rho_0)H(f)]$$

$$= \frac{\partial^2}{\partial x_i \partial x_j} [T_{ij}H(f)] + \frac{\partial}{\partial x_i} [F_i \delta(f)] + \frac{\partial}{\partial t} [Q\delta(f)] \tag{11.110}$$

where

$$T_{ij} = \{\rho u_i u_j + [p - c^2(\rho - \rho_0)]\delta_{ij}\} \quad,$$

$$F_i = -[p\delta_{ij} + \rho u_i(u_j - v_j)] \frac{\partial f}{\partial x_j} \quad, \tag{11.111}$$

$$Q = [\rho_0 v_i + \rho(u_i - v_i)] \frac{\partial f}{\partial x_i} \quad.$$

In an unbounded fluid f cannot be zero (or negative) and Equation (11.110) reduces immediately to Lighthill's Equation (11.51).

In the presence of surfaces, we see that Lighthill's quadrupole sources are essentially unmodified being distributed throughout the fluid, i.e. where $H = 1$. In addition to these sources there are surface concentrations of dipoles and monopoles, the dipole strength density being the surface stress (including an effective Reynolds stress, $\rho u_i(u_j - v_j)$, and the simple source strength density being the rate at which mass is transmitted across unit area of surface.

As Equation (11.110) is valid for all space, including that displaced by the surface, S, we can write down its solution immediately in an analogous manner to the steps that follow Equation (11.53),

$$H(f)(\rho - \rho_0)(\mathbf{x}, t) = \frac{x_i x_j}{x^3} \frac{1}{4\pi c^3} \delta(x - ct) * \frac{\partial^2}{\partial t^2} \{T_{ij}H(f)\}$$
$$x \to \infty$$

$$- \frac{x_i}{x^2} \frac{1}{4\pi c^2} \delta(x - ct) * \frac{\partial}{\partial t} \{F_i \delta(f)\} \tag{11.112}$$

$$+ \frac{1}{x} \frac{1}{4\pi c} \delta(x - ct) * \frac{\partial}{\partial t} \{Q\delta(f)\} \quad.$$

We shall now examine the dipole term in detail to illustrate the development of our final form of this solution,

$$\delta(x - ct) * \frac{\partial}{\partial t} \{F_i \delta(f)\}$$

$$= \frac{\partial}{\partial t} \int_\tau \int_{V(\mathbf{y})} \delta((|\mathbf{x} - \mathbf{y}| - c(t - \tau)) F_i(\mathbf{y}, \tau) \delta[f(\mathbf{y}, t)] d\mathbf{y}\, d\tau \quad,$$

(11.113)

$$= \frac{\partial}{\partial t} \int_\tau \int_{S(f=0)} \delta\{|\mathbf{x} - \mathbf{y}| - c(t - \tau)\} \frac{F_i}{|\nabla f|} d\mathbf{y}\, d\tau \quad.$$

By noting that on S, where $f = 0$,

$$\frac{\partial}{\partial \tau}\{|\mathbf{x} - \mathbf{y}| - c(t - \tau)\} = c\left\{1 - \frac{(x_i - y_i)}{|\mathbf{x} - \mathbf{y}|}\frac{v_i}{c}\right\} \underset{x \to \infty}{\sim} c\left(1 - \frac{v_i x_i}{cx}\right) \quad, \quad (11.114)$$

the τ integration in Equation (11.113) can be carried out immediately to give the value of the integral as

$$\frac{\partial}{\partial t} \int_S \left\{\frac{F_i}{c|\nabla f|\left|1 - \dfrac{v_i x_i}{cx}\right|}\right\}\left(\mathbf{y}, t - \frac{|\mathbf{x} - \mathbf{y}|}{c}\right) d\mathbf{y} \quad.$$

(11.115)

We write the instantaneous outward normal to the surface S at $\mathbf{y}$ as ℓ_i.

$$-\ell_i = \frac{\partial f/\partial y_i}{|\nabla f|} \quad,$$

(11.116)

and Equation (11.112) can then we written in the form

$$H(f)(\rho - \rho_0)(\mathbf{x}, t) \underset{x \to \infty}{-} \frac{x_i x_j}{x^3}\frac{1}{4\pi c^4}\frac{\partial^2}{\partial t^2}\int_{V_f} T_{ij}\left(\mathbf{y}, t - \frac{|\mathbf{x} - \mathbf{y}|}{c}\right) d\mathbf{y}$$

$$+ \frac{x_i}{x^2}\frac{1}{4\pi c^3}\frac{\partial}{\partial t}\int_S \left\{\frac{[\ell_i p + \ell_j \rho u_i(u_j - v_j)]}{\left|1 - \dfrac{v_i x_i}{xc}\right|}\right\}\left(\mathbf{y}, t - \frac{|\mathbf{x} - \mathbf{y}|}{c}\right) d\mathbf{y}$$

(11.117)

$$+ \frac{1}{x}\frac{1}{4\pi c^2}\frac{\partial}{\partial t}\int_S \left\{\ell_i \frac{[\rho v_i + \rho(u_i - v_i)]}{\left|1 - \dfrac{v_i x_i}{xc}\right|}\right\}\left(\mathbf{y}, t - \frac{|\mathbf{x} - \mathbf{y}|}{c}\right) d\mathbf{y} \quad,$$

where V_f is written for the volume actually occupied by fluid, i.e. the volume in which $f > 0$. On the moving surface, $\mathbf{y} = \mathbf{y}^*$, say, is a function of time and, in particular, at the retarded time we have the Döppler–contracted timescale in evidence because then, $\mathbf{y}^*(\tau) = \mathbf{y}\big|_{f=0}$, $\partial \mathbf{y}^*/\partial \tau = \mathbf{v}$ and $\tau = t - |\mathbf{x} - \mathbf{y}|/c$.

$$\frac{\partial y_i^*(\tau)}{\partial t} = v_i \frac{\partial \tau}{\partial t} \Rightarrow \frac{(x_i - y_i)}{|\mathbf{x} - \mathbf{y}|c} \frac{\partial y_i^*}{\partial t} = \frac{(x_i - y_i)}{|\mathbf{x} - \mathbf{y}|c} v_i \frac{\partial \tau}{\partial t}$$

so that,

$$\frac{\partial}{\partial t}\left\{ t - \frac{|\mathbf{x} - \mathbf{y}^*|}{c} \right\} = \left\{ 1 - \frac{v_i x_i}{xc} \right\}^{-1} \;,$$

and

$$\frac{\partial \tau}{\partial t} = \frac{\partial}{\partial t}\left(1 - \frac{|\mathbf{x} - \mathbf{y}^*|}{c} \right) = 1 + \frac{(x_i - y_i)}{|\mathbf{x} - \mathbf{y}|} \frac{\partial y_i^*}{\partial t} \underset{x \to \infty}{=} \frac{x_i v_i}{xc} \frac{\partial \tau}{\partial t} \;,$$

so that,

$$\frac{\partial}{\partial t}\left\{ t - \frac{|\mathbf{x} - \mathbf{y}^*|}{c} \right\} = \left\{ 1 - \frac{v_i x_i}{xc} \right\}^{-1} \;, \tag{11.118}$$

and

$$\frac{\partial y_i^*}{\partial t} = \frac{v_i}{\left\{ 1 - \frac{v_i x_i}{cx} \right\}} \;; \tag{11.119}$$

relations that are readily verified through (11.115).

Equation (11.117) is a general equation representing the exact distant field solution of the full inviscid equations of fluid motion in terms of the turbulence stress tensor T_{ij} and boundary conditions specified on a surface that moves with arbitrary velocity through which the fluid may or may not be passing. Surface acceleration effects can be determined by examining the terms arising from the operator $\partial/\partial t$ acting on the surface velocity $\mathbf{v}$.

There are several interesting cases when the normal velocities of the surface and fluid are not equal (i.e. $\ell_i(u_i - v_i) \neq 0$) so all the surface terms in (11.117) are non–zero. The surface S may be merely a mathematical control surface so that there is no requirement for the surface velocity to be in any way related to the fluid. The surface S could also be a physical surface through which fluid could penetrate, as would be the case for a moving flame front propagating through a combustible mixture. Once conditions at the flame front are specified, the sound field is known through (11.117). A second example is the general area of sound absorbtive surfaces which rely on flow crossing the mean surface for their absorption properties. Aerodynamic sound generation by fluid adjacent to such surfaces could be specified through Equation

(11.117), again with $\ell_i(u_i - v_i) \neq 0$. The sound generated by the vibration of porous flags and sails, flapping with amplitude beyond the bounds of the acoustic model, is also a problem formally solved by (11.117) in the sense that he sound field is determined once conditions are known on the moving surface.

Many sound–producing surfaces are however, perfectly impervious and no fluid can cross them or be absorbed by them.

Then

$$\ell_i(u_i - v_i) = 0 \quad , \tag{11.120}$$

and the surface terms in (11.117) simplify considerably. We shall write ρ_s as the contributions to the distant sound field coming from the surface terms of (11.117) with $\ell_i(u_i - v_i) = 0$. That is the far–field solution of the inhomogeneous wave equation

$$\left\{ \frac{\partial^2}{\partial t^2} - c^2 \nabla^2 \right\} \rho_s = \frac{\partial}{\partial t} \left\{ \rho_0 v_i \delta(f) \frac{\partial f}{\partial x_i} \right\} - \frac{\partial}{\partial x_i} \left\{ p\delta(f) \frac{\partial f}{\partial x_i} \right\}$$

$$\rho_s = \frac{1}{4\pi c^2 x} \frac{\partial}{\partial t} \int_S \ell_i \left\{ \frac{\rho_0 v_i + \frac{x_i}{x} \frac{p}{c}}{\left| 1 - \frac{v_i x_i}{x c} \right|} \right\} \left(\mathbf{y}, t - \frac{|\mathbf{x} - \mathbf{y}|}{c} \right) d\mathbf{y} \quad , \tag{11.121}$$

S being the surface of an impenetrable body moving with arbitrary speed and deforming in an arbitrary way.

The radiation field induced by the surface terms can be quantified in particular cases. One of the most important is that of a surface moving with small–amplitude motion about a fixed position. Then $v/c \ll 1$, the Döppler factor is essentially unity and the surface S can be considered to lie at its mean position. In this approximation, Equation (11.121) gives the surface terms of linear acoustics,

$$\rho_s = \frac{1}{4\pi c^2 x} \frac{\partial}{\partial t} \int_S \ell_i \left(\rho_0 v_i + \frac{x_i}{x} \frac{p}{c} \right) \left(\mathbf{y}, t - \frac{|\mathbf{x} - \mathbf{y}|}{c} \right) d\mathbf{y} \quad . \tag{11.122}$$

If the surface vibration is flow–induced with velocity and length scales U and ℓ respectively, then the frequencies will be of order U/ℓ, and

$$\rho \sim \frac{\rho_0 U^2 \ell}{c^2 x} \quad , \quad \overline{\rho^2} \sim \frac{\rho_0^2 U^4 \ell^2}{c^4 x^2} \quad . \tag{11.123}$$

The proportionality on the fourth power of velocity is characteristic of simple source radiation induced by unsteady flow. The situation is quite different if the integral scales of velocity fluctuation is zero, as it would be if the surface S enclosed a body moving without change of volume. Then the net simple source strength is zero and much more care is necessary in the evaluation of the radiation integrals.

If the surface is fixed and still then it is only the pressure term in (11.122) that induces sound, and if retarded time changes are negligible, as they are if the entire surface is on a scale much smaller than the radiation wavelength, i.e. the surface is compact, then p is typically the fluctuating dynamic head ρU^2, and

$$\rho \sim \rho_0 \frac{U}{c^3} U^2 \frac{\ell}{r}; \quad \overline{\rho^2} \sim \frac{\rho_0{}^2 U^6}{c^6} \frac{\ell^2}{r^2} \quad . \tag{11.124}$$

This dependence on the sixth power of velocity is characteristic of aerodynamic dipoles of compact scale. The most important practical problem of dipole–induced sound centre about lifting bodies in accelerated flight, such as a propeller or aero–engine fan. Then the source velocity v_i is well defined over the interior of the body as well as on the surface, and since the body moves without distortion the rate of strain is zero and, in particular, $\partial v_i / \partial y_i = 0$. The differential form of (11.121) can be rearranged in this case using the fact that

$$\frac{\partial}{\partial t} \left\{ \rho_0 v_i \delta(f) \frac{\partial f}{\partial x_i} \right\} = \frac{\partial}{\partial t} \left\{ \rho_0 v_i \frac{\partial (H-1)}{\partial x_i} \right\} = \rho_0 \frac{\partial^2}{\partial x_i \partial t} \left\{ (H-1) v_i \right\} \quad ,$$

$$= \rho_0 \frac{\partial}{\partial x_i} \left\{ \frac{D}{Dt} - v_j \frac{\partial}{\partial x_j} \right\} \left\{ (H-1) v_i \right\} \quad ,$$

$$= -\rho_0 \frac{\partial}{\partial x_i} \left\{ (1-H) \frac{Dv_i}{Dt} \right\} + \rho_0 \frac{\partial^2}{\partial x_i \partial x_j} \left\{ v_i v_j (1-H) \right\} \quad ,$$

$$\tag{11.125}$$

and integrated to give an alternative exact form for the distant sound field radiated by a body of constant volume, distorting in an otherwise arbitrary fashion, as it moves through the surrounding fluid. All positions $\mathbf{y}$ where $H = 0$ (i.e. $f < 0$) at the retarded time, can contribute to the integral of terms of (11.125) containing the $(1-H)$ factor. They comprise a volume different from the *instantaneous* volume of the body V_0, say, because each instantaneous volume element of the body sweeps through an *emission* volume elongated in the direction of wave emission by the factor $\left| 1 - \dfrac{v_i x_i}{cx} \right|^{-1}$.

The first term in (11.117) is supplemented by ρ_s to give the exact distant field,
as

$$4\pi c^2 x \, (\rho - \rho_0)(\mathbf{x}, t) \underset{x \to \infty}{\sim} \frac{x_i x_j}{x^2} \frac{1}{c^2} \frac{\partial^2}{\partial t^2} \int_\infty \left[H \, T_{ij} + (1 - H)\rho_0 v_i v_j \right] d^3 \mathbf{y}$$

$$+ \frac{x_i}{xc} \frac{1}{c} \frac{\partial}{\partial t} \left\{ \int_{S_{f=0}} \left[\frac{\ell_i p}{\left| 1 - \dfrac{v_i x_i}{cx} \right|} \right] d^3 \mathbf{y} + \int_{V_{f<0}} \rho_0 \left[\frac{Dv_i}{Dt} \right] d^3 \mathbf{y} \right\} \quad , \qquad (11.126)$$

the square brackets $\begin{bmatrix} \ \ \end{bmatrix}$ implying evaluation at the retarded time $\tau = t - |\mathbf{x} - \mathbf{y}|/c$.

It is only in very special cases that this equation simplifies to give a simple algebraic result, the simplest of all being when the body is compact enough that variations in retarded time are negligible and the quadrupoles negligibly weak in comparison with the dipoles. Furthermore, if the body moves as a solid body, the Döppler factor is effectively constant for all sources, and

$$4\pi c^2 (\rho - \rho_0) \underset{x \to \infty}{\sim} \frac{x_i}{xc} \frac{\partial}{\partial t} \left[\frac{F_i + \rho_0 V_0 \frac{Dv_i}{Dt}}{\left| 1 - \dfrac{x_i v_i}{xc} \right|} \right] \quad , \qquad (11.127)$$

$-F_i$ being the force exerted by the flow on the body.

Evidently the surface effects for a compact body are those of a dipole of strength equal to the *excess* of F_i, the force exerted by the fluid on the body, over that which would be required to overcome the rate of change of momentum of the body if it were of density, ρ_0. If there is no excess

$$F_i = \rho_0 \frac{Dv_i}{Dt} V_0 \quad ,$$

the compact body moves with the fluid and is indistinguishable in its dynamic effects from fluid. No sound can thus result from the surface terms and the dipole is seen to have zero strength.

In underwater situations, Mach numbers of practical interest rarely exceed 10^{-2} so that M is a small enough parameter to make most source regions thoroughly compact and to render any quadrupole insignificant in comparison with dipoles (unless, of course, the quadrupole strength is of order M^{-1} relative to the dipole).

However, in aeronautical applications the Mach number is hardly ever small enough to be negligible. The surface corrections to Lighthill's equations are then quite different. Just how different remains at the present time a matter of conjecture. The surface dipole terms are usually assumed to be the main source terms even though it is evident that the quadrupoles are of comparable order when M is not much less than 1. Ffowcs Williams and Hawkings have shown how the sources which dominate at low speed, and are usually (but incorrectly) assumed to remain dominant at higher speeds, actually become negligible compared with other sources as flow Mach number rises to about unity. However, this view has yet to percolate to practical application so that there is as yet little experience by which to judge the usefulness of the more difficult description of the flow once the surface velocities are not very small.

Throughout this section, though we have claimed to be deriving solutions to the equations, the formulae are in fact nothing more than integral identities. To be sure, if we really know T_{ij} together with pressure and velocity at the bounding surfaces, we can certainly use the preceding formulae to determine the radiation field. In aerodynamic applications when the main body of experience gives us data on the surface condition, this is a useful approach. But even there, the approach fails at high enough frequencies because we can no longer rely on the experience gained with steady flows to throw light on the high–frequency problem. Aerofoil theory features the Kutta condition for controlling the flow at a sharp trailing edge. It is viscous effects that ensure the condition is met. However, if the characteristic frequency is significantly in excess of ν/ℓ^2, the inverse of the time taken for viscous effects to act over the source lengthscale ℓ, then the motion is essentially inviscid with no Kutta condition. Normal aerofoil experience is then not particularly helpful to the high–frequency noise production problem, and the issue is better posed as a diffraction effect on known source fields. This is what we shall do now and we shall discover effects which could hardly be deduced by the earlier approach.

11.11 SURFACE EFFECTS AS A PROBLEM IN DIFFRACTION

We have posed the flow–noise problem as that of obtaining the solution to Lighthill's equation

$$\left\{ \frac{\partial^2}{\partial t^2} - c^2 \nabla^2 \right\} (\rho - \rho_0) = \frac{\partial^2 T_{ij}}{\partial x_i \partial x_j} \quad , \tag{11.128}$$

with T_{ij} and the boundary conditions known. The surface values are computable from (11.128) given the surface response equation, so the flow–noise problem is one of establishing the diffraction field induced by a quadrupole distribution near a specified boundary surface. We shall discuss some examples of problems that have been tackled by this scheme, all of which concern either rigid or pressure–release surfaces, the pressure–release surface being assumed to move with "linear" amplitude. The procedure for solving (11.128) is to seek the exact Green function for the problem and to write down the solution directly, as

$$(\rho - \rho_0)(\mathbf{x}, t) = \int_\tau \int_{V_I} \frac{\partial^2 T_{ij}}{\partial y_i \partial y_j} (\mathbf{y}, \tau) G(\mathbf{x}, t; \mathbf{y}, \tau) d\mathbf{y}\, d\tau \quad , \tag{11.129}$$

with G satisfying the equation

$$\left\{ \frac{\partial^2}{\partial t^2} - c^2 \nabla_{\mathbf{x}}^2 \right\} G = \delta(\mathbf{x} - \mathbf{y})\delta(t - \tau) \quad , \tag{11.130}$$

together with the boundary conditions for $(\rho - \rho_0)$.

11.12   THE INFINITE PLANE

Turning first to the problem of multipole sources near an infinite plane, suppose the plane to be rigid and choose coordinates (x_1, x_2, x_3) so that fluid occupies the half space $x_3 > 0$. The Green function G then satisfies Equation (11.130), $x_3 > 0$, together with the boundary condition

$$\frac{\partial G}{\partial x_3} = 0 \quad \text{when} \quad x_3 = 0 \quad , \tag{11.131}$$

and the usual "causality" condition that G be zero for $t < \tau$.

The boundary condition is easily satisfied by adding an "image source" of equal strength at the image point $\mathbf{y}^* = (y_1, y_2, -y_3)$. Thus we have

$$G(\mathbf{x}, t; \mathbf{y}, \tau) = \frac{1}{4\pi c^2 R} \delta(t - \tau - R/c) + \frac{1}{4\pi c^2 R^*} \delta(t - \tau - R^*/c) \quad , \tag{11.132}$$

with

$$R = |\mathbf{x} - \mathbf{y}| \quad \text{and} \quad R^* = |\mathbf{x} - \mathbf{y}^*| \quad .$$

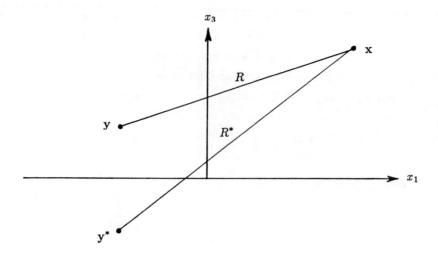

Figure 11.1 Reflection of source in a plane.

Substitution into formula (11.129) reveals that

$$4\pi c^2(\rho - \rho_0)(\mathbf{x}, t) = \int_{V_f} \frac{\partial^2}{\partial y_i \partial y_j} T_{ij}(\mathbf{y}, t - R/c) \frac{d\mathbf{y}}{R}$$

$$+ \int_{V_f} \frac{\partial^2}{\partial y_i \partial y_j} T_{ij}(\mathbf{y}, t - R^*/c) \frac{d\mathbf{y}}{R^*} \quad , \qquad (11.133)$$

the first term of which represents the "incident field". The second term, which accounts for the presence of the scattering boundary, is precisely equivalent to a distribution of quadrupoles throughout the region $x_3 < 0$, each quadrupole T_{ij} in the flow field having an equal image quadrupole at the point of reflection with respect to the plane $x_3 = 0$. Whether such an image flow could exist is quite irrelevant. The effect of the infinite plane is the same as that which would be produced by an imaginary image flow; the quadrupole nature of the scattered field is apparent.

At first sight this result seems to be inconsistent with the general theory according to which the effect of a rigid surface is equivalent to a suitable layer of *dipole* sources, which are much more efficient producers of sound than quadrupoles. The explanation of the paradox is that the dipole elements cancel to leave a less efficient quadrupole.

A very similar analysis results if the infinite plane is a pressure release surface in which case the sign of the second term of (11.132) is simply changed from plus to minus. These results were first presented in the flow–noise context by Powell (1960) and have been generalized to deal with more complicated boundary conditions by Ffowcs Williams (1965). In each case the presence of the boundary causes no significant enhancement of the incident sound field, the effect being essentially some kind of reflection of the original quadrupole source distribution.

In view off this somewhat surprising result, it is natural to enquire whether the presence of *any* scattering surface will have a correspondingly small effect upon the sound field. For smooth finite rigid bodies, the sound field induced is of the dipole type predicted by the general theory but we show below that the presence of a sharp edge can produce large effects, particularly upon multipoles in its immediate vicinity.

In order to deal with these more complicated geometries, we simplify the problem by taking the simple harmonic time dependence $\exp(-i\omega t)$ throughout. This implies no loss of generality since the quadrupole distribution $T_{ij}(\mathbf{x}, t)$ and density fluctuation $(\rho - \rho_0)(\mathbf{x}, t)$ can be written as a superposition of time–periodic functions according to the Fourier transformation

$$T_{ij}(\mathbf{x}, t) = \frac{1}{2\pi} \int_{-\infty}^{\infty} T_{ij}(\mathbf{x}, \omega)e^{-i\omega t}d\omega \tag{11.134}$$

with

$$T_{ij}(\mathbf{x}, \omega) = \int_{-\infty}^{\infty} T_{ij}(\mathbf{x}, t)e^{i\omega t}dt \quad,$$

and similarly for $(\rho - \rho_0)$.

With the time factor $\exp(-i\omega t)$ suppressed, the formula (11.129) for $(\rho - \rho_0)$ reduces to the form

$$(\rho - \rho_0)(\mathbf{x}, \omega) - -\frac{1}{c^2} \int_{V_I} T_{ij}(\mathbf{y}, \omega) \frac{\partial^2}{\partial y_i \partial y_j} G(\mathbf{x}, \mathbf{y}; \omega)d\mathbf{y} \quad, \tag{11.135}$$

where $G(\mathbf{x}, \mathbf{y})$ satisfies

$$(\nabla_{\mathbf{x}}^2 + k^2)G = \delta(\mathbf{x} - \mathbf{y}), \qquad k = \omega/c \quad,$$

with appropriate boundary conditions on the scattering surfaces, and a radiation condition at infinity.

11.13 THE RIGID SPHERE

As a prototype to examine the effect of smooth scattering bodies, use is made of the known results for a rigid sphere of radius a.

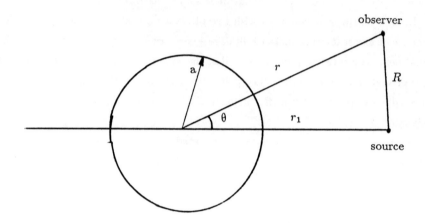

Figure 11.2 A source at $(r_1, 0)$ near a sphere of radius a is observed at (r, θ).
The solution $G(r, \theta)$ due to a unit source at $(r_1, 0)$ is know in the form of an infinite series

$$G(r, \theta) = -\frac{e^{ikR}}{4\pi R} + \frac{ik}{2\pi} \sum_0^\infty \left(n + \frac{1}{2}\right) \frac{j_n'(ka) h_n(kr_1)}{h_n'(ka)} h_n(kr) P_n(\cos \theta) \quad , \quad (11.136)$$

with $R = |\mathbf{r} - \mathbf{r}_1| = (r^2 + r_1^2 - 2rr_1 \cos \theta)^{1/2}$. The first term represents the incident field due to the source alone, and the infinite sum represents the scattered field induced by the presence of the rigid sphere. The form of the far field can be readily calculated for waves whose length is much greater than both a and r_1,

$$G \sim -\frac{e^{ikr}}{4\pi r} e^{-ikr_1 \cos \theta} + \frac{ik}{8\pi} \frac{a^3}{r_1^2} \frac{e^{ikr}}{r} \cos \theta \quad \text{as} \quad r \to \infty, \qquad (11.137)$$

whence the ratio of scattered field against incident field is of order ka^3/r_1^2 for long waves. If the incident field is of the quadrupole type, as in the flow–noise problem, the scattered sound field is larger than that of the incident field. In order to see this clearly, we simply differentiate G twice with respect to r_1, this corresponding to a longitudinal radial quadrupole at $(r_1, 0)$. In this way it is seen that

$$\frac{\partial^2 G}{\partial r_1^2} \sim \frac{k^2 \cos^2 \theta}{4\pi r} e^{ikr} + \frac{3ik}{4\pi} \frac{a^3 \cos \theta}{r_1^4} \frac{e^{ikr}}{r} \quad , \qquad (11.138)$$

in which expression the ratio of scattered field against incident field is of the order a^3/kr_1^4, which is large in the limit under consideration, $ka << 1$, $kr_1 << 1$.

This enhancement of an incident quadrupole field by a rigid sphere is in line with the result predicted by the general theory, and should hold for any smooth finite body.

A simple (monopole) source with time dependence proportional to $\exp(-i\omega t)$ produces density fluctuations of the form

$$\rho \sim \rho_0 = \frac{e^{ikR}}{R} \quad , \qquad k = \omega/c \quad , \tag{11.139}$$

where $R = |\mathbf{x} - \mathbf{y}|$ is the distance from source at $\mathbf{y}$ to the observer at $\mathbf{x}$. By direct differentiation it is seen that the field due to a point dipole of moment μ_i situated at $\mathbf{y}$ is given by

$$(\rho - \rho_0) = -\mu_i \frac{\partial}{\partial x_i} \left(\frac{e^{ikR}}{R} \right) = -\mu_i \frac{\partial R}{\partial x_i} \left(\frac{ik}{R} - \frac{1}{R^2} \right) e^{ikR} \quad . \tag{11.140}$$

At large values of R, this expression takes the form

$$(\rho - \rho_0) \sim -\mu_i \frac{\partial R}{\partial x_i} \, ik \, \frac{e^{ikR}}{R} \quad , \tag{11.141}$$

this "far–field" approximation giving a measure of the energy radiated towards infinity in the form of sound waves. The relative efficiency of source and dipole as radiators of sound is seen from (11.139) and (11.141) to contain the factor k in the dipole term. At low wave numbers k, the source is more efficient than a dipole, this being more efficient than a quadrupole, and so on.

For points very close to the dipole source, on the other hand, the most important term of (11.140) is given by the approximation

$$\rho - \rho_0 \sim \mu_i \frac{\partial R}{\partial x_i} \frac{1}{r^2} \quad , \qquad kR << 1 \quad . \tag{11.142}$$

The field is seen to be very large as $R \to 0$, this indicating the fact that there is a large amount of energy contained within the "near field"; this energy is due to the motion of fluid backwards and forwards from one side of the dipole to the other, to almost the same extent as if the flow were incompressible. The near–field energy of a quadrupole source is even larger, being of order R^{-3}. The transformation of this near–field energy into sound is the key to the explanation of the enhancement of the sound field of quadrupole sources by scattering surfaces. One might anticipate that the presence of a sharp edge within the near field of such a multipole will provide an efficient mechanism for transforming near–field into far–field energy, and this is a well–established fact.

11.14 A SEMI–INFINITE PLANE

For simplicity the scattering surface is taken to be a rigid semi–infinite plane situated in the region $x_2 = 0$, $x_1 > 0$.

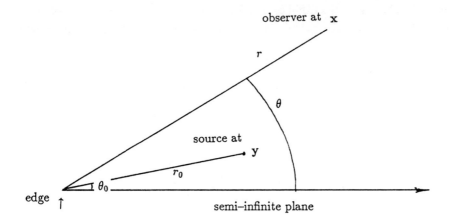

Figure 11.3 Source near the edge of a semi–infinite plane.

Suppose a simple source is situated at the point $\mathbf{y}$, given in polar coordinates by (r_0, θ_0), and an observer is situated at a point $\mathbf{x}(r, \theta)$, in the same plane perpendicular to the scattering half plane. According to the reciprocal theorem, the field is precisely the same as that obtained by interchanging source and observer position. For higher–order multipoles at $\mathbf{y}$ with an observer at $\mathbf{x}$, the problem is equivalent to that of a simple source at $\mathbf{x}$ with an observer measuring the appropriate derivatives at $\mathbf{y}$. In the limit as $r \to \infty$, which determines the far field, it is seen that the source at $\mathbf{x}$ is to be replaced by a plane wave of suitable amplitude, propagating at an angle θ to the scattering half plane. To be precise, if a simple source with incident pressure field

$$p_i = \frac{e^{ik|\mathbf{x}-\mathbf{y}|}}{|\mathbf{x} - \mathbf{y}|}$$

is situated at $\mathbf{x}$, then as $r = x \to \infty$ with $\mathbf{y}$ fixed, we have

$$p_i \sim \frac{e^{ikr}}{r} e^{-ik(y_1 \cos \theta + y_2 \sin \theta)} \quad , \qquad (11.143)$$

which represents a plane wave of amplitude e^{ikr}/r, at angle θ.

Thus we are led to consider the problem of finding the solution at $\mathbf{y}$ due to an incident plane wave, given by

$$p_i = P e^{-ik(y_1 \cos \theta + y_2 \sin \theta)} \quad ,$$

where the amplitude $P = e^{ikr}/R$ is considered a fixed parameter.

It has already been observed that in the vicinity of the edge where there is a singularity, the governing wave equation $(\nabla^2 + k^2)p = 0$ takes the approximate form

$$\nabla^2 p \approx 0 \quad , \qquad k r_0 << 1 \quad ,$$

and the pressure field resembles that appropriate to an incompressible flow. In particular, it follows that

$$\frac{\partial p}{\partial y_1} - i \frac{\partial p}{\partial y_2} \sim P \frac{C(k)}{(y_1 + iy_2)^{1/2}} = P \frac{C(k)}{r_0^{1/2} e^{i\theta_0/2}}, \quad k r_0 \to 0 \quad , \qquad (11.144)$$

where $C(k)$ is an unknown complex constant which depends on the parameter k, and on the angle of incidence θ. For a rigid half plane, the normal derivative $\partial p/\partial n$ is zero on the surface, whence $\partial p/\partial y_2$, is certainly nonsingular, as $r_0 \to 0$ when $\theta_0 = 0$. In order to maintain consistency with the general result (11.144) therefore, the phase of the constant C must be such that the derivatives $\partial p/\partial y_1$, and $\partial p/\partial y_2$ take the forms

$$\frac{\partial p}{\partial y_1} \sim P \frac{A(k)}{r_0^{1/2}} \cos(\theta_0/2) \quad ,$$

and
(11.145)

$$\frac{\partial p}{\partial y_2} \sim P \frac{A(k)}{r_0^{1/2}} \sin(\theta_0/2) \quad ,$$

where A is the same in each expression. Further a dimensional analysis reveals the requirement that the constant $A(k)$ must take the form $A_0 k^{1/2}$, with A_0 depending only on the direction θ of the incident wave. Thus

$$\frac{\partial p}{\partial y_1} \sim P A_0(\theta) \frac{k^{1/2}}{r_0^{1/2}} \cos(\theta_0/2) \quad,$$

and (11.146)

$$\frac{\partial p}{\partial y_2} \sim P A_0(\theta) \frac{k^{1/2}}{r_0^{1/2}} \sin(\theta_0/2) \quad,$$

as $kr_0 \to 0$. Thus a dipole of moment $\mu_i = (\mu_1, \mu_2)$ at a point $\mathbf{y}$ close to the edge induces a pressure disturbance p whose far field takes the form

$$p(r, \theta) \sim A_0(\theta) \frac{k^{1/2}}{r_0^{1/2}} \frac{e^{ikr}}{r} \left\{ \mu_1 \cos(\theta_0/2) + \mu_2 \sin(\theta_0/2) \right\}$$

valid for $kr_0 \ll 1$, $kr \gg 1$.

The field due to higher-order multipoles can be inferred by further differentiation with respect to y_1 and y_2, by means of the identities

$$\frac{\partial}{\partial y_1} = \cos\theta_0 \frac{\partial}{\partial r_0} - \frac{\sin\theta_0}{r_0} \frac{\partial}{\partial \theta_0}; \qquad \frac{\partial}{\partial y_2} = \sin\theta_0 \frac{\partial}{\partial r_0} + \frac{\cos\theta_0}{r_0} \frac{\partial}{\partial \theta_0} \quad.$$

Thus a quadrupole with components T_{11}, T_{12}, T_{22} situated at $\mathbf{y}(r_0, \theta_0)$ has a far field of the form

$$p(r, \theta) \sim A_0(\theta) \frac{k^{1/2}}{2 r_0^{3/2}} \frac{e^{ikr}}{r} \left\{ (T_{22} - T_{11}) \cos\left(\frac{3\theta_0}{2}\right) - T_{12} \sin\left(\frac{3\theta_0}{2}\right) \right\} \quad,$$

(11.147)

and dipoles and quadrupoles with components parallel to the scattering edge are negligible in comparison.

The argument presented above gives the far field for points in the plane containing the source point $\mathbf{y}$. For an observer at a more general position $\mathbf{x}$, the solution can be generated from the special case, by replacing the plane wave by a wave at oblique incidence, i.e.

$$p_i = P \exp\left\{ -ik(y_1 \cos\theta + y_2 \sin\theta)\cos\phi - iky_3 \sin\theta \right\} \quad,$$

the only difference to the result (11.147) being that the wavenumber k is replaced by its component $k \cos\phi$ in the (y_1, y_2) plane. The result for a volume distribution of quadrupoles near the edge is generated by integrating over the source region, which is supposed finite. Thus we have

$$p(r,\theta,\phi) \sim A_0(\theta)k^{1/2}(\cos\phi)^{1/2}\frac{e^{ikr}}{2r}\int_{V_f}\left\{(T_{22}-T_{11})\cos\left(\frac{3\theta_0}{2}\right)\right.$$

$$\left. - T_{12}\sin\left(\frac{3\theta_0}{2}\right)\right\}\frac{d\mathbf{y}}{r_0^{3/2}} \quad, \tag{11.148}$$

as $kr \to \infty$, which gives the far field at a point (r,θ,ϕ) with direction cosines $(\cos\theta\cos\phi)$, $\sin\theta\cos\phi$, $\sin\phi)$ with respect to the source region.

It is remarkable that so much information can be inferred without any need to solve the full boundary–value problem. The r_0– and θ_0–dependence arises entirely from the incompressible nature of the solution near the edge, and the k–dependence is given by dimensional analysis.

An exact analysis is available for this particular problem, where the exact Green function can be expressed in terms of Fresnel integrals. Ffowcs Williams & Hall (1970) have exploited this property to show, in the present notation, that

$$p(r,\theta,\phi) \sim \frac{e^{-i\pi/4}}{(2\pi)^{3/2}}k^{1/2}$$

$$\times \cos\left(\frac{\theta}{2}\right)(\cos\phi)^{1/2}\frac{e^{ikr}}{2r}\int_{V_f}\left\{(T_{22}-T_{11})\cos\left(\frac{3\theta_0}{2}\right) - T_{12}\sin\left(\frac{3\theta_2}{2}\right)\right\}\frac{d\mathbf{y}}{r_0^{3/2}} \quad,$$

which agrees with (11.148) and provides further the θ–dependence.

Finally, to interpret these results within the flow–noise context, an order–of–magnitude argument of the usual kind, with directional dependence ignored, indicates that the presence of the edge modifies the "U^8 law" of Lighthill for unbounded flow. The wavenumber k is replaced by $U/\ell c$, where U is a typical turbulence velocity and ℓ a lengthscale associated with the turbulence. Since the components T_{ij} are themselves proportional to U^2, it is seen that p varies with U according to the formula $p \sim U^{5/2}$, whence the intensity I varies as

$$I \propto U^5 \quad .$$

11.15 SCATTERING BY A WEDGE

A similar analysis can be carried out for a wedge of any angle. If the wedge has exterior angle $(p/q)\pi$, where p/q is a rational number between 1 and 2, then the edge singularity takes the form

$$|\nabla p| \sim \frac{C_0(kr_0)^{q/p}}{r_0} \quad ,$$

in place of (11.146), C_0 being a function of angle only, whence results the intensity law

$$I \propto U^{4+2q/p} \quad . \tag{11.149}$$

In particular, for the half plane we have $p/q = 2$ and recover the fifth power law obtained above. The intensity law for a right–angled corner is obtained by setting $p/q = 3/2$, to get

$$I \propto U^{16/3} \quad .$$

As might be expected, a sharper edge produces more intense scattering.

11.16 SCATTERING BY A RIGID DISC

In view of these results, it is natural to ask whether the enhancement of sound is due essentially to the sharpness of the edge or to the large extent of the scattering surface. It is proposed here that the r_0- and θ_0-dependence is entirely a property of the sharpness of the edge, whilst the important factor $k^{1/2}$ arises on account of the large size of the scatterer.

The simplest problem involving a finite sharp edge is probably that due to a plane wave normally incident on a rigid circular disc.

As before, it is argued that the wave equation for p may be replaced,

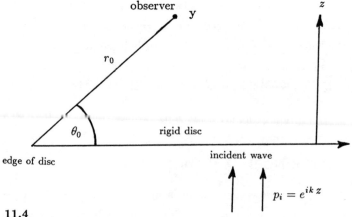

Figure 11.4

for any fixed wavenumber k and fixed radius a, by the Laplace equation for points in the immediate vicinity of the edge. Thus

$$\frac{\partial p}{\partial y_1} \sim \frac{A(k,a)}{r_0^{1/2}} \cos\left(\frac{\theta_0}{2}\right) \quad,$$

and (11.150)

$$\frac{\partial p}{\partial y_2} \sim \frac{A(k,a)}{r_0^{1/2}} \sin\left(\frac{\theta_0}{2}\right)$$

$$kr_0 \to 0 \quad,$$

where $y_1 = r_0\cos\theta_0$ and $y_2 = r_0\sin\theta_0$ are local Cartesian coordinates measured from the edge. The important difference now is that the constant A depends upon *two* parameters k and a. The dimension of A must be $(\text{length})^{-1/2}$ but there are now *two* lengthscales, radius and wavelength, with which this requirement can be met, and A takes a different form according as the waves are short, or long, compared with the radius a.

If the waves are very short (i.e. $ka \gg 1$), then the flow within a few wavelengths will be insensitive to the curvature of the disc, which may consequently be replaced by a semi–infinite plane, to get

$$A(k,a) \sim A_0 k^{1/2} \quad, \quad \text{for} \quad ka \gg 1 \quad, \tag{11.151}$$

and the solution is like that obtained above for a half plane.

For long waves, on the other hand, it is quite inappropriate to neglect the curvature of the edge. In this limit, the sort of analysis suggested by the general theory of Curle (1955) seems the proper point of view, this being essentially to represent the effect of the rigid disc by a system of dipoles, whose strength is determined as if the flow were incompressible. This idea is contained within the general result and shows that the presence of the body is equivalent to a dipole whose strength is proportional to the total force on the body. In particular, the k– dependence is seen to appear in linear form, the constant $A(k,a)$ in (11.150) being proportional to k

$$A(k,a) \sim B_0 k a^{1/2} \quad, \quad \text{for} \quad ka \ll 1 \quad. \tag{11.152}$$

Finally then, we have

$$I \propto U^5 \tag{11.153}$$

for sharp bodies with dimensions much larger than a wavelength, and

$$I \propto U^6 \qquad (11.154)$$

for bodies with dimensions much less than a wavelength.

REFERENCES

Curle, N. (1955). The influence of solid boundaries upon aerodynamic sound. Proc. R. Soc. London A 231:505.

Ffowcs Williams, J.E. (1965). Sound radiation from turbulent boundary layers formed on compliant surfaces. J. Fluid Mech. 22:347.

Ffowcs Williams, J.E. & Hall, L.H. (1970). Aerodynamic sound generation by turbulent flow in the vicinity of a scattering half plane. J. Fluid Mech. 40:657–670.

Powell, A. (1960). Aerodynamic noise and the plane boundary. J. Acoust. Soc. Am. 32:982.

12. VORTEX SOUND

We saw in Chapter 11 that Lighthill's theory of aerodynamic sound identifies the quadrupole term $T_{ij} = \rho u_i u_j + p_{ij} - c^2 \rho' \delta_{ij}$ as the source of sound in an unbounded fluid in nonlinear motion. ρ is the density, $\mathbf{u}$ the particle velocity p_{ij} the compressive stress tensor and c the speed of sound in the distant linearly disturbed fluid. The mean density is denoted by ρ_0, and $\rho' = \rho - \rho_0$ is the density perturbation. It is sometimes convenient to rewrite this quadrupole source in a way which emphasises the dependence of the noise–producing elements of T_{ij} on local vorticity. One advantage of doing this is that vortical regions of the flow are often much more concentrated than the hydrodynamic region over which T_{ij} is nonzero. Moreover, the development of the vorticity field can be described by simple kinematics.

We will begin by considering an unbounded region and then in later sections will go on to investigate some of the effects of scattering surfaces.

12.1 SOUND RADIATION BY A COMPACT TURBULENT EDDY IN AN UNBOUNDED SPACE

An eddy of size ℓ with typical velocity U, generates sound with frequencies of the order U/ℓ. The wavelength of this sound, $\lambda - 0(\ell/M)$, greatly exceeds the eddy size in a low Mach number flow and the sound generation problems can be formulated as a singular perturbation problem. Crow (1970) used the method of matched asymptotic expansions in this way to investigate the sound generated by a turbulent eddy in an isentropic medium. He found that the acoustic source is $\rho_0 \partial^2 v_i v_j / \partial x_i \partial x_j$ to lowest order in the Mach number, where $\mathbf{v}$, the velocity field in a hypothetical incompressible flow, is defined by

$$\mathbf{v} = \operatorname{curl}\left(\int \frac{\omega(\mathbf{y}) d^3 \mathbf{y}}{4\pi |\mathbf{x} - \mathbf{y}|}\right) , \qquad (12.1)$$

with $\boldsymbol{\omega}$ the vorticity vector, $\boldsymbol{\omega} = \operatorname{curl} \mathbf{u}$.

The Green function of the three–dimensional wave equation given in Equation (11.34) can be used to write down the density perturbation generated by the source $\rho_0 \partial^2 v_i v_j / \partial x_i \partial x_j$ in an unbounded fluid,

$$\rho'(\mathbf{x}, t) = \int \rho_0 \frac{\partial^2 v_i v_j}{\partial y_i \partial y_j} \frac{\delta(t - \tau - |\mathbf{x} - \mathbf{y}|/c)}{4\pi c^2 |\mathbf{x} - \mathbf{y}|} d^3 y d\tau \ . \tag{12.2}$$

Howe (1975a) shows how Equation (12.2) may be rearranged to demonstrate the dependence of the sound field on the vorticity. Following his procedure, begin by noting that

$$(\mathbf{v}.\nabla)\mathbf{v} = \boldsymbol{\omega} \times \mathbf{v} + \nabla(\tfrac{1}{2}v^2) \tag{12.3}$$

is a vector identity. The definition (12.1) ensures that $\operatorname{div} \mathbf{v}$ vanishes and hence taking the divergence of Equation (12.3) leads to

$$\frac{\partial^2 v_i v_j}{\partial y_i \partial y_j} = \operatorname{div}(\boldsymbol{\omega} \times \mathbf{v}) + \nabla^2(\tfrac{1}{2}v^2) \ . \tag{12.4}$$

When this is substituted into (12.2), it gives

$$\rho'(\mathbf{x}, t) = \frac{\rho_0}{4\pi c^2} \int \left(\operatorname{div}(\boldsymbol{\omega} \times \mathbf{v}) + \nabla^2 \left(\tfrac{1}{2}v^2 \right) \right) \frac{\delta(t - \tau - |\mathbf{x} - \mathbf{y}|/c)}{|\mathbf{x} - \mathbf{y}|} d^3 y d\tau \ . \tag{12.5}$$

We evidently have a dipole of strength $\rho_0 \operatorname{div}(\boldsymbol{\omega} \times \mathbf{v})$ and an isotropic quadrupole of strength $\tfrac{1}{2}\rho_0 v^2$ integrated with the Green function. In the usual way we can integrate by parts to transfer the derivatives to the Green function,

$$\rho'(\mathbf{x}, t) = \frac{\rho_0}{4\pi c^2} \int \left(-(\boldsymbol{\omega} \times \mathbf{v})_i \frac{\partial}{\partial y_i} \left(\frac{\delta(t - \tau - |\mathbf{x} - \mathbf{y}|/c)}{|\mathbf{x} - \mathbf{y}|} \right) \right.$$

$$\left. + \frac{1}{2} v^2 \frac{\partial^2}{\partial y_i \partial y_i} \left(\frac{\delta(t - \tau - |\mathbf{x} - \mathbf{y}|/c)}{|\mathbf{x} - \mathbf{y}|} \right) \right) d^3 y d\tau \ . \tag{12.6}$$

The Green function depends on $\mathbf{x}$ and $\mathbf{y}$ only in the combination $\mathbf{x} - \mathbf{y}$. Hence

$$\frac{\partial}{\partial y_i} \left(\frac{\delta(t - \tau - |\mathbf{x} - \mathbf{y}|/c)}{|\mathbf{x} - \mathbf{y}|} \right) = -\frac{\partial}{\partial x_i} \left(\frac{\delta(t - \tau - |\mathbf{x} - \mathbf{y}|/c)}{|\mathbf{x} - \mathbf{y}|} \right) ,$$

and Equation (12.6) can be rewritten as

$$\rho'(\mathbf{x}, t) = \frac{\rho_0}{4\pi c^2} \left(\frac{\partial}{\partial x_i} \int (\boldsymbol{\omega} \times \mathbf{v})_i \frac{\delta(t - \tau - |\mathbf{x} - \mathbf{y}|/c)}{|\mathbf{x} - \mathbf{y}|} d^3 y d\tau \right.$$

$$+\frac{\partial^2}{\partial x_i \partial x_i} \int \frac{1}{2}v^2 \frac{\delta(t - \tau - |\mathbf{x} - \mathbf{y}|/c)}{|\mathbf{x} - \mathbf{y}|} \, d^3y d\tau \bigg) .$$

The δ–function may now be used to evaluate the τ–integrals to give

$$\rho'(\mathbf{x}, t) = \frac{\rho_0}{4\pi c^2} \left(\frac{\partial}{\partial x_i} \int \frac{(\boldsymbol{\omega} \times \mathbf{v})_i(\mathbf{y}, t - |\mathbf{x} - \mathbf{y}|/c)}{|\mathbf{x} - \mathbf{y}|} \, d^3y \right.$$

$$\left. + \frac{\partial^2}{\partial x_i \partial x_i} \int \frac{\frac{1}{2}v^2(\mathbf{y}, t - |\mathbf{x} - \mathbf{y}|/c)}{|\mathbf{x} - \mathbf{y}|} \, d^3y \right) . \tag{12.7}$$

The expression for the density perturbation $\rho'(\mathbf{x}, t)$ in (12.7) simplifies further when the observer's position $\mathbf{x}$ is far from the source region, which is, say, near the origin. Then $|\mathbf{x}| \gg |\mathbf{y}|$ and

$$|\mathbf{x} - \mathbf{y}| = (|\mathbf{x}|^2 - 2\mathbf{x} \cdot \mathbf{y} + |\mathbf{y}|^2)^{1/2} \simeq |\mathbf{x}| - \frac{\mathbf{x} \cdot \mathbf{y}}{|\mathbf{x}|} , \tag{12.8a}$$

$$|\mathbf{x} - \mathbf{y}|^{-1} \simeq |\mathbf{x}|^{-1} \tag{12.8b}$$

and

$$\frac{\partial}{\partial x_i} \simeq -\frac{x_i}{|\mathbf{x}|c} \frac{\partial}{\partial t} , \tag{12.8c}$$

(see Equation (11.55)). For $\mathbf{x}$ in the far field,

$$\rho'(\mathbf{x}, t) = \frac{\rho_0}{4\pi c^2 |\mathbf{x}|} \left[-\frac{x_i}{|\mathbf{x}|c} \frac{\partial}{\partial t} \int (\boldsymbol{\omega} \times \mathbf{v})_i \left(\mathbf{y}, t - \frac{|\mathbf{x}|}{c} + \frac{\mathbf{x} \cdot \mathbf{y}}{|\mathbf{x}|c} \right) d^3y \right.$$

$$\left. + \frac{1}{2c^2} \frac{\partial^2}{\partial t^2} \int v^2(\mathbf{y}, t - \frac{|\mathbf{x}|}{c} + \frac{\mathbf{x} \cdot \mathbf{y}}{|\mathbf{x}|c}) \, d^3y \right] . \tag{12.9}$$

The variation in retarded time can be expanded as a Taylor series

$$(\boldsymbol{\omega} \times \mathbf{v})\left(\mathbf{y}, t - \frac{|\mathbf{x}|}{c} + \frac{\mathbf{x} \cdot \mathbf{y}}{|\mathbf{x}|c} \right) = (\boldsymbol{\omega} \times \mathbf{v})(\mathbf{y}, t - \frac{|\mathbf{x}|}{c}) + \frac{\mathbf{x} \cdot \mathbf{y}}{|\mathbf{x}|c} \frac{\partial}{\partial t}(\boldsymbol{\omega} \times \mathbf{v})(\mathbf{y}, t - \frac{|\mathbf{x}|}{c}) + \ldots$$

$$\tag{12.10}$$

We have already noted that, in a low Mach number flow, a turbulent eddy is compact. We would therefore expect the largest contribution to the sound field to be generated by the lowest–order terms in the expansion that lead to nonzero integrals in (12.9).

The total source strength of the isotropic quadrupole, $\frac{1}{2}\rho_0 \int v^2(\mathbf{y}, t)d^3y$, is constant in an inviscid flow because it represents the total incompressible kinetic energy. This is readily seen from the inviscid, incompressible momentum equation

$$\frac{\partial \mathbf{v}}{\partial t} + \mathrm{grad}\left(\frac{p}{\rho_0} + \frac{1}{2}v^2 \right) = -\boldsymbol{\omega} \times \mathbf{v} . \tag{12.11}$$

Taking the scalar product of this equation with $\mathbf{v}$ and noting that div $\mathbf{v} = 0$ leads to

$$\frac{1}{2}\frac{\partial v^2}{\partial t} + \operatorname{div}\left(\mathbf{v}\left(\frac{p}{\rho_0} + \frac{1}{2}v^2\right)\right) = 0 \ . \tag{12.12}$$

Integrating over all space and applying the divergence theorem, the contribution from the second term vanishes because $\mathbf{v}(p/\rho_0 + \frac{1}{2}v^2)$ tends to zero at least as fast as $|\mathbf{y}|^{-3}$ as $|\mathbf{y}| \to \infty$. Hence

$$\frac{\partial}{\partial t}\int v^2(\mathbf{y}, t)d^3\mathbf{y} = 0 \ . \tag{12.13}$$

The first nonzero term in the second integral in Equation (12.9) therefore arises due to the effect of retarded time variations over the source region. But this leads to a contribution to the far–field density fluctuation of order $\rho_0 M^5 \ell/|\mathbf{x}|$, which we will find is negligible in comparison with the leading–order term in low Mach number flow.

We now consider the first integral in (12.9). Using the expansion of the retarded time variation in (12.10) gives

$$-\frac{x_i}{|\mathbf{x}|c}\frac{\partial}{\partial t}\int(\boldsymbol{\omega} \times \mathbf{v})_i\left(\mathbf{y}, t - \frac{|\mathbf{x}|}{c} + \frac{\mathbf{x} \cdot \mathbf{y}}{|\mathbf{x}|c}\right)d^3\mathbf{y}$$

$$= -\frac{x_i}{|\mathbf{x}|c}\frac{\partial}{\partial t}\int(\boldsymbol{\omega} \times \mathbf{v})_i\left(\mathbf{y}, t - \frac{|\mathbf{x}|}{c}\right)d^3\mathbf{y} - \frac{x_i x_j}{(|\mathbf{x}|c)^2}\frac{\partial^2}{\partial t^2}\int y_j(\boldsymbol{\omega} \times \mathbf{v})_i(\mathbf{y}, t - \frac{|\mathbf{x}|}{c})d^3\mathbf{y}. \tag{12.14}$$

The first integral on the right–hand side of (12.14) vanishes identically because the integrand can be expressed through (12.3) as a divergence. The leading term in the far–field density perturbation is therefore

$$\rho'(\mathbf{x}, t) = -\frac{\rho_0}{4\pi c^4|\mathbf{x}|^3}\frac{\partial^2}{\partial t^2}\int(\mathbf{x} \cdot \mathbf{y})(\mathbf{x} \cdot \boldsymbol{\omega} \times \mathbf{v})\left(\mathbf{y}, t - \frac{|\mathbf{x}|}{c}\right)d^3\mathbf{y} \ . \tag{12.15}$$

This is Howe's (1975a) Equation (2.15) with a minor typing mistake corrected. It identifies $\rho_0 \boldsymbol{\omega} \times \mathbf{v}$ as the major aerodynamic source term as suggested by Powell (1964). Lighthill's eighth–power scaling law can be readily recovered from (12.15). But the advantage of this formulation is that it enables exact problems to be solved after integration only over regions where the vorticity is nonzero.

Example. Radiation of sound by an elliptical vortex (Howe 1975a).

Lamb (1932) investigated the incompressible flow field of a cylindrical vortex of slightly elliptical cross–section. Within the ellipse the fluid has uniform vorticity Ω per unit area in a direction perpendicular to the axis of the cylinder. When the ellipse is only slightly deformed from a circle of radius a, Lamb showed that the elliptical cross-section of the cylinder rotates about its axis with angular velocity $\frac{1}{4}\Omega$. The equation of the ellipse can be written as

$$\sigma = a\left(1 + \epsilon \cos(2\psi - \frac{1}{2}\Omega t)\right), \qquad (12.16)$$

where ϵ is a small parameter and (σ, ψ) are polar coordinates. Lamb determined the velocity field within the core

$$\mathbf{v} = \frac{1}{2}\Omega\sigma\left(-\sin\psi - \epsilon\sin(\psi - \frac{1}{2}\Omega t), \quad \cos\psi - \epsilon\cos(\psi - \frac{1}{2}\Omega t), \quad 0\right). \qquad (12.17)$$

Hence, inside the vortex the core, i.e. for $\sigma < a(1 + \epsilon\cos(2\psi - \frac{1}{2}\Omega t))$,

$$\boldsymbol{\omega} \times \mathbf{v} = \frac{1}{2}\Omega^2\sigma\left(-\cos\psi + \epsilon\cos(\psi - \frac{1}{2}\Omega t), \quad -\sin\psi - \epsilon\sin(\psi - \frac{1}{2}\Omega t), \quad 0\right), \quad (12.18)$$

and $\boldsymbol{\omega} \times \mathbf{v}$ is identically zero outside the vortex core.

To evaluate the sound field we will substitute for $\boldsymbol{\omega} \times \mathbf{v}$ from Equation (12.18) into an expression equivalent to (12.15). But in this two–dimensional problem, the source extends over all y_3 and so the source region is certainly not compact in the y_3 direction! The dependence of $|\mathbf{x} - \mathbf{y}|$ on y_3 must be retained and the form equivalent to that in Equation (12.15) is

$$\rho'(\mathbf{x}, t) = -\frac{\rho_0}{4\pi c^4}\frac{\partial^2}{\partial t^2}\int \frac{(x_1 y_1 + x_2 y_2)(\mathbf{x}\cdot\boldsymbol{\omega}\times\mathbf{v})\left(\mathbf{y}, t - (R^2 + y_3^2)^{1/2}/c\right)}{(R^2 + y_3^2)^{3/2}} d^3\mathbf{y} ,$$

$$(12.19)$$

where $R = (x_1^2 + x_2^2)^{1/2}$. When the expression for $\boldsymbol{\omega} \times \mathbf{v}$ in (12.18) is substituted into (12.19), the integrals over σ and ψ can be readily evaluated to obtain

$$\rho'(\mathbf{x}, t) = -\frac{\rho_0\epsilon\Omega^4 a^4 R^2}{128c^4}\int_{-\infty}^{\infty} \frac{1}{(R^2 + y_3^2)^{3/2}} \cos\left[2\theta - \frac{1}{2}\Omega t + \frac{\Omega}{2c}(R^2 + y_3^2)^{1/2}\right] dy_3,$$

$$(12.20)$$

with θ defined by $x_1 = R\cos\theta$, $x_2 = R\sin\theta$. For far–field points at which $\Omega R >> c$, the y_3- integral in (12.20) may be evaluated by changing the integration variable to

y_3/R and using the method of stationary phase described in Chapter 4. This leads to

$$\rho'(\mathbf{x},t) = -\frac{\rho_0 \epsilon \Omega^{7/2} a^{7/2}}{64 c^{7/2}} \left(\frac{\pi a}{R}\right)^{1/2} \cos\left[2\theta + \frac{\pi}{4} - \frac{\Omega}{2}\left(t - \frac{R}{c}\right)\right], \tag{12.21}$$

which has the form expected for the sound field of a two–dimensional eddy. The directivity $\cos[2\theta + \frac{\pi}{4} - \frac{\Omega}{2}(t - \frac{R}{c})]$ is typical of a rotating quadrupole, while the intensity is proportional to M^7, where $M = \Omega a/c$.

12.2 HOWE'S ACOUSTIC ANALOGY

So far we have just considered the sound produced by vortical flows in unbounded space, but we will now go on to discuss an extension to account for the presence of fixed, rigid surfaces. Howe (1975a) reformulated the acoustic analogy with stagnation enthalpy, $B = h + \frac{1}{2}u^2$, as the dependent variable, where h is the enthalpy and u the flow speed. He showed that in the absence of viscous dissipation, the momentum and energy equations can be combined to give

$$\left(\frac{D}{Dt}\left(\frac{1}{c_L^2}\frac{D}{Dt}\right) + \frac{1}{c_L^2}\frac{D\mathbf{u}}{Dt}\cdot\frac{\partial}{\partial\mathbf{x}} - \nabla^2\right) B$$
$$= \operatorname{div}\left(\boldsymbol{\omega} \times \mathbf{u} - T\operatorname{grad}s\right) - \frac{1}{c_L^2}\frac{D\mathbf{u}}{Dt}\cdot\left(\boldsymbol{\omega} \times \mathbf{u} - T\operatorname{grad}s\right), \tag{12.22}$$

where T is the absolute temperature, c_L the local speed of sound and s the entropy per unit mass. Equation (12.22) is exact in the absence of dissipative processes and heat transfer. Howe argues that in the low Mach number limit, it reduces to the simple wave equation

$$\left(\frac{1}{c^2}\frac{\partial^2}{\partial t^2} - \nabla^2\right) B = \operatorname{div}\left(\boldsymbol{\omega} \times \mathbf{v}\right) \tag{12.23}$$

for an isentropic flow. B is then equal to $\int dp/\rho + \frac{1}{2}u^2$.

The surface S bounding the flow is fixed and impenetrable. We see from the momentum equation that

$$\nabla B = \frac{\nabla p}{\rho} + \nabla(\frac{1}{2}u^2) = -\frac{\partial \mathbf{u}}{\partial t} - \boldsymbol{\omega} \times \mathbf{u}.$$

The boundary condition of no normal acceleration on the surface S therefore reduces to

$$\mathbf{n}\cdot\nabla B = 0 \quad \text{on} \quad S, \tag{12.24}$$

provided that the vorticity vanishes on the surface.

A general way of solving Equation (12.23) with the boundary condition (12.24) is to introduce a Green function, $G(\mathbf{y}, \tau | \mathbf{x}, t)$. G is to satisfy

$$\left(\frac{1}{c^2} \frac{\partial^2}{\partial \tau^2} - \frac{\partial^2}{\partial y_i \partial y_i}\right) G = \delta(\mathbf{x} - \mathbf{y}, t - \tau) \tag{12.25}$$

outside the surface S, together with the boundary condition $\mathbf{n}.\nabla\, G = 0$ on S and a radiation condition at infinity. Equations (12.23) and (12.25) can then be combined in the way described in Chapter 11 to show that

$$B(\mathbf{x}, t) = \int_V \operatorname{div}(\boldsymbol{\omega} \times \mathbf{v}) G d^3 \mathbf{y} d\tau , \tag{12.26}$$

where V is the region outside the foreign bodies. In the far field B simplifies to p'/ρ_0 and after integration by parts (12.26) becomes

$$p'(\mathbf{x}, t) = -\rho_0 \int_V (\boldsymbol{\omega} \times \mathbf{v})_i \frac{\partial G}{\partial y_i} d^3 \mathbf{y} d\tau . \tag{12.27}$$

We note that the free–space result in Equation (12.25) can be quickly recovered from this expression. The free–space Green function, which is a solution of (12.15), is given by

$$G(\mathbf{y}, \tau | \mathbf{x}, t) = \frac{\delta(t - \tau - |\mathbf{x} - \mathbf{y}|/c)}{4\pi |\mathbf{x} - \mathbf{y}|} , \tag{12.28}$$

and, with the far–field approximations in (12.8), this reduces to

$$G(\mathbf{y}, \tau | \mathbf{x}, t) = \frac{\delta(t - \tau - |\mathbf{x}|/c + \mathbf{x} \cdot \mathbf{y}/|\mathbf{x}|c)}{4\pi |\mathbf{x}|} . \tag{12.29}$$

Differentiation shows that

$$\frac{\partial}{\partial y_i} G(\mathbf{y}, \tau | \mathbf{x}, t) = \frac{x_i}{4\pi c |\mathbf{x}|^2} \frac{\partial}{\partial t} \delta(t - \tau - |\mathbf{x}|/c + \mathbf{x} \cdot \mathbf{y}/|\mathbf{x}|c) . \tag{12.30}$$

When this is substituted into (12.27), we obtain

$$p'(\mathbf{x}, t) = -\frac{\rho_0}{4\pi c |\mathbf{x}|^2} \frac{\partial}{\partial t} \int \mathbf{x} \cdot \boldsymbol{\omega} \times \mathbf{v} \left(\mathbf{y}, t - \frac{|\mathbf{x}|}{c} + \frac{\mathbf{x} \cdot \mathbf{y}}{|\mathbf{x}|c}\right) d^3 \mathbf{y}. \tag{12.31}$$

Now for the compact source region the retarded time variation can be expanded by the Taylor series in (12.10). Recalling that $\int \boldsymbol{\omega} \times \mathbf{v}(\mathbf{y}, t) d^3 \mathbf{y} = 0$, we find the leading term to be

$$p'(\mathbf{x}, t) = -\frac{\rho_0}{4\pi c^2 |\mathbf{x}|^3} \frac{\partial^2}{\partial t^2} \int (\mathbf{x} \cdot \mathbf{y}) \mathbf{x} \cdot \boldsymbol{\omega} \times \mathbf{v} \left(\mathbf{y}, t - \frac{|\mathbf{x}|}{c}\right) d^3 \mathbf{y} , \tag{12.32}$$

and so recover Equation (12.15).

We have seen that the far–field density fluctuations depend on the aeroacoustic source $\rho_0 \operatorname{div}(\boldsymbol{\omega} \times \mathbf{v})$. However, Equation (12.1) shows that $\mathbf{v}$ itself can be calculated directly from $\boldsymbol{\omega}$. So it might be reasonable to seek a simpler form for the source term: one depending only on the vorticity, $\boldsymbol{\omega}$. That seems even more reasonable when one recalls that the vorticity equation

$$\frac{\partial \boldsymbol{\omega}}{\partial t} + \operatorname{curl}(\boldsymbol{\omega} \times \mathbf{v}) = 0 , \tag{12.33}$$

provides a direct relationship between the nonlinear term $\boldsymbol{\omega} \times \mathbf{v}$ and a term linear in the vorticity. Möhring (1978) shows that the sound field may often be expressed in such a form.

Möhring defines *a vector Green function*, $\mathbf{G}(\mathbf{y}, \tau | \mathbf{x}, t)$, by

$$\operatorname{curl} \mathbf{G} = \operatorname{grad} G . \tag{12.34}$$

If such a Green function exists, it enables Equation (12.27) to be rewritten as

$$p'(\mathbf{x}, t) = -\rho_0 \int_V (\boldsymbol{\omega} \times \mathbf{v}) \cdot \operatorname{curl} \mathbf{G} d^3 y d\tau . \tag{12.35}$$

After integration by parts, this becomes

$$p'(\mathbf{x}, t) = -\rho_0 \int \operatorname{curl}(\boldsymbol{\omega} \times \mathbf{v}) \cdot \mathbf{G} d^3 y d\tau , \tag{12.36}$$

and the vorticity Equation (12.34) shows that this simplifies to

$$p'(\mathbf{x}, t) = \rho_0 \int \frac{\partial \boldsymbol{\omega}}{\partial \tau} \cdot \mathbf{G} d^3 y d\tau .$$

After integration by parts, we obtain

$$\rho'(\mathbf{x}, t) = -\rho_0 \int \boldsymbol{\omega} \cdot \frac{\partial \mathbf{G}}{\partial \tau} d^3 y d\tau ,$$
$$= \rho_0 \frac{\partial}{\partial t} \int \boldsymbol{\omega} \cdot \mathbf{G} d^3 y d\tau , \tag{12.37}$$

since $\mathbf{G}$ depends on τ and t only in the combination $t - \tau$.

The representation for the acoustic density perturbation in (12.37) has a particularly convenient form. Unlike the Powell–Howe formulation in (12.15) it does not contain the flow velocity. Moreover, it depends linearly on the vorticity field.

This means that the contributions from several vortices just add linearly. There is, however, an initial difficulty with this approach. It relies on the existence of the vector Green function $\mathbf{G}$. Taking the divergence of Equation (12.34) and recalling that $\operatorname{div}\operatorname{curl}\mathbf{G} = 0$ shows that such a $\mathbf{G}$ can only exist of $\nabla^2 G = 0$. However, we have from the definition of G in (12.25)

$$\nabla^2 G = \frac{1}{c^2}\frac{\partial^2 G}{\partial \tau^2} - \delta(\mathbf{x} - \mathbf{y},\, t - \tau)\,. \tag{12.38}$$

We must conclude that a vector Green function $\mathbf{G}$ can only exist if the right–hand side of (12.38) vanishes! This δ–function is zero provided $\mathbf{x} \neq \mathbf{y}$. So this term vanishes if $\mathbf{y}$ is in the vortical flow and $\mathbf{x}$ is in the far field. The first term on the right–hand side of (12.38) is nonzero. However, since it involves the factor c^{-2}, it might perhaps vanish to lowest order in the Mach number. The way in which this works in practice is best illustrated by a number of particular examples.

As a first example, we will return to our discussion of the sound generated by vorticity in a low Mach–number flow in unbounded space. Equations (12.15) and (12.32) showed that

$$\rho'(\mathbf{x}, t) = -\frac{\rho_0}{4\pi c^4 |\mathbf{x}|^3}\frac{\partial^2}{\partial t^2}\int (\mathbf{x}\cdot\mathbf{y})\mathbf{x}\cdot\boldsymbol{\omega}\times\mathbf{v}\left(\mathbf{y}, t - \frac{|\mathbf{x}|}{c}\right)d^3\mathbf{y}\,. \tag{12.39}$$

To apply Möhring's method here we would want to find a vector function $\mathbf{G}$ such that $\operatorname{curl}\mathbf{G} = (\mathbf{x}\cdot\mathbf{y})\mathbf{x}$, remembering that the derivatives are, as always, with respect to $\mathbf{y}$. However, $\operatorname{div}((\mathbf{x}\cdot\mathbf{y})\mathbf{x}) = |\mathbf{x}|^2$ and so that is impossible: such a function will not exist. We can, however, find a vector $\mathbf{G}$ such that

$$\operatorname{curl}\mathbf{G} = (\mathbf{x}\cdot\mathbf{y})\mathbf{x} - \frac{1}{3}|\mathbf{x}|^2\mathbf{y}\,, \tag{12.40}$$

since now the divergence of the right–hand side of Equation (12.40) is zero. It is a matter of straightforward algebra to check that

$$\mathbf{G} = \frac{1}{3}(\mathbf{x}\cdot\mathbf{y})\mathbf{x}\times\mathbf{y} \tag{12.41}$$

satisfies (12.40). Substitution for $(\mathbf{x}\cdot\mathbf{y})\mathbf{x}$ in Equation (12.39) from (12.40) and (12.41) shows the far–field density perturbation to be given by

$$\rho'(\mathbf{x}, t) = -\frac{\rho_0}{12\pi c^4 |\mathbf{x}|^3}\frac{\partial^2}{\partial t^2}\left(\int \operatorname{curl}\left[(\mathbf{x}\cdot\mathbf{y})\mathbf{x}\times\mathbf{y}\right]\cdot\boldsymbol{\omega}\times\mathbf{v}\left(\mathbf{y}, t - \frac{|\mathbf{x}|}{c}\right)d^3\mathbf{y}\right.$$
$$\left. + |\mathbf{x}|^2\int \mathbf{y}\cdot\boldsymbol{\omega}\times\mathbf{v}\left(\mathbf{y}, t - \frac{|\mathbf{x}|}{c}\right)d^3\mathbf{y}\right)\,. \tag{12.42}$$

The first integral can be rearranged in the way indicated by the steps leading from Equation (12.35) to (12.37) to show

$$\int \text{curl} \left[(\mathbf{x} \cdot \mathbf{y}) \mathbf{x} \times \mathbf{y} \right] \boldsymbol{\omega} \times \mathbf{v} d^3 \mathbf{y} = -\frac{\partial}{\partial t} \int (\mathbf{x} \cdot \mathbf{y})(\mathbf{x} \times \mathbf{y}) \cdot \boldsymbol{\omega} d^3 \mathbf{y}$$

$$= -\frac{\partial}{\partial t} \int (\mathbf{x} \cdot \mathbf{y}) \mathbf{y} \cdot (\boldsymbol{\omega} \times \mathbf{x}) d^3 \mathbf{y} \ . \tag{12.43}$$

The second integral in Equation (12.42), $\int \mathbf{y} \cdot (\boldsymbol{\omega} \times \mathbf{v}) d^3 \mathbf{y}$, is equal to the total kinetic energy in the fluid (Lamb 1932, p.218) which is constant in an inviscid incompressible flow. The details are as follows. Note from the scalar product of the vector identity (12.3) with $\mathbf{y}$ that

$$\mathbf{y} \cdot (\boldsymbol{\omega} \times \mathbf{v}) = y_i v_j \frac{\partial v_i}{\partial y_j} - y_i \frac{\partial}{\partial y_i} \left(\frac{1}{2} v^2 \right) \ . \tag{12.44}$$

After using div $\mathbf{v} = 0$ and div $\mathbf{y} = 3$, this becomes

$$\mathbf{y} \cdot (\boldsymbol{\omega} \times \mathbf{v}) = \frac{1}{2} v^2 + \text{div} \left[(\mathbf{y} \cdot \mathbf{v}) \mathbf{v} - \frac{1}{2} v^2 \mathbf{y} \right] \ . \tag{12.45}$$

Integrating over all space and applying the divergence theorem to express the second term on the right–hand side of Equation (12.45) as a surface integral leads to

$$\int \mathbf{y} \cdot (\boldsymbol{\omega} \times \mathbf{v}) d^3 \mathbf{y} = \int \frac{1}{2} v^2 d^3 \mathbf{y} \ , \tag{12.46}$$

since $(\mathbf{y} \cdot \mathbf{v}) \mathbf{v} - \frac{1}{2} v^2 \mathbf{y}$ tends to zero faster than $|\mathbf{y}|^{-2}$ as $|\mathbf{y}|$ tends to infinity. In Equation (12.13) we proved that $\int \frac{1}{2} v^2 d^3 \mathbf{y}$ is constant in an unbounded, inviscid incompressible flow. Now we see from (12.46) that $\int \mathbf{y} \cdot (\boldsymbol{\omega} \times \mathbf{v}) d^3 \mathbf{y}$ is also unchanging. It therefore follows that the second integral in Equation (12.42) does not contribute to the far–field density fluctuations, and we have

$$\rho'(\mathbf{x}, t) = \frac{\rho_0}{12 \pi c^4 |\mathbf{x}|^3} \frac{\partial^3}{\partial t^3} \int (\mathbf{x} \cdot \mathbf{y}) \mathbf{y} \cdot (\boldsymbol{\omega} \times \mathbf{x}) d^3 \mathbf{y} \ , \tag{12.47}$$

where $\boldsymbol{\omega}$ is to be evaluated at regarded time $t - |\mathbf{x}|/c$. For a specified compact vorticity field, the integral in (12.47) can be readily calculated to determine the acoustic far field. This form shows that it is only components of vorticity perpendicular to the observer's position vector $\mathbf{x}$ that contribute to the sound field. A vortex element does not radiate in the direction of its axis.

Möhring (1978) used the form in Equation (12.47) to calculate the sound pressured generated by a circular vortex ring. Consider a ring of radius R whose centre is at a position $\mathbf{a}$, and let $\mathbf{n}$ be a unit vector normal to the ring, as shown in Figure 12.1.

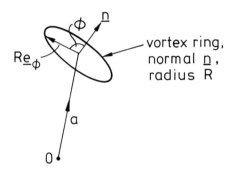

vortex ring, normal $\underline{n}$, radius R

Figure 12.1 The geometry of the ring.

The position of a typical element on the ring can be written as $\mathbf{y}_w = \mathbf{a} + R\mathbf{e}(\phi)$, $0 \le \phi \le 2\pi$, where $\mathbf{e}(\phi)$ denotes a radial unit vector in the ring's plane. The vorticity at $\mathbf{y}_w$ is orthogonal to both $\mathbf{n}$ and $\mathbf{e}$ can can be expressed in the form $\boldsymbol{\omega} = \Gamma \mathbf{n} \times \mathbf{e}\delta(\mathbf{y} - \mathbf{y}_w)$, where the δ–function is a two–dimensional one in the $\mathbf{n}$, $\mathbf{e}$ plane. Then

$$\mathbf{y} \times \boldsymbol{\omega} = \Big\{ (\mathbf{a}\cdot\mathbf{e})\mathbf{n} - (\mathbf{a}\cdot\mathbf{n})\mathbf{e} + R\mathbf{n} \Big\} \Gamma\delta(\mathbf{y} - \mathbf{y}_w) . \tag{12.48}$$

Equation(12.47) involves the integral

$$\int (\mathbf{x}\cdot\mathbf{y})\mathbf{x}\cdot(\mathbf{y} \times \boldsymbol{\omega})d^3\mathbf{y}$$

$$= \Gamma \int (\mathbf{x}\cdot\mathbf{a} + R\mathbf{x}\cdot\mathbf{e})\Big\{ (\mathbf{a}\cdot\mathbf{e})(\mathbf{x}\cdot\mathbf{n}) - (\mathbf{a}\cdot\mathbf{n})(\mathbf{x}\cdot\mathbf{e}) + R\mathbf{x}\cdot\mathbf{n} \Big\} \delta(\mathbf{y} - \mathbf{y}_w)d^3\mathbf{y} . \tag{12.49}$$

Noting that $\int e_i\delta(\mathbf{y} - \mathbf{y}_w)d^3\mathbf{y} = 0$ and $\int e_i e_j \delta(\mathbf{y} - \mathbf{y}_w)d^3\mathbf{y} = \pi R(\delta_{ij} - n_i n_j)$ enables this integral to be evaluated in a straightforward way, and we find

$$\int (\mathbf{x}\cdot\mathbf{y})\mathbf{x}\cdot(\mathbf{y} \times \boldsymbol{\omega})d^3\mathbf{y} = \Gamma R^2 \pi \Big\{ 3(\mathbf{x}\cdot\mathbf{a})(\mathbf{x}\cdot\mathbf{n}) - |\mathbf{x}|^2 \mathbf{a}\cdot\mathbf{n} \Big\} . \tag{12.50}$$

Finally, substitution into Equation (12.47) determines the distant pressure perturbation:

$$p'(\mathbf{x}, t) = \frac{\rho_0}{4c^2|\mathbf{x}|^3} \frac{\partial^3}{\partial t^3} \left[\Gamma R^2 \Big\{ (\mathbf{x}\cdot\mathbf{a})(\mathbf{x}\cdot\mathbf{n}) - \frac{1}{3}|\mathbf{x}|^2 \mathbf{a}\cdot\mathbf{n} \Big\} \right] . \tag{12.51}$$

This simplifies still further if the ring moves in the direction of its normal, $\mathbf{a} = z\mathbf{n}$, say. Then

$$p'(\mathbf{x}, t) = \frac{\rho_0}{4c^2|\mathbf{x}|^3} \frac{\partial^3}{\partial t^3} \left[\Gamma R^2 z \left\{ (\mathbf{x} \cdot \mathbf{n})^2 - \frac{1}{3} |\mathbf{x}|^2 \right\} \right] . \qquad (12.52)$$

The sound field has the directivity of that produced by a quadrupole in the $\mathbf{n} - \mathbf{n}$ direction, together with an isotropic quadrupole. If there are several rings their contributions to the sound pressure just add.

Kambe & Minota (1981 and 1983) investigate the axisymmetric collision of two equal but opposite vortex rings, both theoretically and experimentally. Suppose the common axis of the vortices is the 3–axis, and that one vortex is the image of the other in the plane $y_3 = 0$, as shown in Figure 12.2.

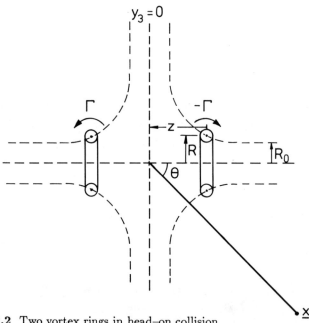

Figure 12.2 Two vortex rings in head–on collision.

It then follows from Equation (12.52) that the far–field pressure perturbation is given by

$$p'(\mathbf{x}, t) = \frac{\rho_0}{4c^2|\mathbf{x}|} \left(\cos^2 \theta - \frac{1}{3} \right) \dddot{Q}(t - |\mathbf{x}|/c) , \qquad (12.53)$$

where $Q(t) = -2\Gamma R^2 z(t)$ and $\cos \theta = x_3/|\mathbf{x}|$.

Equation (12.53) leads to a prediction that the magnitude of p' scales on $\rho_0 U_0^4 R_0/c^2|\mathbf{x}|$, where U_0 is the initial speed of the ring when its radius is R_0. This dependence was confirmed in Kambe & Minota's experiments. They also found that the sign of their measured perturbation changed near $\theta = 55°$. This is consistent with the predicted $\cos^2 \theta - \frac{1}{3}$ directivity since this function changes sign at $\theta = \cos^{-1}(3^{-1/2}) = 54.7°$. However, they note that as the vortex rings collide the inviscid, infinitesimally thin vortex ring model ceases to be valid, and the theory requires extension to include the finite size of the vortex cores and viscosity (Kambe 1986).

12.3 A LINE VORTEX NEAR A SEMI–INFINITE RIGID PLANE

Consider a line vortex with circulation Γ near the edge of a semi–infinite plate which occupies $x_2 = 0$, $x_1 < 0$ as shown in Figure 12.3.

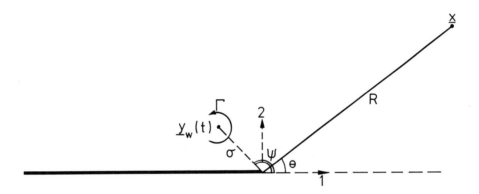

Figure 12.3 A line vortex near the edge of a semi– infinite plate.

This problem has been studied extensively as a model of turbulence passing near an edge (Crighton 1972, Howe 1975a and Möhring 1978). If the vortex position at time t is $\mathbf{y_w}(t)$, the vorticity can be written as

$$\boldsymbol{\omega} = \Gamma \delta(\mathbf{y} - \mathbf{y_w}(t))\mathbf{e_3} , \qquad (12.54)$$

where δ denotes a two–dimensional δ–function in the 1–2 plane and $\mathbf{e_3}$ is a unit vector in the 3–direction.

The Green function is to satisfy

$$\left(\frac{1}{c^2} \frac{\partial^2}{\partial \tau^2} - \frac{\partial^2}{\partial y_i \partial y_i} \right) G(\mathbf{y}, \tau | \mathbf{x}, t) = \delta(\mathbf{x} - \mathbf{y}, t - \tau) , \qquad (12.55)$$

with $\partial G/\partial y_2 = 0$ on $y_2 = 0$, $y_1 < 0$. This is a classical diffraction problem which is amenable to solution by the Wiener–Hopf technique discussed in Chapter 5 (see, for example, Crighton & Leppington 1970). Howe (1975a) rearranges this Green function and shows that for two–dimensional sources close to the edge, with the observer's position $\mathbf{x}$ in the far field, G can be written in the form

$$G(\mathbf{y}, \tau | \mathbf{x}, t) = \frac{\phi(\mathbf{x})\phi(\mathbf{y})}{\pi R} \, \delta \left(t - \tau - \frac{R}{c} \right) , \qquad (12.56)$$

where $R = (x_1^2 + x_2^2)^{1/2}$ and ϕ is the velocity potential for incompressible irrotational flow around a half–plane; $\phi(\mathbf{x}) = R^{1/2} \sin(\theta/2)$ with $\tan\theta = x_2/x_1$.

Möhring's vector Green function, $\mathbf{G}$, is to satisfy (12.34), $\mathrm{curl}\,\mathbf{G} = \mathrm{grad}\,G$. This Green function exists since

$$\frac{\partial^2 G}{\partial y_i \partial y_i} = \frac{\phi(\mathbf{x})}{\pi R} \left(\frac{\partial^2 \phi(\mathbf{y})}{\partial y_i \partial y_i} \right) \delta \left(t - \tau - \frac{R}{c} \right) = 0 .$$

$\mathbf{G}$ can be readily calculated and is given by

$$\mathbf{G}(\mathbf{y}, \tau | \mathbf{x}, t) = \frac{\phi(\mathbf{x})\Psi(\mathbf{y})}{\pi R} \, \delta \left(t - \tau - \frac{R}{c} \right) \mathbf{e}_3 , \qquad (12.57)$$

where $\Psi(\mathbf{y})$ is the stream function for a flow with velocity potential $\phi(\mathbf{y})$. In fact, $\Psi(\mathbf{y}) = -\sigma^{1/2} \cos(\psi/2)$ with $\sigma = (y_1^2 + y_2^2)^{1/2}$, $\tan\psi = y_2/y_1$.

Substitution for $\boldsymbol{\omega}$ and $\mathbf{G}$ from (12.54) and (12.57) respectively into Equation (12.37) gives the far–field pressure perturbation

$$p'(\mathbf{x}, t) = \frac{\rho_0 \Gamma \sin(\theta/2)}{\pi R^{1/2}} \frac{d}{dt} \, \Psi \left[\mathbf{y_w}(t - R/c) \right] . \qquad (12.58)$$

The noise–producing elements of the vortex motion are clearly displayed by Equation (12.58). The vortex generates sound as it crosses the streamlines, $\sigma^{1/2} \cos(\psi/2)$ = constant, of a hypothetical potential flow around the plate. Far from the edge, the vortex moves parallel to these streamlines and is silent. In the vicinity of the edge, the vortex path departs from the potential flow streamlines and the rate at which the vortex crosses these streamlines determines the pressure. The $\sin(\theta/2)$ directivity is typical of a half–plane scattering problem.

Kambe et al. (1985) investigate the three–dimensional version of this flow in which a vortex ring moves past the edge of a semi–infinite plate $x_2 = 0$, $-\infty < x_1 < 0$, $-\infty \leq x_3 \leq \infty$, as shown in Figure 12.4. There they find the calculated directivity to be $\sin(\theta/2)(\sin\phi)^{1/2}$ where $\tan\theta = x_2/x_1$, $\cos\phi = x_3/|\mathbf{x}|$. Their measurements in the plane $x_3 = 0$ are shown in Figure 12.5 and the experimental angular dependence is in excellent agreement with the predicted $\sin(\theta/2)$.

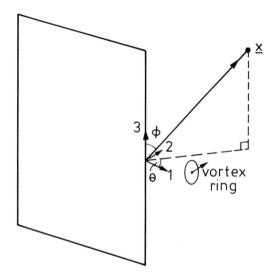

Figure 12.4 The geometry of Kambe et al.'s experiment.

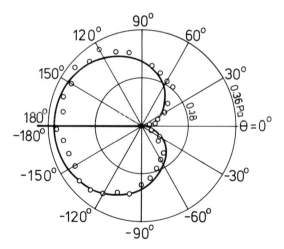

Figure 12.5 The directivity measured by Kambe et al. The circles show data and the curve is $\sin(\theta/2)$ (from Kambe et al. 1985).

12.4 VORTEX SOUND NEAR A CYLINDER

Let us now turn our attention to the generation of sound by vorticity near a fixed rigid cylinder, which is infinitely long with radius a. We introduce a cylindrical coordinate system and write $\mathbf{y} = (\sigma, \psi, y_3)$. Equation (12.27) shows that the distant pressure perturbation is given by

$$p'(\mathbf{x}, t) = -\rho_0 \int_V (\boldsymbol{\omega} \times \mathbf{v})_i \frac{\partial G}{\partial y_i} d^3\mathbf{y}\, d\tau \ , \tag{12.27}$$

where the Green function $G(\mathbf{y}, \tau | \mathbf{x}, t)$ satisfies

$$\left(\frac{1}{c^2} \frac{\partial^2}{\partial \tau^2} - \frac{\partial^2}{\partial y_i \partial y_i} \right) G(\mathbf{y}, \tau | \mathbf{x}, t) = \delta(\mathbf{x} - \mathbf{y}, t - \tau) \ , \tag{12.25}$$

with $\partial G / \partial \sigma = 0$ on $\sigma = a$.

An approximation to $G(\mathbf{y}, \tau | \mathbf{x}, t)$ may be calculated for use with compact fluctuations in vorticity in the way indicated by Howe (1975b). We decompose the Green function into two parts by writing $G = G_i + G_s$, where

$$G_i = \frac{\delta(t - \tau - |\mathbf{x} - \mathbf{y}|/c)}{4\pi |\mathbf{x} - \mathbf{y}|} \tag{12.59}$$

is the solution to (12.25) in unbounded space. For large $|\mathbf{x}|$ and $\mathbf{y}$ near the origin, the far–field approximations in (12.8) can be used to rewrite

$$G_i = \frac{\delta(t - \tau - |\mathbf{x}|/c + \mathbf{x} \cdot \mathbf{y}/|\mathbf{x}|/c)}{4\pi |\mathbf{x}|} \ . \tag{12.60}$$

A Taylor expansion of the δ–function yields

$$G_i = \frac{1}{4\pi |\mathbf{x}|} \delta(t - \tau - |\mathbf{x}|/c) + \frac{\mathbf{x} \cdot \mathbf{y}}{4\pi |\mathbf{x}|^2 c} \delta'(t - \tau - |\mathbf{x}|/c) + \dots \ . \tag{12.61}$$

The second term in this expansion leads to the first nonzero term in $\partial G / \partial y_i$ and if $\mathbf{y}$ is near a cylinder of compact radius, higher–order terms are unnecessary. It is convenient to express $\mathbf{x}$ in the spherical polar coordinates $(|\mathbf{x}|, \theta, \phi)$ illustrated in Figure 12.6.

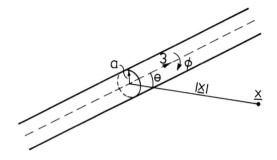

Figure 12.6 The position $\mathbf{x}$ in spherical polar coordinates $(|\mathbf{x}|, \theta, \phi)$.

Then

$$G_i = \frac{1}{4\pi|\mathbf{x}|}\delta(t - \tau - |\mathbf{x}|/c) + \frac{y_3\cos\theta + \sigma\sin\theta\cos(\psi - \phi)}{4\pi|\mathbf{x}|c}\delta'(t - \tau - |\mathbf{x}|/c) + \cdots .$$

$$(12.62)$$

G_i describes the field incident upon the cylinder, while G_s accounts for the scattered or reflected field.

For positions $\mathbf{y}$ close to a compact cylinder, the wave equation (12.25) reduces to $\nabla^2 G = 0$. It is evident that the first two terms in the expansion for G_i satisfy this equation. We need to find the scattered field G_s such that $\nabla^2 G_s = 0$, with $\partial G_s/\partial\sigma = -\partial G_i/\partial\sigma$ on $\sigma = a$ and $G_s \to 0$ as $\sigma \to \infty$. By inspection

$$G_s = \frac{a^2\sin\theta\cos(\psi - \phi)}{4\pi\sigma|\mathbf{x}|c}\delta'(t - \tau - |\mathbf{x}|/c) .$$

$$(12.63)$$

A combination of Equations (12.62) and (12.63) leads to

$$G = \frac{1}{4\pi|\mathbf{x}|}\delta(t - \tau - |\mathbf{x}|/c) + \frac{y_3\cos\theta + (\sigma + a^2/\sigma)\sin\theta\cos(\psi - \phi)}{4\pi|\mathbf{x}|c}\delta'(t - \tau - |\mathbf{x}|/c) + \cdots .$$

$$(12.04)$$

This approximation to G satisfies $\nabla^2 G = 0$, and so a vector Green function $\mathbf{G}(\mathbf{y}, \tau|\mathbf{x}, t)$ exists. Simple algebra shows a solution to curl $\mathbf{G} = \text{grad } G$ to be

$$\mathbf{G} = \frac{1}{4\pi|\mathbf{x}|c}\left(0, \frac{1}{2}\sigma\cos\theta, \left(\sigma - \frac{a^2}{\sigma}\right)\sin\theta\sin(\psi - \phi)\right)\delta'(t - \tau - |\mathbf{x}|/c) \qquad (12.65)$$

in cylindrical polar coordinates. Hence, it follows from Equation (12.37) that

$$p'(\mathbf{x}, t) = \frac{\rho_0\cos\theta}{8\pi|\mathbf{x}|c}\frac{\partial^2}{\partial t^2}\int[\sigma\omega_\psi]\,d^3\mathbf{y} + \frac{\rho_0\sin\theta}{4\pi|\mathbf{x}|c}\frac{\partial^2}{\partial t^2}\int\left[\left(\sigma - \frac{a^2}{\sigma}\right)\sin(\psi - \phi)\omega_3\right]d^3\mathbf{y}.$$

$$(12.66)$$

The first term on the right–hand side of (12.66) describes the sound field of an axial dipole. The strength of the dipole depends on the rate of change of $\int \sigma \omega_\psi d^3\mathbf{y}$, which is the axial impulse of the vorticity (Lamb 1932, p.214). Since an infinitely long cylinder cannot exert an axial force on an inviscid fluid, this component of the impulse of the vortex system is constant and, as we might expect, the axial dipole is identically zero. It therefore follows from (12.66) that the axial component of the vorticity, ω_3, must be nonzero if the vortex sound field is to be of dipole type (Obermeier 1980).

When the vorticity is concentrated into a ring C, with vortex strength Γ, the source term simplifies

$$\int \left(\sigma - \frac{a^2}{\sigma}\right) \sin(\psi - \phi)\omega_3 d^3\mathbf{y} = \Gamma \int_C \left(\sigma - \frac{a^2}{\sigma}\right) \sin(\psi - \phi) d\ell_3$$

$$= \Gamma \int_\Sigma \mathrm{curl}\left\{\left(\sigma - \frac{a^2}{\sigma}\right) \sin(\psi - \phi)\mathbf{e}_3\right\} \cdot d\Sigma ,$$

$$(12.67)$$

where Σ is a surface bounded by the curve C and $\mathbf{e}_3$ is a unit vector in the 3–direction. We can therefore write

$$p'(\mathbf{x}, t) = \frac{\rho_0 \Gamma \sin\theta}{4\pi |\mathbf{x}| c} \frac{\partial^2 \Phi}{\partial t^2}(t - |\mathbf{x}|/c) , \qquad (12.68)$$

where

$$\Phi = \int_\Sigma \mathrm{curl}\left\{\left(\sigma - \frac{a^2}{\sigma}\right) \sin(\psi - \phi)\mathbf{e}_3\right\} \cdot d\Sigma ,$$

$\mathrm{curl}\left\{(\sigma - a^2/\sigma) \sin(\psi - \phi)\mathbf{e}_3\right\}$ is the velocity field in a hypothetical potential flow around the cylinder, in which far from the cylinder there is a unit velocity in the ϕ–direction. The streamlines of such a flow are illustrated in Figure 12.7. The function Φ is the volume flux of this flow through the vortex ring, and we see from Equation (12.68) that the second time derivative of this volume flux generates the far–field pressure (Kambe et al. 1986).

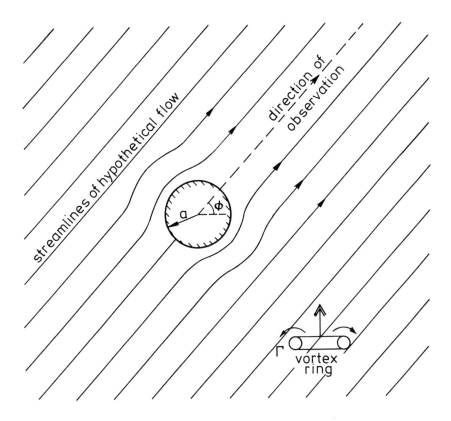

Figure 12.7 Vortex ring near a cylinder.

12.5 VORTEX SOUND NEAR COMPACT BODIES

We will now investigate the generation of sound by vorticity near a rigid compact body with surface S. We can find the Green function $G(\mathbf{y}, \tau | \mathbf{x}, t)$ which satisfies $\partial G / \partial n = 0$ on S in a similar way to that outlined in Section 12.4.

Let us suppose that the rigid body is near the origin, is small in comparison with the wavelength of the sound, and that the source at $\mathbf{y}$ is located well within a wavelength of the body. This means that the flow is essentially incompressible in the vicinity of $\mathbf{y}$. We will take the observer's position $\mathbf{x}$ to be in the far field, and again decompose G into its incident and scattered components by writing $G = G_i + G_s$. G_i is the solution to the wave Equation (12.25) in unbounded space,

$$G_i = \frac{\delta(t - \tau - |\mathbf{x} - \mathbf{y}|/c)}{4\pi |\mathbf{x} - \mathbf{y}|} , \tag{12.69}$$

which for **y** near the origin can be expanded as

$$G_i = \frac{1}{4\pi|\mathbf{x}|}\delta(t - \tau|\mathbf{x}|/c) + \frac{\mathbf{x}\cdot\mathbf{y}}{4\pi|\mathbf{x}|^2 c}\,\delta'(t - \tau - |\mathbf{x}|/c) + \dots . \tag{12.70}$$

The **y**–dependence in G_i is like that in the velocity potential of a uniform flow in the direction **x**. G_s contains the perturbation to this velocity potential induced by the presence of the body, and we can write

$$G(\mathbf{y}, \tau|\mathbf{x}, t) = \frac{1}{4\pi|\mathbf{x}|}\delta(t - \tau - |\mathbf{x}|/c) + \frac{\mathbf{x}\cdot(\mathbf{y} + \boldsymbol{\Phi})}{4\pi|\mathbf{x}|^2 c}\delta'(t - \tau - |\mathbf{x}|/c) ; \tag{12.71}$$

$\Phi_i(\mathbf{y})$ is the scattered velocity potential due to unit flow in i–direction. It satisfies $\nabla^2\Phi_i = 0$, with $\Phi_i \sim 0(|\mathbf{y}|^{-2})$ for $|\mathbf{y}|$ large and $\partial\Phi_i/\partial n = -n_i$ on S. This form for Φ_i ensures that G satisfies the required boundary conditions on S.

We can differentiate the expression for $G(\mathbf{y}, \tau|\mathbf{x}, t)$ in (12.71) and substitute into Equation (12.27). Such a procedure clearly demonstrates the dipole–like directivity of the radiated sound field, but the integrand depends on the fluid velocity **v**. The need to calculate **v** can be circumvented by the use of a vector Green function.

Differentiation of the Green function in (12.71) shows that $\partial G/\partial y_i$ can be written in the form

$$\frac{\partial}{\partial y_i}G(\mathbf{y}, \tau|\mathbf{x}, t) = \frac{\mathbf{U}(\mathbf{y}|\mathbf{x})}{4\pi|\mathbf{x}|c}\,\delta'(t - \tau - |\mathbf{x}|/c) . \tag{12.72}$$

where

$$\mathbf{U}(\mathbf{y}|\mathbf{x}) = \nabla_\mathbf{y}(\mathbf{x}\cdot(\mathbf{y} + \boldsymbol{\Phi}))/|\mathbf{x}| . \tag{12.73}$$

$\mathbf{U}(\mathbf{y}|\mathbf{x})$ is the dimensionless velocity at **y** in a hypothetical, irrotational, incompressible flow field in which the velocity far from the body is in the **x**–direction and of unit magnitude. Since div $\mathbf{U}(\mathbf{y}|\mathbf{x}) = 0$ (where the derivatives are as usual with respect to **y**), it is possible to introduce a vector Green function. Indeed, if $\boldsymbol{\Psi}(\mathbf{y}|\mathbf{x})$ is a vector potential for $\mathbf{U}(\mathbf{y}|\mathbf{x})$, i.e. if $\mathbf{U}(\mathbf{y}|\mathbf{x}) = \operatorname{curl}\boldsymbol{\Psi}(\mathbf{y}|\mathbf{x})$,

$$G(\mathbf{y}, |\mathbf{x}, t) = \frac{\boldsymbol{\Psi}(\mathbf{y}|\mathbf{x})}{4\pi|\mathbf{x}|c}\,\delta'(t - \tau - |\mathbf{x}|/c) . \tag{12.74}$$

Substitution for **G** into Equation (12.37) gives

$$p'(\mathbf{x}, t) = \frac{\rho_0}{4\pi|\mathbf{x}|c}\frac{\partial^2}{\partial t^2}\int \boldsymbol{\omega}\cdot\boldsymbol{\Psi}(t - |\mathbf{x}|/c)d^3\mathbf{y} . \tag{12.75}$$

An example makes the implications of this result clear.

For a sphere of radius a, $\boldsymbol{\Phi} = \frac{1}{2}a^3\mathbf{y}/|\mathbf{y}|^3$. $\mathbf{U}(\mathbf{y}|\mathbf{x})$ has a simple form if we choose to express $\mathbf{y}$ in terms of spherical polar coordinates (σ, θ, ϕ), where θ is the angle between $\mathbf{y}$ and $\mathbf{x}$, and ϕ is the azimuthal angle of $\mathbf{y}$ about the axis $\mathbf{x}$, as shown in Figure 12.8. Then,

$$\mathbf{x}\cdot(\mathbf{y}+\boldsymbol{\Phi})/|\mathbf{x}| = \cos\theta(\sigma + \frac{1}{2}a^3/\sigma^2) \quad .$$

In spherical polar coordinates

$$\mathbf{U}(\mathbf{y}|\mathbf{x}) = \left(\cos\theta(1 - a^3/\sigma^3), \; -\sin\theta(1 + \frac{1}{2}a^3/\sigma^3), 0 \right)$$

and can be expressed as curl $\boldsymbol{\Psi}$ where

$$\boldsymbol{\Psi} = \left(0, 0, \frac{1}{2}\sigma \sin\theta(1 - a^3/\sigma^3) \right) \quad . \tag{12.76}$$

When this is used in (12.75), it leads to

$$p'(\mathbf{x}, t) = \frac{\rho_0}{8\pi|\mathbf{x}|c} \frac{\partial^2}{\partial t^2} \int \omega_\phi \sigma \sin\theta \left(1 - \frac{a^3}{\sigma^3} \right) d^3\mathbf{y} \quad . \tag{12.77}$$

$$= \frac{\rho_0}{4\pi|\mathbf{x}|c} \frac{\partial^2}{\partial t^2} \int \omega_\phi \psi \frac{d^3\mathbf{y}}{\sigma \sin\theta} \quad , \tag{12.78}$$

where ω_ϕ is the azimuthal component of the vorticity and $\psi = \frac{1}{2}\sigma^2 \sin^2\theta(1 - a^3/\sigma^3)$ is the streamfunction for a potential flow around the sphere in the $\mathbf{x}$–direction (Batchelor 1967, p.452).

This result has been obtained by Obermeier (1980) by the method of matched asymptotic expansions. As Obermeier points out, it states that the sound pressure radiated in a direction $\mathbf{x}$ depends on the rate of change in the rate at which the azimuthal component of the vorticity crosses the streamlines of a hypothetical flow around the sphere.

376

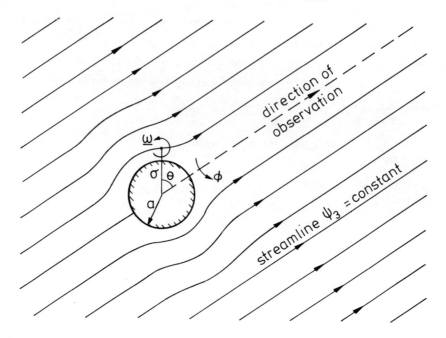

Figure 12.8 Vorticity near a sphere.

REFERENCES

Batchelor, G.K. (1967). An Introduction to Fluid Mechanics. Cambridge University Press.

Crighton, D.G. (1972). Radiation from vortex filament motion near a half plane. J. Fluid Mech. 51:357–362.

Crighton, D.G. & Leppington, F.G. (1970). Scattering of aerodynamic noise by a semi–infinite compliant plate. J. Fluid Mech. 43:721–736.

Crow, S.C. (1970). Aerodynamic sound emission as a singular perturbation problem. Stud. Appl. Maths. 49:21–44.

Howe, M.S. (1975a). Contributions to the theory of aerodynamic sound, with application to excess jet noise and the theory of the flute. J. Fluid Mech. 71:625–673.

Howe, M.S. (1975b). The generation of sound by aerodynamic sources in an inhomogeneous steady flow, J. Fluid Mech. 67:597–610.

Kambe, T. (1986). Acoustic emissions by vortex motions. J. Fluid Mech. 173:643–666.

Kambe, T. & Minota, T. (1981). Sound radiation from vortex systems. J. Sound.

Vib. 74:61–72.

Kambe, T. & Minota, T. (1983). Acoustic wave radiated by head–on collision of two vortex rings. Proc. Roy. Soc. A 386:277–308.

Kambe, T., Minota, T. & Ikushima, Y. (1985). Acoustic wave emitted by a vortex ring passing near the edge of a half–plane. J. Fluid Mech. 155:77–103.

Kambe, T., Minota, T. & Ikushima, Y. (1986). Acoustic waves emitted by vortex–body interaction. Proc. IUTAM Symposium on Aero– and Hydro–Acoustics. Springer–Verlag, pp.21–28.

Lamb, H. (1932). Hydrodynamics. Cambridge University Press.

Möhring, W. (1978). On vortex sound at low Mach number. J. Fluid Mech. 85:685–691.

Obermeier, F. (1980). The influence of solid bodies on low Mach number vortex sound. J. Sound Vib. 72:39–49.

Powell, A. (1964). Theory of vortex sound. J. Acoust. Soc. Amer. 36:177–195.

13. THERMOACOUSTIC SOURCES AND INSTABILITIES

13.1 INTRODUCTION TO THERMOACOUSTIC SOURCES

Thermoacoustics deals with the acoustics of flows in which the variation of entropy plays a significant role. A range of processes are thermoacoustic sources. For example, unsteady combustion, diffusion of heat and mass and turbulent two–phase flows all generate sound. Chapters 11 and 12 demonstrate how the Lighthill theory provides a convenient description of sound generation. There we saw in Equation (11.46), for example, that the Navier–Stokes equation and the equation of mass conservation may be combined to give an inhomogeneous wave equation for the density fluctuations. When investigating thermoacoustic source processes it is convenient to use the pressure perturbation as the dependent variable.

The Lighthill equation (11.46) states that

$$\frac{\partial^2 \rho}{\partial t^2} - c_0^2 \nabla^2 \rho = \frac{\partial^2}{\partial x_i \partial x_j}\left(\rho u_i u_j + \left((p - p_0) - c_0^2(\rho - \rho_0)\right)\delta_{ij} - \tau_{ij}\right) , \qquad (13.1)$$

where ρ is the density, p the pressure, $\mathbf{u}$ the particle velocity, c the sound speed and τ_{ij} the viscous stress tensor. The suffix 0 denotes a mean value in the distant acoustic field. Equation (13.1) can be rearranged to give

$$\frac{1}{c_0^2}\frac{\partial^2 p}{\partial t^2} - \nabla^2 p = \frac{\partial^2}{\partial x_i \partial x_j}\left(\rho u_i u_j - \tau_{ij}\right) - \frac{\partial^2 \rho_e}{\partial t^2} , \qquad (13.2)$$

where ρ_e is the "excess" density

$$\rho_e = \rho - \rho_0 - (p - p_0)/c_0^2 . \qquad (13.3)$$

ρ_e vanishes in the far field but is nonzero in regions where the entropy is significantly different from ambient. Chapters 11 and 12 concentrated on the sound generated by the quadrupole source term $\rho u_i u_j$; the remaining source terms $-\partial^2 \rho_e/\partial t^2 - \partial^2 \tau_{ij}/\partial x_i \partial x_j$ are the topic of this Chapter. They are "thermoacoustic" sources and are nonzero in regions of appreciable irreversible processes. To determine the strength of these sources we need to investigate the thermodynamics of the source region. It is shown in the Appendix 13A that the energy equation for a gas made up of N (possibly reacting) species leads to a relationship of the form

$$\frac{D\rho}{Dt} = \frac{1}{c^2}\frac{Dp}{Dt} + \frac{\alpha}{c_p}\left(\sum_{n=1}^{N}\left.\frac{\partial h}{\partial Y_n}\right|_{\rho,p,Y_m}\rho\frac{DY_n}{DT} + \nabla\cdot\mathbf{q} - \frac{\partial u_i}{\partial x_j}\tau_{ij}\right) ; \qquad (13.4)$$

Y_n is the mass fraction of the nth species, h the enthalpy, $\mathbf{q}$ the heat flux. α is the volumetric expansion coefficient and for an ideal gas is equal to T^{-1}. Conservation of species can be expressed as

$$\rho\frac{DY_n}{Dt} = w_n - \nabla\cdot\mathbf{J}_n , \qquad (13.5)$$

where w_n is the production rate per unit volume of species n by reaction, and $\mathbf{J}_n$ is the flux of species n by diffusion. Equation (13.4) demonstrates that the density of a material particle changes not only due to pressure variations in a compressible fluid, but also because of the expansion caused by heating.

Before the thermodynamic relationship in (13.4) can be used to substitute for $\partial^2 \rho_e/\partial t^2$ in (13.2), one of the partial time derivatives must be replaced by a material derivative. Straightforward algebraic manipulation leads to

$$\frac{\partial\rho_e}{\partial t} = \frac{D\rho_e}{Dt} - \frac{\rho_e}{\rho}\frac{D\rho}{Dt} - \nabla\cdot(\mathbf{u}\rho_e) , \qquad (13.6)$$

where the continuity equation has been used to replace $\nabla\cdot\mathbf{u}$ by $-\rho^{-1}D\rho/Dt$. Once ρ_e in (13.6) is written explicitly as $\rho - \rho_0 - (p-p_0)/c_0^2$, Equation (13.4) can be used to replace $D\rho/Dt$. This leads to

$$\frac{\partial\rho_e}{\partial t} = \frac{\alpha\rho_0}{c_p\rho}\left(\sum_{n=1}^{N}\left.\frac{\partial h}{\partial Y_n}\right|_{\rho,p,Y_m}\rho\frac{DY_n}{Dt} + \nabla\cdot\mathbf{q} - \frac{\partial u_i}{\partial x_j}\tau_{ij}\right) - \nabla\cdot(\mathbf{u}\rho_e)$$

$$\qquad (13.7)$$

$$-\frac{1}{c_0^2}\left(\left(1 - \frac{\rho_0 c_0^2}{\rho c^2}\right)\frac{Dp}{Dt} - \frac{(p-p_0)}{\rho}\frac{D\rho}{Dt}\right) .$$

The inhomogeneous wave equation in (13.2) is therefore equivalent to

$$\frac{1}{c_0^2}\frac{\partial^2 p}{\partial t^2} - \nabla^2 p = -\frac{\partial}{\partial t}\left(\frac{\alpha\rho_0}{c_p\rho}\left(\sum_{n=1}^{N}\frac{\partial h}{\partial Y_n}\Big|_{\rho,p,Y_m}\rho\frac{DY_n}{Dt} + \nabla\cdot\mathbf{q} - \frac{\partial u_i}{\partial x_j}\tau_{ij}\right)\right)$$

$$+ \frac{\partial^2}{\partial x_i\partial x_j}(\rho u_i u_j - \tau_{ij})$$

(13.8)

$$+ \frac{1}{c_0^2}\frac{\partial}{\partial t}\left(\left(1 - \frac{\rho_0 c_0^2}{\rho c^2}\right)\frac{Dp}{Dt} - \frac{(p-p_0)}{\rho}\frac{D\rho}{Dt}\right)$$

$$+ \frac{\partial^2}{\partial x_i\partial t}(u_i\rho_e)\ .$$

We can use the free–space Green function, $\delta(t-\tau-|\mathbf{x}-\mathbf{y}|/c_0)/(4\pi|\mathbf{x}-\mathbf{y}|)$, to write down the solution to this equation. That solution is

$$(p-p_0)(\mathbf{x},t) = -\frac{\partial}{\partial t}\int\left[\frac{\alpha\rho_0}{c_p\rho}\left(\sum_{n=1}^{N}\frac{\partial h}{\partial Y_n}\Big|_{\rho,p,Y_m}\rho\frac{DY_n}{Dt} + \nabla\cdot\mathbf{q} - \frac{\partial u_i}{\partial x_j}\tau_{ij}\right)\right]\frac{d^3\mathbf{y}}{4\pi|\mathbf{x}-\mathbf{y}|}$$

$$+ \frac{\partial^2}{\partial x_i\partial x_j}\int\left[\rho u_i u_j - \tau_{ij}\right]\frac{d^3\mathbf{y}}{4\pi|\mathbf{x}-\mathbf{y}|}$$

$$+ \frac{1}{c_0^2}\frac{\partial}{\partial t}\int\left[\left(1 - \frac{\rho_0 c_0^2}{\rho c^2}\right)\frac{Dp}{Dt} - \frac{(p-p_0)}{\rho}\frac{D\rho}{Dt}\right]\frac{d^3\mathbf{y}}{4\pi|\mathbf{x}-\mathbf{y}|}$$

$$+ \frac{\partial^2}{\partial x_i\partial t}\int\left[u_i\rho_e\right]\frac{d^3\mathbf{y}}{4\pi|\mathbf{x}-\mathbf{y}|}\ .$$

(13.9)

The square brackets denote that the functions they enclose are to be evaluated at a retarded time $t - |\mathbf{x}-\mathbf{y}|/c_0$. When $\mathbf{x}$ is in the far field, the approximations in Equation (12.8) show that the retarded time simplifies to $t - |\mathbf{x}|/c_0 + \mathbf{x}\cdot\mathbf{y}/(|\mathbf{x}|c_0)$, that $|\mathbf{x}-\mathbf{y}|^{-1}$ is approximately $|\mathbf{x}|^{-1}$ and that the derivative $\partial/\partial x_i$ is equivalent to $-x_i/(|\mathbf{x}|c_0)\partial/\partial t$. Then

$$4\pi|\mathbf{x}|(p-p_0)(\mathbf{x},t) = -\frac{\partial}{\partial t}\int\left[\frac{\alpha\rho_0}{c_p\rho}\left(\sum_{n=1}^{N}\frac{\partial h}{\partial Y_n}\bigg|_{\rho,p,Y_m}\rho\frac{DY_n}{Dt} + \nabla\cdot\mathbf{q} - \frac{\partial u_i}{\partial x_j}\tau_{ij}\right)\right]d^3\mathbf{y}$$

$$+\frac{x_i x_j}{|\mathbf{x}|^2 c_0^2}\frac{\partial^2}{\partial t^2}\int\left[\rho u_i u_j - \tau_{ij}\right]d^3\mathbf{y}$$

$$+\frac{1}{c_0^2}\frac{\partial}{\partial t}\int\left[\left(1-\frac{\rho_0 c_0^2}{\rho c^2}\right)\frac{Dp}{Dt} - \frac{(p-p_0)}{\rho}\frac{D\rho}{Dt}\right]d^3\mathbf{y}$$

$$-\frac{x_i}{|\mathbf{x}|c_0}\frac{\partial^2}{\partial t^2}\int\left[u_i\rho_e\right]d^3\mathbf{y}\,.$$

$$(13.10)$$

No approximation (other than $|\mathbf{x}|$ large) has been made in deriving this expression. It conveniently displays the effect of all the thermoacoustic sources, and enables us to review the literature on thermoacoustic source mechanisms within a unified framework. The first term on the right–hand side of Equation (13.10) describes the sound generated by irreversible flow processes and is of monopole type. The second term involves the familiar quadrupole source of Lighthill's jet–noise theory. It is well known that it leads to an acoustic intensity that scales on Mach number to the eighth power. The third term is appreciable if there are regions of unsteady flow with different mean density and sound speed from the ambient fluid. We will see that this term also leads to an M^8 scaling, but with a different coefficient from the jet–noise term. The last term in (13.10) is dipole in nature and describes the effect of momentum changes of density inhomogeneities. We will use this equation to investigate the effects of these various sources.

13.2 COMBUSTION NOISE

In a reacting gas w_n, the rate of production species n, is nonzero and the first term on the right–hand side of Equation (13.10) describes the strong monopole sound of combustion. When combustion occurs unsteadily in a low Mach number flow, this term is far larger than the sound due to the other source mechanisms demonstrated in Equation (13.10) and we can write the distant sound field as

$$4\pi|\mathbf{x}|(p - p_0)(\mathbf{x}, t) = -\frac{\partial}{\partial t} \int \left[\frac{\alpha\rho_0}{c_p\rho} \sum_{n=1}^{N} \frac{\partial h}{\partial Y_n}\bigg|_{\rho,p,Y_m} w_n \right] d^3\mathbf{y} \; . \tag{13.11}$$

For an ideal gas $\alpha/c_p = (\gamma - 1)/c^2$. ρ and c denote the actual density and sound speed in the combustion zone and, since the gas there is hot, differ from the far–field values ρ_0 and c_0. It is shown in the Appendix that if the gas has constant average molecular weight, a good approximation for the combustion of hydrocarbons in air,

$$-\sum_{n=1}^{N} \frac{\partial h}{\partial Y_n}\bigg|_{\rho,p,Y_m} w_n = Q \; , \tag{13.12}$$

where Q is the heat release rate per unit volume. The combustion noise may therefore be rewritten as

$$4\pi|\mathbf{x}|(p - p_0)(\mathbf{x}, t) = \frac{\partial}{\partial t} \int \left[\frac{\rho_0(\gamma - 1)Q}{\rho c^2} \right] d^3\mathbf{y} \; . \tag{13.13}$$

When γ is assumed to be independent of temperature and the combustion takes place at ambient pressure, $\rho c^2 = \gamma p_0 = \rho_0 c_0{}^2$ and this expression for the acoustic pressure perturbation simplifies to

$$4\pi|\mathbf{x}|(p - p_0)(\mathbf{x}, t) = \frac{\gamma - 1}{c_0{}^2} \frac{\partial}{\partial t} \int [Q] d^3\mathbf{y} \; . \tag{13.14}$$

Equation (13.14) predicts that for a compact flame the far–field acoustic pressure perturbation should be proportional to the time derivative of the total rate of heat release evaluated at retarded time. Hurle et al. (1968) confirmed this relationship experimentally by using the light emission from short–lived CH or C_2 free radicals to monitor the rate of combustion. Since their pioneering work, the relationship has been verified using more advanced signal processing (see, for example, Shivashankara et al. 1975).

Equation (13.13) states that the far–field acoustic pressure perturbation is proportional to ρ_0, the density of the surroundings. Let us compare this prediction from the acoustic analogy with the exact solutions of two simple problems.

Examples

(i) Consider combustion occurring with a total rate of heat release $\widehat{Q}e^{-i\omega t}$, concentrated at the origin. The gas within a sphere of radius a is hot and has mean density ρ_1 and sound speed c_1, while the surrounding colder gas has density ρ_0 and sound speed c_0 as shown in Figure 13.1.

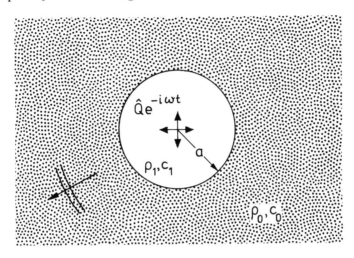

Figure 13.1 Geometry of the simple model problem.

The pressure perturbation is therefore to satisfy the equations

$$\frac{1}{c_1^2}\frac{\partial^2 p}{\partial t^2} - \nabla^2 p = -\frac{i\omega(\gamma-1)}{c_1^2}\,\widehat{Q}e^{-i\omega t}\delta(\mathbf{x}) \quad \text{in } r < a \tag{13.15a}$$

and

$$\frac{1}{c_0^2}\frac{\partial^2 p}{\partial t^2} - \nabla^2 p = 0 \qquad \qquad \text{in } r > a . \tag{13.15b}$$

The general solution to (13.15a) has the form

$$(p-p_0)(r,t) = \left(Ie^{i\omega r/c_1} + Re^{-i\omega r/c_1}\right)\frac{e^{-i\omega t}}{r} \quad \text{in } r \leq a ,$$

and integrating (13.15a) over a small sphere centred on the origin shows that

$$I + R = -\frac{i\omega(\gamma-1)}{4\pi c_1{}^2}\,\widehat{Q} . \tag{13.16}$$

In $r \geq a$ the pressure perturbation can be expressed as

$$(p-p_0)(r,t) = \frac{T}{r}\,e^{-i\omega(t-r/c_0)} . \tag{13.17}$$

It is a matter of straightforward algebra to use Equation (13.16) and conditions of continuity of pressure and particle velocity across $r = a$ to determine the coefficients I, R and T in terms of $\widehat{Q}$. The results simplify when the hot region is compact. Then the solution reduces to

$$(p - p_0)(r, t) = -\frac{\rho_0 i\omega(\gamma - 1)}{4\pi r \rho_1 c_1{}^2} \widehat{Q} e^{-i\omega(t - r/c_0)} \qquad \text{in } r \geq a \quad (13.18a)$$

$$= -\frac{i\omega(\gamma - 1)}{4\pi r c_1{}^2} \widehat{Q} e^{-i\omega t} \left\{ 1 + \frac{r}{a} \left(\frac{\rho_0}{\rho_1} - 1 \right) \right\} \quad \text{in } r \leq a \quad (13.18b)$$

for $\omega a/c_0$ and $\omega a/c_1$ both small in comparison with unity. We see that throughout $r \geq a$ the solution of the exact problem is identical to the far–field pressure predicted from the representation theory (Equation (13.13)). The distant acoustic pressure perturbation is proportional to $\rho_0 \omega(\gamma - 1) \widehat{Q}/\rho_1 c_1^2$. A heavier fluid surrounding the combustion zone increases the radiated acoustic energy by a factor $\rho_0 c_1/\rho_1 c_0$ from the value it would have if the combustion occurred in a space occupied entirely by ρ_1, c_1 fluid. It is interesting to note how large this factor is when the combustion takes place in an air pocket surrounded by steel or water!

(ii) We will now discuss a second but similar example with spherical symmetry, which again has a simple exact solution. Unsteady combustion with a total rate of heat release $\widehat{Q} e^{-i\omega t}$ is distributed uniformly throughout the sphere $r \leq a$. Again the mean density and sound speed have values ρ_1, c_1 within $r < a$ and ρ_0, c_0 in $r > a$. The wave equation in $r < a$ is then

$$\frac{1}{c_1{}^2} \frac{\partial^2 p}{\partial t^2} - \nabla^2 p = -\frac{i\omega(\gamma - 1)3}{4\pi a^3 c_1{}^2} \widehat{Q} e^{-i\omega t} . \qquad (13.19)$$

This is to be solved together with (13.15b) and conditions of continuity of pressure and particle velocity across $r = a$. When the combustion zone is compact, the solution in $r \geq a$ again has the form expressed in (13.18a), thereby verifying again the results obtained from the representation theorem.

We have obtained an expression for the sound generation by combustion noise,

$$(p - p_0)(\mathbf{x}, t) = \frac{1}{4\pi |\mathbf{x}|} \frac{\partial}{\partial t} \int \left[\frac{\rho_0(\gamma - 1)Q}{\rho c^2} \right] d^3\mathbf{y} , \qquad (13.13)$$

and checked it by comparison with two exact solutions. We now go on to compare it with existing theories of combustion noise. Chiu & Summerfield (1974) also find the distant acoustic pressure perturbation to be proportional to $\rho_0/(\rho_f c^2)$. (In their notation ρ_f is the density at the flame.) But since they have expressed the time derivative as a material derivative, their source term appears more complicated. Ffowcs Williams (1982) discusses linear source processes and Equation (13.13) is in agreement with his expression for sound generation by the unsteady addition of heat in a fluid with a different density and sound speed from the far field.

Strahle (1972) relates the sound field to the reaction rate and obtains

$$(p - p_0)(\mathbf{x}, t) = \frac{\gamma - 1}{4\pi|\mathbf{x}|c_1{}^2} \frac{\partial}{\partial t} \int [Q] d^3 \mathbf{y} \tag{13.20}$$

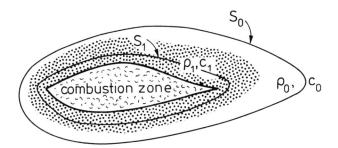

Figure 13.2 The geometry of the flame.

ρ_1 and c_1 are the mean density and sound speed in the hot gas after the combustion, as shown in Figure 13.2. The combustion is considered to be in a perfect gas at constant pressure so that $\rho c^2 = \rho_1 c_1{}^2 = \rho_0 c_0{}^2$. A comparison of (13.13) and (13.20) shows that (13.20) is a factor $c_0{}^2/c_1{}^2 = \rho_1/\rho_0$ smaller than (13.13). This factor is about 1/6 for the combustion of hydrocarbons in air. The difference does not affect the scaling laws derived by Strahle (1972). But, since both Equations (13.13) and (13.20) have been derived from exact representations of the sound field and should be identical, this discrepancy must be resolved.

Strahle (1972) has a monopole source whose strength is the time derivative of

$$\int_V \frac{\partial \rho}{\partial t} d^3 \mathbf{y} \ .$$

The integral is to be evaluated over the source region V in which $\rho - \rho_0$ is appreciably different from $(p - p_0)/c_0{}^2$. From the equation of mass conservation

$$\int_V \frac{\partial \rho}{\partial t} d^3 \mathbf{y} = - \int_S \rho \, \mathbf{u} \cdot d\mathbf{S} \ , \tag{13.21}$$

where S is the surface enclosing the source region V. Strahle takes this bounding surface S to lie in the hot gas, i.e. the surface S_1 in Figure 13.2. On this surface ρ is nearly equal to ρ_1, and so

$$\int_S \rho \, \mathbf{u} \cdot \mathbf{S} \simeq \rho_1 \int_{S_1} \mathbf{u} \cdot d\mathbf{S} \ . \tag{13.22}$$

Strahle goes on to show how $\int \mathbf{u} \cdot d\mathbf{S}$ over any near–field surface enclosing the combustion zone can be related to the heat release rate to obtain the result in (13.20). However, S_1 does not enclose all the acoustic sources. Outside it ρ has strong gradients as its mean value changes from ρ_1 to ρ_0. Acoustic waves travelling through this region move the interface and produce large values of $\partial\rho/\partial t$. To include these sources the bounding surface S should be taken to lie in the ρ_0 fluid i.e. to coincide with surfaces S_0 in Figure 13.2. For this position of S

$$\int_S \rho \mathbf{u} \cdot d\mathbf{S} \simeq \rho_0 \int_{S_0} \mathbf{u} \cdot d\mathbf{S} \ . \tag{13.23}$$

A comparison of Equations (13.22) and (13.23) shows that if all the acoustic sources are to be included, Strahle's result (13.20) should be multiplied by a factor ρ_0/ρ_1. It is then in agreement with Equation (13.13).

Hassan (1974) also obtains an expression for the far–field density perturbation in terms of the heat release rate. His result is

$$(p - p_0)(\mathbf{x}, t) = \frac{\gamma - 1}{4\pi|\mathbf{x}|} \frac{\partial}{\partial t} \int \left[\frac{Q}{c^2}\right] d^3\mathbf{y} \ . \tag{13.24}$$

Comparison with (13.13) shows that it differs from our result by a multiplicative factor ρ/ρ_0. This discrepancy can be resolved quickly because a source term, which is not small, has been neglected by Hassan. He retains $-(\gamma - 1)Q/c^2$ as his main source term, but neglects $\rho_e \nabla \cdot \mathbf{u}$. In the combustion zone, however, $\nabla \cdot \mathbf{u}$ is related to the heat release rate per unit volume by Equation (13.4) and is not negligible;

$$\rho_e \nabla \cdot \mathbf{u} \simeq -\frac{\rho - \rho_0}{\rho} \frac{D\rho}{Dt} \simeq \frac{\rho - \rho_0}{\rho c^2} (\gamma - 1) Q \ .$$

Adding to this term, $-(\gamma - 1)Q/c^2$, retained by Hassan gives a total combustion source $-(\gamma - 1)\rho_0 \, Q/\rho c^2$, a factor ρ_0/ρ larger than Hassan's source and in agreement with our expression.

In Equation (13.13), then, we have an expression for the sound field radiated by unsteady combustion. This has been checked by comparison with the exact solutions of two problems with simple geometries, and we have been able to reconcile apparent differences in the literature of combustion noise.

Our form for the radiated pressure can be used to derive scaling laws for the acoustic power. Let us assume that combustion occurs in a perfect gas at constant pressure. Then $\rho c^2 = \rho_0 c_0{}^2$ and Equation (13.14) is the appropriate form for the distant pressure. The acoustic power can be found by integrating the intensity over a large sphere. For a compact source region, this gives

$$\text{acoustic power} = \frac{\overline{(p - p_0)^2}\, 4\pi |\mathbf{x}|^2}{\rho_0 c_0}$$

(13.25)

$$= \frac{(\gamma - 1)^2}{4\pi \rho_0 c_0{}^5} \int \overline{\frac{\partial Q}{\partial t}(\mathbf{y}, t) \frac{\partial Q}{\partial t}(\mathbf{y} + \boldsymbol{\Delta}, t)}\, d^3 \boldsymbol{\Delta} d^3 \mathbf{y} \; .$$

The overbar denotes a time average. The acoustic power can be converted into an efficiency of sound generation, η, by dividing the heat release rate.

If we consider an open, premixed flame, in which the premixed gas emerges through a circular orifice of diameter D with velocity U and density ρ_0,

$$\int \overline{Q(\mathbf{y}, t) d^3 \mathbf{y}} = \rho_0 \frac{U \pi D^2}{4}\, F \Delta H \; .$$

(13.26)

The combustion has been assumed to be totally efficient. F is the ratio of the mass of fuel to premixed gas, and ΔH is the calorific value of the fuel. Before Equation (13.25) can be used to determine the combustion noise, two questions need to be answered: how should the time derivatives $\partial/\partial t$, and the correlation volume be scaled. There are two main protagonists of scaling, Putnam and Strahle.

Putnam and his colleagues use an extension of Bragg's (1963) expanding monopole theory to determine the combustion noise. But since our expression is exact, we can recover their scaling law by making the same physical assumptions. Giammar & Putnam (1972) view the combustion as being correlated within eddies of size d_L, the burning within each eddy being controlled by the laminar flame speed S_L. Hence the total heat release rate within a correlation volume of $\mathbf{y}$

$$\int \overline{Q(\mathbf{y} + \boldsymbol{\Delta}, t) d^3 \boldsymbol{\Delta}} = \rho_0 S_L d_L^2 F \Delta H \; ,$$

(13.27)

and by comparison with (13.26)

$$d^3 \boldsymbol{\Delta} \sim \frac{4 S_L d_L^2}{\pi U D^2}\, d^3 \mathbf{y} \; .$$

(13.28)

They take

$$\frac{\partial}{\partial t} \sim \frac{u'}{d_L} , \tag{13.29}$$

where u' is a typical turbulent velocity scale. These assumptions lead to

$$\eta = \frac{(\gamma - 1)^2}{4\pi} \left(\frac{u'}{c_0}\right)^2 \frac{S_L}{c_0} \frac{F\Delta H}{c_0{}^2} . \tag{13.30}$$

This is seen to be identical to Giammar & Putnam's result once $(\gamma - 1)F\Delta H/c_0{}^2$ is rewritten as $E - 1$, where E is the expansion ratio. Putnam & Faulkner (1983) show that this efficiency compares well with experiment. However, it has been derived by assuming that the frequency of combustion noise scales as u'/d_L, and experiments show the frequency to be virtually independent of U. Now u' is proportional to U, and Giammar & Putnam try to reconcile their assumed frequency scaling with experiment by arguing, not very convincingly, that d_L is also proportional to U. So although Equation (13.30) gives a reasonable form for the acoustic efficiency, it is based on assumptions of the frequency which are not compatible with observation.

Strahle has developed several scaling laws for combustion noise. Probably the most successful for open, premixed flames is based on an empirical frequency. Strahle (1978) takes

$$d^3\boldsymbol{\Delta} = d^3\mathbf{y} \quad \text{and} \quad \frac{\partial}{\partial t} = \mathrm{St}\,\frac{U}{D} , \tag{13.31}$$

where from curve–fitting to a set of experimental results, St is taken to be proportional to $M^{-0.40}Re^{0.02}Da^{0.39}F^{-1.1}$, with $M = U/c_0$, $Re = \rho_0 U D/\mu$, $Da = \rho_0 S_L^2 D c_p/U\lambda$ and λ is the thermal conductivity. With these assumptions, it follows directly from (13.25) that

$$\eta \propto M^{2.2}\ Re^{0.04}\ Da^{0.78}\ F^{-1.2}\ \frac{\Delta H}{c_p T_0} . \tag{13.32}$$

This is in close agreement with measurements of efficiency from the same set of experiments, which gave

$$\eta \propto M^{2.68}\ Re^{-0.09}\ Da^{0.92}\ F^{-1.26} \tag{13.33}$$

13.3 THE DIFFUSION OF MASS AND HEAT

We have seen that combustion is a strong monopole sound source. The representation theorem (13.10) is in a convenient form to investigate whether there is a monopole source, in the absence of combustion, due to the diffusion of heat and/or species. Monopole sources may lead to an acoustic power output proportional to M^4 and so be important in a low Mach number flow. The effects of heat diffusion have been investigated by Kempton (1976). Morfey (1976) considers the diffusion of both heat and species.

From Equations (13.5) and (13.10), the contribution to the far–field pressure perturbation due to diffusion, $p'_d(\mathbf{x}, t)$, is given by

$$4\pi |\mathbf{x}| p'_d(\mathbf{x}, t) = -\rho_0 \frac{\partial}{\partial t} \int \left[\frac{\alpha}{\rho c_p} \left(\nabla \cdot \mathbf{q} - \sum_{n=1}^{N} \frac{\partial h}{\partial Y_n} \bigg|_{\rho, p, Y_m} \nabla \cdot \mathbf{J}_n \right) \right] d^3 \mathbf{y}. \qquad (13.34)$$

The source involves $\nabla \cdot \mathbf{q}$ and $\nabla \cdot \mathbf{J}_n$, but their integral over the source region at fixed time is zero. We can demonstrate this near cancellation by rewriting Equation (13.34) in the exactly equivalent form,

$$4\pi |\mathbf{x}| p'_d(\mathbf{x}, t) = \rho_0 \frac{\partial}{\partial t} \int \left[\mathbf{q} \cdot \nabla \left(\frac{\alpha}{\rho c_p} \right) - \sum_{n=1}^{N} \mathbf{J}_n \cdot \nabla \left(\frac{\alpha}{\rho c_p} \frac{\partial h}{\partial Y_n} \bigg|_{\rho, p, Y_m} \right) \right] d^3 \mathbf{y}$$

$$(13.35)$$

$$- \rho_0 \frac{\partial^2}{\partial t \partial x_i} \int \left[\frac{\alpha}{\rho c_p} \left(q_i - \sum_{n=1}^{N} \frac{\partial h}{\partial Y_n} \bigg|_{\rho, p, Y_m} J_{ni} \right) \right] d^3 \mathbf{y} .$$

If the very small Dufour effect is neglected, the flux of heat in a gas made up of N species is given by (Williams 1965, p 427)

$$\mathbf{q} = -\lambda \nabla T + \sum_{n=1}^{N} h_n \mathbf{J}_n , \qquad (13.36)$$

where we will use the notation h_n, ρ_n and V_n to denote the partial derivatives

$$\frac{\partial h}{\partial Y_n} \bigg|_{T, p, Y_m}, \frac{\partial \rho}{\partial Y_n} \bigg|_{T, p, Y_m} \quad \text{and} \quad \frac{\partial \rho^{-1}}{\partial Y_n} \bigg|_{T, p, Y_m},$$

respectively. Once $\mathbf{q}$ has been substituted into (13.35), the acoustic pressure perturbation can be simplified by using the chain rule,

$$h_n = \frac{\partial h}{\partial Y_n} \bigg|_{\rho, p, Y_m} + \frac{\partial h}{\partial \rho} \bigg|_{p, Y_n} \rho_n , \qquad (13.37)$$

and the relationship $\partial h/\partial\rho\big|_{p,Y_n} = -c_p/\rho\alpha$. This leads to

$$4\pi|\mathbf{x}|p_d'(\mathbf{x},t) = \rho_0\frac{\partial}{\partial t}\int\left[-\lambda\nabla T\cdot\nabla\left(\frac{\alpha}{\rho c_p}\right) + \sum_{n=1}^{N}\mathbf{J}_n\cdot\left(\nabla V_n - \frac{\alpha\nabla h_n}{\rho c_p}\right)\right]d^3\mathbf{y}$$

$$+ \rho_0\frac{\partial^2}{\partial t\partial x_i}\int\left[\frac{\alpha\lambda}{\rho c_p}\frac{\partial T}{\partial y_i} - \sum_{n=1}^{N}J_{n_i}V_n\right]d^3\mathbf{y}\ .$$

$$(13.38)$$

Equation (13.38) shows that there is only a monopole sound field due to heat diffusion in a low Mach number flow if $\alpha/\rho c_p$ is a function of position. For an ideal gas $\alpha/\rho c_p = (\gamma-1)/\gamma p$. Since fractional changes in p are of order M^2, the existence of a monopole source due to heat diffusion relies on variations in γ. As pointed out by Kempton (1976) and Morfey (1976), this is a very weak effect.

A low Mach number monopole source due to species diffusion depends on variations of V_n or h_n. Morfey discusses how in an ideal gas V_n and h_n are independent of species concentration and can only vary as a function of position due to temperature differences.

In general then, diffusion does not lead to a significant monopole source and its dominant effect is to generate dipole sound. This is described by the second integral in Equation (13.38). When diffusion is significant, these terms lead to a radiated sound power proportional to M^6. Kempton (1976) calculated this sound power exactly in a number of model problems which were amenable to analytical solution.

Equation (13.10) also describes the role of viscosity as a sound source. This is treated comprehensively in Obermeier (1985) and will not be discussed here.

13.4 ACCELERATION OF DENSITY INHOMOGENEITIES

We have seen that diffusion can lead to a dipole–type sound field. A similar field is produced by the change in momentum of compact density inhomogeneities. The acoustic pressure generated by this sound source is described by the last term in Equation (13.10) and we will denote it by $p_a'(\mathbf{x},t)$.

$$4\pi|\mathbf{x}|p_a'(\mathbf{x},t) = -\frac{x_i}{|\mathbf{x}|c_0}\frac{\partial^2}{\partial t^2}\int[u_i\rho_e]d^3\mathbf{y}\ . \tag{13.39}$$

This equation clearly displays the dipole nature of the radiated sound field. It shows that a compact fluid element, even with constant density, radiates sound when it accelerates provided its density is different from ambient — a form of "acoustic *bremsstrahlung*". Such a density inhomogeneity might be due to convected temperature variations downstream of a zone of unsteady combustion, for example. The sound field produced by "hot spots" accelerated through a low Mach number nozzle has been investigated by Ffowcs Williams & Howe (1975). There the situation is more complicated than our free–space result because of unsteady forces on the nozzle walls. Ffowcs Williams & Howe overcome this difficulty by using an exact Green function. Cumpsty & Marble (1977) consider the convection of entropy inhomogeneities through the deaccelerating flow across a turbine, and find it can generate considerable sound energy. From Equation (13.39), the pressure perturbation produced by an accelerating compact density inhomogeneity is proportional to $(\rho - \rho_0)MU^2$. The radiated acoustic power is therefore proportional to M^6.

13.5 TURBULENT TWO–PHASE FLOW

The sound field radiated by turbulence can be changed significantly if that turbulence is within a region of fluid of different density and sound speed from ambient. Variations in ρ and c are particularly dramatic in the case of two–phase flows where, for example, 1% air bubbles by volume in water can cause the sound speed to drop to about 30 m/s, 1/50 of the sound speed in water. These effects have been investigated by Crighton & Ffowcs Williams (1969).

Example We begin by solving a simple problem with an exact solution. Consider an isotropic quadrupole of strength $\rho_1 U^2 e^{-i\omega t}$ concentrated at the origin. The fluid within a sphere of radius a has density ρ_1 and sound speed c_1, while the surrounding fluid has density ρ_0 and sound speed c_0. The pressure perturbation satisfies

$$\frac{1}{c_1^2}\frac{\partial^2 p}{\partial t^2} - \nabla^2 p = \nabla^2(\rho_1 U^2 e^{-i\omega t}\delta(\mathbf{x})) \quad \text{in } r < a \qquad (13.40a)$$

and

$$\frac{1}{c_0^2}\frac{\partial^2 p}{\partial t^2} - \nabla^2 p = 0 \qquad\qquad \text{in } r > a , \qquad (13.40b)$$

with conditions of continuity of pressure and particle velocity across $r = a$. A rearrangement of (13.40a) shows that it an be written as

$$\left(\frac{1}{c_1^2}\frac{\partial^2}{\partial t^2} - \nabla^2\right)\left(p + \rho_1 U^2 e^{-i\omega t}\delta(\mathbf{x})\right) = -\frac{\omega^2}{c_1^2}\rho_1 U^2 e^{-i\omega t}\delta(\mathbf{x}) . \qquad (13.41)$$

We have already discussed the solution to an identical set of equations in Section 13.2. By comparison with (13.15a) and (13.18a), the solution for small $\omega a/c_0$ and $\omega a/c_1$ is

$$(p - p_0)(r, t) = -\frac{\rho_0 \omega^2 U^2}{4\pi r c_1^2} e^{-i\omega(t - r/c_0)} \quad \text{in} \quad r \geq a \ . \tag{13.42}$$

The acoustic power may be found by integrating the intensity over a large sphere and is equal to $\rho_0 \omega^4 U^4/(2^{1/2} 4\pi c_1^4 c_0)$. For comparison with Crighton & Ffowcs Williams's work we convert this into an efficiency by dividing by $\rho_1 U^3 L^2$ where L has dimensions of length. This shows that the efficiency is increased by a multiplicative factor

$$\frac{\rho_0 c_0{}^4}{\rho_1 c_1{}^4} \tag{13.43}$$

over the value it would have if a quadrupole with the same turbulent velocity fluctuations were entirely in ρ_0, c_0 fluid.

For the case of air bubbles in water, ρ_0 is nearly equal to ρ_1 but c_1 can be very much less than c_0. As shown in Chapter 21, the factor $c_0{}^4/c_1{}^4$ may be as large as 10^7 and demonstrates how a small quantity of bubbles can dramatically increase the radiation efficiency of turbulence in water. This predicted increase is entirely in agreement with the work of Crighton & Ffowcs Williams. The other example they consider is that of a dusty gas, for which ρ_1 is much larger than ρ_0, while $\rho_1 c_1{}^2 = \rho_0 c_0{}^2$. Then from (13.43) the efficiency is increased by a factor $c_0{}^2/c_1{}^2$. This is far less augmentation than that predicted by Crighton & Ffowcs Williams (1969), where the corresponding result is $c_0{}^6/c_1{}^6$. We will explain this discrepancy later.

The problem considered here had such a simple geometry than an exact solution was possible. However, in general for an arbitrarily shaped source region, we need to rely on an acoustic analogy of the sort described by Equation (13.10) to determine the radiated sound. Let us consider turbulent flow in a compact region of fluid with properties different from ambient. In the absence of reactions, diffusion and viscosity the representation (13.10) reduces to

$$4\pi|\mathbf{x}|(p-p_0)(\mathbf{x},t) = \frac{\partial}{\partial t}\int\left[\left(\frac{1}{c_0{}^2}-\frac{\rho_0}{\rho c^2}\right)\frac{Dp}{Dt}-\frac{(p-p_0)}{\rho c_0{}^2}\frac{D\rho}{Dt}\right]d^3\mathbf{y}$$

$$+\frac{x_i x_j}{|\mathbf{x}|^2 c_0{}^2}\frac{\partial^2}{\partial t^2}\int[\rho u_i u_j]d^3\mathbf{y} \tag{13.44}$$

$$-\frac{x_i}{|\mathbf{x}|c_0}\frac{\partial^2}{\partial t^2}\int[u_i\rho_e]d^3\mathbf{y}\ .$$

We will rewrite the first term in (13.44) to emphasize the dependence of the dipole strength on density gradients.

In a way equivalent to Equation (13.6), the partial derivative $\partial(u_i\rho_e)/\partial t$ can be related to its material derivative $D(u_i\rho_e)/Dt$

$$\frac{\partial}{\partial t}(u_i\rho_e) = \frac{D}{Dt}(u_i\rho_e) - \frac{u_i\rho_e}{\rho}\frac{D\rho}{Dt} - \frac{\partial}{\partial x_j}(\rho_e u_i u_j)\ . \tag{13.45}$$

The energy equation (13.4) simplifies considerably when there are no chemical reactions and diffusion and viscous effects are negligible. Then

$$\frac{D\rho}{Dt} = \frac{1}{c^2}\frac{Dp}{Dt}\ . \tag{13.46}$$

The momentum equation is just

$$\frac{Du_i}{Dt} = -\frac{1}{\rho}\frac{\partial p}{\partial x_i}\ . \tag{13.47}$$

After substituting (13.46) and (13.47) into (13.45), we obtain

$$\frac{\partial}{\partial t}(u_i\rho_e) = (p-p_0)\frac{\partial}{\partial x_i}\left(\frac{\rho_e}{\rho}\right) - u_i\left(\left(\frac{1}{c_0{}^2}-\frac{\rho_0}{\rho c^2}\right)\frac{Dp}{Dt}-\frac{(p-p_0)}{\rho c_0^2}\frac{D\rho}{Dt}\right)$$

$$-\frac{\partial}{\partial x_i}\left((p-p_0)\frac{\rho_e}{\rho}\right) - \frac{\partial}{\partial x_j}(\rho_e u_i u_j)\ . \tag{13.48}$$

Hence

$$-\frac{x_i}{|\mathbf{x}|c_0}\frac{\partial^2}{\partial t^2}\int [u_i\rho_e]\,d^3\mathbf{y} = -\frac{x_i}{|\mathbf{x}|c_0}\frac{\partial}{\partial t}\int \left[(p-p_0)\frac{\partial}{\partial y_i}\left(\frac{\rho_e}{\rho}\right)\right]d^3\mathbf{y}$$

$$+\frac{x_i}{|\mathbf{x}|c_0}\frac{\partial}{\partial t}\int \left[u_i\left(\left(\frac{1}{c_0{}^2}-\frac{\rho_0}{\rho c^2}\right)\frac{Dp}{Dt}-\frac{(p-p_0)}{\rho c_0^2}\frac{D\rho}{Dt}\right)\right]d^3\mathbf{y}$$

$$+\frac{x_i}{|\mathbf{x}|c_0}\frac{\partial}{\partial t}\int \left[\frac{\partial}{\partial y_i}\left((p-p_0)\frac{\rho_e}{\rho}\right)\right]d^3\mathbf{y}+\frac{x_i}{|\mathbf{x}|c_0}\frac{\partial}{\partial t}\int \left[\frac{\partial}{\partial y_j}(\rho_e u_i u_j)\right]d^3\mathbf{y}\;. \quad (13.49)$$

The last two terms on the right–hand side of Equation (13.49) involve a partial derivative with respect to $\mathbf{y}$. The integrated contributions from these terms are only nonzero because of retarded time variations over the source region. A repetition of the steps between Equations (12.5) and (12.7) shows that these derivatives can be replaced by derivatives with respect to $\mathbf{x}$. Since the far–field $\partial/\partial x_i = -x_i/(|\mathbf{x}|c)\,\partial/\partial t$, Equation (13.49) is equivalent to

$$-\frac{x_i}{|\mathbf{x}|c_0}\frac{\partial^2}{\partial t^2}\int [u_i\rho_e]\,d^3\mathbf{y} = -\frac{x_i}{|\mathbf{x}|c_0}\frac{\partial}{\partial t}\int \left[(p-p_0)\frac{\partial}{\partial y_i}\left(\frac{\rho_e}{\rho}\right)\right]d^3\mathbf{y}$$

$$+\frac{x_i}{|\mathbf{x}|c_0}\frac{\partial}{\partial t}\int \left[u_i\left(\left(\frac{1}{c_0{}^2}-\frac{\rho_0}{\rho c^2}\right)\frac{Dp}{Dt}-\frac{(p-p_0)}{\rho c_0^2}\frac{D\rho}{Dt}\right)\right]d^3\mathbf{y} \quad (13.50)$$

$$-\frac{1}{c_0{}^2}\frac{\partial^2}{\partial t^2}\int \left[(p-p_0)\frac{\rho_e}{\rho}\right]d^3\mathbf{y}-\frac{x_i x_j}{|\mathbf{x}|^2 c_0{}^2}\frac{\partial^2}{\partial t^2}\int [\rho_e u_i u_j]\,d^3\mathbf{y}\qquad.$$

Hence the representation (13.44) can be rewritten in the form

$$4\pi|\mathbf{x}|(p-p_0)(\mathbf{x},t) = \frac{\partial}{\partial t}\int \left[(1+M_r)\left(\left(\frac{1}{c_0{}^2}-\frac{\rho_0}{\rho c^2}\right)\frac{Dp}{Dt}-\frac{(p-p_0)}{\rho c_0^2}\frac{D\rho}{Dt}\right)\right]d^3\mathbf{y}$$

$$+\frac{x_i x_j}{|\mathbf{x}|^2 c_0{}^2}\frac{\partial^2}{\partial t^2}\int [(\rho-\rho_e)u_i u_j]\,d^3\mathbf{y}-\frac{1}{c_0{}^2}\frac{\partial^2}{\partial t^2}\int \left[(p-p_0)\frac{\rho_e}{\rho}\right]d^3\mathbf{y}$$

$$-\frac{x_i}{|\mathbf{x}|c_0}\frac{\partial}{\partial t}\int \left[(p-p_0)\frac{\partial}{\partial y_i}\left(\frac{\rho_e}{\rho}\right)\right]d^3\mathbf{y}\;. \quad (13.51)$$

where $M_r = \mathbf{u}\cdot\mathbf{x}/c_0|\mathbf{x}|$ is the component of the fluid Mach number in the direction of the distant observer. The first three sources in (13.51) are volume terms, while the last involves gradients of density which are only large near the boundary of the two–phase region.

For air bubbles in water

$$\rho = (1 - \beta)\rho_0 + \beta\rho_a . \tag{13.52}$$

β is the volume fraction of bubbles, ρ_0 the density of water and ρ_a the density of air. ρ is nearly equal to ρ_0, but in the bubbly region c is very much less than c_0. The expression (13.51) appears complicated, but only one volume term is important. That is

$$\frac{\partial}{\partial t} \int \left[\frac{1}{c^2} \frac{Dp}{Dt} \right] d^3\mathbf{y} . \tag{13.53}$$

This term involves division by c^2, whereas all the other volume terms are divided by the very much larger $c_0{}^2$.

Following Crighton & Ffowcs Williams, we consider a flow with mean velocity U, turbulent velocity fluctuations of order u' and typical eddy size ℓ, within a bubbly region of linear dimension L. The pressure perturbation is of order $\rho u' U$ and $\partial/\partial t \sim u'/\ell$. The radiated acoustic power can be converted into an efficiency, η, by dividing it by $\rho u' U^2 L^2$, the average working of the fluctuating pressure gradient against the mean flow over the source region. The term in (13.53) leads to

$$\eta = \frac{1}{4\pi} \left(\frac{u'}{U} \right)^5 M^5 \left(\frac{c_0}{\bar{c}} \right)^4 \frac{L}{\ell} . \tag{13.54}$$

The M^5 scaling of efficiency, which is equivalent to M^8 in power, demonstrates that this source is of quadrupole type. The presence of bubbles increases the radiation efficiency by a factor $c_0{}^4/\bar{c}^4$ over the value it would have in pure water.

The radiation efficiency of the density gradient term,

$$-\frac{x_i}{|\mathbf{x}|c_0} \frac{\partial}{\partial t} \int \left[(p - p_0) \frac{\partial}{\partial y_i} \left(\frac{\rho_e}{\rho} \right) \right] d^3\mathbf{y} , \tag{13.55}$$

depends on how abruptly the density varies at the edge of the bubble region. If β varies over the turbulent lengthscale ℓ, so that $|\nabla\beta| \sim \bar{\beta}/\ell$, the term in (13.55) radiates sound with an efficiency

$$\eta = \frac{\bar{\beta}^2}{4\pi} \left(\frac{u'}{U} \right)^3 M^3 \frac{L}{\ell} . \tag{13.56}$$

The Mach number dependence illustrates that this is a dipole–type source. This gradient term has been discussed by Crighton & Ffowcs Williams (1960). It happens to be identically zero in our exact example (13.40) because of the spherical symmetry of that problem, one of the drawbacks of the simple geometry. Whether, for an arbitrarily shaped bubbly region, the dominant source mechanism is described by (13.54) or (13.56) depends on the relative magnitudes of $u'^2 c_0^2/\bar{c}^4$ and $\bar{\beta}^2$. But in either case the bubbles lead to a considerable enhancement of the radiation efficiency.

For dusty gases

$$\rho = \rho_0(1 + f) \quad \text{and} \quad c^2 = \frac{c_0^2}{1 + f} , \tag{13.57}$$

where f, the mass concentration of dust, is assumed to be large in comparison with unity. Since $\rho c^2 = \rho_0 c_0^2$, two terms cancel in the first integrand in Equation (13.51). For a compact dusty region, the volume terms generate sound with efficiency

$$\eta = \frac{1}{4\pi} \left(\frac{u'}{U}\right)^5 M^5 \left(\frac{c_0}{\bar{c}}\right)^2 \frac{L}{\ell} , \tag{13.58}$$

while the boundary term has an efficiency

$$\eta = \frac{1}{4\pi} \left(\frac{u'}{U}\right)^3 M^3 \left(\frac{\bar{c}}{c_0}\right)^2 \frac{L}{\ell} , \tag{13.59}$$

if f varies over the turbulent lengthscale ℓ at the edge of the dust cloud.

The form given in (13.58) is consistent with our exact solution, and means that the dust produces little increase in the radiation efficiency. This augmentation is less than that predicted by Crighton & Ffowcs Williams (1969) by a factor $\bar{c}^4/c_0^4$. There the partial time derivative of $\partial\rho/\partial t - c_0^{-2}\,\partial p/\partial t$ appears as a volume source term and $\partial\rho/\partial t$ is estimated by writing

$$\frac{\partial\rho}{\partial t} = \frac{1}{c^2}\frac{\partial p}{\partial t} , \tag{13.60}$$

so that

$$\frac{\partial\rho}{\partial t} - \frac{1}{c_0^2}\frac{\partial p}{\partial t} = \left(\frac{1}{c^2} - \frac{1}{c_0^2}\right)\frac{\partial p}{\partial t} = \frac{f}{c_0^2}\frac{\partial p}{\partial t} . \tag{13.61}$$

This leads to the quadrupole radiation efficiency being increased by a factor $c_0^6/\bar{c}^6$ due to dust. But since there are considerable mean density gradients, the material density derivative should have been used in (13.60)

$$\frac{\partial\rho}{\partial t} = \frac{1}{c^2}\frac{\partial p}{\partial t} - \mathbf{u}\cdot\nabla\rho . \tag{13.62}$$

The equation of mass conservation enables $\mathbf{u}\cdot\nabla\rho$ to be rearranged as follows:

$$\mathbf{u}\cdot\nabla\rho = \nabla\cdot\left(\mathbf{u}(\rho-\rho_0)\right) + \frac{\rho-\rho_0}{\rho}\frac{D\rho}{Dt}$$

$$= \nabla\cdot\left(\mathbf{u}(\rho-\rho_0)\right) + \frac{\rho-\rho_0}{\rho c^2}\frac{Dp}{Dt}\ . \tag{13.63}$$

When (13.62) and (13.63) are used to substitute for $\partial\rho/\partial t$ in the source term, all the terms involving pressure cancel giving

$$\frac{\partial\rho}{\partial t} - \frac{1}{c_0^2}\frac{\partial p}{\partial t} = -\nabla\cdot\left(\mathbf{u}(\rho-\rho_0)\right)\ . \tag{13.64}$$

After some algebra our scaling law (13.58) for the efficiency of the sound generated by the volume sources can be recovered from this description of the source strength. The effect of dust was exaggerated then in the work of Crighton & Ffowcs Williams (1969) because the poor approximation $\partial\rho/\partial t = c^{-2}\,\partial p/\partial t$ made the source term seem larger than when the correct relationship $D\rho/Dt = c^{-2}\,\partial p/\partial t$ is used.

The acoustic analogy as described by (13.10) can be used to estimate the sound from a compact source region. But when the source is not compact, regarded time variations across the source mean the integrals in (13.10) are hard to estimate. We will now use the model problem with spherical symmetry to investigate the effect of a non–compact two–phase region.

When both $\omega a/c_0$ and $\omega a/c_1$ are large in comparison with unity, the solution to Equation (13.40) is

$$(p-p_0)(r,t) = -\frac{\omega^2\rho_1 U^2 T}{4\pi r c_1^2}\frac{e^{-i\omega(t-(r-a)/c_0-a/c_1)}}{1+Re^{2i\omega a/c_1}}\ , \tag{13.65}$$

in $r \geq a$, where

$$R = \frac{\rho_0 c_0 - \rho_1 c_1}{\rho_0 c_0 + \rho_1 c_1} \quad\text{and}\quad T = \frac{2\rho_0 c_0}{\rho_0 c_0 + \rho_1 c_1}\ .$$

R and T are the pressure reflection and transmission coefficients for sound at normal incidence onto a plane interface separating $\rho_1 c_1$ fluid from $\rho_0 c_0$. For large $\omega a/c_1$, p varies rapidly as a function of frequency, particularly for extreme differences between the two fluids when $|R|$ is approximately unity. So, rather than consider a single frequency, we will investigate the sound field produced by a source with time dependence $g(t)$

$$\frac{1}{c_1^2}\frac{\partial^2 p}{\partial t^2} - \nabla^2 p = \nabla^2(g(t)\delta(\mathbf{x})) \qquad \text{in } r < a \tag{13.66a}$$

$$\frac{1}{c_0^2}\frac{\partial^2 p}{\partial t^2} - \nabla^2 p = 0 \qquad \text{in } r > a\ . \tag{13.66b}$$

Provided $g(t)$ only contains frequencies that are high in comparison with both c_0/a and c_1/a, the solution to (13.66) can be calculated by inverting the solution of the Fourier–transformed problem (13.65). This gives

$$(p - p_0)(r, t) = \frac{T}{4\pi r c_1^2} \int \left\{ -\omega^2 \tilde{g}(\omega) e^{-i\omega(t-(r-a)/c_0 - a/c_1)} \right.$$

(13.67)

$$\left. \times (1 - Re^{2i\omega a/c_1} + R^2 e^{4i\omega a/c_1} - \ldots) \right\} d\omega \ .$$

Here $\tilde{g}(\omega)$ is the Fourier transform of $g(t)$, and the binomial theorem has been used to expand the denominator in (13.65). After integrating term by term, we obtain

$$(p - p_0)(r, t) = \frac{T}{4\pi r c_1^2} \left[\frac{\partial^2}{\partial t^2} g \left(t - \frac{r-a}{c_0} - \frac{a}{c_1} \right) - R \frac{\partial^2}{\partial t^2} g \left(t - \frac{r-a}{c_0} - \frac{3a}{c_1} \right) \right.$$

$$\left. + R^2 \frac{\partial^2}{\partial t^2} g \left(t - \frac{r-a}{c_0} - \frac{5a}{c_1} \right) - \ldots \right] \ .$$

(13.68)

The first term in square brackets in (13.68) describes sound that has travelled a distance a in the c_1 region, $r - a$ in c_0 fluid, and undergone an amplitude of change T when crossing the interface. The sound in the next term has been reflected back once at the $r = a$ interface, travelled an extra distance $2a$ in c_1 fluid and changed phase by π as it passed through the origin. The higher–order terms represent sound waves that have been multiply reflected within $r = a$.

The total acoustic energy, E, may be found by integrating the power output and is equal to $4\pi r^2 \int (p - p_0)^2 dt/\rho_0 c_0$. Substitution for $p - p_0$ from (13.68) shows that

$$E = \frac{T^2}{4\pi \rho_0 c_0 c_1^4} (1 + R^2 + R^4 \ldots) \int_{-\infty}^{\infty} \left(\frac{\partial^2 g}{\partial t^2} \right)^2 dt \ ,$$

(13.69)

if the source g emits for a time less than $2a/c_1$. The series can be summed and substituting for R and T gives

$$E = \frac{1}{4\pi \rho_1 c_1^5} \int_{-\infty}^{\infty} \left(\frac{\partial^2 g}{\partial t^2} \right)^2 dt \ .$$

(13.70)

In this case the turbulent quadrupoles radiate energy as if all the fluid had density ρ_1 and sound speed c_1; the ρ_1, c_1 fluid is so extensive that information that it is bounded does not arrive back at the source during its lifetime. The radiated power is increased by a factor $\rho_1 c_0^5 / \rho_0 c_1^5$ from the value it would have if the same velocity fluctuations occurred in ρ_0, c_0 fluid. This could represent a considerable enhancement due to both air bubbles in water and dust particles in air.

13.6 THERMOACOUSTIC INSTABILITIES

So far in our discussion of the sound generated by thermoacoustic perturbations, we have treated the source as specified. For example, in the consideration of combustion noise in Section 13.2, it was assumed implicitly that the unsteady heat release could be prescribed independently of the sound field. That approach is appropriate when the fluctuations in the rate of heat release are due mainly to turbulence. However, when combustion occurs within an acoustic resonator, the sound field can be so intense that it alters the rate of combustion. Instability is then possible because while the sound waves perturb the combustion, the unsteady combustion generates yet more sound!

Rayleigh (1945) gives a clear physical interpretation of the interchange of energy between sound waves and unsteady heating. He states that the amplitude of a sound wave will increase when heat is added in phase with its pressure perturbation. The addition of heat out of phase with the pressure reduces the amplitude. Chu (1964) adopted a more mathematical approach and his method enables the effects of boundary conditions to be clearly displayed.

Following Chu, let us consider burning within a combustor of volume V, bounded by the surface S, as illustrated in Figure 13.3. For simplicity we will just consider a perfect gas linearly disturbed from rest with no mean heat release or mean flow . However, the analysis may be extended to include mean heat release and mean flow (see Bloxsidge et al. 1988) and the interaction between liquid and gaseous phases (Culick 1988).

For combustion with constant average molecular weight, a combination of Equation (13.12) with the linearized form of (13.4) leads to

$$\frac{\partial \rho'}{\partial t} = \frac{1}{c_0{}^2} \frac{\partial p'}{\partial t} - \frac{\gamma - 1}{c_0{}^2} Q , \qquad (13.71)$$

where Q is the combustion rate per unit volume and ρ' and p' denote the perturbation in density and pressure respectively. ρ_0 and c_0 are the mean values of density and sound speed. The diffusion of heat and species has been neglected in the derivation of (13.71).

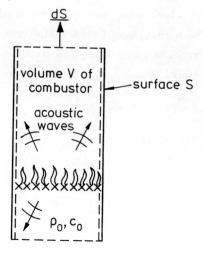

Figure 13.3 Combustion within a resonator.

When the linearized equation of mass conservation

$$\frac{\partial \rho'}{\partial t} + \rho_0 \, \nabla \cdot \mathbf{u} = 0 \; , \tag{13.72}$$

is combined with Equation (13.71), it leads to

$$\frac{1}{c_0^2} \frac{\partial p'}{\partial t} + \rho_0 \, \nabla \cdot \mathbf{u} = \frac{\gamma - 1}{c_0^2} \, Q \; . \tag{13.73}$$

The linearized momentum equation has the usual form

$$\rho_0 \, \frac{\partial u_i}{\partial t} + \frac{\partial p'}{\partial x_i} = \frac{\partial \tau_{ij}}{\partial x_j} \; . \tag{13.74}$$

Equations (13.73) and (13.74) may be combined to form an acoustic energy equation. When Equation (13.73) is multiplied by p'/ρ_0 and added to the product of (13.74) with u_i, it yields, after integration over the volume V,

$$\frac{\partial}{\partial t} \int_V \left(\frac{1}{2} \rho_0 u^2 + \frac{1}{2} \frac{p'^2}{\rho_0 c_0^2} \right) dV = \frac{\gamma - 1}{\rho_0 c_0^2} \int_V p' Q \, dV$$

$$\tag{13.75}$$

$$- \int_S (p' u_j - u_i \tau_{ij}) dS_j - \int_V \frac{\partial u_i}{\partial x_j} \, \tau_{ij} dV \; .$$

The term on the left–hand side of Equation (13.75) is the rate of change of the sum of the kinetic and potential energies within the volume V. The first term of the right–hand side describes the exchange of energy between the combustion and the acoustic waves. As noted by Rayleigh, the acoustic energy tends to be increased when p' and Q are in phase. The surface term accounts for the loss of energy across the bounding surface S, while the last term is just the rate of viscous dissipation.

Equation (13.75) states that disturbances grow if their net energy gain from the combustion is greater than the sum of their energy losses across the boundary and due to dissipation. Therefore the acoustic mode grows in amplitude if

$$\frac{\gamma - 1}{\rho_0 c_0{}^2} \int_V \overline{p'Q} \, dV > \int_S \overline{(u_j p' - u_i \tau_{ij})} \, dS_j + \int_V \overline{\tau_{ij} \frac{\partial u_i}{\partial x_j}} \, dV \,, \qquad (13.76)$$

where the overbar denotes an average over one period of the acoustic oscillation. This is a generalized form of Rayleigh's criterion. When it is satisfied, the combustor has a thermoacoustic instability. Linear waves increase in amplitude until limited by nonlinear effects.

There are many practical devices in which combustion instabilities occur. Rockets, ramjets, aeroengines and gas burners are all susceptible. More than one boiler has been unable to attain its design output because of the onset of damaging oscillations. The occurrence of a combustion instability is invariably detrimental. The oscillations can become so intense that they cause structural damage. Alternatively, they may enhance the heat transfer leading to overheating, or the perturbations may simply become so violent that the flame is extinguished. Two excellent reviews describing different types of combustion oscillations have been given recently (see Culick 1988 and Candel & Poinsot 1988).

The inequality in Equation (13.76) demonstrates how to eliminate a combustion instability. Either the driving term, $\int p'Q \, dV$ should be reduced or the damping in the system should be increased. Classical techniques exploit passive ways of producing this change. They involve the insertion of baffles within the combustor or the connection of quarter–wave tubes or Helmholtz resonators to the walls. These suppression devices are described fully by Putnam (1971). However, at low frequencies tuned resonators are prohibitively large and interest has turned to active ways of stabilizing the system.

One way of stabilizing combustion oscillations is to actively change the boundary conditions so that more acoustic energy is lost on reflection, thereby reversing the inequality in (13.76). This has been demonstrated to be effective both on a laminar flame burning in a duct by Dines (1983), Heckl (1988) and Lang et al. (1987), and on more intense turbulent flames, see Poinsot et al. (1987) and Bloxsidge et al. (1988). However, a more practical means of control is to reverse the inequality (13.76) by inducing combustion out of phase with pressure. Langhorne et al. (1990) showed that combustion oscillations of a $\frac{1}{4}$MW turbulent flame could be controlled by the unsteady addition of extra fuel. A suitably–phased addition of only 3% more fuel reduced the peak in the pressure spectrum due to the combustion instability by some 12dB. Since relatively little unsteady fuel is necessary, the mechanical power requirements of the controller are modest and the system is easy to implement. Active control is emerging as a practical means of eliminating low–frequency thermoacoustic instabilities.

REFERENCES

Bloxsidge, G.J., Dowling, A.P., Hooper, N. & Langhorne, P.J. (1988). Active control of reheat buzz. AIAA J. 26:783–790.

Bragg, S.L. (1963). Combustion noise. J. Inst. of Fuel 36, 12–16.

Candel, S.M. & Poinsot, T.J. (1988). Interactions between acoustics and combustion. Proc. Inst. Acoustics 10:103–153.

Chiu, H.H. & Summerfield, M. (1974). Theory of combustion noise. Acta Astro. 1:967–984.

Chu, B.T. (1964). On the energy transfer to small disturbances in fluid flow. Part I. Acta Mechanics 1:215–234.

Crighton, D.G. & Ffowcs Williams, J.E. (1969). Sound generation by turbulent two–phase flow. J. Fluid Mech. 36:585–603.

Culick, F.E.C. (1988). Combustion in liquid–fuelled propulsion systems — an overview. AGARD–CPP–450.

Cumpsty, N.A. & Marble, F.E. (1977). The interaction of entropy fluctuations with turbine blade rows; a mechanism of turbo–jet engine noise. Proc. Roy. Soc. A. 357:323–344.

Dines, P.J. (1983). Active Control of Flame Noise. PhD Thesis, Cambridge University.

Ffowcs Williams, J.E. (1982). Sound sources in aerodynamics — fact and fiction. AIAA J. 20:307–315.

Ffowcs Williams, J.E. & Howe, M.S. (1975). The generation of sound by density inhomogeneities in low Mach number nozzle flows. J. Fluid Mech. 70:605–622.

Giammar, R.D. & Putnam, A.A. (1972). Combustion roar of premix burners, singly and in pairs. Combustion and Flame 18:435–438.

Hassan, H.A. (1974). Scaling of combustion generated noise. J. Fluid Mech. 66:445–453.

Heckl, M.A. (1988). Active control of the noise from a Rijke tube. J. Sound Vib. 124:117–133.

Hurle, I.R., Price, R.B., Sugden, T.M., & Thomas, A. (1968). Sound emission from open turbulent premixed flames. Proc. Roy Soc. Lond. A 303:409–427.

Kempton, A.J. (1976). Heat diffusion as a source of aerodynamic sound. J. Fluid Mech. 78:1–31.

Lang, W., Poinsot, T. & Candel, S.M. (1987). Active control of combustion instability. Combustion and Flame 70:281–289.

Langhorne, P.J., Dowling, A.P. & Hooper, N. (1990). A practical active control system for combustion instabilities. AIAA J. Prop. and Power 6:324–333.

Morfey, C.L. (1976). Sound radiation due to unsteady dissipation in turbulent flows. J. Sound Vib. 48:95–111.

Obermeier, F. (1985). Aerodynamic sound generation caused by viscous processes. J. Sound Vib. 99:111–120.

Poinsot, T., Bourienne, F., Esposito, E., Candel, S. & Lang, W. (1987). Suppression of combustion instabilities by active control. AIAA-87-1876.

Putnam, A.A. (1971). Combustion–driven Oscillations in Industry. Elsevier (USA).

Putnam, A.A. & Faulkner, L. (1983). An overview of combustion noise. J. Energy 7:458–469.

Rayleigh, J.W.S. (1945). The Theory of Sound. Vol.II. Dover Publications.

Shivashankara, B.N., Strahle, W.C. & Handley, J.C. (1975). Evaluation of combustion noise scaling laws by an optical technique. AIAA J. 13:623–627.

Strahle, W.C. (1972). Some results in combustion generated noise. J. Sound Vib. 23:113–125.

Strahle, W.C. (1978). Combustion noise. Prog. Energy and Combust. Sci. 4:157–176.

Williams, F.A. (1965). Combustion Theory. Addison–Wesley.

13A APPENDIX

The energy equation is

$$\rho \frac{De}{Dt} = -\nabla \cdot \mathbf{q} - p \, \nabla \cdot \mathbf{u} + \frac{\partial u_i}{\partial x_j} \, \tau_{ij} \qquad (13.A1)$$

where e is the internal energy. For a gas made up of N species, Williams (1965) shows that

$$de = T \, ds + \frac{p}{\rho^2} \, d\rho + \sum_{n=1}^{N} \frac{\mu_n}{W_n} \, dY_n \, , \qquad (13.A2)$$

with s entropy, μ_n the chemical potential of species n and W_n its molecular weight. Substituting for e from (13.A2) into (13.A1) and using the mass equation leads to an expression for entropy changes

$$\rho T \frac{Ds}{Dt} = -\nabla \cdot \mathbf{q} + \frac{\partial u_i}{\partial x_j} \tau_{ij} - \sum_{n=1}^{N} \frac{\mu_n}{W_n} \rho \frac{DY_n}{Dt} \, . \qquad (13.A3)$$

Now from the chain rule

$$\frac{Dp}{Dt} = c^2 \frac{D\rho}{Dt} + \left. \frac{\partial p}{\partial s} \right|_{\rho, Y_n} \frac{Ds}{Dt} + \sum_{n=1}^{N} \left. \frac{\partial p}{\partial Y_n} \right|_{s, \rho, Y_m} \frac{DY_n}{Dt} \, . \qquad (13.A4)$$

When (13.A4) is rearranged and (13.A3) used to replace Ds/Dt, it leads to

$$\frac{D\rho}{Dt} = \frac{1}{c^2} \frac{Dp}{Dt} + \left(\nabla \cdot \mathbf{q} - \frac{\partial u_i}{\partial x_j} \, \tau_{ij} \right) \frac{1}{\rho T c^2} \left. \frac{\partial p}{\partial s} \right|_{\rho, Y_n}$$

$$+ \sum_{n=1}^{N} \frac{1}{c^2} \left\{ \frac{1}{T} \frac{\mu_n}{W_n} \left. \frac{\partial p}{\partial s} \right|_{\rho, Y_n} - \left. \frac{\partial p}{\partial Y_n} \right|_{s, \rho, Y_m} \right\} \frac{DY_n}{Dt} \, . \qquad (13.A5)$$

It can be shown by repeated applications of the chain rule and using the enthalpy relationship,

$$dh = T \, ds + \frac{1}{\rho} \, dp + \sum_{n=1}^{N} \frac{\mu_n}{W_n} \, dY_n \, , \qquad (13.A6)$$

that

$$\frac{1}{\rho T c^2} \left. \frac{\partial p}{\partial s} \right|_{\rho, Y_n} = \frac{\alpha}{c_p} \qquad (13.A7)$$

and

$$\frac{1}{c^2} \left\{ \frac{1}{T} \frac{\mu_n}{W_n} \left. \frac{\partial p}{\partial s} \right|_{\rho, Y_n} - \left. \frac{\partial p}{\partial Y_n} \right|_{s, \rho, Y_m} \right\} = \frac{\rho \alpha}{c_p} \left. \frac{\partial h}{\partial Y_n} \right|_{\rho, p, Y_m} \, . \qquad (13.A8)$$

Equation (13.A5) therefore simplifies to

$$\frac{D\rho}{Dt} = \frac{1}{c^2}\frac{Dp}{Dt} + \frac{\alpha}{c_p}\left\{\sum_{n=1}^{N}\left.\frac{\partial h}{\partial Y_n}\right|_{\rho,p,Y_m}\rho\frac{DY_n}{Dt} + \nabla\cdot\mathbf{q} - \frac{\partial u_i}{\partial x_j}\tau_{ij}\right\} . \qquad (13.A9)$$

This is Williams' Equation (4.40) extended to include transport effects and is used in Section 13.1.

From the chain rule

$$\left.\frac{\partial h}{\partial Y_n}\right|_{T,p,Y_m} = \left.\frac{\partial h}{\partial Y_n}\right|_{\rho,p,Y_m} + \left.\frac{\partial h}{\partial \rho}\right|_{p,Y_n}\left.\frac{\partial \rho}{\partial Y_n}\right|_{T,p,Y_m} . \qquad (13.A10)$$

A rearrangement of this expression shows that

$$\sum_{n=1}^{N}\left.\frac{\partial h}{\partial Y_n}\right|_{\rho,p,Y_m}\rho\frac{DY_n}{Dt} = \sum_{n=1}^{N}\left.\frac{\partial h}{\partial Y_n}\right|_{T,p,Y_m}\rho\frac{DY_n}{Dt}$$
$$- \rho\left.\frac{\partial h}{\partial \rho}\right|_{p,Y_n}\left(\sum_{n=1}^{N}\left.\frac{\partial \rho}{\partial Y_n}\right|_{T,p,Y_m}\frac{DY_n}{Dt}\right) . \qquad (13.A11)$$

The sum on the left–hand side of (13.A11) appears in (13.A9). When diffusion of species is neglected, $\rho\, DY_n/Dt = w_n$. Then the first term on the right–hand side of (13.A11) is equal to $-Q$, where Q is the rate heat release per unit volume due to changes in species concentration. The term in brackets in (13.A11) is the change in density at constant p and T due to changes in species concentration. For an ideal gas $p = RT\rho/W_0$, where R is the universal gas constant and W_0 the average molecular weight. If the average molecular weight is constant,

$$\sum_{n=1}^{N}\left.\frac{\partial \rho}{\partial Y_n}\right|_{T,p,Y_m}\frac{DY_n}{Dt} = 0 ,$$

and (A13.11) reduces to

$$-\sum_{n=1}^{N}\left.\frac{\partial h}{\partial Y_n}\right|_{\rho,p,Y_m}\rho\frac{DY_n}{Dt} = Q . \qquad (13.A12)$$

14. EFFECTS OF MOTION ON ACOUSTIC SOURCES

14.1 EFFECTS OF MOTION ON ELEMENTARY SOURCES

An acoustic source in motion radiates a different sound field from that produced when the source is stationary. One effect of source motion is to cause a *Döppler shift* in frequency. This is well known and often observed. For example, the whistle of an approaching train appears to be of higher frequency to a stationary listener than when the train recedes. Another effect of source motion is to change the amplitude of the acoustic field. These effects can be illustrated by considering a moving acoustic monopole which is at a position $\mathbf{x}_s(t)$ at time t. The pressure perturbation satisfies

$$\frac{1}{c^2}\frac{\partial^2 p'}{\partial t^2} - \nabla^2 p' = \frac{\partial}{\partial t}\Big(Q(t)\delta(\mathbf{x}-\mathbf{x}_s(t))\Big) \quad , \tag{14.1}$$

where c is the sound speed. Use of the free–space Green function $\delta(t-\tau-|\mathbf{x}-\mathbf{y}|/c)/(4\pi|\mathbf{x}-\mathbf{y}|)$ enables the solution of this equation to be written down immediately. It is

$$p'(\mathbf{x},t) = \frac{\partial}{\partial t}\int \frac{Q(\tau)\delta(\mathbf{y}-\mathbf{x}_s(\tau))\delta(t-\tau-|\mathbf{x}-\mathbf{y}|/c)}{4\pi|\mathbf{x}-\mathbf{y}|} \, d^3\mathbf{y}\,d\tau \quad ,$$

and the δ–function may be used to evaluate the $\mathbf{y}$–integral giving

$$p'(\mathbf{x},t) = \frac{\partial}{\partial t}\int \frac{Q(\tau)\delta(t-\tau-|\mathbf{x}-\mathbf{x}_s(\tau)|/c)}{4\pi|\mathbf{x}-\mathbf{x}_s(\tau)|} \, d\tau \quad . \tag{14.2}$$

We saw in Chapter 2, Equation (2.36), that

$$\int Q(\tau)\delta(f(\tau))d\tau = \sum_{n=1}^{N} \frac{Q(\tau_n^*)}{|\partial f/\partial \tau|} \quad , \tag{14.3}$$

the sum being taken over all the zeros τ_n^*, $n = 1, \ldots, N$ of $f(\tau)$. Here

$$f(\tau) = t - \tau - |\mathbf{x} - \mathbf{x}_s(\tau)|/c \qquad (14.4)$$

and

$$\frac{df}{d\tau} = -1 + \frac{(x_i - x_{si})}{|\mathbf{x} - \mathbf{x}_s|c} \frac{dx_{si}}{d\tau}$$

$$= -1 + M_r \quad ,$$

where $cM_r = \dfrac{(x_i - x_{si})}{|\mathbf{x} - \mathbf{x}_s|} \dfrac{dx_{si}}{d\tau}$ is the component of the source velocity in the direction of the observer. Evaluation of the remaining integral in (14.2) then leads to

$$p'(\mathbf{x}, t) = \sum_{n=1}^{N} \frac{\partial}{\partial t} \left[\frac{Q(\tau_n^*)}{4\pi r |1 - M_r|} \right] \quad ; \qquad (14.5)$$

τ_n^* is an emission time of sound heard by the observer at time t. At this emission time the source is at $\mathbf{x}_s(\tau_n^*)$ and τ_n^* satisfies

$$c(t - \tau_n^*) = |\mathbf{x} - \mathbf{x}_s(\tau_n^*)| = r \quad , \qquad (14.6)$$

as illustrated in Figure 14.1. If there is more than one solution to this equation, the right–hand side of (14.5) is to be summed over all such τ_n^*.

From (14.6) we note that

$$c\left(1 - \frac{\partial \tau_n^*}{\partial t}\right) = -\frac{(x_i - x_{si})}{|\mathbf{x} - \mathbf{x}_s|} \frac{dx_{si}}{d\tau_n^*} \frac{\partial \tau_n^*}{\partial t} \quad ,$$

i.e.

$$\frac{\partial \tau_n^*}{\partial t} = \frac{1}{1 - M_r} \quad . \qquad (14.7)$$

This illustrates the relationship between the source time τ_n^* and the observer's time t. The *Döppler factor* $(1 - M_r)^{-1}$ is greater than unity for an approaching subsonic source and less than unity for a receding subsonic source and accounts for the frequency shifts commonly heard when noisy vehicles pass by. If M_r is greater than unity, the factor in (14.7) is negative and sounds are heard in reverse order to their emission. This was described in a picturesque way by Lord Rayleigh (1945) who imagined a violinist playing his instrument while approaching at exactly twice the speed of sound. The music is then heard in perfect time and pitch but backwards!

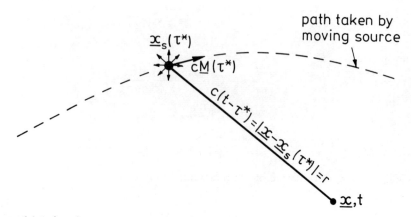

Figure 14.1 An observer at $\mathbf{x}, t$ hears the sound emitted by the source at time τ^*.

Equation (14.5) illustrates how a source in unsteady motion radiates a sound field even if $Q(\tau)$ is constant. For $Q(\tau) = Q_0$, Equation (14.5) becomes

$$p'(\mathbf{x}, t) = \frac{Q_0}{4\pi} \sum_{n=1}^{N} \frac{\partial}{\partial t} \left[\frac{1}{r|1 - M_r|} \right] \quad . \tag{14.8}$$

The relation (14.7) aids in the evaluation of the time derivative and it shows that for large r

$$p'(\mathbf{x}, t) = \frac{Q_0}{4\pi} \sum_{n=1}^{N} \frac{\dot{M}_r}{r|1 - M_r|^3} \quad , \tag{14.9}$$

where the dot denotes differentiation with respect to the argument. When the source is near the origin and moving at a low Mach number, this simplifies to

$$p'(\mathbf{x}, t) = \frac{Q_0\, \mathbf{x} \cdot \dot{\mathbf{U}}}{4\pi |\mathbf{x}|^2 c} \quad . \tag{14.10}$$

The directivity and dependence on the sound speed have the form we would expect from a dipole. This indicates that a nonevolving monopole can generate a dipole sound field when it is in unsteady motion.

The response due to a moving–point dipole can be determined in a similar way. The sound field satisfies the inhomogeneous wave equation

$$\left(\frac{1}{c^2} \frac{\partial^2}{\partial t^2} - \nabla^2 \right) p' = -\frac{\partial}{\partial x_i} \left(F_i(t)\delta(\mathbf{x} - \mathbf{x}_s(t)) \right) \quad , \tag{14.11}$$

and a repetition of the steps between Equations (14.1) and (14.5) shows that the solution is

$$p'(\mathbf{x}, t) = -\sum_{n=1}^{N} \frac{\partial}{\partial x_i} \left[\frac{F_i(\tau_n^*)}{4\pi r |1 - M_r|} \right] . \qquad (14.12)$$

Again, we see that a source in unsteady motion radiates sound even if its source strength is constant. This is one element of propeller noise discussed in Chapter 15, where the steady thrust exerted by a propeller blade generates sound because the blade has a changing velocity as it sweeps out a helical path.

We now restrict our attention to the case where the source moves with a constant velocity. Let us, without loss of generality, assume that the uniformly moving source is travelling in the 1–direction with speed U_1. Then $\mathbf{x}_s(\tau) = \mathbf{U}\tau$, where $\mathbf{U} = (U_1, 0, 0)$. The implicit relationship in (14.6) for the emission time of sound heard at $\mathbf{x}$ at time t simplifies to

$$c(t - \tau^*) = \left((x_1 - U_1\tau^*)^2 + x_2^2 + x_3^2 \right)^{1/2} . \qquad (14.13)$$

This equation can be solved to obtain τ^* explicitly for any observer's position. Squaring (14.13) leads to a quadratic equation for τ^* which has solutions

$$\tau^* = \frac{ct - Mx_1 \pm \left((x_1 - U_1t)^2 + (1 - M^2)(x_2^2 + x_3^2) \right)^{1/2}}{c(1 - M^2)} , \qquad (14.14)$$

with $M = U_1/c$. Spurious roots may have been introduced by squaring and only solutions in (14.14) which have τ^* real and $\tau^* < t$ are solutions of the original Equation (14.13). When the source velocity is subsonic only the solution in (14.14) with the lower choice of sign satisfies these conditions and there is only one value of τ^* satisfying (14.13) for each $(\mathbf{x}, t)$. When the source velocity is supersonic, Equation (14.13) only has a solution for observer positions $\mathbf{x}, t$ inside the Mach cone, illustrated in Figure 14.2, where

$$(1 - M^2)^{1/2}(x_2^2 + x_3^2)^{1/2} < U_1t - x_1 . \qquad (14.15)$$

For positions within this cone, both values of τ^* in (14.14) are real and less than t. The sound emitted at two distinct times is heard simultaneously at $\mathbf{x}$.

410

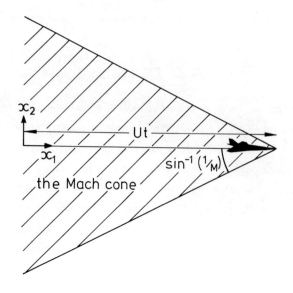

Figure 14.2 The Mach cone.

When a monopole source is in steady motion, the largest contribution to the far–field sound comes from the differentiation of $Q(\tau_n^*)$ in Equation (14.5). This leads to

$$p'(\mathbf{x}, t) = \sum_n \frac{\dot{Q}(\tau_n^*)}{4\pi r|1 - M\cos\theta|^2 \mathrm{sgn}(1 - M\cos\theta)} \quad , \qquad (14.16)$$

where θ is the angle between the observer's position and the 1–axis measured from the source position at emission time and is illustrated in Figure 14.3

$$\cos\theta = (x_1 - x_{s1}(\tau^*))/r \quad . \qquad (14.17)$$

A monopole source at rest emits an omnidirectional sound field. The factor $|1 - M\cos\theta|^{-2}$ in (14.16) means that when the monopole is in subsonic motion the sound in the forward arc is more intense than that in the rear arc. The corresponding far–field form for the dipole sound in (14.12) is

$$p'(\mathbf{x}, t) = \sum_n \frac{\hat{\mathbf{r}} \cdot \dot{\mathbf{F}}(\tau_n^*)}{4\pi r c|1 - M\cos\theta|^2 \mathrm{sgn}(1 - M\cos\theta)} \quad , \qquad (14.18)$$

where $\hat{r}_i(i = 1, 2, 3)$ are the direction cosines of the observer from the emission point.

The variables r and θ are defined in Equations (14.6) and (14.17) respectively. r is the distance between the source and the observer and θ is the angle between the observer's position and the 1–axis. Since they are measured from the position of the source at the time the sound was emitted (see Figure 14.3), they are known as *emission–time coordinates*. It is sometimes convenient to use *reception–time coordinates*, R and Θ, based on the position of the source at reception time t,

$$R = \left((x_1 - U_1 t)^2 + x_2^2 + x_3^2\right)^{1/2} \tag{14.19}$$

and

$$\cos \Theta = (x_1 - U_1 t)/R \quad . \tag{14.20}$$

R and Θ are shown in Figure 14.3

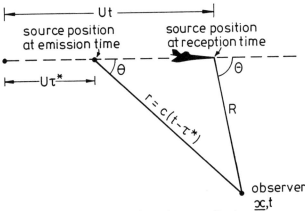

Figure 14.3 Reception and emission–time coordinates.

Reception–time coordinates are just coordinates in a reference frame that moves with the source. They are therefore the natural coordinates in a wind–tunnel geometry in which the source is at rest and the fluid is convected past it with velocity $-U$. R is then the radial distance from the observer to the fixed source, and Θ the angle between the observer's position and the 1–direction measured from the fixed source position in the wind tunnel.

The radical in (14.14) has a simple form when expressed in terms of R and Θ and leads to

$$\tau^* = t - \frac{R}{c(1 - M^2)} \left(M \cos \Theta \mp (1 - M^2 \sin^2 \Theta)^{1/2}\right) \quad . \tag{14.21}$$

When this form for τ^* is substituted into (14.6) and (14.17), it gives expressions relating emission and reception time coordinates. In particular,

$$r(1 - M\cos\theta) = \mp R(1 - M^2 \sin^2\Theta)^{1/2} \tag{14.22}$$

and

$$\cos\theta = \frac{\cos\Theta \mp M(1 - M^2 \sin^2\Theta)^{1/2}}{M\cos\Theta \mp (1 - M^2 \sin^2\Theta)^{1/2}} \quad . \tag{14.23}$$

These relationships can be used to rewrite the far–field pressure in (14.16) and (14.18) in terms of reception–time coordinates.

When the source moves supersonically, the expressions (14.16) and (14.18) for the sound field radiated by point monopoles and dipoles respectively are evidently singular at positions $\mathbf{x}$ such that $M_r = M\cos\theta = 1$. We see from Equation (14.22) that a combination of the condition $M_r = 1$ with the retarded–time Equation (14.13) leads to

$$\sin\Theta = 1/M \quad , \tag{14.24}$$

i.e. the pressure field of a point source is singular on the Mach cone, where

$$U_1 t - x_1 = (M^2 - 1)^{1/2}(x_2^2 + x_3^2)^{1/2} \quad . \tag{14.25}$$

We now go on to consider finite source length which can control the singularity.

Let us consider a monopole moving with constant velocity $\mathbf{U} = (U_1, 0, 0)$ and of length ℓ in the 1–direction, as shown in Figure 14.4. The source strength is expressed conveniently as $q(\eta_1, \tau)\delta(\eta_2)\delta(\eta_3)$, where the $\boldsymbol{\eta}$ coordinate system moves with the velocity $\mathbf{U}$, i.e. $\boldsymbol{\eta} = \mathbf{x} - \mathbf{U}\tau$ and $q(\eta_1, \tau)$ is only nonzero for $0 \leq \eta_1 \leq \ell$.

Equation (14.5) generalizes to

$$\phi(\mathbf{x}, t) = -\int_0^\ell \sum_{n=1}^N \frac{q(\eta_1, \tau_n^*)}{4\pi\rho_0 r|1 - M_r|} d\eta_1 \quad ; \tag{14.26}$$

$\phi(\mathbf{x}, t)$ is the velocity potential, ρ_0 the mean density and $-\rho_0 \partial\phi/\partial t = p'(\mathbf{x}, t)$. Now τ_n^*, which is a solution of

$$c(t - \tau_n^*) = \left((x_1 - \eta_1 - U_1\tau_n^*)^2 + x_2^2 + x_3^2\right)^{1/2} = r \quad , \tag{14.27}$$

is a function of η_1.

Differentiation of (14.27) shows that

$$\frac{\partial\tau_n^*}{\partial\eta_1} = \frac{x_1 - \eta_1 - U_1\tau_n^*}{cr(1 - M_r)} \quad . \tag{14.28}$$

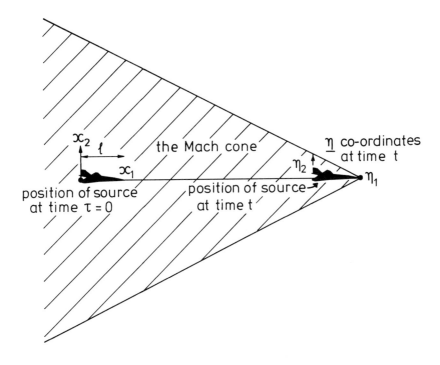

Figure 14.4 A supersonic source of length ℓ.

So as M_r approaches unity, the retarded time τ_n^* varies increasingly rapidly as η_1 varies along the source length. A source of small but nonzero size is appreciably different from a point source near $M_r = 1$. Then, for a source of any nonvanishing extent, retarded time variations over the source region are significant and this makes the integral in Equation (14.26) hard to evaluate. It is more convenient to return to the general representation

$$\phi(\mathbf{x}, t) = -\int \frac{q(\eta_1, \tau)\delta(t - \tau - r/c)}{4\pi \rho_0 r}\, d\eta_1\, d\tau \quad ; \tag{14.29}$$

$r = ((x_1 - \eta_1 - U_1\tau)^2 + x_2^2 + x_3^2)^{1/2}$, and to use the δ–function to evaluate the η_1–integral in (14.29). This leads to

$$\phi(\mathbf{x}, t) = -\int \frac{cq(\eta_1^*, \tau)\, d\tau}{4\pi \rho_0 |x_1 - \eta_1^* - U_1\tau|} \quad , \tag{14.30}$$

where η_1^* is a solution of

$$c(t - \tau) = \left((x_1 - \eta_1^* - U_1\tau)^2 + x_2^2 + x_3^2\right)^{1/2} \quad . \tag{14.31}$$

Let us suppose that the source $q(\eta_1, \tau)$ has a finite lifetime T so that $q(\eta_1, \tau)$ is only nonzero for $0 \leq \eta_1 \leq \ell$, $0 \leq \tau \leq T$. Then, provided $|x_1|$ is large in comparison with both ℓ and $U_1 T$, Equation (14.30) simplifies to

$$\phi(\mathbf{x}, t) = -\frac{c}{4\pi\rho_0 |x_1|} \int_0^T q(\eta_1^*, \tau) d\tau \quad . \tag{14.32}$$

We want to investigate the implications of this result for positions $(\mathbf{x}, t)$ on the Mach cone.

The integrand in (14.32) is only nonzero if η_1^* lies between 0 and ℓ. We see from (14.31) that for large $|\mathbf{x}|$ this requires $|\mathbf{x}| = ct$. The Mach cone condition (14.25) and $|\mathbf{x}| = ct$ combine to show $x_1 = |\mathbf{x}|/M$ and (14.32) can be rewritten as

$$\phi(\mathbf{x}, t) = -\frac{U_1}{4\pi\rho_0 |\mathbf{x}|} \int_0^T q(\eta_1^*, \tau) d\tau \quad . \tag{14.33}$$

It follows directly from the definition of η_1^* in (14.31) that, provided $|\mathbf{x}|$ is large in comparison with $c^2 T^2 / \ell$, η_1^* is effectively zero as τ varies over the interval $(0, T)$. Hence the *Mach wave sound* is given by

$$\phi(\mathbf{x}, t) = -\frac{U_1}{4\pi\rho_0 |\mathbf{x}|} \int_0^T q(0, \tau) d\tau \quad . \tag{14.34}$$

The Mach wave sound heard in the very far field therefore decays inversely with distance. The level involves the integral of the source strength over time. All the sound ever released during the entire life history of the source is heard by the distant observer in one big bang!

So far we have discussed the effect of motion on elementary sources whose strength is specified. However that oversimplifies the problem, because motion also changes the nature of a real source. For example, a stationary small pulsating body is equivalent to a monopole — an injection of mass at a point. This is not the case when the body moves because a moving body changing in size also includes a net unsteady force on the fluid and the sound field is that of a monopole and a coupled dipole.

Various methods have been devised to calculate the sound of a moving source. We will describe these methods and illustrate their strengths and weaknesses by using them to solve the same simple model problem, that of a pulsating sphere in uniform motion. By a change in reference frame, this is equivalent to a stationary sphere in a moving stream. We will assume the Mach number, M, of the oncoming flow to be low. Terms of order M will be retained but M^2 will be neglected in comparison with unity.

Consider a compressible irrotational mean flow specified by a velocity potential $\overline{\phi}(\mathbf{x})$, and suppose that an irrotational disturbance is introduced into the flow, the perturbation in potential being denoted by $\phi(\mathbf{x}, t)$. If the additional term ϕ is small so that only terms linear in it need be retained, the Navier–Stokes equation is greatly simplified. The details are given by Howe (1975). When the Mach number $|\nabla \overline{\phi}|/c$ is small everywhere, the time–independent potential $\overline{\phi}$ satisfies Laplace's equation

$$\nabla^2 \overline{\phi} = 0 \quad , \tag{14.35}$$

with appropriate boundary conditions. In Equation (14.35) terms of order M^2 have been neglected in comparison with unity.

The perturbation potential ϕ satisfies a convected wave equation. It is convected with the velocity of the mean potential $\nabla \overline{\phi}$

$$\left(\frac{\partial}{\partial t} + \frac{\partial \overline{\phi}}{\partial y_i} \frac{\partial}{\partial y_i} \right)^2 \phi - c^2 \nabla^2 \phi = 0 \quad . \tag{14.36}$$

There are several different ways of solving this equation.

14.2 TAYLOR'S TRANSFORMATION

Taylor (1978) solves the convected Equation (14.36) by the introduction of new variables $(\mathbf{Y}, T)$ where

$$\mathbf{Y} = \mathbf{y}, \qquad T = t + \overline{\phi}/c^2 \tag{14.37}$$

and $\phi(\mathbf{y}, t) = \Phi(\mathbf{Y}, T)$.

Then

$$\left. \frac{\partial}{\partial t} \right|_{\mathbf{y}} = \left. \frac{\partial}{\partial T} \right|_{\mathbf{Y}} \quad ,$$

and

$$\left. \frac{\partial}{\partial y_i} \right|_t = \left. \frac{\partial}{\partial Y_i} \right|_T + \frac{1}{c^2} \frac{\partial \overline{\phi}}{\partial y_i} \left. \frac{\partial}{\partial T} \right|_{\mathbf{Y}} \quad .$$

Correct to order M^2, Equation (14.36) becomes

$$\frac{\partial^2 \Phi}{\partial T^2} - c^2 \frac{\partial^2 \Phi}{\partial Y_i \partial Y_i} = 0 \quad , \tag{14.38}$$

i.e. when expressed in terms of the transformed variables $(\mathbf{Y}, T)$, the perturbed velocity potential satisfies a simple wave equation, which for certain geometries can be solved easily.

In order to illustrate the method we will use it to investigate the way in which a mean flow alters the sound field generated by an elementary source, a sphere of variable radius $A(t)$. We will choose $\mathbf{y}$–axes so that the centre of the sphere is at the origin, and far from the origin the mean flow velocity is in the negative y_1–direction with speed U_1, as shown in Figure 14.5. The Mach number $M = U_1/c$ is assumed small enough that M^2 is negligible in comparison with unity, and the pulsation is considered to be linear.

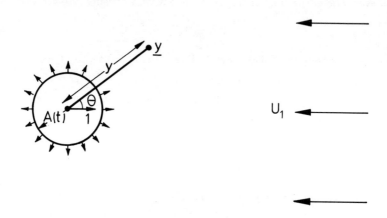

Figure 14.5 A pulsating sphere in a moving stream (the $\mathbf{y}$–reference frame).

The time–independent potential $\overline{\phi}$ satisfies Laplace's equation (14.35), together with a condition of zero normal velocity on the average body surface, a sphere of radius $a(= \overline{A}(t))$

$$\mathbf{n} \cdot \nabla \overline{\phi} = 0 \qquad \text{on} \qquad |\mathbf{y}| = a \quad,$$

where $\mathbf{n}$ is a unit vector in the radial direction. Far from the body there is uniform flow in the –1-direction,

$$\overline{\phi} \to -U_1 y_1 \quad \text{as} \quad |\mathbf{y}| \to \infty \quad,$$

and hence

$$\overline{\phi} = -U_1 \cos \Theta \left(y + \frac{a^3}{2y^2} \right) \quad, \tag{14.00}$$

where $y = |\mathbf{y}|$ and $\cos \Theta = y_1/y$.

On the instantaneous position of the surface of the sphere, the normal fluid velocity must be equal to the normal velocity of the surface. Therefore the boundary condition for ϕ is

$$\mathbf{n} \cdot \nabla \phi = -\mathbf{n} \cdot \nabla \overline{\phi} + \dot{A}(t) \qquad \text{on} \qquad y = A(t) \quad. \tag{14.40}$$

The first term on the right–hand side is proportional to $A - a$, because $\mathbf{n} \cdot \nabla\bar\phi$ is zero on a sphere of radius a. The second term is equal to the normal velocity due to the pulsation. There is a second constraint on ϕ; in the far field it must represent at outward–propagating wave. Substituting for $\bar\phi$ in (14.40) we obtain, after linearization in the perturbation of $A(t)$,

$$\frac{\partial\phi}{\partial y} = 3U_1 \cos\Theta\, \frac{A(t) - a}{a} + \dot A(t) \qquad \text{on} \qquad y = a \ . \tag{14.41}$$

In terms of the Taylor variables $(\mathbf{Y}, T)$, this is

$$n_i \left(\frac{\partial}{\partial Y_i} + \frac{1}{c^2}\frac{\partial\bar\phi}{\partial y_i}\frac{\partial}{\partial T} \right)\phi = \frac{3U_1\cos\Theta}{a}\left(A(T + 3M\cos\Theta a/2c) - a \right)$$

$$+ \dot A(T + 3M\cos\Theta a/2c) \qquad \text{on} \qquad Y = a \ . \tag{14.42}$$

Now $\mathbf{n} \cdot \nabla\bar\phi = 0$ on $Y = a$, and so Equation (14.42) simplifies to

$$\frac{\partial\Phi}{\partial Y} = \frac{3U_1\cos\Theta}{a}\left(A(T + 3M\cos\Theta a/2c) - a \right) + \dot A(T + 3M\cos\Theta a/2c)$$

$$\text{on} \ Y = a \ ; \tag{14.43}$$

M is small and we are neglecting terms of order M^2 in comparison with unity. The variation of time over the surface of the sphere can therefore be expanded in a Taylor series to give

$$\frac{\partial\Phi}{\partial Y} = \frac{3U_1\cos\Theta}{a}\left(A(T) - a \right) + \dot A(T) + \frac{3M\cos\Theta a}{2c}\ddot A(T) \ . \tag{14.44}$$

It remains to solve Equation (14.38) subject to the boundary condition (14.44) and the radiation condition at infinity. For simplicity we now assume that the pulsation is compact, i.e. that the sphere's diameter is much less than a typical wavelength. We make this approximation for convenience only. Solutions can be obtained for a non–compact sphere in a similar way, although of course the algebra and the expression obtained are more complicated (see Taylor 1979). For a compact sphere, the solution is

$$\Phi(\mathbf{Y}, T) = -\frac{\dot A(T - Y/c)a^2}{Y} + \frac{3}{2}U_1 a^2 \frac{\partial}{\partial Y_1}\left(\frac{A(T - Y/c) - a}{Y} \right) \ . \tag{14.45}$$

In the far field where Y is large, Φ simplifies to

$$\Phi(\mathbf{Y}, t) = - \frac{\dot{A}(T - Y/c)a^2}{Y}\left(1 + \frac{3}{2}M\cos\Theta\right) \quad . \tag{14.46}$$

At large distances from the sphere $\bar{\phi} \propto -U_1 y_1$ and correct to order M^2, we can rewrite Equation (14.46) as

$$\phi(\mathbf{y}, t) = - \frac{\dot{A}(t - My_1/c - y/c)a^2}{y(1 - M\cos\Theta)^{3/2}} \quad . \tag{14.47}$$

We see then that the main effect of flow past a stationary vibrating sphere is to modify the amplitude of the radiated potential by the Döppler factor $(1 - M\cos\Theta)^{-3/2}$. This effect is zero at 90° to the flow direction.

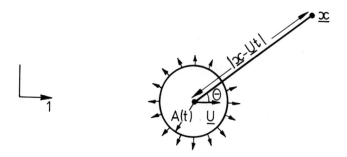

Figure 14.6 A pulsating sphere moving with velocity $\mathbf{U}$ (the $\mathbf{x}$-reference frame).

Let us change to a coordinate system $\mathbf{x} = \mathbf{y} + \mathbf{U}t$ as shown in Figure 14.6. In this frame of reference the flow at infinity is at rest and the sphere moves with a velocity $\mathbf{U}$. In terms of $\mathbf{x}, t$

$$\phi = - \frac{\dot{A}(t - Mx_1/c - |\mathbf{x} - \mathbf{U}t|/c)a^2}{|\mathbf{x} - \mathbf{U}t|(1 - M\cos\Theta)^{3/2}} \tag{14.48}$$

for a compact sphere. Differentiation of ϕ with respect to time at fixed $\mathbf{x}$ gives for large $|\mathbf{x} - \mathbf{U}t|$,

$$\left.\frac{\partial\phi}{\partial t}\right|_{\mathbf{x}} = -(1 + M\cos\Theta)\frac{\ddot{A}(t - Mx_1/c - |\mathbf{x} - \mathbf{U}t|/c)a^2}{|\mathbf{x} - \mathbf{U}t|(1 - M\cos\Theta)^{3/2}} \quad , \tag{14.49}$$

where $M \cos \Theta = M y_1/y = M(x_1 - U_1 t)/|\mathbf{x} - \mathbf{U}t|$, correct to order M. We see from a comparison of Equations (14.48) and (14.49) that

$$\frac{1}{\phi} \left.\frac{\partial \phi}{\partial t}\right|_{\mathbf{x}} = (1 + M \cos \Theta) \frac{\ddot{A}}{A} = \frac{1}{1 - M \cos \Theta} \frac{\ddot{A}}{A} \quad, \tag{14.50}$$

in agreement with the Döppler shift of frequency demonstrated in Equation (14.7).

If Equations (14.22) and (14.23) are used to rewrite the reception–time coordinates $R = |\mathbf{x} - \mathbf{U}t|$ and Θ in terms of emission–time variables r and θ, Equation (14.49) leads to an identical expression for the pressure perturbation due to a moving pulsating sphere to that given in Dowling (1976).

We have seen that Taylor's transformation provides a convenient way of solving the convected wave equation (14.36), and that for simple boundary conditions there is no need to assume that the source is compact.

14.3 HOWE'S METHOD

Howe (1975) has a different method of solving the convected wave equation (14.36). He takes the Fourier transform of this equation and considers

$$\left(-i\omega + \frac{\partial \overline{\phi}}{\partial y_i} \frac{\partial}{\partial y_i}\right)^2 \widetilde{\phi} - c^2 \nabla^2 \widetilde{\phi} = 0 \quad . \tag{14.51}$$

We illustrate Howe's method by solving (14.51) in a region outside a surface S which is impenetrable to the mean flow, i.e. $\mathbf{n} \cdot \nabla \overline{\phi} = 0$ on S. Howe introduces a Green function $G(\mathbf{y}|\mathbf{y}_0)$ which satisfies the *reversed–flow* equation

$$\left(-i\omega - \frac{\partial \overline{\phi}}{\partial y_1} \frac{\partial}{\partial y_1}\right)^2 G - c^2 \nabla^2 G = \delta(\mathbf{y} - \mathbf{y}_0) \tag{14.52}$$

and the boundary condition

$$\mathbf{n} \cdot \nabla G = 0 \quad \text{on} \quad S \quad . \tag{14.53}$$

Equation (14.51) is then multiplied by G and the product of (14.52) and $\widetilde{\phi}$ subtracted from it. Integration of the result over the region outside the surface S yields

$$\widetilde{\phi}(\mathbf{y}_0, \omega) = -c^2 \int_S G \frac{\partial \widetilde{\phi}}{\partial n} d^2 \mathbf{y} \quad, \tag{14.54}$$

where $\mathbf{n}$ is the outward normal to the surface S.

In the pulsating sphere problem, S is the surface $y = a$ and the Fourier transform of (14.41) gives the boundary condition

$$\frac{\partial \widetilde{\phi}}{\partial n} = \left(\frac{3U_1 \cos \Theta}{a} - i\omega \right) \widetilde{A - a} \quad \text{on} \quad y = a \quad . \tag{14.55}$$

Once we have found $G(\mathbf{y}|\mathbf{y}_0)$ we can substitute for $\partial \widetilde{\phi}/\partial n$ in Equation (14.54). Integration will then determine $\widetilde{\phi}$, the Fourier transform of the velocity potential.

Howe (1975) introduces an approximate way of determining $G(\mathbf{y}|\mathbf{y}_0)$ for use with a compact sphere. The method is an extension of the steps between Equations (12.64) and (12.66) to account for the effect of a mean flow. The Green function is decomposed into two parts by writing $G = G_i + G_s$, where G_i describes the field incident upon the sphere, while G_s accounts for the scattered or reflected field. G_i is the solution to the convected wave equation in the absence of the surface S

$$\left(-i\omega + U_1 \frac{\partial}{\partial y_1} \right)^2 G_i - c^2 \nabla^2 G_i = \delta(\mathbf{y} - \mathbf{y}_0) \quad , \tag{14.56}$$

and the solution is

$$G_i = \frac{1}{4\pi c^2 |\mathbf{y} - \mathbf{y}_0|} \exp\left[i\omega(|\mathbf{y} - \mathbf{y}_0| - \mathbf{M} \cdot (\mathbf{y} - \mathbf{y}_0))/c \right] \quad , \tag{14.57}$$

where $\mathbf{M} = (M, 0, 0)$. For $\omega y/c$ small, the exponential can be expanded and if we retain only the first two terms in this expansion, we have for $y_0 = |\mathbf{y}_0| >> y$,

$$G_i \approx \frac{1}{4\pi c^2 y_0} \left(1 - \frac{i\omega \mathbf{y}}{c} \cdot \left(\mathbf{M} + \frac{\mathbf{y}_0}{y_0} \right) \right) \exp\left[i\omega \left(y_0 + \mathbf{M} \cdot \mathbf{y}_0 \right)/c \right] \quad . \tag{14.58}$$

For positions $\mathbf{y}$ close to the cylinder, the convected wave equation (14.52) reduces to $\nabla^2 G = 0$. It is evident that the expansion for G_i in (14.58) satisfies this equation. We need to find the scattered field G_s such that $\nabla^2 G_s = 0$, with $\partial G_s/\partial y = - \partial G_i/\partial y$ on $y = a$ and $G_s \rightarrow 0$ as $y \rightarrow \infty$. By inspection

$$G_s \simeq - \frac{1}{4\pi c^2 y_0} \frac{i\omega \mathbf{y}}{c} \cdot \left(\mathbf{M} + \frac{\mathbf{y}_0}{y_0} \right) \frac{a^3}{2y^3} \exp\left[i\omega \left(y_0 + \mathbf{M} \cdot \mathbf{y}_0 \right)/c \right] \quad . \tag{14.59}$$

The Green function, obtained by adding Equations (14.58) and (14.59), is

$$G \simeq \frac{1}{4\pi c^2 y_0} \left(1 - \frac{i\omega \mathbf{y}}{c} \cdot \left(\mathbf{M} + \frac{\mathbf{y}_0}{y_0} \right) \left(1 + \frac{a^3}{2y^3} \right) \right) \exp\left[i\omega \left(y_0 + \mathbf{M} \cdot \mathbf{y}_0 \right)/c \right] .$$

$$\tag{14.60}$$

The Fourier transform of the velocity perturbation due to a compact pulsating sphere can then be obtained by substituting for G into Equation (14.54). Before doing that we discuss the form of G.

G can be simplified by introducing a new space variable

$$\mathbf{Y} = \mathbf{y}\left(1 + \frac{a^3}{2y^3}\right) \quad . \tag{14.61}$$

In terms of $\mathbf{y}$, again only correct to order $\omega y/c$, G may be rewritten as

$$G \simeq \frac{1}{4\pi c^2 y_0} \exp\left[i\omega\left(|\mathbf{y}_0 - \mathbf{Y}| + \mathbf{M}\cdot(\mathbf{y}_0 - \mathbf{Y})\right)\Big/c\right] \quad . \tag{14.62}$$

Since $y_0 \gg a$, we can rewrite G more symmetrically by introducing

$$\mathbf{Y}_0 = \mathbf{y}_0\left(1 + \frac{a^3}{2y_0^3}\right) \quad .$$

Then

$$G(\mathbf{y}|\mathbf{y}_0) = \frac{1}{4\pi c^2 |\mathbf{Y}_0 - \mathbf{Y}|} \exp\left[i\omega\left(|\mathbf{Y}_0 - \mathbf{Y}| + \mathbf{M}\cdot\left(\mathbf{Y}_0 - \mathbf{Y}\right)\right)\Big/c\right] \quad . \tag{14.63}$$

The effect of the flow around the sphere is expressed by a variable change from $\mathbf{y}$ to $\mathbf{Y}$. Far from the body $\mathbf{y}$ and $\mathbf{Y}$ are equal, but near the body they are quite different. The body's effect on the unsteady flow is equivalent to a distortion of the local three–dimensional Euclidean space.

Howe (1975) comments that the results for sphere can be generalized to an arbitrary compact body. If $\phi_i^*(\mathbf{y})$, $i = 1, 2, 3$, is the perturbation in velocity potential of a uniform flow in the i direction due to the presence of the body, G has the form given in Equation (14.63) with

$$\mathbf{Y} = \mathbf{y} + \boldsymbol{\phi}^*(\mathbf{y}) \quad . \tag{14.64}$$

Rewriting G in this way aids generalization to arbitrarily shaped bodies, but for our present purposes the form in Equation (14.60) is more convenient. When we substitute it into Equation (16.54) and use the boundary conditions in (14.55), we obtain

$$\widetilde{\phi}(\mathbf{y}_o, \omega) = -\frac{1}{4\pi y_0} \int_S \left(\frac{3U_1\cos\Theta}{a} - i\omega\right)\left(1 - \frac{i3\omega}{2c}\left(aM\cos\Theta + \frac{\mathbf{y}\cdot\mathbf{y}_0}{y_0}\right)\right) dS$$
$$\times \widetilde{A - a} \exp\left[i\omega\left(y_0 + \mathbf{M}\cdot\mathbf{y}_0\right)\Big/c\right] \quad .$$

$$\tag{14.65}$$

The integration can be performed in a straightforward way to give

$$\tilde{\phi}(\mathbf{y}_0, \omega) = \frac{i\omega a^2}{y_0} \left(1 + \frac{3}{2} M \cos \Theta_0 \right) \widetilde{A - a} \exp\left[i\omega\left(y_0 + \mathbf{M} \cdot \mathbf{y}_0\right)\Big/c\right] \quad, \qquad (14.66)$$

where $\cos \Theta_0 = y_{01}/y_0$. Hence

$$\phi(\mathbf{y}_0, t) = \frac{1}{2\pi} \int \tilde{\phi}(\mathbf{y}_0, \omega) e^{-i\omega t} dw$$

$$= -\frac{a^2 \dot{A}(t - y_0/c - \mathbf{M} \cdot \mathbf{y}_0/c)}{y_0 (1 - M \cos \Theta_0)^{3/2}} \quad. \qquad (4.67)$$

This is of course the same result as we obtained by Taylor's method. Unlike Taylor's transformation, Howe's technique can only be applied when the source region is compact and then it only gives the distant acoustic field.

14.4 THE METHOD OF MATCHED ASYMPTOTIC EXPANSIONS

Near the body ϕ varies over a lengthscale of order a, but far away compressibility effects are dominant and the relevant measure of length is the wavelength c/ω. For a compact pulsation $\epsilon = \omega a/c << 1$, and these lengthscales become disparate as the compactness ratio ϵ tends to zero. Then the problem for ϕ is amenable to solution by the method of matched asymptotic expansions described in Chapter 6.

The equations for ϕ can be rewritten in terms of nondimensional variables by nondimensionalizing the space coordinate $\mathbf{y}$ with respect to c/ω and time with respect to the timescale of the pulsation $1/\omega$. Then Equation (14.36) becomes

$$\left(\frac{\partial}{\partial t} + \frac{\partial \overline{\phi}}{\partial y_i} \frac{\partial}{\partial y_i}\right)^2 \phi - \nabla^2 \phi = 0 \quad. \qquad (14.68)$$

The same symbols have been used to denote the dimensional and nondimensional variables, but in this equation $\mathbf{y}$ and t are now dimensionless and for the pulsating sphere problem

$$\overline{\phi} = -My_1 \left(1 + \epsilon^3/2y^3\right) \quad. \qquad (14.69)$$

We scale ϕ on ac, and the obvious choice for A is to nondimensionalize it with respect to its mean value a, since A/a is only linearly disturbed from unity. Then the boundary condition (14.41) becomes

$$\frac{\partial \phi}{\partial y} = \frac{3M \cos \Theta}{\epsilon} (A - 1) + \dot{A} \qquad \text{on} \qquad y = \epsilon \quad, \qquad (14.70)$$

where all the parameters are now nondimensional. Equations (14.68) and (14.70) are to be solved subject to a condition that ϕ has outward wave behaviour at infinity.

For small values of ϵ, we can obtain a solution of the form

$$\phi(\mathbf{y}, \epsilon) = \epsilon^\lambda \phi_\lambda(\mathbf{y}, \epsilon) + O(\epsilon^{\lambda+1}) \quad,$$

where λ, the leading order of the far field, is as yet unknown. This outer expansion tacitly assumes that y is at least of order unity, and will not be valid near the body where y is small.

From Equation (14.68) and the form for $\overline{\phi}$ given in (14.69), we see that ϕ_λ satisfies a convected wave equation

$$\left(\frac{\partial}{\partial t} - M \frac{\partial}{\partial y_1}\right)^2 \phi_\lambda - \nabla^2 \phi_\lambda = 0 \quad, \tag{14.71}$$

together with the radiation condition at infinity. Hence

$$\phi^{(\lambda)}(\mathbf{y}, \epsilon) = \epsilon^\lambda \left(\frac{f(t - y - My_1)}{y} + \frac{\partial}{\partial y_i}\left(\frac{f_i(t - y - My_1)}{y}\right) + \ldots\right) \quad, \tag{14.72}$$

where $\phi^{(\lambda)}$ denotes the outer expansion correct to order ϵ^λ. The lowest order of the far field, λ, and the functions $f, f_i, \ldots$, are to be determined from a consideration of the flow in the inner region.

The flow near the body scales on a stretched variable $\mathbf{Y}$, where $\mathbf{Y} = \mathbf{y}/\epsilon$. When the outer expansion is rewritten in terms of the inner coordinate $\mathbf{Y}$

$$\phi^{(\lambda)}(\epsilon\mathbf{Y}, \epsilon) = \epsilon^{\lambda-1} \frac{f(t - \epsilon Y - \epsilon MY_1)}{Y} + \epsilon^{\lambda-2} \frac{\partial}{\partial Y_i}\left(\frac{f_i(t - \epsilon Y - \epsilon MY_1)}{Y}\right) + \ldots \quad,$$
$$\tag{14.73}$$

a new power series in ϵ is obtained. Using the notation introduced in Chapter 6, we denote this series truncated after ϵ_n, by $\phi^{(\lambda,n)}$. Once the flow near the body is known, the functions $f, f_i, \ldots$ can be found by using the Asymptotic Matching Principle in (6.21) to relate the $\phi^{(\lambda,n)}$ to the inner flow variables.

We use the stretched coordinate $\mathbf{Y}$ to describe the flow near the body, and write

$$\phi(\mathbf{y}, \epsilon) \equiv \Phi(\mathbf{Y}, \epsilon) \quad. \tag{14.74}$$

An inner expansion for ϕ can be obtained by expressing Φ in ascending powers of ϵ

$$\Phi(\mathbf{Y}, \epsilon) = \Phi_0 + \epsilon\Phi_1 + O(\epsilon^2) \quad. \tag{14.75}$$

When the convected wave equation (14.68) is rewritten in terms of the variable $\mathbf{Y}$, we find that, correct to order M, $\Phi(\mathbf{Y}, \epsilon)$ is the solution of

$$\frac{\partial^2 \Phi}{\partial Y_i \partial Y_i} - 2\epsilon \frac{\partial \overline{\Phi}}{\partial Y_i} \frac{\partial^2 \Phi}{\partial Y_i \partial t} - \epsilon^2 \frac{\partial^2 \Phi}{\partial t^2} = 0 \quad, \tag{14.76}$$

with

$$\overline{\Phi} = -MY_1 \left(1 + 1/2Y^3 \right) \quad, \tag{14.77}$$

which satisfies the boundary condition

$$\frac{\partial \Phi}{\partial Y} = 3MY_1(A-1) + \epsilon \dot{A} \qquad \text{on a sphere of radius 1} \quad. \tag{14.78}$$

By substituting the power series expansion for Φ into both Equation (14.76) and the boundary condition (14.78), and equating each coefficient of ϵ to zero, we obtain a series of equations

$$\nabla^2 \Phi_0 = 0 \quad \text{with} \quad \frac{\partial \Phi_o}{\partial Y} = 3MY_1(A-1) \quad \text{on} \qquad Y = 1 \tag{14.79a}$$

$$\nabla^2 \Phi_1 = 2 \frac{\partial \overline{\Phi}}{\partial Y_i} \frac{\partial^2 \Phi_0}{\partial Y_i \partial t} \quad \text{with} \quad \frac{\partial \Phi_1}{\partial Y} = \dot{A} \qquad \text{on} \qquad Y = 1 \tag{14.79b}$$

and so on. The solution of (14.79) is

$$\Phi_0 = - \frac{3MY_1}{2Y^3} (A-1) \quad, \tag{14.80a}$$

$$\Phi_1 = - \frac{\dot{A}}{Y} \quad. \tag{14.80b}$$

We denote the inner series terminated after ϵ^n by $\Phi^{(n)}$, i.e. $\Phi^{(0)}(\mathbf{Y}, \epsilon) = \Phi_0$, $\Phi^{(1)}(\mathbf{Y}, \epsilon) = \Phi_0 + \epsilon \Phi_1$. When rewritten in terms of the outer variable $\mathbf{y}$

$$\Phi^{(0)}(\mathbf{y}/\epsilon, \epsilon) = -\epsilon^2 M \frac{3y_1}{2y^3} (A-1) \quad, \tag{14.81a}$$

$$\Phi^{(1)}(\mathbf{y}/\epsilon, \epsilon) = \Phi^{(0)}(\mathbf{y}/\epsilon, \epsilon) - \epsilon^2 \frac{\dot{A}}{y} \quad; \tag{14.81b}$$

a new power series in ϵ is obtained. As in Chapter 6, we define a set of functions $\Phi^{(n,m)}$, where $\Phi^{(n,m)}$ is the expansion of $\Phi^{(n)}(\mathbf{y}/\epsilon, \epsilon)$ terminated after order ϵ^m. This inner solution must match to an outer expansion of the form given in (14.73). The Asymptotic Matching Principle in Equation (6.21) states that

$$\phi^{(\lambda,n)} = \Phi^{(n,\lambda)} \quad, \tag{14.82}$$

where $\phi^{(\lambda,n)}$ has already been defined as the outer expansion correct to order λ, expressed in terms of the inner variables and then terminated after order ϵ^n. It is at once apparent that $\lambda = 2$, and that the quadrupoles and all higher multipoles in the outer expansion must be zero. Then (14.73) becomes

$$\phi^{(2)}(\epsilon\mathbf{Y}, \epsilon) = f_i(t)\frac{\partial}{\partial Y_i}\left(\frac{1}{Y}\right) + \epsilon\left(\frac{f(t)}{Y} - M\dot{f}_i(t)\frac{\partial}{\partial Y_i}\left(\frac{Y_1}{Y}\right)\right) + O(\epsilon^2) \quad . \quad (14.83)$$

When the rule (14.82) is applied to (14.83), we find that

$$f_i(t) = \frac{3M}{2}(A-1)\delta_{i1}$$

and

$$f(t) = -\dot{A} \quad .$$

We have now obtained the far-field solution for ϕ to lowest order in the compactness ratio:

$$\phi(\mathbf{y}, t) = \epsilon^2\left(-\frac{\dot{A}(t - y - My_1)}{y} + 3M\frac{\partial}{\partial y_1}\left(\frac{A(t - y - My_1) - 1}{2y}\right)\right). \quad (14.84)$$

After evaluation of the derivative for large y, this becomes

$$\phi(\mathbf{y}, t) = -\frac{\epsilon^2}{y}\dot{A}\left(t - y - My_1\right)\left(1 + \frac{3}{2}M\cos\Theta\right) \quad , \quad (14.85)$$

or, in dimensional parameters, our previous result

$$\phi(\mathbf{y}, t) = -\frac{a^2\dot{A}(t - (y + My_1)/c)}{y(1 - M\cos\Theta)^{3/2}} \quad . \quad (14.86)$$

14.5 THE LIGHTHILL THEORY

Lighthill's theory of aerodynamic noise provides yet another method of calculating the sound due to a compact pulsating sphere in a mean flow, although the details will not be given here. When applying the Lighthill theory it is convenient to take the $\mathbf{x}$–frame of reference in which the fluid is at rest at infinity and the centre of the sphere has a velocity $\mathbf{U} = (U_1, 0, 0)$. We choose to take a control surface Σ, which also moves with velocity $\mathbf{U}$. Σ is the surface of a sphere of fixed radius b, where b is just large enough to enclose the pulsating sphere at all times. Fluid crosses this control surface.

When the Lighthill theory is applied outside a control surface, it leads to surface monopole and dipole sources in addition to a volume quadrupole source distribution. An expression for the far–field radiated sound was determined in Equation (11.117). Applying that result here, we obtain

$$4\pi|\mathbf{x}|p'(\mathbf{x},t) = \frac{x_i x_j}{|\mathbf{x}|^2 c^2}\frac{\partial^2}{\partial t^2}\int_V\left[\frac{T_{ij}}{1-M_r}\right]d^3\mathbf{y} + \frac{x_i}{|\mathbf{x}|c}\frac{\partial}{\partial t}\int_\Sigma\left[\frac{p'n_i + \rho u_i(u_j - U_j)n_j}{|1-M_r|}\right]d\Sigma$$

$$+ \frac{\partial}{\partial t}\int_\Sigma\left[\frac{\rho_0 u_i + \rho'(u_i - U_i)}{|1-M_r|}\right]n_i d\Sigma \quad ;$$

$$(14.87)$$

T_{ij} is the Lighthill stress tensor, $T_{ij} = \rho u_i u_j + p_{ij} - c^2\rho'\delta_{ij}$, $\mathbf{u}$ the fluid velocity, p_{ij} the compressive stress tensor (in an inviscid fluid $p_{ij} = p'\delta_{ij}$) and ρ the actual fluid density. V is the region outside Σ and $\mathbf{n}$ a unit vector in the direction of the outward normal to Σ. $\mathbf{y}$ is a coordinate measured from the centre of the body, and the square brackets denote that the function they enclose is to be evaluated at retarded time τ^*, where

$$\tau^* = t - |\mathbf{x} - \mathbf{y} - \mathbf{U}\tau^*|/c$$

and in the far–field approximation, $\cos\theta = x_1/|\mathbf{x}|$.

We see that the terms on the right–hand side of Equation (14.87) only involve the flow near the surface of the sphere, and this was determined when considering the inner solution in the method of matched asymptotic expansions. In fact, the nondimensionalization used there is convenient to keep a track of the order of the various terms. When the inner solution is substituted into the integrands on the right–hand side of Equation (14.87), the integrals can be evaluated to yield a form for the pressure in agreement with the previous result, Equation (14.49). Both the monopole and dipole terms in (14.87) are nonzero showing that a moving body producing a mass flux necessarily has a fluctuating force and momentum associated with it. These additional sources are significant and lead to a term in the distant pressure, which is only of order M smaller than the contribution due to the monopole.

The Lighthill theory therefore provides an alternative method to matched asymptotic expansions (MAE) for going from an inner solution to the far field. It is interesting to note that, in a related problem, the sound produced by a rigid sphere of radius a which is undergoing a compact vibration in addition to its mean velocity $\mathbf{U}$, the Lighthill theory requires less information about the inner solution than MAE. Three inner terms need to be determined in MAE, since to apply the matching principle, we need to know Φ_2, the order ϵ^2 term in the inner potential to determine the outer field. But the effect of the derivatives in Lighthill's theory ensures that Φ_2 gives rise to a negligible far field, and only two inner terms need to be calculated in this method. This occurs because of the way in which Φ_2 enters the matching procedure. Although $\epsilon^2\Phi_2$ is small in magnitude near the body, it only decays slowly with distance from the sphere and so in the "matching region" it is comparable with the other terms in the inner expansion, and must be evaluated to obtain the outer field. The Lighthill theory is different. The structure of the Navier–Stokes equation is used to express the distant acoustic field in terms of integrals over the source region. The main sources, the monopole and dipole, depend only on the flow on the surface of the body and there the $\epsilon^2\Phi_2$ term is negligible in comparison with the leading–order terms (Dowling 1978).

REFERENCES

Dowling, A.P. (1976). Convective amplification of real simple sources. J. Fluid Mech. 74:529–546.

Dowling, A.P. (1978). Acoustic Sources in Motion. Ph. D. Thesis, University of Cambridge.

Howe, M.S. (1975). The generation of sound by aerodynamic sources in an inhomogeneous steady flow. J. Fluid Mech. 67:597–610.

Taylor, K. (1978). A transformation of the acoustic equation with implications for wind–tunnel and low–speed flight tests. Proc. R. Soc. Lond. A363:271–281.

Taylor, K. (1979). Acoustic generation by vibrating bodies in homentropic potential flow at low Mach number. J. Sound Vib. 65:125–136.

Rayleigh, J.W.S. (1945). The Theory of Sound, II. Dover Publications.

15. PROPELLER AND HELICOPTER NOISE

15.1 INTRODUCTION

Plans for aircraft powered by high–speed prop fans means that propeller noise can be expected to be an increasingly important aeroacoustic source. Both the near and far acoustic fields of propellers are of interest, with the near field influencing unsteady airframe loading and passenger comfort, and the far field determining the noise level near airports. The propeller is also a major source of noise for underwater vehicles.

The force exerted on the propeller by the airflow can be conveniently separated into two components, a thrust T in the direction of forward motion, and a drag force D opposing the rotation. For an aeroplane propeller it is reasonable to assume in the first instance that the flow coming on to the blade is uniform. Then the thrust is unchanging and the drag is constant in the rotating frame. These steady loads generate harmonic sounds. The broad–band problem is more complicated. Propellers which operate in inhomogeneous flow, for example in a turbulent atmosphere or in the wake of a vessel, support unsteady loads which generate broad–band noise.

We have seen that forces can generate acoustic fields, the sound produced by a force $\mathbf{f}(t)$ whose point of application is in motion with velocity $c\mathbf{M}$ being given by (cf. Chapter 11, Equations (11.115) and (11.127))

$$p'(\mathbf{x}, t) = -\frac{\partial}{\partial x_i}\left[\frac{f_i(\tau)}{4\pi r|1 - M_r|}\right] \quad . \tag{15.1}$$

The square brackets denote that the function they enclose is to be evaluated at retarded time τ. r is the distance between the observer and the source at emission time, and cM_r the component of the source velocity in the direction of the observer. In the far field the derivative $\frac{-\partial}{\partial x_i}$ simplifies to $\frac{x_i}{xc}\frac{\partial}{\partial t}$. Even a steady force in unsteady motion generates an acoustic field. Since the propeller blades are rigid they undergo no volume change and so there is no net monopole source. However, the retarded time variation between mass flux on the two sides of the blade does produce a sound field known as *thickness noise*. These noise sources can be conveniently analyzed within the framework provided by the Ffowcs Williams–Hawkings equation. The density perturbation produced by a rigid surface S in motion is given exactly by (Ffowcs Williams & Hawkings 1969, Equation (5.1))

$$4\pi c^2 \rho'(\mathbf{x}, t) = \frac{\partial^2}{\partial x_i \partial x_j} \int \left[\frac{T_{ij}}{r|1 - M_r|} \right] d\boldsymbol{\eta} - \frac{\partial}{\partial x_i} \int_S \left[\frac{p_{ij} n_j}{r|1 - M_r|} \right] dS(\boldsymbol{\eta})$$
$$+ \frac{\partial}{\partial t} \int_S \left[\frac{\rho_0 u_n}{r|1 - M_r|} \right] dS(\boldsymbol{\eta}) \quad .$$

$$(15.2)$$

The first integral on the right–hand side is over the region outside the rigid blade surfaces S. T_{ij} is the Lighthill's stress tensor $T_{ij} = \rho u_i u_j + p_{ij} - c^2 \rho' \delta_{ij}$, p_{ij} is the compressive stress tensor, and $\mathbf{u}$ the fluid velocity. $\boldsymbol{\eta}$ is a coordinate system, moving with velocity $\mathbf{v} = c\mathbf{M}$. The velocity $\mathbf{v}$ consists of, at most, a translation and rotation and is solenoidal. The square brackets again denote retarded time τ,

$$\tau = t - \left| \mathbf{x} - \boldsymbol{\eta} - \int^t \mathbf{v}(\boldsymbol{\eta}, \tau') d\tau' \right| / c \quad ;$$

$$(15.3)$$

r is the distance between the observer and the source element at emission time, $r = c(t - \tau)$, and $M_r = \mathbf{M} \cdot (\mathbf{x} - \boldsymbol{\eta} - \int^t \mathbf{v}(\boldsymbol{\eta}, \tau') d\tau')/r$.

The quadrupole source term in Equation (15.2) is unimportant for thin low Mach–number blades which only disturb the flow linearly since T_{ij} involves terms quadratic in the fluid perturbation. But for blades with supersonic tip speeds and shock waves there can be a substantial region of the flow with properties different from ambient and the contribution from the quadrupole term is then significant. For thick blades this term is most important for transonic flows, and is negligible for subsonic and fully supersonic flow (Hanson 1979). The second source term in (15.2) is due to the force exerted on the fluid by the blade. This is usually referred to as *loading noise*. The last term in Equation (15.2) is in monopole form and describes the *thickness noise*. This form would be appropriate if the surface were changing in volume but it is not very useful for rigid surfaces. The true nature of the source can be brought out by an exact rearrangement. First note that on the surface S the normal components of the fluid and surface velocities are equal, $u_n = v_n$. The "monopole term" can therefore be rewritten as

$$\frac{\partial}{\partial t} \int_S \left[\frac{\rho_0 v_n}{r|1 - M_r|} \right] dS(\boldsymbol{\eta}) = \frac{\partial}{\partial t} \int \rho_0 v_i \frac{\partial H}{\partial y_i} (f) \, \delta(t - \tau - |\mathbf{x} - \mathbf{y}|/c) \, d\mathbf{y} \, d\tau \quad , \quad (15.4)$$

where, as in Chapter 11, Equation (11.125), H is the Heaviside function and $f(\mathbf{y}, \tau)$ is a function that is positive in the fluid and negative within the regions occupied by blades. The blade convection velocity $\mathbf{v}$ is clearly defined for all points within the blade surfaces and $\mathbf{v}.\nabla H$ can be rearranged as $-\nabla \cdot (\mathbf{v}(1 - H(f)))$ since $\mathbf{v}$ is solenoidal. The monopole term is therefore equal to

$$\frac{\partial}{\partial t} \int \rho_0 \frac{\partial}{\partial y_i}(v_i(1 - H(f))) \delta(t - \tau - |\mathbf{x} - \mathbf{y}|/c) \, d\mathbf{y} \, d\tau \quad,$$

which becomes (cf. the first line, last term of Equation (11.125))

$$\frac{\partial^2}{\partial t \, \partial x_i} \int \rho_0 v_i(1 - H(f)) \delta(t - \tau - |\mathbf{x} - \mathbf{y}|/c) \, d\mathbf{y} \, d\tau \quad,$$

after integration by parts. Finally, changing the integration variable from $\mathbf{y}$ to $\boldsymbol{\eta} = \mathbf{y} - \int^\tau \mathbf{v}(\boldsymbol{\eta}, \tau')d\tau'$, enables the τ–integration to be evaluated. This shows that an equivalent form for Equation (15.2) is

$$4\pi c^2 \rho'(\mathbf{x}, t) = \frac{\partial^2}{\partial x_i dx_j} \int \left[\frac{T_{ij}}{r|1 - M_r|}\right] d\boldsymbol{\eta} - \frac{\partial}{\partial x_i} \int_s \left[\frac{p_{ij}n_j}{r|1 - M_r|}\right] dS(\boldsymbol{\eta})$$
$$+ \frac{\partial^2}{\partial x_i \partial t} \int_{V_0} \left[\frac{\rho_0 v_i}{r|1 - M_r|}\right] d\boldsymbol{\eta} \quad. \tag{15.5}$$

We see from this equivalent form that for compact rigid blade surfaces the last source term in (15.2) is actually a dipole, whose strength is equal to the momentum displaced by the blade, a point already mentioned in Chapter 11, Equation (11.127).

Since the main sound field is periodic, it is sometimes convenient to investigate its structure in the frequency domain, and we begin by considering this spectral decomposition of propeller noise.

15.2 SPECTRAL DECOMPOSITION

As an introductory example we will discuss the hypothetical case of a thin propeller blade with no forward motion generating a steady thrust and torque (drag). There are two consequences of the thin blade assumption; the blades only disturb the flow linearly making the quadrupole source terms in (15.5) negligible except at transonic speeds, and secondly the surface integrals which are at present over both sides of the blade surface can be approximated to a single integral over its mean chord surface. We will use the approach of Hawkings & Lowson (1974) to recover the classical results of Gutin (1938).

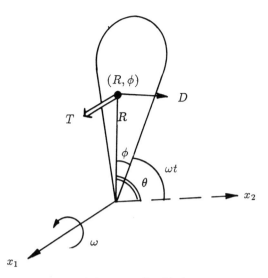

Figure 15.1 The geometry of the propeller blade.

We choose a Cartesian coordinate system such that the 1–axis is parallel to the flow and the mean chord surface is in the 2–3 plane; we will also use a moving cylindrical polar coordinate system R, ϕ, x_1, as shown in Figure 15.1; the blade rotates about the 1–axis with angular frequency ω. For a single blade, Equation (15.5) simplifies to

$$4\pi c^2 \rho'(\mathbf{x}, t) = -\frac{\partial}{\partial x_i} \int \left[\frac{f_i}{r|1 - M_r|} \right] R dR d\phi + \frac{\partial^2}{\partial x_i \partial t} \int \left[\frac{\rho_0 v_i h}{r|1 - M_r|} \right] R dR d\phi \quad ,$$

$$(15.6)$$

where f_i is the force per unit area exerted on the flow by the blade and $h(r, \phi)$ the blade thickness. The source at angle ϕ in the rotating frame is at $\theta = \phi + \omega\tau$ in the absolute frame. Our assumption for this hypothetical steadily–loaded propeller is that:

$$\mathbf{f} = (-T, \ -D\sin\theta, D\cos\theta)$$

$$(15.7)$$

$$\mathbf{v} = (\ 0, \ -\omega R\sin\theta, \omega R\cos\theta)$$

where the thrust T and drag D (both per unit blade area) are functions of the rotating frame coordinates (R, ϕ).

For an observer in the far field, (15.6) simplifies to

$$4\pi c^2 \rho'(\mathbf{x}, t) = \frac{x_i}{x^2 c} \frac{\partial}{\partial t} \int \left[\frac{f_i}{|1 - M_r|} \right] R dR d\phi - \frac{\partial^2}{\partial t^2} \int \left[\frac{\rho_0 v_i h}{|1 - M_r|} \right] R dR d\phi \quad , (15.8)$$

432

and the retarded time τ is the solution of

$$\tau = t - \frac{|x|}{c} + \frac{x_2 R \cos\theta}{xc} + \frac{x_3 R \sin\theta}{xc} \quad ,$$

(15.9)

$$= t - \frac{|x|}{c} + \frac{\sigma R}{xc} \cos(\theta - \Theta) \quad ,$$

where $\Theta = \phi + \omega\tau$ and σ and Θ are defined by

$$x_2 = \sigma \cos\Theta$$

(15.10)

$$x_3 = \sigma \sin\Theta \quad .$$

The sound pressure is harmonic with period $2\pi/\omega$, the amplitude of the nth harmonic $\tilde{p}_n(\mathbf{x})$ being given by

$$\tilde{p}_n(\mathbf{x}) = \omega c^2 \int \rho'(\mathbf{x}, t) e^{in\omega t} dt$$

(15.11)

$$= -\frac{in\omega^2 x_i}{4\pi x^2 c} \int \left[\frac{f_i + \rho_0 in\omega v_i h}{|1 - M_r|} \right] e^{in\omega t} R dR d\phi \, dt \quad ,$$

where the t–integral is over any interval of duration $2\pi/\omega$. We now change the integration variable from t to τ. The time increments are related by $dt = |1 - M_r| d\tau$. With $\mathbf{f}$ and $\mathbf{v}$ written out explicitly as in (15.7),

$$\mathbf{x} \cdot \mathbf{f} = -x_1 T - D\sigma \sin(\theta - \Theta)$$

$$\mathbf{x} \cdot \mathbf{v} = -\omega R\sigma \sin(\theta - \Theta) \quad ,$$

the nth harmonic becomes

$$\tilde{p}_n(\mathbf{x}) = \frac{in\omega^2 e^{in\omega x/c}}{4\pi x^2 c} \int (x_1 T + (D - \rho_0 in\omega^2 Rh)\sigma \sin(\theta - \Theta))$$

(15.12)

$$\times e^{in\omega(\tau - \sigma R \cos(\theta - \Theta)/xc)} R dR d\phi \, d\tau \quad .$$

The τ–integral in Equation (15.12) can be evaluated by using the standard integrals

$$\int_0^{2\pi/\omega} e^{i(n\omega\tau - \alpha\,\cos(\theta-\Theta))}\,d(\omega\tau) = 2\pi J_n(\alpha)e^{-in(\phi-\Theta+\pi/2)} \quad ,$$

and

$$\int_0^{2\pi/\omega} \sin\theta\, e^{i(n\omega\tau - \alpha\,\cos(\theta-\Theta))}\,d(\omega\tau) = -\frac{2\pi n}{\alpha} J_n(\alpha)e^{-in(\phi-\Theta+\pi/2)} \quad .$$

This gives

$$\widetilde{p}_n(\mathbf{x}) = \frac{in\omega e^{in(\omega x/c + \Theta - \pi/2)}}{2xc} \int \left\{ \frac{x_1 T}{x} - \frac{D}{M} + in\omega\rho_0 ch \right\} J_n\left(\frac{nM\sigma}{x}\right) e^{-in\phi}\, R dR d\phi,$$

$$(15.13)$$

where M is the rotational Mach number $\omega R/c$. Equation (15.13) is equivalent to Gutin's (1938) result extended to include the thickness noise which is, however, a small correction at low tip–Mach numbers. The result in Equation (15.13) describes the nth harmonic of the acoustic field generated by a single blade. For a symmetric assembly of B blades the contributions from the individual blades add constructively for harmonics which are integer multiples of B, and all the rest cancel. Hence the sound field consists entirely of harmonics of ωB, the blade–passing frequency, the magnitude of the nth harmonic being given by B times $\widetilde{p}_{nB}(\mathbf{x})$ given in Equation (15.13).

We see from Equation (15.13) that the phase of the nBth harmonic is

$$nB\left(\frac{\omega|x|}{c} + \Theta - \omega t\right) + \quad \text{a constant} \quad .$$

As time increases, surfaces of constant phase propagate towards the observer at speed c and rotate with angular speed ω.

The propeller thrust is usually significantly larger than the drag and, were the section lift/drag ratio $T/D = (Mx_1/x)^{-1}$, then

$$\frac{x_1}{x} \int TJ_n\left(\frac{nM\sigma}{x}\right) e^{in\phi} R dR d\phi - \int \frac{D}{M} J_n\left(\frac{nM\sigma}{x}\right) e^{in\phi} R dR d\phi$$

would vanish. This could lead to a directivity for the loading noise of the form sketched in Figure 15.2.

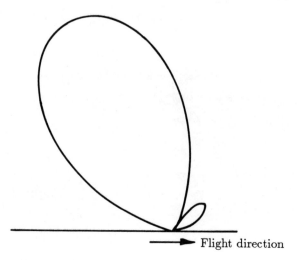

Flight direction

Figure 15.2 A hypothetical directionality pattern of the loading–noise term.

The Bessel function $J_n(nz)$ decays rapidly with increasing n for $0 < z < 1$ but only slowly when z is greater than unity. Hence Equation (15.13) predicts that for low subsonic speeds, the amplitudes of the harmonics decrease rapidly as their order n increases. As the tip speed increases, the higher harmonics become relatively more important. Hawkings & Lowson (1974) point out that Equation (15.13) can also be used to describe the loading and thickness noise of supersonic rotors. Then the acoustic field has a different structure according to whether $M\sigma/x$ is greater or less than unity. In regions where $M\sigma/x$ is greater than unity, the sound field is particularly intense and is made up of many harmonics.

Hanson's (1980a) spectral method is based on a generalization of Equation (15.13), extended to include effects of forward flight and blade sweep. Forward flight with speed cM_F means that the sound heard by a distant observer is Döppler shifted by the factor $|1 - M_F x_1/x|$. The other effect incorporated into Hanson's method is that the mean chord of the blade is no longer assumed to lie in the (x_2, x_3) plane. This enables highly swept thin blades to be considered, the effect of blade displacement being a phase factor describing the variation in retarded time. In a numerical calculation, Hanson considers just the thickness noise and the thrust component of the loading noise for a two–bladed prop fan. He finds that the linear theory predicts the experimental waveshapes with reasonable accuracy (see Figure 15.3). Hanson (1980b) has used this formulation to investigate the effects of blade sweep and twist on the far acoustic field. Peake & Crighton's (1990) version of Hanson's method confirms its general usefulness though it is incorrect in some significant detail.

Flight Mach Number $M_x = 0.32$

Tip rotational Mach Number $M_T = 1.00$

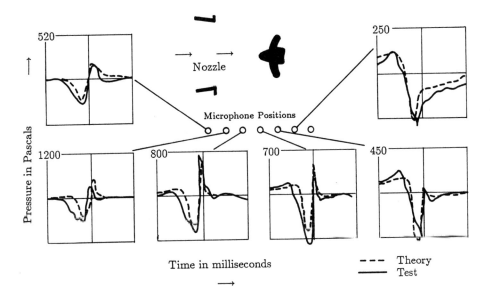

Figure 15.3 Comparison between Hanson's theory and experimental waveforms for a two–blade SR–1 prop fan (from Hanson 1980a).

15.3 TIME DOMAIN METHODS

At high–tip speeds the acoustic field of a propeller involves many spectral components and so it is often more convenient to investigate the time history of the sound
directly. In general, time–domain methods deal with both the near and far field
but all rely on estimates of the surface loading and local value of the stress tensor.
These values are inserted in Equation (15.5) and evaluated numerically. A number
of computer codes for conducting this evaluation have been produced and there are
several examples where theoretical (computed) models of the problem have been
shown to be in excellent agreement with experiment. Schmitz & Yu (1986) find
that the quadrupole plays an essential part at transonic speeds and our Figure 15.4
is a reproduction of their Figure 13, which illustrates this effect very well.

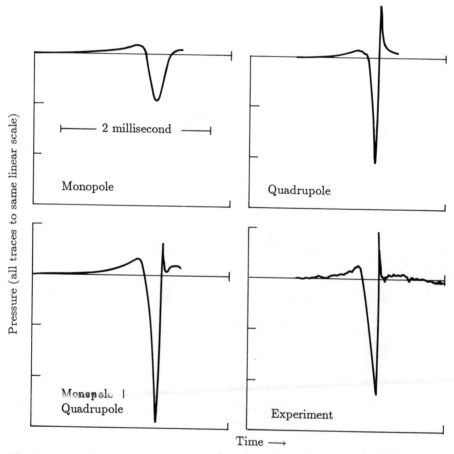

Figure 15.4 Theory–experiment comparison, $M_{\text{tip}} = 0.9$ (from Schmitz & Yu
1986).

At these transonic speeds, shock waves form near the propeller blades and propagate away as the sharp–fronted pulses that are so characteristic of the noise of high–performance helicopters.

Of course in a subject as topical and important as this, there is a rapidly growing and complex literature, and it is often difficult to see the wood from the trees. Some workers have sustained a very comprehensive attack on the problem and Farassat in particular combines the rigorous development of the theory with important practical codes for the rapid evaluation of noise signatures. He has concentrated on the linear terms which he shows are dominant over most of the operational envelope of well–designed propellers and his review of the subject in the (1982) lecture series at the Von Kármán Institute for Fluid Dynamics was an important contribution to the field. Figure 7 of that paper is reproduced here as our Figure 15.5 and illustrates the striking agreement between his computational procedures and measurement on propellers of advanced design. He computes the time history of the pulse radiated by an individual blade and by Fourier transformation obtains the spectrum. Such impressive agreement has spurred him to exploit the theory embedded in the exact Equation (15.2) to form a basis for calculating with it the propeller performance as well as the far radiation field. His review provides evidence that some illusive aspects of high–speed linear aerodynamic theory are better obtained from this equation than they are by more conventional means.

15.4 BROAD–BAND PROPELLER NOISE

The analysis of broad–band noise is of course much more complicated in that it always originates in some form of turbulence which cannot be handled as precisely as the steady loading terms. Inflow turbulence, the boundary–layer turbulence and particularly the unsteady flow in the vicinity of the blade's trailing edge, all contribute significant sources of sound. The tip vortex is usually turbulent and constitutes yet another noise source that is hard to handle analytically. We can do no more in these notes than direct the interested student to the excellent review provided by George & Chan (1984), who give data to indicate how broad–band noise is affected by the main aerodynamic parameters of the rotating blades. Again, the theory is based on Equation (15.2), but supplemented with the techniques of diffraction theory and the established methods for predicting the unsteady response of aerofoils operating in inhomogeneous flow.

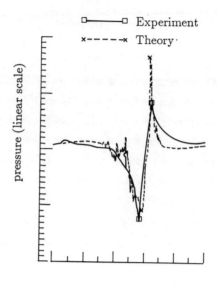

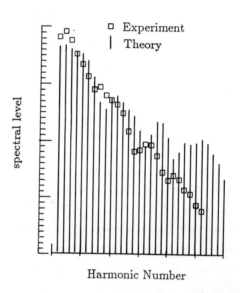

Harmonic Number

Figure 15.5 Comparison of measured and predicted acoustic pressure signature and spectrum for an advanced propeller (from Farassat 1982).

15.5 THE SOUND OF POINT SOURCES IN CIRCULAR MOTION

The general propeller noise problem is very complex and hard to illustrate by exact model problems. Equation (15.2) gives the field in terms of surface integrals, each element of the blade and flow supporting a particular source in rotational motion. It is instructive to examine the fields generated by idealized moving sources because they can be worked out exactly and illustrate well the range of behaviour to be expected but which become confused in the complexity of the mutual interference of the full source array. Li (1986) has documented these properties systematically and we illustrate here with examples of his work just how rich the range of behaviour can be.

The spectral form and the time history of the field radiated by a steady point source in circular motion becomes rich in harmonics, the wave form sharply peaked, as the Mach number approaches unity.

Contra–rotating propellers will have noise sources of a different kind because the blade loading, and therefore source strengths, are periodically modulated by interaction of the rear stage with the flow of the leading blades and vice–versa. The sources then contain a periodic element and the following figures illustrate the behaviour of such sources at various frequency ratios. The frequency ratio is defined as the number of cycles through which the source strength is modulated during each revolution of the propeller blade.

Figures 15.6 – 15 illustrate the waveforms and spectral strengths of sources rotating at various speeds and frequency ratios and Figure 15.16 gives the time histories for fractional frequency ratios, i.e. the propeller makes more than one revolution during a period of its strength modulation cycle, and this condition would be representative of a propeller blade rotating in the wake of another which rotates in the same direction but at a different speed; the time history is intricate but for this simple source is easily interpreted.

Obviously, the coherent but intricate patterns from adjacent sources will interfere to produce a waveform defying simple analytical description. For them the calculational procedures inevitably rest on numerical computational methods, examples of which we have referred to above.

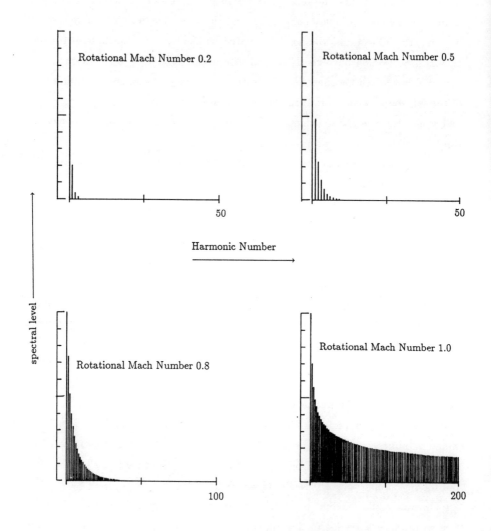

Frequency Ratio = 0

Figure 15.6 Frequency spectra of a steady monopole source at four rotational speeds.

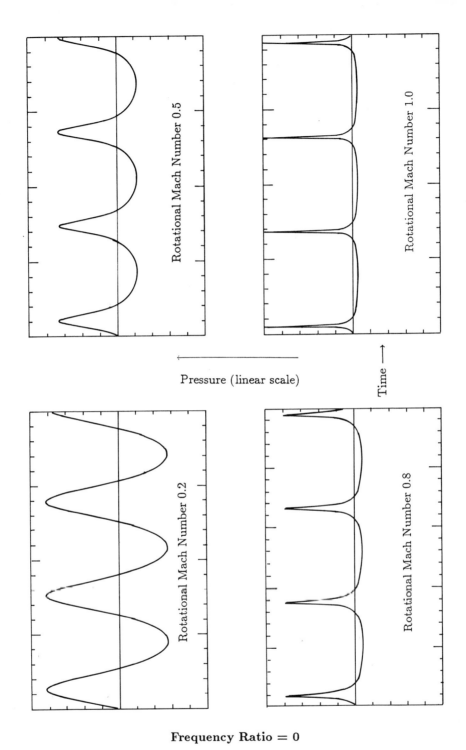

Frequency Ratio = 0

Figure 15.7 Pressure–time history of the sound made by a steady monopole source at four rotational speeds.

442

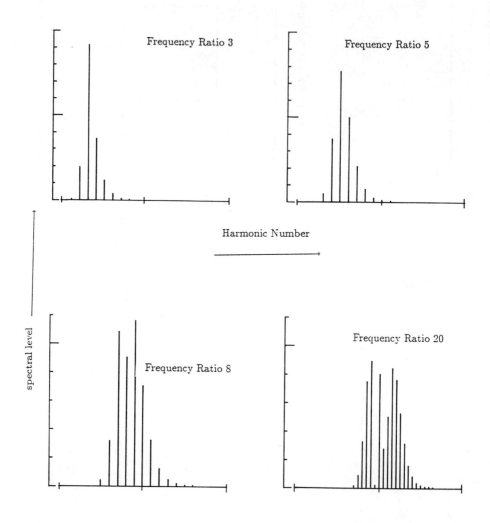

Rotational Mach Number = 0.2

Figure 15.8 Frequency spectra of a rotating harmonic source at four frequency ratios.

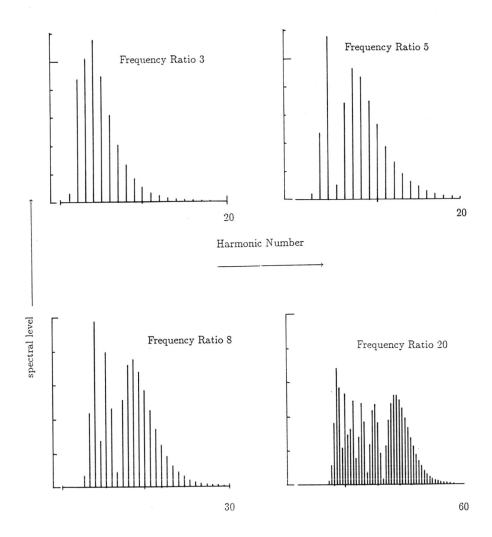

Rotational Mach Number = 0.5

Figure 15.9 Frequency spectra of a rotating harmonic source at four frequency ratios.

444

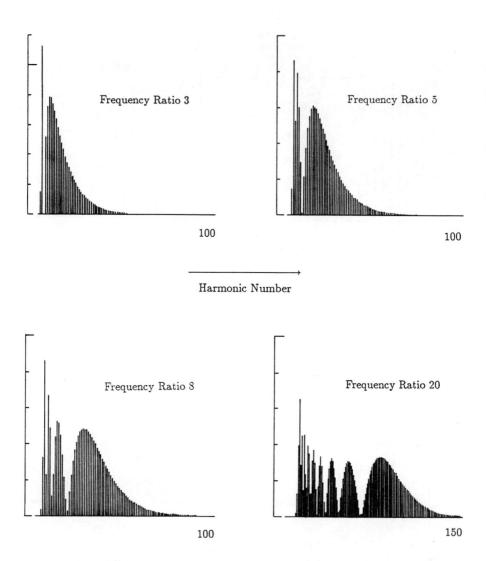

Rotational Mach Number = 0.8

Figure 15.10 Frequency spectra of a rotating harmonic source at four frequency ratios.

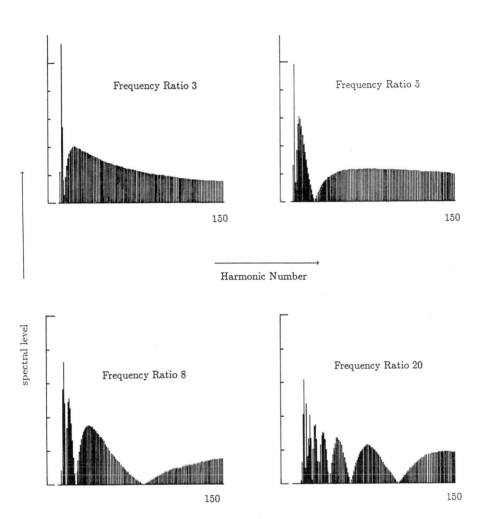

Rotational Mach Number = 1.0

Figure 15.11 Frequency spectra of a rotating harmonic source at four frequency ratios.

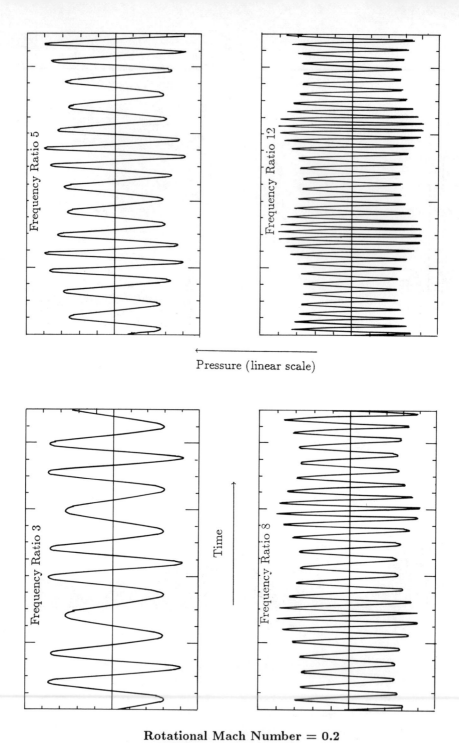

Rotational Mach Number = 0.2

Figure 15.12 Pressure–time history of a rotating harmonic source at four frequency ratios.

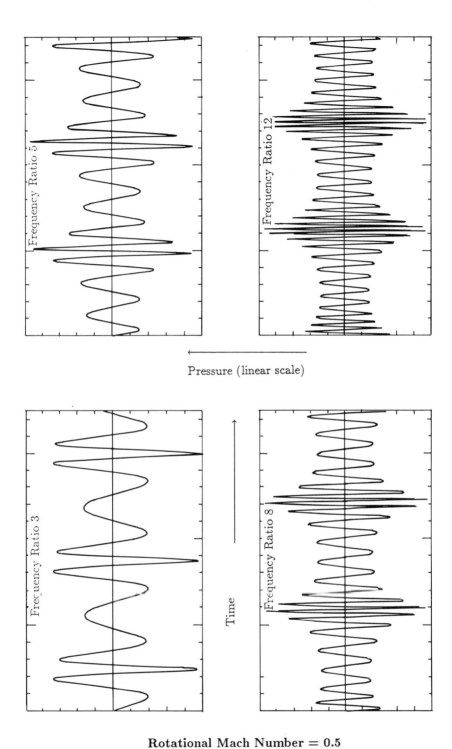

Rotational Mach Number = 0.5

Figure 15.13 Pressure–time history of a rotating harmonic source at four frequency ratios.

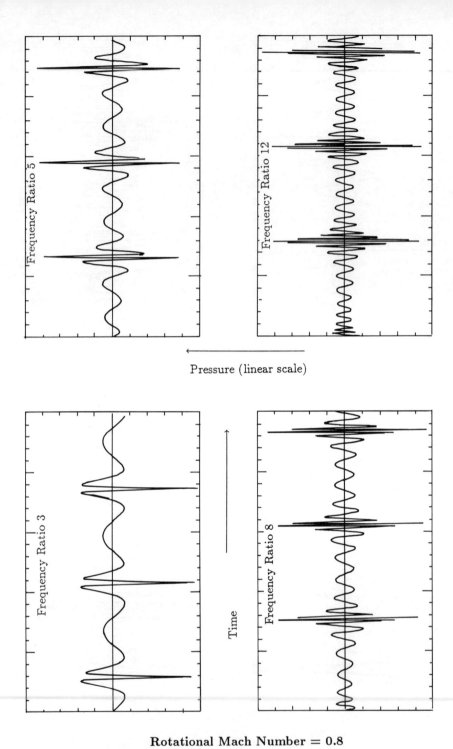

Rotational Mach Number = 0.8

Figure 15.14 Pressure–time history of a rotating harmonic source at four frequency ratios.

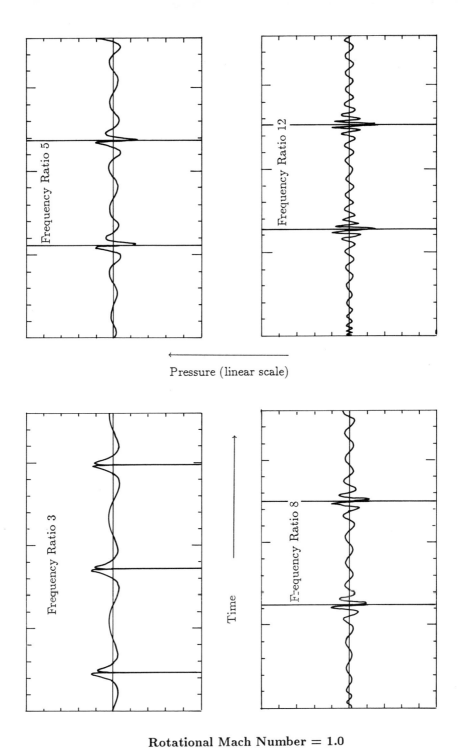

Rotational Mach Number = 1.0

Figure 15.15 Pressure–time history of a rotating harmonic source at four frequency ratios.

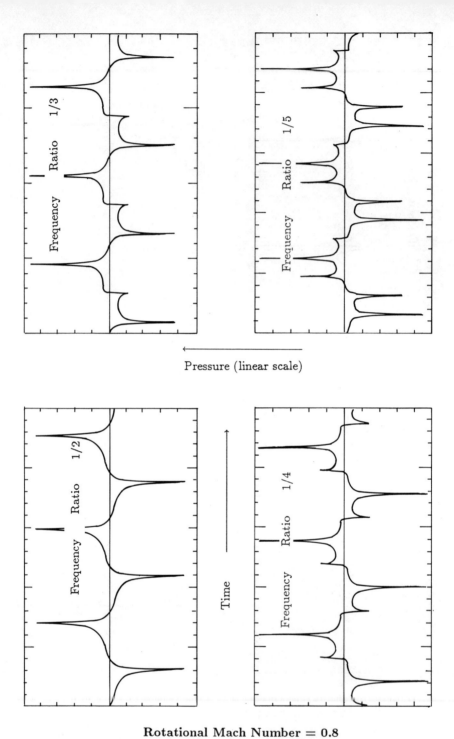

Rotational Mach Number = 0.8

Figure 15.16 Pressure–time history of a rotating harmonic source at four frequency ratios.

REFERENCES

Farassat, F. (1982). Advanced theoretical treatment of propeller noise. Von Kármán Institute for Fluid Dynamics Lecture Series 1982–08.

Ffowcs Williams, J.E. & Hawkings, D.L. (1969). Sound generation by turbulence and surfaces in arbitrarymotion. Phil. Trans. Roy Soc. A 264:321.

George, A.R. & Chan, S.T. (1984). Board Band Noise Analyses. NASA Contract Report 3797.

Gutin, L. (1938). On the sound field of a rotating airscrew. Phys. Z. Sowjetunion A1:57. [English translation NACA Tech. Mem. No. TM1195 (1948).]

Hanson, D.B. & Fink, M.R. (1979). The importance of quadrupole sources in prediction of transonic top speed propeller noise. J. Sound Vib. 62:19.

Hanson, D.B. (1980a). Helicoidal surface theory for harmonic noise of propellers in the far field. AIAA J. 18:1213.

Hanson, D.B. (1980b). Influence of propeller design parameters on far–field harmonic noise in forward flight. AIAA J. 18:1313.

Hawkings, D.L. & Lowson, M.V. (1974). Theory of open supersonic rotor noise. J. Sound Vib. 36:1.

Li, K.M. (1986). Noise of High Speed Propellers — A Prediction Method. Ph.D. Thesis, University of Cambridge.

Peake, N. & Crighton, D.G. (1991). Lighthill quadrupole radiation in supersonic propeller acoustics. J. Fluid Mech. 223:363–382.

Schmitz, F.H. & Yu, Y.H. (1986). Helicopter impulsive noise: theoretical and experimental status. J. Sound Vib. 109:361–422. Sixth European Rotocraft and Powered Lift Aircraft Forum. Paper No.22.

16. FLOW NOISE ON SURFACES

A turbulent boundary–layer flow over a surface generates noise, which is usually referred to as *flow noise*. This Chapter deals with the prediction and properties of the fluctuating surface pressures induced by flow noise and investigates their spectral characteristics. This problem is particularly important in underwater acoustics, where a passive sonar system on a ship or submarine aims to detect and analyze weak sounds emanating from a distant source. However, when the vessel is in motion, the unsteady pressures generated by its turbulent boundary layer can obscure the incoming signal. There are other applications where flow noise is important. For example, the turbulent boundary layer over fuselage is a significant source of high–frequency sound within an aircraft cabin. But in this Chapter we will concentrate on the underwater geometry.

16.1 THE SURFACE PRESSURE SPECTRUM ON A RIGID PLANE WALL

Consider the geometry illustrated in Figure 16.1. The hard surface, $x_3 = 0$, has a turbulent boundary–layer flow over it, and we want to investigate the unsteady surface pressures. Figure 16.2 shows the typical form of $\widetilde{P}_R(k_1, k_2, \omega)$, the wavenumber–frequency decomposition of flow noise on the rigid surface. There are two main peaks. The maximum occurs in the convective régime, which for a turbulent boundary–layer flow in the 1–direction means k_1 of order $-\omega/U_c$ and k_2 small. U_c denotes a typical eddy convection velocity. The second peak is in the vicinity of the acoustic wavenumber where k, the modulus of the vector $\mathbf{k} = (k_1, k_2)$, is equal to ω/c_0. For highly supersonic modes with very low values of k, the surface pressure spectrum becomes "wavenumber white", that is almost independent of k. The level of the pressure spectrum in the convective peak is typically some 40dB larger than near $\mathbf{k} = \mathbf{0}$. Nevertheless spectral elements of the flow noise with sonic and supersonic surface phase speeds are of particular interest because they can mask the incoming signal which travels supersonically over the surface.

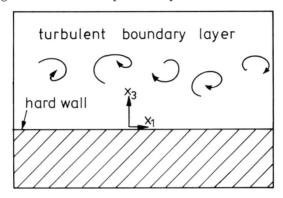

Figure 16.1 A turbulent boundary layer over a hard wall.

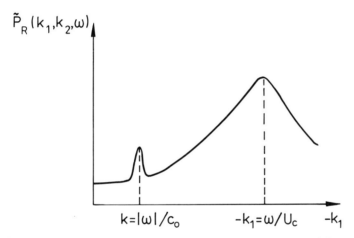

Figure 16.2 The wavenumber frequency decomposition of flow noise on a plane, rigid surface.

The convective region has been considered by Corcos (1963), and is discussed in detail by Willmarth (1975) in a review article. In underwater problems, where flow Mach numbers are so small, the convective spectral components have a low subsonic surface phase speed and are essentially incompressible. It is convenient to introduce $\widehat{P}_R(\boldsymbol{\xi}, \omega)$, the Fourier transform of the cross–correlation of the surface pressure on a rigid surface,

$$\widehat{P}_R(\boldsymbol{\xi}, \omega) = \int \overline{p(\mathbf{x}, t)p(\mathbf{x} + \boldsymbol{\xi}, t + \tau)}e^{i\omega\tau}d\tau \quad . \tag{16.1}$$

Corcos assumes that $\widehat{P}_R(\boldsymbol{\xi}, \omega)$ can be expressed in separable form. He writes

$$\widehat{P}_R(\boldsymbol{\xi}, \omega) = \Phi(\omega) A\left(\frac{\omega \xi_1}{U_c}\right) B\left(\frac{\omega \xi_2}{U_c}\right) e^{i\omega \xi_1 / U_c} \quad . \tag{16.2}$$

U_c is the eddy convection velocity and the nondimensional functions A and B are to be determined from experimental data. The spatial Fourier transform of Equation (16.2) leads directly to the cross–power spectral density

$$\widetilde{P}_R(k_1, k_2, \omega) = \frac{U_c^2}{\omega^2} \Phi(\omega) \widehat{A}\left(1 + \frac{k_1 U_c}{\omega}\right) \widehat{B}\left(\frac{k_2 U_c}{\omega}\right) \quad , \tag{16.3}$$

where $\widehat{A}$ and $\widehat{B}$ are the transforms of A and B,

$$\widehat{A}(\alpha) = \int A(\beta) e^{i\alpha\beta} d\beta \text{ and } \widehat{B}(\alpha) = \int B(\beta) e^{i\alpha\beta} d\beta \quad . \tag{16.4}$$

Blake (1986, Equation (8.49)) fits algebraic expressions to the functions A and B, and shows that the Corcos model provides a good estimate to the cross–power spectral density in the vicinity of the convective peak (see Figure 16.3).

Chase (1987a) makes a number of specific assumptions about the statistics of the velocity field in a turbulent boundary layer to derive an expression for the surface pressure spectrum in an incompressible fluid. He obtains

$$\widetilde{P}_R(\mathbf{k}, \omega) = \frac{\rho^2 u_\tau^3}{\left[K_+^2 + (b\delta)^{-2}\right]^{5/2}} \left\{ C_T k^2 \frac{K_+^2 + (b\delta)^{-2}}{k^2 + (b\delta)^{-2}} + C_M k_1^2 \right\} \quad , \tag{16.5}$$

where ρ is the fluid density, u_τ the friction velocity and δ the boundary–layer thickness

$$K_+^2 = k^2 + (\omega + U_c k_1)^2 / (h u_\tau)^2 \quad . \tag{16.6}$$

The expression in (16.5) clearly involves a number of arbitrary constants. Chase suggests the values

$$C_T = 4.7 \times 10^{-3}, \ C_M = 0.15, \ b = 0.75 \text{ and } h = 3 \quad . \tag{16.7}$$

At the convective peak, $k_1 = -\omega/U_c$, $k_2 = 0$, Equation (16.5) simplifies to

$$\widetilde{P}_R(-\omega/U_c, 0, \omega) = \rho^2 u_\tau^3 \delta^3 \left(\frac{U_c}{\omega\delta}\right)^3 (C_T + C_M) \quad , \tag{16.8}$$

for $\omega\delta/U_c \gg b^{-1}$.

Blake (1986, Figure 8.15) has shown that an earlier version of Chase's empirical expression for the surface pressure spectrum (Chase 1980) agrees with three sets of experimental data (Martini et al. 1984, Jameson 1975 and Martin 1976). Figure 16.3 compares Chase's more recent formula, given in Equation (16.5), with the same data. It is apparent that there is reasonable agreement down to wavenumbers $k_1 \sim -0.4\delta^{*-1}$, where δ^* is the boundary–layer displacement thickness.

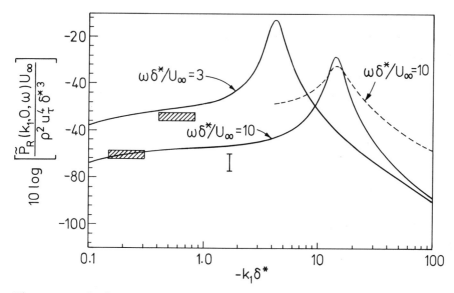

Figure 16.3 Surface pressure spectrum in the convective régime.

▨ experimental data $\quad \omega\delta^*/U_\infty = 3$ ⎫

I experimental data $\quad \omega\delta^*/U_\infty = 10$ ⎬ from Blake (1986, Fig.8.15)

---- Corcos theory ⎪

—— Chase's (1987 formula (Equation16.5)). ⎭

(Note in interpreting Blake's graph we have assumed that $\Omega = \omega\delta^*/U_\infty$. This ensures that the convective peak occurs at $k_1\delta^* = -\omega\delta^*/U_c$.)

In summary, the Corcos theory provides good predictions for the level of the cross–power spectral density of the surface pressure fluctuations in the vicinity of the convective peak. Chase (1987a) has developed an empirical formula which, with suitably chosen constants, fits experimental data for spectral elements with a range of subsonic phase speeds.

However we are primarily interested in spectral elements with supersonic surface phase speeds. These are essentially sound waves generated by turbulence within the boundary layer. Their behaviour is dominated by compressible effects, and their strength can be determined by solving Lighthill equation (Equation (11.46))

$$\frac{\partial^2 \rho'}{\partial t^2} - c_0^2 \nabla^2 \rho' = \frac{\partial^2 T_{ij}}{\partial x_i \partial x_j} \quad ; \qquad (16.9)$$

ρ' denotes the density perturbation and c_0 is the speed of sound in the distant linearly disturbed fluid. Lighthill's quadrupole source, T_{ij}, is equal to $\rho u_i u_j + (p' - c_0^2 \rho')\delta_{ij} - \sigma_{ij}$ where $\mathbf{u}$ is the particle velocity, p' denotes pressure perturbations and σ_{ij} is the viscous stress tensor. Obermeier & Möhring (1984) and Leehey (1988) have given reviews of sound generation by a turbulent boundary–layer flow. We saw in Chapter 11 that the solution of (16.9) in a region V exterior to a surface S leads to surface distributions of monopole and dipole sources. In particular, if S is impenetrable, the solution at a point $(\mathbf{x}, t)$ within V can be written in the form (see for example Goldstein 1976, Equation (3.6))

$$c_0^2 \rho'(\mathbf{x}, t) = \int \int_V \frac{\partial^2 G}{\partial y_i \partial y_j} T_{ij}(\mathbf{y}, \tau) d^3 \mathbf{y} d\tau - \int \int_S \frac{\partial G}{\partial y_i}(p' n_i - \sigma_{ij} n_j) dS d\tau$$
$$+ \int \int_S \frac{\partial G}{\partial \tau} \rho_0 \mathbf{u} \cdot \mathbf{n} dS d\tau \quad , \qquad (16.10)$$

where $\mathbf{n}$ is the outward normal to the surface S. $G(\mathbf{y}, \tau | \mathbf{x}, t)$, a reciprocal Green function of the wave equation, satisfies

$$\frac{1}{c_0^2} \frac{\partial^2 G}{\partial \tau^2} - \nabla^2 G = \delta(\mathbf{x} - \mathbf{y}, t - \tau) \quad \text{in } V \quad , \qquad (16.11)$$

and has inward wave behaviour for large $|\mathbf{y}|$. This does not completely specify $G(\mathbf{y}, \tau | \mathbf{x}, t)$, which is only uniquely determined once a boundary condition is given on S.

In the case of a turbulent boundary layer over an infinite plane wall, S is the surface $x_3 = 0$ and $\mathbf{u} \cdot \mathbf{n} = 0$ on S. Equation (16.10) therefore simplifies to

$$c_0^2 \rho'(\mathbf{x}, t) = \int \int_V \frac{\partial^2 G}{\partial y_i \partial y_j} T_{ij}(\mathbf{y}, \tau) d^3 \mathbf{y} d\tau - \int \int_S \frac{\partial G}{\partial y_3} p'(\mathbf{y}, \tau) dy_1 dy_2 d\tau$$
$$+ \int \int_S \frac{\partial G}{\partial y_i} \sigma_{i3}(\mathbf{y}, \tau) dy_1 dy_2 d\tau \quad . \qquad (16.12)$$

If we use the free–space Green function, $G = \delta(t - \tau - |\mathbf{x} - \mathbf{y}|/c_0)/(4\pi|\mathbf{x} - \mathbf{y}|)$, this becomes

$$c_0^2 \rho'(\mathbf{x}, t) = \frac{\partial^2}{\partial x_i \partial x_j} \int_V \frac{T_{ij}(\mathbf{y}, t - |\mathbf{x} - \mathbf{y}|/c_0)}{4\pi|\mathbf{x} - \mathbf{y}|} d^3 \mathbf{y} + \frac{\partial}{\partial x_3} \int_S \frac{p'(\mathbf{y}, t - |\mathbf{x} - \mathbf{y}|/c_0)}{4\pi|\mathbf{x} - \mathbf{y}|} dy_1 dy_2$$
$$- \frac{\partial}{\partial x_i} \int_S \frac{\sigma_{i3}(\mathbf{y}, t - |\mathbf{x} - \mathbf{y}|/c_0)}{4\pi|\mathbf{x} - \mathbf{y}|} dy_1 dy_2 \quad . \qquad (16.13)$$

The second term on the right–hand side of (16.13) apparently describes the sound field of a normal dipole of strength $p'(\mathbf{y}, \tau)$ distributed over the surface $y_3 = 0$. Compact dipoles are known to be more efficient radiators than compact quadrupoles (Lighthill 1952) and some early discussions of flow noise considered this dipole to be the major source of sound in a turbulent boundary layer. However the surface terms in (16.13) are linear, and so are nonzero even in the sound field. They therefore extend over a large area and the variation of retarded time in the integrand cannot be neglected. This makes the surface integrals in (16.13) hard to estimate. This difficulty was resolved by Powell (1960) who used an image argument to show that in an inviscid fluid the normal surface dipole describes the sound field of the image of the turbulent quadrupole sources reflected in the hard surface.

We can recover Powell's result by using the hard wall Green function G_R in (16.10).

$$G_R(\mathbf{y}, \tau | \mathbf{x}, t) = \frac{\delta(t - \tau - |\mathbf{x} - \mathbf{y}|/c_0)}{4\pi |\mathbf{x} - \mathbf{y}|} + \frac{\delta(t - \tau - |\mathbf{x}^* - \mathbf{y}|/c_0)}{4\pi |\mathbf{x}^* - \mathbf{y}|} \quad , \qquad (16.14)$$

where $\mathbf{x}^*, = (x_1, x_2, -x_3)$, is the image of the position $\mathbf{x}$ in the rigid surface $x_3 = 0$. The normal derivative $\partial G_R / \partial y_3$ vanishes on $y_3 = 0$ and (16.10) becomes

$$c_0^2 \rho'(\mathbf{x}, t) = \frac{\partial^2}{\partial x_i \partial x_j} \int_V \frac{T_{ij}(\mathbf{y}, t - |\mathbf{x} - \mathbf{y}|/c_0)}{4\pi |\mathbf{x} - \mathbf{y}|} d^3\mathbf{y} + \frac{\partial}{\partial x_i^* \partial x_j^*} \int_V \frac{T_{ij}(\mathbf{y}, t - |\mathbf{x}^* - \mathbf{y}|/c_0)}{4\pi |\mathbf{x}^* - \mathbf{y}|} d^3\mathbf{y}$$
$$- \frac{\partial}{\partial x_\alpha} \int_S \frac{\sigma_{\alpha 3}(\mathbf{y}, t - |\mathbf{x} - \mathbf{y}|/c_0)}{2\pi |\mathbf{x} - \mathbf{y}|} dy_1 dy_2 \quad . \qquad (16.15)$$

where α is to be summed over 1 and 2. A comparison of (16.13) and (16.15) shows that

$$\frac{\partial}{\partial x_3} \int \frac{p'(\mathbf{y}, t - |\mathbf{x} - \mathbf{y}|/c_0) - \sigma_{33}(\mathbf{y}, t - |\mathbf{x} - \mathbf{y}|/c_0)}{4\pi |\mathbf{x} - \mathbf{y}|} dy_1 dy_2$$
$$= \frac{\partial^2}{\partial x_i^* \partial x_j^*} \int_V \frac{T_{ij}(\mathbf{y}, t - |\mathbf{x}^* - \mathbf{y}|/c_0)}{4\pi |\mathbf{x}^* - \mathbf{y}|} d^3\mathbf{y} - \frac{\partial}{\partial x_\alpha} \int_S \frac{\sigma_{\alpha 3}(\mathbf{y}, t - |\mathbf{x} - \mathbf{y}|/c_0)}{4\pi |\mathbf{x} - \mathbf{y}|} dy_1 dy_2 \quad .$$

$$(16.16)$$

Evidently there is significant cancellation of the sound field from individual normal surface dipoles, so that their integrated effect is predominantly quadrupole. This illustrates that much care is needed in handling any linear sources!

The last term in Equation (16.16) involves linear in–plane surface dipoles, whose strength depends on the surface shear stress. The contribution of these viscous stresses to the surface pressure fluctuations has long been an area of controversy and is discussed in Section 16.6. In this section we will concentrate on the acoustic field generated by the quadrupole sources and write

$$
c_0^2 \rho'(\mathbf{x}, t) = \frac{\partial^2}{\partial x_i \partial x_j} \int_V \frac{T_{ij}(\mathbf{y}, t - |\mathbf{x} - \mathbf{y}|/c_0)}{4\pi |\mathbf{x} - \mathbf{y}|} d^3\mathbf{y}
$$
$$
+ \frac{\partial}{\partial x_i^* \partial x_j^*} \int_V \frac{T_{ij}(\mathbf{y}, t - |\mathbf{x}^* - \mathbf{y}|/c_0)}{4\pi |\mathbf{x}^* - \mathbf{y}|} d^3\mathbf{y} \quad , \tag{16.17}
$$

or equivalently

$$
c_0^2 \rho'(\mathbf{x}, t) = \int \int_V \frac{\partial^2 G_R}{\partial y_i \partial y_j} T_{ij}(\mathbf{y}, \tau) d^3\mathbf{y} d\tau \quad . \tag{16.18}
$$

We are particularly interested in the spectra of the fluctuations and so it is natural to introduce Fourier transforms,

$$
\widetilde{\rho}(x_3, \mathbf{k}, \omega) = \int \rho'(\mathbf{x}, t) e^{ik_\alpha x_\alpha + i\omega t} d^2\mathbf{x} dt \quad , \tag{16.19}
$$

where α is summed over 1 and 2. The Fourier transform of (16.18) gives

$$
c_0^2 \widetilde{\rho}(x_3, \mathbf{k}, \omega) = \int \int \frac{\partial^2 \widetilde{G}_R}{\partial y_i \partial y_j} T_{ij}(\mathbf{y}, \tau) d^3\mathbf{y} d\tau \quad , \tag{16.20}
$$

where

$$
\widetilde{G}_R = \int G_R(\mathbf{y}, \tau | \mathbf{x}, t) e^{ik_\alpha x_\alpha + i\omega t} d^2\mathbf{x} dt \quad . \tag{16.21}
$$

After substitution from (16.14) into (16.21), the $\mathbf{x}$ and t integrals are standard, and we find (see for example Ffowcs Williams, 1965, Equation (16.7))

$$
\widetilde{G}_R(\mathbf{y}, \tau | x_3, \mathbf{k}, \omega) = -\frac{1}{2\gamma i} e^{i\omega \tau + ik_\alpha y_\alpha} \left\{ e^{i\gamma |y_3 - x_3|} + e^{i\gamma |y_3 + x_3|} \right\} \quad , \tag{16.22}
$$

where $\gamma = (\omega^2/c_0^2 - k^2)^{1/2}$. The root of γ is chosen in such a way that when γ is real it has the same sign as ω, and when γ is purely imaginary $\Im\gamma$ is positive. The form of $\widetilde{G}_R$ simplifies for $x_3 = 0$,

$$
\widetilde{G}_R(\mathbf{y}, \tau | 0, \mathbf{k}, \omega) = -\frac{1}{\gamma i} e^{i(\omega \tau + k_\alpha y_\alpha + \gamma y_3)} \quad . \tag{16.23}
$$

The surface pressure perturbation can be found, by substituting the expression for $\widetilde{G}_R$ given in Equation (16.23) into (16.20), to be

$$
\widetilde{p}(0, \mathbf{k}, \omega) = \frac{d_i d_j}{i\gamma} \int_0^\infty \widetilde{T}_{ij}(y_3, \mathbf{k}, \omega) e^{i\gamma y_3} dy_3 \quad , \tag{16.24}
$$

where $d_\alpha = k_\alpha$, for $\alpha = 1$ or 2, and $d_3 = \gamma$. This result relates the Fourier transform of the wall pressure to the Reynolds stresses in the turbulent boundary layer, and was first derived by Ffowcs Williams (1965).

The power spectral density of the surface pressure, $\widetilde{P}_R(\mathbf{k}, \omega)$, is the Fourier transform of the autocorrelation

$$\widetilde{P}_R(\mathbf{k}, \omega) = \int \overline{p'(x_1, x_2, 0, t) p'(x_1 + \Delta_1, x_2 + \Delta_2, 0, t + \tau)} e^{ik_\alpha \Delta_\alpha + i\omega\tau} d^2 \Delta d\tau \quad ,$$

(16.25)

where the overbar denotes an ensemble average. If the turbulence is statistically stationary and homogeneous in the 1 and 2 directions, $\widetilde{P}_R(\mathbf{k}, \omega)$ is related simply to $\widetilde{p}(0, \mathbf{k}, \omega)$ (see Chapter 3, Equation (3.63)) by

$$\widetilde{P}_R(\mathbf{k}, \omega) = \frac{1}{(2\pi)^3} \int \overline{\widetilde{p}^*(0, \mathbf{k}, \omega) \widetilde{p}(0, \mathbf{k}', \omega')} d^2 \mathbf{k}' d\omega' \quad .$$

(16.26)

The star denotes a complex conjugate. Substitution for $\widetilde{p}(0, \mathbf{k}, \omega)$ from Equation (16.24) gives

$$\widetilde{P}_R(\mathbf{k}, \omega) = \frac{d_i^* d_j^*}{(2\pi)^3 \gamma^*} \int \frac{d_k' d_l'}{\gamma'} \overline{\widetilde{T}_{ij}^*(y_3, \mathbf{k}, \omega) \widetilde{T}_{kl}(y_3', \mathbf{k}', \omega')} \, e^{i(\gamma' y'_3 - \gamma^* y_3)} d^2 \mathbf{k}' d\omega' dy_3 dy_3'.$$

(16.27)

By making use of the homogeneity of the turbulence again, we can write

$$\overline{\widetilde{T}_{ij}^*(y_3, \mathbf{k}, \omega) \widetilde{T}_{kl}(y_3', \mathbf{k}', \omega')} = (2\pi)^3 \delta(\mathbf{k} - \mathbf{k}') \delta(\omega - \omega') T_{ijkl}(y_3, y_3', \mathbf{k}, \omega) \quad , \quad (16.28)$$

where $T_{ijkl}(y_3, y_3', \mathbf{k}, \omega)$ is the cross–power spectral density of the quadrupole source strengths,

$$T_{ijkl}(y_3, y_3', \mathbf{k}, \omega) = \int \overline{T_{ij}(\mathbf{y}, t) T_{kl}(y_1 + \Delta_1, y_2 + \Delta_2, y_3', t + \tau)} e^{ik_\alpha \Delta_\alpha + i\omega\tau} d^2 \Delta d\tau \quad .$$

(16.29)

When Equation (16.28) is used in (16.27), we obtain

$$\widetilde{P}_R(\mathbf{k}, \omega) = \frac{d_i^* d_j^* d_k d_l}{|\gamma|^2} \int T_{ijkl}(y_3, y_3', \mathbf{k}, \omega) e^{i(\gamma y'_3 - \gamma^* y_3)} dy_3 dy_3' \quad .$$

(16.30)

Equation (16.30) is in useful form because, although very little is known about the structure of the turbulent source terms T_{ij}, it is possible to make some reasonable assumptions about the behaviour of their spectral functions, T_{ijkl}. As argued by Bergeron (1973), who uses a matching argument, we expect the source functions

T_{ijkl} to have a Taylor series about $\mathbf{k} = 0$. Hence, in order to investigate the surface pressure in the low wavenumber régime, we can replace $T_{ijkl}(y_3, y_3', \mathbf{k}, \omega)$ in (16.30) by $T_{ijkl}(y_3, y_3', 0, \omega)$. This is equivalent to usual assumption in the Lighthill theory, that at low Mach numbers the source functions can be determined by considering an incompressible flow. Crow (1970) used the method of matched asymptotic expansions to give a mathematical justification of this approximation.

For spectral elements with very low subsonic phase speeds, γ is purely imaginary with a large positive imaginary part. Equation (16.30) then shows that, in this régime, the effect of sources away from the boundary decays exponentially with their distance from the boundary. If, however, the boundary–layer thickness, δ, is small in comparison with $|\gamma|^{-1}$, the exponential terms in the integral in (16.30) are approximately unity, and it is convenient to cast the source terms into non-dimensional form.

The Reynolds stresses, T_{ij}, are proportional to the square of the velocity perturbations, which in turn scale on the friction velocity u_τ. We will nondimensionalize lengths on the boundary–layer displacement thickness, δ^*, and time on δ^*/U_∞, where U_∞ is the free–stream velocity. Hence

$$\int T_{ijkl}(y_3, y_3', 0, \omega) dy_3 dy_3' = \frac{\rho^2 u_\tau^4 \delta^{*5}}{U_\infty} S_{ijkl}\left(\frac{\omega \delta^*}{U_\infty}\right) \quad , \qquad (16.31)$$

where S_{ijkl} is a nondimensional function of $\omega\delta^*/U_\infty$ only. Equation (16.30) becomes

$$\widetilde{P}_R(\mathbf{k}, \omega) = \frac{d_i^* d_j^* d_k d_l}{|\gamma|^2} \frac{\rho^2 u_\tau^4 \delta^{*5}}{U_\infty} S_{ijkl}\left(\frac{\omega \delta^*}{U_\infty}\right) \quad . \qquad (16.32)$$

This result is equivalent to that derived by Ffowcs Williams (1982) and can be used to comment on the limiting forms of the power spectrum of the wall pressure.

For spectral elements with subsonic phase speeds, i.e for spectral components with $|\omega|/c_0 \ll k \ll \delta^{-1}$,

$$\gamma \sim ik \quad , \qquad (16.33)$$

and

$$\widetilde{P}_R(\mathbf{k}, \omega) \sim \frac{d_i^* d_j^* d_k d_l}{k^2} \frac{\rho^2 u_\tau^4 \delta^{*5}}{U_\infty} S_{ijkl}\left(\frac{\omega \delta^*}{U_\infty}\right) \quad . \qquad (16.34)$$

For these wavenumbers, all the components of δ depend only on $\mathbf{k}$ and the power spectral density is proportional to k^2.

For spectral elements with highly supersonic surface phase speeds, $k \ll |\omega|/c_0$,

$$\gamma \sim \omega/c_0 \quad . \qquad (16.35)$$

The predicted surface pressure spectrum simplifies to

$$\widetilde{P}_R(\mathbf{k}, \omega) \sim \frac{\rho^2 u_\tau^4 \delta^{*3} M^2}{U_\infty} \left(\frac{\omega \delta^*}{U_\infty}\right)^2 S_{3333}\left(\frac{\omega \delta^*}{U_\infty}\right) \quad, \tag{16.36}$$

where M is the free–stream Mach number U_∞/c_0. In this limit the wall pressure spectrum becomes independent of wavenumber.

Sevik (1986) collects together experimental data for this highly supersonic régime. He shows that results obtained on a buoyant body in water agree closely with wind–tunnel measurements on both smooth and rough walls, when expressed in the nondimensional form in (16.36). This agreement provides reassurance that the experiments have not been contaminated by extraneous noise sources. Sevik fits the curve

$$\widetilde{P}_R(\mathbf{0}, \omega) \sim 5.6 \, \frac{\rho^2 u_\tau^4 \delta^{*3} M^2}{U_\infty} \left(\frac{\omega \delta^*}{U_\infty}\right)^{-4.5} \tag{16.37}$$

to the experimental data in the frequency range $3 < \omega \delta^*/U_\infty < 30$. This is entirely compatible with the theoretical form in (16.36) and a comparison of the two expressions tells us that

$$S_{3333}(\omega \delta^*/U_\infty) = 5.6(\omega \delta^*/U_\infty)^{-6.5} \quad. \tag{16.38}$$

Since measured pressure spectra are observed to be "wavenumber white", that is practically independent of $\mathbf{k}$, for spectral elements with supersonic phase velocities, we might (after an inspection of (16.32)) expect the other components of $S_{ijkl}(\omega \delta^*/U_\infty)$ to be of the same order.

Chase (1987a) also obtains predictions for hard wall surface pressure spectra in a compressible field. He develops his model from the Lighthill theory by making specific assumptions about the source spectra. In the supersonic régime his expressions simplify considerably and lead to

$$\widetilde{P}_R(\mathbf{0}, \omega) = 1.2 \times 10^{-2} \, \frac{\rho^2 u_\tau^4 M^2 \delta^{*3}}{U_\infty} \left(\frac{u_\tau \delta}{U_\infty \delta^*}\right)^2 \left(\frac{\omega \delta^*}{U_\infty}\right)^{-1} \quad, \tag{16.39}$$

for $\omega \delta^*/U_\infty > 0.1$. It is immediately striking that this prediction has quite a different dependence on frequency from the experimental fit in (16.37). If (following Blake 1986, page 505) we assume that $\delta^*/\delta = 3.6 u_\tau/U_\infty$, Equation (16.39) reduces to

$$\widetilde{P}_R(\mathbf{0}, \omega) = 9.1 \times 10^{-4} \, \frac{\rho^2 u_\tau^4 M^2 \delta^{*3}}{U_\infty} \left(\frac{\omega \delta^*}{U_\infty}\right)^{-1} \quad, \tag{16.40}$$

and we can obtain a numerical comparison between Chase's predictions and the experimental data. This is shown in Figure 16.4. Equation (16.40) tends to underestimate the experimental results by about 20dB near $\omega\delta^*/U_\infty = 3$, and overestimate them by some 14dB near $\omega\delta^*/U_\infty = 30$.

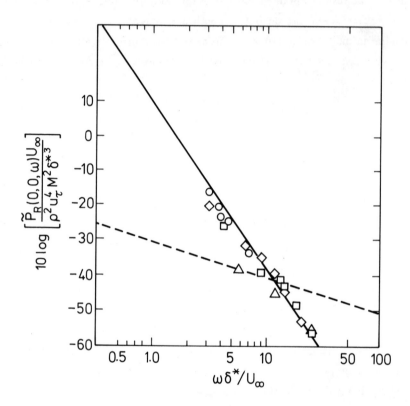

Figure 16.4 Variation of flow noise with nondimensional frequency.

————— Sevik's empirical fit;

━ ━ ━ Chase's model.

The experimental data is from Sevik (1986), 0 buoyant body; ◊ smooth, □ medium rough ($k_s u_\tau/\nu = 50$), △ fully rough ($k_s u_\tau/\nu = 1000$), wind–tunnel data of Geib and Farabee.

We can determine the ratio between the pressure spectral levels near the convective peak and at low wavenumbers from a comparison of (16.8) and (16.37)

$$\frac{\widetilde{P}_R(-\omega/U_c, 0, \omega)}{\widetilde{P}_R(0, 0, \omega)} = \frac{C_T + C_M}{5.6M^2} \frac{U_\infty}{u_\tau} \left(\frac{U_c}{U_\infty}\right)^3 \left(\frac{\omega\delta^*}{U_\infty}\right)^{1.5} . \tag{16.41}$$

This predicts that the level at the convective peak exceeds that for low wavenumbers by 40dB for $\omega\delta^*/U_\infty = 3$ and that it is relatively larger at higher nondimensional frequencies.

Both the theoretical prediction (16.31) and the experimental result (16.37) are proportional to the fourth power of the friction velocity. This is a reminder that the source strength depends strongly on the amplitude of velocity fluctuations. Any device that reduces turbulent velocities is likely to reduce the contribution of the turbulent quadrupoles to flow noise. This opens up exciting new possibilities.

Recent research on drag reduction has lead to the development of "large–eddy break–up devices" or LEBUs (see Nguyen et al. (1984), Bandyopadhyay (1986) or Wilkinson et al. (1987) for reviews of this work). LEBUs are thin splitter plates about a boundary–layer thickness in length, which are positioned at a height between 0.4δ and 0.8δ above the wall. Such devices have been shown to lead to local skin–friction reductions of between 15 and 40%, with some reduction in skin–friction persisting over a downstream distance of $100\delta - 150\delta$. This is accompanied by a corresponding decrease in turbulence intensity, turbulent Reynolds stress and integral lengthscale (Westphal 1986; Bonnet et al. 1987; Coustols et al. 1987).

It is apparent from Equation (16.30) that it is these fluctuating velocities that generate sound in a turbulent boundary layer over a plane wall, and we might hope that an array of LEBUs producing a global reduction in turbulence intensity might lead to a corresponding reduction in noise. Such an argument is clearly an over-simplification. Although the introduction of a LEBU might reduce the strength of the quadrupole sources, the LEBU exerts unsteady forces on the fluid and leads to new dipole sources. These dipoles produce centred waves travelling from the LEBU over the surface with the speed of sound. The installation of LEBUs therefore introduces no additional pressure fluctuations with supersonic surface phase speeds and, for these spectral components, we can expect a reduction in flow noise due to the decrease in the strength of the quadrupole sources, T_{ij}. Encouraging experimental results have been reported by Beeler (1986) and Moller & Leehey (1989). They measured pressure spectra downstream of a LEBU in a turbulent boundary layer and found that the presence of the manipulator reduced the unsteady pressures.

The installation of the LEBU increases the surface pressure spectrum for spectral elements with sonic phase speeds. However there is significant flow noise at these acoustic wavenumbers even without LEBUs. The increase due to the LEBU is not significant at frequencies higher than about $c_0/30\delta$ (Dowling 1989).

The simple theoretical prediction in (16.32) has a nonintegrable singularity for spectral elements with sonic phase speeds, for which $k = |\omega|/c_0$ and $\gamma = 0$. This singularity was investigated by Bergeron (1973) and Ffowcs Williams (1982). It arises due to a two–dimensional form of Olber's paradox, because the turbulent source region is considered to be of infinite extent and the sound field from each source element does not decrease rapidly enough with distance for the integrated effect to be finite. We will use the methods of Ffowcs Williams (1982) to investigate this scale effect.

If we write $q(\mathbf{x}, t) = \partial^2 T_{ij}/\partial x_i \partial x_j$, Equation (16.9) becomes

$$\frac{\partial^2 \rho'}{\partial t^2} - c_0^2 \nabla^2 \rho' = q \quad , \tag{16.42}$$

and the solution to this equation satisfying the hard surface boundary condition is

$$p'(x_1, x_2, 0, t) = \int \frac{q(\mathbf{y}, t - |\mathbf{x} - \mathbf{y}|/c_0)}{2\pi |\mathbf{x} - \mathbf{y}|} d^3 \mathbf{y} \quad . \tag{16.43}$$

In interpreting the singularities it is sufficient to concentrate on $p'_d(\mathbf{x}, t)$, the contribution to $p'(\mathbf{x}, t)$ from distant sources, since the singularities arise due to the integrated effect of these distant sources. As the effect of nearby sources is to be neglected, it will not be possible to deduce the full structure of the pressure distribution as was done in Equation (16.30), but instead the origin of the singularities can be highlighted. For the distant sources within the turbulent wall region $\sigma, = (y_1^2 + y_2^2)^{1/2}$, is much greater than $|\mathbf{x}|$ and

$$|\mathbf{x} - \mathbf{y}| = \sigma - \frac{\mathbf{y} \cdot \mathbf{x}}{\sigma} \quad . \tag{16.44}$$

Hence

$$p'_d(x_1, x_2, 0, t) = \frac{1}{2\pi} \int q\left(\mathbf{y}, t - \frac{\sigma}{c_0} + \frac{\mathbf{y} \cdot \mathbf{x}}{\sigma c_0}\right) \frac{d^3 \mathbf{y}}{\sigma} \quad . \tag{16.45}$$

Similarly a simple change of variable gives

$$p'_d(x_1 + \Delta_1, x_2 + \Delta_2, 0, t + \tau)$$
$$= \frac{1}{2\pi} \int q\left(y_1 + \xi_1, y_2 + \xi_2, y_3', t + \tau - \frac{\sigma}{c_0} + \frac{\mathbf{y} \cdot}{\sigma c_0}(\mathbf{x} + \mathbf{\Delta} - \mathbf{\xi})\right) \frac{d^2 \mathbf{\xi} dy_3'}{\sigma} \quad . \tag{16.46}$$

The autospectrum of the surface pressure, $P_{Rd}(\Delta, \tau)$, can be obtained from the mean value of the product of Equations (16.45) and (16.46)

$$P_{Rd}(\Delta, \tau) = \frac{1}{4\pi^2} \int Q\left(\mathbf{y}, y_3', \boldsymbol{\xi}, \tau + \frac{\mathbf{y} \cdot (\Delta - \boldsymbol{\xi})}{\sigma c_0}\right) \frac{d^2\xi \, dy_3' \, d^3\mathbf{y}}{\sigma^2} \quad , \tag{16.47}$$

where $Q(\mathbf{y}, y_3', \boldsymbol{\xi}, \tau) = \overline{q(\mathbf{y}, t) q(y_1 + \xi_1, y_2 + \xi_2, y_3', t + \tau)}$ is the autocorrelation of q.

The origin of the singularity is now apparent. If the turbulence is homogeneous so that Q is independent of y_1 and y_2, the y_1, y_2 integral in Equation (16.47) is unbounded. However, when we insist that there is a patch of homogeneous turbulence of large but finite extent L, so that Q is independent of y_1, y_2 for $\sigma < L$ and vanishes for $\sigma > L$,

$$P_{Rd}(\Delta, \tau) = \frac{1}{4\pi^2} \int_{\sigma = c_0/\omega}^{L} \int_{\phi=0}^{2\pi} Q\left(y_3, y_3', \boldsymbol{\xi}, \tau + \frac{\mathbf{y} \cdot (\Delta - \boldsymbol{\xi})}{\sigma c_0}\right) dy_3 \, dy_3' \, d^2\xi \, d\phi \, \frac{d\sigma}{\sigma}$$

$$= \frac{\ln(\omega L/c_0)}{4\pi^2} \int_{\phi=0}^{2\pi} Q\left(y_3, y_3', \boldsymbol{\xi}, \tau + \frac{\mathbf{y} \cdot (\Delta - \boldsymbol{\xi})}{\sigma c_0}\right) dy_3 \, dy_3' \, d^2\xi \, d\phi \quad .$$
$$\tag{16.48}$$

The lower limit on σ has been set as c_0/ω (Howe 1987a) as the far–field approximation in (16.45) will only be valid for large σ.

The wall pressure spectrum due to these extensive acoustic sources is given by the Fourier transform of (16.48)

$$\widetilde{P}_{Rd}(\mathbf{k}, \omega) = \int P_{Rd}(\Delta, \tau) e^{ik_\alpha \Delta_\alpha + i\omega\tau} d^2\Delta \, d\tau$$

$$= \ln(\omega L/c_0) \int_{\phi=0}^{2\pi} \widetilde{Q}\left(y_3, y_3', \omega\mathbf{y}/c_0\sigma, \omega\right) \delta(\mathbf{k} - \omega\mathbf{y}/c_0\sigma) dy_3 \, dy_3' \, d\phi \quad ,$$
$$\tag{16.49}$$

where $\widetilde{Q}(y_3, y_3', \mathbf{k}, \omega) = \int Q(y_3, y_3', \boldsymbol{\xi}, \tau) \exp(ik_\alpha \xi_\alpha + i\omega\tau) d^2\xi \, d\tau$ is the power spectral density of q. One of the δ–functions can be used to evaluate the ϕ–integral to give

$$\widetilde{P}_{Rd}(\mathbf{k}, \omega) = 2\ln(\omega L/c_0)\delta(k^2 - \omega^2/c_0^2) \int \widetilde{Q}(y_3, y_3', \mathbf{k}, \omega) dy_3 \, dy_3' \quad . \tag{16.50}$$

This shows that there is a δ–function singularity for spectral elements with sonic surface wave speed. The peak is now integrable. For a small–scale experimental facility, its integrated effect may not be large because it depends on the logarithm of the extent of the turbulent source region.

In terms of the nondimensional function defined in (16.31), we obtain

$$\widetilde{P}_{Rd}(\mathbf{k}, \omega) = \frac{2\rho^2 u_\tau^4 \delta^{*3}}{U_\infty} \ln\left(\frac{\omega L}{c_0}\right) \left(\frac{\omega \delta^*}{c_0}\right)^4 \delta\left((k\delta^*)^2 - (\omega \delta^*/c_0)^2\right) S_{1111}\left(\frac{\omega \delta^*}{U_\infty}\right) ,$$

$$(16.51)$$

in agreement with Ffowcs Williams (1982, Equation (4.18)).

To summarize, we expect the integrated pressure spectrum level near the sonic condition to depend on the geometry or source size for a hard surface. Sevik (1986) reports that experimentally measured levels at $k = \omega/c_0$ are higher than at supersonic wavenumbers, with "most of the energy arriving at grazing incidence from an upstream direction". There is no such anisotropy in Equation (16.51), but it will emerge when we consider mean flow effects in Section 16.4.

16.2 SURFACE CURVATURE

As an introduction the effects of surface curvature, we will investigate flow noise on an infinitely long, rigid cylinder of radius a. Dhanak (1988) gives a full description of the low wavenumber surface pressure spectrum on a circular cylinder and we will summarize his main results. Two limiting cases are of practical importance. The limit $\omega a/c_0 \to \infty$ describes the first–order effects of curvature on a nearly planar array, while the limit of small $\omega a/c_0$ is appropriate for the towed array geometry. In the latter case, the circumferentially–averaged surface pressure is of primary interest.

Sound generation by turbulent boundary–layer flow over the cylinder can be described by Equation (16.10), where S is the surface of the cylinder. Let us suppose that the cylinder axis is in the 1–direction. It is convenient to introduce cylindrical polar coordinates and we write $\mathbf{y} = (\sigma, \phi, y_1), \mathbf{x} = (R, \theta, x_1)$. If we choose the particular Green function that satisfies the boundary condition

$$\frac{\partial G}{\partial \sigma} = 0 \text{ on } \sigma = a , \quad (16.52)$$

Equation (16.10) simplifies to

$$c_0^2 \rho'(\mathbf{x}, t) = \int \int_V \frac{\partial^2 G}{\partial y_i \partial y_j} T_{ij}(\mathbf{y}, \tau) d^3\mathbf{y} \, d\tau . \quad (16.53)$$

We have again neglected the effects of viscous shear stresses on the surface. Their rôle is discussed in Section 16.6.

It is a straightforward matter to evaluate the Green function, $G(\mathbf{y}, \tau | \mathbf{x}, t)$, that satisfies Equation (16.11) and the boundary condition (16.52), by taking Fourier transforms in y_1 and τ and expanding the ϕ dependence in a Fourier series. In particular for a position $\mathbf{x}$ on the cylinder

$$G(\mathbf{y}, \tau | a, \theta, x_1, t) =$$

$$\frac{1}{(2\pi)^3} \int \int \sum_{n=-\infty}^{\infty} F_n(\alpha a) \frac{H_n^{(1)}(\alpha\sigma)}{H_n^{(1)}(\alpha a)} e^{in(\phi-\theta)+ik_1(y_1-x_1)+i\omega(\tau-t)} dk_1 d\omega \quad , \quad (16.54)$$

where $F_n(z) = - H_n^{(1)}(z)/(z H_n^{(1)'}(z))$ and $\alpha = (\omega^2/c_0^2 - k_1^2)^{1/2}$ (see for example, Dhanak 1988, Equation (2.9)). The sign of the square root of α is to be chosen such that it has the same sign as ω when real, with $\Im\alpha$ positive when α is purely imaginary.

After substituting for G in Equation (16.53) and taking Fourier transforms in x_1, t, Fourier series in θ, we obtain

$$\tilde{p}(a, n, k_1, \omega) = a F_n(\alpha a) \int \frac{\partial^2}{\partial y_i \partial y_j} \left(\frac{H_n^{(1)}(\alpha\sigma)}{H_n^{(1)}(\alpha a)} e^{in\phi+ik_1 y_1+i\omega\tau} \right) T_{ij}(\mathbf{y}, \tau) d^3\mathbf{y} d\tau \quad , \tag{16.55}$$

where

$$\tilde{p}(R, n, k_1, \omega) = \int p'(\mathbf{x}, t) e^{in\theta+ik_1 x_1+i\omega t} a \; d\theta dx_1 dt \quad . \tag{16.56}$$

The cross–power spectral density of the surface pressure can be obtained by multiplying $\tilde{p}$ by its complex conjugate. This shows that the surface pressure spectrum is proportional to $|a F_n(\alpha a)|^2$.

In order to investigate the first–order effects of surface curvature on a cylinder of large radius, we need to consider the limit $a \to \infty$, with fixed circumferential wavenumber $k_2 = n/a$. It is convenient to introduce $\gamma = (\omega^2/c_0^2 - k_1^2 - k_2^2)^{1/2}$, then in the limit $|\gamma a| \to \infty$,

$$a F_n(\alpha a) \to -\frac{1}{i\gamma} \quad \text{and} \quad \frac{H_n^{(1)}(\alpha\sigma)}{H_n^{(1)}(\alpha a)} \to e^{i\gamma(\sigma-a)} \tag{16.57}$$

(Dhanak 1988, Equation (3.13)). Equation (16.55) then reduces to the equivalent plane wave result in (16.24).

Dhanak investigates the function $|a F_n(\alpha a)|^2$ in some detail near $\gamma = 0$, and finds that for these wavenumbers even slight curvature affects the surface pressure spectrum. Dhanak shows that $|aF_n(\alpha a)|$ has a maximum near $\gamma = 0$. In particular, for $n = 0$

$$\left|\frac{\omega a}{c_0}F_0(\alpha a)\right|^2 \sim \left(\frac{\omega a}{c_0}\right)^2 \ln^2(|\alpha a|) \quad . \tag{16.58}$$

The surface pressure spectrum induced by axial disturbances is singular but integrable at the sonic condition, $k_1 = \mp\omega/c_0$. As n increases from zero, Dhanak finds that $F_n(\alpha a)$ is always finite, and that its peaks move away from $\alpha = 0$. For large n, $|F_n(\alpha a)|$ has a maximum value of $d_0^{-1}n^{-2/3}$, which occurs near $\gamma^2 = -d_0^2 n^{4/3}a^{-2}$, where $d_0 = 0.837$. The peak width is $0(n^{2/3}a^{-1})$. These considerations cause Dhanak to approximate $|F_n(\alpha a)|$ by

$$\left|\frac{\omega a}{c_0}F_n(\alpha a)\right|^2 \approx \frac{(\omega/c_0)^2}{|\omega^2/c_0^2 - k^2(1-\beta^2)| + \beta^2 k^2} \quad , \tag{16.59}$$

where $k^2 = k_1^2 + n^2/a^2$, $\beta = d_0|\cos\chi|^{2/3}(c_0/\omega a)^{1/3}$ and $\tan\chi = ak_1/n$. This approximation correctly predicts the position, height and width of the peak in $|F_n(\alpha a)|$. Away from the peak,

$$\left|\frac{\omega a}{c_0}F_n(\alpha a)\right|^2 \approx \frac{(\omega/c_0)^2}{|\gamma|^2} \quad , \tag{16.60}$$

the plane result. The first-order effects of surface curvature can therefore be described by replacing $|\gamma|^{-2}$ in Equation (16.32) by the form for a cylinder of large radius $(|\omega^2/c_0^2 - k^2(1-\beta^2)| + \beta^2 k^2)^{-1}$. Hence it follows that $\widetilde{P}_c(\mathbf{k},\omega)$, the pressure spectrum on a cylinder, is given by

$$\widetilde{P}_c(\mathbf{k},\omega) = \frac{d_i^* d_j^* d_k d_l}{|\omega^2/c_0^2 - k^2(1-\beta^2)| + \beta^2 k^2} \frac{\rho^2 u_\tau^4 \delta^{*5}}{U_\infty} S_{ijkl}\left(\frac{\omega\delta^*}{U_\infty}\right) \quad , \tag{16.61}$$

in the limit $a \to \infty$. The expression in (16.61) is equivalent to Dhanak (1988), Equation (4.26). For a cylinder of very large radius whose axis is aligned with the mean flow, we might assume that the source functions $S_{ijkl}(\omega\delta^*/U_\infty)$ have the same form as those for a flat plate.

Howe (1987a) has also investigated the pressure spectrum on a surface with a large but finite radius of curvature. His result corresponding to that given in (16.61) is

$$\widetilde{P}_c(\mathbf{k},\omega) = \frac{d_i^* d_j^* d_k d_l}{|\omega^2/c_0^2 - k^2| + \beta_0^2\omega^2/c_0^2} \frac{\rho^2 u_\tau^4 \delta^{*5}}{U_\infty} S_{ijkl}\left(\frac{\omega\delta^*}{U_\infty}\right) \quad , \tag{16.62}$$

where $\beta_0 = (0.92/0.837)\beta$. As Dhanak points out Howe's expression underestimates the height of the peak by a factor β^2/β_0^2. This difference occurs because Howe erroneously assumes the peak to lie at the acoustic wavenumber $k = \omega/c_0$, rather than its true position $\omega^2/c_0^2 - k^2 = -d_0^2 n^{4/3} a^{-2}$. However, this is not a serious discrepancy. In terms of decibels, it means that the peak has been underestimated by about 1dB.

With this slight reservation about Howe's result, we note that his formulation is more general than that derived by Dhanak. Rather than a circular cylinder, Howe considers an arbitrary rigid wall with principal radii of curvature R_1 and R_2 in orthogonal directions 1 and 2. He finds the surface pressure spectrum to have the form given in (16.62) with

$$\beta_0 = 0.92 \left(\frac{c_0}{\omega k^2}\right)^{1/3} \left(\frac{k_1^2}{R_1} + \frac{k_2^2}{R_2}\right)^{1/3} . \tag{16.63}$$

The contribution from the peak near $k = \omega/c_0$ to the point pressure spectrum can be determined by an integration of Equation (16.61) over wavenumber. Dhanak finds the net contribution to be proportional to $\ln(\omega L_1/c_0)$, where $L_1 = (c_0 a^2/\omega)^{1/3}$. For a patch of turbulence of large but finite extent L over a plane surface we found the corresponding result in (16.51) to be proportional to $\ln(\omega L/c_0)$. Hence we conclude that the surface pressure on a large cylinder is influenced by turbulent sources within a distance L_1. Indeed, as noted by Dhanak, L_1^{-1} is the decay rate of creeping rays around a cylinder.

To summarize, the curvature of a cylinder of large radius a controls the singularity in the surface pressure spectrum that occurs for an unbounded region of turbulence over a flat plate. Surface curvature ensures that the integrated contribution from spectral elements with sonic phase speeds is finite and proportional to $\ln\left((\omega a/c_0)^{2/3}\right)$.

The towed array geometry corresponds to the limit of small $\omega a/c_0$. In this case, the circumferentially–averaged surface pressure is of particular interest. We define $\widehat{p}(a, k_1, \omega)$ to be the Fourier transform of this average surface pressure

$$\widehat{p}(a, k_1, \omega) = \frac{1}{2\pi} \int p'(a, \theta, x_1, t) e^{ik_1 x_1 + i\omega t} dx_1 dt d\theta . \tag{16.64}$$

A comparison with the definition in (16.56) shows that

$$\widehat{p}(a, k_1, \omega) = \frac{1}{2\pi a} \widetilde{p}(a, 0, k_1, \omega) . \tag{16.65}$$

After substitution from (16.55) this becomes

$$\widehat{p}(a, k_1, \omega) = \frac{F_0(\alpha a)}{2\pi H_0^{(1)}(\alpha a)} \int \frac{\partial^2}{\partial y_i \partial y_j} \left(H_0^{(1)}(\alpha \sigma) e^{ik_1 y_1} \right) e^{i\omega \tau} T_{ij}(\mathbf{y}, \tau) d^3\mathbf{y} d\tau \quad .$$

(16.66)

The terms within the brackets are independent of the azimuthal angle ϕ which makes it relatively easy to evaluate the derivatives. This leads to

$$\widehat{p}(a, k_1, \omega) = \frac{F_0(\alpha a)}{2\pi H_0^{(1)}(\alpha a)}$$

$$\times \int \left[\alpha^2 H_0^{(1)''}(\alpha \sigma) T_{\sigma\sigma} + \alpha H_0^{(1)'}(\alpha \sigma) \left(\frac{1}{\sigma} T_{\phi\phi} + 2ik_1 T_{\sigma 1} \right) - k_1^2 H_0^{(1)}(\alpha \sigma) T_{11} \right]$$

$$\times e^{ik_1 y_1 + i\omega \tau} d^3\mathbf{y} d\tau.$$

(16.67)

We note from the definition (16.56) that

$$\int T_{ij}(\mathbf{y}, \tau) e^{ik_1 y_1 + i\omega \tau} \sigma d\phi dy_1 d\tau = \frac{\sigma}{a} \widetilde{T}_{ij}(\sigma, 0, k_1, \omega) \quad .$$

(16.68)

Hence Equation (16.67) simplifies to

$$\widehat{p}(a, k_1, \omega) = \frac{F_0(\alpha a)}{2\pi a H_0^{(1)}(\alpha a)}$$

$$\times \int \left[\alpha^2 H_0^{(1)''}(\alpha \sigma) \widetilde{T}_{\sigma\sigma} + \alpha H_0^{(1)'}(\alpha \sigma) \left(\frac{1}{\sigma} \widetilde{T}_{\phi\phi} + 2ik_1 \widetilde{T}_{\sigma 1} \right) - k_1^2 H_0^{(1)}(\alpha \sigma) \widetilde{T}_{11} \right] \sigma d\sigma \quad .$$

(16.69)

When both the cylinder radius and the boundary–layer thickness are compact, the Hankel functions in (16.69) can be approximated by their small argument asymptotic forms to give

$$\widehat{p}(a, k_1, \omega) = \frac{1}{2\pi a} \int \left[\widetilde{T}_{\sigma\sigma} - \widetilde{T}_{\phi\phi} - 2ik_1 \sigma \widetilde{T}_{\sigma 1} + k_1^2 \sigma^2 \ln(\alpha \sigma) \widetilde{T}_{11} \right] \frac{d\sigma}{\sigma} \quad ,$$

(16.70)

for $\omega a/c_0$ and $|k_1|a$ small in comparison with unity. This result can be recovered from the work of Chase & Noiseux (1982) by neglecting any of their terms that arise due to the mean velocity profile. (Mean flow effects are discussed in Section 16.4.) Ffowcs Williams (private communication) has also derived an expression for the circumferentially–averaged surface pressure. He erroneously omits the contribution from the azimuthal quadrupole, $T_{\phi\phi}$, but otherwise his predicted form is identical to that in Equation (16.70).

Since $k_1\sigma$ is small for all positions within the turbulent boundary layer, the first two terms in the integral in (16.70) have the largest effect on the surface pressure, away from the sonic condition $|k_1| = \omega/c_0$

$$\hat{p}(a, k_1, \omega) \simeq \frac{1}{2\pi a} \int \left[\tilde{T}_{\sigma\sigma} - \tilde{T}_{\phi\phi} \right] \frac{d\sigma}{\sigma} \text{ for } |k_1| \neq \omega/c_0 \quad . \tag{16.71}$$

We note that this term is independent of compressibility. Indeed the effectiveness with which a turbulent source, $\tilde{T}_{ij}(\sigma, 0, k_1, \omega)$, generates surface pressure is independent of both wavenumber and frequency. There is no reason for it to be small at low wavenumber.

The corresponding form for the surface pressure on a planar array is given in Equation (16.24)

$$\tilde{p}(0, \mathbf{k}, \omega) = \frac{d_i d_j}{i\gamma} \int_0^\infty \tilde{T}_{ij}(y_3, \mathbf{k}, \omega) e^{i\gamma y_3} dy_3,$$

where $d_i = k_i$ for $i = 1$ or 2, and $d_3 = \gamma = (\omega^2/c_0^2 - k^2)^{1/2}$. For wavenumbers near the convective peak $|d_i d_j/\gamma| \sim \omega/U_c$ whereas for low wavenumbers it is only of order ω/c_0. This shows that low–wavenumber turbulent sources are relatively ineffective at generating pressure disturbances on a planar array. We see from (16.71) that a line array does not discriminate against the low– wavenumber components of turbulence in a similar way.

The line pressure spectrum can be determined by multiplying (16.70) by its complex conjugate. In Section 16.1 we found the pressure spectrum, due to an unbounded region of turbulence over a planar array, to have a nonintegrable singularity at the sonic condition. We see from (16.70) that the line pressure spectrum is proportional to $(k_1 a)^4 \ln^2 |\omega^2 a^2/c_0^2 - k_1^2 a^2|$ near $|k_1| = \omega/c_0$ (see also Dhanak (1988), Equation (4.7)). This singularity is only logarithmic and is integrable.

If we assume that the cross–power spectral density of the source function can be expanded in a separable form of the type adopted by Corcos (1963) for the surface pressure (see Equation (16.3)), it follows from (16.70) that (Ffowcs Williams, private communication)

$$\hat{P}_c(k_1, \omega) = \rho^2 U_c^3 a^2 \Phi\left(\frac{\omega a}{U_c}\right)$$

$$\times \left(A\left(1 + \frac{U_c k_1}{\omega}\right) + B\left(1 + \frac{U_c k_1}{\omega}\right)(k_1 a)^4 \ln^2 \left| \frac{\omega^2 a^2}{c_0^2} - k_1^2 a^2 \right| \right) \quad ;$$

$$\tag{16.72}$$

U_c is the convection velocity of turbulent eddies, and $A(1 + U_c k_1/\omega)$ has been normalized so that $\int_{-\infty}^{\infty} A(1 + U_c k_1/\omega) d(k_1 U_c/\omega) = 1$. The functions Φ, A and B need to be determined by comparison with experiment.

The most easily measured spectral function on a rigid cylinder is the point pressure spectrum, $\widehat{P}_p(\omega)$. The best data on this seems to be that due to Markowitz (1976) who fits the curve

$$\widehat{P}_p(\omega) = 6\pi \times 10^{-6} \rho^2 U_c^3 a \left[1 + (0.6\omega a/U_c)^2\right]^{-1} \tag{16.73}$$

to his measurements. Of course we are interested in the circumferentially–averaged pressure and not point pressure measurements. Data for the cross–correlation of pressure at different azimuthal angles on the cylinder are needed to obtain information about $\widehat{P}_c(\omega)$ (the power spectral density of the circumferentially–averaged pressure) from (16.73). This cross–correlation has been measured by Willmarth & Yang (1969) who show

$$\widehat{P}_c(\omega) = \widehat{P}_p(\omega) \frac{0.21}{\omega a/U_c} \quad \text{for} \quad \frac{\omega a}{U_c} > 0.25 \quad . \tag{16.74}$$

The form for the pressure spectrum defined by Equations (16.73) and (16.74) is entirely compatible with the separable from in (16.72). Applying the constraint

$$\int_{-\infty}^{\infty} \widehat{P}_c(k_1, \omega) dk_1 = 2\pi \widehat{P}_c(\omega) \quad , \tag{16.75}$$

to equations (16.72) – (16.74) leads to

$$\Phi \left(\frac{\omega a}{U_c} \right) = 2.5\pi^2 \times 10^{-6} \left(\frac{U_c}{\omega a} \right)^2 \left[1 + (0.6\omega a/U_c)^2\right]^{-1} \quad . \tag{16.76}$$

It is evident from (16.72) that at low wavenumbers the pressure spectrum is wavenumber white

$$\widehat{P}_c(k_1, \omega) = \rho^2 U_c^3 a^2 \Phi \left(\frac{\omega a}{U_c} \right) A(1) \quad , \tag{16.77}$$

for $k_1 U_c/\omega << 1$, $|k_1| \neq \omega/c_0$. From a theoretical argument, Chase (1979) chooses

$$\Phi(\omega a/U_c) A(1) = C \left[\alpha^2 + (\omega a/U_c)^2\right]^{-5/2} \quad , \tag{16.78}$$

where α and C are constants. There appears to be no experimental data on the low–wavenumber pressure spectrum published in the open literature. It is therefore impossible to compare the predictions in (16.78) with experiment here.

16.3 FLEXIBLE SURFACES

We have determined the properties of the pressure spectrum induced by turbulence over a hard surface. However, the wavelengths associated with spectral elements with supersonic phase speeds are so long that the fluid loading in water can cause even quite massive surfaces to vibrate. Let us suppose that the surface is homogeneous and that the mechanical properties of the surface lead to a linear relationship between the Fourier transforms of surface displacement and pressure.

The normal velocity, $\mathbf{u} \cdot \mathbf{n}$, in Equation (16.10) is then nonzero and related in some way to the surface pressure perturbation p'. The potentially misleading linear surface terms in (16.10) can be eliminated by using a Green function which satisfies the physical surface condition. This is the approach adopted in Dowling (1983a). However, here we will exploit the work in Sections 16.1 and 16.2 where we determined the surface pressure induced by turbulence over a rigid surface.

The sound generation is described by Equation (16.9)

$$\frac{\partial^2 \rho'}{\partial t^2} - c_0^2 \nabla^2 \rho' = \frac{\partial^2 T_{ij}}{\partial x_i \partial x_j} \quad . \tag{16.79}$$

We want to solve this equation subject to the radiation condition at infinity and a flexible surface boundary condition. It is convenient to decompose the density fluctuations into those that occur near a hard wall, with an additional term to account for surface flexibility. We write

$$\rho'(\mathbf{x}, t) = \rho'_R(\mathbf{x}, t) + \rho'_F(\mathbf{x}, t) \quad , \tag{16.80}$$

where $\rho'_R(\mathbf{x}, t)$ is the density perturbation produced by the same quadrupole sources, T_{ij}, when they are adjacent to a rigid surface, i.e.

$$\frac{\partial^2 \rho'_R}{\partial t^2} - c_0^2 \nabla^2 \rho'_R = \frac{\partial^2 T_{ij}}{\partial x_i \partial x_j} \quad , \tag{16.81}$$

with

$$\mathbf{n} \cdot \nabla \rho'_R = 0 \tag{16.82}$$

on the surface S.

The relationship between $\widetilde{\rho}_R(\mathbf{x}, t)$ and the turbulent sources was determined in Sections 16.1 and 16.2 for plane and cylindrical surfaces respectively. When the expansion (16.80) is substituted into (16.79), it shows that $\rho'_F(\mathbf{x}, t)$ satisfies a homogeneous wave equation

$$\frac{\partial^2 \rho'_F}{\partial t^2} - c_0^2 \nabla^2 \rho'_F = 0 \quad , \tag{16.83}$$

together with a surface boundary condition. We will discuss the solution of these equations for planar and cylindrical geometries separately.

16.3.1 Planar Array

For a planar array, whose mean position is in the plane $x_3 = 0$, let us suppose that the relationship between the surface pressure and the normal displacement, $\xi(x_1, x_2, t)$, can be conveniently expressed in terms of their Fourier transforms,

$$\widetilde{p}(0, \mathbf{k}, \omega) = Z(\mathbf{k}, \omega)\widetilde{\xi}(\mathbf{k}, \omega) \quad . \tag{16.84}$$

Tensioned membranes, bending plates and more complicated multi–layered structures all fall into this category. If the flow near the surface is only linearly disturbed from rest, the linearized momentum equation shows that $\partial\widetilde{p}/\partial x_3 = \rho_0\omega^2\widetilde{\xi}$, where ρ_0 is the mean density. Since $\widetilde{\rho} = \widetilde{p}/c_0^2$, Equation (16.84) can be rewritten to give a surface condition for $\widetilde{\rho}$:

$$\widetilde{\rho}(0, \mathbf{k}, \omega) = \frac{Z}{\rho_0\omega^2} \frac{\partial\widetilde{\rho}}{\partial x_3}(0, \mathbf{k}, \omega) \quad . \tag{16.85}$$

When the expansion (16.80) is substituted into (16.85), it shows that

$$\widetilde{\rho}_F(0, \mathbf{k}, \omega) + \widetilde{\rho}_R(0, \mathbf{k}, \omega) = \frac{Z(\mathbf{k}, \omega)}{\rho_0\omega^2} \frac{\partial\widetilde{\rho}_F}{\partial x_3} \quad . \tag{16.86}$$

The Fourier transform of Equation(16.83) leads to

$$\frac{\partial^2\widetilde{\rho}_F}{\partial x_3^2} + \gamma^2\widetilde{\rho}_F = 0 \quad , \tag{16.87}$$

where $\gamma = (\omega^2/c_0^2 - k^2)^{1/2}$. It is a straightforward matter to solve (16.87) subject to the boundary condition (16.86) and we find that

$$\widetilde{\rho}_F(0, \mathbf{k}, \omega) = -\frac{\widetilde{\rho}_R(0, \mathbf{k}, \omega)}{1 - i\gamma Z/\rho_0\omega^2} \quad . \tag{16.88}$$

This leads directly to the surface pressure fluctuation

$$\widetilde{p}(0, \mathbf{k}, \omega) = c_0^2 \left(\widetilde{\rho}_R(0, \mathbf{k}, \omega) + \widetilde{\rho}_F(0, \mathbf{k}, \omega)\right) \tag{16.89}$$

$$= \frac{i\gamma}{i\gamma - \rho_0\omega^2/Z}\widetilde{p}_R(0, \mathbf{k}, \omega) \quad . \tag{16.90}$$

The power spectral density of the surface pressure, $\widetilde{P}(\mathbf{k}, \omega)$, can be calculated by multiplying (16.90) by its complex conjugate. This shows that

$$\widetilde{P}(\mathbf{k}, \omega) = \frac{|\gamma|^2}{|i\gamma - \rho_0 \omega^2 / Z|^2} \widetilde{P}_R(\mathbf{k}, \omega) \quad ; \tag{16.91}$$

$\widetilde{P}_R(\mathbf{k}, \omega)$ is "rigid–wall" or blocked pressure spectrum. It describes the surface pressure induced on a rigid wall by the same turbulent quadrupoles. The multiplying factor $|\gamma|^2 / |i\gamma - \rho_0 \omega^2 / Z|^2$ evidently describes the difference between radiation from sources near flexible and rigid walls. After substitution for $\widetilde{P}_R(\mathbf{k}, \omega)$ from (16.30), we obtain

$$\widetilde{P}(\mathbf{k}, \omega) = \frac{d_i^* d_j^* d_k d_l}{|i\gamma - \rho_0 \omega^2 / Z|^2} \int T_{ijkl}(y_3, y_3', \mathbf{k}, \omega) e^{i(\gamma y_3' - \gamma^* y_3)} dy_3 dy_3' \quad . \tag{16.92}$$

This relates the wall pressure spectrum to source terms which depend on the cross–correlation of the Reynolds stresses. The factor $d_i d_j^* d_k d_l / |i\gamma - \rho_0 \omega^2 / Z|^2$ describes how the turbulent flow radiates sound within the boundary layer over the flexible surface. The influence of a large plane homogeneous flexible boundary is simply to "reflect" the sound generated by the turbulence (Ffowcs Williams 1965). The product $d_i^* d_j^* d_k d_l$ describes the propagation of different directional elements, but the main structure of the surface pressure spectrum comes from the $|i\gamma - \rho_0 \omega^2 / Z|^{-2}$ term. We will investigate how this factor varies for different surfaces of practical interest. Many of the surfaces we have in mind are massive (5cm–thick steel plates, for example). While long wavelength sound waves can deform them ($k \sim \omega / c_0$), they are effectively rigid to turbulence with its shorter wavelengths ($k_1 \sim -\omega / U_c$). In this case, we might expect the cross–correlation of the turbulent sources, T_{ij}, to be uninfluenced by surface flexibility. Replacing the integral in (16.92) by its form for a rigid surface as given in (16.31) leads to

$$\widetilde{P}(\mathbf{k}, \omega) = \frac{d_i^* d_j^* d_k d_l}{|i\gamma - \rho_0 \omega^2 / Z|^2} \frac{\rho^2 u_\tau^4 \delta^{*5}}{U_\infty} S_{ijkl}\left(\frac{\omega \delta^*}{U_\infty}\right) \quad , \tag{16.93}$$

provided the boundary–layer thickness is small in comparison with $|\gamma|^{-1}$.

As an introductory example we will consider a bending plate backed by a void. For such a surface

$$Z(\mathbf{k}, \omega) = m\omega^2 - Bk^4, \quad , \tag{16.94}$$

where m is the mass of the plate per unit area and $B = Ed^3 / 12(1 - \nu^2)$, E is Young's modulus, ν is Poisson's ratio and d is the plate thickness. In the absence of fluid loading, flexural waves of frequency ω propagate in the plate with a phase speed V equal to $(B\omega^2/m)^{1/4}$. At the coincidence frequency ω_c, V is equal to the sound speed c_0. At frequencies above coincidence V is supersonic and at lower frequencies V is subsonic.

Equation (16.93) shows that the surface pressure spectrum under an infinite extent of turbulence is given by

$$\widetilde{P}(\mathbf{k},\omega) = d_i^* d_j^* d_k d_l |F(\mathbf{k},\omega)|^2 \frac{\rho^2 u_\tau^4 \delta^{*5}}{U_\infty} S_{ijkl}\left(\frac{\omega\delta^*}{U_\infty}\right) \quad , \tag{16.95}$$

where

$$F(\mathbf{k},\omega) = \frac{m\omega^2 - Bk^4}{\rho_0\omega^2 - i\gamma(m\omega^2 - Bk^4)} \quad . \tag{16.96}$$

An investigation of the form of $F(\mathbf{k},\omega)$ will determine the influence of the wall impedance on the surface pressure spectrum. The structure of $F(\mathbf{k},\omega)$ depends on N, the Mach number of the in vacuo–bending wave speed, V/c_0, and on the fluid loading factor $\rho_0 c_0/m\omega$, which is essentially the ratio between the mass of fluid within a wavelength of the surface and the mass of the plate. Plots of $F(\mathbf{k},\omega)$ are given in Figure 16.5 for positive ω.

The function $F(\mathbf{k},\omega)$ can also be investigated analytically. For heavy plates and highly supersonic spectral elements $F(\mathbf{k},\omega) \sim ic_0/\omega$. The pressure spectrum is then identical to the hard surface result in (16.32). It is also apparent from expression (16.96) that $F(\mathbf{k},\omega)$ (and hence the pressure spectrum) vanishes for spectral elements whose phase speeds are equal to V. The pressure spectrum has a singularity at $\kappa(\omega)$, a zero of $\rho_0\omega^2 - i\gamma(m\omega^2 - Bk^4)$. For heavy plates $\kappa(\omega)$ can be determined iteratively by an expansion in powers of ρ_0. This shows that there is a singularity in the pressure spectrum near $k = \omega/V$ for $V < c_0$ and near $k = \omega/c_0$ if $V > c_0$ (with the phase speed ω/k just subsonic). These predictions for a heavy plate are confirmed by the plots in Figure 16.5.

Figure 16.5 illustrates the variation of $20\log_{10}|\omega F(\mathbf{k},\omega/c_0)|$ with nondimensional wavenumber kc_0/ω. The plate parameters used in these calculations correspond to a 5cm–thick steel plate in water at a frequency of (a) 3kHz and (b) 6kHz. In Figure 16.5a, V is subsonic and the singularity in the pressure spectrum occurs for spectral elements with phase speeds nearly equal to the bending wave speed in vacuo. In Figure 16.5b, V is supersonic and the singularity in the pressure spectrum occurs for modes with nearly sonic phase speeds. These singularities are double poles and are stronger than those found for sonic spectral elements on a hard surface. Just as in the hard surface case they arise due to a scale effect, because the source region has been assumed to be infinite.

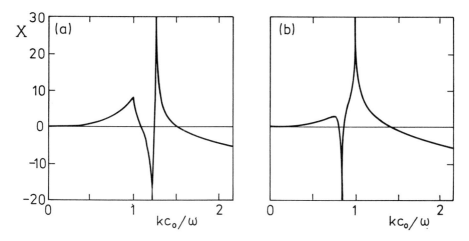

Figure 16.5 Plots of $X = 20\log_{10}|\omega F(\mathbf{k}, \omega)/c_0|$ versus nondimensional wavenumber for a bending plate with (a) $m\omega/\rho_0 c_0 = 5$, $N = 0.8$; (b) $m\omega/\rho_0 c_0 = 10$, $N = 1.2$.

Since the peaks in the surface pressure spectrum are dominated by the influence of distant sources, the first step in understanding them is to determine the response far from a source adjacent to a bending plate. $\kappa(\omega)$, the zeros of $\rho_0\omega^2 - i\gamma(m\omega^2 - Bk^4)$, are the wavenumbers of free modes of the plate–fluid system. These modes are excited by a source near the surface and travel subsonically over the plate. Since they have a subsonic surface speed they are evanescent in the fluid. All the energy in these modes therefore remains "trapped" within a disc near the surface and energy conservation suggests that the pressure disturbances associated with the modes should decay like $r^{-1/2}$, where r is the distance from a source near the surface. An asymptotic evaluation of the response due to a point source shows this to be true. This decay rate is slower than that for sound near a hard surface and this leads to stronger singularities in the pressure spectrum. Using the same techniques as in Section 16.1 to investigate the effect of a finite patch of turbulence shows that there is a δ–function singularity for spectral elements with wavenumber $\kappa(\omega)$. The strength of the δ–function grows linearly with L, the extent of the turbulent source region. The details are given in Dowling (1983a). Although consideration of a finite turbulent source region makes the peaks in the surface pressure spectrum integrable, for reasonably sized source regions, these singularities still have a considerable effect. The integrated pressure level due to the singularities can be significantly larger than the levels elsewhere in the spectrum. Therefore we look for other means of limiting the pressure field.

Howe (1979) investigated the influence of the fluid's viscosity on the pressure spectrum on the hard surface, but found that it only had a small effect. For a flexible surface the wall provides another source of dissipation. The damping in the bending plate will be modelled by giving Young's modulus a small imaginary part. Then $Z(\mathbf{k}, \omega) = m\omega^2 - Bk^4(1 - i\eta)$, where the damping factor η has the same sign as ω. Plots of $|F(\mathbf{k}, \omega)|$ with this surface condition are given in Figure 16.6. As one might expect the damping has most effect on the pressure spectrum for spectral elements whose phase speed is nearly equal to V, the speed of bending waves in the unloaded plate. The pressure spectrum no longer vanishes for spectral components with phase speeds equal to V as it did for the undamped plate. Also the singularity that occurred near V for frequencies below coincidence is reduced to a reasonable level by dissipation in the plate. The damping has very little effect on the singularity for nearly sonic elements at frequencies above coincidence, essentially because little plate vibration is involved in these glancing modes.

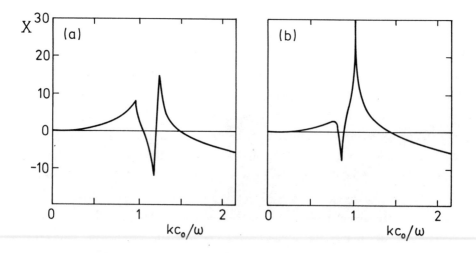

Figure 16.6 Plots of $X = 20\log_{10}|\omega F(\mathbf{k}, \omega)/c_0|$ versus nondimensional wavenumber for a damped bending plate with $\eta = 0.05$ and (a) $m\omega/\rho_0 c_0 = 5$, $N = 0.8$; (b) $m\omega/\rho_0 c_0 = 10$, $N = 1.2$.

There is some experimental evidence that certain surface coatings can also reduce the surface pressure fluctuations due to flow noise. We will model such a coating by a fluid layer of thickness T, with density ρ_s and sound speed c_s. A sketch of the coating is given in Figure 16.7. The impedance of this composite surface can be found in a straightforward way (Maidanik 1984) leading to $Z(\mathbf{k}, \omega)$. Equation (16.95) then gives the pressure spectrum on the surface of the coating. In fact the pressure spectrum on the bending plate, $\widetilde{P}_B(\mathbf{k}, \omega)$, is of more practical interest since it can be compared directly with the pressure on the uncoated plate. The relationship between the pressure spectra on these two surfaces can easily be determined and

$$\widetilde{P}_B(\mathbf{k}, \omega) = d_i^* d_j^* d_k d_l |F_B(\mathbf{k}, \omega)|^2 \frac{\rho^2 u_\tau^4 \delta^{*5}}{U_\infty} S_{ijkl} \left(\frac{\omega \delta^*}{U_\infty} \right) \quad , \tag{16.97}$$

where $|F_B(\mathbf{k}, \omega)|$ is shown in Figures 16.8 and 16.9.

The coating can have a significant effect on the pressure spectrum for spectral components with sonic phase speeds. If the coating is sufficiently thick and c_s is greater than c_0, it is possible for the coating to control the singularity that would occur for sonic modes in an uncoated plate whenever V is supersonic. The graphs in Figure 16.8 are for a 4cm–thick coating with $c_s = 1.5c_0$ and $\rho_s = 1.5\rho_0$ on a 5cm–thick steel plate at frequencies of (a) 3kHz and (b) 6kHz. A comparison of Figures 16.5 and 16.8 shows that the coating can eliminate the singularity in the spectrum for frequencies above coincidence, but that it has little effect at frequencies below coincidence.

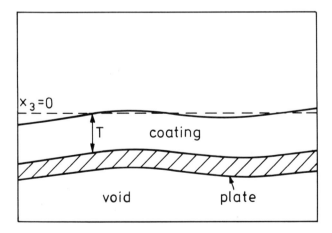

Figure 16.7 The plate and coating.

480

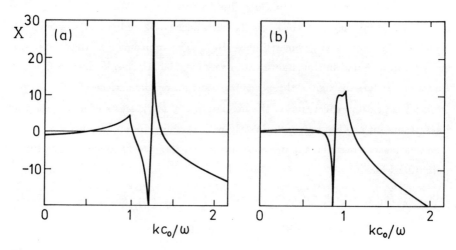

Figure 16.8 Plots of $X = 20\log_{10}|\omega F_B(\mathbf{k},\omega)/c_0|$ versus nondimensional wavenumber for a coated plate with $c_s = 1.5c_0$, $\rho_s = 1.5\rho_0$ and (a) $m\omega/\rho_0 c_0 = 5$, $N = 0.8$; (b) $m\omega/\rho_0 c_0 = 10$, $N = 1.2$.

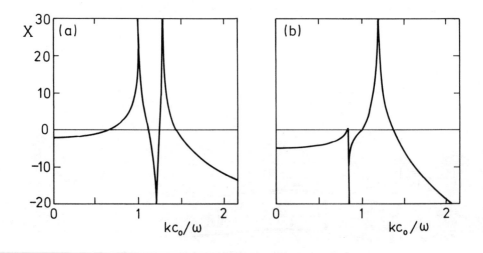

Figure 16.9 Plots of $X = 20\log_{10}|\omega F_B(\mathbf{k},\omega)/c_0|$ versus nondimensional wavenumber for a coated plate with $c_s = 0.75c_0$, $\rho_s = 0.75\rho_0$ and (a) $m\omega/\rho_0 c = 5$, $N = 0.8$; (b) $m\omega/\rho_0 c_0 = 10$, $N = 1.2$.

A low sound speed coating with $c_s < c_0$ has an adverse effect on the pressure spectrum generated by the turbulent flow. It can produce a singularity at approximately sonic phase velocities for frequencies below coincidence, where for an uncoated plate the pressure spectrum is not particularly large. Figure 16.9 shows plots of $|F_B(\mathbf{k}, \omega)|$ for a 5cm–thick steel plate covered by a 4cm–thick coating layer with $c_s = 0.75c_0$ and $\rho_s = 0.75\rho_0$, again at frequencies of 3kHz and 6kHz. A comparison of Figure 16.5a and Figure 16.9a shows that the low–speed coating has introduced a new singularity not present in the uncoated plate.

The theory can be extended to investigate more complicated surfaces. If the wall is treated more realistically as a thin elastic slab rather than a bending plate, the wall pressure spectrum has an additional narrow band peak or "spike" (Dowling 1983b). This occurs because admitting that the elastic medium can support general elastic deformations introduces a new mode of vibration which is not present in the simple bending plate theory. In addition to the flexural waves displayed by bending plate theory, the elastic region can support a symmetric mode which propagates with the compressive wave speed $\sqrt{E/\rho_m(1 - \nu^2)}$, where E is Young's modulus, ν is Poisson's ratio and ρ_m is the density of the elastic medium. Dowling (1983b) considers a thin slab with identical fluid in contact with both faces. The surface pressure spectrum under the plate is found to have a local maximum for spectral elements whose phase speeds are equal to either the compressive or the flexural wave speeds.

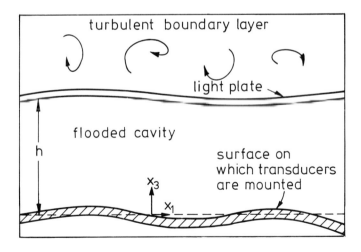

Figure 16.10 A sonar dome.

The effect of placing a sonar "dome" over a bending plate can be investigated in a similar way. A sonar dome consists of a light surface and a flooded cavity of height h over a massive plate (see Figure 16.10). The surface impedance of the plate and flooded cavity can be calculated in a straightforward way (Maidanik 1968). The flow noise can then be predicted using the result in Equation (16.93) for a general flexible surface. If the covering surface is sufficiently light it has little effect on pressure perturbations passing through it and the main effect of the dome arrangement is to ensure that the turbulent sources are held away from the massive surface on which the pressure spectrum is measured. We see from (16.92) that the surface pressure spectrum for spectral elements with subsonic phase speeds decays exponentially with vertical distance between the surface and the sources. The amplitudes of the elements with supersonic phase speeds are not reduced in this way. The domed sonar system can therefore be used to discriminate against the subsonic modes. It dramatically reduces the convective peak in the surface pressure spectrum without a significant change to propagating acoustic waves, thereby increasing the signal-to-noise ratio.

16.3.2 Towed Arrays

In Section 16.2 we determined the circumferentially–averaged pressure spectrum on a rigid cylinder. That work can be extended to determine the line spectrum of axial pressure within a towed array.

We will consider the array to be a fluid–filled flexible cylinder of mean radius a. The thin walls of the cylinder have thickness d and negligible mass. They are made of an elastic material with Young's modulus E, damping η and Poisson ratio ν. The array is filled with fluid with the same mean density and sound speed as the surrounding fluid.

Let us introduce polar coordinates (R, θ, x_1) and denote the displacement of the cylinder walls from radius a by $\xi(\theta, x_1, t)$. It is convenient to take Fourier transforms in x_1 and t and expand the θ–dependence as a Fourier series

$$\widetilde{\xi}(n, k_1, \omega) = \int \xi(\theta, x_1, t) e^{in\theta + ik_1 x_1 + i\omega t} a\, d\theta\, dx_1\, dt \quad . \tag{16.98}$$

Axial pressure disturbances are only coupled to circumferentially–averaged fluctuations and so we are particularly interested in the $n = 0$ mode.

Within the cylinder, pressure fluctuations are finite and satisfy a homogeneous wave equation. We can therefore write

$$\tilde{p}(R, 0, k_1, \omega) = S J_0(\alpha R) \text{ for } R \leq a \quad , \tag{16.99}$$

where $\alpha = (\omega^2/c_0^2 - k_1^2)^{1/2}$.

Outside the cylinder, we decompose fluctuations in the way indicated by Equations (16.80) – (16.83) and expand

$$p'(\mathbf{x}, t) = p'_R(\mathbf{x}, t) + p'_F(\mathbf{x}, t) \quad ; \tag{16.100}$$

$p'_R(\mathbf{x}, t), = c_0^2 \rho'_R(\mathbf{x}, t)$, is the pressure field induced by quadrupoles T_{ij} adjacent to a rigid cylinder and was determined in Section 16.2. We see from Equation (16.83) that $p'_F(\mathbf{x}, t)$ satisfies a homogeneous wave equation and this, together with the radiation condition at infinity, leads to

$$\tilde{p}_F(R, 0, k_1, \omega) = T H_0^{(1)}(\alpha R) \qquad \text{for} \qquad R \geq a \quad . \tag{16.101}$$

The constants S and T in Equations (16.99) and (16.101) can be determined from the surface conditions.

The normal velocity must be continuous across the surface $R = a$. Hence

$$\tilde{\xi}(0, k_1, \omega) = \frac{S\alpha}{\rho_0 \omega^2} J_0'(\alpha a) = \frac{T\alpha}{\rho_0 \omega^2} H_0^{(1)'}(\alpha a) \quad , \tag{16.102}$$

where we have used Equation (16.82) to say that $\partial \tilde{p}_R / \partial R = 0$ on $R = a$. The equation of motion of a thin curved plate (see for example Bolotin 1963, pp.274–279) is

$$E(1 - i\eta)(1 + \beta^4 k_1^4)\tilde{\xi}(0, k_1, \omega) = - \frac{a^2}{d} [\tilde{p}(R, 0, k_1, \omega)]_{a^-}^{a^+} \quad , \tag{16.103}$$

where $\beta = \left(a^2 d^2/12(1 - \nu^2)\right)^{1/4}$. S and T can be calculated by substituting Equations (16.99) – (16.102) into (16.103). This shows that

$$S = \frac{\tilde{p}_R(a, 0, k_1, \omega)}{(1 + \beta^4 k_1^4) J_1(\alpha a) E(1 - i\eta) d\alpha/(\rho_0 \omega^2 a^2) - i2/(\pi \alpha a H_1^{(1)}(\alpha a))} \quad . \tag{16.104}$$

The axial pressure then follows directly from (16.99) and is given by

$$\tilde{p}(0, 0, k_1, \omega) = \frac{\tilde{p}_R(a, 0, k_1, \omega)}{(1 + \beta^4 k_1^4) J_1(\alpha a) E(1 - i\eta) d\alpha/(\rho_0 \omega^2 a^2) - i2/(\pi \alpha a H_1^{(1)}(\alpha a))} \quad ; \tag{16.105}$$

$\tilde{p}(0, 0, k_1, \omega)$ is the Fourier transform of the axial pressure induced within a towed array by its turbulent boundary layer. $\tilde{p}_R(a, 0, k_1, \omega)$ is the Fourier transform of the circumferentially–averaged surface pressure that would be produced on a rigid cylinder of radius a by the same turbulent quadrupole sources. The transfer function between these two pressures has singularities which appear as additional peaks in the towed–array line spectra.

The transfer function simplifies considerably when both $|k_1|$ and ω/c_0 are small in comparison with a^{-1}. For realistic geometries, this ensures that $\beta^4 k_1^4$ is negligible in comparison with unity. Moreover the Bessel functions can be replaced by their small argument asymptotic forms to give

$$\tilde{p}(0, 0, k_1, \omega) = \frac{\tilde{p}_R(a, 0, k_1, \omega)}{1 + c_b^2(1 - i\eta)\alpha^2/\omega^2} \quad , \tag{16.106}$$

where $c_b = (Ed/2\rho_0 a)^{1/2}$. If the damping η is small the axial pressure evidently has a strong peak whenever

$$1 + c_b^2 \frac{\alpha^2}{\omega^2} = 0 \quad . \tag{16.107}$$

After substituting $\alpha = (\omega^2/c_0^2 - k_1^2)^{1/2}$, we find that the roots of (16.107) are at

$$k_1^2 = \frac{\omega^2}{c_0^2} + \frac{\omega^2}{c_b^2} \quad . \tag{16.108}$$

For $c_b \ll c_0$, this simplifies to $|k_1| \sim \omega/c_b$. The axial pressure spectrum has a peak at the bulge wave speed c_b.

Near the convective peak, $\alpha \sim i\omega/U_c$ and (16.106) reduces to

$$\tilde{p}(0, 0, -\omega/U_c, \omega) \sim -\frac{U_c^2}{c_b^2(1 - i\eta)} \tilde{p}_R(a, 0, -\omega/U_c, \omega) \text{ for } \frac{\omega a}{U_c} \ll 1 \quad . \tag{16.109}$$

Since the ratio U_c/c_b is generally small, the transmission of flow noise from the circumference to the axis of a towed array is poor near the convective peak.

When $\omega a/U_c$ is large, the large argument forms for the Bessel functions can be used in (16.105) near the convective peak. This leads to

$$\tilde{p}(0, 0, -\omega/U_c, \omega) \sim -\frac{U_c^2}{c_b^2(1 - i\eta)} \left(\frac{\pi}{2}\right)^{1/2} \left(\frac{\omega a}{U_c}\right)^{3/2} \frac{e^{-\omega a/U_c}}{1 + \beta^4 \omega^4/U_c^4} \tilde{p}(a, 0, -\omega/U_c, \omega) \tag{16.110}$$

for $1 \ll \omega a/U_c \ll c_b^2/U_c^2$. Again transmission is poor near the convective peak.

The line spectrum of axial pressure within a towed array can be determined by multiplying (16.105) by its complex conjugate. We obtain

$$\widehat{P}(k_1,\omega) = \frac{\widehat{P}_R(k_1,\omega)}{\left|(1 + \beta^4 k_1^4)J_1(\alpha a)E(1 - i\eta)d\alpha/(\rho_0\omega^2 a^2) - i2/(\pi\alpha a H_1^{(1)}(\alpha a))\right|^2} \quad ; $$

(16.111)

$\widehat{P}_R(k_1,\omega)$ is the circumferentially–averaged line pressure spectrum induced on a rigid cylinder by the fluctuating Reynolds stresses, T_{ij}. It is related to these sources in the way indicated in Equation (16.70). If we assume that the cross–power spectral density of these source functions can be expanded in separable form, $\widehat{P}_R(k_1,\omega)$ is described by (16.72).

As we noted in Section 16.2, Equation (16.77) implies that $\widehat{P}_R(k_1,\omega)$ becomes wavenumber white at low wavenumbers, with

$$\widehat{P}_R(k_1,\omega) = \rho^2 U_c^3 a^2 \Phi(\omega a/U_c)A(1) \quad , $$

(16.112)

for $|k_1| << \omega/U_c$. We have already noted that the denominator in (16.111) simplifies for this wavenumber limit, and combining (16.106), (16.111) and (16.112) leads to

$$\widehat{P}(k_1,\omega) = \frac{\rho^2 U_c^3 a^2 \Phi(\omega a/U_c)A(1)}{|1 + c_b^2(1 - i\eta)\alpha^2/\omega^2|^2} \quad . $$

(16.113)

This expression describes the influence of array compliance on the low wavenumber components of flow noise. For $k_1 \sim 0$, the denominator $\sim |1 + c_b^2(1 - i\eta)/c_0^2|^2$, and is approximately unity. Highly supersonic modes are unaffected by array flexibility. Equation (16.113) predicts that the line pressure spectrum, $\widehat{P}(k_1,\omega)$, has a peak near $\omega = |k_1|c_b$. The height and width of this peak depend on the value of the damping factor η.

16.4 MEAN FLOW EFFECTS

So far we have discussed the boundary–layer flow as if it just consisted of turbulence. In practice there is also a mean flow $\mathbf{U}$ parallel to the surface. We will discuss the effects of this mean flow by considering a turbulent boundary–layer over a plane homogeneous flexible wall which is linearly disturbed from its rest position $x_3 = 0$. Then $\mathbf{U} = (U(x_3), 0, 0)$ and it is appropriate to introduce $\mathbf{u}'$, the difference between the fluid velocity and the mean flow $\mathbf{U}$,

$$\mathbf{u}' = \mathbf{u} - \mathbf{U} \quad . $$

(16.114)

The Lighthill sources in Equation (16.9) can be expressed in terms of $\mathbf{u}'$ to give

$$\left(\frac{\partial^2}{\partial t^2} - c_0^2 \nabla^2\right) \rho' = \frac{\partial^2}{\partial x_i \partial x_j} \left(\rho u_i' u_j' + \rho U_i u_j' + \rho u_i' U_j + \rho U_i U_j + (p' - c_0^2 \rho')\delta_{ij} - \sigma_{ij}\right)$$

(16.115)

The source terms $\rho u_i' u_j'$ are quadratic in the velocity fluctuations. They are negligible in a weak sound field and are only significant where there is turbulence. In contrast the terms $\rho U_i u_j' + \rho u_i' U_j$ depend linearly on $\mathbf{u}'$. They are nonzero even for a sound wave propagating through the region and should not be regarded as a source. They describe propagation effects and occur because the wave operator on the left–hand side of (16.115) does not describe the propagation of sound correctly when there is a mean flow. A more complicated operator is needed to ensure that the source terms are quadratic in the velocity fluctuations. That has been derived by Lilley (1971). We follow Chase & Noiseux (1982) who show that the equations of mass and momentum conservation can be combined into the form

$$\frac{\partial^2}{\partial x_3^2} \left(\frac{\tilde{\rho} + \tilde{T}_{33}/c_0^2}{\omega + U k_1}\right) + f \frac{\tilde{\rho} + \tilde{T}_{33}/c_0^2}{\omega + U k_1}$$

$$= \frac{1}{c_0^2} \left\{\frac{k_\alpha k_\beta \tilde{T}_{\alpha\beta} + \gamma^2 \tilde{T}_{33}}{\omega + U k_1} + 2 i k_\alpha \frac{\partial}{\partial x_3} \left(\frac{\tilde{T}_{\alpha 3}}{\omega + U k_1}\right)\right\} \quad .$$

(16.116)

In this expression $T_{ij} = \rho u_i' u_j' + (p' - c_0^2 \rho')\delta_{ij} - \sigma_{ij}$, α and β are to be summed over 1 and 2 and the tilde again denotes a Fourier transform,

$$f(x_3, \mathbf{k}, \omega) = \frac{(\omega + U k_1)^2}{c_1^2} - k_1^2 - k_2^2 + \frac{d^2 U}{dx_3^2} \frac{k_1}{\omega + U k_1} - 2 \left(\frac{dU}{dx_3}\right)^2 \frac{k_1^2}{(\omega + U k_1)^2}$$

(16.117)

and $\gamma = ((\omega + U k_1)^2 / c_0^2 - k^2)^{1/2}$, with the sign of the square root chosen so that γ has the same sign as ω when γ is real and $\Im\gamma$ is positive when γ is purely imaginary. In the case of no mean flow Equation (16.116) reduces to the Fourier transform of (16.9).

Outside the boundary layer the fluid is only linearly disturbed from the free stream velocity $(U_\infty, 0, 0)$ and $T_{ij} = 0$. Since disturbances must either decay at large x_3 or be outward propagating sound waves,

$$\frac{\partial \tilde{\rho}}{\partial x_3} = i \gamma_\infty \tilde{\rho} \qquad \text{for large positive } x_3 \quad ;$$

(16.118)

$\gamma_\infty, = ((\omega + U_\infty k_1)^2 / c_0^2 - k^2)^{1/2}$, is the limit of γ as x_3 tends to infinity.

We will assume that the impedance is uniform over the whole surface $x_3 = 0$ and that the relationship between the surface pressure and displacement can be conveniently expressed in terms of their Fourier transforms

$$\widetilde{p}(0, \mathbf{k}, \omega) = Z(\mathbf{k}, \omega)\widetilde{\xi}(\mathbf{k}, \omega) \quad . \tag{16.119}$$

Very near the surface the flow is only linearly disturbed and $\widetilde{p} = c_0^2\widetilde{\rho}$. The linearized 3–component of the momentum equation and the surface condition (16.119) then give

$$\rho_0\omega^2\widetilde{\rho}(0, \mathbf{k}, \omega) = Z(\mathbf{k}, \omega)\partial\widetilde{\rho}/\partial x_3 \quad , \tag{16.120}$$

where ρ_0 is the mean density. We will again neglect any surface shear stress.

The ordinary differential Equation (16.116) is to be solved subject to the boundary conditions (16.118) and (16.120). Standard Green function techniques make it easy to determine the solution. It is convenient to introduce a function $E(x_3)$, which satisfies the homogeneous equation

$$\frac{d^2 E}{dx_3^2} + fE = 0 \quad , \tag{16.121}$$

together with the boundary conditions

$$\frac{dE}{dx_3} \to i\gamma_\infty E \text{ as } x_3 \to \infty \quad , \tag{16.122}$$

and

$$E = 1 \text{ on } x_3 = 0 \quad . \tag{16.123}$$

In terms of E the solution to (16.116) is

$$\widetilde{p}(0, \mathbf{k}, \omega) = \frac{\omega}{E'(0) + U'(0)k_1/\omega - \rho_0\omega^2/Z}$$

$$\times \int \left\{ E(y_3)\frac{k_\alpha k_\beta\widetilde{T}_{\alpha\beta} + \gamma^2\widetilde{T}_{33}}{\omega + Uk_1} - \frac{\partial E}{\partial y_3}\frac{2ik_\alpha\widetilde{T}_{\alpha3}}{\omega + Uk_1} \right\} dy_3 \quad ; \tag{16.124}$$

(see Chase & Noiseux (1982) or Dowling (1986) for the derivation of this relationship).

The function $E(x_3)$ can be evaluated for a particular mean–flow profile by a numerical solution of (16.121). The results in Figure 16.11 were obtained for a tanh–velocity profile $U(x_3) = U_\infty \tanh(x_3/\Delta)$, where Δ is a measure of the boundary-layer height. But in fact the dominant flow effect does not depend on the details of the boundary–layer profile. That can be seen from an asymptotic solution for $E(x_3)$, valid for low values of the flow Mach number $M = U_\infty/c_0$ and low wavenumbers i.e. $k \sim \omega/c_0$.

We seek an expansion for E as a series in the powers of the Mach number, M. To lowest order $E(x_3) = \exp(i\gamma_\infty x_3)$. More algebra is required to obtain the $0(M)$ term, but it can be shown (Dowling 1986) that

$$E'(0) = i\gamma_\infty(1 + MB) - U'(0)k_1/\omega - 2k_1\omega M\delta^*/c_0 \quad ; \tag{16.125}$$

δ^* is the boundary–layer displacement thickness, defined by $\delta^* = \int_0^\infty (1 - U(x_3)/U_\infty)dx_3$, B is an order one function. It is evaluated in Dowling (1986), but its precise form does not concern us here.

Substitution for $E(x_3)$ in the representation (16.124) gives

$$\widetilde{p}(0, \mathbf{k}, \omega) = \frac{d_i d_j}{E'(0) + U'(0)k_1/\omega - \rho_0\omega^2/Z} \int \widetilde{T}_{ij}(y_3, \mathbf{k}, \omega)e^{i\gamma_\infty y_3} dy_3 \quad ; \tag{16.126}$$

$d_\alpha = k_\alpha$ for $\alpha = 1$ or 2 and $d_3 = \gamma_\infty$. The terms in the numerator cannot all vanish simultaneously and so we have only kept the lowest–order terms there. Higher–order terms are evaluated by Chase & Noiseux (1982). The leading–order term in the denominator can vanish and so terms of order M must be retained there. These terms are neglected by Chase & Noiseux.

The wall pressure spectrum, $\widetilde{P}_M(\mathbf{k}, \omega)$, can be derived from the Fourier transform of the surface pressure in the way outlined in Section 16.1, and is given by

$$\widetilde{P}_M(\mathbf{k}, \omega) = \frac{d_i^* d_j^* d_k d_l}{|E'(0) + U'(0)k_1/\omega - \rho_0\omega^2/Z|^2} \int T_{ijkl}(y_3, y_3', \mathbf{k}, \omega)e^{i(\gamma_\infty y'_3 - \gamma_\infty^* y_3)} y_3 dy_3'. \tag{16.127}$$

In the low Mach number limit, Equation (16.125) shows that

$$E'(0) + U'(0)k_1/\omega - \rho_0\omega^2/Z(\mathbf{k}\ \omega) = i\gamma_\infty(1 + MB) - 2\omega k_1\delta^* M/c_0 - \rho_0\omega^2/Z(\mathbf{k}, \omega). \tag{16.128}$$

We will begin by discussing results for a hard surface.

16.4.1 A Hard Surface

The normal displacement always vanishes on a hard surface and so $Z(\mathbf{k}, \omega)$ is infinite. The denominator in (16.127) then simplifies to

$$E'(0) + U'(0)k_1/\omega = i\gamma_\infty(1 + MB) - 2\omega k_1 \delta^* M/c_0 \quad . \tag{16.129}$$

If we neglect terms of order $Mk_1\delta^*$ in (16.129), the right–hand side reduces to $i\gamma_\infty$, the uniform mean flow result. This vanishes for spectral elements with sonic phase speeds. A uniform mean flow has no effect other than to produce a Döppler shift in frequency. Haj Hariri & Akylas (1985a) use a matching argument to investigate the effect of a mean–flow profile. They neglect terms of order $M\omega\delta^*/c_0$ and so, as predicted by (16.128), just obtain the uniform flow result for low wavenumbers. The variation in mean–flow velocity leads to the additional terms in (16.129). The profile has most effect on the pressure spectrum near γ_∞, where the leading–order term vanishes. Then $i\gamma_\infty MB$ is negligible since both γ_∞ and M are small, and the dominant effect of the mean–flow profile is contained in the term $-2\omega k_1\delta^* M/c_0$. This is independent of the details of the boundary layer, depending only on its integrated effect through the flow Mach number U_∞/c_0, and the displacement thickness, δ^*.

When the effect of the mean–flow profile is included, the denominator in (16.127) is small near $\gamma_\infty \approx 0$, but it can only vanish when γ_∞ is purely imaginary. Then the first term on the right–hand side of (16.129) is real and negative, and we see from Equation (16.129) that $E'(0) + U'(0)k_1/\omega$ can vanish only if ω and k_1 have opposite signs, i.e. for downstream–propagating modes. For upstream–propagating modes $E'(0) + U'(0)k_1/\omega$ is always nonzero, having a minimum modulus of $2\omega^2\delta^* M/c_0^2$ at $\gamma_\infty = 0$. This result was first obtained by Dowling (1983c), who crudely modelled the boundary layer by a vortex sheet above a hard wall. The analysis we have summarized here is from Dowling (1986) and extends the earlier result to an arbitrary mean–flow profile. It shows that the equivalent vortex sheet is to be placed a boundary–layer displacement thickness above the wall. Reutov & Rybushkina (1986) reached the same conclusion independently by a matching argument.

The pressure spectrum therefore has a double pole near $\gamma_\infty = 0$ for downstream–propagating spectral elements, but it is always finite for upstream–propagating elements. This was verified by numerical calculations for a tanh–velocity profile, $U_\infty \tanh(x_3/\Delta)$, and the results are shown in Figure 16.11. The plot clearly demonstrates that the predicted pressure spectrum is larger for downstream–propagating modes with sonic phase speeds than it is for upstream–propagating modes. Integration shows that for the tanh–velocity profile $\delta^* = \Delta \ln 2$, and the height of the peak for upstream–propagating modes in Figure 16.11 agrees with $-20\log_{10}(2\omega\delta^* M/c_0)$, the level predicted from the low Mach number asymptotic theory.

Sevik (1986) noted that measurements of pressure spectral elements with sonic phase speeds indicate that most of the energy is propagating downstream in agreement with our predictions.

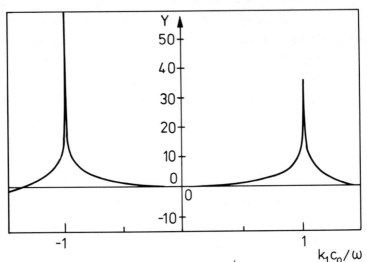

Figure 16.11 Plot of $Y = 20\log_{10}\left|\dfrac{\omega/c_0}{E'(0) + U'(0)k_1/\omega}\right|$ versus nondimensional wavenumber for flow over a rigid wall with $M = 0.01$, $\omega\delta^*/c_0 = 1$, $k_2 = 0$.

In Section 16.1 we found that when the mean–flow profile is neglected, the pressure spectrum has a simple pole for spectral elements with sonic phase speeds. It was possible to interpret the occurrence of these singularities as due to a form of Olber's paradox: the turbulent source region had been assumed to be of infinite extent and the sound from distant sources did not decay sufficiently rapidly with distance for their integrated effect to be finite. A mean–flow profile enhances the singularity for downstream–propagating modes, but eliminates it for upstream–propagating modes. Again, this has a simple physical interpretation.

When there is a mean–flow profile, there are downstream– propagating free modes of the system in which the waves travel supersonically (i.e. faster than local sound waves) within the slowly moving fluid in the boundary layer, but subsonically within the moving fluid outside it. The energy in these modes therefore remains "trapped" near the wall and conservation of energy then suggests that downstream of a source the pressure disturbance will only decay with the inverse square root of the distance. Upstream of a source, pressure fluctuations decay much more rapidly. The downstream–propagating modes decay more slowly with distance from the source than those in a uniform stream and account for the stronger singularity in the pressure spectrum under an infinite region of turbulence.

16.4.2 Flexible Surfaces

The effect of surface flexibility is to produce an extra term $-\rho_0 \omega^2 / Z(\mathbf{k}, \omega)$ in the denominator of (16.127). The relative importance of the mean–flow profile and the surface flexibility is described by the ratio of the two terms $2\omega k_1 \delta^* M / c_0$ and $\rho_0 \omega^2 / Z(\mathbf{k}, \omega)$. For realistic boundary–layer thicknesses over bending plates of practical thicknesses and properties in water, the effect of the mean flow profile is small in comparison with the surface flexibility.

We noted in Section 16.3 that the surface impedance for a sonar dome can be calculated in a straightforward way. Equation (16.127) then leads to a formal prediction for the effects of a mean–flow profile over a dome. The presence of a dome is found to greatly enhance the influence of the mean flow. This result has a simple physical interpretation.

Let us consider an elementary model of a sonar dome. We will suppose that the covering plate is so light that it is completely transparent to sound waves. The main effects of the dome are therefore to produce a stand–off distance h between the turbulent sources and the backing plate and to introduce an additional layer of height h in which the mean velocity is zero. The term δ^* in (16.128) arises from the integrated velocity defect $\int_0^\infty (1 - U(x_3)/U_\infty) dx_3$. The effect of this simplified dome is to increase δ^* to $\delta^* + h$, i.e.

$$E'(0){+}U'(0)k_1/\omega{-}\rho_0\omega^2/Z(\mathbf{k},\omega){=}i\gamma_\infty(1{+}MB'){-}2\omega k_1(\delta^*{+}h)M/c_0{-}\rho_0\omega^2/Z_p(\mathbf{k},\omega),$$

$$(16.130)$$

for a dome of height h over a backing plate of impedance $iZ_p(\mathbf{k},\omega)/\omega$. We should compare $2\omega k_1(\delta^* + h)M/c_0$ with $\rho_0\omega^2/Z_p(\mathbf{k},\omega)$ to assess the relative importance of the mean flow and surface flexibility. Often the backing plate is so massive that it is effectively rigid. As a particular example, let us consider the addition of a dome of nondimensional height $\omega h/c_0 = 5$ to the case shown in Figure 16.11. This reduces the maximum for upstream–propagating elements to a level which is only about 20dB larger than the value for highly supersonic modes.

When a dome is present, we would expect the pressure spectrum to be very much larger for downstream–propagating modes with sonic phase speeds than it is for upstream–propagating modes. Roebuck & Richardson (1981, private communication) have observed this in underwater experiments.

16.5 SCATTERING FROM THE CONVECTIVE PEAK TO LOW WAVENUMBERS

The treatment of flexible surfaces in Section 16.3 was restricted to homogeneous surfaces. Then we found that the flexible–wall pressure spectrum was influenced only by the same wavenumber and frequency components of the turbulent quadrupole sources. This is illustrated explicitly in Equations (16.91) and (16.105), where the source strength has been expressed in terms of the pressure it induces on a rigid wall. Any inhomogeneity in the surface properties provides a mechanism at which wavenumber conversion can take place. In Section 16.1 we noted that the rigid–wall pressure spectrum is typically 40dB larger near the convective peak ($k_1 \sim -\omega/U_c$) than its value for low wavenumbers. In this section we investigate whether scattering from the convective peak is a significant contributor to the low–wavenumber pressure spectrum. Various scattering mechanisms have been investigated in the literature. These include the supports and boundaries of a flexible surface, surface roughness and the nonlinear motion of compliant coatings. We will discuss the effects of these various mechanisms separately.

16.5.1 Supports and Boundaries

As an introductory example let us consider the geometry illustrated in Figure 16.12.

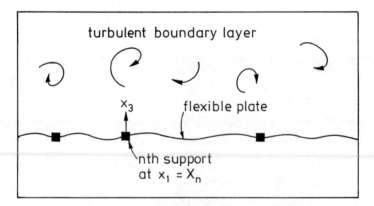

Figure 16.12 A bending plate constrained by line supports.

A plate has bending stiffness B and mass/unit area m and its mean position is the $x_3 = 0$ plane. This plate is constrained by N line supports or ribs with their axes in the 2–direction at positions $x_1 = X_n$, $n = 1, \cdots, N$. The supports exert localized forces on the surface, whose equation of motion can be written in the form

$$-m \frac{\partial^2 \xi}{\partial t^2} - B \left(\frac{\partial^2}{\partial x_1^2} + \frac{\partial^2}{\partial x_2^2} \right)^2 \xi - p'(x_1, x_2, 0, t) = \sum_{n=1}^{N} \sum_{m=0}^{3} A_{mn}(x_2, t) \frac{\partial^m}{\partial x_1^m} \delta(x_1 - X_n) \quad .$$

(16.131)

The δ–functions describe forces, couples and higher order moments exerted on the plate by the supports. The functions $A_{mn}(x_2, t)$ are to be determined in the course of solution. The restriction $m \leq 3$ ensures that the displacement remains finite at the supports (Howe 1986).

As in Section 16.3 we will again decompose the fluctuations into those that are produced by the same turbulent quadrupoles near a hard wall, and a correction for surface flexibility. We write

$$p'(\mathbf{x}, t) = p'_R(\mathbf{x}, t) + p'_F(\mathbf{x}, t) \quad ,$$

(16.132)

in a way analogous to the decomposition in (16.80). $p'_R(\mathbf{x}, t)$ satisfies Lighthill's equation, together with the hard wall boundary condition $\partial p'_R / \partial x_3 = 0$ on $x_3 = 0$, and was determined in Section 16.1. $p'_F(\mathbf{x}, t)$ satisfies a homogeneous wave equation, the radiation condition at infinity and a surface boundary condition.

After taking Fourier transforms in x_1, x_2 and t, we find that

$$\widetilde{p}_F(x_3, \mathbf{k}, \omega) = D e^{i\gamma x_3} \quad ,$$

(16.133)

where $\gamma = (\omega^2 / c_0^2 - k^2)^{1/2}$. D can be determined in terms of p'_R and A_{mn} by combining the equation of motion of the surface (16.131), with the linearized momentum equation,

$$\rho_0 \frac{\partial^2 \xi}{\partial t^2} = -\frac{\partial p'}{\partial x_3} \quad .$$

(16.134)

This leads to

$$\widetilde{p}(x_3, \mathbf{k}, \omega) = \widetilde{p}_R(x_3, \mathbf{k}, \omega) - e^{i\gamma x_3} \left\{ \frac{\widetilde{p}_R(0, \mathbf{k}, \omega) + \Sigma \Sigma \widehat{A}_{mn}(k_2, \omega)(-ik_1)^m e^{-ik_1 X_n}}{1 - i\gamma(m\omega^2 - Bk^4)/\rho_0 \omega^2} \right\} ,$$

(16.135)

where

$$\widehat{A}_{mn}(k_2, \omega) = \int A_{mn}(x_2, t) e^{ik_2 x_2 + i\omega t} dx_2 dt \quad .$$

(16.136)

$p'(\mathbf{x}, t)$ can be found in terms of the (as yet unknown) functions $\widehat{A}_{mn}(k_2, \omega)$ by inverting the Fourier transform. The surface displacement follows from Equation (16.134)

$$\xi(x_1, x_2, t) = -\frac{1}{(2\pi)^3} \int \left\{ \frac{\widetilde{p}_R(0, \mathbf{k}, \omega) + \Sigma\Sigma\widehat{A}_{mn}(k_2, \omega)(-ik_1)^m e^{-ik_1 X_n}}{\rho_0\omega^2 - i\gamma(m\omega^2 - Bk^4)} \right\}$$

$$\times\; i\gamma e^{-ik_1 x_1 - ik_2 x_2 - i\omega t} dk_1\, dk_2\, d\omega \quad.$$

(16.137)

The functions $\widehat{A}_{mn}(k_2, \omega)$ account for the wavenumber scattering. Through these terms $\widetilde{p}(0, \mathbf{k}, \omega)$ is related to $\widetilde{p}_R(0, \mathbf{k}, \omega)$, $k_1 \neq K_1$. The $\widehat{A}_{mn}(k_2, \omega)$ must be determined from the boundary conditions at the supports.

If, for example, the plate is simply supported at its junctions with the ribs, then

$$\xi(x_1, x_2, t) = \frac{\partial^2\xi}{\partial x_1^2}(x_1, x_2, t) = 0 \text{ for all } x_2, \text{ as } x_1 \to X_n \mp 0, n = 1, \cdots, N \quad.$$

(16.138)

On the other hand, for rigidly clamped edges,

$$\xi(x_1, x_2, t) = \frac{\partial\xi}{\partial x_1}(x_1, x_2, t) = 0 \text{ for all } x_2, \text{ as } x_1 \to X_n \mp 0, n = 1, \cdots, N \quad.$$

(16.139)

The four boundary conditions at each rib in either (16.138) or (16.139) are sufficient, in principle, to express the $4N$ unknowns $\widehat{A}_{mn}(k_2, \omega)$ in terms of $\widetilde{p}_R(0, \mathbf{k}, \omega)$. The flexible–surface pressure spectrum then follows from (16.135).

Howe (1986 & 1988a) uses this procedure to investigate a turbulent boundary–layer flow over a bending plate with two line supports a distance a apart. He considers both simply supported and clamped joints, together with "free" boundaries in which the ribs impose little restraint on the plate and

$$\frac{\partial^2\xi}{\partial x_1^2}(x_1, x_2, t) = \frac{\partial^3\xi}{\partial x_1^3}(x_1, x_2, t) = 0 \text{ for all } x_2, \text{ as } x_1 \to X_n \mp 0, n = 1, 2, \cdots, N \quad.$$

(16.140)

In this case the normal displacement, $\xi(x_1, x_2, t)$, is allowed to be discontinuous at the ribs!

For a steel plate in water with practical supports and $\omega a/c_0 \gg 1$, Howe finds that the principal source of radiation is due to the excitation of flexural modes of the panel by the low–wavenumber components of the turbulent sources and their diffraction at the supports. In almost all cases the diffraction of the convective peak leads to terms which are very much smaller. The exception is for the case of "free" boundaries when the plate displacement is allowed to be discontinuous at the supports. Howe found that this extreme boundary condition could produce significant wavenumber conversion.

In Howe's geometry fluid loading enables much of the bending wave energy to be transmitted across the supports. Indeed there is almost perfect transmission at frequencies above coincidence (Howe 1986) and this limits the influence of the finite length of panel. To highlight the effects of finite panel size Dowling (1990) investigated flow noise on a bending plate inserted in an otherwise rigid baffle.

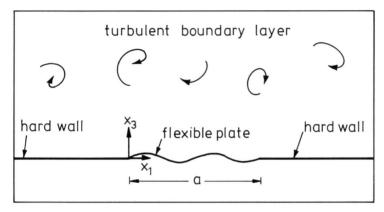

Figure 16.13 A turbulent boundary layer over a plate of finite length.

Dowling's geometry is illustrated in Figure 16.13. The plate has a streamwise length a and infinite width, and is mounted in the plane wall $x_3 = 0$. The rest of the wall is rigid. The plate boundaries at $x_1 = 0$ and $x_1 = a$ are simply supported, so that

$$\xi(x_1, x_2, t) = \frac{\partial^2 \xi}{\partial x_1^2}(x_1, x_2, t) = 0 \text{ at } x_1 = 0 \quad \text{and} \quad a \quad . \tag{16.141}$$

The plate is excited by a turbulent boundary layer and Dowling finds an expression for the power spectral density of the surface pressures. She finds that, for a 5cm–thick steel plate in water at reasonable frequencies, scattering from the convective peak into low wavenumbers leads to a contribution to the surface pressure spectrum which is lower than that produced by low wavenumber sources.

The physical interpretation of this result is that the plate is effectively rigid at the convective wavenumber ω/U_c. At this condition the wavelength is so short that bending stiffness ensures negligible plate deflection. Since the pressure field near the convective peak drives little plate vibration, little energy is scattered at the junctions between the plate and the hard wall.

In summary, for 5cm–thick steel plates in water at reasonable frequencies, wavenumber conversion from the convective peak does not appear to be a significant source of low wavenumber pressure fluctuations. The situation is less clear for GRP domes and this is a current area of research.

So far we have only discussed end effects for planar arrays. Wavenumber conversion can also occur at the end caps of towed array modules, where convective modes are scattered into bulge waves. However surface damping ensures that the scattered bulge waves decay as they propagate away from the end. Away from the end caps, direct excitation from low wavenumber turbulence is larger than the scattered field at all but low frequencies.

16.5.2 Surface Roughness

Continual exposure to sea water may spoil the surface finish causing an initially smooth surface to become rough. Near the convective peak, Chase's empirical model in Equation (16.5) continues to describe the pressure spectrum on a rough surface. Changes in the friction velocity, u_τ, and the boundary–layer thickness, δ, account for the effects of surface roughness. The effects of surface roughness on the low wavenumber elements of flow noise are more complicated. Indeed roughness elements provide inhomogeneities at which scattering from the convective peak to low wavenumbers can take place. Howe has investigated this in a series of papers (1984a and 1988b). In his approach all the quadrupole sources, T_{ij}, are assumed to lie above the highest roughness elements, i.e. the flow is taken to be only linearly disturbed within a roughness height of the surface. In addition the turbulent sources are assumed to be uninfluenced by the surface roughness. This is a significant restriction. It may be appropriate at low frequencies, $\omega\delta^*/U_\infty << 1$, where the wall pressure spectrum may be expected to be controlled by large–scale velocity fluctuations. But for $\omega\delta^*/U_\infty \gtrsim 1$, local separation and vortex shedding from the roughness will alter the source strength T_{ij} (Obermeier & Möhring 1984, Reutov & Rybushkina 1988). Having made these approximations, Howe finds that the influence of the surface roughness may be described as a diffraction problem.

We will introduce the general idea of the diffraction of flow noise by considering turbulence near a single hemispherical boss on a hard surface (see Figure 16.14).

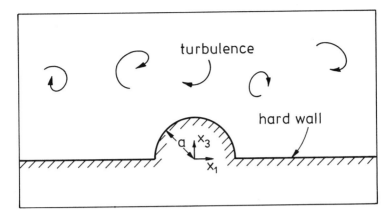

Figure 16.14 A hemispherical boss on the rigid surface $x_3 = 0$.

The surface of the boss is rigid and so the normal fluid velocity vanishes on it. For a Green function $G(\mathbf{y}, \tau | \mathbf{x}, t)$ that satisfies

$$\frac{\partial G}{\partial n} = 0 \tag{16.142}$$

on both the hemisphere and $y_3 = 0$, Equation (16.10) simplifies to

$$c_0^2 \rho'(\mathbf{x}, t) = \int \int_V \frac{\partial^2 G}{\partial y_i \partial y_j} T_{ij}(\mathbf{y}, \tau) d^3 y d\tau \quad . \tag{16.143}$$

The technique developed by Howe (1975) enables the Green function to be written down immediately for use with sources for which $|\mathbf{y}| << |\mathbf{x}|$ and $\omega |\mathbf{y}|/c_0 << 1$. It is (see Howe 1975, Equation (16.84))

$$G(\mathbf{y}, \tau | \mathbf{x}, t) = \frac{1}{4\pi |\mathbf{x} - \mathbf{Y}|} \delta \left(t - \tau - \frac{|\mathbf{x} - \mathbf{Y}|}{c_0} \right) + \frac{1}{4\pi |\mathbf{x}^* - \mathbf{Y}|} \delta \left(t - \tau - \frac{|\mathbf{x}^* - \mathbf{Y}|}{c_0} \right) ,$$
$$\tag{16.144}$$

where $\mathbf{x}^*$ is the image of $\mathbf{x}$ in the plane wall, $\mathbf{x}^* = (x_1, x_2, -x_3)$ and

$$\mathbf{Y} = \mathbf{y} \left(1 + \frac{a^3}{2|\mathbf{y}|^3} \right) \quad . \tag{16.145}$$

Differentiation of (16.144) shows that for $|\mathbf{y}| \sim a$

$$\frac{\partial^2 G}{\partial y_i \partial y_j} = 0 \left(\frac{1}{|\mathbf{x}| a c_0} \delta' \left(t - \tau - \frac{|\mathbf{x} - \mathbf{Y}|}{c_0} \right) \right) \quad . \tag{16.146}$$

It then follows from (16.143) that quadrupoles near the hemispherical boss produce a sound field characteristic of a dipole source. Essentially this occurs because the quadrupole field exerts unsteady forces on the compact boss and so radiates dipole sound.

Howe (1984a) extends the result in (16.144) to determine an approximate Green function that satisfies the boundary condition, $\partial G/\partial n = 0$, on an infinite number of hemispherical bosses of radius a, distributed over the surface $y_3 = 0$. He writes $G(\mathbf{y}, \tau|\mathbf{x}, t)$ in the form given in (16.144) where now for small x_3

$$Y_\alpha = y_\alpha + \frac{\mu a^3}{2} \sum_m \frac{y_\alpha - \xi_{m\alpha}}{|\mathbf{y} - \boldsymbol{\xi}_m|^3}, \qquad \alpha = 1, 2$$

$$Y_3 = y_3 \quad ; \tag{16.147}$$

$\boldsymbol{\xi}_m$ is the centre of the mth boss. The factor μ is to take approximate account of interactions between neighbouring bosses. It is not appreciably different from unity.

Equation (16.143) can be readily integrated by parts, since in Howe's approximation $T_{ij}(\mathbf{y}, \tau)$ vanishes throughout $0 \le y_3 \le a$. This leads to

$$p'(\mathbf{x}, t) = \int \int G(\mathbf{y}, \tau|\mathbf{x}, t) q(\mathbf{y}, \tau) d^3\mathbf{y} d\tau \tag{16.148}$$

where as in (16.42) we have written $q(\mathbf{y}, \tau) = \partial^2 T_{ij}/\partial y_i \partial y_j$. Following Howe (1988b) we will evaluate the pressure on a plane surface above the bosses, $x_3 = a$. After substituting for $G(\mathbf{y}, \tau|\mathbf{x}, t)$ from (16.144) into (16.148) and taking Fourier transforms x_1, x_2 and t, we find

$$\tilde{p}(a, \mathbf{k}, \omega) = -\frac{\cos(\gamma a)}{4\pi^2 i\gamma} \int e^{i(\gamma y_3 + \mathbf{k}\cdot\mathbf{Y} - \mathbf{K}\cdot\mathbf{y})} \tilde{q}(y_3, \mathbf{K}, \omega) d^3\mathbf{y} d^2\mathbf{K} \quad , \tag{16.149}$$

where $\gamma = (\omega^2/c_0^2 - k^2)^{1/2}$. If ka is small in comparison with unity $e^{i\mathbf{k}\cdot\mathbf{Y}}$ can be expanded. It follows from (16.147) that

$$e^{i\mathbf{k}\cdot\mathbf{Y}} = e^{i\mathbf{k}\cdot\mathbf{y}} \left(1 + \frac{1}{2} i\mu a^3 \sum_m \frac{\mathbf{k}\cdot(\mathbf{y} - \boldsymbol{\xi}_m)}{|\mathbf{y} - \boldsymbol{\xi}_m|^3} \right) \quad . \tag{16.150}$$

This enables (16.149) to be rewritten as

$$\tilde{p}(a, \mathbf{k}, \omega) = \tilde{p}_R(a, \mathbf{k}, \omega) + \tilde{p}_{SC}(a, \mathbf{k}, \omega) \quad , \tag{16.151}$$

where

$$\tilde{p}_R(a, \mathbf{k}, \omega) = -\frac{\cos(\gamma a)}{4\pi^2 i\gamma} \int e^{i\gamma y_3 + i(\mathbf{k} - \mathbf{K})\cdot\mathbf{y}} \tilde{q}(y_3, \mathbf{K}, \omega) d^3\mathbf{y} d^2\mathbf{K} \quad . \tag{16.152}$$

Now $\int e^{i(\mathbf{k}-\mathbf{K})\cdot\mathbf{y}}d^2\mathbf{y} = (2\pi)^2\delta(\mathbf{k}-\mathbf{K})$ and so (16.152) simplifies to

$$\widetilde{p}_R(a,\mathbf{k},\omega) = -\frac{\cos(\gamma a)}{i\gamma}\int e^{i\gamma y_3}\widetilde{q}(y_3,\mathbf{k},\omega)dy_3 \quad ; \qquad (16.153)$$

$\widetilde{p}_R(a,\mathbf{k},\omega)$ describes the pressure produced by sources $q(\mathbf{y},\tau)$ near a smooth, rigid surface. It is clear that $\widetilde{p}_R(a,\mathbf{k},\omega)$ is only influenced by elements of the source with the same wavenumber and frequency. $\widetilde{p}_{SC}(a,\mathbf{k},\omega)$ arises from the influence of the second term in (16.150).

$$\widetilde{p}_{SC}(a,\mathbf{k},\omega) =$$

$$-\frac{\cos(\gamma a)\mu a^3}{8\pi^2\gamma}\int e^{i\gamma y_3+i(\mathbf{k}-\mathbf{K})\cdot\mathbf{y}}\sum_m \frac{\mathbf{k}\cdot(\mathbf{y}-\boldsymbol{\xi}_m)}{|\mathbf{y}-\boldsymbol{\xi}_m|^3}\widetilde{q}(y_3,\mathbf{K},\omega)d^3\mathbf{y}d^2\mathbf{K} \quad . \qquad (16.154)$$

As Howe (1988b) points out the y_1 and y_2 integrals are standard and lead to

$$\widetilde{p}_{SC}(a,\mathbf{k},\omega) =$$

$$\frac{i\cos(\gamma a)\mu a^3}{4\pi\gamma}\int e^{i(\gamma-|\mathbf{k}-\mathbf{K}|)y_3}\frac{\mathbf{k}\cdot(\mathbf{K}-\mathbf{k})}{|\mathbf{K}-\mathbf{k}|}\sum_m e^{i(\mathbf{K}-\mathbf{k})\cdot\boldsymbol{\xi}_m}\widetilde{q}(y_3,\mathbf{K},\omega)dy_3d^2\mathbf{K} \quad ; \qquad (16.155)$$

$\widetilde{p}_{SC}(a,\mathbf{k},\omega)$ describes the field scattered by the roughness elements. It is evident that it involves wavenumber conversion because $\widetilde{p}_{SC}(a,\mathbf{k},\omega)$ is related to $\widetilde{q}(y_3,\mathbf{K},\omega)$ for $\mathbf{K}\neq\mathbf{k}$. Now $\widetilde{q}(y_3,\mathbf{K},\omega)$ is largest near the convective peak. Hence the main contribution to the integral comes from near $\mathbf{K}\sim(-\omega/U_c,0)$. Then for low wavenumbers, $\mathbf{k}$,

$$\frac{\mathbf{k}\cdot(\mathbf{K}-\mathbf{k})}{|\mathbf{K}-\mathbf{k}|}\sim -k_1 \quad . \qquad (16.156)$$

The scattered pressure in (16.155) is proportional to k_1 and evidently vanishes when $k_1 = 0$. We saw in Figure 16.4 that the data for zero wavenumber collapses onto a universal curve for a range of roughness heights provided that the source functions are scaled appropriately. This indicates that diffraction is not a significant contributor to flow noise at zero wavenumber. However we cannot infer from this data that scattering is unimportant for all low wavenumbers, because it is evident from Equations (16.155) and (16.156) that $k_1 = 0$ is a special case.

Howe (1988b) considers an average of N bosses/unit area, randomly distributed over the surface $x_3 = 0$, and producing a roughness density $\sigma = N\pi a^2$. He uses Equations (16.153) and (16.155) to derive expressions for the direct and scattered pressure fluctuations respectively. In Howe's analysis the cross–correlation of the source function $q(\mathbf{x}, t)$ is determined from Chase's (1987a) empirical model. We saw in Figures 16.3 and 16.4 that Chase's model gives a good fit to experimental data near the convective peak, but that it has quite a different frequency dependence from measurements at low wavenumbers. Howe investigates the case $M = 0.005$, $a/\delta = 0.01$, $\sigma = 0.2$ in detail. He finds that, at frequencies such that $\omega\delta/U_\infty \leq 10$, the scattered field is negligible in comparison with the direct field. The comparison between Chase's model and experiment in Figure 16.4 shows that Chase's model underpredicts the measured zero wavenumber pressure spectrum in this frequency range. Hence we can conclude from Howe's work and Figure 16.4, that for a Mach number of 0.005 and $\omega\delta/U_\infty \leq 10$, the pressure field scattered from roughness elements is negligible in comparison with that excited directly by the low wavenumber components of turbulence. At higher mean flow Mach numbers the scattered field will be even less important. For frequencies ω, such that $\omega\delta/U_\infty > 10$, Howe finds that his model predicts that the pressure field scattered from a specified turbulent source can exceed the direct field. However this result is less convincing because it is precisely in this frequency range that we would expect the turbulent sources to be modified by separation and vortex shedding from the roughness elements. In a related problem Howe (1976) shows that the effect of vortex shedding from the trailing edge of an aerofoil is to effectively cancel the scattered sound field of a convecting eddy. Vortex shedding may have a similar effect on the pressure field scattered by the hemispherical bosses, but is not included in Howe's (1988b) analysis.

Experimental data is needed to clarify the position. In particular, investigation of the wavenumber dependence of the surface pressure spectrum on a rough surface would indicate whether or not the scattered field is significant. At low wavenumbers, $|k_1| < \omega/c_0$, the pressure field excited directly by low wavenumber turbulent sources is predicted to be wavenumber white (see Equation (16.153)), while the scattered pressure spectrum is proportional to k_1^2 (see Equations (16.155) and (16.156)). We noted in the introduction that the low wavenumber pressure spectrum measured on a smooth, rigid wall is independent of k_1, for $|k_1| < \omega/c_0$. If the same were true on a rough wall, it would indicate that scattering from roughness elements is not important. Conversely a strong dependence on k_1^2 would illustrate the dominance of a scattered field.

16.5.3 Compliant Coatings

In Section 16.3, we found that a homogeneous compliant coating in linear motion simply changes the surface impedance and can have a beneficial effect on the wall pressure spectrum (compare Figures 16.5 and 16.8). However, if the deflection of the surface is sufficiently large that nonlinear effects are important, deformation of the coating can provide a wavenumber conversion mechanism.

Howe (1984b) models a coating by a fluid layer with zero shear wave speed. He investigates scattering from turbulent sources near the convective peak to low wavenumbers. He says the scattered field must dominate at low enough free–stream Mach numbers, but gives no indication of its importance in a practical example.

Chase (1987b) discredits Howe's approach. He considers an elastic wall with a low but nonzero shear wave speed, c_t. Howe's zero shear wave speed model is inappropriate for modes near the convective peak with very small phase speeds. For $c_t > U_\infty$, Chase finds that the scattered pressure field is negligible in comparison with the measured low wavenumber pressure spectrum. Indeed for $c_t = 5U_\infty$, it is an amazing 129dB below the measured level.

Finally we note that care should be taken when deciding how to fix a compliant coating to a surface. Howe (1987b) considers a coating held by an embedded strut transverse to the flow direction. In Howe (1988c) scattering from a transverse seam is considered. Howe concludes that in both cases the scattered field is significant. However he again treats the coating as a fluid layer with zero shear wave speed and so underestimates its rigidity for modes near the convective peak. This may mean that he has overestimated the scattering from these inhomogeneities. If the strut or seam is aligned parallel rather than transverse to the mean flow, it cannot scatter pressure fluctuations from the convective régime to low wavenumbers.

16.6 THE RÔLE OF SURFACE SHEAR STRESS

In Equation (16.15) we derived a representation for the sound generated by turbulence in a boundary–layer flow over the rigid surface $x_3 = 0$

$$c_0^2 \rho'(\mathbf{x}, t) = \frac{\partial^2}{\partial x_i \partial x_j} \int_V \frac{T_{ij}(\mathbf{y}, t - |\mathbf{x} - \mathbf{y}|/c_0)}{4\pi |\mathbf{x} - \mathbf{y}|} d^3\mathbf{y} + \frac{\partial^2}{\partial x_i^* \partial x_j^*} \int_V \frac{T_{ij}(\mathbf{y}, t - |\mathbf{x}^* - \mathbf{y}|/c_0)}{4\pi |\mathbf{x}^* - \mathbf{y}|} d^3\mathbf{y}$$

$$- \frac{\partial}{\partial x_\alpha} \int_S \frac{\sigma_{\alpha 3}(y, t - |\mathbf{x} - \mathbf{y}|/c_0)}{2\pi |\mathbf{x} - \mathbf{y}|} dy_1 \, dy_2 \quad ,$$

$$(16.15)$$

where $\mathbf{x}^* = (x_1, x_2, -x_3)$ and α is summed over 1 and 2. The low wavenumber components of surface pressure generated by the two quadrupole terms were evaluated in Section 16.1. We will now consider the contribution from the surface dipole.

This problem was addressed by Landahl (1975), who determined the order of magnitude of the integrated fluctuating shear stress induced by bursting in a turbulent boundary layer and used it to estimate the dipole term in (16.15). He concluded that the dipole radiation could be expected to be much stronger than the quadrupole sound.

If we adopt Landahl's approach we could investigate the cross–power spectral density of the pressure field generated by the surface shear stress. Let us write

$$p_d'(\mathbf{x}, t) = -\frac{\partial}{\partial x_\alpha} \int_S \frac{\sigma_{\alpha 3}(\mathbf{y}, t - |\mathbf{x} - \mathbf{y}|/c_0)}{2\pi |\mathbf{x} - \mathbf{y}|} dy_1 \, dy_2 \quad . \tag{16.157}$$

This is equivalent to

$$p_d'(\mathbf{x}, t) = \int \frac{\partial G_R}{\partial y_\alpha} \sigma_{\alpha 3}(\mathbf{y}, \tau) dy_1 dy_2 d\tau \quad , \tag{16.158}$$

where $G_R(\mathbf{y}, \tau | \mathbf{x}, t)$ is given in (16.14). After taking Fourier transforms in x_1, x_2 and t and using Equation (16.22), we find that the surface pressure fluctuations are described by

$$\tilde{p}_d(0, \mathbf{k}, \omega) = -\frac{k_\alpha}{\gamma} \tilde{\sigma}_{\alpha 3}(0, \mathbf{k}, \omega) \quad ; \tag{16.159}$$

$\gamma = (\omega^2/c_0 - k^2)^{1/2}$, while $\tilde{p}_d(x_3, \mathbf{k}, \omega)$ and $\tilde{\sigma}_{\alpha 3}(x_3, \mathbf{k}, \omega)$ are the Fourier transforms of $p_d'(\mathbf{x}, t)$ and $\sigma_{\alpha 3}(\mathbf{x}, t)$ respectively.

The power spectral density of the dipole surface pressure can be determined by multiplying (16.159) by its complex conjugate. This leads to

$$\tilde{P}_d(\mathbf{k}, \omega) = \frac{k_\alpha k_\beta}{|\gamma|^2} \tilde{\Sigma}_{\alpha \beta}(\mathbf{k}, \omega) \quad , \tag{16.160}$$

where $\tilde{\Sigma}_{\alpha \beta}(\mathbf{k}, \omega)$ is the cross–power spectral density of the surface shear stress

$$\tilde{\Sigma}_{\alpha \beta}(\mathbf{k}, \omega) = \int \overline{\sigma_{\alpha 3}(y_1, y_2, 0, t) \sigma_{\beta 3}(y_1 + \Delta_1, y_2 + \Delta_2, 0, t + \tau)} e^{ik_\alpha \Delta_\alpha + i\omega \tau} d^2 \Delta d\tau \quad . \tag{16.161}$$

The spirit of Landahl's approach is to assume that $\tilde{\Sigma}_{\alpha \beta}(\mathbf{k}, \omega)$ is produced by turbulent fluctuations. It would then be independent of the speed of sound for a low Mach number flow. Hence

$$\tilde{\Sigma}_{\alpha \beta}(\mathbf{k}, \omega) \approx \tilde{\Sigma}_{\alpha \beta}(0, \omega) \quad \text{for} \quad k \le \omega/c_0 \quad . \tag{16.162}$$

Throughout the low–wavenumber range, Equation (16.160) then predicts that

$$\widetilde{P}_d(\mathbf{k},\omega) = \frac{k_\alpha k_\beta}{|\gamma|^2}\widetilde{\Sigma}_{\alpha\beta}(\mathbf{0},\omega) \quad . \tag{16.163}$$

This dipole surface pressure spectrum vanishes at zero wavenumber and is proportional to k^2 at low wavenumbers. This is at variance with experimental observations which show the pressure spectrum on a hard smooth surface to be virtually independent of $\mathbf{k}$ at low wavenumbers (as sketched in Figure 16.2). This contradiction is resolved by Howe (1979).

Howe notes that the shear stress σ_{ij} is nonzero even for a sound wave propagating through the region. In our notation this means that there are important elements of $\Sigma_{\alpha\beta}(\mathbf{k},\omega)$ that depend on compressibility, and that $\Sigma_{\alpha\beta}(\mathbf{k},\omega)$ varies in some way with $\mathbf{k}$ throughout the low wavenumber régime. Equations (16.162) and (16.163) are therefore wrong, and Equation (16.160) is useless as a predictor of surface pressure until $\Sigma_{\alpha\beta}(\mathbf{k},\omega)$ is known. In practice fluctuations in pressure and shear stress are coupled and need to be determined together.

Howe argues that all perturbations in the shear stress at wavenumbers $k \leq \omega/c_0$ correspond to acoustic modes, and that these terms should not be regarded as acoustic sources. Rather they describe propagation effects, illustrating how turbulence modifies the propagation of acoustic waves in the vicinity of a hard wall. This supposition is similar to the one we made in Section 16.4, where we noted that when there is a mean flow the standard Lighthill quadrupole source contains terms linear in the velocity fluctuation (see Equation (16.115)). We argued that these linear terms describe propagation effects and should be included in the operator as in Equation (16.116).

When the viscous stress terms are taken to the left-hand side of Lighthill's equation and determined in conjunction with the pressure fluctuations, Howe finds the effect of the wall shear stress to be unimportant at almost all wavenumbers. The exception is for spectral elements with sonic phase speeds. When viscosity was neglected we found the surface pressure spectrum under an unbounded region of turbulence to be infinite at this condition. The inclusion of shear stress fluctuations controls this singularity. Howe finds that for $k \sim \omega/c_0$

$$\widetilde{P}(\mathbf{k},\omega) \sim \widetilde{P}(\mathbf{0},\omega)M^{-2}F(Re_{\delta^*},\omega\delta^*/U_\infty) \quad , \tag{16.164}$$

F varies between 600 and 1.8×10^5 for $10^4 \leq Re_{\delta^*} \leq 10^7$, $1 \leq \omega\delta^*/U_\infty \leq 10$. For a Mach number of 0.01, Equation (16.164) predicts the pressure spectrum at the sonic condition to be between 68 and 93dB higher than the level at zero wavenumber. In practice, finite boundary–layer extent, surface curvature, surface flexibility and damping, as discussed in Sections 16.1, 16.2, 16.3, are likely to have a greater influence on the level at the sonic condition.

Howe's analysis is for acoustic waves whose wavelengths are long in comparison with the boundary–layer height. Haj Hariri & Akylas (1985b) relax this restriction. They find that then, as well as resolving the singularities in the pressure spectrum, the shear stresses also modify the source terms. However they conclude that, in typical applications, the additional sources lead to surface pressures that are negligible in comparison with those generated by the quadrupole sources.

In summary both current theories and experimental observations indicate that, for all practical purposes, the influence of the wall shear stresses on the surface pressure spectrum can be neglected. This statement is in contradiction to statements made in a recent review (Leehey 1989).

16.7 SUMMARY

Flow noise on rigid plane and curved surfaces has been reviewed in Sections 16.1 and 16.2, including theoretical approaches and such data as are available. Surface flexibility has been discussed in Section 16.3 and leads to additional peaks in the low wavenumber pressure spectrum on both planar and towed arrays. At the low Mach numbers in underwater applications the mean–flow profile is generally less important than surface flexibility. The exception is for turbulent flow over a sonar dome. A dome of height h introduces an additional layer of fluid in which the mean velocity is zero. This enhances the effect of the mean flow (see Section 16.4).

Various scattering mechanisms have been investigated in Section 16.5. Scattering from supports, ribs and the boundaries of steel plates in water at reasonable frequencies has been found not to be a significant source of low wavenumber pressure fluctuations away from the edges. That is the principal source of low wavenumber pressure fluctuations is excitation by the low wavenumber components of the turbulent sources, rather than diffraction of the convective peak.

The effect of surface roughness on the pressure spectrum near the convective peak can be accounted for by a simple scaling on the friction velocity u_τ and the boundary–layer height δ both of which change with roughness (see Equation (16.5)). The same is true at zero wavenumber as shown by Figure 16.4. The situation is less clear for low but nonzero wavenumbers, $0 < |k_1| < \omega/c_0$, and, as discussed in Section 16.5, experimental data is required to see whether scattering of the convective peak by roughness elements is a significant source of low wavenumber pressure fluctuations.

Finally in Section 16.6 we found the surface shear stress not to be an important source of acoustic pressure fluctuations. The inclusion of these viscous terms into theoretical analyses controls the singularity which would otherwise occur in the pressure spectrum on a rigid, plane surface for spectral elements with sonic surface phase speeds, under an unbounded region of turbulence. In practice, finite boundary–layer extent, surface curvature, surface flexibility and damping, as discussed in Sections 16.1, 16.2 and 16.3, are likely to have a more important effect on the level at the sonic condition.

REFERENCES

Bandyopadhyay, P.R. (1986). Review — mean flow in turbulent boundary layers disturbed to alter skin friction. Trans. ASME:J. Fluids Engng. 108:127–140.

Beeler, G.B. (1986). Turbulent boundary–layer wall pressure fluctuations downstream of a tandem LEBU. AIAA J. 24:689–691.

Bergeron, R.F. (1973). Aerodynamic sound and the low–wavenumber wall-pressure spectrum of nearly incompressible boundary-layer turbulence. J. Acoust. Soc. Amer. 54:123–133.

Blake, W.K. (1986). Mechanics of Flow–Induced Sound and Vibration . Volume II, Academic Press.

Bolotin, V. V. (1963). Non–Conservative Problems of the Theory of Elastic Stability, Pergamon Press.

Bonnet, J.P., Delville, J. & Lemay, J. (1987). Study of LEBUs modified turbulent boundary layer by use of passive temperature contamination. Proc. Turbulent Drag Reduction by Passive Means , R. Aero. Soc. 1:45–68.

Chase, D.M. (1979). Modeling the wavenumber–frequency spectral density of wall pressure in turbulent flow on a cylinder. Chase Inc., Tech. Memo. No.9, Boston.

Chase, D.M. (1980). Modeling the wavevector–frequency spectrum of turbulent boundary layer wall pressure. J. Sound Vib. 70:29–67.

Chase, D.M. (1987a). The character of the turbulent wall pressure spectrum at subconvective wavenumbers and a suggested comprehensive model. J. Sound Vib. 112:125–147.

506

Chase, D.M. (1987b). The estimated level of low–wavenumber pressure generated by non–linear interaction of a compliment wall with a turbulent boundary layer. J. Sound Vib. 116:25–32.

Chase, D.M. & Noiseux, C.F. (1982). Turbulent wall pressure at low wavenumbers: relation to nonlinear sources in planar and cylindrical flow. J. Acoust. Soc. Amer. 72:975–982.

Corcos, G.M. (1963). Resolution of pressure in turbulence. J. Acoust. Soc. Amer. 35:192–199.

Coustols, E., Cousteix, J. & Belanger, J. (1987). Drag reduction on riblet surfaces and through outer layer manipulators. Proc. Turbulent Drag Reduction by Passive Means. R. Aero Soc. 2:250–289.

Crow, S.C. (1970). Aerodynamic sound emission as a singular perturbation problem. Studies in Appl. Maths. 49:21–40.

Dhanak, M.R. (1988). Turbulent boundary layer on circular cylinder: the low–wavenumber surface pressure spectrum due to a low–Mach number flow. J. Fluid Mech. 191:443–464.

Dowling, A.P. (1983a). The low wavenumber wall pressure spectrum on a flexible surface. J. Sound Vib. 88:11–25.

Dowling, A.P. (1983b). Sound generation by turbulence near an elastic wall. J. Sound Vib. 90:309–324.

Dowling, A.P. (1983c). Flow–acoustic interaction near a flexible wall. J. Fluid Mech. 128:181–198.

Dowling, A.P. (1986). Mean flow effects on the low–wavenumber pressure spectrum on a flexible surface. Trans. ASME J. Fluids Engng. 108:104–108.

Dowling, A.P. (1989). The effect of large–eddy breakup devies on flow noise. J. Fluid Mech. 208:193–223.

Dowling, A.P. (1990). Flow noise. A.B. Wood Medal Lecture, Proc. I.O.A., 12: 33–71.

Ffowcs Williams, J E (1965). Boundary–layer pressure and the Corcos model: a development to incorporate low–wavenumber constraints. J. Fluid Mech. 22:507–519.

Ffowcs Williams, J E (1982). Surface pressure fluctuations induced by boundary–layer flow at finite Mach numbers. J. Fluid Mech. 125:9–25.

Goldstein, M.E. (1976). Aeroacoustics. McGraw Hill.

Haj Hariri, H. & Akylas, T. R. (1985a). Mean–flow effects on the low–wavenumber wall–pressure spectrum of a turbulent boundary layer over a compliant surface. J. Acoust. Soc. Amer. 77:1840–1844.

Haj Hariri, H. & Akylas, T.R. (1985b). The wall shear–stress contribution to boundary–layer noise. Phys. Fluids 28:2727–2729.

Howe, M.S. (1975). The generation of sound by aerodynamic sources in an inhomogeneous steady flow. J. Fluid Mech. 67:597–610.

Howe, M.S. (1976). The influence of vortex shedding on the generation of sound by convected turbulence. J. Fluid Mech. 76:711–740.

Howe, M.S. (1979). The rôle of surface shear stress fluctuations in the generation of boundary–layer noise. J. Sound Vib. 65:159–164.

Howe, M.S. (1984a). On the generation of sound by turbulent boundary–layer flow over a rough wall. Proc. R. Soc. Lond. A, 395:247–263.

Howe, M.S. (1984b). On the production of sound by turbulent boundary–layer flow over a compliant coating. IMA J. Applied Maths. 33:189–203.

Howe, M.S. (1986). Attenuation and diffraction of bending waves at gaps in fluid loaded plates. IMA J. Applied Maths. 36:247–262.

Howe, M.S. (1987a). On the structure of the turbulent boundary–layer wall pressure spectrum in the vicinity of the acoustic wavenumber. Proc. R. Soc. Lond. A412:389–401.

Howe, M.S. (1987b). Production of sound by turbulent flow over an embedded strut. IMA J. Applied Maths. 39:99–120.

Howe, M.S. (1988a). Diffraction radiation produced by turbulent boundary–layer excitation of a panel. J.Sound Vib. 121:47–65.

Howe, M.S. (1988b). The turbulent boundary–layer rough–wall pressure spectrum at acoustic and subconvective wavenumbers. Proc. R. Soc. Lond. A415:141–161.

Howe, M.S. (1988c). Diffraction of flow noise by a flexible seam. J. Sound Vib. 125:291–304.

Jameson, P.W. (1975). Measurement of the low wavenumber component of turbulent boundary–layer pressure spectral density. Proc. 4th Symp. Turb. Liq.:192–200.

Landahl, M.T. (1975). Wave mechanics of boundary–layer turbulence and noise. J. Acoust. Soc. Am. 57:824–831.

Leehey, P. (1988). Structural excitation by a turbulent boundary layer: an overview.

ASME J. Vib., Acoust., Stress and Reliability in Design 110:220–225.

Leehey, P. (1989). Dynamic wall pressure measurements. Advances in Fluid Mechanics Measurements Lecture Notes in Engineering 45, ed. M.Gal–ek–Hak,: Springer–Verlag.

Lighthill, M.J. (1952). On sound generated aerodynamically. I. General theory. Proc. R. Soc. Lond. A211:564–587.

Lilley, G.M. (1971). Sound generation in shear flow turbulence. Fluid Dynamics Trans. (Poland) 6:405–420.

Maidanik, G. (1968). Domed sonar system. J. Acoust. Soc. Amer. 44:113–124.

Maidanik, G. (1984). Modification caused by compliant layers and blankets in the pressure field induced on a boundary. ASME J. Vib., Acoust., Stress and Reliability in Design 106:369–375.

Markowitz, A.E. (1976). Turbulent boundary layer wall pressure fluctuations and wall acceleration measurements on a long flexible wall cylinder towed at sea. NUSC, TR5305:New London.

Martin, N.C. (1976). Wavenumber Filtering by Mechanical Structures. Ph.D. Thesis, MIT Cambridge, Mass.

Martini, K.F., Leehey, P. & Moeller, M. (1984). Comparison of techniques to measure the low wavenumber spectrum of a turbulent boundary layer. MIT Acoustics and Vibration Laboratory Rep. 92828–1: Cambridge, Mass.

Moller, J.C. & Leehey, P. (1989). Measurement of wall shear and wall pressure downstream of a honeycomb boundary layer manipulator. MIT Acoustics and Vibration Laboratory. Rep. No. 97457–3: Cambridge, Mass.

Nguyen, V.D., Dickinson, J., Jean, Y., Chalifour, Y., Anderson, J., Lemay, J., Haeberle, D. & Larose, G. (1984). Some experimental observations of the law of the wall behind large–eddy breakup devices using servo–controlled skin friction balances. AIAA–84–0346.

Obermeier, F. & Möhring, W. (1984). Aerodynamic sound generation by turbulent boundary–layer flows along solid and compliant walls. Z. Flugwiss. Weltraumforsch 8:181–192.

Powell, A. (1960). Aerodynamic noise and the plane boundary. J. Acoust. Soc. Amer. 32:982–990.

Reutov, V.P. & Rybushkina, G.V. (1986). Formation of the acoustic pressure spectrum in a turbulent boundary–layer flow on a plate. Sov. Phys. Acoust. 32:217–221.

Sevik, M.M. (1986). Topics in hydroacoustics. IUTAM Symp. on Aero– and

Hydro– Acoustics, Lyon, France, Springer–Verlag:285-308.

Westphal, R.V. (1986). Skin friction and Reynolds stress measurements for a turbulent boundary layer following manipulation using flat plates. AIAA–86–0283.

Willmarth, W.W. (1975). Pressure fluctuations beneath turbulent boundary layers. Ann. Rev. Fluid Mech. 7:13–38.

Willmarth, W.W. & Yang, C. (1969). Wall pressure fluctuations beneath an axially symmetric turbulent boundary layer on a cylinder. ONA Project 02149:University of Michigan.

Wilkinson, S.P., Anders, J.B., Lazos, B.S., & Bushnell, D.M. (1987). Turbulent drag reduction research at NASA Langley — progress and plans. Proc. Turbulent Drag Reduction by Passive Means. R. Aero. Soc. 1:1–32.

17. FLUID–LOADING INTERACTION WITH VIBRATING SURFACES

17.1 INTRODUCTION

When an elastic structure in contact with a fluid vibrates, unsteady pressures are generated in the fluid and exert a surface loading on the structure additional to that which may be exerted by mechanical excitation or by the surface pressures of a boundary layer formed over a perfectly rigid structure. The additional loading is referred to as *"fluid loading"*; it is a highly nonlocal loading, not necessarily connected with fluid compressibility or bulk fluid motion, which is crucial in determining many aspects of most structural acoustics problems in one part of parameter space or another. When the fluid loading is important (which, as we shall see, it is for some aspects of the acoustic or vibration fields even when the fluid density is small compared with that of the structure material) one is faced with a *fully–coupled* problem, involving transverse waves (and occasionally longitudinal waves) on the structure and their nonlocal coupling to either incompressible pressure fluctuations or compression waves in the fluid. Recently there has been considerable success in analytical studies of these problems, leading to simple and explicit descriptions of the structural and acoustic fields and the identification of *resonance* and other significant conditions, even under conditions of "heavy" fluid loading. The basic point of these notes is to illustrate in the simplest problem the rôles played by the five essential competing mechanisms for plane elastic plates under static fluid loading, to illustrate how the balances between them are different for different aspects of the system response and in different parameter ranges, and to give a general asymptotic method which provides a powerful approach to fluid–loaded structure problems.

17.2 THE DIFFERENT RÔLES OF FLUID LOADING

Our aim here is to show that fluid loading cannot be characterized once and for all as either "light" or "heavy" even at a fixed frequency, in a given configuration and for fixed values of the material constants for the fluid and the structure. We exemplify the different rôles which fluid loading may have by looking at the familiar problem of a locally excited plane structure under fluid loading. This is really the canonical problem of the theory of fluid–loaded structures. It has been recently re–examined in some detail (see Junger & Feit (1986) and Crighton (1988)) and shown to contain a great richness of physical effects. A proper understanding of the much briefer treatment below is an essential prerequisite to the study of more realistic structures which are finite in extent, or otherwise inhomogeneous (e.g. because they possess rib supports, or have discontinuities in material properties).

Assume that we have a thin elastic plate, of specific mass m, bending stiffness B, with no internal losses, lying in the plane $y = 0$, with static compressible fluid of density ρ_0 and sound speed c_0 in $y > 0$ and a vacuum in $y < 0$. Transverse waves on the structure, coupled to longitudinal compression waves in the fluid, are excited by a time–harmonic *line* force applied to the plate along the z–axis, the force $F_0 \exp(-i\omega t)$ acting in the y–direction and generating two–dimensional fluid motion in the (x, y) plane and a one–dimensional motion along the plate (see Figure 17.1).

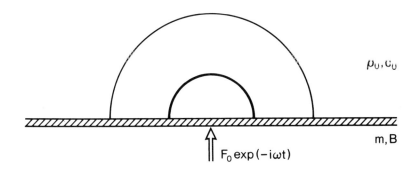

Figure 17.1 The configuration in the line–force excitation problem of fluid–loading theory.

Introduce

$$k_0 = \omega/c_0 \qquad \text{acoustic wavenumber at frequency } \omega;$$

$$k_p = (m\omega^2/B)^{\frac{1}{4}} \qquad \text{free plate wavenumber in vacuum;}$$

$$\hat{\mu} = \rho_0/m \qquad \text{with dimensions of wavenumber;}$$

$$\phi(x,y) \qquad \text{potential in fluid;}$$

$$p(x,y) \qquad \text{pressure in fluid;}$$

$$v(x) \qquad \text{plate velocity in } y\text{–direction;}$$

$$r = (x^2 + y^2)^{1/2} \qquad \text{cylindrical range from excitation;}$$

$$\theta = \tan^{-1}(y/|x|) \qquad \text{angle from surface of plate } (0 \leq \theta \leq \pi/2) \ .$$

Then the line–force–driven plate problem is defined by

$$\left(\frac{\partial^2}{\partial x^2} + \frac{\partial^2}{\partial y^2} + k_0^2\right)\phi = 0 \quad,$$

$$p = \rho_0 i\omega\phi \quad,$$

$$\frac{\partial \phi}{\partial y}(x,0) = v(x) \quad, \tag{17.1}$$

$$\left(B\frac{\partial^4}{\partial x^4} - m\omega^2\right)\left(\frac{v}{-i\omega}\right) = F_0\delta(x) - p(x,0) \quad,$$

with conditions of exponential decay or outgoing radiation at infinity. Define Fourier Transforms by

$$\tilde{\phi}(k,y) = \int_{-\infty}^{+\infty} \phi(x,y)\exp(ikx)dx$$

$$\phi(x,y) = (1/2\pi)\int_C \tilde{\phi}(k,y)\exp(-ikx)dk, \text{ etc. }, \tag{17.2}$$

where C will be specified a little later. Then

$$\left(\frac{\partial^2}{\partial y^2} - \gamma^2\right)\widetilde{\phi}(k,y) = 0 \quad,$$

$$\widetilde{p}(k,y) = \rho_0 i\omega\widetilde{\phi}(k,y) \quad,$$

$$\frac{\partial\widetilde{\phi}(k,0)}{\partial y} = \widetilde{v}(k) \quad,$$

$$B\left(k^4 - k_p^4\right)\left(\frac{\widetilde{v}(k)}{-i\omega}\right) = F_0 - \widetilde{p}(k,0) \quad,$$

(17.3)

and for the solution of the first of these we take

$$\widetilde{\phi}(k,y) = \widetilde{\phi}(k,0)\exp(-\gamma y) \quad,$$
(17.4)

where $\gamma = +(k^2 - k_0^2)^{1/2}$ for $|k| > k_0$ and $\gamma = -i(k_0^2 - k^2)^{1/2}$ for $|k| < k_0$ (so that $\exp(-\gamma y - i\omega t) = \exp\{i(k_0^2 - k^2)^{1/2}y - i\omega t\}$ then represents outgoing waves in the y–direction for the "supersonic" wavenumber components ($|k| < k_0$) and exponential decay for the "subsonic" components ($|k| > k_0$)). The function γ with this behaviour on the real axis can be analytically continued into the complex k–plane provided the branch cut from $+k_0$ goes to infinity in the first quadrant and that from $-k_0$ to infinity in the third. No further particularization of the location of the branch cuts is needed, though some choices may be more *convenient* than others; we take the cuts to be vertical lines from $\pm k_0$ to $\pm k_0 \pm i\infty$.

One then easily gets Fourier integral solutions in which, in the inversion integrals, we replace k by $-k$ for convenience, and find

$$\phi(x,y) = \frac{i\omega F_0}{2\pi B}\int_C \frac{\exp\{ik|x| - \gamma y\}dk}{(k^4 - k_p^4)\gamma - \mu k_p^4} \quad,$$
(17.5)

$$v(x) = -\frac{i\omega F_0}{2\pi B}\int_C \frac{\gamma\exp(ik|x|)dk}{(k^4 - k_p^4)\gamma - \mu k_p^4} \quad,$$
(17.6)

in which γ is as previously defined (an even function of k). The integration path C must be taken as the real axis, with indentations just above any singularities of the integrand on the negative real axis and just below any on the positive. This follows from considering a dissipation mechanism, which is equivalent to making $\Im\omega > 0$ and thereby displaces singularities on the negative real axis downwards, those on the positive real axis upwards. (A much more elaborate consideration of "causality" requirements is needed when there is mean flow past the structure and the system can support *instability* waves. For static fluid these causality requirements are met by the above choice of C.)

Now this simple model problem already involves the competition between five distinct physical attributes represented by the five terms occurring in the denominators of (17.5) and (17.6):

$$(k^4 - k_p^4) \quad (k^2 - k_0^2)^{1/2} - \mu k_p^4 \ ,$$

$$\text{I} \quad \text{II} \quad \text{III} \quad \text{IV} \quad \quad \text{V} \ , \tag{17.7}$$

and these may be referred to as

> I: structural stiffness;
> II: structural inertia;
> III: fluid pressures; $\qquad\qquad$ (17.8)
> IV: fluid compressibility;
> V: fluid inertia.

The terms involve, in addition to the current wavenumber k, the other three wave-number-like quantities

$$k_0 \ , \quad k_p \text{ and } \mu \quad ,$$

and this means that any dimensionless field quantity, Ψ say, can be expressed as

$$\Psi \equiv \Psi\{k_0|\mathbf{x}| \text{ or } k_p|\mathbf{x}| \ , \ \hat{\mathbf{x}} \ , \ \mu/k_0 \ , \ k_0/k_p\} \quad , \tag{17.9}$$

that is, as a function of range made dimensionless with k_0^{-1} or k_p^{-1} as seems most appropriate, of angle (specified by the unit vector $\hat{\mathbf{x}}$) and of the two ratios $k_0 : k_p : \mu$. We have indicated a common choice, namely

$$\alpha = \mu/k_0 = \rho_0 c_0/m\omega \quad , \tag{17.10}$$

and

$$M = k_0/k_p = c_p/c_0 \quad , \tag{17.11}$$

for these ratios, where we can call α a fluid–loading parameter, and M a phase Mach number, the ratio of phase speed of structural waves *in a vacuum* ($c_p(\omega) = (B\omega^2/m)^{\frac{1}{4}}$) to the sound speed. Note that both α and M involve the frequency and the material constants for the structure and the fluid. In terms of frequency alone,

$$M = \Omega^{1/2} = (\omega/\omega_g)^{1/2} \quad , \tag{17.12}$$

where

$$\omega_g = (mc_0^4/B)^{1/2} \tag{17.13}$$

is a reference frequency, known as the *"coincidence"* or *"critical"* frequency and defined by the condition

$$c_p(\omega_g) = c_0 \quad . \tag{17.14}$$

(There is no implication here that the free wave phase speed in the presence of the fluid is equal to c_0 at ω_g or, indeed, at any other real frequency; we stress that ω_g is a *convenient reference frequency*, defined by the structural properties in the absence of fluid, and by the sound speed.)

To illustrate the scaling law (17.9), consider the acoustic far field. As $k_0 r \to \infty$ (what this means will be discussed further in a moment) we may use the method of stationary phase, which filters out of the integral (17.5) the dominant contribution, from a single wavenumber component $k = k_0 \cos\theta, \gamma = -ik_0 \sin\theta$, corresponding to propagation in the observer direction $\hat{\mathbf{x}} = (\cos\theta, \sin\theta)$ of a locally plane acoustic wave. The method gives the result

$$\int_C F(k)e^{ikx - \gamma y}dk \sim \left(\frac{2k_0\pi}{r}\right)^{1/2} \sin\theta e^{ik_0 r - \pi i/4} F(k_0 \cos\theta) \quad , \tag{17.15}$$

which, when applied to (17.5), yields a far–field pressure

$$\left(\frac{p}{\rho_0 c_0^2}\right) \sim \left(\frac{F_0}{mc_0^2}\right) \frac{\exp(i\tilde{r} + \pi i/4)}{(2\pi\tilde{r})^{1/2}} D(\theta, M, \alpha) \quad , \tag{17.16}$$

with

$$D(\theta, M, \alpha) = \left[\frac{\sin\theta}{(1 - M^4 \cos^4\theta)\sin\theta + i\alpha}\right] \quad . \tag{17.17}$$

Here $(p/\rho_0 c_0^2)$ and (F_0/mc_0^2) are dimensionless pressure and force, and $\tilde{r} = k_0 r$ is dimensionless range.

Suppose that we had ignored fluid loading in the calculation of the surface motion, and had then calculated the sound field which would be produced by that calculated surface motion. Such a procedure has often been adopted in dealing with semi–infinite or finite plates. Then we would delete the term $-p(x, 0)$ from the last of (17.1) but leave ϕ determined in terms of the now decoupled field $v(x)$ as before. This would lead to a far field identical with (17.16) and (17.17) except that $D(\theta, M, \alpha)$ would be replaced by $D_0(\theta, M) \equiv D(\theta, M, \alpha = 0)$. In a gross sense, then, as far as the very distant acoustic field is concerned, fluid–loading effects are small if $\alpha << 1$ and large if $\alpha >> 1$, and in general the transition between the two extremes is smooth and unexceptional, so that we can replace $<<$ by $\lesssim$ ("less than about"), etc. Heavy fluid loading thus seems to correspond, for *given* structure and fluid media, to low frequencies, and it is common to define a "null frequency" ω_n by

$$\alpha(\omega_n) \equiv \frac{\rho_0 c_0}{m\omega_n} = 1 \quad , \tag{17.18}$$

and to regard fluid loading as heavy or light according as $\omega \lesssim \omega_n$ or $\omega \gtrsim \omega_n$.

That characterization is not valid for all aspects of the fluid–structure response, however. For the structural response, the "natural" variables are $v/(\omega F_0/Bk_p^3)$, $x_0 = k_p|x|$ and $\zeta = k/k_p$, which give, from (17.6),

$$\left(\frac{v}{\omega F_0/Bk_p^3}\right) = -\frac{i}{2\pi} \int_C \frac{(\zeta^2 - M^2)^{1/2} \exp(i\zeta x_0)d\zeta}{(\zeta^4 - 1)(\zeta^2 - M^2)^{1/2} - \alpha M} \quad , \tag{17.19}$$

but now, although $\alpha = 0$ again corresponds to the vacuum dynamics with fluid loading ignored, it is clear that fluid loading is now "light" or "heavy" according as $\alpha M \lesssim 1$ or $\alpha M \gtrsim 1$. In terms of null frequency and coincidence frequency, the heavy fluid–loading condition for structural response is

$$\omega \lesssim \omega_n \left(\frac{\omega_n}{\omega_g}\right) \quad , \tag{17.20}$$

corresponding to frequencies far below ω_n. Indeed, fluid loading remains light at the null frequency, if it is the structural response that is under discussion (see again later). Thus the condition for heavy fluid loading certainly depends on whether one looks at the acoustic field or structural response.

Even if we stay with the acoustic field, fluid loading is very much a function of *angle* and also of *range*. As far as angle is concerned, the limit $D(\theta, M, \alpha) \to D_0(\theta, M)$ is not approached uniformly in two important angular sectors.

(i) If θ is small, $\theta = O(\alpha)$ or smaller, then

$$D(\theta, M, \alpha) \sim \frac{\theta}{(1 - M^4)\theta + i\alpha} \quad , \tag{17.21}$$

which is not approximated by $D_0(\theta, M) \sim (1 - M^4)^{-1}$; in particular, while $D_0 \to (1 - M^4)^{-1}$ as $\theta \to 0$, $D \to 0$ like $(\theta/i\alpha)$. The near–grazing–incidence field is controlled by fluid loading no matter how small α may be, and fluid loading changes what is (for $M < 1$) an almost uniform monopole directivity $(1 - M^4 \cos^4 \theta)^{-1}$ into a small–angle field of essentially dipole type, like $\sin \theta/(i\alpha)$. The dipole directivity guarantees a vanishing leading order (in r) acoustic field on the structure itself —

a phenomenon well known in the optical and radar analogues as the Lloyd's Mirror effect.

(ii) Consider frequencies "above coincidence". Then $M > 1$, and a "Mach" or "coincidence" angle θ_M can be defined as

$$\theta_M = \text{arc}\cos M^{-1} \quad . \tag{17.22}$$

Evidently now the α term in (17.17) must be retained for θ near to θ_M, and specifically for $(\theta - \theta_M) = O(\alpha)$. Accordingly we write $\theta = \theta_M + \alpha\theta'$, and then the proper approximation to D as $\alpha \to 0$ for $O(1)$ values of θ' is

$$D(\theta, M, \alpha) \sim \left(\frac{1}{\alpha}\right)\left[\frac{\sin\theta_M}{4M\theta'\sin^2\theta_M + i}\right] \quad . \tag{17.23}$$

Observe that $D = O(1)$ for θ not close to θ_M, and that $\alpha << 1$ when $\omega >> \omega_g$ (for steel in water, $\alpha = 0.133$ when $\omega = \omega_g$). Therefore, (17.23) shows that fluid–loading effects control (a) the large $O(\alpha^{-1})$ amplitude of the field in the coincidence beams near $\theta = \theta_M$ (whereas the field elsewhere, except for small angles $\theta = O(\alpha)$, is not significantly affected by fluid loading when $\omega >> \omega_g$); and (b), the width $|\theta - \theta_M| = O(\alpha)$ of the coincidence beams (whereas the directivity function D is elsewhere controlled by the parameter M rather than α). See Figure 17.2 for a sketch of the directivity functions for frequencies below and above coincidence.

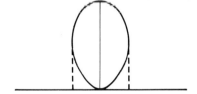
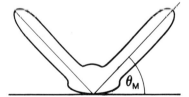

Figure 17.2 Polar plots of $|D(\theta, M, \alpha)|$ versus θ for (a) frequencies below coincidence, $M < 1$, and (b) frequencies above coincidence, $M > 1$, with $\alpha << 1$ in both cases. The dotted line in (a) corresponds to the variation of $|D_0(\theta, M)|$.

Turning to range dependence, it turns out from more detailed analysis of the coincidence beams that the condition for the validity of (17.16) is

$$k_0 r \gg 1 \qquad \text{for} \quad |\theta - \theta_M| \gtrsim \alpha \quad ,$$

but

$$k_0 r \gg \alpha^{-2} \qquad \text{for} \quad |\theta - \theta_M| \lesssim \alpha \quad ,$$

implying that while the outgoing cylindrical wave structure is fully developed away from the coincidence beams at ranges of a wavelength or so, it is not developed within them until ranges of α^{-2} wavelengths (perhaps 100 wavelengths or more). The explanation for this comes from looking at the surface response which, for $M > 1$, is dominated by a structural wave with a complex wavenumber whose real part is close to k_p and whose (positive) imaginary part is small and proportional to the fluid–loading parameter α. The wave travels supersonically over the surface, decaying slowly as it goes, with an attenuation length of order α^{-1} wavelengths; the implied energy loss is, of course, in the form of radiation loss to the intense coincidence beams, and we have a "leaky wave" which launches its energy to the radiation field over a length $\ell \sim (k_0 \alpha)^{-1}$ of the structure. The plate motion is then equivalent to a large projector or antenna, for which the Rayleigh distance d is given by $d \sim k_0 \ell^2 \sim k_0 (k_0 \alpha)^{-2}$. At ranges $k_0^{-1} \ll r \ll d$, or $1 \ll k_0 r \ll \alpha^{-2}$, we have a collimated non–spreading plane beam, which in fact comprises just the plane wavefronts generated by the supersonic wavy wall patterns. At ranges around d, diffraction begins to spread the beam, with the usual mathematical description in terms of Fresnel functions. Then for $r \gg d$ the diffraction is complete and we have a cylindrically spreading field — though still a highly directive one. In this case, fluid–loading effects control the radiation loss, and hence the release of structural energy to the fluid, and thereby the diffraction process in the distant acoustic field.

At frequencies below coincidence there is a much more complicated dependence on range. If one takes (17.6) for the surface response and deforms the path C upwards onto the edges of the vertical cut from k_0 to $k_0 + i\infty$, one gets an expression for $v(x)$ as a sum of residue terms — which constitute what may be called the *surface wave field* — and of a branch line integral which defines the *acoustic field* v_a,

$$v_a = -\left(\frac{i\omega F_0}{2\pi B}\right) 2\mu k_p^4 \exp(ik_0|x|) \int_0^\infty \frac{i\gamma}{P(k)} \exp(-u|x|)du \quad ,$$

where

$$P(k) \equiv (k^4 - k_p^4)^2 (k^2 - k_0^2) - \mu^2 k_p^8 \quad , \tag{17.24}$$

and the integral is up the right side of the cut with $k = k_0 + iu$. As $|x| \to \infty$ the integral is dominated by the region near $k = k_0$, and

$$v_a \sim -\left(\frac{i\omega F}{2\pi B}\right) 2\mu k_p^4 \exp(ik_0|x|) \frac{i(2k_0)^{1/2}}{(-\mu^2 k_p^8)} \int_0^\infty (iu)^{1/2} e^{-u|x|} du$$

$$= O\left(|x|^{-3/2}\right) \quad . \tag{17.25}$$

This is in accord with the Lloyd's Mirror requirement — the coefficient of the "leading–order" term $\exp(ik_0 r)/r^{1/2}$ vanishes and the first non–zero term is $O(|x|^{-3/2})$, and its magnitude is controlled by the $(-\mu^2 k_p^8)$ term, which arises from fluid loading (specifically, from fluid inertia).

If one adopted the approach of ignoring the fluid loading in calculating the plate response, $P(k)$ would lack the $(-\mu^2 k_p^8)$ term, and in consequence the integrand would be proportional to $u^{-1/2} \exp(-u|x|)$, giving $v_a = O(|x|^{-1/2})$. A much more detailed analysis shows that this latter result does indeed hold for an extensive range of $|x|$; if the parameter α is small, as it is at frequencies somewhat less than ω_g for most common situations, then the $|x|^{-1/2}$ result holds for $1 << k_0|x| << \alpha^{-2}$, but it cannot hold to arbitrarily large $|x|$ where the $\mu^2 k_p^8$ term is all–important. A transition takes place around $k_0|x| = \alpha^{-2}$, described by error or Fresnel functions, and for $k_0|x| >> \alpha^{-2}$ the Lloyd's Mirror $|x|^{-3/2}$ decay applies. One might again try to interpret the α^{-2} scaling in terms of some Rayleigh length, but we have not found a convincing explanation this way of why a theory based on neglect of fluid loading works for $k_0|x| << \alpha^{-2}$, and why it takes fluid–loading effects a distance $>> \alpha^{-2}k_0^{-1}$ to enforce the $|x|^{-3/2}$ decay. It is clear, nonetheless, that the significance of fluid–loading effects is very much a function of range, angle and of the physical quantity under discussion. However, as both α and M involve the frequency and the material parameters, no comprehensive theoretical treatment can be achieved of even the simplest problem (at least through analytical studies which might lead to clear understanding of the dominant balances and how they change) on the basis of a characterization of fluid–loading effects by the values of α, M. A far more useful choice of parameters is made in Section 17.3.

17.3 INTRINSIC FLUID–LOADING PARAMETER

Fluid–loading problems are very hard to solve exactly, and for geometries and configurations of practical interest it is essential to be able to make useful simplifying approximations. Various approximations have been devised, some of them "rational", but many have no status as rational theories and are simply expedient measures to get a tractable problem. Of the rational approximations, most deal with light or heavy fluid loading by taking a limit $\alpha \to 0$ or $\alpha \to \infty$, but with other parameters, such as $M = k_0/k_p$, held fixed. To achieve this one has to imagine the limit $\alpha \to 0$ or $\alpha \to \infty$ implemented by considering a sequence of fluid media, all of the same sound speed but of decreasing or increasing density, respectively, and dealing with the same structure and operating frequency throughout. Such an approach does not correspond to the situation with which one is usually faced in practice. There one usually has the same structure and fluid media (steel and water in marine applications, aluminium and air in most aeronautical cases) and imagines the severity of fluid loading to change because of a change in structure thickness or, much more commonly, in the frequency of interest. Variations of M with ω are often at least as significant as those of α, and the two approaches lead to quite different results.

What is needed then is a parameter which measures fluid loading in a way independent of the values of plate thickness and frequency, so that variations in those quantities can be entirely accounted for by variations in M alone. Such an *intrinsic fluid–loading parameter* does indeed exist for thin elastic plates in static fluid; it is

$$\epsilon = \frac{\rho_0 c_0}{m\omega_g}$$

$$= \alpha\Omega = \alpha M^2 \quad .$$

(17.26)

In terms of Young's modulus E, the plate material density ρ_p and Poisson's ratio σ, we have

$$\epsilon = \frac{\rho_0}{\rho_p} \left[\frac{E}{12c_0^2(1 - \sigma^2)\rho_p} \right]^{1/2} ,$$

(17.27)

and for double–sided fluid loading by the same fluid the value of ϵ is to be doubled. This parameter seems first to have been introduced by Nayak (1970), but its systematic use is much more recent. Apart from the usefulness of ϵ as a parameter which is the same for all plates of a given material embedded in a given fluid, it is, further, fortunate that the values of ϵ for cases of greatest interest are small;

for aluminium in air, $\epsilon = 0.00213$, while for steel in water $\epsilon = 0.133$. Even for aluminium in water (as is used in some simulations of aircraft fuel tank vibrations) the value of ϵ is not large; $\epsilon = 0.3903$.

At least for the common steel/water and aluminium/air problems, we therefore advocate not only the presentation of results in terms of ϵ and M rather than α and M, but the systematic exploitation of the smallness of ϵ in conjunction with singular perturbation techniques (matched asymptotic expansions and multiple scaling methods). This is not to say that all fluid loading is "light", but that the "fluid loading at coincidence", as measured by ϵ, is light. Taking $\epsilon \ll 1$ in no way restricts the *actual* fluid loading at any particular plate thickness and frequency (as measured, for example, by α or αM as in Section 17.2) to small values. Indeed, the size of M (or Ω) *relative* to ϵ will determine the severity of fluid–loading effects in a given context and will provide a much finer measure of those effects than could otherwise be obtained. It will also be possible, at least in principle, to systematically cover *all* frequencies, *all* ranges from a given point on the structure, and indeed *all* values of any other parameter (angle, rib spacing, plate size, ...) occurring in a fluid loaded structure problem — as has been demonstrated in recent papers. We next give a simple illustration of this approach at work.

17.4 THE FREE WAVES ON A FLUID–LOADED PLATE

This issue has been examined at some length elsewhere (see the references cited) and here we look just at the *unique* subsonic wave which exists under all conditions (within the model of Section 17.2). The wavenumber for this wave has $k > k_0$ and is a zero of the denominator of (17.5), namely, of

$$(\zeta^4 - 1)(\zeta^2 - M^2)^{1/2} - \frac{\epsilon}{M} = 0 \quad , \tag{17.28}$$

in terms of ϵ, M and $\zeta = k/k_p$. Suppose first that $M < 1$, and order M with respect to ϵ by writing $M = \epsilon^\beta M_\beta$. Then look for a solution of (17.28) with M_β fixed, i.e. solve

$$(\zeta^4 - 1)(\zeta^2 - \epsilon^{2\beta} M_\beta^2)^{1/2} - \frac{\epsilon^{(1-\beta)}}{M_\beta} = 0 \quad . \tag{17.29}$$

Clearly there is a root near $\zeta = 1$ (i.e. near $k = k_p$) provided $0 \le \beta < 1$, and therefore in this context fluid–loading effects provide only a small correction to the vacuum dynamics unless $M = O(\epsilon)$ or smaller. For $\beta = 0$ we find that

$$\zeta = 1 + \frac{\epsilon}{4M(1 - M^2)^{1/2}} + O(\epsilon^2) \quad , \tag{17.30}$$

and for $0 < \beta < 1$

$$\zeta = 1 + \frac{\epsilon^{(1-\beta)}}{4M_\beta} + O(\epsilon^{2-2\beta}) \quad . \qquad (17.31)$$

Observe that at the null frequency we have precisely $\beta = 1/2$, $M_\beta = 1$ and

$$\zeta = 1 + \frac{\epsilon^{1/2}}{4} + O(\epsilon) \quad , \qquad (17.32)$$

confirming the earlier statement that at the null frequency fluid–loading effects on the structural response are small (they vanish as $\epsilon \to 0$, though not very fast, as $\epsilon^{1/2}$).

A significant shift in the free wavenumber occurs only when $\beta \geq 1$, and $\beta = 1$ is the significant ("distinguished") value. Here we have (exactly)

$$(\zeta^4 - 1)(\zeta^2 - \epsilon^2 M_1^2)^{1/2} - \frac{1}{M_1} = 0 \quad , \qquad (17.33)$$

and to leading order ζ satisfies

$$(\zeta^4 - 1)|\zeta| - \frac{1}{M_1} = 0 \quad , \qquad (17.34)$$

and is not necessarily close to 1 since M_1 is a general $O(1)$ quantity. We call $M = O(\epsilon)$ the *frequency range of significant fluid loading* (again, *in this particular context only*), and $M = O(\epsilon^\beta)$ with $0 \leq \beta < 1$ corresponds to frequencies at which fluid loading is light.

Observe that in significant fluid loading the only mechanism which can be neglected is that of fluid compressibility (IV); mechanisms I, II, III and V are comparably important in determining the solution of (17.34). In contrast, for light fluid loading, ζ is close to 1 and the balance is essentially that of vacuum dynamics, between I and II alone.

To understand problems of significant fluid loading one should really concentrate on $M = O(\epsilon)$; however, even in this simple problem the results are not sufficiently tractable; ζ satisfies a quintic equation about which general results can be proved but no explicit solution given. It is therefore advantageous sometimes to go to $\beta > 1$, i.e. $M/\epsilon \to 0$ as $\epsilon \to 0$, or $M \ll \epsilon$, where a further approximation can be made and substantially greater tractability gained.

For $\beta > 1$ put

$$\zeta = \epsilon^{(1-\beta)/5}\xi \quad , \qquad (17.35)$$

and let $\epsilon \to 0$ with ξ and M_β fixed. Then

$$\xi^4 |\xi| - \frac{1}{M_\beta} = 0 \quad , \tag{17.36}$$

the relevant solution of which is

$$\xi = M_\beta^{-1/5} \quad . \tag{17.37}$$

In this frequency range (more than a *decade* below the null frequency for steel in water) the dominant mechanisms are I, III and V — structural stiffness, fluid pressure and fluid inertia. Structural inertia has become negligible at the low frequencies represented by $M \ll \epsilon$. We call this the régime of *heavy fluid loading*. It appears from a number of calculations which have now been made that the results for heavy fluid loading are qualitatively and quantitatively similar to those for significant fluid loading, and this makes the simplification of heavy fluid loading even more useful. The essential point of that simplification lies not so much in the explicit solution (17.37) but rather in the fact that the whole dispersion function (17.28) — a function of ϵ and M independently — can, for $\beta > 1$, be expressed as a multiple of a universal dispersion function

$$X^4 |X| - 1 \quad , \tag{17.38}$$

with

$$X = M_\beta^{\frac{1}{5}} \xi = M^{\frac{1}{5}} \epsilon^{-\frac{1}{5}} \zeta \quad ,$$

and the properties of (17.38) can be computed (if necessary) once and for all.

For $M > 1$ we have to distinguish the subsonic surface wave mode from the supersonic "leaky" wave. Details are given in some of the references below, as are applications of the asymptotic method sketched above to semi–infinite, finite and periodically inhomogeneous geometries.

REFERENCES

Cremer, L., Heckl, M. & Ungar, E.E. (1973). Structure–Borne Sound. Springer–Verlag, Berlin.

Crighton, D.G. (1989). The 1988 Rayleigh Medal Lecture: fluid loading — the interaction between sound and vibration. J. Sound Vib. 133:1–27.

Junger, M.C. & Feit, D. (1986). Sound, Structures, and Their Interaction. (2nd ed.) M.I.T. Press, Cambridge, Massachusetts.

Nayak, P.R. (1970). Line admittance of infinite isotropic fluid–loaded plates. J. Acoust. Soc. Am. 47:191–201.

PART III
Wave Modification

18.   SCATTERING AND DIFFRACTION

18.1 BASIC EQUATIONS

It can be shown that for small perturbations of a compressible fluid from rest, the pressure fluctuation p, density fluctuation ρ and velocity potential ϕ satisfy a scalar wave equation

$$(\nabla^2 - c^{-2}\partial^2/\partial t^2)\phi(\mathbf{x}, t) = q(\mathbf{x}, t) , \qquad (18.1)$$

where c is the wave speed and q represents a distribution of sound sources. The three variables, p, ρ and ϕ are related through the formulae

$$p = c^2\rho = -\rho_0 \partial\phi/\partial t , \qquad (18.2)$$

where ρ_0 is the mean fluid density.

If the forcing term $q(\mathbf{x}, t)$ and the boundary conditions happen to be simple harmonic in time, with angular frequency ω, then the linear nature of (18.1) ensures that ϕ will have similar time dependence. Thus if

$$q(\mathbf{x}, t) = \Re\left\{q_0(\mathbf{x})e^{-i\omega t}\right\} , \quad \phi(\mathbf{x}, t) = \Re\left\{\phi_0(\mathbf{x})e^{-i\omega t}\right\} , \qquad (18.3)$$

and the wave equation (18.1) becomes

$$(\nabla^2 + k^2)\phi_0(\mathbf{x}) = q_0(\mathbf{x}) , \qquad (18.4)$$

where $k = \omega/c$ is the acoustic wavenumber, and (18.4) is sometimes called the "Helmholtz equation" or "reduced wave equation".

In one dimension, with no sources present, the solutions are simply

$$\phi_0(x) = Ae^{\pm ikx} ; \qquad (18.5)$$

hence

$$\phi = \Re\left\{A \exp\left[-i\omega(t \mp x/c)\right]\right\} .$$

This corresponds to plane waves travelling in the positive or negative x–direction according as the $-$sign or $+$sign is taken in (18.5). Evidently the wavelength λ is given by $2\pi/k$, since this parameter added to x leaves (18.5) unchanged.

In general any disturbance $q(\mathbf{x}, t)$, and hence $\phi(\mathbf{x}, t)$, can be expressed as a superposition of time–harmonic functions by Fourier transformation in time (provided that $q(\mathbf{x}, t)$ behaves suitably as $t \to \pm\infty$). Thus

$$q(\mathbf{x}, t) = \frac{1}{2\pi} \int_{-\infty}^{\infty} q(\mathbf{x}, \omega) e^{-i\omega t} d\omega \quad, \tag{18.6}$$

where

$$q(\mathbf{x}, \omega) = \int_{-\infty}^{\infty} q(x, t) e^{i\omega t} dt \quad, \tag{18.7}$$

and similarly for ϕ. If the physical problem is subject to a "causality" condition — that q and ϕ vanish for $t < 0$ — then the functions $q(\mathbf{x}, \omega)$ and $\phi(\mathbf{x}, \omega)$ must be *analytic* in the complex variable ω in some upper half–plane $\Im \omega > \alpha$.

Fundamental Source Solution

The fundamental Green's function for the Helmholtz equation (18.4) with $q_0(\mathbf{x}) = \delta(\mathbf{x})$ can be derived either from first principles or else as a special case of the general wave equation. A detailed calculation has already been given in Chapter 2 for the Green's function appropriate to the general wave equation

$$\left(\nabla^2 - c^{-2}\partial^2/\partial t^2\right) G(\mathbf{x}, \boldsymbol{\xi}; t) = \delta(\mathbf{x} - \boldsymbol{\xi})\delta(t) \,, \tag{18.8}$$

and the solution subject to the causality condition was shown to be

$$G(\mathbf{x}, \boldsymbol{\xi}; t) = -(4\pi R)^{-1}\delta(t - R/c), \quad R = |\mathbf{x} - \boldsymbol{\xi}| \quad.$$

Equation (18.8) is of course a special case of (18.1) with $q = \delta(\mathbf{x} - \boldsymbol{\xi})\delta(t)$, whence (18.7) gives

$$q(\mathbf{x}, \omega) = \delta(\mathbf{x} - \boldsymbol{\xi}) \quad. \tag{18.9}$$

It follows that the solution of the equation

$$(\nabla^2 + k^2)\phi(\mathbf{x}; \boldsymbol{\xi}; k) = \delta(\mathbf{x} - \boldsymbol{\xi}) \tag{18.10}$$

is

$$\phi = -\int_{-\infty}^{\infty} \frac{\delta(t - R/c)}{4\pi R} e^{i\omega t} dt = -\frac{e^{ikR}}{4\pi R} \quad. \tag{18.11}$$

This is therefore the Green's function for the Helmholtz equation, corresponding the the "causality" requirement. With this radially symmetric problem, one could proceed from first principles by seeking a solution of (18.10) in the form $\phi = \phi(R)$. This would yield two possible solutions, namely $\phi_1 = -(4\pi R)^{-1}e^{+ikR}$ and $\phi_2 = -(4\pi R)^{-1}e^{-ikR}$. With the time factor $\exp(-iwt)$ included, it is seen that ϕ_2 corresponds to inward travelling waves that can not be made compatible with the causality condition. The relevant solution ϕ_1 corresponds to outward travelling waves. In such time–harmonic problems, the causality condition is replaced by a boundary condition at infinity: waves travel outwards towards infinity.

Boundary Condition at Infinity

Other singular solutions of the Helmholtz equation can be generated by differentiating (18.11) with respect to x, y or z. Such solutions satisfy (18.4) with $q = 0$ except at the origin. Each solution has the property that its far field is of the form

$$\phi \sim F(\theta, \lambda)\, \frac{e^{ikr}}{kr} \quad \text{as} \quad r \to \infty , \tag{18.12}$$

where (r, θ, λ) are polar coordinates related to any convenient origin $r = 0$. The wavefronts, or surfaces of constant phase, are spherical and the *amplitude at infinity* $F(\theta, \lambda)$ will in general vary with angular position. It is reasonable to expect that (and it can be shown that) any solution generated by finite radiating or scattering bodies in three dimensions has the far–field form (18.12). The determination of the far–field amplitude $F(\theta, \lambda)$ is often the primary concern of an investigation. An equivalent form of the radiation condition (18.12) is

$$r\left(\frac{\partial \phi}{\partial r} - ik\phi\right) \to 0 \quad \text{as} \quad r \to \infty ,$$

the "Sommerfeld radiation condition". It rejects solutions of the form $Fr^{-1}e^{-ikr}$ which correspond to incoming waves at infinity.

Two dimensions: in two dimensions the Green's function satisfies

$$(\nabla^2 + k^2)G = \delta(\mathbf{x} - \boldsymbol{\xi}) ,$$

and can be calculated from (18.11) by integrating along the z–axis. Thus

$$\begin{aligned}
G(x, y; \xi, \eta) &= -\frac{1}{4\pi} \int_{-\infty}^{\infty} \frac{\exp(ik(r^2 + z^2)^{1/2})}{(r^2 + z^2)^{1/2}}\, dz \\
&= \frac{1}{4i}\left\{J_0(kr) + iY_0(kr)\right\} \equiv \frac{1}{4i}\, H_0^{(1)}(kr) ,
\end{aligned} \tag{18.13}$$

where $r = |\mathbf{x} - \boldsymbol{\xi}|$ and J_0, Y_0 denote Bessel functions of order zero; $H_0^{(1)} = J_0 + iY_0$ is called the Hankel function of the first kind.

As $r \to \infty$, $G \sim$ constant times $r^{-1/2}e^{ikr}$. In two dimensions the general far–field form equivalent to (18.12) is

$$\phi \sim F(\theta)(kr)^{-1/2}e^{ikr} . \tag{18.14}$$

18.2 EXACT SOLUTIONS

For simple geometries, such as planes, spheres, elliptic cylinders and wedges, exact solutions can often be calculated, typically in the form of infinite series or integrals. Quite apart from their immediate interest, exact solutions are useful in motivating and checking approximate methods. A few typical cases are listed below, to illustrate some of the methods involved.

I. Infinite plane with piston

A plane piston S, set in an infinite rigid baffle S_0, vibrates with outward normal velocity $V_1(\mathbf{x}, t)$. Without loss of generality the velocity V_1 and the fluid potential $\Phi(\mathbf{x}, t)$ may be assumed to be simple harmonic in time; thus

$$\Phi(\mathbf{x}, t) = \Re\left\{\phi(\mathbf{x})e^{-iwt}\right\}, \quad V_1(\mathbf{x}, t) = \Re\left\{V(\mathbf{x})e^{-iwt}\right\} .$$

More general time dependence can in principle be analyzed by Fourier transformation in t. The governing equations for $\phi(\mathbf{x})$ are

$$(\nabla^2 + k^2)\phi = 0 \quad \text{in the fluid} \quad x_3 > 0 \ , \tag{18.15}$$

$$\left. \begin{array}{ll} \partial\phi/\partial x_3 = V & \text{on the piston } S \quad x_3 = 0 \ , \\ \partial\phi/\partial x_3 = 0 & \text{on the baffle } S_0 \quad x_3 = 0 \ , \end{array} \right\} \tag{18.16}$$

on choosing coordinates so that the baffle and piston are at $x_3 = 0$. In addition there is an outgoing wave condition (18.12).

This is one of the few problems in which the "exact Green's function" $G(\mathbf{x}, \boldsymbol{\xi})$ can be found in simple form. This function is specified by

$$(\nabla^2 + k^2)G(\mathbf{x}, \boldsymbol{\xi}) = \delta(\mathbf{x} - \boldsymbol{\xi}), \quad \xi_3 > 0 \ ,$$

with

$$\partial G/\partial x_3 = 0 \quad \text{at} \quad x_3 = 0,$$

$$\tag{18.17}$$

and a radiation condition at infinity. Physically, G represents the potential at $\mathbf{x}$ due to a point source at $\boldsymbol{\xi}$ in the presence of a rigid wall at $x_3 = 0$. Now it has been shown (18.11) that the function $-(4\pi R)^{-1} \exp(ikR)$, $R = |\mathbf{x} - \boldsymbol{\xi}|$, satisfies the homogeneous Helmholtz equation except for a source singularity at $\mathbf{x} = \boldsymbol{\xi}$, and it remains to satisfy the boundary condition (18.17) at $x_3 = 0$. This is readily achieved by simply adding an "image source" at the point $(\xi_1, \xi_2, -\xi_3)$; thus

$$G(\mathbf{x}, \boldsymbol{\xi}) = -(4\pi R)^{-1} \exp(ikR) - (4\pi R_1)^{-1} \exp(ikR_1) \quad , \tag{18.18}$$

where

$$R^2 = (x_1 - \xi_1)^2 + (x_2 - \xi_2)^2 + (x_3 - \xi_3)^2 \quad ,$$

and

$$R_1^2 = (x_1 - \xi_1)^2 + (x_2 - \xi_2)^2 + (x_3 + \xi_3)^2 \quad .$$

It is easy to verify that $\partial G / \partial x_3 = 0$ at $x_3 = 0$. The potential $\phi(\mathbf{x})$ can now be written down explicitly in terms of G. For an application of Green's formula to the fuctions $\phi(\mathbf{x})$ and $G(\mathbf{x}, \boldsymbol{\xi})$ gives

$$
\begin{aligned}
\phi(\boldsymbol{\xi}) &= \int_{S+S_0} G(\mathbf{x}, \boldsymbol{\xi}) \, \frac{\partial \phi}{\partial n} \, (\mathbf{x}) dx \\
&= -\frac{1}{2\pi} \int_S V(\mathbf{x}) \, \frac{e^{ikR}}{R} \, dx \quad ,
\end{aligned}
\tag{18.19}
$$

evaluated over the piston S, with $R^2 = (x_1 - \xi_1)^2 + (x_2 - \xi_2)^2 + \xi_3^2$. Clearly (18.19) gives an exact solution for ϕ in terms of the prescribed velocity distribution V on the piston.

High–Frequency Limit

An interesting case arises when the wavenumber ka is large (where a is the piston dimension). Assume firstly that $V(x_1, x_2)$ is independent of k. On account of the oscillatory term $\exp(ikR)$ in the integral (18.19), the integrand varies rapidly as $\mathbf{x}$ varies across S and this destructive interference implies a relatively small value for the potential. According to Kelvin's method of stationary phase (see Chapter 4), the most significant contribution to the integral will usually come from the vicinity of the *stationary* point of R where this function varies most slowly as a function of $\mathbf{x}$ [Note that there may be equally important contributions if the field point $\boldsymbol{\xi}$ is such that the phase R is *constant* round part of the boundary S.] Consider a fixed point $\boldsymbol{\xi}$. Then the stationary point of R is obviously at $x_1 = \xi_1$, $x_2 = \xi_2$. In this vicinity

$$V(\mathbf{x}) \sim V(\xi_1, \xi_2) \quad ,$$

and

$$R = \left\{ (x_1-\xi_1)^2 + (x_2-\xi_2)^2 + \xi_3^2 \right\}^{1/2} = \xi_3 \left\{ 1 + \frac{(x_1-\xi_1)^2}{2\xi_3^2} + \frac{(x_2-\xi_2)^2}{2\xi_3^2} + \ldots \right\} . \quad (18.20)$$

The stationary phase method then gives the contribution

$$\phi(\boldsymbol{\xi}) \sim -\frac{1}{2\pi} \frac{V(\xi_1, \xi_2)}{\xi_3} \int_{-\infty}^{\infty} dx_1 \int_{-\infty}^{\infty} dx_2 \exp \left\{ ik \left(\xi_3 + \frac{(x_1-\xi_1)^2}{2\xi_3} + \frac{(x_2-\xi_2)^2}{2\xi_3} \right) \right\}$$

$$= -\frac{1}{2\pi} \frac{V(\xi_1, \xi_2) e^{ik\xi_3}}{\xi_3} \int_{-\infty}^{\infty} \exp \left(\frac{ik(x_1-\xi_1)^2}{2\xi_3} \right) dx_1 \int_{-\infty}^{\infty} \exp \left(\frac{ik(x_2-\xi_2)^2}{2\xi_3} \right) dx_2 .$$

The two integrals are equal to each other and have the value

$$\int_{-\infty}^{\infty} \exp \frac{ikx^2}{2\xi_3} dx = \left(\frac{2\xi_3}{k} \right)^{1/2} e^{i\pi/4} \pi^{1/2}$$

(see Chapter 4). Finally

$$\phi(\xi_1, \xi_2, \xi_3) \sim -\frac{1}{2\pi} \frac{V(\xi_1, \xi_2)}{\xi_3} e^{ik\xi_3} \left(\frac{2\xi_3}{k} \right) \pi e^{i\pi/2} \quad ,$$

i.e.

$$\phi \sim -(i/k)V(\xi_1, \xi_2) e^{ik\xi_3} \quad \text{as} \quad k \to \infty \quad , \quad (18.21)$$

provided $\boldsymbol{\xi}$ lies within the "beam" which is a cylinder of cross–section S and perpendicular to the plane $x_3=0$. If $\boldsymbol{\xi}$ lies outside this beam, the point of stationary phase is excluded from the range of integration and the potential is of lower order $(k^{-3/2})$ as $k \to \infty$. The result (18.21) is for fixed ξ_3 and is not uniformly valid in ξ_3. If that distance is large, for example, the approximation following (18.20) is inadequate and a different procedure is required.

Beams in directions other than the perpendicular from the baffle can be produced by choosing the velocity V to depend suitably on wavenumber. If $V = V_0 \exp(ikx_2\sin\theta)$ for example ($\theta > 0$), then the rapidly oscillating term in the integral (18.19) is of the form $\exp(ik(R + x_2\sin\theta))$ and the phase function $R + x_2\sin\theta$ is stationary when the line from $\mathbf{x}$ to $\boldsymbol{\xi}$ is at an angle θ to the x_3–axis.

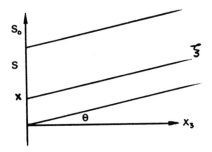

Figure 18.1

Thus the beam has its axis in the (x_2, x_3) plane at an angle θ to the normal.

II. (i) Hard sphere

The potential is calculated here for a point source in the presence of a hard sphere. This is the exact Green's function for the more general problem of arbitrary incident field and arbitrary velocity distribution on the sphere.

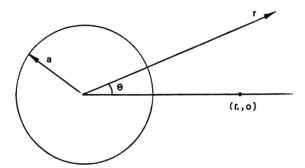

Figure 18.2

Suppose the source is at distance r_1 from the centre of a sphere of radius $a < r_1$, and choose spherical polar coordinates (r, θ, λ) so that the polar angle θ is zero at the source point. By symmetry the potential $\phi(r, \theta)$ is independent of λ and the source is taken to be simple harmonic in time. The potential ϕ is specified by the equations

$$(\nabla^2 + k^2)\phi = \delta(\mathbf{r} - \mathbf{r}_1), \quad r > a \quad , \tag{18.22}$$

$$\partial\phi/\partial r = 0, \qquad r = a \quad , \tag{18.23}$$

and a radiation condition at infinity.

Now the homogeneous equation

$$(\nabla^2 + k^2)\phi = 0$$

has separable solutions of the form

$$h_n^{(1)}(kr)P_n(\cos\theta) \quad \text{and} \quad h_n^{(2)}(kr)P_n(\cos\theta) \quad,$$

where n is an integer, to ensure regular solutions at $\theta = 0$ and $\theta = \pi$, P_n is the Legendre polynomial of degree n, and $h_n^{(1)}$ and $h_n^{(2)}$ are spherical Hankel functions

$$h_n^{(1)}(x) = j_n(x) + iy_n(x) \equiv \left(\frac{\pi}{2x}\right)^{1/2}\left\{J_{n+\frac{1}{2}}(x) + iY_{n+\frac{1}{2}}(x)\right\} \quad,$$

and $h_n^{(2)}(x) = j_n(x) - iy_n(x)$ is the complex conjugate of $h_n^{(1)}(x)$ if x is real. Spherical Bessel functions are of simple form since Bessel functions of order $(n + \frac{1}{2})$ can be expressed in terms of trigonometric functions. In fact

$$h_0^{(1)}(x) = -i\,\frac{e^{ix}}{x}, \quad h_0^{(2)}(x) = +i\,\frac{e^{-ix}}{x} \quad,$$

$$h_1^{(1)}(x) = -h_0^{(1)'}(x) = -(i/x^2 + 1/x)e^{ix} \quad.$$

In order to satisfy the radiation condition at infinity, Hankel functions of the second kind are rejected and henceforth h_n will be used to mean $h_n{}^{(1)}$.

The first few Legendre polynomials are given by

$$P_0(c) = 1, \quad P_1(c) = c, \quad P_2(c) = \tfrac{1}{2}(3c^2 - 1) \quad.$$

The Legendre polynomials $P_n(c)$ have a useful orthogonality property, namely

$$\int_{-1}^{1} P_m(c)P_n(c)dc = \begin{cases} (n + \tfrac{1}{2})^{-1} & \text{if } m = n \\ \\ 0 & \text{if } m \neq n \end{cases} \quad. \tag{18.24}$$

The solution for ϕ is now constructed in the form

$$\phi = -(4\pi R)^{-1}e^{ikR} + \sum_{n}^{\infty} A_n h_n(kr)P_n(\cos\theta) \quad,$$

$$= \phi_i + \phi_s \quad,$$

which is seen to have the correct source singularity as $R = |\mathbf{r} - \mathbf{r}_1| \to 0$, and satisfies the Helmholtz equation and radiation condition. It remains to choose the constants A_n to satisfy the boundary condition (18.23). To this end, the incident potential $\phi_i = -(4\pi R)^{-1}e^{ikR}$ is itself written as a series in Legendre polynomials and spherical Bessel functions, according to the addition theorem (Watson 1966)

$$-\frac{e^{ikR}}{4\pi R} = \frac{-ik}{2\pi} \sum_0^\infty (n + \tfrac{1}{2})j_n(kr)h_n(kr_1)P_n(\cos\theta) \quad \text{if} \quad r < r_1 \quad,$$

with r and r_1 interchanged for $r > r_1$. Setting $\partial\phi/\partial r = 0$ at $r = a$ leads to the conclusion that

$$A_n = \frac{ik}{2\pi} \frac{j_n'(ka)h_n(kr_1)}{h_n'(ka)} (n + \tfrac{1}{2}), \quad \text{whence}$$

$$\phi = -\frac{e^{ikR}}{4\pi R} + \frac{ik}{2\pi} \sum_0^\infty (n + \tfrac{1}{2}) \frac{j_n'(ka)h_n(kr_1)}{h_n'(ka)} h_n(kr)P_n(\cos\theta) \,. \quad (18.25)$$

The infinite series for the scattered potential ϕ_s can be shown to converge for $r > a$, as expected, but the convergence is slow when k is large. The sum can be used directly to estimate ϕ when ka and kr are small. [For large wavenumber k, it is found convenient to recast (18.25) in the form of a contour integral according to the "Watson Transformation" (Watson 1919).] As $x \to \infty$, the Hankel function $h_n(x)$ has the form

$$h_n(x) \sim -(i/x)\exp(ix - \tfrac{1}{2}in\pi) \quad . \quad (18.26)$$

If we let $kr \to \infty$ in the sum (18.25), the scattered potential is seen to have the limiting form

$$\phi_s \sim F(\theta)(kr)^{-1}e^{ikr}$$

with

$$F(\theta) = \frac{k}{2\pi} \sum_0^\infty (n + \tfrac{1}{2})e^{-in\pi/2} \frac{j_n'(ka)h_n(kr_1)}{h_n'(ka)} P_n(\cos\theta) \quad . \quad (18.27)$$

The limiting process requires some justification, since the estimate (18.26) is valid for fixed n as $x \to \infty$, whereas n runs to infinity in the sum. A more careful analysis, based on an integral identity for ϕ_s in terms of the Green's function $G = (-4\pi R)^{-1}\exp(ikR)$ confirms that (18.27) is correct.

II. (ii) Vibrating sphere

A similar calculation can be made for the axially symmetric radiation problem in which a sound field of potential $Re\{\phi(r,\theta)\exp(-i\omega t)\}$ is generated by the outward normal velocity $Re\{V(\theta)\exp(-i\omega t)\}$ at a spherical surface $r = a$, where (r,θ) denote spherical polar coordinates. If the vibrations have small amplitude, the boundary condition $\partial\phi/\partial r = V$ can be applied at the mean position $r = a$. Hence $\phi(r, \theta)$ is subject to the equations

$$(\nabla^2 + k^2)\phi = 0, \quad r > a \quad,$$

$$\partial\phi/\partial r = V(\theta), \quad r = a \quad,$$

and an outgoing wave condition at infinity.

Pose the solution

$$\phi = \sum_0^\infty A_n h_n(kr) P_n(\cos\theta) \quad,$$

which satisfies the Helmholtz equation and outward wave condition. It remains to satisfy the boundary condition at $r = a$, whence

$$\sum_0^\infty A_n k\, h'_n(ka)\, P_n(\cos\theta) = V(\theta) \quad.$$

The given profile $V(\theta)$ can be expressed as a Fourier–type series

$$V(\theta) = \sum \beta_n P_n(\cos\theta) \;,$$

with β_n found, using the orthogonality relations (18.24), to be

$$\beta_n = (n + \tfrac{1}{2}) \int_0^\pi V(\theta) \sin\theta\, P_n(\cos\theta) d\theta \quad.$$

Thus the condition at $r = a$ gives $A_n = \{k\, h'_n(ka)\}^{-1} \beta_n$ and the solution for ϕ is complete.

A simple estimate follows in the high–frequency limit $ka \gg 1$ (whence $kr \gg 1$ as $r \geq a$). Use of the asymptotic form (18.26) for h_n leads to the result

$$\phi \sim (-i/k)(a/r)e^{ik(r-a)} \sum_0^\infty \beta_n P_n(\cos\theta) \quad,$$

i.e.

$$\phi \sim (-i/k)(a/r)V(\theta)\, \exp\{ik(r-a)\} \quad \text{as} \quad ka \to \infty \quad. \tag{18.28}$$

This result could also be deduced using the ray theory arguments developed later in this Chapter.

III. Two–dimensional waveguide

A compressible fluid is contained within a pair of parallel plates $y = 0$ and $y = a$, on which surfaces the potential takes the value zero (i.e. the planes are "acoustically soft"). Vibrations are forced by a line source of strength $e^{-i\omega t}$ at (x_1, y_1). Thus the velocity potential $G(\mathbf{x}, \mathbf{x}_1)$ satisfies

$$(\nabla^2 + k^2)G = \delta(x - x_1)\delta(y - y_1), \quad 0 < y_1 < a, \quad 0 \le y \le a \quad , \tag{18.29}$$

$$G = 0 \quad \text{at} \quad y = 0 \quad \text{and} \quad y = a \quad , \tag{18.30}$$

and a radiation condition as $|x| \to \infty$.

Note that the solution will be unbounded if k has one of the "resonant" values $n\pi/a$. For then the standing–wave solution $\sin ky$ satisfies the homogeneous Helmholtz equation and the boundary condition (18.30) without any forcing. It is assumed henceforth that $k \neq n\pi/a$.

The solution will be constructed in terms of solutions of the *homogeneous* Helmholtz equation with conditions (18.30). Separation of variables leads to solutions of the form

$$\left(A \sin \alpha y + B \cos \alpha y \right) \exp\left\{ \pm i(k^2 - \alpha^2)^{1/2} x \right\} \quad ,$$

where α is an arbitrary separation constant to be determined from the boundary conditions (18.29), which yield $B = 0$ and $\alpha = \alpha_n \equiv n\pi/a$. Thus one writes

$$G = \sum_1^\infty A_n \sin \alpha_n y \, \exp\left\{ +i(k^2 - \alpha_n^2)^{1/2}|x - x_1| \right\}, \quad \alpha_n = \frac{n\pi}{a} \quad , \tag{18.31}$$

which satisfies the wave equation (18.29) for $\mathbf{x} \neq \mathbf{x}_1$, the boundary conditions (18.30) and the radiation condition. For values of α_n greater than k, the square root $(k^2 - \alpha_n^2)^{1/2}$ is interpreted as $+i(\alpha_n^2 - k^2)^{1/2}$ in order that the potential is not exponentially large as $|x| \to \infty$. It remains to choose A_n to give the correct singularity at (x_1, y_1). Thus

$$(\nabla^2 + k^2)G \equiv \sum_1^\infty A_n \sin \alpha_n y \left(\frac{d^2}{dx^2} + k^2 - \alpha_n^2 \right) \exp\left\{ i\left(k^2 - \alpha_n^2\right)^{1/2}|x - x_1| \right\}$$

$$= \delta(x - x_1)\,\delta(y - y_1) \quad .$$

Multiply by $\sin \alpha_m y \equiv \sin(m\pi y/a)$ and integrate from 0 to a to get

$$\frac{a}{2} A_m \left(\frac{d^2}{dx^2} + k^2 - \alpha_m^2 \right) \exp\left\{ i\left(k^2 - \alpha_m^2\right)^{1/2}|x - x_1| \right\} = \sin \alpha_m y_1\, \delta(x - x_1) \quad ,$$

which can be integrated across $x = x_1$ to get $A_n = (ai)^{-1} (k^2 - \alpha_n^2)^{-1/2} \sin \alpha_n y_1$. Thus

$$G = -\frac{i}{a} \sum_1^\infty \frac{\sin \alpha_n y \, \sin \alpha_n y_1}{(k^2 - \alpha_n^2)^{1/2}} \exp\left\{ i \left(k^2 - \alpha_n^2\right)^{1/2} |x - x_1| \right\} \quad . \qquad (18.32)$$

For large $|x - x_1|$ some of the modes are exponentially small, so that

$$G \sim -\frac{i}{a} \sum_{n < ka/\pi} \frac{\sin \alpha_n y \, \sin \alpha_n y_1}{(k^2 - \alpha_n^2)^{1/2}} \exp\left\{ i \left(k^2 - \alpha_n^2\right)^{1/2} |x - x_1| \right\} \quad .$$

IV. Semi–infinite hard screen

The solution for a plane wave incident upon a semi–infinite screen has been given in Chapter 5 as an example of the Wiener–Hopf technique.

18.3 APPROXIMATE SOLUTIONS

As it is not usually possible to obtain exact solutions to diffraction and scattering problems unless the geometry of the scattering surface is particularly simple, it is natural to seek approximate solutions when a parameter of the problem takes limiting values (e.g. $ka \to 0$ where a is a typical obstacle dimension). One can proceed via the differential equations or else from an integral equation formulation such as was described in Chapter 2. The following problems are typical.

I. Low–frequency expansion: series in k.

As the wavenumber $k \to 0$, the Helmholtz equation formally reduces to Laplace's equation, and as a first approximation for a particular problem it is typical that the corresponding *static* problem ($k = 0$) has to be solved. Higher–order corrections involve solving the inhomogeneous Laplace equation. Typical prototype wave functions are

$$e^{ikx} = 1 + ikx - \frac{k^2}{2} x^2 + \dots \quad , \quad kx << 1 \quad ;$$

$$H_1^{(1)}(kr) \cos \theta = \frac{2i}{\pi} \cos \theta \left\{ \frac{1}{kr} - \frac{1}{2} kr \log kr + \dots \right\} , \quad kr << 1 \quad .$$

Two useful points are clear from these elementary results. Firstly, the approximations are not uniformly valid as $k \to 0$ since they obviously require kx and kr to be small: consequently there is some difficulty in assigning boundary conditions at "infinity" to individual terms in the expansions. Secondly, the expansions are not necessarily power series in k, since log k terms might be present: at the outset of a given problem it is not clear what the correct form of an expansion should be. Both these difficulties can be surmounted by the use of the method of matched expansions, and several examples are given in detail in Chapter 6. Another approach, to be followed here, involves the use of integral equations and is illustrated by the problem of scattering of plane waves by a soft sphere of radius a, with ka small.

Scattering by a soft sphere

This problem is described in detail by Morse & Feshbach (1953). A plane wave of incident potential $\phi_i = \exp(ikz)$ is scattered by a sphere of radius a on which $\phi = 0$. The governing equations are

$$(\nabla^2 + k^2)\phi = 0, \qquad r > a \quad , \tag{18.33}$$

$$\phi = 0, \qquad r = a \quad , \tag{18.34}$$

$$\phi - \phi_i \sim f(\theta)(kr)^{-1} e^{ikr} \quad \text{as} \quad r \to \infty \quad , \tag{18.35}$$

where θ is the polar angle measured from the axis $z = 0$. An application of Green's theorem to the scattered potential $\phi - \phi_i$ and the elementary source potential $-(4\pi R)^{-1} \exp(ikR)$ leads to the integral identity

$$\phi(\mathbf{x}) = e^{ikz} - \frac{1}{4\pi} \int_s R^{-1} e^{ikR} \frac{\partial \phi}{\partial n} \, d\xi \tag{18.36}$$

for $\mathbf{x}$ outside the sphere S, where $R = |\mathbf{x} - \boldsymbol{\xi}|$. Now $\exp(ikz)$ and $\exp(ikR)$ can each be expanded as a regular power series in k, valid for $|kr| << 1$. A balance of terms in the identity (18.36) indicates that ϕ has a similar expansion; thus we pose the approximation

$$\phi(\mathbf{x}) = \phi_0(\mathbf{x}) + ik\phi_1(\mathbf{x}) + \frac{(ik)^2}{2!} \phi_2(\mathbf{x}) + \dots \quad , \quad kr << 1 \quad . \tag{18.37}$$

Substitution of (18.37) into the differential equation (18.33) and surface condition (18.34) then requires

$$\nabla^2 \phi_0 = 0, \quad \nabla^2 \phi_1 = 0, \quad \nabla^2 \phi_n = n(n-1)\phi_{n-2}, \quad n \geq 2 \quad , \tag{18.38}$$

and

$$\phi_0 = 0, \quad \phi_1 = 0, \quad \phi_n = 0, \quad r = a \quad . \tag{18.39}$$

There remains the problem of determining conditions at "infinity" for the constituent potentials ϕ_n. Since (18.37) holds only for $kr \ll 1$, we can not substitute (18.37) into (18.35) and hope to get sensible information. The missing conditions will be inferred by appealing again to identity (18.36). For on substituting (18.37) into (18.36) and equating like terms in k, we find

$$\phi_0 = 1 - \frac{1}{4\pi} \int_s \frac{1}{R} \frac{\partial \phi_0}{\partial n} \, d\xi$$

$$\phi_1 = z - \frac{1}{4\pi} \int_s \left\{ \frac{1}{R} \frac{\partial \phi_1}{\partial n} + \frac{\partial \phi_0}{\partial n} \right\} d\xi$$

$$\phi_2 = z^2 - \frac{1}{4\pi} \int_s \left\{ \frac{1}{R} \frac{\partial \phi_2}{\partial n} + 2\frac{\partial \phi_1}{\partial n} + R\frac{\partial \phi_0}{\partial n} \right\} d\xi \quad ,$$

and so on. These are the integral forms for a series of static problems for $\phi_0, \phi_1, \phi_2, \ldots$ and uniquely determine the functions. We shall use them simply to extract boundary conditions at infinity by letting $r \to \infty$, to get

$$\phi_0 \to 1$$

$$\phi_1 \to z - \frac{1}{4\pi} \int_s (\partial \phi_0 / \partial n) \, d\xi$$

$$\phi_2 \to z^2 - \frac{r}{4\pi} \int \frac{\partial \phi_0}{\partial n} \, d\xi + \frac{1}{4\pi} \int a \cos \alpha \, \frac{\partial \phi_0}{\partial n} \, d\xi - \frac{2}{4\pi} \int \frac{\partial \phi_1}{\partial n} \, d\xi \quad ,$$

where α is the angle shown in the diagram and $\cos \alpha = (ar)^{-1} \xi \cdot \mathbf{x}$.

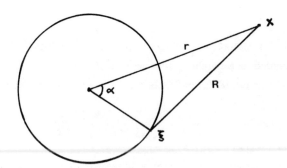

Figure 18.3

To calculate $\phi_0, \phi_1, \phi_2 \ldots$ recall that the axially symmetric solutions of Laplace's equation in spherical polars are $r^n P_n(\cos \theta)$ and $r^{-n-1} P_n(\cos \theta)$, where P_n denotes the Legendre polynomial of degree n.

(i) Calculation of ϕ_0 : This potential is specified by the conditions $\nabla^2 \phi_0 = 0$, $\phi_0 = 0$ when $r = a$ and $\phi_0 \to 1$ as $r \to \infty$. The solution is

$$\phi_0 = 1 - a/r \quad .$$

(ii) Calculation of ϕ_1 : Here $\nabla^2 \phi_1 = 0$, $\phi_1 = 0$ when $r = a$ and

$$\phi_1 \sim z - \frac{1}{4\pi} \int_s \frac{\partial \phi_0}{\partial n} \, d\xi = r \cos\theta - a \quad \text{as} \quad r \to \infty \quad .$$

The solution is

$$\phi_1 = \cos\theta \, (r - a^3/r^2) - a(1 - a/r) \quad ,$$

having noted that the relevant harmonic functions are $r \cos\theta$, $r^{-2}\cos\theta$, 1, $1/r$.

(iii) Calculation of ϕ_2: The governing differential equation is (see (18.38))

$$\nabla^2 \phi_2 = 2\phi_0 = 2 - 2a/r \quad , \tag{18.40}$$

with $\phi_2 = 0$ at $r = a$, and

$$\phi_2 \sim z^2 - \frac{r}{4\pi} \int_s \frac{\partial \phi_0}{\partial n} \, d\xi + \frac{1}{4\pi} \int_s a \cos\alpha \, \frac{\partial \phi_0}{\partial n} \, d\xi - \frac{2}{4\pi} \int_s \frac{\partial \phi_1}{\partial n} \, d\xi \quad ,$$

$$= r^2 \cos^2\theta - ar + 0 + 2a^2 \quad \text{as} \quad r \to \infty \quad .$$

Now (18.40) has a particular solution $\phi_2 = \frac{1}{3} r^2 - ar$, to which we add the harmonic functions $(3\cos^2\theta - 1)(Ar^2 + Br^{-3}) + \cos\theta(Cr + D/r^2) + E + F/r$ noting that $P_2(\cos\theta) = \frac{1}{2}(3\cos^2\theta - 1)$. Thus

$$\phi_2 = \frac{1}{3}(3\cos^2\theta - 1)\left(r^2 - \frac{a^5}{r^3}\right) + 2a^2 - \frac{4}{3}\frac{a^3}{r} + \frac{1}{3}r^2 - ar$$

on accommodating the boundary conditions at infinity and at $r = a$.

(iv) Amplitude at infinity: The approximation (18.37) does not give the far–field solution, and for this we turn once again to the useful identity (18.36) which gives

$$\phi_s \sim -\frac{e^{ikr}}{4\pi r} \int_s \frac{\partial \phi}{\partial n} \, e^{-ik a \cos\alpha} d\xi \quad \text{as} \quad r \to \infty \quad .$$

Thus

$$f(\theta) = -\frac{k}{4\pi} \int_s e^{-ik\,a\,\cos\alpha} \frac{\partial\phi}{\partial n}\, d\xi \quad, \tag{18.41}$$

which is an *exact* expression for the far–field amplitude in terms of the distribution $\partial\phi/\partial n$ on S. This latter function is given approximately by (18.37): on substituting into (18.41) and evaluating the integrals, we find

$$f(\theta) = -ka + ik^2 a^2 + \frac{1}{3}\,k^3 a^3(2 - 3\cos\theta) + \cdots \quad. \tag{18.42}$$

II. High-frequency approximations: ray theory

As a prototype for the type of expansion appropriate for high frequencies, consider the source potential

$$\phi = -(4\pi r)^{-1} e^{ikr} \quad.$$

The first point to note is that ϕ can *not* be simply expanded in inverse powers of k, when $k \to \infty$. The potential consists of a slowly varying amplitude factor $-(4\pi r)^{-1}$ together with a rapidly oscillating term e^{ikr}. The wavefronts are the surfaces of constant phase, $r = $ constant. The potential varies rapidly in the direction perpendicular to the wavefronts but not tangential to the wavefronts. Similar remarks hold for the plane–wave potential

$$\phi = A e^{ikx} \quad,$$

where the wavefronts are given by $x = $ constant.

For a general solution of the Helmholtz equation it is argued that within a small region of space, for large k (i.e. small wavelength) there is but little variation in conditions over a distance of one wavelength. It is expected then that the potential will behave locally like a plane wave, i.e.

$$\phi \sim A(\mathbf{x}, k) e^{iku(\mathbf{x})} \quad \text{as} \quad k \to \infty \quad, \tag{18.43}$$

where the amplitude A and phase u are slowly varying functions. More generally, one writes

$$\phi \sim \exp\{iku(\mathbf{x})\}\, \Sigma\, k^{-n} A_n(\mathbf{x}) \quad, \tag{18.44}$$

this being the direct analogue of the WKB method for ordinary differential equations. The functions $A_n(\mathbf{x})$ and $u(\mathbf{x})$ are found by substituting the series (18.44) into the equations for ϕ and equating powers of k. It may sometimes be preferable to express u as an inverse series $u \sim u_0(x) + k^{-1}u_1(\mathbf{x}) + k^{-2}u_2(\mathbf{x}) + \ldots$, which may produce a more uniformly valid approximation, but this will not be pursued here. The present approach amounts to absorbing the terms $\exp\{iu_1 + ik^{-1}u_2 + \ldots\}$ into $A(\mathbf{x}, k)$.

Example: Radiation from a vibrating circular cylinder

A circular cylinder of radius a vibrates with prescribed normal velocity $V(\theta)\exp(-i\omega t)$. The governing equations are

$$(\nabla^2 + k^2)\phi = 0, \quad r \geq a \ ,$$

$$\partial\phi/\partial r = V(\theta), \quad r = a \ ,$$

$$\phi \sim f(\theta)(kr)^{-1/2}e^{ikr} \quad \text{as} \quad r \to \infty \ .$$

Put $\phi(\mathbf{x}; k) = A(\mathbf{x}; k)\exp(iku(\mathbf{x}))$ with $A \sim A_0 + k^{-1}A_1 + k^{-2}A_2 + \ldots$. Note that

$$\frac{\partial\phi}{\partial x_i} = \left(\frac{\partial A}{\partial x_i} + ikA\frac{\partial u}{\partial x_i}\right)e^{iku}$$

and

$$\frac{\partial^2\phi}{\partial x_i^2} = \left(\frac{\partial^2 A}{\partial x_i^2} + ikA\frac{\partial^2 u}{\partial x_i^2} + 2ik\frac{\partial A}{\partial x_i}\frac{\partial u}{\partial x_i} - k^2 A\left(\frac{\partial u}{\partial x_i}\right)^2\right)e^{iku} \quad .$$

It follows that the Helmholtz equation takes the form

$$\nabla^2 A + ikA\nabla^2 u + 2ik\nabla A \cdot \nabla u + k^2 A(1 - (\nabla u)^2) = 0 \quad . \tag{18.45}$$

The boundary condition becomes

$$\frac{\partial A}{\partial r} + ikA\frac{\partial u}{\partial r} = V(\theta) \quad \text{and} \quad u = 0 \quad \text{at} \quad r = a \ , \tag{18.46}$$

and

$$u \sim r \quad \text{as} \quad r \to \infty \quad . \tag{18.47}$$

Equating powers of k^2 in (18.45) gives

$$(\nabla u)^2 = 1 \quad ,$$

and use of the boundary conditions on u requires

$$u = r - a \quad . \tag{18.48}$$

Thus (18.46) reduces to

$$\nabla^2 A + ikr^{-1}A + 2ik\,\partial A/\partial r = 0 \quad .$$

Equating powers of k in the differential equation (18.45) and boundary condition (18.46) leads to the result that $A_0 \equiv 0$ and

$$2\,\partial A_1/\partial r + A_1/r = 0 \quad \text{with} \quad A_1 = -iV(\theta) \quad \text{when} \quad r = a \quad .$$

The solution is $A_1 = -i(a/r)^{1/2}V(\theta)$ and the leading term for ϕ is

$$\phi \sim -\frac{i}{k}\left(\frac{a}{r}\right)^{1/2}V(\theta)e^{ik(r-a)} \quad . \tag{18.49}$$

Higher–order terms A_n are given inductively by

$$2\frac{\partial A_n}{\partial r} + \frac{1}{r}A_n = i\nabla^2 A_{n-1}, \quad \text{with} \quad A_n = i\frac{\partial A_{n-1}}{\partial r} \quad \text{at} \quad r = a \quad .$$

The above method is plausible, but not rigorous since the series (18.44) is usually not convergent. It is assumed, however, that the series is the asymptotic expansion for $\phi(\mathbf{x}; k)$. The leading terms do coincide with results obtained rigorously from integral equations and for particular special cases like the circular cylinder and infinite plane. The method can be used successfully for any convex bodies. The leading term is expected to give a good approximation except near singularities (near a source, for example) and near "shadow" boundaries. A full account of geometrical acoustics has been developed by Keller (1953).

18.4 MATCHED ASYMPTOTIC EXPANSIONS: DUCT PROBLEM

Several examples have been given on the method in Chapter 6, which gave a detailed description of the procedure in a variety of commonly occurring types of expansion.

The method is described here for the relatively simple two–dimensional problem of an incident plane wave $\phi_i = \exp(ikx')$ travelling down a rigid duct from $x' = -\infty$.

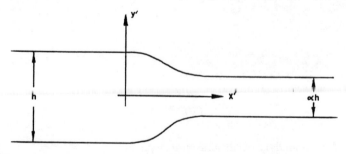

Figure 18.4

The duct width changes from h to αh, where α is of order unity and may be greater than unity or less than unity. It is assumed that the transition from h to αh takes place over a distance x' that is of order h, and an approximate solution is sought for small values of $\epsilon = kh$; that is, the wavelength is much greater than the duct width. The velocity potential $\phi'(x', y')$ is subject to the equations

$$(\partial^2/\partial x'^2 + \partial^2/\partial y'^2 + k^2)\phi' = 0 \quad \text{in the fluid} \quad , \tag{18.50}$$

$$\partial\phi'/\partial n = 0 \quad \text{on the duct walls} \quad , \tag{18.51}$$

$$\phi' \sim T \exp(ikx') \quad \text{as} \quad x' \to +\infty \quad , \tag{18.52}$$

$$\phi' \sim \exp(ikx') + R \exp(-ikx') \quad \text{as} \quad x' \to -\infty \quad . \tag{18.53}$$

At distances $|\mathbf{x}'|$ much less than the wavelength $2\pi/k$, we anticipate that the wave equation will reduce to Laplace's equation and the flow in this "inner region" close to $x' = 0$ will be quasi–static and solved by conformal transformation, for example. In this region then, the significant lengthscale is the duct width h, so we define inner variables X and Y by the relations

$$X = x'/h, \quad Y = y'/h, \quad \phi = \Phi(X, Y) \quad , \tag{18.54}$$

in terms of which the governing equations (18.50) and (18.51) become

$$(\nabla^2 + \epsilon^2)\Phi = 0, \quad \text{with} \quad \partial\Phi/\partial n = 0 \quad \text{on the duct} \quad . \tag{18.55}$$

In the limit $\epsilon \to 0$, the equation reduces to Laplace's equation as expected. Boundary conditions as $X \to \pm\infty$ have to be inferred by matching with appropriate outer approximations.

For in the "outer regions" $x' \gg h$ and $x' \ll -h$, the wave motion will be essentially one dimensional, since all other disturbances in a parallel–walled duct decrease exponentially as $\exp(-\text{constant } |x'|/h)$ and are negligible when $|x'| \gg h$. Thus on defining outer variables

$$x = kx' \quad \text{and} \quad y = ky' \quad , \tag{18.56}$$

the appropriate approximations are

$$\phi_R \sim T e^{ix} \quad \text{in the right outer region} \quad x \gg \epsilon \quad , \tag{18.57}$$

$$\phi_L \sim e^{ix} + R e^{-ix} \quad \text{in the left outer region} \quad x \ll -\epsilon \quad . \tag{18.58}$$

The coefficients R and T remain to be found and have the asymptotic forms

$$T \sim T_0 + \ldots \quad \text{and} \quad R \sim R_0 + \ldots \quad \text{as} \quad \epsilon \to 0 \quad .$$

In the notation of Chapter 6 the zeroth order right outer approximation is

$$\phi_R^{(0)} = T_0 e^{ix} \quad .$$

On rewriting in terms of the inner variable $X = x/\epsilon$ and letting $\epsilon \to 0$ with X fixed, this gives an inner limit of the outer approximation; thus

$$\phi_R^{(0)} = T_0 e^{i\epsilon X} = T_0 (1 + i\epsilon X - \tfrac{1}{2} \epsilon^2 X^2 + \ldots) \quad .$$

In particular, on truncating at order ϵ,

$$\phi_R^{(0,1)} = T_0 (1 + i\epsilon X) \quad .$$

Similarly $\phi_L^{(0,1)} = (1 + R_0) + i\epsilon X (1 - R_0)$ and the matching principles $\phi_R^{(0,1)} = \Phi^{(1,0)} (X \to +\infty)$, $\phi_L^{(0,1)} = \Phi^{(1,0)} (X \to -\infty)$ suggests the inner development

$$\Phi \sim \Phi^{(1)} = \Phi_0 + \epsilon \Phi_1 \quad . \tag{18.59}$$

Substitution into (18.55) shows that Φ_0 and Φ_1 are both harmonic, with zero normal derivatives on the duct, and with conditions

$$\Phi_0 \sim T_0, \quad \Phi_1 \sim i T_0 X \quad \text{as} \quad X \to +\infty \quad , \tag{18.60}$$

$$\Phi_0 \sim (1 + R_0), \quad \Phi_1 \sim i(1 - R_0)X \quad \text{as} \quad X \to -\infty \quad . \tag{18.61}$$

The problem for Φ_0 has the trivial solution $\Phi_0 = \text{constant}$, and in particular

$$T_0 = (1 + R_0) \quad .$$

The problem for Φ_1 corresponds to a steady incompressible flow down a duct that changes from width unity to α as X changes from $-\infty$ to $+\infty$.

Define the particular solution $\Phi_p(X, Y)$ which has far–field behaviour

$$\Phi_p = X + o(1) \quad \text{as} \quad X \to +\infty \quad . \tag{18.62}$$

It follows that

$$\Phi_p \sim \alpha X + K + o(1) \quad \text{as} \quad X \to -\infty \quad , \tag{18.63}$$

where the coefficient α is necessary to ensure continuity of mass flow: the constant K is *not* arbitrary but is a definite constant that depends on the local geometry. If the geometry consists of straight lines K can be calculated using a Schwarz–Christoffel transformation; here we can consider K to be known in principle. Noting the boundary condition (18.60), Φ_1 can be expressed as

$$\Phi_1 = iT_0\Phi_p + A \quad , \tag{18.64}$$

where A has to be determined by matching. The formulae (18.63) and (18.61) combine with (18.64) to show that

$$\alpha T_0 = 1 - R_0 \quad ;$$

hence

$$R_0 = \frac{1-\alpha}{1+\alpha} \quad \text{and} \quad T_0 = \frac{2}{1+\alpha} \quad . \tag{18.65}$$

Evidently T and R depend only on the duct width ratio α, to leading order, and are independent of the precise geometry: the leading term (18.65) can be obtained by a much simpler argument (see Ffowcs Williams & Dowling 1983). An improved estimate for R and T requires consideration of the next order inner expansion. The form of (18.63), together with (18.59) suggests the developments

$$\phi_R \sim \phi_R^{(1)} = (T_0 + \epsilon T_1)e^{ix}, \quad \phi_L \sim \phi_L^{(1)} = e^{ix} + (R_0 + \epsilon R_1)e^{-ix} \tag{18.66}$$

and hence

$$\Phi \sim \Phi^{(2)} = \Phi_0 + \epsilon\Phi_1 + \epsilon^2\Phi_2 \quad . \tag{18.67}$$

In order to accomplish the matching requirements $\phi^{(1,2)} = \Phi^{(2,1)}$ at both left and right regions, note that

$$\phi_R^{(1,2)} - T_0 + \epsilon(T_1 + iT_0X) + \epsilon^2(iT_1X - \tfrac{1}{2}T_0X^2) \tag{18.68}$$

and

$$\phi_L^{(1,2)} = (1+R_0) + \epsilon(R_1 + iX - iR_0X) + \epsilon^2(-iR_1X - \tfrac{1}{2}X^2 - \tfrac{1}{2}R_0X^2) \quad . \tag{18.69}$$

The coefficients of 1, ϵ and ϵ^2 in expression (18.68) and (18.69) provide boundary conditions as X tends respectively to $+\infty$ and $-\infty$ for the three potentials Φ_0, Φ_1 and Φ_2. In particular, the behaviour of Φ_1 given by (18.62) — (18.64) shows that $T_1 = A$ and $R_1 = iT_0K + A$; on eliminating A

$$R_1 - T_1 = iT_0K \quad , \tag{18.70}$$

where K is known in principle and depends on the geometry. We need one more relation between R_1 and T_1, to be obtained from the inner potential Φ_2. Substitution of expansion (18.67) into the differential equation (18.55) gives

$$\nabla^2 \Phi_2 = -\Phi_0 = -T_0 \quad ,$$

with $\partial \Phi_2 / \partial n = 0$ on the duct walls. Conditions at infinity come from matching: specifically, formulae (18.68) and (18.69) show that

$$\begin{aligned}
\Phi_2 &\sim -\tfrac{1}{2} T_0 X^2 + i T_1 X \quad \text{as} \quad X \to +\infty \quad , \\
\Phi_2 &\sim -\tfrac{1}{2}(1 + R_0) X^2 - i R_1 X \quad \text{as} \quad X \to -\infty \quad ,
\end{aligned} \tag{18.71}$$

and since $(1 + R_0) = T_0$ the quadratic term is equal to $-\tfrac{1}{2} T_0 X^2$ at both ends. Some crucial information can be inferred without detailed solution of the problem for Φ_2, by appealing to the integral identity

$$\int \nabla^2 \Phi_2 \, dS = \int (\partial \Phi_2 / \partial n) \, ds \tag{18.72}$$

applied to the region S inside the duct and bounded by distant control surfaces at $X = \pm N$, where N is large enough to allow the distant forms (18.71) to be used. Now $\nabla^2 \Phi_2 = -T_0$ is constant so the left–hand side of the identity (18.72) is equal to $-T_0$ times the area

$$= -T_0(N\alpha + N + \Delta) \quad ,$$

where Δ is the shaded area in Figure 18.5. As for the right–hand side of (18.72), there is no contribution from the duct walls where $\partial \Phi_2 / \partial n = 0$. Hence on using (18.71), the identity (18.72) gives

$$i T_1 \alpha + i R_1 = -T_0 \Delta \quad . \tag{18.73}$$

Solving for T_1 and R_1 between (18.70) and (18.73), and using the result (18.65) for T_0, we find

$$T_1 = \frac{2i(\Delta - K)}{(\alpha + 1)^2} \quad \text{and} \quad R_1 = \frac{2i(\Delta + K\alpha)}{(\alpha + 1)^2} \quad . \tag{18.74}$$

Evidently the correction terms ϵT_1 and ϵR_1 to the elementary results (18.65) depend on the local geometry through the area parameter Δ and through the number K that is defined implicitly (see (18.63)) by the potential function Φ_p.

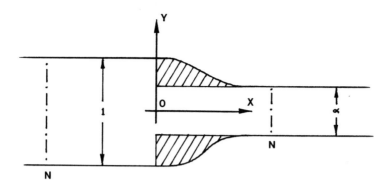

Figure 18.5

REFERENCES

Dowling, A.P. & Ffowcs Williams, J.E. (1983). Sound and Sources of Sound. Ellis Horwood, Chichester.

Keller, J.B. (1953). Proc. Symp. Microwave Optics. McGill University.

Morse, P.M. & Feshbach, H. (1953). Methods of Theoretical Physics, Part I. McGraw–Hill, New York.

Watson, G.N. (1919). The diffraction of electric waves by the earth. Proc. Roy. Soc. Lond. A95:83–99.

Watson, G.N. (1966). A Treatise on the Theory of Bessel Functions. Cambridge University Press.

19. INVERSE SCATTERING

19.1 INTRODUCTION

A typical scattering problem in acoustics concerns the effect on an incident sound field, such as a time–periodic plane wave of velocity potential

$$\phi_i = \Re\left\{\exp\{ik\boldsymbol{w}\cdot\boldsymbol{x} - \omega t\}\right\} \quad , \tag{19.1}$$

by the introduction of a scattering body. The incident field might alternatively be due to a source or a distribution of sources.

The total velocity potential will be written as $\Re\{\phi(\boldsymbol{x})\exp(-i\omega t)\}$ and the time factor $\exp(-i\omega t)$ will henceforth be suppressed. It is sometimes convenient to express ϕ as a sum

$$\phi = \phi_i + \phi_s \quad , \tag{19.2}$$

in which the "scattered" potential ϕ_s accounts for the presence of the scattering body. It satisfies the Helmholtz wave equation

$$(\nabla^2 + k^2)\phi_s = 0 \tag{19.3}$$

outside the scattering surface S, where $k = \omega/c$, $c = $ sound speed; ∇^2 denotes the two– or three–dimensional Laplacian operator, as appropriate.

On S, a fairly general boundary condition has the form

$$\frac{\partial \phi}{\partial n} - i\gamma k\phi = 0 \quad , \tag{19.4}$$

where γ is a given smooth function. We shall concentrate on the two special cases of a *hard* surface ($\gamma = 0$) on which $\partial\phi/\partial n = 0$ and the *soft* surface ($\gamma = \infty$) on which $\phi = 0$.

Finally there is a radiation condition at infinity, to ensure that scattered waves travel outwards; this condition can be expressed as

$$r(\partial \phi_s/\partial r - ik\phi_s) \to 0 \quad \text{as} \quad r = |x| \to \infty$$

for the three–dimensional case, and

$$r^{1/2}(\partial \phi_s/\partial r - ik\phi_s) \to 0 \quad \text{as} \quad r \to \infty$$

in two dimensions. These conditions are equivalent to requiring the distant fields ϕ_s to be of the form (see Chapter 18)

$$\phi_s \sim (kr)^{-1} e^{ikr} f(\eta) \quad \text{in three dimensions} \tag{19.5}$$

$$\phi_s \sim (kr)^{-1/2} e^{ikr} f(\eta) \quad \text{in two dimensions} \;, \tag{19.6}$$

where $\eta = x/|x|$ is the direction in which $r \to \infty$. In either case, f is called the "far–field amplitude" or the "directivity pattern". It obviously depends on the wavenumber k and on the direction w of the plane wave (or on the nature of any other incident field); thus one should write

$$f = f(w, \eta, k) \;, \tag{19.7}$$

but the arguments w and k will sometimes be suppressed where there is no risk of ambiguity. For a given scatterer, with its associated boundary condition, f is determinate in principle. Its evaluation is often the main aim of a calculation.

Inverse Problem

Inverse scattering poses the problem the other way around. That is to say, given an incident field and given the directivity pattern f (at a given frequency, say), find the shape of the scatterer S. An excellent comprehensive review is given by Sleeman (1982). The present chapter will discuss just a few relatively simple results.

We start with some preliminary results in two dimensions. Separation of variables leads to the conclusion that a sum

$$\phi_s = \sum_{-\infty}^{\infty} i^n A_n \, H_n^{(1)}(kr) \, e^{in\theta} \tag{19.8}$$

satisfies the Helmholtz Equation (19.3) and radiation condition at infinity, provided the A_n are such that (19.8) converges. Here $H_n^{(1)}(kr) = J_n(kr) + iY_n(kr)$ is the

Hankel function of the first kind, and the factor i^n has been inserted simply for later algebraic convenience.

Now if the cross section S of our two–dimensional scatterer is finite in extent, a general expansion theory for wave functions ensures that ϕ_s can be expanded in the form (19.8), where the coefficients A_n are determined by S. The sum (19.8) will converge outside some circle $r > a$ where a is such that S lies within the circle $r = a$. In particular, if we formally let $r \to \infty$ in the sum and use the known asymptotic behaviour

$$H_n^{(1)}(kr) \sim (2/\pi kr)^{1/2} \exp\{ikr - in\,\pi/2 - i\pi/4\} \quad ,$$

we get

$$\phi_s \sim (kr)^{-1/2} e^{ikr} f(\theta) \quad \text{as} \quad kr \to \infty \quad ,$$

where

$$f(\theta) = (2/\pi)^{1/2}\, e^{-i\pi/4} \sum_{-\infty}^{\infty} A_n\, e^{in\,\theta} \tag{19.9}$$

is the directivity pattern. If we know $f(\theta)$ at a given wavenumber k, then the Fourier coefficients A_n are known in principle and can be computed. Thus the scattered potential ϕ_s is given at once by formula (19.8) for r greater than some value a (which is as yet unknown). The sum (19.8) will usually diverge inside this circle and therefore fails to represent ϕ_s. It has to be continued analytically in order to obtain values for ϕ_s inside $r = a$. The process is similar to that of the analytic continuation of functions of a complex variable, and the same terminology is used in the present context.

The corresponding analysis for finite three–dimensional scatterers is similar. In this case ϕ_s has the representation

$$\phi_s = \sum_{n,m} A_{nm} \, \exp\left(\frac{1}{2}(n+1)\pi i\right) h_n^{(1)}(kr)\, S_{nm}(\boldsymbol{\eta}), \qquad r > a \quad , \tag{19.10}$$

where $h_n^{(1)}$ is a spherical Bessel function, S_{nm} denotes a spherical (surface) harmonic, and $\boldsymbol{\eta}$ $\boldsymbol{x}/|\boldsymbol{x}|$, a is such that the scatterer lies within the sphere $r = a$. The far field amplitude is obtained from the asymptotic formula $h_n(kr) \sim (kr)^{-1} \exp\{ikr - \frac{1}{2}(n+1)\pi i\}$, whence

$$f(\boldsymbol{\eta}) = \sum_{n,m} A_{nm}\, S_{nm}(\boldsymbol{\eta}) \quad . \tag{19.11}$$

Knowledge of $f(\boldsymbol{\eta})$ determines the coefficients A_{nm}, whence (19.10) gives ϕ_s for $r > a$ with interior values obtained by analytic continuation.

19.2 METHOD OF IMBRIALE & MITTRA

This method uses the above analysis to provide a constructive method for finding the shape of a two–dimensional soft scatterer as follows (for details, see Imbriale & Mittra 1970; Mittra 1973).

Suppose firstly that S is known from the outset to be convex. It is given that $\phi = \phi_i + \phi_s$ vanishes on some convex curve to be found. The far–field amplitude $f(\theta)$ is given, with respect to some origin O. Without loss of generality we may take the coordinate system to be such that the incident plane wave propagates parallel to the x–axis; thus

$$\phi_{inc} = e^{ikx} \quad .$$

Given the amplitude $f(\theta)$, one can compute the Fourier coefficients A_n of formula (19.9); to be exact, a large finite number of them can be calculated. Hence

$$f(\theta) \approx (2/\pi)^{1/2} e^{-i\pi/4} \sum_{-N}^{N} A_n\, e^{in\,\theta}, \quad 0 < \theta < 2\pi \quad , \tag{19.12}$$

where N is large enough to ensure a good approximation to $f(\theta)$. In most of what follows, the limits $-N$ and N will not be written explicitly.

According to (19.8) and (19.2), the total potential for $r > a$ (with a as yet unknown) has the representation

$$\phi = e^{ikr\cos\theta} + \Sigma\, i^n A_n\, H_n^{(1)}(kr) e^{in\,\theta} \quad , \tag{19.13}$$

where the A_n are now known. If this expression is computed for a succession of decreasing values of r, there will eventually be a point on a circle of radius a_0 at which $\phi = 0$, this being one point of the scatterer S.

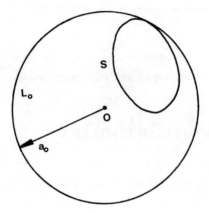

Figure 19.1

At this stage we know that S is inside the circle $L_0(r = a_0)$ and we know one point where S touches L_0. It is not possible to take smaller values of r to find other points of S, since the sum (19.13) will generally diverge inside L_0. Other points of S are now found by shifting the origin to a different position O_1, given by $(r, \theta) = (r_1, \theta_1)$, say. Let (r', θ') denote polar coordinates with respect to the new origin. Since $r' \sim r - r_1 \cos(\theta - \theta_1)$ and $\theta' \sim \theta$ as $r \to \infty$, the new directivity pattern $f'(\theta)$ relative to the origin O_1, is

$$f'(\theta) = f(\theta) \exp\{ikr_1 \cos(\theta - \theta_1)\}$$

$$= (2/\pi)^{1/2} e^{-i\pi/4} \sum_{-\infty}^{\infty} A'_m e^{im\theta} \quad .$$

The A'_m are related to A_n by the formula

$$\sum_{-\infty}^{\infty} A'_m e^{im\theta} = \sum_{-\infty}^{\infty} A_n e^{in\theta} \exp\left(ikr_1 \cos(\theta - \theta_1)\right)$$

$$= \sum_{-\infty}^{\infty} A_n e^{in\theta} \sum_{-\infty}^{\infty} J_p(kr_1) i^p e^{ip(\theta - \theta_1)} \quad .$$

Hence

$$A'_m = \sum_{-\infty}^{\infty} A_n i^{(m-n)} J_{m-n}(kr_1) \exp\left(-i(m - n)\theta_1\right) \quad ,$$

which can be computed from the known values of A_n. Thus a large finite number of the A'_m are known, and we have a different representation

$$\phi_s = \Sigma\, A'_m\, i^m H^{(1)}_m(kr')e^{im\,\theta'}, \quad r' > a_1 \quad , \tag{19.14}$$

which will converge outside a different circle $L_1(r' = a_1)$ centred at O_1, where L_1 contains S. In particular, by computing $\phi = \phi_i + \phi_s$ for successive smaller values of r' until ϕ vanishes at some point with $r' = a_1$, we determine L_1 and another point of contact with S (see Figure 19.2). The process can be continued successively to limit further the possible region occupied by S and to find further points on S.

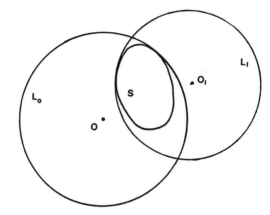

Figure 19.2

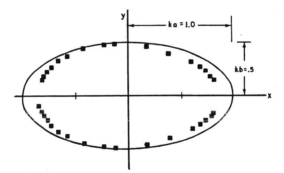

Figure 19.3 (©1970 IEEE)

Figure 19.3, taken from Imbriale & Mittra (1970), shows their results for this procedure applied to scattering data for an elliptic cylinder $x^2/a^2 + y^2/b^2 = 1$, with $ka = 1$, $kb = 1/2$. In practice, the amplitude function $f(\theta)$ may not be known for all θ, but just for a limited range $\theta_1 < \theta < \theta_2$. To deal with this, Imbriale & Mittra suggest that $f(\theta)$ be approximated by a sum $P_N(\theta) = \sum\limits_{-N}^{N} \overline{A}_n\, e^{in\,\theta}$, with coefficients

$\overline{A_n}$ chosen to minimize the error indicator

$$\int_{\theta_1}^{\theta_2} |f(\theta) - P_N(\theta)|^2 \, d\theta \quad .$$

Their results are good for the example of a circle if $\theta_1 = -135°$, $\theta_2 = +135°$, and are reasonable even in the limited range $\theta_1 = -60°$, $\theta_2 = +60°$.

It is clear that the process described above will not determine S if there are several scatterers or if S is not convex. For example, if S consists of two circles, the procedure would provide only the information that S lies within the shaded region of Figure 19.4.

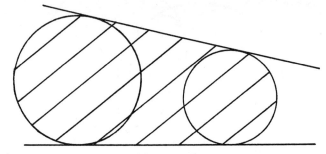

Figure 19.4

Imbriale & Mittra (1970) provide a straightforward extension to deal with such cases. This is based on the fact that any solution of the Helmholtz equation that is regular at and near some origin O_1, can be expanded in the form

$$\phi_s = \sum_{-\infty}^{\infty} C_n \, J_n(kr') \, e^{in \, \theta'}, \quad r' < b_1 \quad , \tag{19.15}$$

where b_1 is the distance from O_1 to the nearest singularity. In the present context, this ensures that (19.15) converges *within* the circle K_1 that just touches S. The difference between the exterior expansion (19.8) and the interior expansion (19.15) is that $H_n^{(1)}(kr)$ satisfies the radiation condition but is singular at $r' = 0$, whilst $J_n(kr')$ is regular at $r' = 0$ but does not correspond to outward–travelling waves at infinity.

Thus Equation (19.15) is valid inside the circle K_1. The coefficients C_n can be obtained in terms of A_n, provided there is an overlap of the inside of K_1 and the outside of L_0; this is certainly ensured if O_1 is chosen to be outside L_0. Comparing the two expansions (19.8) and (19.15), and using an addition theorem for Bessel functions (Watson 1941), namely

$$H_n^{(1)}(kr)\,e^{in(\theta-\theta_1)} = \sum_{m=-\infty}^{\infty} H_{n-m}^{(1)}(kr_1)\,J_m(kr')\,e^{im(\theta'-\theta_1)} \quad ,$$

one finds

$$C_m = \sum_{n=-\infty}^{\infty} i^n\,A_n\,H_{n-m}^{(1)}(kr_1)\,\exp\!\left(i(n-m)\theta_1\right) \quad . \tag{19.16}$$

Thus the C_m are known in principle (and a large finite number are known approximately in practice), whence the representation (19.15) can be considered as being known. Now we take the increasing values of r' until $\phi_i + \phi_s$ vanishes at some point $r' = b_1$, $\theta' = \theta_1'$. This ensures that S lies *outside* the circle $K_1(r' = b_1)$ and gives a point of contact with K_1. The process can be extended by choosing another origin O_2 given by $(r', \theta') = (r_2, \theta_2)$ choosing O_2 to lie inside K_1. If (r'', θ'') denote polar coordinates with respect to O_2, then we seek to continue the limited representation (19.15) by re–expanding about O_2, viz.

$$\phi_s = \sum_{-\infty}^{\infty} C_n''\,J_n(kr'')\,e^{in\,\theta''}, \qquad r'' < b_2 \quad , \tag{19.17}$$

for some value b_2 (as yet unknown). A comparison of expressions (19.15) and (19.17), both valid in some overlap region, and the addition theorem

$$J_n(kr')\,e^{in(\theta'-\theta_2)} = \sum_{m=-\infty}^{\infty} J_{n-m}(kr_2)\,J_m(kr'')\,e^{im(\theta''-\theta_2)}$$

leads to the result

$$C_m'' = \sum_{n=-\infty}^{\infty} C_n\,J_{n-m}(kr_2)\,e^{-i(n-m)\theta_2} \quad , \tag{19.18}$$

and the representation (19.17) can be regarded as known. Taking successively increasing values for r'' until $\phi_i + \phi_s$ vanishes at some point on a circle K_2, we obtain another point of S and also the information that S is outside K_2. The process can be repeated indefinitely in principle to get a series of circles $K_1, K_2, \ldots$ outside of which S must lie (Figure 19.5).

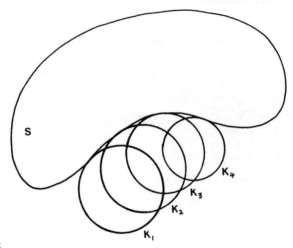

Figure 19.5

There are formidable numerical problems associated with this technique, since sums like Equations (19.15) — (19.18) have to be truncated; and C_n in Equation (19.16), for example, is sensitive to the values A_m at large m. A full discussion is given by Imbriale & Mittra (1970) who give results for a pair of circles of radii a and separation $2b$ between their centres, with $ka = 1$ and $kb = 2.5$.

The methods described above seem to give reasonable reconstructions for modest values of k times a characteristic obstacle dimension d. But such procedures, based on analytical continuation ideas, are inherently ill–posed and subject to numerical instabilities. The following approach is an attractive alternative.

19.3 OPTIMIZATION METHOD

The method described in Colton & Monk (1987), for the scattering from an acoustically soft surface S, recasts the problem as one in optimization. It overcomes the numerical difficulties associated with the analytical continuation procedures described above, and runs briefly as follows.

Suppose that a plane wave, with potential given by (19.1), is incident upon a scattering surface S that encloses the origin. The total potential $\phi = \phi_i + \phi_s$ satisfies the Helmholtz equation (19.3) and the boundary condition

$$\phi(\boldsymbol{x}) = 0, \quad \text{for} \quad \boldsymbol{x} \quad \text{on} \quad S \ . \tag{19.19}$$

It is assumed firstly that k^2 is not one of the interior eignevalues of the interior Dirichlet problem. It is also assumed that the scatterer is "starlike"; that is, the surface S can be represented in the form

$$x = r_s(\eta)\eta \quad,$$

where η is the unit vector in direction x and r_s is single valued.

Given the far–field amplitude function $f(w, \eta, k)$, defined by Equation (19.5) at fixed k, the problem is to determine the function $r_s(\eta)$ that specifies the scattering surface S. Colton & Monk (1987) relate the amplitude $f(w, \eta, k)$ to a function $\psi(x, k)$ that corresponds to the (scattered) potential inside S induced by a point source at the origin: thus

$$(\nabla^2 + k^2)\psi(x) = 0, \qquad x \text{ inside } S \quad, \tag{19.20}$$

with

$$\psi(x) = (1/4\pi r_s)\exp(ikr_s), \quad x \text{ on } S \quad. \tag{19.21}$$

Now an application of Green's formula to the potential $\phi(x)$ and the Green's function $-(4\pi R)^{-1}\exp(ikR)$, with $R = |x - y|$, in the region outside S, leads to the identity

$$\phi(y) = \phi_i(y) - \frac{1}{4\pi}\int_S R^{-1}e^{ikR}\frac{\partial\phi}{\partial n}(x)\,dx \quad,$$

where n denotes the outward normal from S. It follows that the far–field amplitude f has the representation

$$f(w, \eta, k) = -(k/4\pi)\int_S \frac{\partial\phi}{\partial n}(x)\exp(-ik\eta \cdot x)\,dx \quad. \tag{19.22}$$

Now define S_1 to be the sphere of unit radius and centre at the origin. Identity (19.22) can be multiplied by a function $g(\eta)$ (to be chosen) and integrated with respect to η over the unit sphere S_1 to get

$$-(4\pi/k)\int_{S_1} f(w, \eta, k)g(\eta)\,d\eta = \int_S \frac{\partial\phi}{\partial n}(x)\,\psi(x)\,dx \quad, \tag{19.23}$$

where

$$\psi(x) = \int_{S_1} g(\eta)\exp(-ik\eta \cdot x)\,d\eta \quad, \tag{19.24}$$

satisfies the Helmholtz equation if the kernel function $g(\eta)$ is suitably smooth. Functions of the form (19.24) are called Herglotz wave functions, with Herglotz kernel $g(\eta)$. It is now supposed that the domain inside S is such that the interior potential $\psi(x)$ defined by Equations (19.21) and (19.22) can so be represented.

In this case the integral (19.23) is now shown to have the value unity. For, from (19.23) and (19.21)

$$-(4\pi/k)\int_{S_1} f(\boldsymbol{w},\boldsymbol{\eta},k)g(\boldsymbol{\eta})\,d\boldsymbol{\eta} = \frac{1}{4\pi}\int_S \frac{\partial\phi}{\partial n}(\boldsymbol{x})\,\frac{e^{ikr}}{r}\,d\boldsymbol{x}$$

$$= \frac{1}{4\pi}\int_S \left\{\frac{\partial\phi}{\partial n}\frac{e^{ikr}}{r} - \phi\frac{\partial}{\partial n}\left(\frac{e^{ikr}}{r}\right)\right\}d\boldsymbol{x} \quad,$$

where $r = |\boldsymbol{x}|$, on using (19.19),

$$= \frac{1}{4\pi}\int_S \left\{\frac{\partial\phi_i}{\partial n}\frac{e^{ikr}}{r} - \phi_i\frac{\partial}{\partial n}\left(\frac{e^{ikr}}{r}\right)\right\}d\boldsymbol{x} \quad. \tag{19.25}$$

The last step follows from the fact that the integral

$$I = \int_{S'}\left\{G\frac{\partial\phi_s}{\partial n} - \phi_s\frac{\partial G}{\partial n}\right\}d\boldsymbol{x}, \quad \text{with} \quad G = -(4\pi r)^{-1}e^{ikr} \quad,$$

is invariant with respect to any surface S' on or outside S, by virtue of Green's formula applied to ϕ_s and G; taking S' to be a sphere of large radius R_0, one finds that $I \to 0$ as $R_0 \to \infty$, hence that $I \equiv 0$, when use is made of the radiation condition satisfied by ϕ_s. Finally, the integral (19.25) is seen to have the value $\phi_i(0) = 1$, from Green's formula applied to ϕ_i and G, with $(\nabla^2 + k^2)G = \delta(\boldsymbol{x})$. Thus

$$-(4\pi/k)\int_{S_1} f(\boldsymbol{w},\boldsymbol{\eta},k)\,g(\boldsymbol{\eta})\,d\boldsymbol{\eta} = 1 \quad, \tag{19.26}$$

for all incidence directions $\boldsymbol{w}$.

The problem is now specified by the two identities (19.26) and (19.21) to determine g, then r_s. Thus Colton & Monk (1987) formulate the optimization problem to minimize

$$\sum_{n=1}^N \left|\int_{S_1}(4\pi/k)f(\boldsymbol{w}_n,\boldsymbol{\eta},k)\,g(\boldsymbol{\eta})\,d\boldsymbol{\eta} + 1\right|^2 \quad, \tag{19.27}$$

with respect to $g(\boldsymbol{\eta})$ from a suitable function class. Given g, hence ψ from Equation (19.24), there is a second optimization problem to minimize

$$\int_{S_1}\left|\psi(r_s(\boldsymbol{\eta})\boldsymbol{\eta}) - (4\pi r_s)^{-1}e^{ikr_s}\right|^2 d\boldsymbol{\eta} \quad, \tag{19.28}$$

with respect to $r_s(\boldsymbol{\eta})$ from a suitable function class. The estimate for r_s gives an approximation to the surface S.

This optimization procedure remedies the difficulties that are associated with the analytic continuation method described earlier. Important details are given in the original paper of Colton & Monk (1987). They also give results for several axially symmetric problems, using trial functions (involving associated Legendre polynomials) in the form of Fourier series in the azimuthal angle. Their results give excellent reconstructions for a variant of shapes, such as the oblate spheroid, "peanut" shape and "acorn" shape.

19.4 HIGH–FREQUENCY LIMIT

Suppose that a plane wave of potential

$$\phi_i = \exp(ik\mathbf{w}\cdot\mathbf{x})$$

propagates in the direction $\mathbf{w}$ and is incident upon an acoustically soft surface S (whose exact location is as yet unknown).

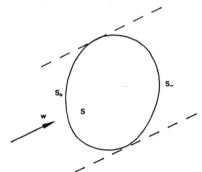

Figure 19.6

Let $\mathbf{x}_1$ be any point in the fluid region outside S, and

$$G(\mathbf{x};\mathbf{x}_1) = -(4\pi R)^{-1}\exp(ikR), \qquad R = |\mathbf{x} - \mathbf{x}_1| \tag{19.29}$$

is the Green's function that satisfies $(\nabla^2 + k^2)G = \delta(\mathbf{x} - \mathbf{x}_1)$ in three–dimensional space. Application of Green's formula to the pair of wave functions $\phi_s(\mathbf{x})$ and $G(\mathbf{x};\mathbf{x}_1)$ leads to the identity

$$\phi_s(\mathbf{x}_1) = \int_s \left(G\,\frac{\partial\phi_s}{\partial n} - \phi_s\,\frac{\partial G}{\partial n} \right) d\mathbf{x} \quad,$$

where n is the outward normal from S. In particular, as $r_1 = |\mathbf{x}_1| \to \infty$ with $\mathbf{x}_1/r_1 = \boldsymbol{\eta}$, $R \sim r_1 - \mathbf{x}\cdot\boldsymbol{\eta}$ and $\partial R/\partial n \sim -\mathbf{n}\cdot\boldsymbol{\eta}$, so

$$\phi_s \sim \frac{-e^{ikr_1}}{4\pi r_1}\int_s e^{-ik\boldsymbol{\eta}\cdot\mathbf{x}} \left(\frac{\partial\phi_s}{\partial n} + ik\boldsymbol{\eta}\cdot\mathbf{n}\,\phi_s \right) d\mathbf{x} \quad,$$

i.e.

$$f(\boldsymbol{\eta}, \boldsymbol{w}) = \frac{-k}{4\pi} \int_s \left(\frac{\partial \phi_s}{\partial n} + ik\boldsymbol{\eta} \cdot \boldsymbol{n} \phi_s \right) e^{-ik\boldsymbol{\eta} \cdot \boldsymbol{x}} \, d\boldsymbol{x} \quad . \qquad (19.30)$$

Now for a soft scatterer, $\phi_s = -\phi_i = -\exp(ik\boldsymbol{w} \cdot \boldsymbol{x})$ on S, but $\partial \phi_s / \partial n$ is not known. In the high–frequency limit ($k \to \infty$), the Kirchoff approximation states that $\partial \phi / \partial n$ can be calculated on the "illuminated" side S_+ of S (see Figure 19.6) as if each element δS of S_+ were part of an infinite tangent plane, and that $\partial \phi / \partial n \approx 0$ on the shadow side S_-. Thus

$$\frac{\partial \phi}{\partial n} \approx 2 \frac{\partial \phi_i}{\partial n} \quad \text{on} \quad S_+, \qquad \frac{\partial \phi}{\partial n} \approx 0 \quad \text{on} \quad S_- \quad ,$$

i.e.

$$\frac{\partial \phi_s}{\partial n} \approx \pm \frac{\partial \phi_i}{\partial n} = \pm ik\boldsymbol{n} \cdot \boldsymbol{w} \, \exp(ik\boldsymbol{w} \cdot \boldsymbol{x}) \quad \text{on} \quad S_\pm \quad .$$

Thus (19.30) becomes

$$f(\boldsymbol{\eta}, \boldsymbol{w}) \approx \frac{ik^2}{4\pi} \int_{s_+} \boldsymbol{n} \cdot (\boldsymbol{\eta} - \boldsymbol{w}) \, \exp\Big(ik(\boldsymbol{w} - \boldsymbol{\eta}) \cdot \boldsymbol{x} \Big) \, d\boldsymbol{x}$$

$$+ \frac{ik^2}{4\pi} \int_{s_-} \boldsymbol{n} \cdot (\boldsymbol{\eta} + \boldsymbol{w}) \, \exp\Big(ik(\boldsymbol{w} - \boldsymbol{\eta}) \cdot \boldsymbol{x} \Big) \, d\boldsymbol{x} \quad .$$

In particular, the amplitude in the back–scattered direction ($\boldsymbol{\eta} = -\boldsymbol{w}$) is

$$f(-\boldsymbol{w}, \boldsymbol{w}) \sim \frac{-ik^2}{2\pi} \int_{s_+} \boldsymbol{n} \cdot \boldsymbol{w} \, \exp\Big(2ik\boldsymbol{w} \cdot \boldsymbol{x} \Big) \, d\boldsymbol{x} \quad .$$

If S is irradiated by the plane wave $\exp(-ik\boldsymbol{x} \cdot \boldsymbol{w})$ and we again measure f in the back–scattered direction, we get

$$f(\boldsymbol{w}, -\boldsymbol{w}) \sim \frac{ik^2}{2\pi} \int_{s_-} \boldsymbol{n} \cdot \boldsymbol{w} \, \exp\Big(-2ik\boldsymbol{w} \cdot \boldsymbol{x} \Big) \, d\boldsymbol{x} \quad .$$

Thus if we define the function Q as

$$Q \equiv f(-\boldsymbol{w}, \boldsymbol{w}) + \overline{f(\boldsymbol{w}, -\boldsymbol{w})} \quad , \qquad (19.31)$$

where the overbar denotes a complex conjugate, then

$$Q \sim -\frac{ik^2}{2\pi} \int_s \boldsymbol{n} \cdot \boldsymbol{w} \, \exp\Big(2ik\boldsymbol{w} \cdot \boldsymbol{x} \Big) \, d\boldsymbol{x} \quad ,$$

where Q is known in principle, since the amplitude function f is considered to be known. The divergence theorem may be used to recast the right–hand side as a volume integral

$$Q \sim \frac{k^3}{\pi} \int_V e^{2\,ik\boldsymbol{w}\cdot\boldsymbol{x}}\, d\boldsymbol{x} \quad , \tag{19.32}$$

where V is the volume inside S. Now we introduce the characteristic function γ associated with the scatterer S, as

$$\gamma(\boldsymbol{x}) = \begin{cases} 1 & \text{if } \boldsymbol{x} \text{ is inside } S \\ 0 & \text{if } \boldsymbol{x} \text{ is outside } S \quad . \end{cases} \tag{19.33}$$

Then

$$\Gamma(\boldsymbol{k}) = \frac{\pi Q}{k^3} = \int \gamma(\boldsymbol{x})\, e^{2ik\cdot\boldsymbol{x}}\, d\boldsymbol{x} \quad , \tag{19.34}$$

where $\boldsymbol{k} = k\boldsymbol{w}$. Evidently Q is like a Fourier transform of the characteristic function γ. Since Q is known, it is tempting to infer that γ (hence S) can be found by a Fourier inversion of formula (19.34), thus

$$\gamma = \frac{1}{\pi^3} \int \Gamma(\boldsymbol{k})\, e^{-2\,i\boldsymbol{k}\cdot\boldsymbol{x}}\, d\boldsymbol{k} \quad .$$

The snag is that we know $\Gamma(\boldsymbol{k})$ only for large k, while the inversion integral needs Γ for all k. One way around this difficulty (see Prosser 1977) is to take Γ as given by (19.34) for large k, and to define Γ in any smooth manner for small and medium values of k. Then its Fourier inverse will not be precisely γ but will have the same singularity, namely a unit jump at the surface S.

Prosser also points out another way of linking the directivity pattern f with the geometry of the scatterer S. For a convex soft scatterer S, the method of geometrical acoustics leads to a result for the far–field amplitude $f(\boldsymbol{\eta}, \boldsymbol{w})$ in which the modulus and phase are related, respectively, to quantities $K(\boldsymbol{\eta} - \boldsymbol{w})$ and $P(\boldsymbol{\eta} - \boldsymbol{w})$. Here K is the Gaussian curvature (i.e. the inverse product of the two principal radii of curvature) of S at the unique "specular point" whose normal $\boldsymbol{n}$ is parallel to $\boldsymbol{\eta} - \boldsymbol{w}$, and P is the distance from the origin to the tangent plane through the specular point. Since f is taken to be known (at high frequencies, say), we can regard K and P as known functions of the normal $\boldsymbol{n}$. The conclusion is tantalizing. For knowledge of the Gaussian curvature at each point of a convex surface is known to determine S *in principle* — this is Minkowski's problem, solved by Nirenberg (1953), whose proof is unfortunately non–constructive. To date there is no systematic method of constructing S from the knowledge of its Gaussian curvature.

Alternatively, knowledge of $P(n)$ determines a convex surface S, where n is the normal direction. For S is the envelope of the family of planes with normal direction n and distance $P(n)$ from the origin. A large number of such planes would give a bounding polyhedron, but the severe practical difficulty is that P comes from the *phase* of expression $f(\eta, w)$ and is difficult to measure with any accuracy at large k.

REFERENCES

Colton, D. & Monk, P. (1987). The numerical solution of the three–dimensional inverse scattering problem for time–harmonic acoustic waves. Siam J. Sci. Stat. Comput. 8:278– 291.

Imbriale, W.A. & Mittra, R. (1970). The two dimensional inverse scattering problem. IEEE Trans. Antennas & Propag. AP-18:633–642.

Mittra, R. (Editor) (1973). Computer Techniques for Electromagnetics. Pergamon, Oxford, New York, Toronto.

Nirenberg, L. (1953). The Weyl and Minkowski problems in differential geometry in the large. Comm. Pure Appl. Math. 6:337–394.

Prosser, R.T. (1977). Can one see the shape of a surface? Amer. Math. Monthly 84:259–270.

Sleeman, B.D. (1982). The inverse problem of acoustic scattering. IMA J. Appl. Math. 29:113–142.

Watson, G.N. (1941). A Treatise on the Theory of Bessel Functions. Cambridge University Press.

20. RESONATORS

20.1 INTRODUCTION

There does not seem to be a simple statement of the type "A resonator is defined as
..."; therefore a few examples may show what is meant in acoustics by resonators.
Examples of single degree of freedom resonators are shown in Figure 20.1. Multi
degree of freedom resonators are shown in Figure 20.2. Continua that act as acoustic
resonators are the air columns in an organ pipe or flute, finite strings, rods, beams,
rings, shafts, plates, bells, any air volume in a room, etc. surrounded by non–
absorbing boundaries.

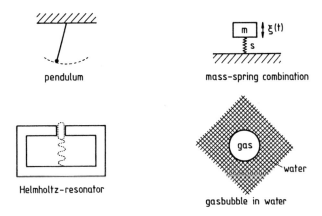

Figure 20.1 Examples of single degree of freedom resonators.

In all these cases it is typical that mechanical energy is stored as kinetic energy
or as potential energy and there is an exchange of energy. Another common feature
is the existence of special frequencies, the so–called resonance frequencies, which can
be excited much easier than others and which determine the ringing or reverberation
of a system when no more energy is transferred to the system from an outside source.

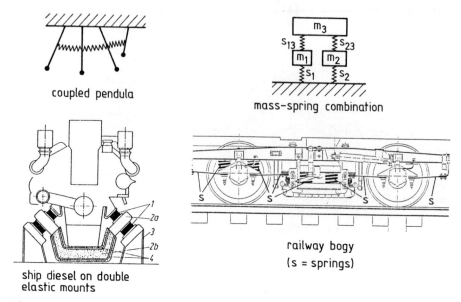

coupled pendula

mass-spring combination

ship diesel on double
elastic mounts

railway bogy
(s = springs)

Figure 20.2 Examples of multi degree of freedom resonators.

Apart from the lumped or distributed masses and springs which are necessary to form resonators, there are always mechanisms which absorb energy. These so–called dampers or dashpots which transform mechanical energy into heat are responsible for the fact that

- all resonances die out sooner or later if outside energy is no longer added;
- at the resonance frequencies the amplitudes remain finite, and
- the resonances are not absolutely sharp but smeared out over a certain frequency region which increases with damping.

20.2 SINGLE DEGREE OF FREEDOM SYSTEMS

20.1.1 Equation of Motion Without Damping

The simplest resonator is a mass m sitting on a spring with stiffness s (see Figure 20.1). The equation of motion is

$$m \frac{d^2\xi(t)}{dt^2} + s\xi(t) = F(t) \,. \tag{20.1}$$

Here $F(t)$ is the outside force acting on the mass and $\xi(t)$ is the motion of the mass.

As was shown in Chapter 2, the Green's function of (20.1) is

$$G(t, \tau) = \frac{H(t - \tau)}{m\omega_0} \sin \omega_0(t - \tau) , \qquad (20.2)$$

where

$$\omega_0 = \sqrt{s/m} \qquad (20.3)$$

is the resonance frequency of the system. Using the convolution formula (see Chapter 2), the general solution of (20.1) is

$$\xi(t) = F * G = \int_{-\infty}^{+\infty} F(\tau) \frac{H(t - \tau)}{m\omega_0} \sin \omega_0(t - \tau) d\tau$$
$$= \int_{-\infty}^{t} \frac{F(\tau)}{m\omega_0} \sin \omega_0(t - \tau) d\tau . \qquad (20.4)$$

In physical terms Equation (20.4) states that the excitation $F(\tau)$ can be considered as a sequence of minute pulses of momentum $F(\tau)d\tau$. Because of linearity the motions caused by these pulses are added up.

An alternative method of solving Equation(20.1) is to use Fourier transforms. In this case the solution is written as

$$\xi(t) = \frac{1}{2\pi} \int_{-\infty}^{+\infty} \frac{\widetilde{F}(\omega)}{s - m\omega^2} e^{-i\omega t} d\omega , \qquad (20.5)$$

with $\widetilde{F}(\omega)$ being the Fourier transform of $F(t)$

$$\widetilde{F}(\omega) = \int_{-\infty}^{+\infty} F(t) e^{+i\omega t} dt . \qquad (20.6)$$

In physical terms, Equation (20.6) shows that the motion consists of a sum of purely harmonic vibrations, which are obtained by a simple multiplication when the excitation is also expressed as a sum of purely harmonic terms.

Example. A purely sinusoidal excitation starts at $t = 0$

$$F(t) = \begin{cases} 0 & \text{for } t < 0; \\ A \cos \omega_f t & \text{for } t > 0 . \end{cases} \qquad (20.7)$$

Equation (20.4) gives

$$\xi(t) = \frac{A}{m\omega_0} \int_0^t \cos \omega_f \tau \sin \omega_0(t - \tau) d\tau$$
$$= \frac{-A}{m(\omega_f + \omega_0)} \sin\left(\frac{\omega_f + \omega_0}{2} t\right) \frac{\sin \Delta_f t}{\Delta_f} \qquad (20.8)$$

with $\Delta_f = (\omega_f - \omega_0)/2$.

This example shows that there is an amplitude modulation with frequency $(\omega_f - \omega_0)/2$. For $\Delta_f = 0$; i.e. when exciting frequency and resonance frequency are exactly equal the motion increases proportionally with time t because in this case

$$\lim_{\Delta_f \to 0} \frac{\sin \Delta_f t}{\Delta_f t} = 1.$$

20.2.2 Influence of Damping

According to (20.8), the amplitude of a sinusoidally–driven oscillator would increase indefinitely for $\omega_0 = \omega_f$ and would constantly show amplitude modulations (beats) if $\omega_0 \neq \omega_f$, i.e. a steady state would never be reached. This result does not agree with experience. Better agreement with real–life resonators can be reached if energy losses are introduced. The easiest way to do this is to include a dashpot which adds a velocity proportional force to (20.1),

$$m \frac{d^2 \xi(t)}{dt^2} + r \frac{d\xi(t)}{dt} + s\xi(t) = F(t) . \tag{20.9}$$

The Green's function now is

$$G(t, \tau) = \frac{H(t - \tau)}{m \omega_R} e^{-\delta(t-\tau)} \sin \omega_R (t - \tau) \tag{20.10}$$

with

$$\delta = \frac{r}{2m}, \quad \omega_R^2 = \frac{s}{m} - \delta^2 = {\omega_0}^2 - \delta^2 .$$

Making a convolution of (20.10) with (20.7) one can see that a steady state is reached (i.e. the influence of the starting point $t = 0$ disappears). It also turns out that this occurs after a time which is proportional to $1/\delta$. Any beating phenomenon also dies out.

If one works with Fourier transforms, the general solution of (20.9) is

$$\xi(t) = \frac{1}{2\pi} \int_{-\infty}^{+\infty} \frac{\widetilde{F}(\omega)}{s - i\omega r - \omega^2 m} e^{-i\omega t} d\omega . \tag{20.11}$$

Very often the absolute value of the integrand in this formula is called the spectrum of the motion,

$$|\widetilde{\xi}(\omega)| = \left| \frac{\widetilde{F}(\omega)}{s - i\omega r - \omega^2 m} \right| . \tag{20.12}$$

A conceptual difficulty associated with (20.12) is due to the fact that $\widetilde{F}(\omega)$ is the result of an integration from $-\infty$ to $+\infty$. Since this is too long for an average experimenter, the integration is made only over finite times. This causes difficulties at the beginning and end of each measurement. These difficulties disappear within a time proportional to $1/\delta$.

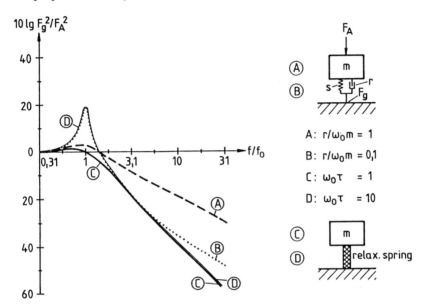

Figure 20.3 Frequency response of a damped oscillator.

A frequency response for a damped oscillator is shown in Figure 20.3 (curves A and B). It should be borne in mind that such curves are obtained only if the frequency sweep is made so slowly that for each frequency a steady state is reached.

Equation (20.9) is a good representation of a damped oscillator if there is actually a velocity proportional damping device *parallel* to the spring. In other cases, e.g. when material losses in the spring (e.g. rubber) are the dominant loss mechanisms, (20.9) may be misleading when applied indiscriminately. A better method for characterising lossy materials is the relaxation model which goes back to Boltzmann (1876). In this model it is assumed that the force $F(t)$ at time t, which is necessary to cause a certain displacement $\xi(t)$, depends not only on $\xi(t)$ but also on the displacements $\xi(t - \Delta)$ at previous times $t - \Delta$. Or, in mathematical terms,

$$F_s(t) = s_1 \xi(t) - \int_0^\infty \xi(t - \Delta)\phi(\Delta)d\Delta \ . \tag{20.13}$$

Here $\phi(\Delta)$ is the memory function, which in its most simple form is

$$\phi(\Delta) = \frac{s_2}{\tau} e^{-\Delta/\tau} \, . \qquad (20.14)$$

Here τ is the so–called relaxation time. Other forms of (20.14), especially sums containing several relaxation times, may also appear. Combining (20.14) and (20.13) and making a few rearrangements we obtain, when we introduce this expression into the equation of motion for a simple degree of freedom system,

$$m\frac{d^2\xi(t)}{dt^2} + s_1\xi(t) - \frac{s_2}{\tau}\int_{-\infty}^{t} \xi(\Delta)e^{-(t-\Delta)/\tau}d\Delta = F(t) \, . \qquad (20.15)$$

A Fourier transform of (20.15) yields for the displacement spectrum

$$\widetilde{\xi}(\omega) = \frac{\widetilde{F}(\omega)}{\left(s_1 - \dfrac{s_2}{1 + \omega^2\tau^2}\right) - \omega^2 m - i\dfrac{\omega\tau s_2}{1 + \omega^2\tau^2}} \, . \qquad (20.16)$$

Comparison with (20.12) shows that the term corresponding to the stiffness increases from $s_1 - s_2$ to s_1 as the frequency increases, with the maximum change around $\omega\tau \approx 1$. The imaginary term describing the damping is also frequency dependent; the maximum is achieved near $\omega\tau = 1$. High polymers such as rubber show just this behaviour. The transmissibility calculated from (20.16) is shown for two examples in Figure 20.3. (curves C and D). Note that viscous damping and relaxation damping lead to rather different results at frequencies far above the resonance.

A third method to describe damping is the so–called loss factor η. It is defined as the imaginary part of the stiffness (or any other similar quantity such as Young's modulus or shear modulus). Using the loss factor, the equation of motion for a single mass resonator would be

$$m\frac{d^2\xi(t)}{dt^2} + s(1 - i\eta)\xi(t) = F(t) \, . \qquad (20.17)$$

As it stands, (20.17) does not make much sense, because there is no such thing as an imaginary modulus or an imaginary displacement, which would follow from (20.17). If one works with pure tones, however, or with combinations of pure tones, one can use the convention

$$\xi(t) = \Re\left\{\sum_{n=1}^{N} \widetilde{\xi}(\omega_n)e^{-i\omega_n t}\right\} \, , \qquad (20.18)$$

and a similar expression for $F(t)$. One then obtains for the spectrum

$$\tilde{\xi}(\omega_n) = \frac{\tilde{F}(\omega_n)}{s(1 - i\eta) - \omega_n^2 m} \ .$$

(20.19)

This type of approach is very convenient as long as one is interested in spectra, because one can use the same calculation as in the undamped case, the only difference being that s is replaced by $s(1 - i\eta)$. Very often the operator $\Re\{\ldots\}$ and even the factor $e^{-i\omega_n t}$ are omitted for the sake of brevity. This can be done without any problems as long as one works in the frequency domain. Transformation into the time domain may cause problems when the concept of loss factor or complex modulus is used. To remind the reader of this fact, (20.18) was written as a sum and not as an integral. Comparison of (20.12) with (20.19) shows that $s\eta$ corresponds to ωr.

20.2.3 Energy Considerations

If (20.9) is multiplied by the velocity $\dot{\xi} = d\xi/dt$, the term on the right–hand side is the power P that is transmitted from the outside force into the system. Thus we have

$$m\ddot{\xi}\dot{\xi} + s\xi\dot{\xi} + r\dot{\xi}^2 = F\dot{\xi} = P_i \ .$$

This can be rearranged to give

$$\frac{d}{dt}\left(\frac{1}{2}m\dot{\xi}^2 + \frac{1}{2}s\xi^2\right) + r\dot{\xi}^2 = P_i \ .$$

(20.20)

This equation shows that the transmitted power P serves to change the sum of kinetic and potential energy and to feed the power $r\dot{\xi}^2$ into the damper. In the steady–state case, $d/dt = 0$ and therefore

$$\overline{P}_i = r\overline{\dot{\xi}}^2 = \frac{s\eta}{\omega}\overline{\dot{\xi}}^2 \approx m\omega\eta\overline{\dot{\xi}}^2 = m\omega\eta\overline{v}^2 \ .$$

(20.21)

The overbar indicates averaging over many periods. The frequency which appears here is the resonance frequency, because it is mainly around this frequency where power transfer occurs. To see this, the following example is considered.

Example. When the excitation is given by

$$F(t) = \Re\left\{\tilde{F}_n e^{-i\omega_n t}\right\} \ ,$$

the steady–state response is

$$\xi(t) = \Re \left\{ \frac{\widetilde{F}_n e^{-i\omega_n t}}{s(1-i\eta) - \omega_n^2 m} \right\} ; v(t) = \Re \left\{ \frac{-i\omega_n \widetilde{F}_n e^{-i\omega_n t}}{s - \omega_n^2 m - is\eta} \right\} .$$

Time averaging yields

$$\begin{aligned}
\overline{P} &= \frac{\omega_n s \eta}{(s - \omega_n^2 m)^2 + s^2 \eta^2} \left(F'^2 + F''^2 \right) \\
&= \frac{\omega_n s \eta}{(s - \omega_n^2 m)^2 + s^2 \eta^2} |\widetilde{F}_n|^2 = \frac{s\eta}{\omega_n} |\widetilde{v}_n|^2 .
\end{aligned}$$

$$(20.22)$$

Here the abbreviation $\widetilde{F}_n = F' + jF''$ was used. it can be seen that for a given F_n the maximum power transfer occurs when $s - \omega_n^2 m = 0$. Away from the resonance the power transfer increases with the losses.

20.2.4 Helmholtz Resonators

There are many cases where the device for storing kinetic energy is not as well defined as in a simple mass spring system. A typical example is the so–called Helmholtz resonator, or the F–holes in a violin, or the holes in perforated absorbing lining.

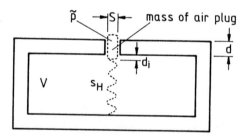

Figure 20.4 Helmholtz resonator.

For the configuration shown in Figure 20.4, the spring is the air volume. From the ideal gas law in the adiabatic case, it follows that volume change ΔV causes a pressure change Δp which are related by

$$\frac{\Delta p}{\Delta V} = \kappa \frac{p_{\text{stat}}}{V} ;$$

p_{stat} = static pressure, V = total resonator volume, κ = ratio of specific heats (in air $\kappa \approx 1.4$). If we assume that the volume change is due to the air plug of area

S and displacement ξ, we find $\Delta V = -S\xi$ (negative sign because outward motion is positive). The pressure increase is related to the force acting from the resonator volume to the air plug by $F_r = S\Delta p$. Thus we have

$$F_r = \xi \kappa S^2 p_{\text{stat}}/V = s_H \xi . \qquad (20.23)$$

We can now write the balancing equation for the forces acting on the air plug of mass $m = \rho S(d + d_i)$ as

$$m\ddot{\xi} + S_H\xi + F_M = F_0 . \qquad (20.24)$$

Here F_0 is an external force, and F_M the force acting from the medium outside the resonator onto the mass plug. d is the thickness of the resonator neck and d_i an internal end correction which will be discussed later. Going to the frequency domain and replacing the displacements by the velocities ($\tilde{v} = i\omega\tilde{\xi}$), we find

$$\left(-i\omega m + \frac{s_H}{-i\omega}\right)\tilde{v} + \tilde{F}_M = \tilde{F}_0 . \qquad (20.25)$$

The restriction to single frequencies is useful because F_M is not a very simple function of the velocity of the air plug. If $\tilde{p}$ is the sound pressure right over the resonator hole, we have

$$\tilde{F}_M = \int \tilde{p}dS .$$

Assuming that $\tilde{v}$ has the same value over the entire area of the neck, we can write

$$\tilde{F}_M = \tilde{v}\int \frac{\tilde{p}}{\tilde{v}}dS = \tilde{v}Z_S . \qquad (20.26)$$

Z_S is the impedance which is seen by a plug or a piston when it radiates into the surrounding medium.

In most cases Z_S is difficult to find analytically; therefore we look for an approximate solution. If we assume that the air plug in the resonator behaves like a circular piston of radius a and area $S = \pi a^2$ sitting in a very large baffle, the sound pressure at any point is given by Rayleigh's formula

$$\tilde{p} = \frac{-i\omega\rho}{2\pi}\int_S \tilde{v}(S)\frac{e^{ikr}}{r}dS . \qquad (20.27)$$

This formula is obtained by assuming that each element dS with volume flow $\tilde{v}(S)dS$ acts as a monopole source. In our case $\tilde{v}(S) = \tilde{v}$ is constant, therefore $\tilde{p}/\tilde{v}$ can be calculated for each point over the vibrating surface. The next step is to integrate $\tilde{p}/\tilde{v}$ over S. The result of this operation is (see Morse 1948):

$$Z_S = \pi a^2 \rho c (A' - iA'')$$

$$A' = 1 - \frac{2}{ka} J_1(2ka) \to A' \approx k^2 a^2/2 \qquad \text{for } ka \ll 1 \qquad (20.28a)$$

$$A'' = \frac{4}{\pi} \int_0^{\pi/2} \sin(2ka \cos\alpha) \sin^2\alpha \, d\alpha \to A'' \approx 8ka/3\pi \text{ for } ka \ll 1.$$

It is interesting to note that on the surface of a small hemisphere with area S vibrating in its zero'th mode (breathing), the ratio between pressure and surface velocity is

$$\frac{\tilde{p}}{\tilde{v}} = \rho c \left(\frac{k^2 s}{2\pi} - ik\sqrt{\frac{S}{2\pi}} \right) \quad \text{for } k\frac{S}{2\pi} \ll 1 \,, \qquad (20.28b)$$

which would give a very similar value for the radiation impedance.

If (20.28) and (20.26) are introduced into (20.25), the result is

$$\left(-i\omega m + \frac{s_H}{-i\omega} + \rho c S A'(-i\rho c S A'') \right) \tilde{v} = \tilde{F}_0$$

or

$$\left[-i\omega(m + \rho c S A''/\omega) + \rho c S A' - \frac{s_H}{i\omega} \right] \tilde{v} = \tilde{F}_0 \,. \qquad (20.29)$$

Since A' and A'' are always positive, (20.29) represents a single degree of freedom resonator with mass $m + \rho c S A''/\omega$, damping $\rho c S A'$ and stiffness s_H. In the most interesting case of small holes, we have for the damping term

$$\rho c S k^2 a^2/2 \,.$$

This is a so–called radiation damping, because it describes the power which is lost from the resonator by radiation into the surrounding medium. For $ka \ll 1$ the mass can be written as

$$m + \rho c S \frac{8ka}{3\pi\omega} = m + \rho S \frac{8a}{3\pi} - \rho S \left(d + d_i + \frac{8a}{3\pi} \right) \,.$$

We see that the influence of the medium outside the resonator has, apart from the radiation damping, the same effect as an increase of the air plug thickness from d to $d + 8a/3\pi$. If we assume that the same increase is also on the inside of the resonator, i.e. $d_i \approx 8a/3\pi$, we have finally for the effective mass: $\qquad m = \rho S(d + 2d_{\mathrm{corr}})$

for the resonance frequency: $\omega_0^2 = \dfrac{s_H}{m} = \dfrac{\kappa S^2 p_{\mathrm{stat}}}{Vm} = \dfrac{c^2 S}{V(d + 2d_{\mathrm{corr}})}$

$$(20.30)$$

for the decay constant: $\qquad \delta = \dfrac{\rho c S k^2 a^2}{2 \cdot 2m} = \dfrac{\omega^2 a^2}{4c(d + 2d_{\mathrm{corr}})} \quad ,$

with $\qquad d_{\mathrm{corr}} = 8a/3\pi \approx 0.8a$. In the last formula $8/3\pi$ has been set to 0.8 because the inner correction seems to be somewhat smaller than the outer one.

For non–circular holes, a reasonable agreement with measured data is obtained if, as an effective radius, $a_{\mathrm{eff}} \approx 1.06 S^{3/4} U^{-1/2}$ is taken (S = area, U = perimeter of the hole) (see Mechel 1989). In the case of slit of width b,

$$d_{\mathrm{corr}} \approx \frac{b}{\pi} \ln \frac{2\pi}{kb} - 0.107$$

should be used. When the Helmholtz resonator is not mounted in a very large baffle, but is small compared with the wavelength, the correction term is $0.61a < d_{\mathrm{corr}} < 0.8a$. Finally, if many resonators are mounted in a quadratic array with spacing ℓ, the value

$$d_{\mathrm{corr}} = 0.79a \left[1 - 1.47\sqrt{\pi a^2/\ell^2} + 0.47(\pi a^2/\ell^2)^{3/2} \right]$$

should be used.

20.2.5 Pulsating Bubbles in Water

A single air or gas bubble in water is also a resonator. Here the gas volume is a spring, which according to (20.23) has the stiffness

$$s_B = \frac{\kappa S^2 p_{\mathrm{stat}}}{V} . \tag{20.31}$$

For the effect of surface tension which acts as a second spring, see Chapter 21. The bubble is assumed to be spherical, thus $S/V = 3/a$. The effect of the surrounding water is again represented by the ratio of pressure to surface velocity of the bubble. Assuming monopole radiation from the pulsating bubble, we get

$$p(r, t) = \widetilde{p}e^{-i\omega t} = \frac{Q}{4\pi r}e^{+ikr}e^{-i\omega t}$$

$$v(r, t) = \widetilde{v}e^{-i\omega t} = \frac{-Q}{4\pi i\omega\rho}\left(\frac{1}{r^2} - \frac{ik}{r}\right)e^{ikr}e^{-i\omega t} .$$

$$(20.32)$$

For $r = a$, this gives

$$\widetilde{p} = \widetilde{v} \, \frac{-i\omega\rho a}{1 - ika} \, . \tag{20.33}$$

Thus the balancing equation for the forces on the bubble surface are similar to (20.25) and (20.29)

$$- \left(i\omega \frac{\rho a S}{1 + k^2 a^2} - \frac{\omega\rho k a^2 S}{1 + k^2 a^2} + \frac{s_B}{i\omega} \right) \widetilde{v} = \widetilde{F}_0 \, . \tag{20.34}$$

So we finally have in the most interesting case of $ka \ll 1$

for the effective mass: $\qquad m = \rho S a = 4\pi\rho a^3$

for the resonance frequency: $\omega_0^2 = \dfrac{s_B}{m} = 3\dfrac{\rho_B}{\rho} \dfrac{c_B^2}{a^2}$

$$\tag{20.35}$$

for the decay constant: $\qquad \delta = \dfrac{\omega k a}{2} \, .$

Here the formula $c_B^2 = \kappa p_{\text{stat}}/\rho_B$ was used. ρ_B = density of the gas in the bubble, c_B = its speed of sound.

20.2.6 Contact Resonances

For Helmholtz resonators and gas bubbles in water, the effect of the surrounding medium is to effectively increase the vibrating mass and cause some additional damping. This is typical for the so–called radiation loading by a medium without shear stiffness, e.g. air or water. The situation is different when the surrounding medium has also a finite shear stiffness, as is the case for any solid medium. Under these circumstances, the influence of the surrounding medium may also have stiffness character. The most common example in this respect is contact resonance. This occurs when a solid body rests on an elastic medium forming a resonating system where the mass of the solid body is the effective mass, the local elasticity of the medium is the stiffness and radiation of shear and compressional waves contributes to the damping.

If the contact area is independent of the static load (e.g. punch), the local elasticity (i.e. contact stiffness) is

$$s_c = \pi \, \frac{Ga}{1 - \mu} \, . \tag{20.36}$$

If the contact area changes substantially with the static load, which occurs for the contact of curved bodies, the Hertzian theory holds (Johnson 1985). It yields for the linearized contact stiffness

$$s_c = \frac{3}{2} \left(\frac{16}{9} R F_{st} E^{*2} \right)^{1/3}$$

$$\frac{1}{R} = \frac{1}{R_1} + \frac{1}{R_2}; \quad \frac{1}{E^*} = \frac{1 - \mu_1^2}{E_1} + \frac{1 - \mu_2^2}{E_2};$$

(20.37)

G = shear modulus, μ = Poisson's ratio, R_1, R_2 = radius of curvature of the contacting bodies, E_1, E_2 = Young's moduli, F_{st} = static load.

20.2.7 Scattering Cross Section of a Resonator

A small Helmholtz resonator, which is exposed to a plane sound wave of pressure amplitude $\tilde{p}_i$, has a surprisingly large scattering cross section when the exciting frequency and the resonance frequency coincide. To see this one makes use of the fact that according to (20.10) and (20.30) the radiation damping can be presented by

$$r = \rho c S k^2 a^2 / 2 .$$

Thus the power lost, i.e. radiated, by the resonator is, according to (20.21),

$$p = r |\tilde{v}|^2 / 2 = \rho c S k^2 a^2 |\tilde{v}|^2 / 4 .$$

(20.38)

From (20.29) we find

$$\tilde{F}_0 \approx 2 \tilde{p}_i S = -\tilde{v} [i\omega(m + \rho c S A'' / \omega) - \rho c S A' + S_H / i\omega] .$$

In this equation it is assumed that the pressure driving the Helmholtz resonator is, because of the reflection, twice the incoming pressure. Combining the equations gives, with the approximations for A' and A''

$$P = \frac{8 P_i}{k^2 a^2} \frac{1}{1 + 4 \left(\dfrac{d + 2 d_{corr}}{ka} \right)^2 (1 - \omega_0^2 / \omega^2)^2} .$$

(20.39)

Here $P_i = |\tilde{p}_i|^2 S / 2 \rho c$ is the acoustic power impinging on the resonator hole. Since $ka << 1$, it is easily possible that $P >> P_i$. This is especially true for $\omega = \omega_0$, where $P / P_i = 2 \lambda^2 / \pi^2 a^2$ (λ = wavelength). Sometimes this result is expressed by stating that the scattering cross section of a resonator at the resonance frequency is approximately λ^2.

20.2.8 Rayleigh's Principle for Determining the Fundamental Resonance Frequency

According to Temple & Bickley (1933), "... the characteristic feature of Rayleigh's method ... is its emphasis on energy rather than on force. The equations of motion of the system studied play a secondary part, the primary rôle being played by the energy equation. ... What is novel about Rayleigh's work is the conversion of the energy method into a practical tool of rapid, approximate calculation ... . The energy method of determining the frequencies of vibration is an immediate consequence of the principle that, in simple–periodic motion, the mean values of the total kinetic and potential energies of the vibrating system are equal".

A simple example using a mass spring system immediately shows how the principle works. The kinetic and potential energies are

$$
\begin{aligned}
E_{\mathrm{kin}} &= \frac{1}{2}m\dot{\xi}^2 = \frac{1}{2}m\omega_0^2\widetilde{\xi}^2\cos^2\omega_0 t \\
E_{\mathrm{pot}} &= \frac{1}{2}s\xi^2 = \frac{1}{2}s\widetilde{\xi}^2\sin^2\omega_0 t \ .
\end{aligned}
\tag{20.40}
$$

Taking the time averages gives

$$
\overline{E_{\mathrm{kin}}} = \frac{1}{4}m\omega_0^2\widetilde{\xi}^2; \quad \overline{E_{\mathrm{pot}}} = \frac{1}{4}s\widetilde{\xi}^2 \ .
$$

Equating these two expressions results in

$$
\omega_0^2 = s/m \ .
$$

A not–so–simple example is a mass spring system where the mass of the spring is not negligible. If ρ_f is the density of the spring, S_F its cross section, and ℓ_F its length,

$$
\begin{aligned}
E_{\mathrm{kin}} &= \frac{1}{2}m\omega_0^2\widetilde{\xi}^2\cos^2\omega_0 t + \frac{1}{2}\int_0^{\ell_F}\rho_F S_F\omega_0^2\left(\widetilde{\xi}\frac{x}{\ell_F}\right)^2 dx\cos^2\omega_0 t \\
&= \frac{1}{2}m\omega_0^2\widetilde{\xi}^2\cos^2\omega_0 t + \frac{1}{2}\frac{\rho_F S_F \ell_F}{3}\omega_0^2\widetilde{\xi}^2\cos^2\omega_0 t \ .
\end{aligned}
\tag{20.41}
$$

Here it was assumed that $\widetilde{\xi}$ increases linearly from $x = 0$ to $x = \ell_F$.

Equating the mean value of (20.41) with the mean value of the potential energy, as given by (20.40), yields

$$\omega_0^2 = \frac{s}{m + \frac{1}{3}\rho_F S_F \ell_F} \ .$$

As another example let us consider a finite plate of density ρ_p, thickness h_p, length ℓ and Young's module E_p, which is mounted in a distance d in front of a rigid wall. Neglecting the mass of the air, the kinetic energy of this "panel absorber" is

$$E_{\text{kin}} = \frac{1}{2}\rho_p h_p \int_0^\ell \omega_0^2 \widetilde{\xi}(x)^2 dx \cos^2 \omega_0 t \ .$$

The potential energy in the bending of the plate is

$$E_{1,\text{pot}} = \frac{1}{2} \frac{E_p h_p^3}{12} \int_0^\ell \left(\frac{d^2 \widetilde{\xi}(x)}{dx^2} \right)^2 dx \sin^2 \omega_0 t \ . \tag{20.42}$$

The potential energy in the air is

$$E_{2,\text{pot}} = \frac{1}{2} \frac{\rho c^2}{d} \int_0^\ell \widetilde{\xi}(x)^2 dx \sin^2 \omega_0 t \ ;$$

ρ = density of air, c = speed of sound, ρc^2 = compressibility.

The resonance frequency is, according to Rayleigh's principle, given by

$$\omega_0^2 = \omega_1^2 + \omega_2^2 \tag{20.43}$$

with

$$\omega_2^2 = \frac{\rho c^2}{d \rho p h_P}$$

$$\omega_1^2 = \frac{E_P h_P^2}{12 \rho_P} \int_0^\ell \left(\frac{d\widetilde{\xi}(x)}{dx^2} \right)^2 dx / \int_0^\ell \widetilde{\xi}(x)^2 dx \tag{20.44}$$

In many cases ω_1^2 will be known, because it is the resonance frequency of the plate under the given boundary condition without the influence of the air backing (see e.g. Leissa 1969).

If ω_1^2 has to be calculated, it is necessary to make a reasonable assumption on the displacement distribution $\xi(x)$ over the plate area. Fortunately, the final result is not very sensitive to the actual choice of $\xi(x)$ as long as the boundary conditions are fulfilled. (This is one of the great advantages of Rayleigh's principle.) If, for example, the plate is clamped at $x = 0$ and $x = \ell$, the boundary conditions are

$$\widetilde{\xi}(0) = \widetilde{\xi}(\ell) = \frac{\partial \widetilde{\xi}(0)}{\partial x} = \frac{\partial \widetilde{\xi}(\ell)}{\partial x} = 0 \ .$$

A possible function which fulfills the boundary conditions is

$$\widetilde{\xi}(x) = 1 - 2\left(\frac{2x}{\ell} - 1\right)^2 + \left(\frac{2x}{\ell} - 1\right)^4 \ . \tag{20.45}$$

Introduced into (20.44), this gives

$$\omega_1^2 = 504\frac{E_P h_P^2}{12\rho_p \ell^4} \ .$$

The correct value for the fundamental frequency of a clamped plate would have the factor 501 instead of 504. This is very close in spite of the fact that (20.45) is not the exact equation for the displacement.

It is worthwhile to note that Rayleigh's principle always gives an upper bound for the fundamental frequency; e.g. for a circular plate of radius a with clamped edges, Rayleigh's principle, using the approximation $\xi(r) = (a^3 - 3ar^2 + 2r^3)$, gives $\omega_1^2 = 105\, E_p h_p^2 / 12\rho_p a^4$; the correct formula has a factor 104.36 instead.

20.3 MULTI DEGREE OF FREEDOM SYSTEMS

20.3.1 Equations of Motion

Multi degree of freedom systems are described by a system of differential equations. They can be obtained by calculating the forces that act on each element of the system. An alternative, and in the author's mind more convenient, way for finding the equation of motion is the application of Lagrange's equation of the second kind. This very fundamental equation of theoretical mechanics uses as the most important quantity the so-called Lagrangian $L(v_n, \xi_n)$, which is the difference between the total kinetic energy E_{kin} and the total potential energy E_{pot}. For non dissipative systems with N degrees of freedom, Lagrange's equation is

$$\frac{d}{dt}\frac{\partial L(v_n, \xi_n)}{\partial v_n} - \frac{\partial L(v_n, \xi_n)}{\partial \xi_n} = 0 \qquad n = 1, 2 \ldots N \ . \tag{20.46}$$

Here v_n is the velocity and ξ_n the displacement of the n=-th element. If there is some dissipation, which is identical or can be approximated by viscous dampers, there is an additional term

$$\frac{\partial E_{\text{diss}}}{\partial v_n} \,,$$

and if there are outside forces F_n which can be derived from a potential function Q, we have on the right side an additional term

$$\frac{\partial Q}{\partial \xi_n} = F_n \,. \tag{20.47}$$

In most acoustic applications the amplitudes are small and therefore the energies are quadratic functions of the displacements

$$E_{\text{kin}} = \frac{1}{2} \sum_{n=1}^{N} m_n v_n^2 + \frac{1}{2} \sum_{\nu=1}^{N} \sum_{\mu=\nu+1}^{N} m_{\mu\nu}(v_\mu - v_\nu)^2$$

$$E_{\text{pot}} = \frac{1}{2} \sum_{n=1}^{N} s_n \xi_n^2 + \frac{1}{2} \sum_{\nu=1}^{N} \sum_{\mu=\nu+1}^{N} s_{\mu\nu}(\xi_\mu - \xi_\nu)^2 \tag{20.48}$$

$$E_{\text{diss}} = \frac{1}{2} \sum_{n=1}^{N} r_n v_n^2 + \frac{1}{2} \sum_{\nu=1}^{N} \sum_{\mu=\nu+1}^{N} r_{\mu\nu}(v_\mu - v_\nu)^2 \,.$$

Obviously the energy stored in the spring $s_{\mu\nu}$ or dissipated in the dashpot $r_{\mu\nu}$ is independent of μ and ν. Therefore

$$s_{\mu\nu} = s_{\nu\mu} \text{ and } r_{\mu\nu} = r_{\nu\mu}, \text{ also } m_{\mu\nu} = m_{\nu\mu} \,. \tag{20.49}$$

If (20.48) is introduced in (20.46) which is modified by the dissipation and by outside forces, one finds

$$(M)\left(\frac{d^2\xi}{dt^2}\right) + (R)\left(\frac{d\xi}{dt}\right) + (K)(\xi) = (F) \,. \tag{20.50}$$

Here $v = d\xi/dt$ is used. $(M), (K), (R)$ are the mass, stiffness and damping matrices, $(v), (\xi)$ are the velocity and displacement vectors.

This system of linear differential equations can be solved as usual by assuming the $e^{-i\omega t}$ dependence

$$\left[-\omega^2(M) - i\omega(R) + (K)\right](\tilde{\xi}e^{-i\omega t}) = (\tilde{F}e^{-i\omega t}) \,. \tag{20.51}$$

Because of the symmetry relations (20.49), the matrices $(M), (R), (K)$ are also symmetric. In a less compact form the system of linear equations (20.51) as obtained from (20.48) for the usual case $m_{\mu\nu} = 0$, would look as follows

$$a_{11}\tilde{\xi}_1 + a_{12}\tilde{\xi}_2 + a_{13}\tilde{\xi}_3 + \ldots = \tilde{F}_1$$
$$a_{12}\tilde{\xi}_1 + a_{22}\tilde{\xi}_2 + a_{23}\tilde{\xi}_3 + \ldots = \tilde{F}_2$$
$$a_{13}\tilde{\xi}_1 + a_{23}\tilde{\xi}_2 + a_{33}\tilde{\xi}_3 + \ldots = \tilde{F}_3$$
$$\quad " \qquad\qquad " \qquad\qquad "$$

(20.52)

with

$$a_{nn} = -\omega^2 m_n - i\omega\left(r_n + \sum_{\nu=1}^{N} r_{\nu n}\right) + s_n + \sum_{\nu=1}^{N} s_{\nu n}$$

$$a_{n\mu} = i\omega r_{n\mu} - s_{n\mu} \qquad\qquad \text{for } n \neq \mu .$$

An essential feature of (20.51) and (20.52) is the symmetry of the matrix. It is a consequence of the quadratic form of the energy expressions. It can also be considered as an example of Newton's actio = reactio law.

20.3.2 Reciprocity and Conjugate Energies

If all outside forces except $\tilde{F}_{1\alpha}$ are zero, the displacement amplitude $\tilde{\xi}_{2\alpha}$ of the second mass is obtained from (20.51) or (20.52) by applying the rules of linear algebra

$$\tilde{\xi}_{2\alpha} = -\tilde{F}_{1\alpha} \begin{vmatrix} a_{12} & a_{23} & a_{24}\ldots \\ a_{13} & a_{33} & a_{34}\ldots \\ " & " & " \\ " & " & " \end{vmatrix} \Big/ \text{Det} \quad .$$

Similarly, if all outside forces except $\tilde{F}_{2\beta}$ vanish, we find for the displacement amplitude of the mass m_1

$$\tilde{\xi}_{1\beta} = -\tilde{F}_{2\beta} \begin{vmatrix} a_{12} & a_{13} & a_{14}\ldots \\ a_{23} & a_{33} & a_{34}\ldots \\ " & " & " \\ " & " & " \end{vmatrix} \Big/ \text{Det} \quad .$$

Det is the determinant of the matrix; the vertical lines also indicate determinants. Dividing the two expressions, we find

$$\tilde{\xi}_{2\alpha}\tilde{F}_{2\beta} = \tilde{F}_{1\alpha}\tilde{\xi}_{1\beta} .$$

(20.53)

Since the masses can be numbered in any way, (20.53) expresses the principle of reciprocity. It states, if in an experiment α a force $\widetilde{F}_{\mu\alpha}$ acting at mass m_μ causes a displacement $\widetilde{\xi}_{\nu\alpha}$ at mass m_ν, a force $\widetilde{F}_{\nu\beta}$ acting at mass m_ν causes a displacement $\widetilde{\xi}_{\mu\beta}$ at mass m_μ in such a way that the relation $\widetilde{F}_{\mu\alpha}\,\widetilde{\xi}_{\mu\beta} = F_{\nu\alpha}\,\widetilde{\xi}_{\nu\beta}$ holds. If the motions are expressed in terms of angles ϕ, the corresponding relation is ($M =$ moment)

$$\widetilde{M}_{\mu\alpha}\widetilde{\phi}_{\mu\beta} = \widetilde{M}_{\nu\alpha}\widetilde{\phi}_{\nu\beta} \, .$$

Similar expressions are possible if other coordinates are used, but in any case the product which is equal has to have the dimension of an energy or a mechanical power.

The same type of argument that led to (20.53) can also be used to prove Heaviside's law of conjugate energies. It says the following: if in an experiment α the forces $\widetilde{F}_{\nu\alpha}$ acting at positions ν cause the displacements $\widetilde{\xi}_{\mu\alpha}$ at the different positions $\mu \neq \nu$, then in another experiment β the forces $\widetilde{F}_{\mu\beta}$ acting at positions μ cause the displacements $\widetilde{\xi}_{\nu\beta}$ at positions ν in such a way that the following relation for the conjugate energies (i.e. force from one experiment multiplied by the displacement from the other one at the same point) holds (see also Figure 20.5),

$$\sum \widetilde{\xi}_{\nu\alpha}\widetilde{F}_{\nu\beta} = \sum \widetilde{\xi}_{\mu\alpha}\widetilde{F}_{\mu\beta} \, . \tag{20.54}$$

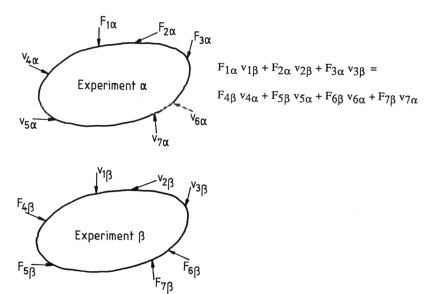

Figure 20.5 Law of conjugate energies (example).

20.3.3 Eigenfrequencies and Modes (Modal Expansion)

The equations of motion for multi degree of freedom systems (20.50) and (20.51) are very similar to the equation for a mass–spring–dashpot system; therefore one expects similar phenomena to occur. One such phenomenon is resonance, i.e. very high amplitudes in the lightly damped case when the exciting frequency is close to a resonance. As (20.53) etc. show, the resonance condition is

$$\text{Det} = \alpha_0 + \alpha_1 \omega^2 + \alpha_2 \omega^4 \ldots \alpha_{N-1} \omega^{2(N-1)} + \alpha_N \omega^{2N} = 0 . \tag{20.55}$$

Here the determinant is written as a polynomial in ω^2. There are a maximum of N values of ω^2 (the roots of (20.55)) which make the determinant vanish; therefore there are at most N resonances. The roots of (20.55) may be complex; thus the general solution of (20.51) for vanishing excitation is

$$\xi_n(t) = \sum_{\nu=1}^{N} (a_{n\nu} e^{-i\omega_\nu t} + b_{n\nu} e^{i\omega_\nu t}) e^{-\delta_\nu t} . \tag{20.56}$$

$\omega_\nu - i\delta_\nu$ and $-\omega_\nu - i\delta_\nu$ are the complex roots of (20.51), the coefficients $a_{n\nu}$ and $b_{n\nu}$ depend on the initial conditions. It is seen that the free motion of a multi degree of freedom system consists of many decaying harmonic vibrations which may give rise to rather complicated vibration patterns. When no damping is present (r_n and $r_{\mu\nu}$ are zero) the decay constants δ_ν vanish and all the roots of (20.55) are real.

The following discussion concerning the normal modes (eigenmodes) is restricted to systems without damping. But the functions obtained this way may also be used when the damping is small and when the roots of (20.55) have small imaginary parts. If the damping is so high that in the frequency response of a system the individual resonances can no longer be identified separately, difficulties may arise; they are responsible for the problems associated with the modal analysis of complicated systems with high damping.

Without proof, the following properties of normal frequencies and normal modes are listed.

(a) Without damping and outside excitation ($R = 0, F = 0$), Equation (20.51) becomes an eigenvalue equation

$$\left[-\omega^2 (M) + (K) \right] (\widetilde{\xi}_E) = 0 \text{ or } \omega^2 (M)(\widetilde{\xi}_E) = (K)(\widetilde{\xi}_E) . \tag{20.57}$$

The roots of ω_ν^2 of the determinant obtained from (20.55) are the eigenvalues or normal frequencies. They are all real and positive since (M) and (K) are real and symmetric. If some of the roots of (20.55) should be equal, we assume that a very small change in (M) and (K) is made which has no practical influence on the vibration behaviour but causes a minute difference between the eigenvalues. (There are more formal and more exact methods to approach the problem of equal roots but this so–called degeneracy problem would lead too far here.)

(b) For each value of ω_ν^2 with $1 \leq \nu \leq N$ there is one amplitude vector $(\tilde{\xi}_{\nu E})$ with components $\tilde{\xi}_{1\nu E}, \tilde{\xi}_{2\nu E}, \tilde{\xi}_{3\nu E} \dots$ which solves (20.57). These so–called eigenvectors are completely determined by (20.57) except for a constant factor.

(c) The eigenvectors are orthogonal, i.e. the free constant factor just mentioned can be chosen in such a way that

$$(\tilde{\xi}_{\mu E})^t (M)(\tilde{\xi}_{\nu E}) = \delta_{\nu\mu} , \qquad (20.58)$$

and consequently

$$(\tilde{\xi}_{\mu E})^t (K)(\tilde{\xi}_{\nu E}) = \delta_{\nu\mu}\omega_\nu^2 ;$$

$\delta_{\mu\nu}$ = Kronecker's symbol.

(d) Since the eigenvectors form a complete set, any amplitude distribution for an excitation by outside forces with a fixed angular frequency ω and amplitudes $\tilde{F}_1, \tilde{F}_2 \dots \tilde{F}_N$ can be described as

$$\tilde{\xi}_n = \sum_{\nu=1}^{N} \alpha_\nu \tilde{\xi}_{n\nu E} , \qquad (20.59)$$

where the coefficients α_ν are given by

$$\alpha_\nu = \frac{1}{\omega_\nu^2 - \omega^2} \sum_{\mu=1}^{N} \tilde{F}_\mu \tilde{\xi}_{\mu\nu E} \qquad (20.60)$$

Proofs for these very important formulae can be found, e.g. in Courant & Hilbert (1931). Since in most cases α_ν becomes very small, provided that $\omega_\nu^2 \gg \omega^2$ (i.e. when the exciting frequency is well below the ν–th resonance frequency), it may be sufficient to extend the sum in (20.59) to a value of ν which is far below the maximum value N. This is especially true for FEM calculations, where N may be many thousands and where the interesting frequencies may be below the tenth resonance. In such a case, a maximum value of ν around twenty or thirty quite often is sufficient. This way tremendous decrease in computing time is achieved.

(e) The modal expansion equations (20.59) and (20.60) hold in good approximation when a small damping is present. All that has to be done in this case is to replace ω_ν^2 by its complex value obtained from (20.55). The eigenvectors are assumed unchanged by the damping.

20.4 CONTINUOUS, RESONATING SYSTEMS

20.4.1 Equations of Motion and Modal Expansion

Resonators consisting of finite continuous media can be treated in much the same was a multi degree of freedom systems. The first step is to establish a partial differential equation or a system of such equations which describe the motion. These equations can be found from conservation laws or from Hamilton's principle.

Typically, an equation obtained this way has the following form

$$L\{p(x,t)\} = F(x,t) . \tag{20.61}$$

Here p is the field quantity of interest, $L\{\ldots\}$ is a linear differential operator, and F a representation of the force which is driving the system from outside. If we restrict ourselves to purely harmonic motion, (20.61) usually takes the form

$$L\{\widetilde{p}(x)\} - \omega^2 m\widetilde{p}(x) = \widetilde{F}(x) . \tag{20.62}$$

The continuous (no–source) version of (20.62) has non–trivial solutions only at certain eigenfrequencies ω_ν. At these frequencies the solution is the ν–th normal mode, which is called $\phi_\nu(x)$, and which is given by

$$L\{\phi_\nu(x)\} - \omega_\nu^2 m\phi_\nu(x) = 0 . \tag{20.63}$$

The conditions at the boundaries of the finite continuous system enter through the values of ω_ν and ϕ_ν, because all ϕ_ν must fulfill the boundary conditions. If there is no damping or other loss mechanism, ω_ν is real and the normal modes are orthogonal, i.e.

$$\int \phi_\nu(x)m\phi_\mu(x)dx = \delta_{\nu\mu} . \tag{20.64}$$

The integral has to be taken over the entire structure, x stands for x_1 in one dimension and for x_1, x_2 or x_1, x_2, x_3 in two or three dimensions. it can be shown that the ϕ_ν form a complete set of orthogonal functions; therefore any other function can be expressed as a linear combination of the ϕ_ν. In particular,

$$\widetilde{F}(x) = \sum_{\nu=1}^{\infty} F_\nu \phi_\nu(x)m \tag{20.65}$$

with

$$F_\nu = \int \widetilde{F}(x)\phi_\nu(x)dx \ . \tag{20.66}$$

This can be proved by multiplying (20.65) with $\phi_\mu(x)$, integrating over x, and applying (20.64).

Quite often in the literature the quantity m, which is the mass per unit length, area or volume, is assumed to be constant. In such cases m does not appear in (20.64) and (20.66) but in some other way. The final result is unaffected by that. The expansion formula can also be applied to $\widetilde{p}(x)$ in (20.62), giving

$$\widetilde{p}(x) = \sum_{\nu=1}^{\infty} p_\nu \phi_\nu(x). \tag{20.67}$$

Introducing this and (20.65) into (20.62) results in

$$\sum_{\nu=1}^{\infty} p_\nu \left[L\{\phi_\nu(x)\} - \omega^2 m\phi_\nu(x) \right] = \sum_{\nu=1}^{\infty} F_\nu \phi_\nu(x)m \ . \tag{20.68}$$

The linear differential operator $L\{\}$ can be inserted from (20.63) giving

$$\sum_{\nu=1}^{\infty} p_\nu \left[m\omega_\nu^2 - m\omega^2 \right]\phi_\nu(x) = \sum_{\nu=1}^{\infty} F_\nu \phi_\nu(x)m \ . \tag{20.69}$$

Multiplication with $\phi_\mu(x)$ and integration yields, because of (20.64),

$$p_\nu = \frac{F_\nu}{\omega_\nu^2 - \omega^2}; \ \widetilde{p}(x) = \sum_{\nu=1}^{N} \frac{\phi_\nu(x)}{\omega_\nu^2 - \omega^2} \int \widetilde{F}(x)\phi_\nu(x)dx \ . \tag{20.70}$$

Equation (20.70) is the solution of (20.62) provided the eigenfunctions $\phi_\nu(x)$ and the eigenfrequencies are known. The result is very similar to the method for the solution of multi degree of freedom systems described briefly at the end of the last section with the eigenfunctions $\phi_\nu(x)$ corresponding to the eigenvectors $\widetilde{\xi}_{n\nu E}$. The main difference is that in the continuous case, there is an infinite number of modes.

Strictly speaking (20.70) holds only if no energy losses are present. But if the damping is not too high, one can, without loss of accuracy, assume that the normal modes are not affected by the damping and that the eigenfrequencies have a small imaginary part. The calculations become especially simple if the concept of loss factor η (see (20.17), (20.21)) is used. All that has to be done is to replace ω_ν^2 by $\omega_\nu^2(1 - i\eta)$.

20.4.2 Examples of Resonances in Continua

Equation (20.70) is very useful for general problems or for those where average solutions are sufficient (see Statistical Energy Analysis). For specific problems, such as finding the vibration pattern of an engine under certain running conditions, (20.70) hides the real problem which lies in finding the eigenfrequencies ω_ν and mode shapes ϕ_ν for a given structure under certain boundary conditions. Nowadays FEM is very much used in this field, i.e. a continuous structure is modelled as a multi degree of freedom system and then it is a matter of taste whether the resulting mode shapes are called eigenvectors or eigenfunctions.

There are, however, a few simple cases for which ω_ν and ϕ_ν can be expressed explicitly. Some of them are given here because knowing such functions and using them for estimates is very helpful in getting a good understanding of the vibrations of continua. The list shows only two– and three–dimensional examples because one–dimensional problems will be treated in detail in the next section.

Example: (the eigenfunctions are not normalized to unity; in this way the expressions are a little shorter)

(a) Membrane with mass per unit area ρh and tension T; clamped, i.e. zero displacement at boundary

- rectangular, with sides ℓ_1 and ℓ_2:

$$\phi_\nu(x_1, x_2) = \sin \frac{\nu_1 \pi x}{\ell_1} \sin \frac{\nu_2 \pi x_2}{\ell_2}$$

$$\omega_\nu^2 = \pi^2 \frac{T}{\rho h} \left[\left(\frac{\nu_1}{\ell_1} \right)^2 + \left(\frac{\nu_2}{\ell_2} \right)^2 \right] ; \quad \nu_1, \nu_2 = 1, 2, 3 \ldots \tag{20.71}$$

- circular, with radius a

$$\phi_\nu(r, \theta) = J_{\nu_1}(\pi \gamma'_{\nu_1, \nu_2} r/a) \cos \nu_1 \theta$$

$$\omega_\nu^2 = \pi^2 \frac{T}{\rho h} (\gamma'_{\nu_1, \nu_2}/a)^2$$

$$\gamma'_{0,1} \approx 0.765; \quad \gamma'_{1,1} \approx 1.22; \quad \gamma'_{2,1} \approx 1.635; \gamma'_{0,2} \approx 1.757 \tag{20.72}$$

$$\gamma'_{\nu_1, \nu_2} \approx \frac{\nu_1}{2} + \nu_2 - \frac{1}{4} \quad \text{for } \nu_1 \text{ and } \nu_2 \text{ large .}$$

(b) Flat space filled with gas or liquid with speed of sound c; rigid walls
- rectangular, with sides ℓ_1, ℓ_2

see membrane, replace $T/\rho h$ by c^2

- circular, with radius a

see membrane, replace $T/\rho h$ by c^2 and $\gamma'_{\nu 1, \nu 2}$ by $\gamma_{\nu 1, \nu 2}$

$$\gamma_{1,0} \approx 0.586; \gamma_{2,0} \approx 0.972; \gamma_{0,1} \approx 1.22$$

$$\gamma_{3,0} \approx 1.34 \; ; \gamma_{4,0} \approx 1.693; \gamma_{1,1} \approx 1.697 \; .$$

(c) Plate in bending motion, $D = Eh^3/12(1-\mu^2) = $ bending stiffness, $E = $ Young's modulus, $\mu = $ Poisson's ratio.

- rectangular, simply supported, i.e. zero displacement and moment along boundaries:

$$\phi_\nu(x_1, x_2) = \sin \frac{\nu_1 \pi x_1}{\ell_1} \sin \frac{\nu_2 \pi x_2}{\ell_2}$$

$$\omega_\nu^2 = \pi^4 \frac{D}{\rho h} \left[\left(\frac{\nu_1}{\ell_1} \right)^2 + \left(\frac{\nu_2}{\ell_2} \right)^2 \right]^2 \; . \tag{20.73}$$

(d) Ring of radius a and thickness h:

$$\phi_\nu = \cos \nu \theta$$

$$\omega_0 = \sqrt{E/\rho}/a$$

$$\omega_\nu^2 = \frac{E}{\rho} \frac{h^2}{12a} \frac{\nu^2 (\nu^2 - 1)^2}{\nu^2 + 1}$$

and $\tag{20.74}$

$$\omega_\nu^2 = \frac{E}{\rho} \frac{1}{a^2} \{1 + \nu^2\}$$

$$\omega_\nu^2 = \frac{E}{\rho} \frac{1}{a^2} \frac{1-\mu}{2} \nu^2 \; .$$

(e) Cylinder of radius a and length ℓ; simply supported:

$$\phi_\nu(z, \theta) = \cos \nu_1 \theta \sin \frac{\nu_2 \pi z}{\ell}$$

$$\omega_\nu^2 \approx \frac{E(\nu_2 \pi a/\ell)^4}{\rho a^2 \left[(\nu_2 \pi a/\ell)^2 + \nu_1^2 \right]^2}$$

$$+ \frac{Eh^2}{\rho 12 a^4} \left\{ \left[(\nu_2 \pi a/\ell)^2 + \nu_1^2 \right]^2 - \frac{\nu_1^2 (4 - \mu) - 2 - \mu}{2(1 - \mu)} \right\} \tag{20.75}$$

and

$$\omega_\nu^2 \approx \frac{1}{a^2} \frac{E}{\rho} \left[1 + \nu_1^2 + \left(\frac{\nu_2 \pi a}{\ell} \right)^2 \right]$$

$$\omega_\nu^2 \approx \frac{1}{a^2} \frac{E}{\rho} \frac{1-\mu}{2} \left[\nu_1^2 + \left(\frac{\nu_2 \pi a}{\ell} \right)^2 \right] \; .$$

(f) Room filled with gas or liquid with speed of sound c; rigid walls:

- rectangular with sides ℓ_1, ℓ_2, ℓ_3

$$\phi_\nu(x_1, x_2, x_3) = \cos\frac{\nu_1 \pi x_1}{\ell_1} \cos\frac{\nu_2 \pi x_2}{\ell_2} \cos\frac{\nu_3 \pi x_3}{\ell_3}$$

$$\omega_n^2 = \pi^2 c^2 \left[\left(\frac{\nu_2}{\ell_1}\right)^2 + \left(\frac{\nu_2}{\ell_2}\right)^2 + \left(\frac{\nu_3}{\ell_3}\right)^2 \right],$$

(20.76)

- cylindrical with radius a and length ℓ

$$\phi_\nu(r, z, \theta) = J_{\nu 1}\left(\frac{\omega_\nu r}{a}\right) \cos \nu_2 \theta \cos(\nu_3 \pi z / \ell)$$

$$\omega_\nu^2 = \pi^2 c^2 \left[\left(\frac{\gamma_{\nu 1, \nu 2}}{a}\right)^2 + \left(\frac{\nu_3}{\ell_3}\right)^2 \right]$$

(20.77)

$\gamma_{\nu 1, \nu 2}$ see flat space .

20.4.3 Resonances in One–Dimensional Continua

Resonances in one dimension can usually be calculated for any boundary condition, because there are only two waves (possibly additional near fields) which propagate in opposite direction. All that has to be done is to adjust the amplitudes of these waves in such a way that the the boundary conditions are fulfilled; the resonance is then determined as a frequency for which the amplitude becomes very large.

Another way of finding resonances is to apply the principle of "closed wave trains". The idea behind this method (Cremer & Heckl 1988) is to add up the phase shifts which a wave experiences during a "round trip". Resonances are then defined by "closed trains", i.e. when the phase shifts add up to a multiple of 2π.

A possible way of expressing the sound waves in a pipe or duct is

- pressure

$$p_1(x) = A e^{ik_1 x_1} + B e^{-ik_1 x_1}$$

$$p_2(x) = C e^{ik_2 x_2} + D e^{-ik_2 x_2}$$

- velocity

(20.70)

$$v_1(x) = \frac{A}{\rho_1 c_1} e^{ik_1 x_1} - \frac{B}{\rho_1 c_1} e^{-ik_1 x_1}$$

$$v_2(x) = \frac{C}{\rho_2 c_1} e^{ik_2 x_2} - \frac{D}{\rho_2 c_2} e^{-ik_2 x_2} .$$

Here $\rho_1, c_1, k_1 = \omega/c_1$ are density, speed of sound and wavenumber in section 1 of the pipe (see Figure 20.6) and $\rho_2, c_2, k_2 = \omega/c_2$ are the corresponding quantities in section 2. For waves on strings or longitudinal or torsional waves in rods, the same relations hold basically.

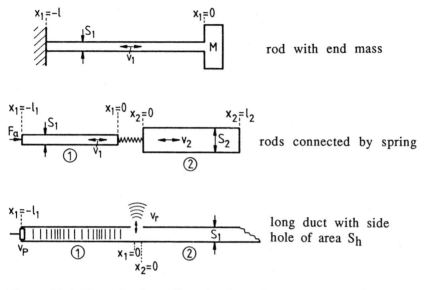

rod with end mass

rods connected by spring

long duct with side hole of area S_h

Figure 20.6 Example of one–dimensional resonators.

In each equation the first term represents the wave travelling in positive direction, the second term the wave in negative direction. The extension to more than two sections is obvious. If section 2 is infinite or very long and damped, D is zero. The unknown factors can be found by applying the boundary conditions at $x_1 = 0, x_2 = \ell_2$ and the two boundary conditions at $x_1 = \ell_1$ or $x_2 = 0$.

Examples.

• rod with mass M at one end (see Figure 20.6)
  boundary conditions:

$$C = D = 0$$

$$v_1(-\ell) = 0$$

$$S_1 p_1(0) = -i\omega M v_1(0)$$

(inertial force from mass and force at end of rod are equal).

Characteristic equation for resonances:

$$\frac{\omega M}{\rho_1 c_1 S_1}\, tgk_1\ell = 1$$

$$\text{or}\quad \left(\frac{\omega\ell}{c_1}\right)^2 + \frac{1}{3}\left(\frac{\omega\ell}{c_1}\right)^4 + \frac{2}{15}\left(\frac{\omega\ell}{c_1}\right)^6 \dots = \frac{\rho_1 S_1 \ell}{M}.$$

(20.79)

- rod with exciting force F_a at $x_1 = -\ell_1$, spring of stiffness s connecting the two sections, free end at $x_1 = \ell_1$ (see Figure 20.6)

  boundary conditions

$$Ae^{-ik_1\ell_1} + Be^{ik_1\ell_1} = F_a/S_1$$

$$Ce^{ik_2\ell_2} + De^{-ik_2\ell_2} = 0$$

$$S_1 p_1(0) = S_2 p_2(0); \quad S_1(A+B) = S_2(C+D)$$

$$\frac{-s}{i\omega}\left[v_1(0) - v_2(0)\right] = S_1 p_1(0); \quad \frac{-s}{i\omega}\left[\frac{A-B}{\rho_1 c_1} - \frac{C-D}{\rho_2 c_2}\right] = S_1(A+B).$$

The first two equations are obvious; the third states that forces are transmitted without change through a massless spring, the fourth equation is Hooke's law; it relates the displacement difference $(v_1 - v_2)/(-i\omega)$ to the acting force $p_1 S_1$. Carrying out the calculations yields

$$A = \frac{F_a}{-2iS_1}\,\frac{-i\alpha + \epsilon}{\alpha\cos k_1\ell_1 + \epsilon\sin k_1\ell_1}$$

$$B = \frac{F_a}{-2iS_1}\,\frac{i\alpha - \epsilon}{\alpha\cos k_1\ell_1 + \epsilon\sin k_1\ell_1}$$

$$D = \frac{-i}{tgk_2\ell_2}\,\frac{S_1}{S_2}(A+B); \quad C = De^{-2ik_2\ell_2}$$

$$\alpha = \frac{s}{\omega\rho_1 c_1 S_1}; \quad \epsilon = 1 - \frac{s}{\omega\rho_2 c_2 S_2 tgk_2\ell_2}.$$

(20.80)

- very long pipe with driving piston and side hole (see Figure 20.6)

  boundary conditions:

$$D = 0;$$

$$v_1(-\ell_1) = v_p; \quad Ae^{-ik_1\ell_1} - Be^{ik_1\ell_1} = v_p\rho_1 c_1$$

$$p_1(0) = p_2(0); \quad A + B = C;$$

$$Z_r\left[v_1(0) - v_2(0)\right] = p_1(0); \quad \frac{Z_r}{\rho_1 c_1}(A - B - C) = A + B.$$

The first two equations are obvious; the third one states that at the hole there is no jump in pressure; in the fourth equation Z_r is the radiation impedance (see Equation (20.28)); it relates the volume flow through the hole (i.e. the difference of the volume flow on both sides of the hole) to the pressure at this point

$$Z_r = \frac{p_1}{v_r} = \frac{p_1}{S_h(v_1 - v_2)/S_1} \approx \frac{S_1}{S_h}(\rho c k^2 a^2 - i\omega\rho a) \ .$$

The last part of this equation is obtained by assuming that the hole acts as a monopole radiator of area $S_h = 4\pi a^2$. After a few calculations, we find

$$\begin{aligned}
&A = v_p\rho_1 c_1(\rho_1 c_1 + 2Z_r)/2\beta; \quad B = -v_p\rho_1 c_1 \cdot \rho_1 c_1/2\beta \\
&C = v_p\rho_1 c_1 Z_r/2\beta; \quad \beta = (\rho c + Z_r)\cos k_1\ell_1 + jZ_r \sin k_1\ell_1 \ .
\end{aligned} \tag{20.81}$$

Typically Z_r is small compared with ρc; therefore the resonance condition $|\beta|$ = Min is close to the relation $\cos k_1\ell_1 = 0$ which holds for a pipe closed on one end and open at the other (the position of the hole).

If there are more than two sections in a one–dimensional system, (20.78) can be extended in a straightforward way. There is nothing basically new to be done, but obviously the calculation effort becomes bigger.

Sometimes it is useful to writer (20.78) in terms of a transfer matrix for each section

$$\begin{aligned}
p(\ell) &= \quad p(0)\cos k\ell - j\rho c v(0)\sin k\ell \\
v(\ell) &= -\frac{j}{\rho c}p(0)\sin k\ell + v(0)\cos k\ell \ .
\end{aligned} \tag{20.82}$$

Here $p(0), v(0)$ and $p(\ell), v(\ell)$ are the field quantities at the beginning and the end of a section. A combination of sections obviously results in matrix multiplication. The method underlying (20.82) is used extensively in muffler design (see e.g. Munjal 1987).

One–dimensional combinations of beams in bending motion can be treated in a rather similar fashion, because the near fields which play a rôle here can also be expressed by exponential functions. There are four field quantities which have to be considered; therefore four linear equations are needed for each section. If v is the velocity, w the angular velocity, M the moment, F the force of the bending motion, these equations are

$$
\begin{pmatrix} v(x) \\ w(x) \\ M(x) \\ F(x) \end{pmatrix} = \begin{pmatrix} 1, & 1, & 1, & 1 \\ ik_B, & -ik_B, & -k_B, & k_B \\ \dfrac{-Dk_B^2}{i\omega}, & \dfrac{-Dk_B^2}{i\omega}, & \dfrac{Dk_B^2}{i\omega}, & \dfrac{Dk_B^2}{i\omega} \\ -\dfrac{Dk_B^3}{\omega}, & \dfrac{Dk_B^3}{\omega}, & \dfrac{Dk_B^3}{i\omega}, & -\dfrac{Dk_B^3}{i\omega} \end{pmatrix} \begin{pmatrix} Ae^{ik_B x} \\ Be^{-ik_B x} \\ Ce^{-k_B x} \\ De^{-k_B(\ell-x)} \end{pmatrix}
\tag{20.83}
$$

Here $k_B^4 = \omega^2 \rho S / D$ is the bending wavenumber, D the bending rigidity, S the cross–section. The unknown factors A, B, C, D which appear here have to be found for each section from the boundary conditions. It would be beyond the scope of these notes to work out examples for (20.83). Computers are usually used to solve the systems of linear equations which describe connected beams and rods.

REFERENCES

Boltzmann, L. (1876). Ann d. Physik, Erg. Bd. 7:624–654.

Courant, R. & Hilbert, D. (1931). Methoden der mathematischen Physik Bd.1. Springer–Verlag, Berlin, Chapter 1.

Cremer, L., Heckl, M. & Ungar, E.E. (1988). Structure Borne Sound. Springer– Verlag, Berlin, Chapter II.4a.

Johnson, K.L. (1985). Contact Mechanics. Cambridge University Press, Chapter 4.2.

Leissa, A.W. (1969). Vibration of Paltes. NASA SP–160.

Mechel, F.P. (1989). Schallabsorber Band I. S. Hirzel Verlag Stuttgart, Kap.9.9.

Morse, P.M. (1948). Vibration and Sound. McGraw Hill Book Company, Chapter VII.28.

Munjal, M.L. (1987). Acoustics of ducts and mufflers. J. Wiley New York, Chapters 1 and 2.

Temple, G. & Bickley, W.G. (1933). Rayleigh's Principle and its Application to Engineering. Dover Publications.

21. BUBBLES

21.1 INTRODUCTION

Bubbly liquids, even with small volume concentrations of air bubbles, have very unusual hydrodynamic and acoustic behaviour. Extreme near–field pressures can be developed from the highly nonlinear oscillation of bubbles in an unsteady environment, and there is a renewal of interest in this topic because of the chaotic dynamical behaviour that is now known to occur in many single degree of freedom systems with nonlinearity and subject to periodic forcing. As far as the acoustic properties are concerned, small concentrations of bubbles by volume can lead to sound speeds (at low frequencies) which are much lower even than the pure air sound speed, and which can be lower by a factor of 10 or more than the pure water speed. At higher frequencies the sound speed can become even lower, its real part being essentially zero for a range of frequencies around the "typical" bubble resonance frequency and then, for still higher frequencies, being *greater* than the pure water sound speed. Dissipation is highly significant at all but the lower frequencies, and especially significant around the bubble resonance frequencies. These features have stimulated much work on sound propagation in bubbly liquids, the scattering and transmission of sound by clouds of bubbly liquid (such as boundary layers and wakes) and the propagation of finite–amplitude sound pulses (which have a shock wave and/or soliton structure which is not possible in pure water).

In this Chapter we discuss first the dynamics of a single bubble, giving the nonlinear Rayleigh equation, the linear bubble resonance frequency and the forced response on and off resonance. Compressibility effects are included where appropriate, along with a mention of dissipative mechanisms. Then we look at a distribution of bubbles, obtaining first the low–frequency sound speed from elementary continuum principles and discussing second, on the basis of a model involving a sparse distribution of identical bubbles, the dispersive effects associated with incorporation of the inertia effects in the near fields of individual bubbles. Some implications of these results are given as regards linear acoustics, and then the nonlinear equation for propagation of a low–frequency finite–amplitude pulse is quoted and its shock wave and soliton types of solution are mentioned.

21.2 MOTION OF A SINGLE BUBBLE

Consider a single bubble at rest apart from radial, spherically symmetric oscillations, in infinite otherwise homogeneous fluid. Assume first that the fluid is strictly incompressible, inviscid and non–(heat)–conducting; we assume nevertheless that the fluctuations in the bubble are isothermal rather than adiabatic, on the basis that an adiabatic compression of the whole bubble–plus–fluid system produces only a small temperature rise in the liquid and any higher temperatures produced in the gas bubble are rapidly reduced by conduction across the boundary of the bubble.

If r is the distance from the bubble, whose radius at time t is $R(t)$, then the potential is the monopole solution of Laplace's equation with $\partial\phi/\partial r = \dot{R}$ on $r = R$, namely

$$\phi = \frac{-R^2\dot{R}}{r} \quad . \tag{21.1}$$

The pressure $p(r, t)$ is found from Bernoulli's integral

$$\frac{\partial\phi}{\partial t} + \frac{p}{\rho_\ell} + \frac{1}{2}\left(\frac{\partial\phi}{\partial r}\right)^2 = \frac{p_\infty}{\rho_\ell} \tag{21.2}$$

to be

$$p = p_\infty + \rho_\ell\left\{\frac{R^2\ddot{R}}{r} + \frac{2R\dot{R}^2}{r} - \frac{R^4\dot{R}^2}{r^4}\right\} \tag{21.3}$$

where p_∞ is the ambient pressure and ρ_ℓ is the uniform liquid density. Now if p_g is the (uniform) pressure inside the bubble and T the surface tension,

$$p_g = p(R) + \frac{2T}{R} \quad , \tag{21.4}$$

while if the gas behaves isothermally

$$p_g R^3 = p_0 a^3 \tag{21.5}$$

where p_0 and a are the pressure and radius with no motion, so that

$$p_0 = p_\infty + \frac{2T}{a} \quad . \tag{21.6}$$

These equations lead to an equation, first found by Rayleigh, for oscillation of a bubble of equilibrium radius a in fluid at pressure p_∞;

$$R\ddot{R} + \frac{3}{2}\dot{R}^2 + \frac{p_\infty}{\rho_\ell}\left(1 - \frac{a^3}{R^3}\right) + \frac{2T}{\rho_\ell R}\left(1 - \frac{a^2}{R^2}\right) = 0 \quad . \tag{21.7}$$

There has been much analytical and numerical work done recently on this equation and generalizations of it, particularly with regard to its nonlinear aspects, and for the case when the right side contains a forcing term $-(p'_\infty/\rho_\ell)\sin\omega t$ corresponding to excitation of the bubble by an incident sound wave of amplitude p'_∞ and frequency ω.

For small oscillations we put $R = a + S(t)$ and linearize to get

$$\ddot{S} + \omega_0^2 S = 0 \tag{21.8}$$

where the bubble resonance frequency, often known as the Minnaert frequency, is given by

$$\omega_0 = \left(\frac{3p_\infty}{\rho_\ell a^2} + \frac{4T}{\rho_\ell a^3}\right)^{1/2} \quad . \tag{21.9}$$

Resonant oscillations of bubbles of sizes often found in brooks and streams fall in the audible frequency range, and contribute most of the "babbling" noise of brooks. The effect of the surface tension term here and elsewhere is small unless the bubble size is less than about 10^{-5}m, and will henceforth be neglected.

If the ambient pressure is itself a function of time, $p_\infty + p'_\infty(t)$, the equation for forced (linear) oscillations is

$$\ddot{S} + \omega_0^2 S = -\frac{p'_\infty}{\rho_\ell a} \tag{21.10}$$

and at frequencies much less than the bubble resonance frequency the response is simply

$$S = -\frac{p'_\infty}{\rho_\ell a \omega_0^2} \quad , \tag{21.11}$$

or

$$S = -\frac{1}{3}\left(\frac{p'_\infty}{p_\infty}\right)a \tag{21.12}$$

when surface tension can be ignored.

The general inclusion of compressibility effects is complicated and we shall deal just with the linear problem. Then the potential satisfies

$$(\nabla^2 + k_0^2)\phi = 0 \tag{21.13}$$

if we assume a time dependence $\exp(-i\omega t)$ for all fluctuating quantities, and write $k_0 = \omega/c_0$. The appropriate spherically symmetric radiating solution is

$$\phi = r^{-1} A \exp(ik_0 r) ,$$

the condition at the linearized bubble wall $r = a$ giving

$$\frac{\partial \phi}{\partial r} = \left(-\frac{A}{a^2} + \frac{ik_0 A}{a} \right) e^{ik_0 a} = -i\omega S$$

if the bubble radius is $R(t) = a + S(t)$ and $S(t) = S e^{-i\omega t}$. Calculating the pressure at $r = a$ from ϕ and equating it to the gas pressure obtained by linearizing (21.5) then gives the expression

$$\omega^2 = \frac{3 p_\infty}{\rho_\ell a^2} (1 - ik_0 a) \tag{21.14}$$

for free oscillations, surface tension being here neglected.

The radius is always small compared with the acoustic wavelength in pure water, and hence $k_0 a$ is small; then if we write $\omega = \omega_1 + i\omega_2$ we get

$$\omega_1^2 = \left(\frac{3 p_\infty}{\rho_\ell a^2} \right) \quad \text{and} \quad \omega_2 = -\omega_1 \left(\frac{k_0 a}{2} \right) \tag{21.15}$$

with a negative sign corresponding to *radiation damping*. For forced oscillations, with $p'_\infty(t) = p'_\infty e^{-i\omega t}$, the response $S e^{-i\omega t}$ is given by

$$\left[-\omega^2 + \omega_1^2 - 2i\delta\omega \right] S = -\frac{p'_\infty}{\rho_\ell a} \tag{21.16}$$

where the damping coefficient is

$$\delta = \omega_1 \left(\frac{k_0 a}{2} \right) \tag{21.17}$$

and ensures a finite response at resonance, $\omega = \omega_1$. At frequencies below resonance we find

$$\phi = \left(\frac{p'_\infty}{\rho_\ell i\omega} \right) \left(\frac{a}{r} \right) e^{ik_0 r}$$

so that the pressure on the bubble surface is just p'_∞ itself. This pressure is then scattered to infinity with a (uniform) monopole directivity and provides a powerful mechanism by which the field of an acoustically inefficient source can be enhanced. Suppose, for example, that the bubble lies within the near field of a turbulent eddy, of lengthscale L and velocity scale $U \ll c_0$. Then the order of magnitude of p'_∞ is $\rho_\ell U^2$ and the pressure scattered to infinity from the forced oscillations of the bubble is

$$p \sim \rho_\ell U^2 (a/r)$$

whereas the radiation from the eddy itself is of quadrupole nature, with

$$p \sim \rho_\ell U^2 (L/r) M^2$$

where $M = U/c_0$ is the Mach number. The field scattered by the bubble is, in order of magnitude, that of free turbulence with a lengthscale $L \sim M^{-2} a$, vastly greater than the size of the scattering object.

The damping factor δ in (21.16) is in practice not usually dominated by radiation damping, which is the only form of dissipation so far considered. There are three other "elementary" forms of dissipation (along with other more esoteric mechanisms associated with surface chemistry). These are

(i) viscous dissipation associated with the radial motion in the neighbourhood of the bubble,

(ii) viscous dissipation arising from the relative motion between bubble and surrounding fluid (if a massless sphere moves with velocity $U(t)$ through fluid which, far away, has velocity $V(t)$, the force on the sphere is

$$-\frac{2}{3}\pi a^3 \rho_\ell \left(\frac{dU}{dt} - \frac{dV}{dt} \right) + \frac{4}{3}\pi a^3 \rho_\ell \frac{dV}{dt} \ ,$$

the first pair of which represent the relative acceleration with "virtual mass" $(2/3)\pi a^3 \rho_\ell$, the last a buoyancy term arising from the ambient pressure gradient into which the sphere is placed — and the force must be zero, so that

$$\frac{dU}{dt} = 3\frac{dV}{dt} \tag{21.18}$$

and there is boundary–layer drag in this relative motion),

(iii) thermal conduction across the wall of the bubble and the thermal dissipation of energy.

Assessments of the relative importance of these four dissipation mechanisms have been made; none of the four is always negligible, but the relative importance of each depends rather sensitively and differently on the frequency and on the actual distribution of bubble sizes. There are circumstances of practical relevance in which each of the four mechanisms dominates the other three.

21.3 SOUND SPEED IN BUBBLY LIQUID — LOW FREQUENCIES

Let ρ_ℓ, ρ_g be the liquid and gas densities in a *suspension* of gas bubbles in liquid; by *suspension* we mean that there is no relative motion between bubbles and gas. We assume that the bubbles are small and sparsely distributed, but sufficiently densely distributed on some macroscopic scale that they can be regarded as determining the properties of a single phase *continuum* of density ρ.

Let α be the volume concentration of bubbles, i.e. the fraction of unit volume of mixture occupied by gas. Then the continuum density is

$$\rho = (1 - \alpha)\rho_\ell + \alpha\rho_g \qquad (21.19)$$

and because of the assumption of no relative motion, the mass of gas in unit *mass* of mixture must be constant; the volume of unit mass of mixture is $1/\rho$, the fraction of this occupied by bubbles is α/ρ and the mass of gas in this fraction is $\alpha\rho_g/\rho$, and hence

$$\frac{\alpha\rho_g}{\rho} = \text{constant} \qquad . \qquad (21.20)$$

Suppose now that equilibrium conditions prevail, of uniform pressure everywhere within the bubbles and the liquid. This will be a proper assumption at very low frequencies — which turns out to mean frequencies far below the resonance frequency ω_0. Then we can calculate the sound speed c in the mixture from its definition $c^2 = dp/d\rho$, using (21.19) and (21.20) and the definitions $c_\ell^2 = dp/d\rho_\ell$, $c_g^2 = dp/d\rho_g$ of the liquid and gas sound speeds. We have, from (21.19),

$$\frac{d\rho}{dp} = \frac{(1 - \alpha)}{c_\ell^2} + \frac{\alpha}{c_g^2} + (\rho_g - \rho_\ell)\frac{d\alpha}{dp} \qquad . \qquad (21.21)$$

Now from (21.20)

$$\alpha\rho_g = \left(\frac{\alpha\rho_g}{\rho}\right)\rho$$

where the factor $(\alpha\rho_g/\rho)$ is constant, so that

$$\rho_g\frac{d\alpha}{dp} + \frac{\alpha}{c_g^2} = \left(\frac{\alpha\rho_g}{\rho}\right)\frac{d\rho}{dp} ,$$

which gives

$$\frac{d\alpha}{dp} = \left(\frac{\alpha}{(1-\alpha)\rho_\ell + \alpha\rho_g}\right)\frac{d\rho}{dp} - \frac{\alpha}{\rho_g c_g^2} \quad . \tag{21.22}$$

Substitution of this in (21.21) gives, on reduction,

$$\frac{1}{c^2} = \frac{d\rho}{dp} = \frac{(1-\alpha)^2}{c_\ell^2} + \frac{\alpha^2}{c_g^2} + \alpha(1-\alpha)\left\{\frac{\rho_\ell^2 c_\ell^2 + \rho_g^2 c_g^2}{\rho_\ell \rho_g c_\ell^2 c_g^2}\right\} \quad . \tag{21.23}$$

As

$$\alpha \to 0$$

$$c \to c_\ell \, ,$$

and as

$$\alpha \to 1$$

$$c \to c_g \quad .$$

However, the transition from c_ℓ to c_g is not monotonic, and for quite modest values of α, c drops to values not only much less than c_ℓ, but indeed much less than c_g, before recovering from below to the pure gas value c_g as $\alpha \to 1$.

In (21.23) we can approximate the last term by

$$\frac{\alpha(1-\alpha)\rho_\ell}{\rho_g c_g^2}$$

and if this is then compared with either of the first two terms, it is seen to dominate unless α is very close to 0 or to 1. Thus

$$c^2 \approx \frac{\rho_g c_g^2}{\alpha(1-\alpha)\rho_\ell}$$

where, if isothermal conditions hold in the bubbles, $c_g^2 = p_0/\rho_g$ and if adiabatic conditions hold, then $c_g^2 = \gamma p_0/\rho_g$ with γ the specific heat ratio. Then

$$c^2 \approx \frac{\gamma p_0}{\alpha(1-\alpha)\rho_\ell} \quad , \tag{21.24}$$

with γ taken as 1 for isothermal behaviour.

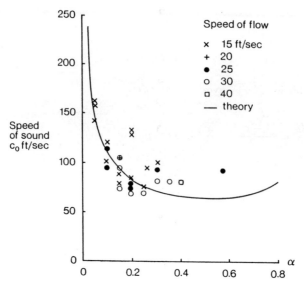

Figure 21.1 Theory and measurement (from Campbell & Pitcher 1954) for low–frequency sound speed as a function of volume concentration α. Speed of flow (water tunnel with gas bubble injection at a grid): ×15 ft/sec, + 20, • 25, ∘ 30, □ 40. Line: theory represented by (21.24).

Note the minimum value of c as a function of α for given p_0 and ρ_ℓ; this occurs at $\alpha = 0.5$, corresponding to closely packed bubbles, and gives, for water at one bar, $c = 20$ m/sec. The theoretical result (21.24) used in Figure 21.1 is based on the assumption of *isothermal* bubble oscillations, which seems to be the more realistic assumption. Note also that $c \ll c_\ell$, and even $< c_g$, for quite small values of α; if $\alpha = 1\%$ at $p_0 = 1$ bar, $c = 100$ m/s.

21.4 SOUND SPEED IN BUBBLY LIQUID — DISPERSIVE EFFECTS

At all but the lowest frequencies the equilibrium assumptions used above no longer hold, and in particular the pressures in the bubbles and in the surrounding fluid cannot be taken as equal because of the inertia of the fluid near each bubble. This introduces *dispersion*, the frequency dependence of the sound speed. We consider *linear* bubble oscillations, in the absence of surface tension and dissipation, in which case the mean pressure p_0 is the same in bubbles and liquid.

We introduce the following:

mixture density	$\rho = \rho_0 + \rho'$
pressure in liquid	$p = p_0 + p'$
liquid density	$\rho_\ell = \rho_{\ell 0} + \rho'_\ell$
pressure in gas	$p_g = p_0 + p'_g$
bubble radius	$R = R_0 + R'$
(identical bubbles)	
number density (number of bubbles per unit volume of mixture)	$n = n_0 + n'$
liquid and gas velocity	$u = u'$
volume concentration	$\alpha = \alpha_0 + \alpha'$

$$(21.25)$$

The continuity equation expressing mass conservation for the mixture is (for one–dimensional flow)

$$\frac{\partial \rho}{\partial t} + u \frac{\partial \rho}{\partial x} + \rho \frac{\partial u}{\partial x} = 0 , \qquad (21.26)$$

and the corresponding equation for conservation of bubbles is

$$\frac{\partial n}{\partial t} + u \frac{\partial n}{\partial x} + n \frac{\partial u}{\partial x} = 0 , \qquad (21.27)$$

while the Euler equation for the mixture is

$$\rho \left(\frac{\partial u}{\partial t} + u \frac{\partial u}{\partial x} \right) + \frac{\partial p}{\partial x} = 0 . \qquad (21.28)$$

The linearized versions of these are

$$\frac{\partial p'}{\partial t} + \rho_0 \frac{\partial u'}{\partial x} = 0 , \qquad (21.29)$$

$$\frac{\partial n'}{\partial t} + n_0 \frac{\partial u'}{\partial x} = 0 , \qquad (21.30)$$

$$\rho_0 \frac{\partial u'}{\partial t} + \frac{\partial p'}{\partial x} = 0 . \qquad (21.31)$$

If we neglect the gas contribution $\alpha \rho_g$ to ρ then we have $\rho = (1 - \alpha)\rho_\ell$, which gives $\rho_0 = (1 - \alpha_0)\rho_{\ell 0}$ and

$$\rho' = (1 - \alpha_0)\rho'_\ell - \frac{\alpha' \rho_0}{(1 - \alpha_0)} . \qquad (21.32)$$

The fluctuation ρ'_ℓ in liquid density can be related simply to p' through the liquid sound speed,

$$p' = c^2_\ell \rho'_\ell \quad , \tag{21.33}$$

and the linearized version of the Rayleigh equation is

$$p'_g - p' = \frac{\rho_0}{(1 - \alpha_0)} \ddot{R}' \quad , \tag{21.34}$$

where the pressure p in the Rayleigh equation has been taken to be the mixture pressure p'. Then the relation $\alpha = (4/3)\pi R^3 n$ gives

$$\alpha_0 = \frac{4}{3}\pi R^3_0 n_0$$

and

$$\alpha' = 4\pi R^2_0 n_0 R' + \frac{4}{3}\pi R^3_0 n' \quad . \tag{21.35}$$

Finally, if isothermal conditions prevail in the bubbles then $p_g R^3 = p_0 R^3_0$, or in linearized form,

$$p'_g R^3_0 - p_0 3 R^2_0 R' = 0 \quad . \tag{21.36}$$

In (21.29)–(21.36) we have a set of eight homogeneous equations for the eight fluctuations defined in (21.25). They are equivalent to a higher–order wave equation for any one of them. To derive this it is simplest to assume a variation $\exp(ikx - i\omega t)$ for each quantity. Then, after much algebra, one finds, quite simply,

$$\frac{k^2}{\omega^2} = \frac{(1 - \alpha_0)^2}{c^2_\ell} + \frac{1}{c^2_0(1 - \frac{\omega^2}{\omega^2_0})} \quad , \tag{21.37}$$

where now c_0 refers to the low–frequency sound speed as defined by (21.24) with $\gamma = 1$, and ω_0 is the bubble resonance frequency,

$$\omega_0 = \left(\frac{3p_0}{\rho_{\ell 0} R^2_0} \right)^{1/2} \quad .$$

The left side of this equation is the inverse square of the phase speed $c_p(\omega)$ at frequency ω ($c_p(\omega) = \omega/k$). For $\omega \ll \omega_0$ the second term dominates on the right, and we get $c_p(\omega) = c_0$, recovering the result of Section 21.3. We see also that $c_p = 0$ at $\omega = \omega_0$, and remains zero from ω_0 to a higher frequency given approximately (for $\alpha_0 \ll 1$) by

$$\omega \sim \omega_0 \left(\frac{c_\ell}{c_0} \right) \quad . \tag{21.38}$$

For frequencies just above this value, c_p is very large, and for much higher values of ω, c_p remains larger than c_ℓ, approaching the value $c_\ell/(1-\alpha_0)$ as $\omega \to \infty$. We observe that while at low frequencies the effect of the bubbles is to make the mixture less stiff than the water, the bubbles actually make the mixture *more stiff* than pure water when driven above their resonance frequency. Thus while for $\omega \ll \omega_0$ we have essentially the stiffness all in the gas and the inertia all in the liquid, for $\omega \gg \omega_0$ we have again the inertia all in the liquid, but now with increased liquid stiffness because of the bubbles.

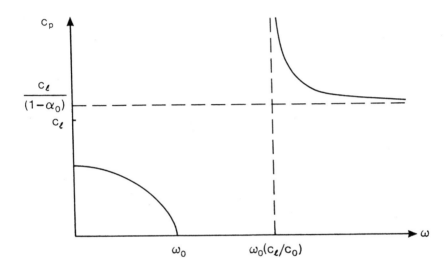

Figure 21.2 Schematic plot of the variation of the phase speed $c_p(\omega)$ (taken as real) defined by (21.37) as a function of frequency ω.

Naturally, the introduction of dissipation removes the singular behaviour seen in Figure 21.2, though the trends illustrated there have certainly been seen in experiments — in particular (in the same experiment with a fixed concentration but with varying frequency) the initial decrease in frequency (below a value c_0, itself well below the pure liquid sound speed), followed by a rapid increase to values much higher than the pure liquid sound speed (to values as high as 3000m/sec), followed in turn by a decrease towards the pure liquid sound speed, have all been observed. If dissipation is modelled in the bubble response equation by the inclusion of a factor

δ as in (21.16), the modification to (21.37) is easily found to be

$$\frac{k^2}{\omega^2} = \frac{(1-\alpha_0)^2}{c_\ell^2} + \frac{1}{c_0^2\{(1-\frac{\omega^2}{\omega_0^2}) - 2i\delta\frac{\omega}{\omega_0}\}} \quad,$$

and the form of variation of the real part of ω/k is as in Figure 21.3 below. The imaginary part of the phase speed, i.e. the attenuation rate, peaks in the vicinity of the bubble resonance frequency — and hence attenuation measurements can be used to get information on the distribution of bubble sizes in the ocean.

The striking properties of the variation of sound speed with frequency and with concentration have led to many attempts to control sound propagation away from propellers on marine vessels by the use of "bubble screens". These, however, have advantages and disadvantages, because while the dissipation may be large near resonance conditions, it is not large elsewhere and then the large ρc impedance-mismatch can both reduce and amplify incident sound waves, depending upon the screen thickness. There is further the point (Crighton & Ffowcs Williams 1969) that the acoustic output of turbulent bubbly flow exceeds that of the same flow in pure liquid by the large factor $(c_\ell/c_0)^4$ (this applying to turbulent fluctuations at frequencies well below typical bubble resonance frequencies — which are of order 1 kHz for bubbles with $R_0 = 1$ mm at one bar pressure). Thus if the bubble screen is allowed to interact with the turbulent boundary layer or wake, the enhanced sound generation by the turbulent flow may offset any benefit derived from the reflecting properties associated with the large jump in sound speed at the interface between bubbly and clear liquid.

21.5 NONLINEAR WAVES IN BUBBLY LIQUID

The low–frequency (long wave) behaviour of (21.37) is the same in form as that of the linear version of the so–called Boussinesq equation which governs weakly dispersive surface gravity waves on shallow water. One might expect that a nonlinear theory of pressure waves in bubbly liquids would lead to equations similar to the nonlinear Boussinesq equations — which are known to possess soliton solutions. Indeed, the Boussinesq equations reduce, for waves travelling in one direction only, to the KdV equation in which soliton behaviour was first observed. One–dimensional weakly nonlinear compression waves of long wavelength in bubbly liquid do indeed satisfy the KdV equation, in the form

$$\frac{\partial p}{\partial t} + c_0\frac{\partial p}{\partial x} + \frac{c_0}{p_0}p\frac{\partial p}{\partial x} + \frac{c_0^3}{2\omega_0^2}\frac{\partial^3 p}{\partial x^3} = 0 \quad, \tag{21.39}$$

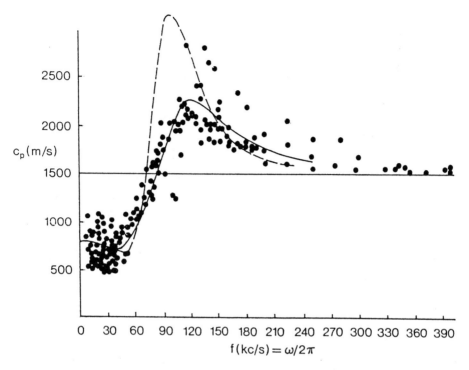

Figure 21.3(a) Sound speed in gas bubble–water mixture as a function of frequency. $R_0 = 1.2 \times 10^{-4}$m $\pm 4.10^{-5}$m, $\alpha = 2 \times 10^{-4}$. The solid curve represents the theoretical results, with $\delta = 0.5$, for the actual bubble distribution. The broken curve would hold if all the bubbles had the same size $R_0 = 1.1 \times 10^{-4}$m. The dots represent measured values (Fox, Curley & Larson 1955).

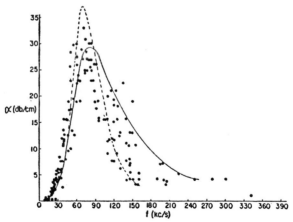

Figure 21.3(b) Attenuation of pressure waves in a water–air bubble mixture. (Further details are the same as in Figure 21.3a above.)

608

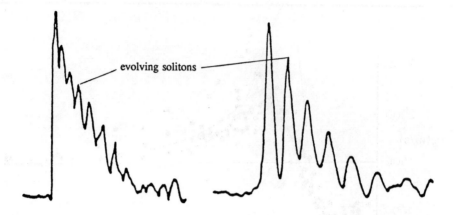

Figure 21.4 Evolution of solitons from a triangular initial pressure disturbance in a bubbly liquid (from Roelofsen 1981.)

where p is the pressure perturbation (Van Wijngaarden 1972). The existence of soliton–type pulses in bubbly liquids has been seen experimentally; see the sketch in Figure 21.4 above, and the reference quoted.

In some applications, the effect of dissipation is important, and then the left side of (21.39) acquires a term

$$-\frac{\delta c_0^2}{2\omega_0}\frac{\partial^2 p}{\partial x^2} \qquad (21.40)$$

and the equation becomes the Burgers–KdV equation (BKdV). The BKdV cannot have soliton solutions (indeed it cannot be integrated using any of the schemes currently known) and instead it has shock wave solutions giving a transition from p_1 to $p_2 > p_1$. If the dispersive term in (21.39) is not large enough compared with the dissipative term, the transition is essentially like the shock wave in a pure liquid or gas and given by a hyperbolic tangent solution to the Burgers equation. If the dispersion is large enough, the shock transition develops oscillations on the downstream side (cf. Figure 21.5), the structure of which is well predicted by theory based on the BKdV equation.

In summary, the linear and nonlinear theory of sound propagation in bubbly liquids is well developed, shows a range of quite remarkable effects, and gives predictions in quite reasonable accord with careful experiments.

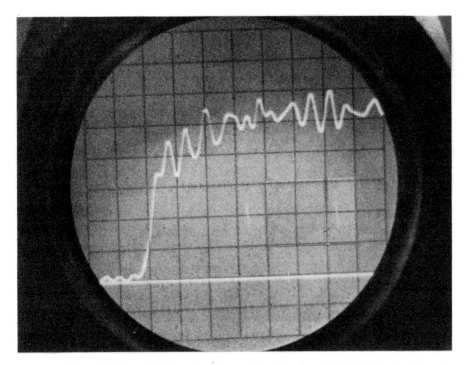

Figure 21.5 Shock transitions in (a) pure liquid or gas, (b) bubbly liquid with sufficient dispersion (from Noordzij 1973).

REFERENCES

Batchelor, G.K. (1969). Compression waves in a suspension of gas bubbles in liquid. Fluid Dynamics Transactions (Poland) 4:425–445.

Campbell, I.J. & Pitcher, A.S. (1954). Flow of air–water mixtures. Symposium at Admiralty Research Laboratory, Teddington, Middlesex.

Crighton, D.G. & Ffowcs Williams, J.E. (1969). Sound generation by turbulent two–phase flow. J. Fluid Mech. 36:585 603.

Fox, F.E., Curley, S.R. & Larson, G.S. (1955). Phase velocity and absorption measurements in water containing air bubbles. J. Acoust. Soc. Am. 27:534–539.

Noordzij, L. (1973). Shock Waves in Mixtures of Liquid and Air. Doctoral Thesis, Technological University, Twente, Netherlands.

Roelofsen, P. (1981). Solitons in Liquid/Bubble Mixtures. Master's Thesis, Technological University, Twente, Netherlands.

Wijngaarden, L. van (1972). One–dimensional flow of liquids containing small gas bubbles. Ann. Rev. Fluid Mech. 4:369–396.

22. REVERBERATION

22.1 INTRODUCTION

When someone speaks in a large empty room, speech intelligibility is rather poor, because the sound which is just emitted is mixed with the reverberating sound that was emitted a few tenths of a seconds (or even seconds) earlier. When someone strikes a bell or a similar lightly–damped mechanical system, a sound can be heard for many seconds because it takes rather a long time until the vibratory energy in the system is transferred into heat or radiated.

When a strong sound pulse is emitted under water, one may get back a small reverberant signal which is caused by the sound reflection (echos) at the sea surface, the sea bottom, or by obstacles in the water (e.g. fish). The time it takes for a reverberation to decay may be milliseconds or less at high frequencies; it may be of the order of a few second in large rooms, or it may be many minutes when a small explosion in the arctic sea is observed. The discussion of the common features of such reverberation phenomena is the topic of this Chapter.

22.2 DECAY OF RESONANCES IN SYSTEMS WITH FEW MODES

The response of a simple resonator to a sudden pulse of delta–function type (e.g. a hammer blow on a mass spring system) is given by the Green's function (see Equation (20.10)). If the amplitude is squared and averaged over a time which is much shorter than the decay time but longer than the period of the resonance frequency, we get

$$\upsilon^2(t) = \upsilon_0^2 e^{-2\delta t},$$

(22.1)

Since squaring and averaging is done by many common instruments, this decay can be observed quite easily and the decay constant δ can be measured. Very often the decay is expressed by a reverberation time T, which is defined as the time necessary for an amplitude to decrease by a factor of 1000, i.e. 60 dB. The relation between decay constant and reverberation time, as well as with other quantities, is given in Table 22.1. The quantity Λ which appears in the Table is the logarithmic decay; it is defined as

$$\Lambda = \ln(v_n/v_{n+1}) \ .$$

Here v_n and v_{n+1} are the amplitudes at two times which are separated by one period; usually the maxima of two adjacent periods are taken. Λ is a useful quantity when the damping is so high that averaging over one period, which is underlying (22.1), does no longer make sense.

Table 22.1 Equations relating damping parameters at resonance.

$$\delta = r/2m \ ; \ \omega_R^2 = \frac{s}{m} - \delta^2 \ ; \ \delta = 6.9007/T \ ;$$

$$\eta \approx r/\omega_R m \ ; \ \eta = 13.815/\omega_R T \ ; \ \eta = \Lambda/\pi \ ;$$

$$\eta = 1/Q \ ; \ \eta = \tan\phi \ ; \ \eta = b/\omega_R \ .$$

$s =$ stiffness, $m =$ mass, $r =$ damping (see Equation (20.9)), $T =$ reverberation time, $\eta =$ loss factor, $\omega_R =$ resonance frequency, $\Lambda =$ logarithmic decrement, $Q =$ quality factor, $\phi =$ phase shift between force and displacement, $b =$ bandwidth of resonance (distance between 3 dB down points in radians).

When reverberation measurements are made by recording the decay of the motion after a continuous excitation is suddenly switched off, some problems may arise because the type of continuous signal has an influence on decay. If the driving signal has its main frequency content far away from the resonance, there is hardly any reverberation, because the decay of energy can take place only near the resonance (i.e. the solution of the homogeneous vibration equation). If the driving signal has a very strong component at the resonance frequency, the decay follows Equation (22.1). Strange results may occur when the driving signal has its strongest component near but not exactly at the resonance; in this case there may be beating phenomena similar to those that are found when a resonator is excited by a single frequency source near the resonance (see Equation (20.8)). The exponential decay given by (22.1) can be completely disturbed by this beating.

When a multi degree of freedom resonator is excited by a pulse, the decay is again determined by the Green's function which in this case is of the form

$$v_n(t) = \sum_{\nu=1}^{N} v_{n\nu} e^{-\delta_\nu t} \sin \omega_\nu t \ . \tag{22.2}$$

Here v_n is the velocity of the n–th element, δ_ν and ω_ν are the decay constants and resonance frequencies of the modes, and $v_{n\nu}$ are the modal amplitudes at position n. The values of $v_{n\nu}$ are different when different elements of the system are excited; they also depend on the type of excitation.

When a system has only a few resonances which are well separated (i.e. the spacing between resonances is large compared with the bandwidth caused by damping), the modes in (22.2) can be separated by frequency filtering. This way a multi–modal system becomes almost a system of independent single degree of freedom resonators and can be treated this way. The situation is much more complicated when the frequency spacing is such that the motion at one frequency consists of several modes of comparable amplitude. In such a case the reverberation curves consist of a sum of exponential functions and if the δ_ν are different, one gets different decay curves for different points of excitation or points of measurement, i.e. for different values of the amplitudes $v_{n\nu}$. If measurements are made at many points (e.g. the excitation points are chosen in such a way that only one mode is excited), it is possible to distinguish modes due to their spatial distribution and find individual decay constants. This is one of the problems in modal analysis.

22.3 REVERBERATION IN SYSTEMS WITH MANY MODES (e.g. LARGE ROOMS)

22.3.1 Modal Approach

As long as the damping is small, the vibration behaviour of any system can be described as a sum of modes; thus (22.2) is valid. There are, however, two difficulties;

(a) the number of modes can be so large that they can no longer be handled individually. As examples we mention an air–filled space of 50 m³ which as approximately 53000 modes below 1000 Hz and a 0.5 mm aluminium plate of $2 \ m^2$ which has approximately 2800 modes below 2000 Hz. For comparison we have to keep in mind that a reverberation time of $T = 1$ seconds corresponds (according to Table 22.1) to a bandwidth of 13.8 radian or 2.2 Hz;

(b) in the equation describing the motion of damped systems, the losses are assumed to be distributed over the whole system. This is especially true when we use the concept of loss factor, which is based on the assumption that the materials have inner losses that can be expressed by a complex modulus. In reality, however, the main losses are very often concentrated, e.g. as sound absorbing linings in rooms or as friction–damped interfaces in complicated structures.

The difficulty associated with the great number of modes does not cause any problems as long as one does not attempt to distinguish individual modes with their decay constants. In most cases it is even possible to assume that the decay constants δ_ν within a certain frequency range are equal; therefore (22.2) can be approximated to

$$v_n(t) \approx e^{-\delta t} \sum_{\nu=1}^{N} v_{n\nu} \sin \omega_\nu t \ . \tag{22.3}$$

In the average this again gives (22.1). It should be borne in mind, however, that this is an approximation and therefore it might well be that, in actual cases, the reverberation is not exponential, i.e. it does not follow a straight line on a dB versus time plot, and that the result depends on the type of excitation and the measurement position. Typically, reverberation curves are "hanging through", because the modes with long reverberation time which may not have been fully excited, determine the long–term decay. Apart from a gross deviation from the exponential decay, reverberation measurements in multi–modal systems show quasi–random fluctuations of the decay level (see Figure 22.1a). The uncertainties caused this way could be avoided by averaging over a great many curves. Fortunately, there is a more elegant method called "the method of integrated impulse response" (Schröder 1965; Kuttruff 1973).

To prove the underlying equations we assume that a room or other system is excited by a random noise signal $r(t)$. If this signal is switched off at $t = 0$, the decaying signal is given by the convolution

$$p(t) = \int_{-\infty}^{0} r(\tau)g(t - \tau)d\tau = \int_{t}^{\infty} g(\tau)r(t - \tau)d\tau \ . \tag{22.4}$$

In this relation, which holds for $t \geq 0$, the function $g(\tau)$ is the Green's function or impulse response function at the point of measurement for the source point of interest. Squaring and averaging gives

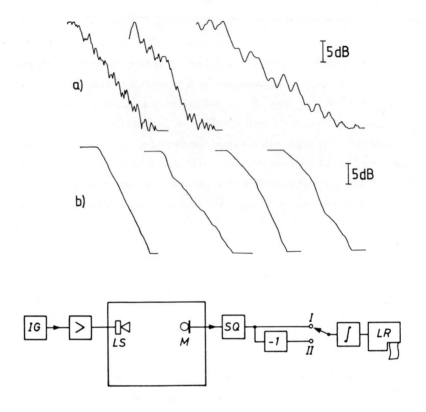

IG = random noise generator
SQ = squaring device
LR = level recorder
∫ = integrator e.g. capacitor
-1 = multiplier with minus one
switch position I the signal is integrated and stored,
switch position II exactly the same excitation is used, the signal is integrated, subtracted from the stored signal, the difference is recorded.

Figure 22.1 Experimentally obtained reverberation curves.

(a) conventional method; (b) integrated impulse response.

$$< p^2(t) > \ = \ < \int_t^\infty g(\tau)r(t-\tau)d\tau \int_t^\infty g(z)r(t-\tau)dz >$$

$$= \int \int g(\tau)g(z) < r(t-\tau)r(t-z) > d\tau dz$$

$$= A \int \int g(\tau)g(z)\delta(\tau-z)d\tau dz = A \int_t^\infty g(\tau)^2 d\tau \qquad (22.5)$$

$$= A \int_0^\infty g(\tau)^2 d\tau - A \int_0^t g(\tau)^2 d\tau \ .$$

The average here is taken over many realizations of the exciting random noise signal $r(t)$. The delta function appears because $< r(t-\tau)r(t-z) >$ is the autocorrelation function which for purely white noise is known to be of delta type (See Chapter 3). The rest of (22.5) is obvious; the last expression is shown here only because it is sometimes more convenient for measurement purposes. The advantage of (22.5) is that the continuously integrated impulse response gives much clearer reverberation curves than the impulse response itself (see Figure 22.1).

22.3.2   Reverberation as a Sequence of Reflections

22.3.2.1 One–dimensional approach

The influence of partially reflecting walls on the reverberation can easily be calculated for the sound travelling back and forth in a pipe of finite length as shown in Figure 22.2.

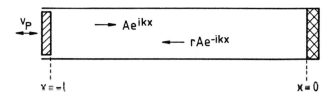

Figure 22.2 Pipe with partial reflector at $x = 0$.

If at the end $x = -\ell$ there is a rigid piston with driving velocity v_p, and if at the end $x = 0$ there is an absorbing device with reflection coefficient r, the sound field is given by (the common time factor $e^{-i\omega t}$ is omitted)

$$p = A \left(e^{ikx} + re^{-ikx} \right)$$

$$v = \frac{A}{\rho c} \left(e^{ikx} - re^{-ikx} \right) \ . \qquad (22.6)$$

Here $k = \omega/c$ is the wavenumber. The unknown factor A is obtained by applying the boundary condition $v(-\ell) = v_p$ at the perfectly reflecting end. This way we get

$$p = \frac{v_p \rho c e^{ik\ell}}{1 - r e^{2ik\ell}} \left(e^{ikx} + r e^{-ikx} \right) . \tag{22.7}$$

Writing the reflection coefficient as an absolute value and a phase shift ϕ,

$$r = |r| e^{-i\phi} = e^{-\gamma} e^{-i\phi} \tag{22.8}$$

we get

$$p = \frac{v_p \rho c e^{-\gamma/2} e^{ik\ell'}}{1 - e^{-\gamma} e^{2ikg}} \left(e^{\gamma/2} e^{ikx'} + e^{-\gamma/2} e^{-ikx'} \right) ; \tag{22.9}$$

with $g = \ell - \phi/2k$; $x' = x + \phi/2k$.

If there were perfect reflection $|r| = 1$ at $x = 0$ and instead some losses during propagation, i.e. if the wavenumber would consist of a real part k' and an imaginary part k''

$$\mathbf{k} = k' + ik'' ;$$

the wavefield would be

$$p = \frac{v_p \rho c e^{-k''\ell} e^{ik'\ell}}{1 - e^{-2k''\ell} e^{2ik'\ell}} \left(e^{-k''x} e^{ik'x} + e^{k''x} e^{-ik'x} \right) . \tag{22.10}$$

Comparing (22.9) with (22.10), we see that they are very similar, provided we set

$$|r| = e^{-\gamma} e^{-2k''\ell} . \tag{22.11}$$

The remaining differences show up only in the space dependence of the functions:
- the phase shift at the reflection has the same effect as a change in origin by $\phi/2k$;
- if the losses are small, i.e. $\gamma = 2k''\ell \ll 1$, there is practically no difference between (22.9) and (22.10); if, however, there is a large damping, i.e. $\gamma = 2k''\ell \gg 1$, then (22.9) gives a non–decaying wave $\sim e^{ik(x+\ell)}$ which is absorbed completely at $x = 0$, whereas Equation (22.10) gives a continuously decaying wave $\sim e^{-k''(\ell+x)} e^{ik'(\ell+x)}$ which disappears before it reaches the other end.

In the following, we restrict ourselves to the small damping case $\gamma = 2k''\ell \ll 1$. Introducing the absorption coefficient

$$\alpha = 1 - |r|^2 = 1 - e^{-4k''\ell} , \tag{22.12}$$

we can make a simple approximation for small damping

$$\alpha \approx 4k''\ell . \tag{22.13}$$

This equation states that a concentrated absorbing device with $\alpha \ll 1$ can be treated the same way as a (distributed) lossy medium with the imaginary wavenumber

$$k'' = \alpha/2\ell_m . \tag{22.14}$$

Here ℓ_m is the distance travelled by the wave between two reflections. In our case $\ell_m = 2\ell$.

There are two important practical consequences of absorption in a room. One is the influence on reverberation time, the other is the influence on the average sound pressure.

- The reverberation time in a one–dimensional room with absorption coefficient α and mean wave travel distance ℓ_m can be found directly from (22.14) because, by introducing the loss factor η, we can write

$$\mathbf{k} = k' - ik'' = \omega\sqrt{\frac{\rho}{K(1 - i\eta)}} \approx \omega\sqrt{\frac{\rho}{K}}(1 + i\eta/2) = \frac{\omega}{c}(1 + i\eta/2) . \tag{22.15}$$

Thus

$$\frac{\omega}{2c}\eta = \frac{\alpha}{2\ell_m} \quad \text{or} \quad T = 13.815 \, \ell_m/\alpha c . \tag{22.16a}$$

If losses during propagation due to a loss factor η_M of the medium, and losses at the boundaries due to an absorption coefficient α giving an equivalent loss factor $\eta_A = \alpha c/\omega\ell_m$ exist simultaneously, then we would have to write

$$\eta = \eta_M + \eta_A \quad \text{and} \quad T = 13.81/\omega\eta . \tag{22.16b}$$

The relation between η and T was taken from Table 22.1.

- The average sound pressure can be found by squaring and integrating (22.9) over x. The result is

$$\left|\overline{p^2}\right| = \frac{v_p^2 \rho^2 c^2 (1 + |r|^2)}{1 - 2|r| \cos 2k\ell' + r^2} . \tag{22.17}$$

This expression has maxima at $k\ell' = \nu\pi$; i.e. at the room resonances. If we assume that $k\ell'$ is large and the excitation is due to a signal covering a large bandwidth, we can integrate over the frequency (i.e. k) and get

$$\left|\overline{\overline{p^2}}\right| = v_p^2 \rho^2 c^2 \frac{1 + |r|^2}{1 - |r|^2} = v_p^2 \rho^2 c^2 \frac{2 - \alpha}{\alpha} \approx \frac{v_p^2 \rho^2 c^2}{\alpha} 2 . \tag{22.18}$$

Up to now only the wave approach has been used to describe reverberation, but it has become already quite clear that reverberation can also be understood as a sequence of reflections. To elaborate this point further we consider a very long pipe which is excited by a sound pulse (Figure 22.3).

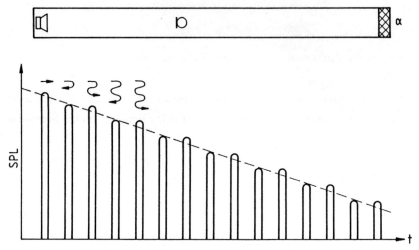

Figure 22.3 Decay of sound pulse (wavepacket) in a room.

This pulse travels without change through the pipe, gets reflected and travels back with an amplitude reduced by a factor r, etc. If we observe the sound signal at a certain position, we obviously get a sequence of decaying pulses. After n reflections the amplitude or sound pressure level (SPL) are given by

$$|p_n^2| = |p_0^2|\,|r^2|^n = |p_0|^2\,|r^2|^{tc/\ell_m} \tag{22.19}$$

or

$$L_n = 10\,\lg\left|\frac{p_n}{p_0}\right|^2 = \frac{tc}{\ell_m}10\,\lg|r^2|$$

$$= \frac{tc}{\ell_m}4.34\ln(1-\alpha) \approx -4.34\frac{\alpha tc}{\ell_m}\quad.$$

For the 60 dB decay we get, in agreement with (22.16) for the reverberation time,

$$T = 13.815\ell_m/\alpha c\quad. \tag{22.20}$$

In (22.19) we used the fact that up to the n–th reflection the time $t = n\ell_m/c$ has passed.

22.3.3.2 Statistical approach for two and three dimensions

In two– or three–dimensional spaces with small damping we can use the same arguments as in the last paragraph, i.e. we can assume that the absorption along boundaries can be "smeared out" by introducing an imaginary part of the wavenumber
(see (22.14)) and that the reverberation can be seen as a sequence of reflections
leading to (22.20). Thus the remaining question is: How large is ℓ_m? To answer
this question we assume first that all boundaries have the same average absorption
coefficient α_m.

In order to find the mean free path ℓ_m in two dimensions, we follow Kosten
(1960) and divide the area S into a large number of parallel strips of width dB (see
Figure 22.4).

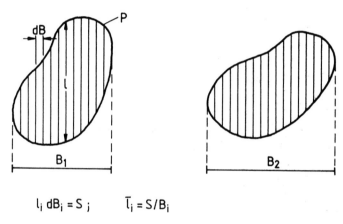

$$l_i\, dB_i = S \; ; \qquad \overline{l}_i = S/B_i$$

Figure 22.4 Derivation of mean free path.

This way we find

$$\overline{\ell}_1 = S/B_1$$

where B_1 is the projection of the contour. Next we rotate the area a little and get
new values for $\overline{\ell}_2$ and B_2. The mean free path is found by averaging the ℓ_i over all
angles of rotation. Here we have to take into account that longer projections count
more because there are more parallel strips of width dB; therefore we have to use
as statistical weight the factor $B_i/\overline{B}$, where $\overline{B}$ is the average over all B_i. This way
we get, for the mean free path,

$$\ell_m = \sum \overline{\ell}_i \frac{B_i}{B} = S/\overline{B} \quad .$$

Thus the final result is

$$\ell_m = \pi \, S/P \quad . \tag{22.21}$$

In three dimensions the argument is very similar. The only difference is that now $\overline{\cos \phi_i} = 1/2$, giving

$$\ell_m = 4 \, V/S \quad . \tag{22.22}$$

Here V is the volume of the room and S its surface. It is shown by Kosten (1960) that the expressions (22.21), (2.22) are also valid if the spaces have re–entrant parts.

Numerical calculations of mean–free paths for many rooms showed that (22.22) agrees within 1% with the exact values and is rather independent of room shape.

Introducing (22.21), (22.22) into (22.16a.b), we find, for the reverberation time in two dimensions,

$$T = 13.815/\omega\eta_M + \alpha_m cP/\pi\omega S) \quad , \tag{22.23}$$

and in three dimensions

$$T = \frac{55.26}{c} \frac{1}{4k\eta_M + \alpha_m S/V} \rightarrow \frac{0.163 \, V}{4k\eta_M V + \alpha_m S} \quad . \tag{22.24}$$

In air–filled spaces at not too high frequencies the term η_M can often be neglected giving the famous Sabine's formula with the constant 0.163 which holds for air at normal temperatures and when all quantities are SI–units.

22.4 EXAMPLES OF SOUND ABSORBERS

22.4.1 Porous Absorbers

In almost all sound absorbers for airborne sound, the actual transfer of sound energy into heat takes place in the viscous boundary layer close to the extremely large surfaces of porous materials. In most cases the motion of the skeleton of the porous material can be neglected; thus for sound–absorbing purposes, such materials (fibre glass, foam with open pores, etc.) can be described the same way as any other homogeneous medium. The linearized equations expressing continuity of mass and momentum are

$$\frac{\sigma}{\rho_0 c_0^2} \frac{\partial p}{\partial t} + \frac{\partial v}{\partial x} = 0$$

$$\frac{\rho_0}{\sigma} \frac{\partial v}{\partial t} + rv + \frac{\partial p}{\partial x} = 0 \quad . \tag{22.25}$$

Here ρ_0, c_0 are density and speed of sound in the surrounding gas, σ is the porosity, i.e. $1 - \sigma$ is the fraction of volume occupied by the fibres, etc. (usually $0.9 <$

$\sigma < 0.98$), and r is the flow resistance per unit thickness. Sometimes the equations contain a "structure factor". But since this is a fudge factor, which is only used to make the data fit when the absorber has a very complicated structure with many "pockets", it is omitted here. Applying the usual harmonic time dependence, we find that plane waves in such a material are only possible if the wavenumber is given by

$$k_p^2 = k_0^2(1 + i\sigma r/\omega \rho_0) \quad ; \qquad (22.26)$$

$k_0 = $ wavenumber in the surrounding gas. For the wave impedance of plane waves, we find from (22.25)

$$Z_p = \frac{p}{v} = \frac{\rho_0 c_0}{\sigma}\sqrt{1 + i\sigma r/\omega \rho_0} \quad . \qquad (22.27)$$

Equations (22.26), (22.27) fully describe the acoustic behaviour of a porous absorber. They especially show that sound waves decay rather fast in such an absorber because the imaginary part of the wavenumber is quite high; they also show that there is some sound reflection at the surface of a porous absorber because $Z_p \neq \rho_0 c_0$. Typical values of the flow resistance are $10^4 < r < 4 \cdot 10^4$ (see Mechel 1989).

Absorption coefficients of porous absorbers can be calculated in a straightforward (but sometimes tedious) way be performing the following steps (see also Figure 22.5):

(a) find the boundary conditions at the different interfaces of the absorber;

(b) represent the sound field in the porous absorber by

$$p_p(x) = A e^{+ik_p x} + B e^{-ik_p x}$$

$$v_p(x) = \frac{1}{Z_p}\left(A e^{+ik_p x} - B e^{-ik_p x}\right) \quad ; \qquad (22.28)$$

(c) calculate the impedance Z_{tot} at the surface of the whole combination;

(d) find the reflection coefficient form

$$r = \frac{Z_{tot} - \rho_0 c_0}{Z_{tot} + \rho_0 c_0} \quad ; \qquad (22.29)$$

(e) calculate

$$\alpha = 1 - |r|^2 \quad . \qquad (22.30)$$

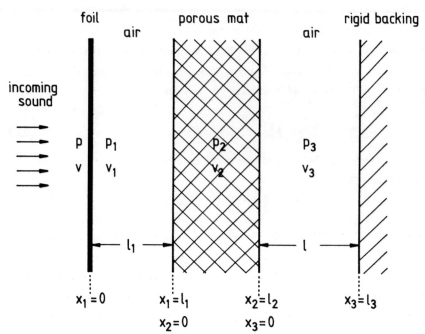

Figure 22.5 Example of absorbing lining.

Example:

Porous layers covered by thin plate or foil.

In the case shown in Figure 22.5 the pressures and velocities in the porous material are given by (22.28). For the air spaces, similar equations hold; the only difference is that k_p is replaced by k_0 and Z_p by $\rho_0 c_0$. The boundary conditions are

$$v_3(\ell_3) = 0; p_2(\ell_2) = p_3(0); v_2(\ell_2) = v_3(0); p_1(\ell_1) = p_2(0);$$
$$v_1(\ell_1) = v_2(0); v = v_1; p - p_1(0) = i\omega m v \quad . \tag{22.31}$$

Here m is the mass per unit area of the foil. There are enough boundary conditions to find $Z_{\text{tot}} = p/v$ and then to continue with (22.29), (22.30). If, instead of the foil there were a perforated plate of thickness ℓ_0, hole spacing d and hole diameter a, its effect is similar to a mass (see Chapter 20.2.4); all that has to be done is to replace m by $\rho_0(\ell_0 + 0.8a)d^2/\pi^2 a^2)$, where $d^2/\pi a^2$ is the perforation ratio.

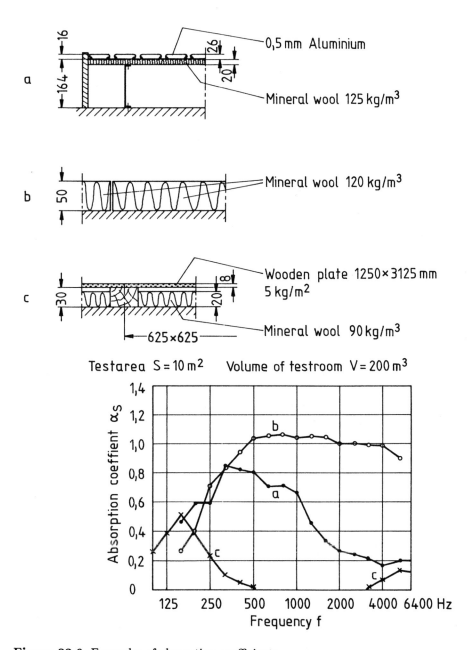

Figure 22.6 Examples of absorption coefficients.

If the foil is replaced by a thin curtain with total flow resistance $r\ell_0$, the flow resistance and the mass act "in parallel"; therefore the last equation in (22.31) would have to be replaced by

$$\frac{v}{p - p_1(0)} = \frac{1}{-i\omega m} + \frac{1}{r\ell_0} \quad . \tag{22.32}$$

For normal incidence of the incoming sound, the method just described gives reasonable agreement with measured data. If the sound comes from all directions, the formulae (22.25) through (22.31) still show which frequency ranges are absorbed predominantly, but the calculated values of α may be different from experimental data. Figure 22.6 shows three typical experimental results. The measurements were made in a reverberation chamber with sound coming from all directions.

22.4.2 Sound Absorption by Vibration Damping

Porous absorbers, though very common, are not the only means of sound absorption. Since a few years there have been rather effective airborne sound absorbers without any porous material (and without using active control). They work on the principle that sound energy is first transformed into vibratory mechanical energy and then this energy is transferred into heat by using materials with high loss factors. Sound absorbing linings in water tanks also work on this principle; rubber–like materials which are excited in one or more highly–damped resonances are generally used for this purpose.

Figure 22.7 shows sketches of such absorbers. In reality, this type of absorber looks more complicated with more layers, different materials, more complicated geometry, etc. but the basic feature is the same. It consists of vibration amplitude enhancement by one or more resonances and by consequent energy absorption in the lossy material.

The mechanism of the absorption can be readily understood when we represent the system by a single mass–spring resonator with resonance frequency ω_0 and a loss factor η. In this case, the impedance seen by the incoming sound wave is (see Chapter 20.2.2)

$$\mathbf{Z}_{\text{tot}} = -i\omega m \left[1 - \frac{\omega_0^2}{\omega^2}(1 - i\eta) \right] = Z_t' - iZ_t'' \quad . \tag{22.33}$$

Here m is the vibratory mass per unit area of the system. Introducing this into (22.29), (22.30) gives

$$\alpha = \frac{4\rho_0 c_0 Z_t'}{(Z_t' + \rho_0 c_0)^2 Z_t''^2} \quad .$$ (22.34)

The maximum of this function is at $Z_t'' = 0$, i.e. at the resonance frequency of the mass–spring system. High values of α can be achieved, especially when $Z_t' = \rho_0 c_0$, but the bandwidth can be small. This is why in practical applications multi–resonance systems are used.

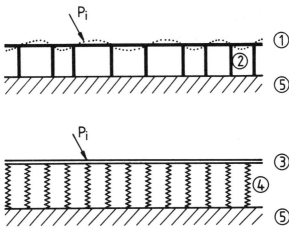

① vibrating skin with high loss factor

② support to cause resonances of the skin sections

③ cover or membrane

④ springs with high internal damping

⑤ rigid backing

Figure 22.7 Idealized examples of absorbers incorporating vibrating materials with high inner losses (e.g. high polymers).

22.5 SOUND FIELDS IN LARGE ROOMS

In Section 22.3.2.2 the formulae for reverberation were derived under the assumption that all surfaces are covered by the same sound–absorbing material with the coefficient α_m, which was assumed to be small. In reality, the distribution of absorbing material is quite irregular. Unfortunately, it is almost impossible to derive exact equations for the sound field in large rooms with several patches of highly absorbing material. Therefore, approximate methods have been developed. One of these is the "ray–tracing approach", which is based on geometrical acoustics. The general idea is to calculate many reflections of the "sound particles" which are emitted by the source. Absorption, and to a certain degree scattering, can also be taken into account in the calculations. In principle, it is possible to find in this matter the impulse response and also the steady–state response of an irregularly–shaped room for each source receiver combination.

The second method is the "image–source approach", where wall reflections are replaced by image sources. Again, impulse response and steady state can be found. The computational effort of the method can be very high, e.g. in a 15000 m^3 room consisting of 30 walls, the number of images that might contribute to the sound field within the first 0.4 seconds is of the order 10^{14}. For this reason, the method of ray–tracing and image sources has been combined (see Vian & van Maercke 1985, Krokstadt & Sorsdahl 1968 and Vorländer 1989 for more information on this problem).

These methods would, however, lead too far here; therefore we "solve" the problem by assuming that the effect of absorbing patches of area S_i and absorption coefficient α_i can be smeared out over the whole surface, i.e.

$$\alpha_m S = \sum \alpha_i S_i \quad . \tag{22.35}$$

This expression is then inserted into (22.24)

Obviously (22.35) relies on the assumption that each absorbing patch is hit by the soundwaves with equal likelihood. Thus the reverberation formula is correct only if the sound field is uniform and consists of waves travelling in all directions with the same probability. There are many cases —rooms with reflecting side walls and absorbing ceilings, sound fields in the vicinity of sources or sinks (absorbers, open windows), etc. — where these conditions are not fulfilled, but there remain many practical situations for which the simple reverberation theory can be applied. For such cases it is even possible to derive some additional results (Schröder 1954, 1962).

- If a room is big, the average distance between two modal frequencies, which according to Equation (8.21) is $\Delta\omega = 2\pi c^3/\omega^2 V$, can be much smaller than the bandwidth of a resonance for which Table 22.1 gives $b = 13.8/T$. Thus if

$f \gg c^3 T/27.6V$, we can be sure that many modes are excited even if the source should radiate only pure tones. Under this condition it is reasonable to assume that the real part p_r and the imaginary part p_i of the sound pressure amplitude $\widetilde{p} = p_r - ip_i$ at each point are random variables with a Gaussian distribution. Thus

$$w(p_r) = \frac{1}{\sqrt{2\pi\overline{p_r^2}}}e^{-p_r^2/2\overline{p_r^2}}$$

$$w(p_i) = \frac{1}{\sqrt{2\pi\overline{p_i^2}}}e^{-p_i^2/2\overline{p_i^2}} . \tag{22.36}$$

Here $w(\ldots)$ is the probability density.

- Applying the laws of probability theory, one finds from (22.36) that for pure tone excitation of a large room, the frequency response is a random function which on a dB–scale has a mean fluctuation of 11 dB.

- If $p^2(\omega)$ is the (rather irregular) frequency response of the squared pressure at any point, the average spacing between "level crossings" (points where $p^2(\omega) = \overline{p^2}$) is given by (see Rice 1944)

$$\Delta\omega_s = \pi\overline{p^2}/\sqrt{\left[\frac{dp^2(\omega)}{d\omega}\right]^2} . \tag{22.37}$$

If we apply modal expansion to a room, we find

$$p(\omega) = \sum \frac{A_i}{i\,[\omega^2 - \omega_i^2(1 - i\eta)]} = \sum \frac{A_i}{\eta\omega_i^2}\,\frac{1}{1 - i\dfrac{\omega_i^2 - \omega^2}{\omega_i^2\eta}}$$

$$\approx \frac{A_i T}{13.8\omega_i}\,\frac{1}{1 - i(\omega_i - \omega)T/0.9} . \tag{22.38}$$

Here we again applied $\eta = 1.38/T\omega_i$ and made the approximation $\omega_i^2 - \omega^2 = (\omega_i - \omega)(\omega_i + \omega) \approx 2\omega_i(\omega_i - \omega)$.

If now all the operations are made that are necessary to calculate (22.37), one finds

$$\Delta\omega_s = \pi\sqrt{\frac{2\pi}{3}}\,\frac{6.9}{T} \quad \text{or} \quad \Delta f_s \approx 5/T . \tag{22.39}$$

In going from (22.38) to (22.39) all cross products are neglected because we assume randomness.

- If the sound pressure amplitudes at two points in a diffuse sound field are correlated, the correlation function follows a

$$\frac{\sin k_0 r}{k_0 r} \tag{22.40}$$

law, where r is the distance between the measuring points (Cook et al. 1955).

- Other interesting results on the statistics of energy distribution and energy flux can be found for diffuse sound fields (Ebeling 1984).

22.6 REVERBERATION CAUSED BY MANY SCATTERERS

Reverberation can also be observed in unbounded spaces, which do not show resonant behaviour. A typical example in this respect is the reverberation in the sea where sound is scattered at obstacles and irregular boundaries give rise to a fairly long ringing after the sound source has been switched off. The main difference between reverberation caused by the reflections at the walls of a closed room and that caused by the scattering in an essentially unbounded space lies in the time dependence. Normal reverberation shows an exponential decay where pressure p (or sound pressure level L) are given by

$$p^2 \sim e^{-6.91t/T}; \quad L = \text{const} - 60t/T . \tag{22.41}$$

On the contrary, reverberation by scattering follows a power law of the type

$$p^2 \sim A t^{-n}; \quad L = \text{const} - 1 - n \ell g t . \tag{22.42}$$

The exponent n has been observed in the sea to be (Chapman 1967)

$n \approx 2$ when the scatters are distributed in a uniform way;

$n \approx 3$ when there are layers of isotropic scatters;

$n \approx 4$ when the reverberation is due mainly to surface scattering.

For the special case of a pistol shot in the woods the decay, according to Kuttruff (1967), is given by

$$p^2 \sim t^{-3/2} e^{-\beta t} . \tag{22.43}$$

This shows a power–law decay with $n = 1.5$ and an exponential decay due to the energy losses during propagation.

An estimate of the reverberation due to scattering can be obtained by calculating the one–dimensional sound propagation in the presence of many scatterers, i.e. in an opaque medium. We assume that there are n scatters per unit length and that each of them scatters a fraction Q of the incoming sound. The mean free path is then

$$\ell_S = 1/nQ \ . \tag{22.44}$$

Next we consider a large number of sound pulses propagating with speed c in both directions. Let $\rho_+(x,t)$ and $\rho_-(x,t)$ be the density of pulses in positive or negative direction. Following Kuttruff (1967), we then write

$$\rho_\pm(x \pm c\Delta t, t + \Delta t) = \rho_\pm(x,t) - \frac{c\Delta t}{\ell_S}\rho_\pm(x,t) + \frac{c\Delta t}{\ell_S}\rho_\mp(x,t) \ . \tag{22.45}$$

The three terms on the right–hand side are: undisturbed propagation, scattering losses of the primary component, scattering gains from the opposite component. Making a Taylor expansion of the left–hand side of (22.45) and comparing it with the right–hand side, gives

$$\frac{1}{c}\frac{\partial \rho_+}{\partial t} = -\frac{\partial \rho_+}{\partial x} - \frac{1}{\ell_S}(\rho_+ - \rho_-)$$
$$\frac{1}{c}\frac{\partial \rho_-}{\partial t} = \frac{\partial \rho_-}{\partial x} + \frac{1}{\ell_S}(\rho_+ - \rho_-) \quad . \tag{22.46}$$

Adding and subtracting both equations and eliminating the $(\rho_+ - \rho_-)$ terms gives

$$c^2 \frac{\partial^2 \rho}{\partial x^2} = \frac{\partial^2 \rho}{\partial t^2} + \frac{2c}{\ell_S}\frac{\partial \rho}{\partial t} \ . \tag{22.47}$$

Here $\rho = \rho_+ + \rho_-$. Equation (22.47) is the so–called "telegraphist's equation" for which solutions are known. For the special case of a delta–function excitation

$$\rho(x,0) = 2k\delta(x); \quad [\partial\rho/\partial t]_{t=0} = 0 \ ,$$

one can show that

$$\rho(x,t) \approx k \sqrt{\frac{2}{\pi\ell_S ct}}e^{-x^2/\ell_S ct} \text{ (for } ct >> x \text{ and } x > \ell_S) \tag{22.48}$$

is an asymptotic solution. (To show this insert (22.48) in (22.47) and keep only terms in $1/t^{3/2}$ and $1/t^{5/2}$). Equation (22.48) is also a solution of the diffusion equation which is obtained when the second time derivative in (22.47) is omitted. It can be seen that in the one–dimensional case, the exponential term vanishes as t becomes large; therefore the reverberation follows a $t^{-1/2}$ power law.

In three dimensions exact calculations are more involved. But if we again neglect the second time derivative in (22.47) and replace the second space derivative by its analogon in spherical coordinates, we get

$$c^2 \frac{1}{r} \frac{\partial}{\partial r} \left(r^2 \frac{\partial \rho}{\partial r} \right) = \frac{2c}{\ell_S} \frac{\partial \rho}{\partial t} \ . \tag{22.49}$$

Introducing

$$\rho = K t^\alpha e^{-\beta r^2 t^B}$$

and comparing coefficients in t, yields

$$\rho(r, t) \sim t^{-3/2} e^{-3r^2/4\ell_S ct} \ . \tag{22.50}$$

This expression agrees reasonably well with measurements in air. As well as (22.48), it shows that reverberation caused by scattering causes a decay which follows a power law in time.

REFERENCES

Chapman, R.P. (1967). Sound Scattering the the Ocean. In: Underwater Acoustics, Vol.2 (ed. V.M. Albers). Plenum Press, Chapter 9.

Cook, R.K., Waterhouse, R.V., Behrendt, R.V., Edelmann, R.D. & Rhompson, M.C. (1955). J. Acoust. Soc. Amer. 27:1072.

Ebeling, K.J. (1984). Statistical Properties of Random Wave Fields. In: Physical Acoustics (eds. Mason & Thurston), Vol.XVII, Academic Press, Chapter 4.

Kosten, C.W. (1960). Acustica 10:245.

Krokstadt, A., Strom, S. & Sorsdahl, S. (1968). J. Sound Vib. 8:118.

Kuttruff, H. (1967). Acustica 18:131.

Kuttruff, H. (1973). Room Acoustics. Applied Science Publishers London. Chapter VIII.

Mechel, F. (1989). Schallabsorber. Hirzel Verlag Stuttgart.

Rice, S.O. (1944). Bell Syste. Tech. J. 23:282.

Schröder, M.R. (1954). Acustica 4:594.

Schröder, M.R. (1962). J. Acoust. Soc. Amer. 34:1819.

Schröder, M.R. (1965). J. Acoust. Soc. Amer. 37:409.

Vian, J.P. & van Maercke, D. (1985). Proceedings of the Institute of Acoustics, Cambridge.

Vorländer, H.M. (1989). J. Acoust. Soc. Amer. 86:172.

23. SOLITONS

23.1 INTRODUCTION

The solitary wave (John Scott Russell's "Great Wave of Translation") was discovered in 1834 by Russell during investigations of the Edinburgh Union Canal. In the shallow water wave problem the solitary wave is a single hump, of large lengthscale, travelling without change of form. Its existence was predicted theoretically by Rayleigh and Boussinesq in the 1870's and then definitively by Korteweg and de Vries in 1895 in the paper in which the now–famous KdV equation was first derived. The wave of permanent form can exist only because of a delicate balance between the linear mechanism of dispersion (produced partly because of finite–depth effects and partly because of surface tension) and the nonlinear one by which the higher parts of a water wave travel more quickly than the lower. Dispersion allows the different frequency or wavelength components of a disturbance to travel away at different speeds, so that a concentrated elevation would in time disperse into an oscillatory wave train along which the wavelength changes slowly and continuously. The nonlinear mechanism in isolation leads to overturning of the wave, as the higher parts overtake the lower. The dispersive and nonlinear mechanisms precisely balance in the solitary wave, and the amplitude, lengthscale and propagation speed are all related; in a certain nondimensional set of coordinates moving with the speed of propagation of waves of infinitesimal amplitude the solitary wave solution of KdV is

$$\eta(x, t) = 2a^2 \, \text{sech}^2 \left[a(x - 4a^2 t) \right] \tag{23.1}$$

so that if the lengthscale a^{-1} is given, the amplitude is $2a^2$ and the speed $V = 4a^2$ — narrow pulses are high and travel fast.

Russell performed a remarkable series of laboratory experiments in which he obtained the speed–amplitude relation above and also, apparently, observed the amazing way in which two KdV solitary waves interact. This most important observation lay unnoticed for 130 years, and little more was done on the KdV equation or its solitary wave solutions (or those of other nonlinear wave equations or evolution equations). The solitary waves were regarded as very special solutions, in which the necessary nonlinearity–dispersion balance could only exceptionally be created and which, once established, could presumably be easily upset.

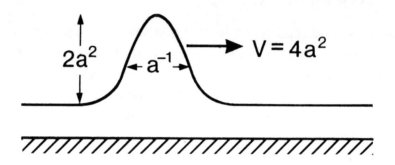

Figure 23.1 The KdV solitary wave.

In 1955, however, Fermi, Pasta and Ulam (FPU) were studying the nonlinear interactions between a chain of masses connected by nonlinear springs and observed, in numerical experiments, that an initial state did not get scrambled by these interactions but, rather, recurred again and again (FPU–recurrence). To understand this phenomenon, Kruskal and Zabusky derived a continuum model for the FPU oscillator chain, and found, at a certain level of approximation, the KdV equation. Examining numerical solutions of KdV, they observed the wave–steepening tendency of nonlinearity, followed by the development of oscillations in the waveform as dispersion began to affect the shorter scales associated with the steep parts of the wave. The interactions between the oscillations were surprising; the oscillations corresponded to the production of a series of KdV solitary waves, the larger of which overtook the smaller, leading to a complicated interaction. But eventually the large separated out ahead of the small, leaving each of the solitary waves with precisely the amplitude, shape, scale and speed that it had before the interaction. The only evidence that an interaction had taken place lay in the phase of each wave; each

each wave suffered a phase shift in the interaction, and that is all.

The implications of this were far–reaching; solitary waves of the KdV type had an identity which, once established, was permanent. Because of their particle–like properties, Zabusky and Kruskal (1965) named the solitary wave solutions of KdV "solitons" (actually, "solitrons" was proposed, but conflicted with the name of a TV company in the USA). We refer to *any* solution $\eta = f(x - ct)$, for fixed c, and such that $\eta \to$ constants as $x - ct \to \pm\infty$, as a *"solitary wave"* solution of a nonlinear wave evolution equation, and reserve the word *"soliton"* for systems in which two solitary waves interact with at most a phase shift following the interaction. Many physical systems support solitary waves; a good number of interest in a wide range of contexts have now been found in which the soliton property holds. Further, theoretical work in nonlinear waves has been revolutionized by the invention by Kruskal and his colleagues at Princeton (Gardner, Greene and Miura, principally) of a new mathematical technique, now called the *Inverse Spectral* (or *Scattering*) *Transform* (IST). This technique is a kind of Fourier Transform method for certain *nonlinear problems*, and it leads, in principle, to a complete solution of the initial-value problem (given $\eta(x, t{=}0)$ find $\eta(x, t > 0)$) for those problems. The most important feature of this general solution is that almost any $\eta(x, 0)$ will generate a finite number of solitons in $\eta(x, t)$ and these solitons are then permanent and dominate the solution in most regions of space as time increases.

Developments of IST, in particular to the range of nonlinear wave equations to which it can be applied, are continuing, as are discoveries of new physical and biological systems in which soliton–type behaviour can be observed (though perhaps not yet analysed within current limitations of IST and related methods). The following notes are intended to provide some motivation and familiarity with the ideas in their simplest context. Fortunately, that context — the KdV equation, is the one of most relevance to acoustics; it describes the evolution of pressure pulses in one of the few circumstances and media in which acoustic waves are subject to appreciable dispersion. That medium is the bubbly liquid; another example of acoustic importance involves dispersion associated with a duct or wave–guide, and here the nonlinear–dispersive modulation of a duct mode leads to a different nonlinear wave equation, the Nonlinear Schrödinger equation (NLS). This also has soliton solutions, but the mathematical apparatus for NLS is much more complicated than for KdV and will not be discussed here. An excellent introduction to the whole field is given by Drazin & Johnson (1989), while there are numerous advanced texts listed in the References.

23.2 WATER WAVES; LINEAR THEORY

Consider a water layer of uniform depth h undergoing *small* $2D$ motion in an (x, y) plane, with y vertical. There is a plane bottom at $y = -h$ and a free surface (of constant, atmospheric, pressure) at $y = 0$ in the undisturbed state. T is the surface tension divided by the water density ρ, g is gravity. The motion in the fluid is irrotational and for the potential $\phi(x, y, t)$ and surface elevation $\eta(x, t)$ we can look for solutions

$$\phi = \Phi(y) \exp(ikx - i\omega t) ,$$
$$y = d \exp(ikx - i\omega t) . \tag{23.2}$$

Solving Laplace's equation and imposing the conditions at $y = 0$ and $y = -h$ leads, in a familiar way, to the *linear dispersion relation* between ω and k,

$$\omega^2 = (gk + Tk^3) \tanh kh . \tag{23.3}$$

For wavelengths long compared with h we can take

$$\tanh kh \simeq kh - \frac{1}{3}(kh)^3 ,$$

and then

$$\omega^2 \simeq c_0^2 k^2 + k^4 \left(Th - \frac{1}{3}gh^3 \right) ,$$

or

$$\omega = \pm \left[c_0 k + k^3 \frac{(Th - \frac{1}{3}gh^3)}{2c_0} + O(k^5) \right] . \tag{23.4}$$

Here $c_0 = (gh)^{1/2}$ is the phase speed ω/k of very long waves ($kh \to 0$). The k^3 term is positive if $h < 0.5$cm (for an air–water free surface) and negative for greater depths; it is a first approximation to dispersion due to finite–depth effects and to surface tension.

Consider just waves propagating in the positive x–direction; then we need the $+$ sign, and to $O(k^3)$,

$$\omega = c_0 k + \beta k^3 , \qquad \text{say,} \tag{23.5}$$

The *phase velocity* (propagation speed of particular phases of the waves, such as the crests and troughs) is

$$c_p(k) = \frac{\omega}{k} = c_0 + \beta k^2 , \tag{23.6}$$

while the *group velocity* (propagation speed of a wave packet of dominant wavelength $2\pi/k$, or the propagation speed for energy at wavelength $2\pi/k$) is

$$c_g(k) = \frac{d\omega}{dk} = c_0 + 3\beta k^2 \ . \tag{23.7}$$

Relative to axes travelling at speed c_0, waves with larger k travel faster than those of smaller k, and the short waves disperse out farthest. Relative to the convected axes we have $c_g = 3c_p$, so that the packet travels faster than any individual wave phase in it, and wave crests and troughs must be continually created at the head of the packet and annihilated at the rear.

The dispersion relation (23.5) is equivalent to a partial differential equation for $\eta(x,t)$: it is

$$\eta_t + c_0\eta_x - \beta\eta_{xxx} = 0 \ , \tag{23.8}$$

the linearized form of the KdV equation.

23.3 FOURIER TRANSFORM SOLUTION OF INITIAL–VALUE PROBLEM

For comparison with the IST method for the nonlinear KdV it is worth putting down the FT solution for the linear KdV (in the form obtained by replacing $x - c_0 t$ by a new coordinate x)

$$\eta_t = \beta\eta_{xxx} \quad , \tag{23.9}$$

with

$$\eta(x,0) = \eta_0(x) \ .$$

The evolution of $\eta(x,t)$ with t is complicated; in the FT method we transform $\eta(x,t)$ to an equivalent $\widetilde{\eta}(k,t)$ whose time evolution is much simpler. Starting from $\eta(x,0)$ we calculate $\widetilde{\eta}(k,0)$ by *direct FT*, then $\widetilde{\eta}(k,t)$ from the *simple evolution* in k–space, then $\eta(x,t)$ by *inverse FT*.

Define

$$\widetilde{\eta}(k,t) = \int_{-\infty}^{+\infty} \eta(x,t)e^{-ikx}dx \ .$$

Then the FT of η_{xxx} is

$$\int_{-\infty}^{+\infty} \eta_{xxx}e^{-ikx}dx = (ik)^3\widetilde{\eta}(k,t) \ ,$$

after three integrations by parts and assuming $\eta, \eta_x, \eta_{xx}, \to 0$ as $|x| \to \infty$. Applying FT to the linear KdV gives

$$\frac{\partial \widetilde{\eta}}{\partial t}(k, t) = -i\beta k^3 \widetilde{\eta}(k, t) \; , \tag{23.10}$$

a simple ordinary differential equation for the evolution in k–space. The solution is

$$\widetilde{\eta}(k, t) = \widetilde{\eta}(k, 0) \exp[-i\omega(k)t] \tag{23.11}$$

where

$$\omega(k) = \beta k^3 \; ,$$

as in (23.5). Application of the inverse transform gives

$$\eta(x, t) = \frac{1}{2\pi} \int_{-\infty}^{+\infty} \widetilde{\eta}(k, 0) \; e^{ikx - i\omega(k)t} dk \; , \tag{23.12}$$

as the exact solution of the general initial–value problem.

It is not often possible to explicitly evaluate integrals of this kind; even the simplest $\eta(x, 0)$ $(\eta(x, 0) = \delta(x) \Rightarrow \widetilde{\eta}(k, 0) = 1)$ leads to Airy functions. But the behaviour at large times can be found by stationary phase methods. The upshot is that for $\beta > 0$, $t \to +\infty$ and $x/t > 0$ fixed, there is an oscillatory wavetrain — whatever $\widetilde{\eta}(k, 0)$ — with amplitude decaying as $t^{-1/2}$, while if $x/t < 0$ there is much more rapid decay (at least as fast as t^{-1}). If $\beta < 0$ oscillations are seen in $x < 0$, as one would expect from the signs of the phase and group velocities, βk^2 and $3\beta k^2$ respectively. Similar features will emerge for the nonlinear problem — together with a totally new feature, the emergence of *solitons*.

The essence of the FT procedure is illustrated in the diagram:

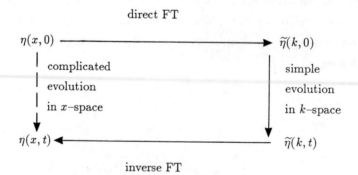

23.4 WEAKLY NONLINEAR THEORY

In the nondispersive limit $\beta \to 0$, we have $\omega = c_0 k$ or equivalently, $\eta_t + c_0\eta_x = 0$. Waves of all scales travel at speed c_0, and the solution of the initial–value problem becomes $\eta(x,t) = \eta_0(x - c_0 t)$ — representing translation at speed c_0 of the original profile $\eta_0(x)$ without change of shape. Now if the amplitude of $\eta_0(x)$ is increased, it becomes important to retain some nonlinear terms. In the present problem, nonlinear terms arise in the relation between the pressure and the potential, and they also arise because the free surface condition is really to be applied at $y = \eta(x,t)$, this leading to an infinite series of nonlinear terms when functions evaluated at $y = \eta$ are evaluated at $y = 0$ by Taylor–series expansions.

In weakly nonlinear theory one retains the dominant nonlinear terms (when all functions are expanded in a power series in some amplitude parameter); these are usually, but by no means always, quadratic. At this level of approximation the effect of nonlinearity is often to cause different parts of a waveform to propagate at different speeds, so that the constant c_0 is replaced by $c(\eta)$ where $c(\eta)$ has an expansion usually of the form $c(\eta) = c_0 + \alpha\eta + \ldots$. The weakly nonlinear but nondispersive evolution equation is then of the form

$$\eta_t + c_0\eta_x + \alpha\eta\eta_x = 0 \tag{23.13}$$

where in the commonest cases (ordinary gas dynamics and water waves) $\alpha > 0$, though there are physical circumstances in which $\alpha < 0$ (and others in which $\alpha = 0$, perhaps because of some symmetry condition, and then the lowest–order nonlinear term is cubic, proportional to $\eta^2\eta_x$).

The solution to (23.13) with $\eta(x,0) = \eta_0(x)$ (often referred to as a *simple wave* or *Riemann wave*) is

$$\eta = \eta_0\{x - (c_0 + \alpha\eta)t\}$$

or, parametrically,

$$\eta = \eta_0(\xi) \tag{23.14}$$

where $\xi(x,t)$ is defined implicitly by

$$\xi = x - c_0 t - \alpha t\eta_0(\xi)$$

and is simply a phase variable with a nonlinear distortion term. Now the initial profile η_0 is not propagated without change of shape — though the values taken by $\eta(x,t)$ are precisely the values which were present in the initial signal $\eta_0(x)$, so that the propagation process is one of *pure distortion*. The effect of the nonlinear term is to cause a steepening of backward sloping parts of the waveform, and ultimately to a triple–valued waveform which, though it may appear to represent a breaking water wave, is unacceptable physically. A typical sequence of waveforms for this type of quadratic nonlinearity is shown below:

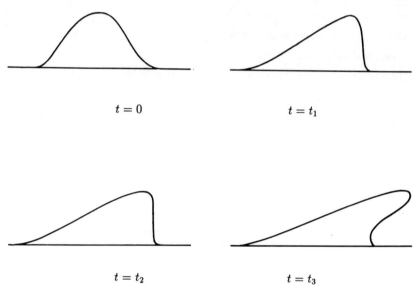

$$t = 0 \qquad\qquad\qquad\qquad t = t_1$$

$$t = t_2 \qquad\qquad\qquad\qquad t = t_3$$

Figure 23.2 Progressive distortion of a simple wave. The waveform is seen from a frame moving at speed c_0, so that the waveform would remain static in linear theory. Nonlinear distortion is seen at $t = t_1$, leading at $t = t_2$ to an infinite gradient at one point, and at times $t_3 > t_2$ to a triple–valued waveform.

Surface gravity waves do indeed have this type of nonlinear deformation when the wavelengths are long compared with the water depth. The value of α is found to be

$$\alpha = \frac{3}{2}\frac{c_0}{h} \, .$$

Note that in weakly nonlinear theory the nonlinear terms are always *locally* small; the ratio of $\alpha\eta\eta_x$ to $c_0\eta_x$ is of order (η/h) which is supposed always small. However the structure of the nonlinear terms is such that after sufficiently long times they produce a dramatic change in the wave shape; if η_m is the maximum value of η_0 and L a measure of the length of η_0, a dramatic change, possibly even breaking, will happen after a time of order $(L/\eta_m)(h/c_0)$. Before such dramatic changes can take place, other mechanisms, initially negligible, will have been brought into play. Sometimes these relate to dissipation associated with viscous forces on the high gradient parts of the waveform, and in that case a balance is ultimately achieved between convective nonlinearity and viscous dissipation in what are called *shock waves*. If dissipation is not significant, dispersion usually is, because the nondispersive approximation cannot usually be sustained for the high wavenumber components representing the steeply sloping parts of the wave. Then the balance achieved — for the water wave problem — is that of Russell's solitary wave, or soliton.

23.5 THE KORTEWEG–DE VRIES EQUATION

If both weak (quadratic) nonlinearity and weak ($\omega \sim k^3$) dispersion are allowed, an equation for η can be derived which combines the equations of the two previous sections; it is the *KdV equation*

$$\eta_t + c_0\eta_x + \alpha\eta\eta_x = \beta\eta_{xxx} \ , \tag{23.15}$$

with c_0, α, β as already defined. The translation $x' = x - c_0 t$ removes the $c_0\eta_x$ term and gives

$$\eta_t + \alpha\eta\eta_{x'} = \beta\eta_{x'x'x'}$$

and putting $\eta = k_1\eta*$, $x' = k_2 x*$, $t' = k_3 t*$ allows, with appropriate choice of k_1, k_2, k_3, any nonzero coefficients to be placed in front of the three terms. Apart from the obvious choice (dropping the $*$)

$$\eta_t + \eta\eta_x = \eta_{xxx} \ ,$$

popular (and more convenient) choices are

$$\eta_t + 6\eta\eta_x + \eta_{xxx} = 0$$

and

$$\eta_t - 6\eta\eta_x + \eta_{xxx} = 0 \ . \tag{23.16}$$

We shall work with (23.16), which is standard. The variable η is then *minus* the surface elevation.

A *travelling wave solution* of permanent form is $\eta = f(x - Vt)$ for some constant V. Put this in (23.16) with $\xi = x - Vt$; then f must satisfy

$$-Vf' - 6ff' + f''' = 0$$

or

$$-Vf - 3f^2 + f'' = A \text{ (constant)} .$$

This equation can be integrated further in terms of elliptic functions, and in particular there is a periodic elliptic function solution (representing a periodic wave of finite amplitude) $f = \operatorname{cn}^2(\xi|m)$ for some parameter m. Waves of this kind are called *cnoidal waves*.

If we look for a *localized solution*, such that f and all its derivatives vanish at $\xi = \pm\infty$, then $A = 0$, and one multiplies by f' and integrates again to get

$$-\frac{Vf^2}{2} - f^3 + \frac{1}{2}f'^2 = 0 .$$

Write $V = 4a^2$ and then

$$\pm\frac{df}{f(2a^2 + f)^{1/2}} = 2^{1/2}\,d\xi .$$

To integrate the left side, put $f = -2a^2 \operatorname{sech}^2\theta$; then

$$\theta = \pm a(\xi - \xi_0) ,$$

$$f = -2a^2 \operatorname{sech}^2 [a(x - 4a^2 t - \xi_0)] . \tag{23.17}$$

Here a and ξ_0 are arbitrary constants. Equation (23.17) is the solitary wave or *single–soliton solution* to KdV. Note that all localized f have a real, and therefore $f < 0$. Solitary waves are necessarily waves of *elevation*. This obviously nonlinear property was seen by Scott Russell, who failed to produce, experimentally, any solitary wave of depression; attempts led, instead, to the production of an oscillatory, lengthening, decaying wave train. IST provides the mathematical reason for this.

23.6 INVERSE SPECTRAL TRANSFORM

We want to solve the initial–value problem for KdV, in the form

$$u_t - 6uu_x + u_{xxx} = 0$$
$$u(x,0) = u_0(x) \tag{23.18}$$

with $u_0(x) \to 0$ rapidly as $|x| \to \infty$. Gardener, Greene, Kruskal and Miura in 1967 discovered how this can be solved in terms of three associated linear problems. Again, as with FT's, the idea is to find a set of quantities, analogous to $\widetilde{u}(k,t)$, whose time evolution is simple. For the KdV, these quantities are the *spectral data* for the *Schrödinger equation* (of quantum mechanics).

23.7 DIRECT SPECTRAL OR SCATTERING PROBLEM

Given a function $v(x)$ such that $v(x) \to 0$ as $|x| \to \infty$, consider the eigenvalue (spectral) problem defined by the *Schrödinger equation*

$$-\Psi_{xx} + v(x)\Psi = \lambda\Psi \quad . \tag{23.19}$$

Spectral theory tells us that if $v(x)$ vanishes rapidly enough as $|x| \to \infty$, then for a certain discrete *finite* set of real negative λ, $\lambda_n = -K_n^2, 0 < K_1 < K_2 < \dots < K_N$, (23.19) has a solution such that $\Psi = \Psi_n \to 0$ as $|x| \to \infty$ and $\int_{-\infty}^{+\infty} \Psi_n^2 dx$ is finite. The λ_n (or equivalently the K_n) are called the *discrete eigenvalues*. For $\lambda = \lambda_n$ we normalize the Ψ_n (which can be taken as real) so that $\int_{-\infty}^{+\infty} \Psi_n^2 dx = 1$. Then as $x \to \to +\infty$, $\Psi_n \sim c_n \exp(-K_n x)$ for some definite coefficients c_n. It turns out that the squares $\gamma_n = c_n^2$ are most significant, and we call the γ_n the *normalization coefficients*.

For positive λ there are no solutions with square–integrable eigenfunctions; all the solutions are oscillatory at infinity, like $\exp(\pm i\lambda^{1/2} x)$. For each positive λ we write $\lambda = k^2$ and choose the corresponding solution of (23.19) so that

$$\Psi \sim e^{-ikx} + b(k)e^{ikx} \qquad \text{as } x \to +\infty$$

$$\Psi \sim a(k)e^{-ikx} \qquad \text{as } x \to -\infty . \tag{23.20}$$

If we imagine a time factor $e^{-i\omega\tau}$ ($\tau = $ time for this discussion), $e^{-ikx-i\omega\tau}$ represents a wave incident from $x = +\infty$, travelling in the negative x–direction towards a "potential well" with potential $v(x)$ which scatters the wavetrain. Then $b(k)$ is the *reflexion coefficient* and $a(k)$ *the transmission coefficient*.

Then the *spectral data* for the potential $v(x)$ are the set of quantities

$$S = \left\{ K_n; \gamma_n; b(k) \text{ for } -\infty < k < \infty \right\} , \tag{23.21}$$

i.e. the discrete eigenvalues, the normalization coefficients and the reflexion coefficient $b(k)$.

23.8 TIME EVOLUTION OF THE SPECTRAL DATA

Now suppose $v(x) = u(x,t)$ is a function of a parameter t. We ask "how will the scattering or spectral data S change as t changes, and what constraint does this place on the development with t of $u(x,t)$?" Here we introduce the idea of an *isospectral problem* — one in which the discrete eigenvalues λ_n are independent of t. It turns out that if λ_n is an eigenvalue for (23.19) with $v = u(x,t)$, then $d\lambda_n/dt = 0$ if u satisfies

$$u_t - 6uu_x + u_{xxx} = 0 \ ,$$

the KdV equation; solutions of the KdV equation have the property that the Schrödinger eigenvalues to which they give rise at each t are independent of t.

To prove this we have to postulate an equation for the development with t of the eigenfunction Ψ (dropping the suffix n). We try

$$\Psi_t = A\Psi + B\Psi_x \tag{23.22}$$

where A, B are scalar functions (which may depend on $u(x,t)$ and on λ but not on Ψ). This is the most general linear local relation; any linear dependence on Ψ_{xx}, Ψ_{xxx} etc. could be eliminated by use of (23.19). Next we impose the *integrability condition*, that Ψ_{xxt} from (23.19) should equal Ψ_{txx} from (23.22). With some trial and error (a systematic method has been devised) we find that if

$$A = -u_x, \ B = 4\lambda + 2u \ , \tag{23.23}$$

then the integrability condition reduces to

$$\left[u_t - 6uu_x + u_{xxx} - \lambda_t \right] \Psi = 0 \ . \tag{23.24}$$

Hence if u satisfies KdV, then the t–evolution of the discrete eigenvalues is simple,

$$\lambda(t) = \text{constant} \ . \tag{23.25}$$

Examination of (23.22) at $x = +\infty$ gives

$$\Psi_t \sim -4K_n^2 \Psi_x$$

and with $\Psi \sim c_n(t) \exp(-K_n x)$ this gives, with use of (23.25), a linear differential equation

$$\frac{dc_n}{dt} = 4K_n^3 c_n \ ,$$

or

$$\frac{d\gamma_n}{dt} = 8K_n^3 \gamma_n \; , \tag{23.26a}$$

for the normalization coefficients.

For the continuous spectrum, $\lambda = k^2$, (23.23) is not general enough and we find that we need $A = -u_x + 4ik^3$. Then taking $x \to +\infty$ we get

$$\Psi_t \sim 4ik^3 \Psi + 4k^2 \Psi_x$$

which with use of the first of (23.20) gives

$$\frac{db(k,t)}{dt} = 8ik^3 b(k,t) \tag{23.26b}$$

$$\left(\text{while } x \to -\infty \text{ gives } \frac{da(k,t)}{dt} = 0 \right) .$$

Therefore the evolved spectral data are

$$K_n(t) = K_n(0) \quad , \tag{23.27a}$$

$$\gamma_n(t) = \gamma_n(0) \exp(8K_n^3 t) \quad , \tag{23.27b}$$

$$b(k,t) = b(k,0) \exp(8ik^3 t) . \tag{23.27c}$$

Take now $u_0(x)$ and solve (23.19) for the scattering or spectral data at $t = 0$, i.e. calculate $\{K_n(0), \gamma_n(0), b(k,0)\}$ — a direct spectral problem. Then follow the evolution of the spectral data, using (23.25) and (23.26); we obtain simple expressions for the spectral data $S(t)$ given $S(0)$, while $u(x,t)$ evolves according to KdV. Finally, use the methods of *inverse scattering* or *spectral theory* to recover $u(x,t)$ from $S(t)$.

23.9 INVERSE SPECTRAL PROBLEM

The final problem here — inverse scattering — has been thoroughly worked out for the Schrödinger equation. We quote only the result. From the spectral data, compute

$$F(x) = \frac{1}{2\pi} \int_{-\infty}^{+\infty} b(k)e^{ikx}dk + \sum_{n=1}^{N} \gamma_n e^{-K_n x} . \tag{23.28}$$

Then solve, for $y > x$, the Gel'fand–Levitan–Marchenko (GLM) linear integral equation

$$K(x,y) + F(x+y) + \int_{x}^{\infty} K(x,z)F(y+z)dz = 0 \tag{23.29}$$

for the function $K(x,y)$. Then the scattering potential u is given by

$$u = -2\frac{d}{dx}K(x,x) \tag{23.30}$$

(t in all this being merely a parameter).

Though vastly more complicated, the essence of the IST follows closely the ideas behind the FT, as the diagram below shows.

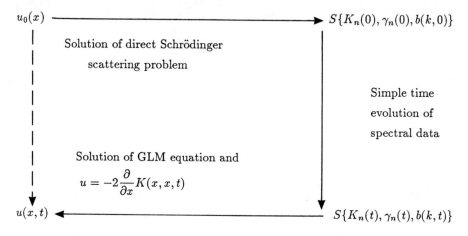

$u_0(x) \longrightarrow S\{K_n(0), \gamma_n(0), b(k,0)\}$

Solution of direct Schrödinger scattering problem

Simple time evolution of spectral data

Solution of GLM equation and
$$u = -2\frac{\partial}{\partial x}K(x,x,t)$$

$u(x,t) \longleftarrow S\{K_n(t), \gamma_n(t), b(k,t)\}$

This programme can be carried through in all detail only for certain convenient special $u_0(x)$ — in particular for $u_0(x) = \mu\,\mathrm{sech}^2\nu x$. For these potentials the spectral data can be calculated in terms of Legendre functions, and the most vital result is that the potentials are reflexionless, $b \equiv 0$, when $\nu = 1$ and $\mu = -N(N+1)$ for integer N. In this case, and essentially only in this case, is an exact solution of the GLM equation possible. It is then solvable by elementary methods (separation of variables) and the solution for u can be written as

$$u(x,t) = -2\frac{\partial^2}{\partial x^2}\ln\left|\Delta\right| \tag{23.31}$$

where $|\Delta|$ is the determinant of the $N \times N$ matrix

$$\Delta_{mn} = \delta_{mn} +$$

$$\frac{(\gamma_m\gamma_n)^{1/2}}{(K_m + K_n)}\exp\left\{-(K_m + K_n)x + 4(K_m^3 + K_n^3)t\right\} \tag{23.32}$$

in which γ_m, K_m come from the scattering problem *at* $t = 0$, and the eigenvalues are in fact $K_n = n$. This solution is the N–soliton solution to KdV; for $N = 1$ it reduces to the form (23.1) and for $N = 2$ it takes the form of a quotient of two linear combinations of hyperbolic cosines,

$$u(x,t) = -12 \, \frac{3 + 4\cosh(2x - 8t) + \cosh(4x - 64t)}{\left\{3\cosh(x - 28t) + \cosh(3x - 36t)\right\}^2} \ . \tag{23.33}$$

The N–soliton solution can be interpreted as follows; as $t \to -\infty$ we have N widely separated solitons each corresponding to one discrete eigenvalue of $u_0(x)$ and, if $K_1 < K_2 < ... < K_N$, arranged so that the highest soliton is farthest to the left, etc (note from (23.32) that the propagation speed of the soliton K_n is proportional to K_n^2). Then as t increases, the solitons overtake and interact with each other, but as $t \to +\infty$ they separate out in reverse order, with the highest now farthest to the right. The only trace of the interaction is a phase shift, or fixed displacement, of each soliton (each phase shift in general different) from the position it would have reached without interaction. Observe that, regardless of the sign of β, the solitons in the present coordinates always propagate to the right.

If the potential is not reflexionless there is also a contribution from the continuous spectrum (i.e. from $b(k,t)$) which generally prevents exact solution of the GLM equation. However, it has been proved that the contribution of the continuous spectrum to $u(x,t)$ consists of a decaying oscillating dispersive wavetrain (which may indeed propagate away in the direction *opposite* to that of the solitons) — very much as in linear theory. Indeed, as the amplitude of $u_0(x)$ becomes small, it can be shown that there are no longer any discrete eigenvalues and that the GLM equation with only the continuous contribution to F in (23.28) reduces to the Fourier inversion problem of linear theory. We see that the solitons are an inherently nonlinear phenomenon, that they dominate *any* solution after a sufficiently long time, and that they retain their identity permanently. Properties of this kind are of interest in a vast range of problems in nonlinear science (e.g. as models for a continuum representation of elementary particles; in nonlinear information transmission lines, whether man–made or nerve axons; in the propagation of large amplitude optical and acoustical pulses; as a description of the Great Red Spot of Jupiter; in geophysical and astrophysical fluid dynamics; etc.).

The method (IST) originally developed for KdV was in the early 1970's quickly extended to cover several other nonlinear evolution equations of physical interest — in particular,

$$u_t \pm 6u^2 u_x + u_{xxx} = 0 \qquad \text{(mKdV — modified KdV)}$$

$$iu_t = u_{xx} \pm |u|^2 u \qquad \text{(NLS — nonlinear Schrödinger) (23.34)}$$

$$u_{xt} = \sin u \qquad \text{(SG — sine–Gordon)} \quad .$$

Further extensions (all using the *isospectral* idea) have now led to a list of more than 40 essentially different equations of physical significance (and infinitely many, generated by the same procedure, which have not so far occurred in practice). There have been extensions to cover waves travelling in opposite directions, and limited success so far with extensions to two or three space dimensions. For example, the CKdV (cylindrical KdV)

$$u_t - 6uu_x + \frac{u}{2t} + u_{xxx} = 0 \qquad (23.35)$$

has soliton–like solutions and can be "solved" in a sense by a version of IST but it has been proved that no version of IST as presently understood can solve the SKdV (spherical KdV)

$$u_t - 6uu_x + \frac{u}{t} + u_{xxx} = 0 \; . \qquad (23.36)$$

A serious restriction is that none of the equations so far solved by IST is permitted to contain dissipative mechanisms (except, in a highly degenerate way, the famous Burgers equation $u_t + uu_x = \epsilon u_{xx}$), and that is almost certain to remain a permanent restriction.

The simplest acoustic medium sustaining soliton behaviour is the bubbly liquid, where it has been shown that at low frequencies the KdV equation itself is the right model. If dissipative effects need to be included, the model becomes the Burgers–Korteweg–de Vries equation (BKdV)

$$u_t - 6uu_x = \epsilon u_{xx} - u_{xxx} \qquad (23.37)$$

which is not solvable by IST and does not have strict soliton solutions. However, the dissipative term may sometimes be treated as small, and then one can deal with KdV solitons whose amplitude, scale and velocity change slowly with time under the action of weak dissipation. At higher frequencies a more complicated model equation (a nonlinear Klein–Gordon equation) can be derived which has, according to numerical solutions, no soliton properties, and no IST has yet been found for it.

In any event, the discovery of IST and the soliton in the last 25 years has represented a breakthrough in qualitative and quantitative understanding of nonlinear physics, with every indication of further possibilities for significant extension of the ideas and methods.

REFERENCES: BOOKS AND REVIEW ARTICLES

Ablowitz, M.J., Kaup, D.J., Newell, A.C. & Segur, H. (1974). The Inverse Scattering Transform — Fourier Analysis for Nonlinear Problems. Studies in Applied Math. 53:249–315.

Ablowitz, M.J. & Segur, H. (1981). Solitons and the Inverse Scattering Transform. SIAM, Philadelphia.

Bullough, R.K. & Caudrey, P.J. (eds.) (1980). Solitons. Springer–Verlag, Berlin.

Calogero, F. (ed.) (1978). Nonlinear Evolution Equations Solvable by the Spectral Transform. Pitman, London.

Dodd, R.K., Eilbeck, J.C., Gibbon, J.D. & Morris, H.C. (1982). Solitons and Nonlinear Wave Equations. Academic Press, London.

Drazin, P.G. & Johnson, R.S. (1989). Solitons: An Introduction. Cambridge University Press, Cambridge.

Fordy, A.P. (ed.) (1990). Soliton Theory: A Survey of Results. Manchester University Press, Manchester.

Lamb, G.K. (1980). Elements of Soliton Theory. Wiley, New York.

Lonngren, K. & Scott, A.C. (eds.) (1978). Solitons in Action. Academic Press, New York.

Newell, A.C. (1985). Solitons in Mathematics and Physics. SIAM, Philadelphia.

Scott, A.C., Chu, F.Y.F. & McLaughlin, D.W. (1973). The Soliton — a New Concept in Applied Science. Proc. IEEE 61:1443–1483.

Whitham, G.B. (1974). Linear and Nonlinear Waves. Wiley, New York.

24. NONLINEAR ACOUSTICS

24.1 INTRODUCTION

Nonlinear acoustics is often said to be the study of the effects associated with the small but finite amplitude of a sound wave. Certainly there are *local* effects associated with finite–amplitude sound waves, such as the production of *steady* quantities from what appears to be a purely oscillatory motion with zero mean values. Steady streaming flow in the axial direction generated by large amplitude piston motion is a common example, but phenomena like cavitation in an oscillatory sound field provide other examples of steady effects. However, a more important aspect of nonlinear acoustics is associated with the large distortion of a sound wave which is produced over a sufficiently large time or distance by the accumulation of effects corresponding to arbitrarily small local nonlinearities. This distortion is often severe over distances of practical interest, and if it is, then the amplitude decay due to geometrical constraints and due to attenuation mechanisms, and the directivity pattern, will be governed by laws quite different from those of linear acoustics. Another aspect of nonlinear acoustics of particular interest in underwater applications involves the so–called "parametric array", in which the nonlinear interaction, over a large range, of two primary sound beams of comparable frequency, generates a low–frequency field at the difference frequency. The interaction region acts as a virtual source of large spatial extent, and therefore the low–frequency beam is narrow, highly directional, and with very low–level side–lobes.

In these notes we try first to find the basic effects of nonlinearity on a small–amplitude sound wave in the absence of dissipation and spreading, showing the local linear relations and the cumulative nonlinear ones leading to shock formation. Then we include thermoviscous dissipation, and examine the shock structure and the decay of an initially sinusoidal signal. We shall see that the amplitude of the sinusoidal signal which eventually again emerges at large ranges is *independent* of the initial amplitude — the phenomenon of *amplitude saturation*. We shall also describe equations which incorporate the effects of geometrical spreading.

24.2 LINEAR LOCAL BEHAVIOUR, NONLINEAR CUMULATIVE BEHAVIOUR

Take one–dimensional non–dissipative motion; the Euler equations are

$$\rho\left(\frac{\partial u}{\partial t} + u\frac{\partial u}{\partial x}\right) = -\frac{\partial p}{\partial x} \quad ,$$

$$\frac{\partial \rho}{\partial t} + \rho\frac{\partial u}{\partial x} + u\frac{\partial \rho}{\partial x} = 0 \quad , \tag{24.1}$$

$$(p/p_0) = (\rho/\rho_0)^\gamma \quad ,$$

where γ is the specific heat ratio for a gas, and an empirically fitted constant for condensed media ($\gamma \sim 7$ for water). In a sound wave of amplitude U, frequency ω and wavenumber k_0 we have pressure fluctuations of order $p - p_0 \sim \rho_0 c_0 U$, density fluctuations $\rho - \rho_0 \sim (p - p_0)/c_0^2 \sim \rho_0 M$, where $c_0 = (\gamma p_0/\rho_0)^{1/2}$ is the sound speed and M the Mach number U/c_0. Therefore, to assess the effects of small nonlinearity ($M << 1$) we take the following dimensionless variables, which are $O(1)$ in a small–amplitude sound wave:

$$\bar{t} = \omega t, \; \bar{x} = k_0 x, \; \bar{u} = u/U \quad ,$$

$$\bar{p} = (p - p_0)/\rho_0 c_0 U, \; \bar{\rho} = (\rho - \rho_0)/M\rho_0 \quad . \tag{24.2}$$

If we insert these in (24.1) we get, exactly,

$$(1 + M\bar{\rho})\left(\frac{\partial \bar{u}}{\partial \bar{t}} + M\bar{u}\frac{\partial \bar{u}}{\partial \bar{x}}\right) = -\frac{\partial \bar{p}}{\partial \bar{x}} \quad ,$$

$$\frac{\partial \bar{\rho}}{\partial \bar{t}} + \frac{\partial \bar{u}}{\partial \bar{x}} + M\frac{\partial}{\partial \bar{x}}(\bar{\rho}\bar{u}) = 0 \quad , \tag{24.3}$$

$$(1 + M\gamma\bar{p}) = (1 + M\bar{\rho})^\gamma \quad .$$

Now we make an expansion in the small parameter M,

$$u = u_0 + Mu_1 + \ldots$$

$$p = p_0 + Mp_1 + \ldots \tag{24.4}$$

$$\rho = \rho_0 + M\rho_1 + \ldots \quad ,$$

where we have dropped the overbars on all variables, and where now p_0, ρ_0 are not to be confused with the earlier p_0, ρ_0 which defined the ambient state. At $O(M^0)$ we have

$$\frac{\partial u_0}{\partial t} + \frac{\partial p_0}{\partial x} = 0 \; ,$$

$$\frac{\partial \rho_0}{\partial t} + \frac{\partial u_0}{\partial x} = 0 \; , \tag{24.5}$$

$$p_0 - \rho_0 = 0 \; ,$$

which are the equations of ordinary linear non–dissipative acoustics. They give

$$\left(\frac{\partial^2}{\partial t^2} - \frac{\partial^2}{\partial x^2} \right) (u_0, p_0, \rho_0) = 0 \; , \tag{24.6}$$

so that

$$u_0 = p_0 = \rho_0 = f(t - x) \; , \tag{24.7}$$

if we restrict ourselves to a wave propagating in the positive x–direction only.

To find the first nonlinear corrections we look at $O(M)$:

$$\frac{\partial u_1}{\partial t} + \frac{\partial p_1}{\partial x} = -u_0 \frac{\partial u_0}{\partial x} - \rho_0 \frac{\partial u_0}{\partial t} \; ,$$

$$\frac{\partial \rho_1}{\partial t} + \frac{\partial u_1}{\partial x} = -\rho_0 \frac{\partial u_0}{\partial x} - u_0 \frac{\partial \rho_0}{\partial x} \; , \tag{24.8}$$

$$p_1 - \rho_1 = \left(\frac{\gamma - 1}{2} \right) \rho_0^2 \; .$$

If we write $\theta = t - x$ (or $\omega(t - x/c_0)$ in the original variables, so that θ is the phase variable of linear acoustics) and $f'(\theta)$ for $df/d\theta = \partial f/\partial t = -\partial f/\partial x$, we find

$$\frac{\partial^2 u_1}{\partial t^2} - \frac{\partial^2 u_1}{\partial x^2} = \left(1 + \frac{\gamma - 1}{2} \right) (f^2)'' \; . \tag{24.9}$$

The nonlinear forcing on the right side arises from two apparently distinct terms in the Euler equations; $(f^2)''$ from the nonlinear convective acceleration $u\partial u/\partial x$, $\frac{\gamma-1}{2}(f^2)''$ from the nonlinearity of the p–ρ relationship. The latter is numerically more important for water, the former for air.

The general solution of (24.9) comprises a complementary function (which for waves travelling in the positive x-direction only can be absorbed into (24.7) if required) plus a particular integral which we can guess is of the form

$$u_1 = tg(t - x)$$

provided

$$2g'(\theta) = \left(\frac{\gamma + 1}{2}\right)(f^2)'' \quad , \tag{24.10}$$

or

$$g(\theta) = \frac{\gamma + 1}{2}f(\theta)f'(\theta) \quad .$$

Thus

$$u = f(t - x) + M\left(\frac{\gamma + 1}{2}\right)tf(t - x)f'(t - x) + O(M^2) \quad . \tag{24.11}$$

Alternatively, $u_1 = xh(t - x)$ gives

$$u = f(t - x) + M\left(\frac{\gamma + 1}{2}\right)xf(t - x)f'(t - x) + O(M^2) \quad , \tag{24.12}$$

which is equivalent to (24.11).

We shall consider functions in the form (24.12), functions of a linear phase variable θ and of range x, as is appropriate for the solution of boundary–value or *signalling* problems in which we seek $u(x, t)$ given the value $u(x = 0, t)$, with $0 \leq x < \infty$. The location $x = 0$ might be an approximation to the position of a moving piston on which u is controlled, or $u(0, t)$ might be a measured time–history of u at $x = 0$. As a case of common interest, suppose, in physical variables, that $u = U \sin \omega t$ at $x = 0$; then here $f(\theta) = \sin \theta$, and if another term, $O(M^2)$ is calculated much as above, the solution (24.12) can be written either as a series expansion in $M x \sin \theta$,

$$u = \sin \theta + \left(\frac{\gamma + 1}{2}\right)M x \sin \theta \cos \theta - \frac{1}{2}\left(\frac{\gamma + 1}{2}\right)^2 M^2 x^2 \sin^3 \theta + O(M^3) \quad ,$$

or as a Fourier series in θ, with range–dependent coefficients,

$$u = \sin\theta \left\{ 1 - \frac{3}{32}(\gamma+1)^2 M^2 x^2 + O(M^4 x^4) \right\}$$

$$+ \sin 2\theta \left\{ \left(\frac{\gamma+1}{4} \right) Mx + O(M^3 x^3) \right\} \qquad (24.13)$$

$$+ \sin 3\theta \left\{ \frac{(\gamma+1)^2}{32} M^2 x^2 + O(M^4 x^4) \right\} + \text{.......} \quad .$$

This shows how the nonlinear terms produce *harmonics* whose amplitudes (for $n \geq 2$) initially grow with range, the amplitude of the first harmonic (fundamental) initially decreasing with range in order to supply energy for pumping to higher frequencies. Recall that the system is free of dissipation, and indeed it can be shown that to the order indicated in (24.13) the energy lost from the first harmonic just balances that supplied to the second, the third and higher harmonics containing negligible energy to this order of approximation. (In order to verify this one has to check that the mean flux $< (p - p_0)u >$ is independent of x to the appropriate order, and $p - p_0$ *cannot* be simply taken as $\rho_0 c_0 u$ as in linear theory; instead $p - p_0$ must be calculated to $O(M^2)$ from the known form for u in (24.13).)

The series (24.12) or (24.13) is an asymptotic representation for $u(x,t)$ as $M \to 0$ for any fixed interval $0 \leq x \leq x_1$ say. However, it contains *secular terms* which make the series nonuniform for large x; the secular terms are those which involve powers of x in addition to factors involving $f(t - x)$ and its derivatives. They make the ratio of successive terms in the expansion of order (Mx), and render the expansion useless for $x = O(M^{-1})$ and larger. In order to derive a description of the effect of wave amplitude on wave development for $x = O(M^{-1})$ we use the well–established Method of Multiple Scales. (Note that we are *not* looking, at these large x, at a small nonlinear correction to the basic wave $u \simeq f(t - x)$; for when $x = O(M^{-1})$, $u \simeq f(t - x)$ is *not* a valid approximation. In other words, linear acoustics does not give even a valid first approximation, however small the wave amplitude, at sufficiently large times or distances, and unless dissipation or some other mechanism intervenes, the long–time behaviour of any acoustic wave, no matter how weak, is inherently nonlinear, as we shall see.)

For multiple scales, we regard t, x and $X = Mx$ as independent variables, so that $\partial/\partial x \to \partial/\partial x + M\partial/\partial X$, and expand in the form

$$u = u_0(x, t, X) + Mu_1(x, t, X) + \text{....} \quad . \qquad (24.14)$$

We shall choose the dependence of u_0 on X to ensure that no secular terms now arise in u_1, and consequently (Mu_1/u_0) will be uniformly bounded for all x, t, X, giving us a uniformly valid first approximation $u \simeq u_0(x, t, X)$. Substituting in

$$(1 + M\rho)\left[\frac{\partial u}{\partial t} + Mu\left(\frac{\partial}{\partial x} + M\frac{\partial}{\partial X}\right)u\right] + \left(\frac{\partial}{\partial x} + M\frac{\partial}{\partial X}\right)p = 0 \quad,$$

$$\frac{\partial \rho}{\partial t} + (1 + M\rho)\left(\frac{\partial}{\partial x} + M\frac{\partial}{\partial X}\right)u + Mu\left(\frac{\partial}{\partial x} + M\frac{\partial}{\partial X}\right)\rho = 0 \quad, \tag{24.15}$$

$$p - \rho = \frac{\gamma - 1}{2}M\rho^2 + \cdots \quad,$$

gives, at $O(M^0)$, the system (24.5) again, with the dependence on X not yet specified. Confining attention to waves travelling in the positive x–direction, and introducing $\theta = t - x$, we have

$$u_0 = p_0 = \rho_0 = q(\theta, X) \quad, \tag{24.16}$$

say. At $O(M)$ we find

$$\frac{\partial u_1}{\partial t} + \frac{\partial p_1}{\partial x} = -\frac{\partial q}{\partial X} \quad,$$

$$\frac{\partial \rho_1}{\partial t} + \frac{\partial u_1}{\partial x} = 2q\frac{\partial q}{\partial \theta} - \frac{\partial q}{\partial X} \quad, \tag{24.17}$$

$$p_1 - \rho_1 = \left(\frac{\gamma - 1}{2}\right)q^2 \quad.$$

Elimination gives

$$\frac{\partial^2 u_1}{\partial t^2} - \frac{\partial^2 u_1}{\partial x^2} = \left(1 + \frac{\gamma - 1}{2}\right)\frac{\partial^2}{\partial \theta^2}(q^2) - 2\frac{\partial^2 q}{\partial \theta \partial X} \quad, \tag{24.18}$$

where the first term on the r.h.s corresponds to that on the right of (24.9) and led before to secular terms. All terms on the right of (24.18) lead to particular solutions proportional to x times functions of (θ, X), and these will render (24.14) invalid for large x unless we choose the right side of (24.18) to vanish identically. Thus

$$u \simeq q(\theta, X) \tag{24.19}$$

is a uniform first approximation provided $q(\theta, X)$ satisfies the nonlinear equation

$$\frac{\partial q}{\partial X} - \left(\frac{\gamma + 1}{2}\right)q\frac{\partial q}{\partial \theta} = 0 \quad, \tag{24.20}$$

where we have integrated once with respect to θ and ignored any constant of integration. Equation (24.20) — known as the *simple wave* equation — is to be solved with (in general) q at $x = X = 0$ (i.e. $q(t, 0)$) set equal to the specified value of u at $x = 0$. The equation shows that the evolution is inherently nonlinear for large ranges; if the amplitude parameter $M = U/c_0$ is made smaller, the straightforward series (24.12) — representing small nonlinear corrections to a linear field — remains valid for a greater range of x, but when $x = O(M^{-1})$ no description short of the fully nonlinear form (24.20) can hold. The form (24.20) may *not* be solved by series expansions unless all terms in the series are calculated and somehow summed, for (24.20) and its boundary condition contain no small expansion parameter, as it is expressly assumed that q, X and θ are all $O(1)$.

24.3 CHARACTERISTIC SOLUTION FOR SIMPLE WAVES

Writing $Z = (\frac{\gamma+1}{2})X$, and taking $u(x = 0, t) = f(t)$, we have to solve

$$\frac{\partial q}{\partial Z} - q\frac{\partial q}{\partial \theta} = 0 \quad , \tag{24.21}$$

with $q = f(\theta)$ at $Z = 0$. Equation (24.21) is a hyperbolic equation equivalent to an ordinary differential equation along paths in a (Z, θ) plane. Along any given curve, for which θ is some definite function of Z, we have

$$\frac{dq}{dZ} = \frac{\partial q}{\partial Z} + \left(\frac{d\theta}{dZ}\right)\frac{\partial q}{\partial \theta} \quad ,$$

and so (24.21) states that

$$\frac{dq}{dZ} = 0 \text{ on the } characteristic \text{ curves } \Gamma : \frac{d\theta}{dZ} = -q \quad . \tag{24.22}$$

Thus, on any particular Γ, q is a constant; but this then shows that Γ is a straight line,

$$\Gamma : \theta = -qZ + \phi \quad ,$$

where we identify Γ by the point ϕ on the θ–axis through which Γ passes at $Z = 0$. The value of q on this Γ is the value q has at the point $Z = 0$, $\theta = \phi$, namely $f(\phi)$. Hence, $q = f(\phi)$ where $\theta = -qZ + \phi$, gives an implicit description of the solution — or in other words

$$q = f(\phi) \quad , \tag{24.23}$$

where f is the initial function or initial wave shape (at $x = 0$) and $\phi(Z, \theta)$ is defined implicitly by

$$\phi = \theta + Z f(\phi) \quad . \tag{24.24}$$

ϕ is a nonlinear phase variable, and the solution for ϕ in terms of θ, Z is only one–valued, in general, for Z less than some finite value. For Z less than this value the solution (24.23) represents simply a *pure distortion* of the original waveform; the only values of q present at range Z are those present at $Z = 0$, but a given value of f is found at a fixed value of ϕ rather than at a fixed value of the phase θ.

Equations (24.23) and (24.24) state that

$$q = f(\theta + qZ) \quad , \tag{24.25}$$

and therefore if q is plotted as a function of θ at successive ranges Z, a given value of q moves to the *left* at a *"speed"* q, so that a sine–wave would distort as shown below.

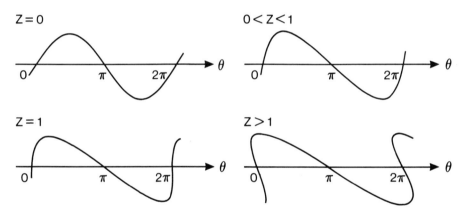

Figure 24.1 Nonlinear distortion, with increasing range Z, of a simple wave.

These solutions and diagrams are perfectly consistent with the "forward breaking" often shown in books and papers. If we revert to dimensional variables u, x, t it is shown in books on gas dynamics that in a right–running simple wave, the theory of Riemann Invariants gives the equation

$$\frac{\partial u}{\partial t} + \left(\frac{\gamma + 1}{2} u + c_0 \right) \frac{\partial u}{\partial x} = 0 \quad , \tag{24.26}$$

which shows that a given value of u (a given "wavelet") travels at speed $c_0 + (\frac{\gamma+1}{2})u$, so that points where $u = 0$ travel at the linear acoustic speed c_0, values of $u > 0$

travelling faster than this and $u < 0$ travelling slower, so that crests overtake troughs and the wave breaks forward as shown below.

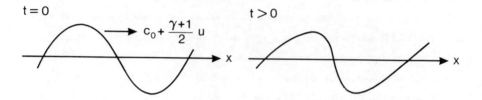

Figure 24.2 Nonlinear distortion, with time, of a simple wave.

To obtain (24.21) from (24.26) we argue that *locally* $u << c_0$ and hence that *locally* $\partial u/\partial t + c_0 \partial u/\partial x \simeq 0$. Hence

$$\frac{\partial u}{\partial t} + c_0 \frac{\partial u}{\partial x} + \left(\frac{\gamma + 1}{2}\right) u \frac{\partial u}{\partial x}$$

(24.27)

$$\sim \frac{\partial u}{\partial t} + c_0 \frac{\partial u}{\partial x} - \left(\frac{\gamma + 1}{2}\right) \frac{u}{c_0} \frac{\partial u}{\partial t} \quad,$$

and if we now take x and $\tau = t - x/c_0$ as independent variables, then

$$\frac{\partial}{\partial t} \to \frac{\partial}{\partial \tau} \quad, \quad \frac{\partial}{\partial x} \to \frac{\partial}{\partial x} - \frac{1}{c_0} \frac{\partial}{\partial \tau} \quad,$$

so that

$$\frac{\partial u}{\partial x} - \frac{\gamma + 1}{2c_0^2} u \frac{\partial u}{\partial \tau} = 0 \quad .$$

Finally write $u = Uq$, $Z = (\frac{\gamma+1}{2})Mk_0 x$, $\theta = \omega\tau$ and we get (24.21), which is much more appropriate for boundary–value problems than is (24.26), which is suitable for initial–value problems with $u(x, t = 0)$ specified.

Note lastly that if $u_0 - q$, is found by solving (24.21), then the (dimensionless) pressure and density fluctuations are given by (24.16) — i.e by the *linear laws*. The essential idea is that at every point *local linear theory holds, but the cumulative long–range effects are always nonlinear.*

24.4 THE FUBINI SOLUTION

A famous explicit solution of (24.21) is that given by Fubini (in 1935). If

$$\frac{\partial q}{\partial Z} - q\frac{\partial q}{\partial \theta} = 0, \quad q(0,\theta) = \sin\theta \ ,$$

then

$$q(Z,\theta) = 2\sum_{n=1}^{\infty} \frac{J_n(nZ)}{nZ}\sin n\theta \ . \tag{24.28}$$

To prove this observe that the differential equation preserves the parity and periodicity of the initial function, and hence

$$q(Z,\theta) = \sum_{n=1}^{\infty} A_n(Z)\sin n\theta \ ,$$

where

$$
\begin{aligned}
A_n(Z) &= \frac{2}{\pi}\int_0^{\pi} q(Z,\theta)\sin n\theta d\theta \\
&= \frac{2}{\pi}\int_0^{\pi} \sin\theta\sin(n\phi - nZ\sin\phi)(1 - Z\cos\phi)d\phi \\
&= \frac{2}{\pi n}\int_0^{\pi} \cos\phi\cos(n\phi - nZ\sin\phi)d\phi \\
&= -\frac{2}{\pi nZ}\int_0^{\pi}(1 - Z\cos\phi)\cos(n\phi - nZ\sin\phi)d\phi \\
&\quad + \frac{2}{\pi nZ}\int_0^{\pi}\cos(n\phi - nZ\sin\phi)d\phi \\
&= \frac{2}{\pi nZ}\int_0^{\pi}\cos(n\phi - nZ\sin\phi)d\phi \\
&= \frac{2}{nZ}J_n(nZ) \ ,
\end{aligned}
$$

by a standard integral expression. If the J_n are expanded in power series we retrieve and extend the series expansions for the Fourier coefficients which were obtained earlier by direct perturbation methods.

The Fubini form of solution holds only for $0 \le Z < 1$; for $Z > 1$ the $(\phi - \theta)$ relation is not (1–1) and the integral over θ cannot be transformed into one over ϕ without modification. Consequently the large–Z behaviour of the initially sinusoidal waveform cannot be obtained from the Fubini representation.

24.5 MULTI–VALUED WAVEFORMS AND SHOCK WAVES

The transformation from θ to $\phi, \phi = \theta + Zf(\phi)$, is one–valued for $Z = 0$ and therefore for sufficiently small Z. But

$$\frac{\partial \phi}{\partial \theta} = (1 - Zf')^{-1}$$

shows that the transformation will not be (1–1) once Z exceeds the first value at which $1 - Zf'(\theta) = 0$ for some θ, i.e. for $Z > Z_*$ where

$$Z_* = \left(\frac{1}{\max f'}\right). \tag{24.29}$$

This shows that if f' is *anywhere* positive a triple–valued region will be produced in the solution $q = f(\phi)$, the region first arising at the points θ in the phase where $f'(\theta)$ has its maximum value (so only a wave of pure expansion, f' everywhere negative, does not lead ultimately to a triple–valued waveform).

The triple–valued solution is unacceptable; and in reality (in media for which dissipation rather than dispersion significantly affects short–scale disturbances) a narrow region of rapid change is produced, as shown below.

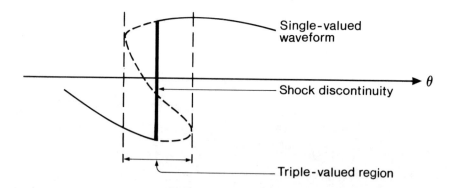

Figure 24.3 Insertion of a shock discontinuity in a triple–valued simple wave.

The thickness of the transition region is, initially at any rate, small compared with the wavelength, and can be idealized (for limited purposes) as a discontinuity; it is a *weak shock wave* (weak because the wave amplitude U was assumed to be everywhere small compared with c_0). The question that needs to be answered, if we are going to deal with solutions consisting of continuous arcs separated by weak shock discontinuities, is "where, in the range of θ for which there is a triple–valued waveform at any given Z, is the shock to be placed, and how does its phase location change with Z?"

The answer to this can be obtained in two ways. One can argue that the differential equation for q is a restricted form, for continuous waves, of an integral conservation principle

$$\frac{d}{dZ}\int_A^B q\,d\theta = \left[\frac{1}{2}q^2\right]_A^B \quad , \tag{24.30}$$

and the principle must continue to hold (essentially because in the present model it represents conservation of both mass and momentum) if the wave is discontinuous. If there is a shock discontinuity at $\theta = \theta_s(Z)$ and q_+, q_- are the values of q at θ_s+ and θ_s-, then

$$\frac{d}{dZ}\int_A^B q\,d\theta = \int_A^B \frac{\partial q}{\partial Z}\,d\theta + (q_- - q_+)\frac{d\theta_s}{dZ} \quad ,$$

while

$$\left[\frac{1}{2}q^2\right]_A^B = \frac{1}{2}(q_B - q_A)(q_B + q_A) \quad .$$

Let $B \to \theta_s+, A \to \theta_s-$; then the integral term vanishes, and

$$\frac{d\theta_s}{dZ} = -\frac{1}{2}(q_+ + q_-) \quad . \tag{24.31}$$

Thus the shock moves, in the direction of decreasing θ, at a "speed" which is the average of the speeds q_+, q_- on either side of the shock. If this relation is transformed to physical variables and coordinates it says that the shock propagation velocity U_s (in the direction of increasing x) is given by

$$U_s = \frac{1}{2}c_0 + \frac{\gamma+1}{4}(U_+ + U_-) \quad , \tag{24.32}$$

where $U_\pm$ are the values of the (physical) velocity $u(x,t)$ on either side of the shock. The shock velocity is the mean of the "wavelet" velocities $c_0 + \frac{\gamma+1}{2}u$ on either side of the shock. In graphical terms this means that the shock is located in the phase so as to cut off equal areas of the multi–valued waves on either side, as in the diagram below.

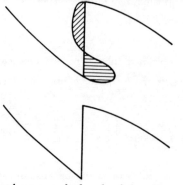

triple–valued wave; equal areas.

composite wave including shock.

Figure 24.4 Equal–areas rule for shock location.

The results (24.32) and (24.31), or the "equal areas" rule, hold only for systems with quadratic nonlinearity, as in (24.30). In the more general case, if (24.30) can be replaced by

$$\frac{d}{dZ}\int_A^B q\,d\theta = \left[Q(q)\right]_A^B \quad,$$

where $Q(q)$ may loosely be called the flux of q, then

$$\frac{d\theta_s}{dZ} = -\frac{Q(q_+) - Q(q_-)}{(q_+ - q_-)} \quad, \tag{24.33}$$

and the equal areas rule again holds, but the shock propagation speed is not simply the mean of the wavelet speeds on the two sides.

"*Weak shock theory*" consists of solving nondissipative wave equations, such as (24.21), and, for $Z > Z_*$, fitting shock discontinuities in according to the equal areas rule or the analytical prescription (24.33). Note that, for quadratic nonlinearity and the parameter $(\gamma + 1)/2$, or its equivalent, *positive*, only compression shocks are formed; as time (or θ) increases, discontinuities are only ever needed to increase the value of q (i.e. of $u_0 = p_0 = \rho_0$). There are fluids (certain hydrocarbons) in which $(\gamma + 1)/2 < 0$ always, and others in which $(\gamma + 1)/2 = 0$ along a curve in some thermodynamic space, and then expansion shocks only, or both expansion and compression shocks, will be produced.

The second way of locating the shocks — and a more general and fundamental way — is to bring in another mechanism to eliminate the triple–valued feature, and then determine where the shocks are located as the new mechanism is allowed to decrease in "size". This will be done in the next section.

24.6 THERMOVISCOUS DIFFUSION — THE BURGERS EQUATION

Suppose now that the fluid has viscosity μ, but that heat conduction can be ignored and that viscous dissipation of energy is small enough for the uniform entropy assumption to be maintained. Then the only change to (24.1) is the addition to the right of the first equation of a term

$$\frac{4}{3}\mu\frac{\partial^2 u}{\partial x^2} \quad,$$

which means that the first of (24.3) acquires a term

$$\frac{4}{3}\left(\frac{\nu k_0^2}{\omega}\right)\frac{\partial^2 \bar{u}}{\partial \bar{x}^2}$$

on the right, where $\nu = \mu/\rho_0$ (and μ is a weak function of temperature, under normal conditions, and can be taken as constant). The factor $\nu k_0^2/\omega = \nu k_0/c_0$ is an inverse Reynolds number, based on sound speed c_0 and inverse wavenumber k_0^{-1}.

Suppose now that $(\nu k_0^2/\omega)$ is comparable with M, and write $\frac{4}{3}\nu k_0^2/\omega = \alpha M$, so that the first of (24.3) is now

$$(1 + M\bar{\rho})\left(\frac{\partial \bar{u}}{\partial \bar{t}} + M\bar{u}\frac{\partial \bar{u}}{\partial \bar{x}}\right) + \frac{\partial \bar{p}}{\partial \bar{x}} = \alpha M\frac{\partial^2 \bar{u}}{\partial \bar{x}^2} \quad.$$

Then, proceeding as before with the multiple scales expansion (24.14), the only changes are that the first of (24.17) gets an extra term

$$\alpha\frac{\partial^2 u_0}{\partial x^2} = \alpha\frac{\partial^2 q}{\partial \theta^2}$$

on the right, while (24.18) gets an extra term

$$\alpha\frac{\partial^3 q}{\partial \theta^3} \quad.$$

Thus elimination of secular terms allows the long–range evolution, under comparable weak nonlinearity and viscous diffusion, to be determined by

$$\frac{\partial q}{\partial X} - \left(\frac{\gamma+1}{2}\right)q\frac{\partial q}{\partial \theta} = \alpha\frac{\partial^2 q}{\partial \theta^2} \quad,$$

or

$$\frac{\partial q}{\partial Z} - q\frac{\partial q}{\partial \theta} = \epsilon\frac{\partial^2 q}{\partial \theta^2} \quad , \tag{24.34}$$

where Z is as before, and

$$\epsilon = \frac{4}{3}\left(\frac{\nu k_0^2}{\omega}\right)\left(\frac{1}{M}\right)\left(\frac{2}{\gamma+1}\right) \quad . \tag{24.35}$$

Equation (24.34) is the famous *Burgers' equation*, the canonical equation representing a competition between linear evolution, quadratic nonlinearity and viscous diffusion. Although it was derived above on the basis that $\epsilon = O(1)$ it can be shown to be valid for both large and small values of ϵ, and the main interest is in solutions of (24.34) as $\epsilon \to 0$. Equation (24.34) continues to hold if heat conduction is included, but ϵ then acquires a comparable additional contribution involving the thermal diffusivity.

Equation (24.34) is remarkable among all nonlinear parabolic equations in that it is the only such equation to have an exact linearization and an exact general solution for arbitrary initial conditions. The linearization is the *Hopf–Cole transformation*. Start by writing (24.34) in the form

$$q_Z = \left[\epsilon q_\theta + \frac{1}{2}q^2\right]_\theta \quad ,$$

which guarantees the existence of a potential Ψ such that

$$q = \Psi_\theta, \quad \epsilon q_\theta + \frac{1}{2}q^2 = \Psi_Z \quad ,$$

so that Ψ satisfies

$$\Psi_Z = \epsilon\Psi_{\theta\theta} + \frac{1}{2}\Psi_\theta^2 \quad , \tag{24.36}$$

which is known as the Potential Burgers Equation. Now put

$$\Psi = A\ln\psi \quad ,$$

and we find that the nonlinear terms in (24.36) cancel if we choose $A = 2\epsilon$. Thus

$$q = 2\epsilon\frac{\partial}{\partial \theta}\ln\psi \tag{24.37}$$

linearizes Burgers equation to the heat conduction equation

$$\psi_Z = \epsilon\psi_{\theta\theta} \quad . \tag{24.38}$$

From any solution to (24.38) we can generate a solution to (24.34), while if we are given $q(0,\theta)$, (24.37) determines $\psi(0,\theta)$, the initial–value problem solution to (24.38) determines $\psi(Z,\theta)$, and $q(Z,\theta)$ is then given by (24.37) again.

The general solution for $q(Z,\theta)$ obtained in this way has been examined in detail and (see Lighthill (1956), Whitham (1974) and Rudenko & Soluyan (1977)) it has been proved that as $\epsilon\to0$ solutions tend to the solutions of the simple wave equation (24.21), with near–discontinuities at locations predicted as in Section 24.5 by weak shock theory. This proof extends to long ranges at which nonlinear distortion is so severe that shock waves have been set up, but it does *not* cover very long ranges at which the shock waves have thickened appreciably and begun to "interfere" with the lossless portions of the wave. This very–long–range behaviour will be discussed in a moment.

24.7 SHOCK–WAVE STRUCTURE

The lossless simple wave equation (24.21) is an "outer" approximation to the Burgers equation, with Z,θ held fixed in the limit $\epsilon \to 0$. An "inner" approximation provides a description of the narrow region — the interior of the shock — in which q increases rapidly from q_- to q_+, say. We take coordinates located in the vicinity of the apparent discontinuity, and suspect the shock width to be $O(\epsilon)$; a suitable inner coordinate is therefore

$$\Theta = \frac{\theta - \theta_s(Z)}{\epsilon} \quad, \tag{24.39}$$

in terms of which (24.34) reads

$$\frac{\partial q}{\partial Z} - \frac{\theta'_s}{\epsilon}\frac{\partial q}{\partial \Theta} - \frac{1}{\epsilon}q\frac{\partial q}{d\Theta} = \frac{1}{\epsilon}\frac{\partial^2 q}{\partial \Theta^2} \quad. \tag{24.40}$$

Expanding q as $q_0 + \epsilon q_1 + ...$ gives

$$-(q_0 + \theta'_s)\frac{\partial q_0}{\partial \Theta} = \frac{\partial^2 q_0}{\partial \Theta^2} \quad, \tag{24.41}$$

an ordinary differential equation, Z appearing only parametrically.

The equation can be integrated, with the result

$$q_0 = -\theta'_s + \bar{q}_0(Z)\tanh\left(\frac{\bar{q}_0(Z)(\Theta - \Theta_0(Z))}{2}\right) \quad, \tag{24.42}$$

where $\bar{q}_0, \Theta_0$ are arbitrary constants of integration with respect to Θ, and therefore functions of Z. To match with an outer solution, substitute (24.39) and let $\epsilon \to 0$ with $\theta > \theta_s(Z)$. Then $q \sim -\theta'_s + \bar{q}_0(Z)$ which must equal the limit q_+ of the outer solution as $\theta \to \theta_s$ from the right. Hence

$$-\theta'_s + \bar{q}_0(Z) = q_+ \quad,$$

and similarly

$$-\theta'_s - \bar{q}_0(Z) = q_- \quad.$$

This gives

$$\theta'_s = -\frac{1}{2}(q_+ + q_-) \quad, \tag{24.43}$$

in agreement with (24.31), for the shock propagation speed, while

$$\bar{q}_0(Z) = \frac{1}{2}(q_+ - q_-) \quad. \tag{24.44}$$

Hence, in terms of q_+ and q_-, the inner shock solution is

$$q_0 = \frac{1}{2}(q_+ + q_-) + \frac{1}{2}(q_+ - q_-)\tanh\left(\frac{(q_+ - q_-)(\Theta - \Theta_0(Z))}{4}\right) \quad. \tag{24.45}$$

The function Θ_0 is not determined by this leading–order matching. It locates the shock centre, in the sense that $q_0 = \frac{1}{2}(q_+ + q_-)$ when $\Theta = \Theta_0(Z)$, and q_0 is antisymmetric about this point. In some cases symmetry dictates a particular value of $\Theta_0(Z)$. For example, for the sine wave the shocks first form at $Z = 1$ at phases $\theta = 2n\pi$, and the odd symmetry implies that the shocks remain permanently located there — so that if we take one period, $-\pi < \theta \le \pi$ say, the shock is at $\theta = 0$, is antisymmetric about $\theta = 0$, and hence $\Theta_0(Z) = 0$ for all Z. In other problems $\Theta_0(Z)$ has to be determined by higher–order matching, and it is often found not merely to be nonzero, but to become unbounded as $Z \to \infty$. This is one of several possible reasons why weak shock theory cannot remain valid for large Z; the shocks refuse to stay in the neighbourhood of the weak shock location, and viscous effects cause them to drift far through the waveform. Corrections to (24.45) can also be obtained, writing $q = q_0 + \epsilon q_1 + \dots$ with Θ as independent variable. The expression for q_1 is very complicated, but shows that in many cases ϵq_1 will not remain small compared with q_0 when Z is large. This arises because q_1 contains terms arising from $\partial q_\pm / \partial Z$, representing the rate of change of the flow just outside the shocks, and in many cases this rate of change is sufficiently rapid that the balance between nonlinearity and diffusion expressed in (24.41) cannot be maintained. A "steady" *Taylor shock solution* like (24.45) is then no longer possible, and weak shock theory again fails. Weak shock theory in fact fails at large ranges in a variety of different ways, depending on the initial function $f(\theta)$ and on whether other effects (relaxation, geometrical spreading, ...) are included. See Nimmo & Crighton (1986) for a general classification of the ultimate decay of nonlinear acoustic waves under the influence of geometrical spreading effects, quadratic nonlinearity and viscous diffusion.

24.8 THE FAY SOLUTION

In 1931 R.D. Fay gave a solution for a periodic nonlinear wave subject to thermoviscous diffusion. He did not deal with Burgers equation as such, but instead approximated the complete equations at various stages. The result he obtained turns out actually to be an exact solution of Burgers equation. It is *not* the solution subject to an initial sinusoidal boundary value, $q(Z=0, \theta) = \sin\theta$ but, surprisingly, it *is* the asymptotic form which that other exact solution takes, as $\epsilon \to 0$, for all ranges Z greater than 1 (the range at which shocks first form).

The Fay solution of

$$\frac{\partial q}{\partial Z} - q\frac{\partial q}{\partial \theta} = \epsilon\frac{\partial^2 q}{\partial \theta^2} \tag{24.46}$$

is

$$q = 2\epsilon \sum_{n=1}^{\infty} \frac{\sin n\theta}{\sinh(n\epsilon Z)} \quad . \tag{24.47}$$

If Z is $O(1)$ and the profile not rapidly changing, so that large values of n are irrelevant, then we may put $\sinh(n\epsilon Z) \sim (n\epsilon Z)$ and

$$q \sim \frac{2}{Z} \sum_{n=1}^{\infty} \frac{\sin n\theta}{n} \quad , \tag{24.48}$$

which is the Fourier series representation of the sawtooth wave shown below in Figure 24.5. Alternatively

$$
\begin{aligned}
q &\sim \left(\frac{\pi - \theta}{Z}\right) && \text{for } 0 < \theta < \pi \\
&\sim -\left(\frac{\pi + \theta}{Z}\right) && \text{for } -\pi < \theta < 0
\end{aligned}
\tag{24.49}
$$

with periodic continuation to other θ.

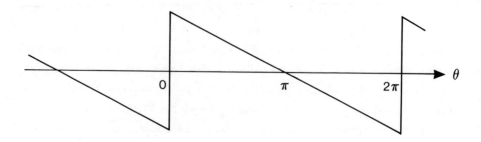

Figure 24.5 The periodic sawtooth wave produced by nonlinear distortion of a sinusoidal wave, for $1 << Z << \epsilon^{-1}$.

The emergence of this *lossless* sawtooth wave can be predicted from the original lossless solution

$$q = \sin\phi, \phi = \theta + Z\sin\phi \quad .$$

As Z increases through 1 the nonlinear distortion produces almost straight line segments, with shock discontinuities at $\theta = \pm 2n\pi$ whose amplitudes are determined by $q_+ = -q_- = \sin\phi_+$, where $\phi_+ = Z\sin\phi_+$. The half–amplitude q_+ starts at 0 for $Z = 1$, increases to a maximum value of 1 when $Z = \pi/2$ (at which point the shock gives a transition from the highest point $q = +1$ to the lowest, $q = -1$, on the wave) and then decreases monotonically. For $Z >> 1$, ϕ_+ must be near π, $\phi_+ = \pi - \phi'$ say and then $\pi \sim Z\phi'$ so that $q_+ \sim \pi/Z$. This provides the shock amplitude $2\pi/Z$ for $Z >> 1$ and agrees with the jump in the Fay solution (24.49). That solution can therefore be regarded as asymptotic for $1 << Z << \epsilon^{-1}$.

In the vicinity of $\theta = 2n\pi$ there is a rapid transition from π/Z to $-\pi/Z$ and the Fay solution must be approximated in another way which does not involve Fourier series. It is found that an alternative asymptotic representation of the Fay solution, for $-\pi < \theta < +\pi$, is

$$q = \frac{1}{Z}\left\{-\theta + \pi\tanh\frac{\pi\theta}{2\epsilon Z}\right\} \quad . \tag{24.50}$$

Amazingly, this result of asymptotic approximation of the exact Fay solution (itself the result of asymptotic approximation to another exact solution found from the Hopf–Cole transformation) is again an exact solution of the full Burgers equation. It is called the Khokhlov solution. If $Z = O(1)$ and θ not as small as $O(\epsilon)$ the tanh is effectively $+1$ for $\theta > 0$ and -1 for $\theta < 0$, so that we regain the sawtooth form

(24.49). If $Z = O(1)$ and $\theta = O(\epsilon)$, $\theta = \epsilon\Theta$, say, then

$$q \sim \frac{\pi}{Z} \tanh\left(\frac{\Theta}{2(\pi/Z)^{-1}}\right) \quad , \qquad (24.51)$$

which is of exactly the form in (24.45), with $\Theta_0(Z) = 0$ by symmetry and with $q_+ = -q_- = \pi/Z$. Thus (24.51) describes a transition in a thin shock wave of thickness $O(\epsilon)$ around $\theta = 0$, from q_- to q_+.

We observe, however, that the effective shock thickness is not simply $O(\epsilon)$, but $O(\epsilon Z)$, and so the shocks gradually thicken and begin to interfere with the main body of the wave. Although (24.50) is an exact solution to Burgers equation, we cannot use it to analyse the subsequent behaviour because it is not periodic. We must therefore revert to the Fay solution in Fourier series form,

$$q = 2\epsilon \sum_{n=1}^{\infty} \frac{\sin n\theta}{\sinh(n\tilde{Z})} \quad , \qquad (24.52)$$

where $\tilde{Z} = \epsilon Z$ is assumed $O(1)$, and this describes the way in which the shocks thicken and "weak shock theory" breaks down. A point of special importance here is that the left side is u/U and the right is simply proportional to $1/U$ if $\tilde{Z} = O(1)$. Therefore at these ranges, $Z = O(\epsilon^{-1})$, u is independent of the initial amplitude U and we have *amplitude saturation*. This is shown most clearly by letting $\epsilon Z >> 1$, in which case the wave reverts to a single fundamental Fourier component,

$$q \sim 4\epsilon \exp(-\epsilon Z) \sin\theta \quad , \qquad (24.53)$$

or in physical variables,

$$u \sim \left(\frac{4}{\gamma+1}\right)\left(\frac{4}{3}\frac{\omega\nu}{c_0}\right) \exp\left\{-\frac{4}{3}\left(\frac{\nu k_0^2}{\omega}\right)(k_0 x)\right\} \sin\omega\left(t - \frac{x}{c_0}\right) \quad . \qquad (24.54)$$

Here we have included viscosity only and ignored thermal conductivity. The exponential factor in (24.54) is simply the decay factor of linear acoustics, so that after an initial period of wave steepening (described by the Fubini solution) followed by shock formation and intensification, then shock thickening (described by the Fay solution), the wave subsides under linear laws, but with an effective initial amplitude

$$"U" = \left(\frac{4}{3}\frac{\omega\nu}{c_0}\right)\left(\frac{4}{\gamma+1}\right) \quad ,$$

determined by these nonlinear processes, and independent of the actual initial amplitude U. The decay phase described by (24.53) is referred to as *"old–age"*. It is of considerable interest to determine the circumstances under which old–age linear decay sets in, and whether amplitude saturation is found or not. There are many different possibilities, depending on the details of the initial profile, the attenuation mechanisms included, the type of nonlinearity (e.g. cubic for shear waves in solids) and other linear energy–conserving mechanisms which may be included.

24.9 EFFECTS OF AREA CHANGE

The effects of wavefront area spreading (or contraction) might be expected to compete strongly with (or strongly augment) nonlinear distortion. Let x be a coordinate in the propagation direction along a *horn* or *ray–tube* of area $A(x)$; the area is supposed to change slowly on the acoustic wavelength scale, so that $k_0 L \gg 1$ where $L \sim \frac{d}{dx} \ln A(x)$. Then the effect of this linear energy–conserving mechanism can be included in a modified Equation (24.26),

$$\frac{\partial u}{\partial t} + \left(\frac{\gamma + 1}{2} u + c_0 \right) \frac{\partial u}{\partial x} + \frac{u c_0}{2} \frac{d}{dx} \ln A(x) = 0 \; , \qquad (24.55)$$

or, in terms of x and $\tau = t - x/c_0$,

$$\frac{\partial u}{\partial x} + \frac{1}{2} u \frac{d}{dx} \ln A(x) - \frac{\gamma + 1}{2c_0^2} u \frac{\partial u}{\partial \tau} = 0 \quad . \qquad (24.56)$$

If viscous effects are included then

$$\frac{\partial u}{\partial x} + \frac{1}{2} u \frac{d}{dx} \ln A(x) - \frac{\gamma + 1}{2c_0^2} u \frac{\partial u}{\partial \tau} = \beta \frac{\partial^2 u}{\partial \tau^2} \quad , \qquad (24.57)$$

for some appropriate β. We refer to this as a *generalized Burgers equation*, the generalization involving a linear term with a coefficient dependent on x (independent of x in the case of an exponential horn). The lossless form (24.56) can always be solved exactly (though implicitly). With viscous effects included it has been proved that (24.57) cannot be linearized by any generalization of the Hopf–Cole transformation *except* when $A(\tau) - $ constant. Approximate and numerical solutions are therefore much needed.

The first two terms of (24.57) indicate that $A^{1/2} u$ is conserved in linear nondissipative acoustics, so define $v = A^{1/2} u$ and then

$$\frac{\partial v}{\partial x} - \frac{\gamma + 1}{2c_0^2} A^{-1/2} v \frac{\partial v}{\partial \tau} = \beta \frac{\partial^2 v}{\partial \tau^2} \quad . \qquad (24.58)$$

Next define

$$\frac{dZ}{dx} = A^{-1/2}(x), Z = \int_{x_0}^{x} A^{-1/2}(x')dx' \quad , \tag{24.59}$$

and then

$$\frac{\partial v}{\partial Z} - \frac{\gamma+1}{2c_0^2} v \frac{\partial v}{\partial \tau} = \beta A^{1/2}(x) \frac{\partial^2 v}{\partial \tau^2} \quad , \tag{24.60}$$

which may be put in the dimensionless form

$$\frac{\partial q}{\partial Z} - q \frac{\partial q}{\partial \theta} = \epsilon G(Z) \frac{\partial^2 q}{\partial \theta^2} \quad , \tag{24.61}$$

with $q = f(\theta)$ at $Z = 0$. This shows how the lossless version ($\epsilon = 0$) can always be solved, but a form $f(\theta)$ which produces a shocked wave in one–dimensional plane flow will not necessarily do so here. Suppose f produces a shock first at $Z = Z_*$ in plane flow. Then if $x = +\infty$ implies either that the corresponding Z for (24.61) is also $+\infty$, or some *finite* value greater than Z_*, then a shock will be formed in the horn flow. For example, for spherical waves $A \sim x^2$ and so $Z = x_0 \ln(x/x_0)$ where x_0 is the radius at which we prescribe u. If initially we have $u = U \sin \omega t$ at $x = x_0$, a shock will be produced at $Z = 1$ and phase $\theta = 2n\pi$, where

$$Z = \frac{\gamma+1}{2c_0^2} \omega U x_0 \ln\left(\frac{x}{x_0}\right) \quad , \tag{24.62}$$

i.e. at

$$x = x_0 \exp\left\{\left(\frac{2}{\gamma+1}\right)\left(\frac{c_0}{\omega x_0}\right)\left(\frac{c_0}{U}\right)\right\} \quad , \tag{24.63}$$

a much greater distance than the shock–formation range

$$x = x_0 \left(1 + \left(\frac{2}{\gamma+1}\right)\left(\frac{c_0}{\omega x_0}\right)\left(\frac{c_0}{U}\right)\right) \quad , \tag{24.64}$$

for plane flow unless the "spreading factor" $(\omega x_0/c_0)$ is much greater than the Mach number inverse c_0/U, in which case (24.63) and (24.64) are asymptotically equal.

For cylindrically spreading waves the transformation corresponding to (24.62) is

$$Z = \frac{\gamma+1}{2c_0^2} \omega U x_0 \left| 1 - \left(\frac{x}{x_0}\right)^{1/2} \right| \quad , \tag{24.65}$$

from which the shock formation range can again be found. For these cases the "effective viscosity" $\epsilon G(Z)$ is a strong function of range Z; for cylindrical flow

$G(Z) \sim Z$ and for spherical flow $G(Z) \sim \exp Z$. These viscous terms are needed at first only in the shocks, which are initially thin and in them $G(Z)$ may be taken as constant — and in that case we have a version of "weak shock theory" for converging or diverging waves. However, the failure of weak shock theory for large ranges is very much affected by the variation of $G(Z)$ and the initial profile. See Nimmo & Crighton (1986) for analysis of this failure and classification of the ultimate behaviour of the waves.

REFERENCES

Lighthill, M.J. (1956). Viscosity Effects in Sound Waves of Finite Amplitude. In: Surveys in Mechanics (eds. G.K. Batchelor & R.M. Davies). Cambridge University Press, pp.250–351.

Nimmo, J.J.C. & Crighton D.G. (1986). Geometrical and diffusive effects in nonlinear propagation over long ranges. Phil. Trans. R. Soc. Lond. A.320:1–35.

Rudenko, O.V. & Soluyan, S.I. (1977). Theoretical Foundations of Nonlinear Acoustics. English translation by R.T. Beyer, Consultants Bureau, New York.

Whitham, G.B. (1974). Linear and Nonlinear Waves. Wiley–Interscience, New York.

25. CHAOTIC DYNAMICS AND APPLICATIONS IN ACOUSTICS

25.1 INTRODUCTION

Nearly thirty years ago E.N. Lorenz, working in meteorology at the Massachusetts Institute of Technology, showed that a very simple three–component nonlinear dynamical system could rapidly develop unimaginable complexity, making the solutions effectively chaotic, random and unpredictable, despite the availability of a simple deterministic prescription (three ordinary differential equations) for the evolution of the system. Lorenz's famous equations are

$$\dot{x} = -\sigma x + \sigma y \ ,$$

$$\dot{y} = rx - y - zx \ , \tag{25.1}$$

$$\dot{z} = xy - bz$$

where σ, r and b are constants, and x, y, z are three dependent variables, representing a drastic truncation of an infinite set of coupled nonlinear differential equations for the amplitudes of Fourier modes in a model of convection due to buoyancy forces in the atmosphere. Chaotic unpredictable behaviour results quickly if these equations are integrated numerically from, say, $x(0) = z(0) = 0, y(0) = 1$, provided (σ, r, b) lie in appropriate regions of parameter space. Figure 25.1 (from Bender & Orszag 1978) shows $y(t)$ for the case (a) ($\sigma = 3$, $r = 17$, $b = 1$) in which there is regular behaviour and (x, y, z) approaches a stable equilibrium point $(-4, -4, 16)$ (this corresponding physically to steady periodic convection cells) and for the case (b) ($\sigma = 3$, $r = 26$, $b = 1$) in which there is irregular behaviour. If case (b) is recomputed one finds that the first few excursions are exactly repeated, but that after quite a small number (~ 10 say) of excursions, deviations appear and are magnified so rapidly that after say 15 units of t there is no resemblance between

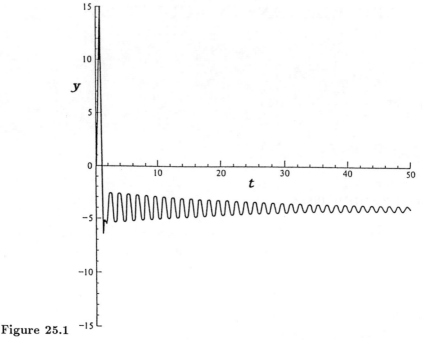

Figure 25.1

(a) Evolution of $y(t)$ in the Lorenz model, with $r = 17$, from $x(0) = z(0) = 0, y(0) = 1$. There is slow, regular, oscillatory approach to the stable critical point $(-4, -4, 16)$.

the two evolutions starting from apparently identical initial conditions and with apparently identical parameter values. In case (b) there are *unstable* equilibrium points at $(5, 5, 25)$ and $(5, -5, 25)$ and the system trajectory winds around one of these for some time, then is repelled and wanders around the other, making haphazard jumps from one region to the other after random numbers of turns around the equilibrium points, as shown in the (x, y) projection of case (b) in Figure 25.1(c) (from Bender & Orszag 1978). There is no question of the uniqueness theorem for differential equations being violated, simply that small errors (in this case computer round–off) may be exponentially magnified and rapidly accumulate. Such behaviour (apparently known to Poincaré 90 years ago) is called *deterministic chaos*, and has been found now in very many simple nonlinear models involving ordinary differential equations, partial differential equations and recurrence and iteration processes.

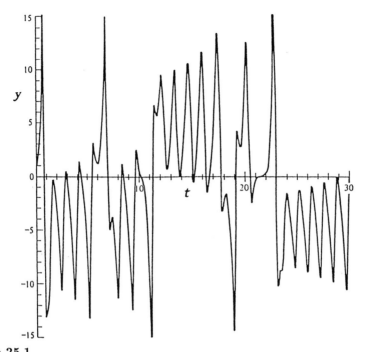

Figure 25.1

(b) Evolution of $y(t)$ in the Lorenz model, with $r = 26$, from $x(0) = z(0) = 0$, $y(0) = 1$. Behaviour is intermittent and irregular.

For ordinary differential equations, a phase space of at least three dimensions is needed, for in two dimensions the possible motions can be classified as approach to a fixed equilibrium point or to a limit cycle (periodic motion), with no possibility of chaos. Here we emphasize that all this refers to systems with some dissipation, which do have some definite asymptotic behaviour rather than — as for dissipationless Hamiltonian systems — wandering everywhere, with uniform probability, over the available phase space. Thus there are one or more *attractors* in the phase space which the solution point $(x, y, z, ...)$ approaches as $t \to \infty$; in two dimensions this may be a point attractor, or a periodic attractor, but in three dimensions we have the possibility of *strange attractors* which may be thought of loosely as continuous but infinitely–many-sheeted surfaces occupying (in three dimensions) a volume of dimension greater than two but less than three. The chaos comes from lack of knowledge of where *along* the attractor the system point is at any time.

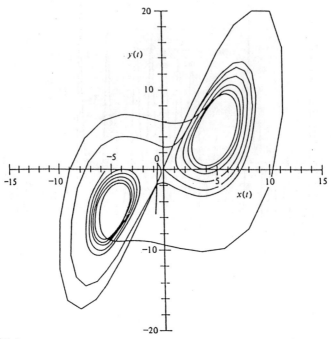

Figure 25.1

(c) A plot of $y(t)$ versus $x(t)$ in the Lorenz model; conditions as in Figure 25.1(b).

Acoustics provides a good physical field of application for many of the ideas developed recently in dynamical systems theory, in the phenomena of cavitation and nonlinear bubble oscillations. Taking the case of the unforced oscillations of a single bubble, we have in the simplest model the Rayleigh equation

$$R\ddot{R} + \frac{3}{2}\dot{R}^2 + \frac{p_\infty}{\rho_\ell}\left(1 - \frac{R_0^3}{R^3}\right) = 0 \tag{25.2}$$

where p_∞ is the ambient pressure and R the instantaneous radius. Although this system is nonlinear, it cannot exhibit chaos, for it occupies a two–dimensional phase space $(R, \dot{R})$, has an energy integral and, in fact, is a completely integrable system. If, however, the bubble is subject to a single–frequency forcing, leading to the equation

$$R\ddot{R} + \frac{3}{2}\dot{R}^2 + \left(1 - \frac{R_0^3}{R^3}\right)\frac{p_\infty}{\rho_\ell}(1 + A\sin\omega t) = 0 \ , \tag{25.3}$$

then chaos will in general result for certain ranges of the parameters involved, because the equation can be regarded as the three–component autonomous system

$$R\frac{dV}{d\tau} + \frac{3}{2}V^2 + \left(1 - \frac{R_0^3}{R^3}\right)\frac{p_\infty}{\rho_\ell}(1 + A\sin\omega\tau) = 0 \ ,$$

$$\frac{dR}{d\tau} = V \ , \qquad\qquad (25.4)$$

$$\frac{d\tau}{dt} = 1 \ .$$

Computer studies of more elaborate versions of this model have been made in Lauterborn (1986), as have measurements by Lauterborn and his colleagues of the pressure fluctuations in a field of cavitation bubbles subjected to forcing of single frequency ω and varying amplitude A. Lauterborn (1986) in fact provides a very good introduction to many aspects of chaotic dynamics, with a more recent article (Lauterborn & Holzfuss 1991) specifically on chaos in acoustics, but other scenarios for transition from regular to chaotic behaviour have also been observed in other physical and mathematical systems and will be noted briefly below. There is an immense body of mathematical theory relating to deterministic chaos, but only rarely are there proofs that particular physical systems satisfy all the conditions required for the mathematical theorems. Instead, there are experimental or computational determinations of strange attractors and quantities that characterize them (the attractor dimension, and exponents giving the rate at which the attractor is approached) but nothing to prove that the measured or computed attractors are genuinely chaotic and that they might not turn out to be periodic attractors of very long period if the measurements or computation could be carried far enough. The question of practical interest is, of course, whether an attractor appears "strange" and chaotic over timescales of practical interest.

25.2 SENSITIVITY TO INITIAL CONDITIONS

The extreme sensitivity of a chaotic system to minute changes in initial conditions is often found surprising, and one should try to understand it in a simple problem. The essentials are contained in the "sieve" mapping of the unit interval onto itself, as sketched in Figure 25.2. The point "A" might be taken initially to have binary representation 0.110000...1, while "B" equals 0.10111... . The initial points differ in their second binary digit, but after two applications of the sieve mapping, differ in their first binary digit, and so are widely separated. From this point on the precise location of the points is determined by increasingly high digits in the binary representations,

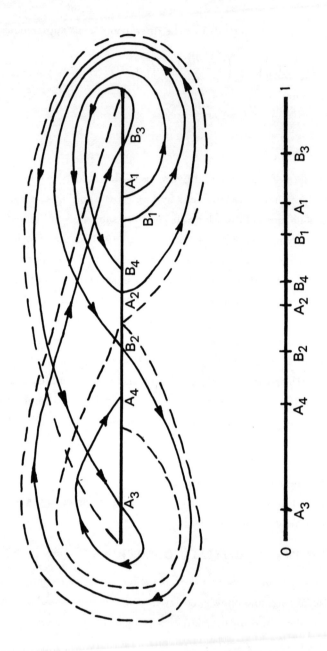

Figure 25.2 The "sieve" mapping of the unit interval, illustrating the rapid divergence of two points "A" and "B" initially close together.

and even if one had 16–bit precision initially, then after 16 iterations of the mapping all that information would be lost and the location of the points would be completely random and unpredictable, determined in fact by errors and noise at a level below the initial precision. Thus points initially differing in their Nth binary digit separate by $O(1)$ after N iterations, and the number of loops performed around 0 or 1 thereafter has all the properties of a random sequence, $LLRLLLRRLRRLLL...RLRRLL...$. The location of the image point for some given initial point is then the interval $(0, 1)$ covered randomly with uniform probability.

The critical assertion now is that all quadratically nonlinear iterative processes (mappings, ordinary differential equations, partial differential equations, difference equations,...) are *generically equivalent* to this sieve mapping, and hence in a typical nonlinear evolving dynamical system, initial errors must be expected to accumulate exponentially with an order–1 rate, so that over each natural time–step for the process, at least one binary digit of initial information is lost.

25.3 THE PERIOD–DOUBLING ROUTE TO CHAOS

Chaotic response is not found everywhere in the parameter space of some system, but instead there is a series of transitions, or "bifurcations", from one type of behaviour to another as one goes from one region of parameter space to another; if one imagines increasing the amplitude of external forcing of a nonlinear oscillator, for example, then there is a definite scenario, a sequence of bifurcations, taking one from regular periodic motion at the lowest (linear) forcing levels to chaos at some higher level. Quite a number of different scenarios have been discovered in different physical and mathematical systems. Perhaps the best known and most intensively studied is the Feigenbaum sequence of period–doubling bifurcations.

Imagine a definite mechanical system, such as a spherically symmetrically oscillating bubble. At the lowest levels of some external acoustic forcing of fixed frequency, the bubble responds in a linear periodic oscillation at the forcing frequency. As the amplitude increases, the response remains periodic, but the fundamental period is slightly distorted by the nonlinearity, and weak integral harmonics are generated. In the phase space (R, V) of Equation (25.4), the response is a limit cycle, a closed curve described once in the fundamental period, and increasingly

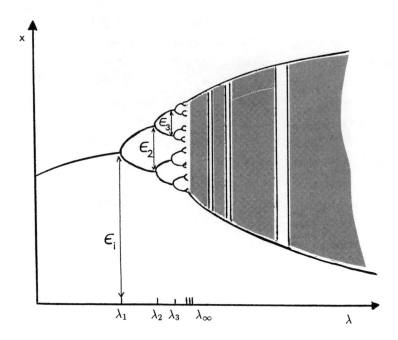

Figure 25.3 Sketch of a bifurcation diagram to illustrate period–doubling to chaos, and defining the parameters λ_i of Equation (25.6) and ϵ_i of Equation (25.7).

distorted from the linear elliptical shape as the amplitude increases. The first period–doubling bifurcation occurs at some finite forcing amplitude, and the limit cycle splits into a double loop which is covered in twice the fundamental period. In the frequency spectrum of the response, a spike appears at half the fundamental frequency, together with sum and difference frequency combinations formed from this subharmonic and the fundamental and all integral harmonics. At the next bifurcation, the period doubles again and a second subharmonic, at one quarter of the fundamental frequency, appears together with combination tones. Further period doublings occur increasingly often, and an infinite number of period doublings occurs before the forcing amplitude has reached some critical value, and for greater amplitudes the response has infinite period — i.e. is aperiodic — and the frequency spectrum is continuous, even though the forcing spectrum has only a single discrete tone.

Feigenbaum studied this scenario in detail for the quadratic iteration scheme

$$x_{n+1} = 4\lambda x_n(1 - x_n) \tag{25.5}$$

which maps the interval $[0, 1]$ onto itself, λ being an external control parameter. The sequence of parameter values at which the subharmonic bifurcations occurred was found to approach a geometric sequence, and

$$\frac{\lambda_i - \lambda_{i-1}}{\lambda_{i+1} - \lambda_i} \to \delta = 4.6692 \quad \text{as} \quad i \to \infty \ , \tag{25.6}$$

while the amplitudes of the response at the subharmonic frequencies (see Figure 25.3) were found to scale according to

$$\frac{\epsilon_i}{\epsilon_{i+1}} \to \alpha = 2.5029 \quad \text{as} \quad i \to \infty \ . \tag{25.7}$$

It has since been found that this behaviour occurs in many other systems, defined in some cases by other iteration maps, or by a set of ordinary differential equations, and also in experiments relating to systems actually governed by partial differential equations, and the same universal Feigenbaum constants are found for the bifurcation parameter ratios and for the amplitude scalings.

Some evidence in acoustics for the subharmonic route to chaos is found in the experiments of Lauterborn (1986) and Lauterborn & Parlitz (1988). The following figures (Figures 25.4—25.9) give some examples from Lauterborn's experiments, in which a cavitation bubble field was irradiated by sound of fixed frequency and an amplitude increasing linearly with time. The sonograms or "visible speech plots" show clearly the production of energy in the subharmonics, but 1/3 and 1/5 subharmonics and their combination tones are also observed because of a very complicated nonlinear resonance structure for the bubble oscillations which is not present in the simpler models to which the Feigenbaum scenario applies. The subharmonic route was also observed in numerical integrations of a much more complicated form of the Rayleigh equation for oscillations of a single bubble under external forcing. The presence of subharmonics in the response of bubbles had been observed experimentally for many years before its explanation through the Feigenbaum mechanism. More important is the fact that the spectrum becomes broad–band and noisy, with no significant tonal elements at and beyond some finite forcing amplitude of a single–frequency excitation.

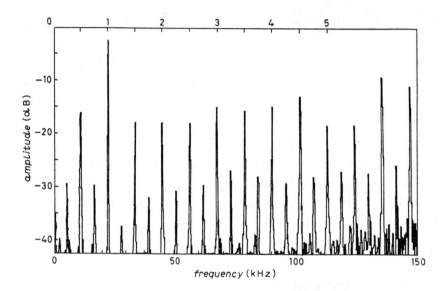

Figure 25.4

Acoustic–cavitation bubble field sound emission spectrum. Driving frequency is $f_a = 23.56$ kHz, and the spectrum has been taken at 26 V after linear increase of the voltage from 0 V to this value in 160 ms. Spectral lines at $\frac{1}{2} f_a$ and $\frac{1}{4} f_a$ and their (odd) harmonics are present (from Lauterborn 1986).

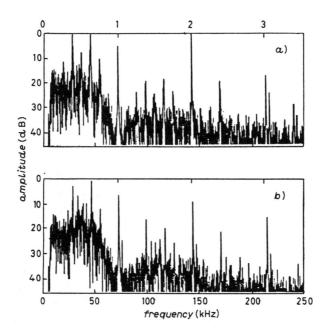

Figure 25.5

Acoustic–cavitation bubble field sound emission spectrum. Driving frequency is $f_a = 71$ kHZ. The spectra have been taken after linear increase of the voltage at the transducer to 110 V in 240 ms (a) and 119 V in 259 ms (b). Spectral lines at $\frac{1}{8} f_a$ and $\frac{1}{16} f_a$ (especially above 100 kHz); (b) are present together with their (odd) harmonics (measurements done by J. Holzfuss) (from Lauterborn 1986).

682

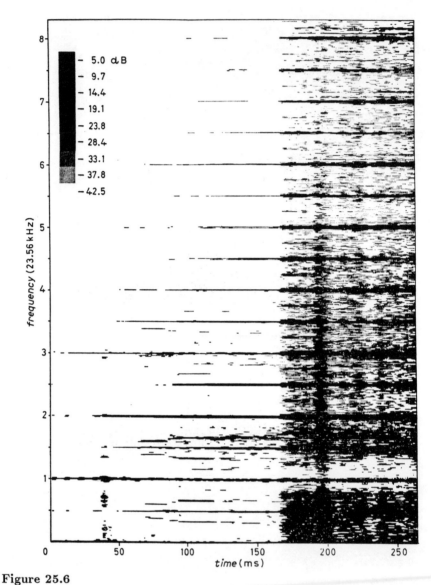

Figure 25.6

Spectral bifurcation diagram from an experiment with linearly increasing voltage (0 to 60 V in 262 ms) at the pieoelectric cylinder producing the cavitation bubble field. Driving frequency is $f_a = 23.56$ kHz. 370 spectra are plotted *vs.* time. Sampling rate is 500 kHz (measurements done by E. Cramer) (from Lauterborn 1986).

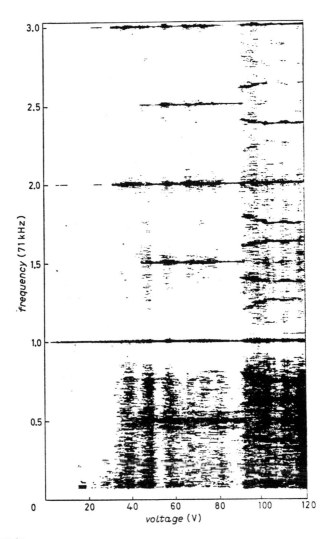

Figure 25.7

Spectral bifurcation diagram from an experiment with linearly increasing voltage
(0 to 120 V in 262 ms) at the pieoelectric cylinder producing the cavitation bubble
field. Driving frequency is $f_a = 71$ kHz. 990 spectra are plotted *vs.* time. Sampling
rate is 500 kHz. Note the appearance of $\frac{1}{5} f_a$ together with its harmonics and
its sliding to $\frac{1}{8} f_a$ with its harmonics (measurements done by J. Holzfuss) (from
Lauterborn 1986).

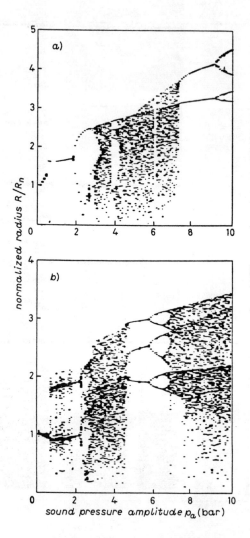

Figure 25.8

Bifurcation diagrams of a bubble with a radius at rest of $R_n = 100$ μm (a) and $R_n = 230$ μm (b). Frequency of the driving sound field is $f_a = 23.56$ kHz. The sound-pressure amplitude p_a is calculated for values in steps of 0.1 bar whereby the same starting point ($R = R_n$, $\dot{R} = 0$) is taken for all calculations at different p_a. The scattered points that do not follow the obvious bifurcation pattern are due to transients that have not yet died out (result of computer–time limitations; calculations done by E. Suchla; from Lauterborn 1986).

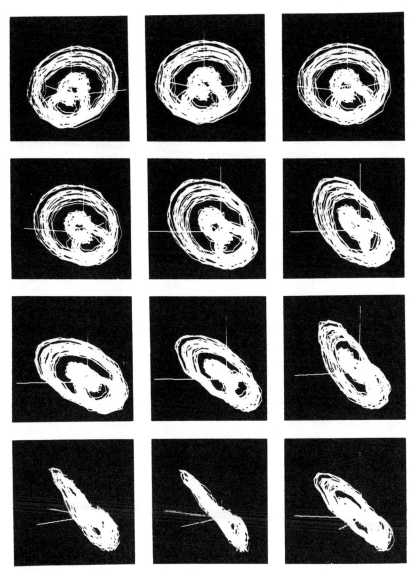

Figure 25.9

Phase–space analysis of acoustic–cavitation noise. The piezoelectric cylinder is driven at 43 V at a frequency of $f_a = 23.56$ kHz. Sampling rate is 1 MHz. The trajectory constructed in a three–dimensional phase space by plotting three measured pressure values $(p(t), p(t-T), p(t-2T))$ with $T = 5$ μs is rotated by 15° about the vertical axis in the figures between consecutive frames. The object found may be said to be a strange attractor (measurements done by J. Holzfuss) (from Lauterborn 1986).

25.4 OTHER ROUTES TO CHAOS

Landau Scenario

The first proposition for a sequence of bifurcations at increasing values of some parameter R is due to Landau. For definiteness, consider Taylor–Couette flow between concentric cylinders, the inner of which rotates steadily, R being a dimensionless rotation rate or Taylor number. In the Landau scenario one expects that for $R < R_c$ the system has a unique fixed point attractor, which simply means that the fluid is in steady motion with no variation along the axis (except in thin boundary layers on the top and bottom ends and the side wall). At $R = R_c$ several fixed point attractors appear, each of these corresponding to a possible motion and each corresponding to different variations along the axis — i.e. to different Taylor vortex states, as shown in Figure 25.10. Such steady Taylor vortex flow persists until $R = R_{c_1}$ at which time "wavy vortex flow" sets in. The multiple attractors are no longer points, but limit cycles, each representing a periodic wavy vortex flow, generally with different frequencies. The system can be represented by the motion of a system point along a given limit cycle, or torus T^1. At $R = R_{c_2}$, a bifurcation occurs to quasi–periodic flow, in which there are two frequencies, not usually commensurable (i.e. rationally related) and the system point moves on the surface of a two–dimensional torus T^2, the path on the torus never closing if the frequencies are incommensurable. Landau imagined that more and more bifurcations of this kind would occur, so that at some large value of R one would have a quasi–periodic motion with many basic incommensurable frequencies, the nonlinear interactions between which would give a frequency spectrum with very many lines and a very complicated behaviour of the signal in real time, but actually corresponding to motion on a torus in a many–dimensional space. Such quasi–periodic motion cannot be described as "turbulence", because it does not display the essential sensitivity to initial conditions, and although the first two or three bifurcations of the Landau scenario have been observed in various systems, there is usually then an abrupt transition to chaos, for reasons suggested below.

Newhouse–Ruelle–Takens scenario

In 1978, Newhouse, Ruelle & Takens proved a theorem to the effect that in every neighbourhood of a vector field on a torus T^m there exists a strange attractor if m is equal to three or greater. This means that quasi–periodic motion on a three–

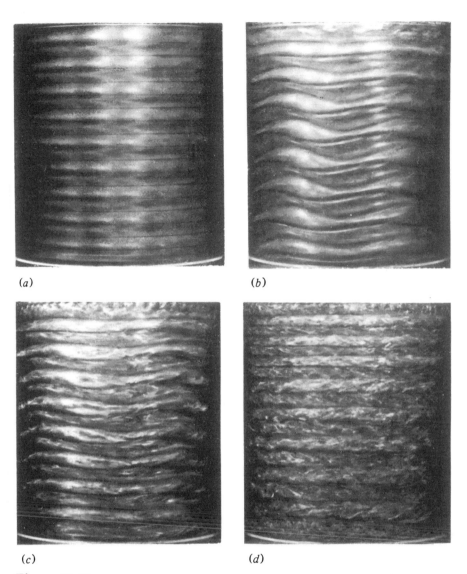

(a) (b)

(c) (d)

Figure 25.10

(a) $R/R_c = 1.1$. Time–independent Taylor vortex flow with 18 vortices. The flows at the upper and lower fluid surfaces are inward. The vertical bars are fiducials separated by $10°$ angles. (b) $R/R_c = 6.0$. (c) $R/R_c = 16.0$. (d) $R/R_c = 23.5$. Figures (b) and (c) illustrate wavy vortex flow (with four waves around the annulus), while in (d) the waves have disappeared. In (b), (c), and (d) there are 17 vortices, and the flow is outward at the upper surface and inward at the lower surface (from Fenstermacher et al. 1979).

688

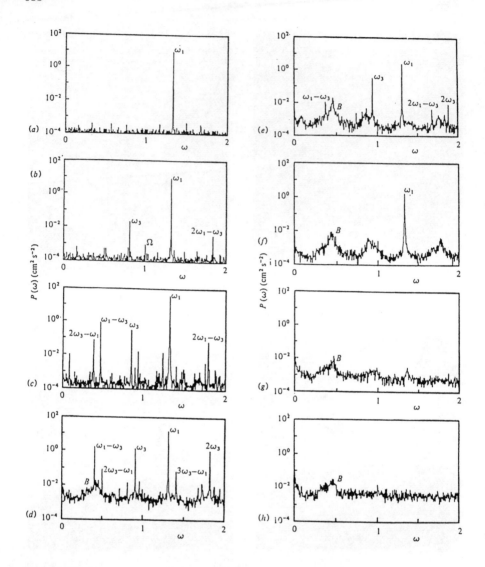

Figure 25.11

Taylor–vortex flow; spectrum of velocity at increasing Taylor number. Bifurcations to (a) wavy vortex flow on T^1; (b) quasi–periodic motion on T^2 (followed by development of combination frequencies $n_1\omega_1 \pm n_3\omega_3$); (g) broad–band spectrum. No significant energy was detected at harmonics other than $n_1\omega_1 \pm n_3\omega_3$ (from Fenstermacher et al. 1979).

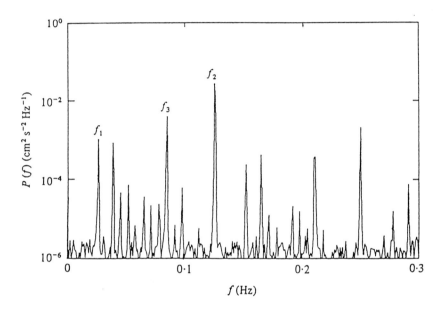

Figure 25.12

Motion with three incommensurate frequencies on T^3 observed in thermal convection experiments by Gollub & Benson (1980).

Spectrum showing presence of three incommensurate frequencies (route IIId), $R/R_c = 42.3$. All peaks are linear combinations of the three frequencies f_1, f_2, and f_3.

dimensional torus is structurally unstable, and would not be observable, the system immediately going on to the nearby chaotic strange attractor. This theorem certainly provides an explanation for some observations in Taylor–vortex flow, where three–frequency motion is not observed and instead the two–frequency motion is replaced by a broad–band spectrum (see Figure 25.11). Such an abrupt bifurcation from the first steps of the Landau scenario directly into chaos, with a continuous

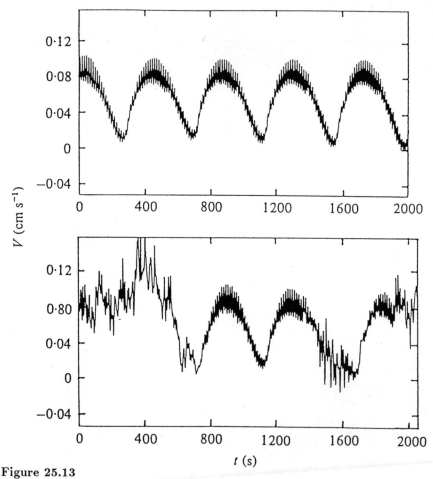

Figure 25.13

Quasi–periodic motion on T^2, with intermittent noise (from Gollub & Benson 1980). Velocity records showing the presence of intermittent noise (route IV). In (a), $R/R_c = 100.4$, the fluid is quasi–periodic with frequencies $f_1 = 0.00233$ and $f_2 = 0.07563$ Hz. In (b), at $R/R_c = 102.8$, intervals of quasi–periodic motion are interspersed between very noisy intervals.

broad–band spectrum, has been seen in a number of systems — but so also has regular three–frequency motion which does not immediately bifurcate to chaos (see Figure 25.12). The reasons why the system can stay on the torus T^3 despite the existence of the chaotic attractor arbitrarily close by seem not to be understood at the present time.

Intermittency Scenario

This scenario starts by following the Landau sequence to quasi-periodic motion on T^2. A typical signal as a function of time is shown in Figure 25.13. As the parameter R increases, the signal alternates randomly between patches of quasi-periodic motion with two frequencies, and bursts of intermittent noisy behaviour. At first the lengths of the noisy periods are small, but as R increases, they also increase, eventually to replace the quasi–periodic motion everywhere. Systems with this type of transition to chaos have also been intensively studied, mathematically and experimentally.

Other Transitions

The scenarios mentioned above have been seen in different physical systems, and indeed have been seen in the same system (the Taylor–vortex flow configuration) in different regions of the parameter space corresponding to the Taylor number, the ratio of gap width to inner cylinder radius, and the aspect ratio or ratio of cylinder length to outer cylinder diameter. Other transitions have been observed in this configuration also, and in other systems. In some cases there is no sequence of bifurcations at all. For example, if the inner cylinder in the Taylor configuration is fixed, and the outer one rotated from rest, there is an immediate bifurcation to fully chaotic flow. For low–aspect ratio configurations, where only a small number (no more than 4) of Taylor vortices is seen after the first bifurcation for rotation of the inner cylinder, a transition mechanism called "Shilnikov homoclinicity" has been observed, for which there is mathematical theory. The essential idea is that there is a homoclinic point, a fixed point of saddle kind, through which pass stable and unstable manifolds in the phase space, and a possible motion is along a homoclinic orbit of infinite period going out from the fixed point at $t = -\infty$ along the unstable manifold and returning along the stable manifold at $t = +\infty$. Such a motion is regular, and indeed it corresponds to the travelling wave solutions — shock waves or solitons — of partial differential equations of certain kinds. Shilnikov proved, however, that chaotic orbits exist near homoclinic points. Chaos attributed to the Shilnikov mechanism has been seen in Taylor–vortex flows with low–aspect ratio cylinders. See the velocity traces in Figure 25.14 and the projections into three-dimensional space of the corresponding strange attractors, as given in the Figure 25.14.

The point to be emphasized is that the much–publicized period–doubling route to chaos is simply one of many possibilities, though one clearly seen in many experiments which also confirm, to reasonable accuracy, the theoretical predictions for the two Feigenbaum constants. In many cases, however, even where the subharmonic route to chaos is observed, it is not simply through period doublings. In the case of the cavitation bubble experiments, for example, other subharmonics at one–third or one–fifth of the fundamental frequency are observed, these probably being connected, in the computer experiments on the Rayleigh equation, with various nonlinear resonances. In a physical experiment there are of course linear and nonlinear resonances in non–symmetric modes which provide a further resonant superstructure for the basic quadratically nonlinear period–doubling bifurcations.

25.5 CHARACTERIZATION OF CHAOS

In many cases standard methods of data analysis can be used to suggest that a sequence of bifurcations to chaos is taking place in some experiment, and to identify which scenario is being followed. For example, a direct plot of the time series of the measured signal (pressure $p(t)$ at a fixed point, in the bubble cavitation experiments) may indicate the presence of intermittency, or of the Shilnikov mechanism, or may indicate a direct transition to chaos. See, for example, the time–series traces in Figures 25.13 and 25.14. The quasi–periodic route (stemming from a series of Hopf bifurcations from one limit cycle to another) can be detected through the Fourier spectrum of the signal, as can the Feigenbaum period–doubling mechanism (see Figures 25.12 and 25.15). Various new tools have been devised also to characterize the number of dimensions which are needed to encapsulate the essential dynamical behaviour and to suggest the geometrical form of these attracting sets which the system approaches and to quantify the rate of approach, and these will be described briefly below.

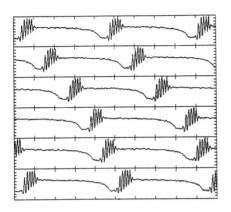

(a) Time series of radial velocity component of the flow measured at mid–gap of the annulus in the regular régime.

(b) "Phase portrait" reconstructed from the time series of (a).

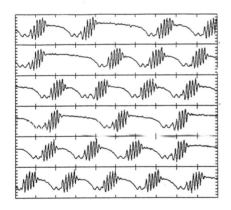
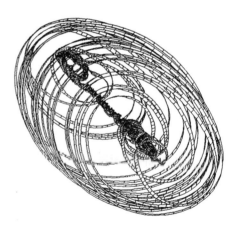

(c) Time series of radial velocity component of the flow measured at mid–gap of the annulus in the irregular régime.

(d) "Phase portrait" reconstructed from the time series of (c).

Figure 25.14 Shilnikov mechanism for transition in Taylor vortex flow (from Mullin & Price 1989).

694

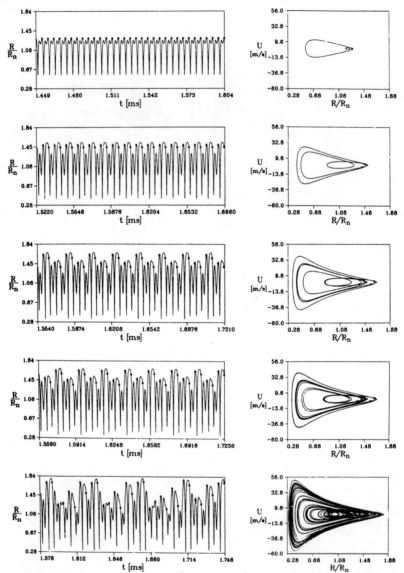

Figure 25.15 Period–doubling route to chaos demonstrated by the attractors for a bubble oscillator. Left column: radius–time solution curves; middle–left column: trajectories in state space; middle–right column: Poincaré section plots; right column: power spectra. Radius of bubble at rest $R_n = 10$ μm, sound–pressure amplitude 90 kPa (0.9 bar), driving frequency 1st row: 207 kHz; 2nd row: 197 kHz; 3rd row: 193 kHz; 4th row: 192.5 kHz; 5th row: 190 kHz (from Lauterborn & Parlitz 1988).

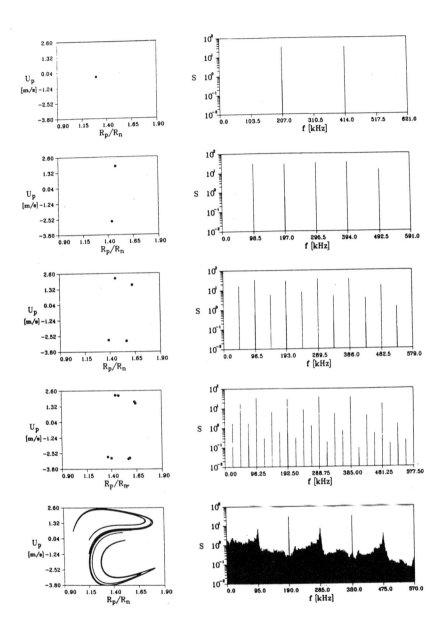

Figure 25.15 continued (from Lauterborn & Parlitz 1988).

Phase Space

The name "phase space" is borrowed from Hamiltonian dynamics, and there it is necessarily of even dimension $2N$, formed by the N generalized coordinates and associated N generalized momenta. Here we are concerned with more general time–dependent systems, where the number of variables needed to define the instantaneous state of the system may be odd or even — and where we assume the system to be autonomous (i.e. with coefficients not explicitly dependent upon the time) by adopting the device of introducing a new dependent variable τ and associated evolution equation $d\tau/dt = 1$.

In many systems, the governing equations are infinite–dimensional partial differential equations, but the question is whether there are a few important degrees of freedom which define a small finite–dimensional phase space in which much of the interesting dynamics takes place. An attempt to see whether this is possible can be made on the basis of an "embedding theorem" which suggests that one tries to project the dynamics onto a phase space spanned by the values of the measured signal $p(t)$ and its values at a set of delayed time instants $p(t-T), p(t-2T), p(t-3T)...$. The simplest case to try is a three–dimensional space, but if the system point clearly fills that space, then one has to try higher dimensions. Remarkably, the pressure measured at an arbitrary point in a field of cavitation bubbles seems to define a dynamical system in a phase space of dimension three — and indeed, the attractor (almost certainly a "strange attractor") to which the system point is attracted occupies a subspace of dimension less than three. Figure 25.9 shows this attractor, as measured by Lauterborn and Holzfuss, with rotation of the phase space about one axis by 15° from one frame to the next. The attractor occupies a disk–like region, but is actually infinitely–many–sheeted normal to the disk, not filling a three–dimensional volume, but occupying a volume much larger than a surface. In fact the "dimension" of this attractor has been found to be about 2.5. What is not known at present is how to deduce a set of three ordinary differential equations for the phase–space coordinates for which Figure 25.9 shows the underlying attractor, but much work is being done on this at present, as it constitutes a central problem of prediction. One would like, from measurements (or data taken from the Stock Exchange, as a very different example) to construct the "current" attractor for some system on the basis of recent data, and from this to deduce "current" underlying differential equations, which could then be used at least for short–term prediction. Then, from the prediction, and possibly from further measured data, one could up–date the attractor, refine the differential equations, and continue the prediction for the next small step.

It is also the case that at present the features, qualitative and quantitative, of an attractor cannot be immediately deduced from the governing dynamical equations, and quantities such as the dimension have to be obtained from long–time integration of the equations.

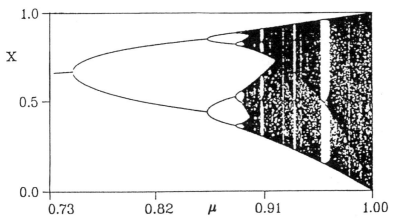

Figure 25.16 The bifurcation diagram of the logistic parabola in the form $X_{n+1} = 4\mu X_n(1 - X_n)$ (from Lauterborn & Parlitz 1988).

Bifurcation Diagrams

In a bifurcation diagram, transitions through a regular sequence of bifurcations are clearly seen when one plots, as a function of an increasing control parameter, the attractors of the system. To display things on paper, a projection is used and only one coordinate of the attractor is plotted against the control parameter. This is completely adequate to characterize the bifurcations of one–dimensional maps like the logistic iteration of Equation (25.5), the bifurcation diagram for which is shown below in Figure 25.16. We see the beginnings of the period–doubling sequence, with very rapid accumulation to chaos with, inside the chaotic range, bands in which the reverse period–halving takes place, together with other bands of periodic oscillation of other periods — e.g. period 3. Lauterborn and his colleagues have extended this idea to that of a "spectral bifurcation diagram", in which one plots the frequency spectrum of the attractor against the control parameter. Such diagrams are the "sonograms" referred to earlier, in which one has to use white–grey–black scales as in Figures 25.6 and 25.7, or full colour, to show how the character of the spectrum changes with the magnitude of the control parameter.

Poincaré Sections and Return Maps

A considerable simplification can be achieved, especially for three–dimensional systems, by taking sections by a Poincaré plane (at least locally) in the three–dimensional phase space. Then the continuous dynamics along the surface of the attractor are converted to a dynamical iteration process in the Poincaré plane with some functional relationship between the coordinates of a point of intersection with the plane and the coordinates of the next one. Typically, a strange attractor appears in a section as a many–branched curve, such as that shown below in Figure 25.17 for the cavitation bubble attractor. Self–similarity is apparent here; if the region enclosed by the box is magnified, one finds exactly the same banded structure at finer and finer scales. Strange attractors are believed to be continuous along the attractor surface, but normal to the surface to have a "Cantor–like" structure. (The classic example for formation of a Cantor set involves taking the unit interval $[0, 1]$ and removing the middle third, and then again removing the middle third from each of the two intervals left, and so on (always keeping the end points).)

If the dynamics can be captured by a Poincaré plane, then the process is effectively described by a nonlinear mapping of the plane onto itself, a topic of long study by pure mathematicians. Things become (relatively) even simpler if the two–dimensional mapping can be reduced to a one–dimensional one, either through use of a coordinate along the attractor or because the attractor is essentially made up of parts of a single curve. Again, this possibility is realized in the bubble cavitation problem. There, in some particular circumstances, the attractor in the Poincaré section of the three–dimensional phase space takes the form of eight linear segments. (The phase–space coordinates are the instantaneous radius of a bubble and its corresponding velocity, and the time, and so the Poincaré section amounts to measurement of the radius and velocity at a fixed phase of the forcing oscillation.) This illustrates the co–existence of periodicity and chaos. As the bubble motion proceeds, the system point cuts through the Poincaré section first on segment 1, then on segment 2, then on 3, etc., repeating the cycle in a periodic fashion with an overall period 8 times that of the forcing frequency. The chaos enters through the fact that the location along a given segment is random, though of course in principle deterministically predicted by the complete equations of motion. The Fourier spectrum of the motion shows broad–band chaotic response, but with strong spectral lines at a frequency 1/8th the driving frequency and at all integral harmonics of that 1/8th subharmonic (see Figure 25.18).

The mappings in the plane are called Poincaré maps or first return maps, and the one–dimensional maps may be called reduced Poincaré maps, or attractor maps.

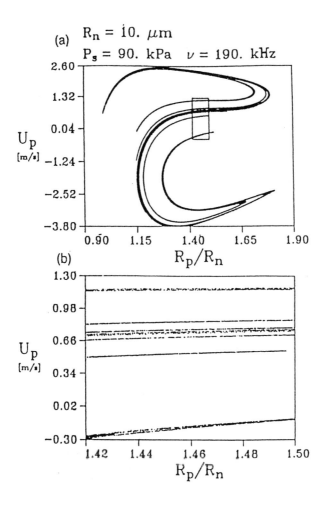

Figure 25.17 A strange bubble attractor in a Poincaré section plane. (a) Total view; (b) exploded view that indicates the self–similarity of the band structure. The coordinates in the Poincaré section are the bubble radius R_p and radial velocity U_p at a fixed phase of the periodic forcing (from Lauterborn & Parlitz 1988).

Attractor Dimensions

Attractors determined by measurement or computation seem to be many–sheeted non–space–filling objects, of non–integer dimension. An infinity of ways has been found to attach a "dimension" to such objects, of which the first, and most widely used in pure mathematics, is the Hausdorff dimension, or external measure, while the best known and easiest to use in experiment or computation is the correlation dimension associated with the names of Grassberger and Procaccia. For the Hausdorff dimension, one covers the set of points forming the attractor by cubes of side r in the appropriate dimension and writes $M(r)$ for the smallest number of such cubes which will cover the point set. Then one defines

$$d_0 = \lim_{r \to 0} \left[\ln M(r) / \ln(1/r) \right] \tag{25.8}$$

which naturally gives the values $0, 1, 2, 3$, for point, line, area and volume sets respectively.

For the correlation dimension, one supposes that N points of the attractor are given from computation or measurement, and asks how many pairs of them, labelled defining the correlation sum by

$$C(r) = \lim_{N \to \infty} \frac{1}{N^2} \sum_{\substack{i,j=1 \\ i \neq j}}^{N} H(r - |\mathbf{r_i} - \mathbf{r_j}|) \tag{25.9}$$

where H denotes the Heaviside unit function, and then defining the correlation dimension by

$$d_2 = \lim_{r \to 0} \left[\ln C(r) / \ln(1/r) \right] \quad . \tag{25.10}$$

In applications, a plot of $\ln C(r)$ against $\ln r$ shows a linear portion only over a restricted range of r. Taking r too small must not be done because the finite precision available in computation or experiment makes the points on the attractor space–filling at small r, and consequently the attractor dimension appears (wrongly) to be that of the phase space itself. Nevertheless, calculation of the correlation dimension is now feasible from real data, and fractal dimensions around 2.5 have been calculated for numerous attractors in a three–dimensional phase space (and a fractal dimension of, say, 1.3 for the attractor section by the Poincaré plane). Related quantities such as the "topological entropy" can also be calculated, but again there is no way of estimating the dimension or the entropy from some set of model equations without extensive computation.

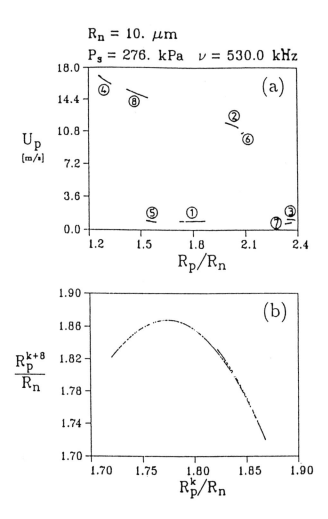

Figure 25.18 Period–8 chaotic attractor of a bubble oscillator in a Poincaré section plane (a) and a subharmonic attractor map of order 8 (b). the encircled numbers in (a) indicate the succession of the section points (from Lauterborn & Parlitz 1988).

Lyapunov Exponents

When describing the motion of a system near an equilibrium point, one can talk of a set of "characteristic exponents" μ_i which give the (exponential) rate of approach of the system to or away from the equilibrium point. The same idea is useful to assess the rate at which two trajectories diverge in the flow towards a strange attractor, or indeed any other kind of attractor. The corresponding exponents μ_i are called Lyapunov exponents, and may be defined in a global fashion, for the whole of one trajectory, or may be used in a local sense and constructed from local data. If the dynamics takes place in an N–dimensional space, then one can define, locally or globally, N Lyapunov exponents, and if the largest of these is positive, neighbouring trajectories will separate exponentially rapidly. The largest (positive) Lyapunov exponent thus quantifies the idea of "sensitivity to initial conditions", which is the essence of deterministic chaos. Determination of the Lyapunov exponents has been carried out, from real data, in a number of experimental situations and the difficulties associated with this are fully described in the references given below.

25.6 SUMMARY

Some of the early papers in the field of deterministic chaos spoke in the 1970's of addressing the "turbulence problem" with completely new tools. It is now recognized that deterministic chaos studies have rather little to do with the problem of predicting the averaged or typical features of fully–developed turbulence. For such turbulence at large Reynolds number, a very large number of degrees of freedom (phase–space coordinates) are needed — at least of the order $(\mathrm{Re})^{9/4}$, where Re is the Reynolds number, itself of order 10^6 or more in many applications of interest. There is not much hope for description by a relatively small finite subset of these, although recent developments in the theory of "inertial manifolds" have shown rigorously that for certain systems (including two–dimensional Navier–Stokes flows) there is indeed a finite–dimensional inertial manifold to the motion on which all other degrees of freedom are "slaved". There is no certainty that these results can be extended to three–dimensional flows. Nevertheless, recent studies also indicate that some important observed features, such as the "bursting" phenomenon in boundary layers, may indeed be simulated by a nonlinear dynamical system of quite moderate size (fewer than ten coordinates, say).

What the dynamical systems ideas have done is to show a large number of quite different routes for transition from regular to chaotic behaviour in simple fluid–mechanical closed systems, such as the Taylor vortex flow and the flow associated with thermal convection in a box. The acoustic cavitation experiments also provide a good example of a fluid–mechanical field in which dynamical systems ideas have made a startling impact, and there are many nonlinear mechanical, electrical and chemical oscillating systems in which there has been similar success in explaining and predicting strange aperiodic behaviour which was hitherto incapable of explanation. Deterministic chaos theory has also changed attitudes in many areas of physics and engineering, through the recognition that to have chaos it is not necessary to have the vast number of interacting degrees of freedom that one has in the turbulence problem; a quadratic map of the real axis onto itself, or three coupled ordinary differential equations with a nonlinear term, provide very simple examples giving extraordinarily complex and unpredictable evolution with time. There is, further, not a single route to irregular behaviour in these systems, but many different routes, in different parts of parameter space, for the same system. Some of these routes can be studied through very simple models, and "universal" scaling behaviour found. Extreme sensitivity to variations in initial conditions — and to the parameter values of the system — is an essential feature of evolving nonlinear systems. The process by which deviations accumulate is often talked of, both loosely and in rigorous mathematical terms, through the analogy of the "baker's transformation", involving the stretching of a sheet of dough, then folding of it back onto itself, then repeated stretching and folding. Even highly accurate initial data, by any normal scientific standards, quickly becomes worthless under such a transformation, and "prediction", as might be desirable for weather forecasting or the movement of stock prices, becomes impossible even over quite short times. More reliable prediction schemes do, however, seem to be possible, based on dynamical systems ideas which tell one how a system might possibly be modelled and how the model might be up–dated in the light of its own predictions and of other new data at later times.

REFERENCES

Barenblatt, G.I., Iooss, G. & Joseph, D.D. (eds.) (1983). Nonlinear Dynamics and Turbulence. Pitman, Boston.

Bender, C.M. & Orszag, S.A. (1978). Advanced Mathematical Methods for Scientists and Engineers. McGraw–Hill, New York.

Bergé, P, Pomeau, Y. & Vidal, C. (1986). Order Within Chaos. Wiley, New York.

Cvitanović, P. (1984). Universality in Chaos (a collection of reprints). Adam Hilger, Bristol.

Fenstermacher, P.R., Swinney, H.L. & Gollub, J.P. (1979). Dynamical instabilities and the transition to chaotic Taylor–vortex flow. J. Fluid Mech. 94:103–128.

Gollub, J.P. & Benson, S.V. (1980). Many routes to turbulent convection. J. Fluid Mech. 100:449–470.

Guckenheimer, J. & Holmes, P. (1983). Nonlinear Oscillations, Dynamical Systems and Bifurcations of Vector Fields. Springer–Verlag, New York.

Hao Bai–Lin (ed.) (1984). Chaos (a collection of reprints). World Scientific, Singapore.

Lauterborn, W (1986). Acoustic Turbulence. In: Frontiers in Physical Acoustics (ed. D. Sette), pp.124–144, North–Holland, Amsterdam.

Lauterborn, W. & Holzfuss, J. (1991). Acoustic chaos. Int. J. Bifurcation & Chaos 1:13–26.

Lauterborn, W. & Parlitz, U. (1988). Methods of chaos physics and their application in acoustics. J. Acoust. Soc. Am. 84:1975–1993.

Moon, F.C. (1987). Chaotic Vibrations. Wiley, New York.

Mullin, T. & Price, T.J. (1989). An experimental observation of chaos arising from the interaction of steady and time–dependent flows. Nature 340:294–296.

Thompson, J.M.T. & Stewart, H.B. (1986). Nonlinear Dynamics and Chaos. Wiley, Chichester.

Wiggins, S. (1988). Global Bifurcations and Chaos. Springer–Verlag, New York.

26. ANTI–SOUND

26.1 INTRODUCTION

Anti–Sound is simply a catchy name given to a sound field deliberately created to interfere destructively with an undesirable sound, the linear strictly anti–phase superposition of the two fields amounting everywhere to zero disturbance. Silence might then be regarded as the superposition of *sound* and *anti–sound*, but that view appeals only when silence is maintained by opposing sources, individually active but collectively silent, a silence that is easily broken by obstructing the field of either the primary or the interfering source.

Sources that naturally interfere destructively are common. The diaphragm of an unmounted loudspeaker makes sound by compressing air as it advances on one side while expanding it on the other; the diaphragm does not vary in volume so neither does the air have to accommodate any global volumetric change. What sound is generated results from the imperfect (destructive) interference between the opposing compression and expansion fields, the imperfection being due to the fact that sound takes longer to travel to the observer from the more distant parts. Loudspeaker practice is to mount the moving diaphragm in such a way that the source from the rear is baffled from earshot; then only half of the destructively interfering array is heard and heard much louder than before.

The term *anti–sound* is not appropriate for such passively arranged destructive fields. They are usually known as *multipoles*, generated by more than one source (pole), with an efficiency and strength absolutely determined by the totality of sources, which are collectively less efficient than any single member of the set. If they are not collectively weaker, then there is no significant destructive interference and their field is not so usefully thought of as multipole. In fact, waves from a source array may interfere constructively in some directions (Figure 26.1), thereby

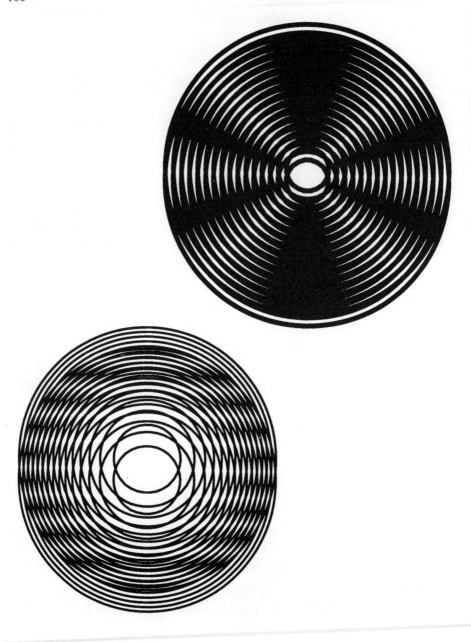

Figure 26.1 Constructive and destructive interference forms beam patterns.

forming columnated beams, and one could arrange to minimize such effects by appropriately phasing other interfering sources.

There is no difference between the sound of a naturally occurring phase–related multiple source system and that created artificially with control (electronic and acoustic apparatus) but the latter can be changed and adapted at the "flick of a switch" as it were. This brings new versatility and scope to practical acoustics. That is the reason it deserves a distinct name; anti–sound is no different from normal sound other than the fact that it is *very precisely* manufactured with strength, phase and spatial characteristics to fill some perceived need — usually the suppression of a noise. *Anti–sound* is synonymous with *active control of sound*. The technology needed to first determine the characteristics of the offending noise and then to duplicate and launch its inverse, is one that obviously includes the facility of audio–reproduction at an unusual level of sophistication.

The subject is an interdisciplinary blend of acoustics, control engineering and electronics, all applied initially to acoustic fields, fields that are small–amplitude vibrations of an elastic medium. Vibration of structures falls into the same category, and the ideas are equally applicable to both. The subject is actually much wider than acoustics in that it is potentially applicable to any weak perturbation of a system, even when the perturbation is an instability mode whose growth would normally change the global characteristics of the system. Fluid flows, for example, are often unstable to weak disturbances that grow into turbulence or large eddying motions which are rarely desirable. Sometimes the early instability phase is recognizable as a linear wave system, which can be modified by a device arranged to create the opposite of those disturbance waves; in this aspect of the subject these are still the very early days.

At first sight the technique seems rather improbable. The silencing of machines by loudspeaker emissions carefully arranged to phase match the primary field is an activity that attracts both the subject's enthusiasts and the critical eye of the hardened sceptic, conditioned to believe in the inevitability of some noises that are prime targets for control by anti–sound. It is generally the low–frequency noise problems that have proved most difficult to counter by conventional means. Sound absorbing materials are not good at low frequencies and it is the low–frequency sounds that propagate for large distances without atmospheric attenuation. Neither are they much obstructed by containing walls. Until active control became an option, the main technique for the containment of low–frequency sound was added mass, the lower the frequency the bigger the necessary noise–controlling mass. That sort of cure has obvious drawbacks and limitations, especially where the noise is to be controlled on a flying aircraft.

But it is at the very low frequencies that are so difficult to control by passive methods that the active alternative is most feasible. Active schemes require the intelligent processing of information in time to effect a change before the monitored field has passed. That is obviously easiest at low frequency. For that reason, it is the low–frequency noises that have provided the impetus for large–scale practical demonstrations. An active sound absorber was first demonstrated on a gas turbine in 1980. The bottom octave that was the subject of control was attenuated by some fifteen decibels. Seventy-two large loudspeakers powered by about a kilowatt of controlled signal eliminated the low–frequency rumbling from the gas turbine. The subject is a real one and might before long come into general use.

26.2 ONE–DIMENSIONAL WAVES

Linear waves are particularly amenable to active control because it is possible to predict their subsequent behaviour from an "upstream" observation in past time. The knowledge that the future behaviour in some region is a linear function of a previously observed state guarantees that its extinction by anti–sound is feasible in that region. Practice has shown that technology is already adequate for theoretically optimal configurations to be closely achievable. The control information for the generation of anti–sound can be acquired "upstream" in advance of the time it is needed. This greatly extends the stability margin of the system and is the principal reason for why there are so many different ways of implementing anti–sound on simple wavefields.

We illustrate the subject here by taking a particular example that is simple enough to be analyzed exactly and for which all the characteristics of the model can be seen in complete detail. It is through such definite examples that intuitive feel for the subject is developed, an important point in a subject where Rayleigh's principle that "energies" are additive is frequently viewed as a foundation stone. Rayleigh's principle rests on the mutual incoherence of most sounds; that cannot be the case with anti–sound and that is why the subject sometimes appears to run counter to common experience and intuition.

We consider a one–dimensional sound wave governed by the wave equation

$$\frac{\partial^2 p}{\partial t^2} - c^2 \frac{\partial^2 p}{\partial x^2} = 0 \quad , \tag{26.1}$$

having nontrivial wave solutions

$$p = f(x + ct) + g(x - ct) \quad , \tag{26.2}$$

f and g being any function of their arguments.

A point source of strength $\frac{\partial q}{\partial t}(t)$ at $x = 0$ generates the outgoing wave $\frac{1}{2c}$ $q(t - |x|/c)$ and a point dipole of strength $h(t)$ at $x = 0$ generates the outgoing wave $\frac{-\text{sgn}\, x}{2c^2} h(t - |x|/c)$. These two sources together generate, according to the equation

$$\frac{\partial^2 p}{\partial t^2} - c^2 \frac{\partial^2 p}{\partial x^2} = \frac{\partial}{\partial t}\Big(q(t)\delta(x)\Big) + \frac{\partial}{\partial x}\Big(h(t)\delta(x)\Big) \quad , \tag{26.3}$$

the combined outgoing field

$$p = \frac{1}{2c}\left\{ q\left(t - \frac{|x|}{c}\right) - \frac{\text{sgn}\, x}{c} h\left(t - \frac{|x|}{c}\right) \right\} \quad , \tag{26.4}$$

which, in any particular case, represents a wave biased either to the left or to the right. For example, when h is chosen to equal $-cq$, then only a right–going wave is launched by the combined sources, and

$$p(x, t) = \frac{H(x)q(t - \frac{|x|}{c})}{c} \quad . \tag{26.5}$$

This kind of source combination could be added to a wave of strength q, incident on the origin from the left, to annihilate it without trace. Equation (26.5) is then the prescription of the anti–sound field, which is superposed on the sound to induce silence on the right of the origin.

Waves travelling in a particular direction can be distinguished from others by a similar wave–detection strategy, and this will be useful in alerting the controller of the wave which is shortly to arrive, and inform it of what waveform the anti–source should produce. The general field is described by Equation (26.2). If the combination measure of that field represented by $\frac{1}{2}\left\{ c \frac{\partial p}{\partial x} - \frac{\partial p}{\partial t} \right\} = m(t)$, is taken at $x = -L$, then

$$m(t) = \frac{1}{2}\left(c \frac{\partial p}{\partial x} - \frac{\partial p}{\partial t} \right) = -\frac{\partial g}{\partial t}(-L - ct) = -c \frac{\partial g}{\partial x}(-L - ct) \quad , \tag{26.6}$$

which gives immediately the strengths of the sources that are to be activated at $x = 0$ time L/c later to eliminate the right–travelling wave in $x > 0$

$$\left. \begin{array}{l} \dfrac{\partial g}{\partial t}(t) = -m(t - L/c) \\[2mm] h(t) = -cm(t - L/c) \end{array} \right\} \quad . \tag{26.7}$$

This illustrates important characteristics of practical tools in anti–sound technology. The need to measure signals accurately, to delay them in time and to combine linearly multiple signals with prescribed time delays is paramount. That is exactly the kind of thing that is done well with digital processors. The practice is, of course, more intricate than this example implies. Pressure gradient microphones are not usually as effective as simple pressure–measuring elements, and pressure gradients tending to be estimated by differencing two microphone signals; differencing emphasizes errors.

The combination measure

$$\left\{p(x,t) - p(x+L,\ t-L/c)\right\} = g(x-ct) - g(x+2L-ct) \quad , \qquad (26.8)$$

is another way of identifying the right–travelling component of the wave. From this measure the signal g can be extracted by passing the measured signal through a filter with input/output transfer function $(1 + e^{-2i\omega L/c})^{-1}$. Pre–filtering will be necessary to avoid components at frequencies $\frac{1}{2}\, n\pi c/L$, n being an odd integer, frequencies at which no information is gathered by this two–element method of measurement because of the undamped filter resonances.

In principle then, one–dimensional waves are easily recognized, launched with appropriate directional bias and phase to effect mutual annihilation.

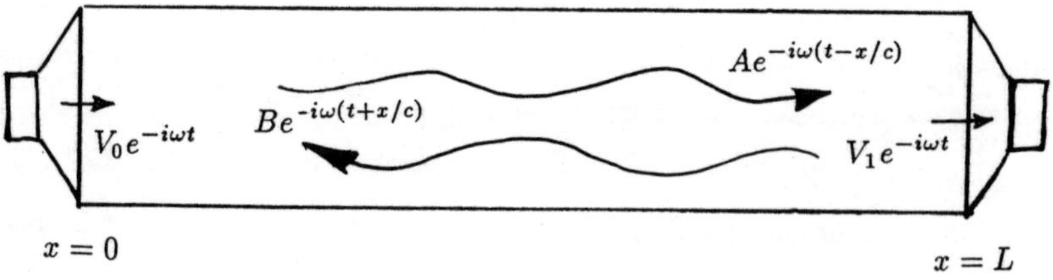

$$\overrightarrow{I} = \left\{ \frac{|A|^2 - |B|^2}{2} \right\} \rho_0 c$$

$$= \frac{g(V_0 v_1^*)}{2\, \sin \omega L/c}$$

Figure 26.2 One–dimensional waves in an acoustic duct.

26.3 ONE–DIMENSIONAL WAVES BETWEEN TWO COHERENT SOURCES

Suppose that two loudspeakers face each other at opposite ends of a duct of length L, a situation illustrated at Figure 26.2. Suppose that the loudspeaker at $x = 0$ is driven to generate an axial velocity $V_0 e^{-i\omega t}$ while that at $x = L$ maintains there a velocity $V_1 e^{-i\omega t}$. The (one–dimensional) velocity in the duct will be the superposition of two waves of frequency ω, one travelling "upstream" and the other "downstream".

$$u(x,t) = Ae^{-i\omega(t-x/c)} + Be^{-i\omega(t+x/c)} \qquad (26.9)$$

from which the boundary condition gives

$$V_0 = A + B \quad \text{and} \quad V_1 = Ae^{i\omega L/c} + Be^{-i\omega L/c} \quad , \qquad (26.10)$$

so that

$$A = \frac{V_1 - V_0 e^{-i\omega L/c}}{2i \sin \omega L/c}$$

and

$$B = \frac{V_0 e^{i\omega L/c} - V_1}{2i \sin \omega L/c} \quad . \qquad (26.11)$$

Note that $A = B$ if $V_0 = 0$, a condition that expresses the fact that a rigid termination to the wavefield at $x = 0$ corresponds to a reflector in which the velocity perturbations in the opposite travelling waves cancel. Similarly, $V_1 = 0$ corresponds to the same conditions imposed at a distance L "downstream".

The sound pressure in the positive–going wave is $\rho_0 c$ times the velocity perturbation in that wave, while the pressure–velocity ratio in the negative propagating wave is $-\rho_0 c$,

$$p(x,t) = \rho_0 c \left\{ Ae^{-i\omega(t-x/c)} - Be^{-i\omega(t+x/c)} \right\} \quad . \qquad (26.12)$$

Now it is clear that the wave generated by the loudspeaker at $x = 0$ is *not* determined by that loudspeaker alone. It is critically dependent on what the other loudspeaker is doing. For example, if V_1 were maintained at a value $V_0 e^{-i\omega L/c}$, then A would be zero and only "upstream"–travelling waves would exist. In that event, the power flow would be from the loudspeaker at $x = L$ to the loudspeaker at $x = 0$ through which the wave would be absorbed and its energy consumed. Alternatively, with the $x = 0$ loudspeaker still maintained at the same prescribed velocity but with the $x = L$ loudspeaker driven with velocity amplitude $V_0 e^{i\omega L/c}$, then B would be zero and the only wave would be that travelling from $x = 0$ to $x = L$ and that would also be the direction of energy flow.

The mean value of the energy flux in the positive x direction, the intensity I, is the mean value of the pressure–velocity product

$$I = \frac{1}{2} \rho_0 c \left\{ |A|^2 - |B|^2 \right\} \quad , \tag{26.13}$$

which is independent of x. All the power flows from one loudspeaker to the other.

The *direction* of power flow depends on which of the forward– or backward–travelling elements is bigger. That is solely determined by the relative *phases* of V_0 and V_1, because, if

$$V_0 = |V_0| e^{i\theta_0} \quad \text{and} \quad V_1 = |V_1| e^{i\theta_1} \quad , \tag{26.14}$$

$$\left.\begin{aligned} A &= \frac{|V_1| e^{i\theta_1} - |V_0| e^{i\theta_0} e^{-i\omega L/c}}{2i \sin \omega L/c} \quad , \\[2em] B &= \frac{|V_0| e^{i\theta_0} e^{i\omega L/c} - |V_1| e^{i\theta_1}}{2i \sin \omega L/c} \quad , \end{aligned}\right\} \tag{26.15}$$

so that

$$|A|^2 - |B|^2 = \frac{|V_0|\,|V_1| \sin(\theta_1 - \theta_0)}{\sin \omega L/c} \tag{26.16}$$

and

$$I = \frac{1}{2} \rho_0 c |V_0|\,|V_1| \frac{\sin(\theta_1 - \theta_0)}{\sin \omega L/c} \quad . \tag{26.17}$$

If V_0 and V_1 are in phase there is no power flow in the sound field (unless $\sin \omega L/c = 0$, which are the special cases corresponding to the free models of the duct). Power flows from $x = 0$ to $x = L$ when $\sin \omega L/c \sin(\theta_1 - \theta_0)$ is positive. It is reversed when the product is negative.

The two loudspeakers are an integral number of half wavelengths apart when $\sin \omega L/c$ vanishes. Then any nontrivial sound field cannot be statistically stationary in time. Damping would hold finite the response, but in the absence of damping the field grows with time.

For example,

$$u = Ae^{-i\omega_n(t-x/c)} + (V_0 - A)e^{-i\omega_n(t+x/c)}$$

$$+ (V_0 + V_1)\left(\frac{ct}{2L} - \frac{x}{2L}\right)e^{-i\omega_n(t-x/c)} - (V_0 + V_1)\left(\frac{ct}{2L} + \frac{x}{2L}\right)e^{-i\omega_n(t+x/c)} \quad , \tag{26.18}$$

is a solution when $\omega_n L/c = \pi$, and this is so for any value of A; the sound is undetermined to within any free mode of system. The power flow for this system is affected by the presence of the free mode which changes the pressure at the loudspeakers without affecting their velocities. By choosing A appropriately, power can be made to flow in the positive or negative direction of x even though the loudspeaker motion is exactly prescribed, but together the loudspeakers supply energy to the wave at a rate increasing in proportion to time; the "mean" energy in the inter–loudspeaker length is increasing with the square of time at this condition of undampened resonance.

26.4 ONE–DIMENSIONAL ANTI–SOUND

Consider now a one–dimensional sound wave incident on $x = 0$ from $x < 0$, which propagates to $x + L$ where it is partly reflected, partly transmitted and, possibly, partly amplified or absorbed. A single loudspeaker at $x = 0$ is used to effect the active control, the object of which is to minimize the wave activity in $x > 0$. Sound disappearing up the tube in the negative x direction is partly reflected back to supplement whatever sound was generated in the principal source area. The situation is illustrated in Figure 26.3. There are many possible strategies for actively controlling this system and we describe here one of the easiest to implement.

Our objective is to minimize the sound at 2 where the pressure is $p_2(t)$, using the signal produced by the microphone at 1 where it measures the pressure $p_1(t)$. Unit signal at frequency ω into the control loudspeaker produces pressure T_{01} at microphone 1 and T_{02} at microphone 2. These transfer functions would be measured, ideally with the system running in its operational condition, measured by injecting random noise into the controller (a noise which is thus uncorrelated with the machine's own noise field) and correlating the pressures measured with microphones 1 and 2 with the control signal. Also required is the information relating the pressure measured at microphone 1 to that at microphone 2 in the absence of the controller. The transfer function T_{12} that relates these is easily measured.

714

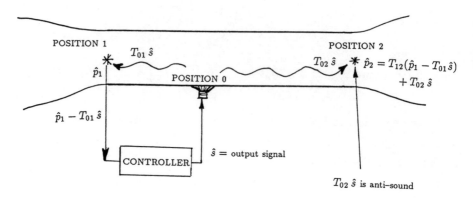

Figure 26.3 Control of sound at one point from observations at another.

The controller must now be driven with a signal $\widehat{s}$ (at frequency ω) such that the pressure $\widehat{p}_2$ is zero. That part of the pressure at 1 which is not produced by the controller is

$$\widehat{p}_1 - T_{01}\,\widehat{s} \quad,$$

and this is linearly related to that part of the pressure at 2 which is *not* due to the controller. The controller–induced pressure at 2, $\widehat{s}\,T_{02}$ is thus to be the opposite of the sound that is there in the controller's absence,

$$(\widehat{p}_1 - T_{01}\,\widehat{s}\,)T_{12} \quad,$$

i.e.

$$\widehat{s}\,T_{02} = (-\widehat{p}_1 + T_{01}\,\widehat{s}\,)T_{12}$$

$$\widehat{s} = \frac{\widehat{p}_1 T_{12}}{(T_{01}T_{12} - T_{02})} = T\widehat{p}_1 \quad . \tag{26.19}$$

A control filter that produces this value of $\widehat{s}$ from the measured $\widehat{p}_1$, will consequently maintain silence at station 2.

In practice, of course, the controller could only model approximately the real thing, and errors would degrade the system's performance. Two kinds of errors are distinguishable. Firstly, a detector measuring $\widehat{p}_1$ would be contaminated by noise

$\widehat{n}$. Secondly, the implementation of the filter characteristic would be imperfect, producing a signal T_a times its input rather than T,

$$\widehat{s} = T_a(\widehat{p}_1 + \widehat{n}) \quad . \tag{26.20}$$

The incoming field at 1 is the existing field $\widehat{p}_1$ minus the controller–induced pressure, $T_{01}\widehat{s}$, and this is related to the (uncontrolled) pressure at 2, $\widehat{p}_{2i}$ by the transfer function T_{12}. The incident pressure at 2 is therefore

$$\widehat{p}_{2i} = (\widehat{p}_1 - T_{01}\widehat{s})T_{12} = \widehat{p}_1(1 - T_{01}T_a)T_{12} - T_{01}T_aT_{12}\widehat{n} \quad . \tag{26.21}$$

The controller produces at 2, the additional field

$$T_{02}\widehat{s} = T_{02}T_a(\widehat{p}_1 + \widehat{n}) \quad . \tag{26.22}$$

This can be expressed in terms of the incident field at 2 by using Equation (26.21) to give the controller–induced anti–sound field at 2 as

$$T_{02}\widehat{s} = \frac{T_{02}T_a}{T_{12}} \frac{\widehat{p}_{2i}}{(1 - T_{01}T_a)} + \frac{T_{02}T_a\widehat{n}}{(1 - T_{01}T_a)} \quad . \tag{26.23}$$

The sum of the incident and controller–induced pressure at 2 is then

$$\widehat{p}_{2i} + T_{02}\widehat{s} = \frac{\left\{1 - T_a\left(\dfrac{T_{12}T_{01} - T_{02}}{T_{12}}\right)\right\}\widehat{p}_{2i} + T_{02}T_a\widehat{n}}{(1 - T_{01}T_a)} \tag{26.24}$$

$$= \frac{1 - T_a/T}{1 - T_{01}T_a}\widehat{p}_{2i} + \frac{T_{02}T_a\widehat{n}}{1 - T_{01}T_a} \quad .$$

With a perfect noiseless control system, then, there would be silence at 2 when the controller characteristic $T_a = T$. But errors in the characteristic and noise degrade the system.

Good attenuative performance not only demands a good controller ($T_a = T$) but also requires that the signal entering the controller ($\widehat{p}_1 + \widehat{n}$) contains as little as possible of the controller–induced contamination at station 1. That contamination is $T_{01}T_a(\widehat{p}_1 + \widehat{n})$. The difference $(1 - T_{01}T_a)$ therefore represents the fraction of the controller signal that has not already been through the controller. If this is small, then noise and system error are accentuated and the attenuative performance deteriorates according to Equation (26.24). It is always important that the detection system should be made as immune as possible from contamination by the controller output. The controller gets irritable when handling the same information more than once and shows it by exaggerating the effects of the transfer function's interactions and noise!

Noise not only degrades the attenuative performance of a controller, it also alters the controller characteristics for which optimal control is achieved. Another way of writing Equation (26.24) for the controlled sound at 2, $\widehat{p}_c$, say, is

$$\widehat{p}_c = \widehat{p}_{2i} + T_{02}\widehat{s} = (1+Z)\widehat{p}_{2i} + Z\,\widehat{n}\,T_{12}$$

(26.25)

$$\text{where} \qquad Z = \frac{T_{02}T_a/T_{12}}{1 - T_{01}T_a} \quad,$$

and the optimal controller would minimize the variance $\overline{|\widehat{p}_c|^2}$ of this residual

$$\overline{|\widehat{p}_c|^2} = |1+Z|^2\,\overline{|\widehat{p}_{2i}|^2} + |Z|^2\,\overline{|\widehat{n}|^2}\,|T_{12}|^2 \quad.$$

(26.26)

It is clear that an optimum value of Z exists and that its optimum value depends on the signal–to–noise ratio. Bringing $-Z$ close to unity to minimize the first term inevitably involves an emphasis of the noise. That optimum condition is most easily seen by writing Equation (26.26) in the form

$$\overline{|\widehat{p}_c|^2} = \frac{\overline{|\widehat{n}|^2}\,|T_{12}|^2}{1 + \dfrac{\overline{|\widehat{n}|^2}\,|T_{12}|^2}{\overline{|\widehat{p}_{2i}|^2}}} + \left|Z + \frac{1}{1 + \dfrac{\overline{|\widehat{n}|^2}\,|T_{12}|^2}{\overline{|\widehat{p}_{2i}|^2}}}\right|^2 \left\{\overline{|\widehat{p}_{2i}|^2} + \overline{|\widehat{n}|^2}\,|T_{12}|^2\right\} \quad. \; (26.27)$$

Noise in the controller reduced the "gain", Z, for which optimal performance is obtained.

$$Z_{\text{opt}} = \frac{T_{02}T_a/T_{12}}{1 - T_{01}T_a} = \frac{-1}{1 + \dfrac{\overline{|\widehat{n}|^2}\,|T_{12}|^2}{\overline{|\widehat{p}_{2i}|^2}}}$$

(26.28)

is the optimal controller setting and that corresponds to a controller sound level at station 2 of

$$\overline{|\widehat{p}_c|^2_{\text{min}}} = \frac{\dfrac{\overline{|\widehat{n}|^2}\,|T_{12}|^2}{\overline{|\widehat{p}_{2i}|^2}}}{1 + \dfrac{\overline{|\widehat{n}|^2}\,|T_{12}|^2}{\overline{|\widehat{p}_{2i}|^2}}}\;\overline{|\widehat{p}_{2i}|^2} \quad.$$

(26.29)

The attenuation factor, the ratio of the best controlled sound level to that existing prior to control is the reciprocal of *one plus the signal–to–noise ratio*.

The "signal" in this respect is the pressure at the detector in the absence of control, $\widehat{p}_{2i}/T_{12}$, and the "noise" is the corruption of that signal in the input to the controller

$$S/N = signal\text{-}to\text{-}noise\ ratio = \frac{\overline{|\widehat{p}_{2i}|^2}}{|T_{12}|^2\ \overline{|\widehat{n}|^2}} \quad . \tag{26.30}$$

The maximum achievable attenuation is, therefore, from (26.29), equal to

$$\frac{\overline{|\widehat{p}_c|^2_{\min}}}{\overline{|\widehat{p}_{2i}|^2}} = \frac{1}{1 + S/N} \quad . \tag{26.31}$$

When the controller noise level is high, $Z_{\rm opt}$ is small and only very little attenuation is possible. But when the signal–to–noise ratio is high, $Z_{\rm opt}$ is close to -1 and the sound level at 2 can then be reduced to a level nearly as low as the low controller noise itself. The incident sound level and the controller noise bound the optimal performance of the anti–sound system.

These performance bounds can be established by measurement before the controller is designed, because the signal–to–noise ratio is actually a measure of the coherence between the detector signal and the sound at the station to be controlled

$$S/N = \frac{\gamma^2}{1 - \gamma^2} \quad ; \tag{26.32}$$

γ being the modulus of the coherence function, and the maximum achievable attenuation is then

$$\frac{\overline{|\widehat{p}_c|^2_{\min}}}{\overline{|\widehat{p}_{2i}|^2}} = 1 - \gamma^2 \quad . \tag{26.33}$$

The coherence between the detector signal and the controlled sound level can be a useful measure of the controller's performance, because when working optimally that coherence is zero. There is no more useful information to be extracted from the detector about the form of the sound at 2 and there can be no further attenuative improvement. We define the modulus of that detector/monitor coherence when the controller is operating by γ_c

$$\gamma_c^2 = \frac{\left\{\overline{|\widehat{s}\ \widehat{p}_c|}\right\}^2}{\overline{|\widehat{s}|^2}\ \overline{|\widehat{p}_c|^2}} \quad , \tag{26.34}$$

which can be evaluated from Equations (26.23) and (26.25)

$$\hat{s} = \frac{Z\,\hat{p}_{2i} + T_{12}Z\,\hat{n}}{T_{02}} \quad , \quad ,$$

$$\hat{p}_c = (1 + Z)\hat{p}_{2i} + T_{12}Z\,\hat{n} \quad ,$$

$$\overline{|\hat{s}\,\hat{p}_c|} = |Z + Z^2|\,\overline{|p_{2i}|^2} + |T_{12}|^2\,|Z|^2\,\overline{|\hat{n}|^2} \Big/ T_{02} \quad , \tag{26.35}$$

$$\overline{|\hat{s}|^2} = |Z|^2\,\overline{|p_{2i}|^2} + T_{12}^2\,|Z|^2\,\overline{|\hat{n}|^2} \Big/ T_{02}^2 \quad , \tag{26.36}$$

$$\overline{|p_c|^2} = |1 + Z|^2\,\overline{|p_{2i}|^2} + |T_{12}|^2\,|Z|^2\,\overline{|\hat{n}|^2} \quad , \tag{26.37}$$

$$\gamma_c^2 = \frac{\left\{ |Z + Z^2| + \frac{|Z|^2}{S/N} \right\}^2}{\left\{ |Z|^2 + \frac{|Z|^2}{S/N} \right\}\left\{ |1 + Z|^2 + \frac{|Z|^2}{S/N} \right\}} \quad , \tag{26.38}$$

$$= 0 \quad \text{when} \quad Z = Z_{\text{opt}} = \frac{-S/N}{1 + S/N} \quad \text{from} \quad (26.28) \quad .$$

The question of whether a stable controller can be constructed with the optimal value of Z is one that depends on the detailed coherence between the detector and primary sound fields. A perfect noiseless system can, in principle, be defined which will guarantee absolute silence at station 2. Furthermore, silence at station 2 implies silence everywhere in $x > 0$, no matter what the details of the acoustic space beyond, so that in practice the sound field has been reflected back into the $x < 0$ region from whence it came. The sound pressure at $x = 0$ is thus maintained at zero by the controller, and since the loudspeaker then operates with *no* pressure change, it is energetically *neutral*. All the acoustic energy coming from $x < 0$ is thus returned by the controller (regardless of the precise geometry in $x > 0$) to be either absorbed or re–reflected back. If it is all re reflected back (from $x < 0$), or indeed if it is amplified, then the system is unstable. But if there is some loss of energy on reflection, less will reappear than was returned by the controller, and a bounded response is assured. The energy is then eventually consumed by the dissipative element in $x < 0$.

The design of the filter characteristics may be much easier with some field measures than others; since it is only the right–travelling wave that excites the system in $x > 0$, then there would be a clear advantage in a measuring strategy that only responds to that rightward wave.

As an interesting aside, one notes that, in this one–dimensional case, the maximum possible acoustic power that can be extracted by the single control loudspeaker is exactly half the power in the incident plane wave. The incident wave of pressure p_i is supplemented by the anti–sound pressure p_a, which is a wave travelling to the right in $x > 0$ and to the left in $x < 0$.

The energy flux incident on the control loudspeaker is

$$\frac{\overline{p_i^2}}{\rho_0 c} \quad ,$$

while that leaving it is

$$\frac{\overline{(p_i + p_a)^2}}{\rho_0 c}$$

to the right, and

$$\frac{\overline{p_a^2}}{\rho_0 c}$$

to the left. The difference between incoming and outgoing energy flux is the power absorbed P_a, say, per unit area of the wave

$$P_a = \frac{\overline{p_i^2}}{\rho_0 c} - \frac{\overline{(p_i + p_a)^2}}{\rho_0 c} - \frac{\overline{p_a^2}}{\rho_0 c}$$

(26.39)

$$= \frac{1}{2} \frac{\overline{p_i^2}}{\rho_0 c} - \frac{\overline{(p_i + 2p_a)^2}}{2\rho_0 c} \quad .$$

Maximum power is absorbed by the loudspeaker when p_a is set equal to $-\frac{1}{2} p_i$ in $x > 0$, at which condition exactly half the incident wave power is extracted.

26.5 THE OPTIMAL PHYSICALLY REALIZABLE CONTROLLER

We consider now the question of how realizable is the optimal performance which we have shown to depend on the signal–to–noise ratio, i.e. the coherence between the detector signal and the primary sound field. We have to incorporate a further crucial constraint arising from the fact that the controller cannot possibly anticipate its input. The transfer function through the controller must be both stable and causal.

The sound to be controlled is, in the absence of the control, given by

$$p(t) = \frac{1}{2\pi} \int_{-\infty}^{\infty} \hat{p}(\omega)e^{-i\omega t}d\omega \quad , \tag{26.40}$$

and this is to be supplemented by the anti–sound signal

$$p_a(t) = \frac{1}{2\pi} \int_{-\infty}^{\infty} \hat{p}_a(\omega)e^{-i\omega t}d\omega \quad , \tag{26.41}$$

which is produced linearly and causally by the control system from a signal $s(t)$ monitored at some convenient place by a detector,

$$s(t) = \frac{1}{2\pi} \int_{-\infty}^{\infty} \hat{s}(\omega)e^{-i\omega t}d\omega \quad . \tag{26.42}$$

The controller produces the anti–sound p_a by performing a linear operation on its input signal s, an operation we can represent by the symbol L

$$Lp_a = s \quad , \tag{26.43}$$

$$LI = \delta \quad , \tag{26.44}$$

is the particular case when the input signal is impulsive; I the impulse response of the control system, must be zero until the impulse has occurred. According to Section 2.4 then,

$$p_a = \delta * p_a = p_a * LI = Lp_a * I = s * I \tag{26.45}$$

gives the anti–sound as a convolution of the controller input and the impulse response,

$$p_a(t) = \int_{-\infty}^{\infty} I(\tau)s(t-\tau)d\tau = \int_{0}^{\infty} I(\tau)s(t-\tau)d\tau \quad , \tag{26.46}$$

because

$$I(\tau) = 0 \quad \text{for all} \quad \tau < 0 \quad .$$

The Fourier transform of the impulse response is the controller's transfer function T_{ps}^{+}, a plus function because it is the transform of a function that exists only in positive time

$$T_{ps}^+(\omega) = \int_{-\infty}^{\infty} I(t)e^{i\omega t}dt = \int_0^{\infty} I(t)e^{i\omega t}dt \quad , \tag{26.47}$$

$$I(t) = \frac{1}{2\pi} \int_{-\infty}^{\infty} T_{ps}^+(\omega)e^{-i\omega t}d\omega = 0 \quad \text{for} \quad t < 0; \tag{26.48}$$

$T_{ps}^+(\omega)$ is an analytic function in the upper half of the complex ω plane and must therefore correspond to a stable controller. This can be checked by evaluating Equation (26.48) by Cauchy's theorem, the contour being closed in the upper half plane when t is negative.

The convolution theorem gives the Fourier transform of Equation (26.46) as

$$\widehat{p}_a(\omega) = T_{ps}^+(\omega)\,\widehat{s}(\omega) \quad , \tag{26.49}$$

which is the alternative Fourier space description of Equation (26.43) that defines the controller characteristic.

This anti–sound combines with the primary sound field to leave the residual (sum) pressure fluctuation

$$(p + p_a)(t) = \frac{1}{2\pi} \int_{-\infty}^{\infty} (\widehat{p} + T_{ps}^+\,\widehat{s})(\omega)e^{-i\omega t}d\omega \quad , \tag{26.50}$$

whose variance is

$$\overline{(p + p_a)^2} = \frac{1}{(2\pi)^2} \int\int_{-\infty}^{\infty} \overline{(\widehat{p} + T_{ps}^+\,\widehat{s})(\omega)(\widehat{p} + T_{ps}^+\,\widehat{s})(\alpha)}\; e^{-i(\omega+\alpha)t}d\omega\,d\alpha \quad . \tag{26.51}$$

The mean products of Fourier transforms are related to the cross–spectral densities, as follows:

$$\overline{\widehat{p}(\omega)\widehat{p}(\alpha)} = 2\pi\widehat{R}_{pp}(\alpha)\delta(\omega + \alpha) \quad ,$$

$$\overline{\widehat{p}(\omega)\widehat{s}(\alpha)} = 2\pi\widehat{R}_{sp}(\omega)\delta(\omega + \alpha) \quad ,$$

$$\tag{26.52}$$

$$\overline{\widehat{p}(\alpha)\widehat{s}(\omega)} = 2\pi\widehat{R}_{sp}(-\omega)\delta(\omega + \alpha) \quad ,$$

$$\overline{\widehat{s}(\omega)\widehat{s}(\alpha)} = 2\pi\widehat{R}_{ss}(\alpha)\delta(\omega + \alpha) \quad ,$$

so that Equation (26.51) becomes

$$\overline{(p+p_a)^2} = \frac{1}{2\pi} \int_{-\infty}^{\infty} \left\{ \widehat{R}_{pp}(\omega) + T_{ps}^+(\omega)\widehat{R}_{sp}(-\omega) + T_{ps}^+(-\omega)\widehat{R}_{sp}(\omega) \right.$$

$$\left. + T_{ps}^+(\omega)T_{ps}^+(-\omega)\widehat{R}_{ss}(\omega) \right\} d\omega \quad .$$

(26.53)

This equation simplifies by Wiener's technique of spectral factorization and the use of Parseval's theorem,

$$\int_{-\infty}^{\infty} x(t)y(t)dt = \frac{1}{2\pi} \int_{-\infty}^{\infty} \widehat{x}(\omega)\widehat{y}(-\omega)d\omega \quad ,$$

which shows that if $\widehat{x}$ is a "plus" function, and $\widehat{y}$ a "minus" function, being Fourier transforms of functions zero for negative and positive time, respectively, then the product $x(t)y(t)$ is always zero, and

$$\int_{-\infty}^{\infty} \widehat{x}^+(\omega)\widehat{y}^-(-\omega)d\omega = 0 = \int_{-\infty}^{\infty} \widehat{x}^+(-\omega)\widehat{y}^-(\omega)d\omega \quad . \tag{26.54}$$

The factorization technique allows Equation (26.53) to be re-expressed as the sum of squared quantities from which the minimum is obvious by inspection. The factorization involves writing the terms in Equation (26.53) as products of "plus" and "minus" functions, and eliminating those elements shown by Equation (26.54) to integrate to zero.

The product of two "plus" functions is itself a "plus" function; similarly with the "minus" functions,

$$\frac{1}{2\pi} \int_{-\infty}^{\infty} \widehat{x}^+(\omega)\widehat{y}^+(\omega)e^{-i\omega t}d\omega = \int_{-\infty}^{\infty} y(\tau)x(t-\tau)d\tau \quad ,$$

$$= \int_{0}^{\infty} y(\tau)x(t-\tau)d\tau \quad \text{because} \quad y(\tau) = 0; \quad t < 0$$

$$= 0; \quad t < 0 \quad .$$

(26.55)

We make use of the fact that the auto-spectrum of the control signal $\widehat{R}_{ss}(\omega)$ is an even non-negative function of ω; it is usually a rational function of ω^2, and may be factorized into the product of nonzero "plus" and "minus" functions,

$$\widehat{R}_{ss}(\omega) = \widehat{R}_{ss}(-\omega) = R^+(\omega)R^-(\omega) = R^+(-\omega)R^-(-\omega) \quad ,$$

i.e.

$$R^-(\omega) = R^+(-\omega)$$

and

$$\widehat{R}_{ss}(\omega) = R^+(\omega)R^+(-\omega) \quad . \tag{26.56}$$

This representation of $\widehat{R}_{ss}(\omega)$ allows Equation (26.53) to be written in the form

$$\overline{(p+p_a)^2} = \frac{1}{2\pi} \int_{-\infty}^{\infty} \left\{ \widehat{R}_{pp}(\omega) + T_{ps}^+(\omega)R^+(\omega)\frac{\widehat{R}_{sp}(-\omega)}{R^-(-\omega)} + T_{ps}^+(-\omega)R^+(-\omega)\frac{\widehat{R}_{sp}(\omega)}{R^-(\omega)} \right.$$

$$\left. + T_{ps}^+(\omega)R^+(\omega)T_{ps}^+(-\omega)R^+(-\omega) \right\} d\omega \quad . \tag{26.57}$$

Equation (26.54) shows that only the "plus" elements of a function survive multiplication by the complex conjugate of a "plus" function and integration over frequency, so that it is useful to express functions as the sum of "plus" and "minus" elements. In particular, when

$$\frac{\widehat{R}_{sp}(\omega)}{R^-(\omega)} = \left[\frac{\widehat{R}_{sp}(\omega)}{R^-(\omega)}\right]^+ + \left[\frac{\widehat{R}_{sp}(\omega)}{R^-(\omega)}\right]^- \tag{26.58}$$

is used in (26.57) and the "plus"–"minus" elimination implied by (26.54) is made, Equation (26.57) becomes, exactly, equal to

$$\overline{(p+p_a)^2} = \frac{1}{2\pi} \int_{-\infty}^{\infty} \left\{ \widehat{R}_{pp} - \left|\left[\frac{\widehat{R}_{sp}}{R^-}\right]^+\right|^2 + \left|T_{ps}^+R^+ + \left[\frac{\widehat{R}_{sp}}{R^-}\right]^+\right|^2 \right\} d\omega \quad . \tag{26.59}$$

This powerful statement of the residual field's variance shows at once
(a) that the optimal controller has the transfer function

$$T_{ps}^+ = \frac{-1}{R^+}\left[\frac{\widehat{R}_{sp}}{R^-}\right]^+ \quad , \tag{26.60}$$

(b) that the optimal transfer function, being a plus function, is physically realizable and stable, and

(c) that the minimum possible variance of the sound under control will be

$$\overline{(p+p_a)_{\min}^2} = \frac{1}{2\pi} \int_{-\infty}^{\infty} \left\{ \widehat{R}_{pp} - \left|\left[\frac{\widehat{R}_{sp}}{R^-}\right]^+\right|^2 \right\} d\omega \quad . \tag{26.61}$$

The power spectral density of the sound is optimally reduced by a factor $(1 - \gamma^2_+)$, γ^+ being the modulus of the "causal part" of the coherence function,

$$\gamma^2_+ = \frac{1}{\widehat{R}_{pp}} \left| \left[\frac{\widehat{R}_{sp}}{R^-} \right]^+ \right|^2 = \left| T^+_{ps} \right|^2 \frac{\widehat{R}_{ss}}{\widehat{R}_{pp}} \leqslant 1 \quad . \tag{26.62}$$

The causality constraint qualifies and restricts the realizability of the $(1 - \gamma^2)$ attenuation (26.33) we had previously shown to be an upper performance bound.

The foregoing optimal solution is that corresponding to the Wiener filter and which is usually described through the Wiener–Hopf equation. That equation can be obtained by differentiating Equation (26.53) with respect to the real part of T^+_{ps} and requiring that gradient to be zero. The absolute minimum also corresponds to a zero of the derivative of Equation (26.53) with respect to the imaginary part of T^+_{ps} (keeping the real part constant). Addition of those two equations produces the Wiener–Hopf equation

$$\int_{-\infty}^{\infty} \left\{ \widehat{R}_{sp}(\omega) + T^+_{ps}(\omega)\widehat{R}_{ss}(\omega) \right\} d\omega = 0 \quad , \tag{26.63}$$

an equation that is usually posed in the form: find a function T^+_{ps} analytic in the upper half plane, such that

$$\widehat{R}_{sp}(\omega) + T^+_{sp}(\omega)\widehat{R}_{ss}(\omega) = \widehat{Q}(\omega), \quad \text{say} \quad , \tag{26.64}$$

is analytic in the lower half plane, a condition that guarantees the required condition (26.63) by application of a Cauchy's theorem on a contour closed in the lower half plane.

The Fourier transform of Equation (26.64)

$$Q(t) = \frac{1}{2\pi} \int_{-\infty}^{\infty} \widehat{Q}(\omega)e^{-i\omega t} dt = 0 \quad \text{for} \quad t > 0 \quad , \tag{26.65}$$

i.e.

$$R_{sp}(t) + \int_0^{\infty} R_{ss}(t - \tau)I(\tau)d\tau = 0 \quad \text{for} \quad t > 0 \quad , \tag{26.66}$$

provides the alternative statement of the Wiener–Hopf equation, the solution of which is (26.60).

It is of course much easier to write down these symbols than to know what they mean! The power of Wiener's result is bought at a high price in understanding the detailed implications of the "plus" and "minus" qualifications in Equation (26.60). The fact that the causality constraint does degrade performance follows from this result, and it is possible to calculate precisely what the various functions are once R_{sp} and R_{ss} are known, in practice by direct measurement.

INDEX